Advances in Intelligent and Soft Computing 159

Editor-in-Chief

Prof. Janusz Kacprzyk
Systems Research Institute
Polish Academy of Sciences
ul. Newelska 6
01-447 Warsaw
Poland
E-mail: kacprzyk@ibspan.waw.pl

For further volumes:
http://www.springer.com/series/4240

David Jin and Sally Lin (Eds.)

Advances in Future Computer and Control Systems

Volume 1

Springer

Editors
Prof. David Jin
International Science & Education
Researcher Association
Wuhan
China

Prof. Sally Lin
International Science & Education
Researcher Association
Guang Zhou
China

ISSN 1867-5662 e-ISSN 1867-5670
ISBN 978-3-642-29386-3 e-ISBN 978-3-642-29387-0
DOI 10.1007/978-3-642-29387-0
Springer Heidelberg New York Dordrecht London

Library of Congress Control Number: 2012934955

Printed on acid-free paper

Springer is part of Springer Science+Business Media (www.springer.com)

Preface

In the proceeding of FCCS2012, you can learn much more knowledge about Future Computer and Control Systems all around the world. The main role of the proceeding is to be used as an exchange pillar for researchers who are working in the mentioned field. In order to meet high standard of Springer, the organization committee has made their efforts to do the following things. Firstly, poor quality paper has been refused after reviewing course by anonymous referee experts. Secondly, periodically review meetings have been held around the reviewers about five times for exchanging reviewing suggestions. Finally, the conference organization had several preliminary sessions before the conference. Through efforts of different people and departments, the conference will be successful and fruitful.

During the organization course, we have got help from different people, different departments, different institutions. Here, we would like to show our first sincere thanks to publishers of Springer, AISC series for their kind and enthusiastic help and best support for our conference.

In a word, it is the different team efforts that they make our conference be successful on April 21–22, 2012, Changsha, China. We hope that all of participants can give us good suggestions to improve our working efficiency and service in the future. And we also hope to get your supporting all the way. Next year, In 2013, we look forward to seeing all of you at FCCS2013.

February 2012

FCCS2012 Committee

Preface

[illegible]

Committee

Honor Chairs

Prof. Chen Bin	Beijing Normal University, China
Prof. Hu Chen	Peking University, China
Chunhua Tan	Beijing Normal University, China
Helen Zhang	University of Munich, China

Program Committee Chairs

Xiong Huang	International Science & Education Researcher Association, China
LiDing	International Science & Education Researcher Association, China
Zhihua Xu	International Science & Education Researcher Association, China

Organizing Chair

ZongMing Tu	Beijing Gireida Education Co.Ltd, China
Jijun Wang	Beijing Spon Technology Research Institution, China
Quanxiang	Beijing Prophet Science and Education Research Center, China

Publication Chair

Song Lin	International Science & Education Researcher Association, China
Xionghuang	International Science & Education Researcher Association, China

International Committees

Sally Wang	Beijing normal university, China
LiLi	Dongguan University of Technology, China
BingXiao	Anhui university, China
Z.L. Wang	Wuhan university, China
Moon Seho	Hoseo University, Korea
Kongel Arearak	Suranaree University of Technology, Thailand
Zhihua Xu	International Science & Education Researcher Association, China

Co-sponsored by

International Science & Education Researcher Association, China
VIP Information Conference Center, China
Beijing Gireda Research Center, China

Reviewers of FCCS2012

Z.P. Lv	Huazhong University of Science and Technology
Q. Huang	Huazhong University of Science and Technology
Helen Li	Yangtze University
Sara He	Wuhan Textile University
Jack Ma	Wuhan Textile University
George Liu	Huaxia College Wuhan Polytechnic University
Hanley Wang	Wuchang University of Technology
Diana Yu	Huazhong University of Science and Technology
Anna Tian	Wuchang University of Technology
Fitch Chen	Zhongshan University
David Bai	Nanjing University of Technology
Y. Li	South China Normal University
Harry Song	Guangzhou Univeristy
Lida Cai	Jinan University
Kelly Huang	Jinan University
Zelle Guo	Guangzhou Medical College
Gelen Huang	Guangzhou University
David Miao	Tongji University
Charles Wei	Nanjing University of Technology
Carl Wu	Jiangsu University of Science and Technology
Senon Gao	Jiangsu University of Science and Technology
X.H Zhan	Nanjing University of Aeronautics
Tab Li	Dalian University of Technology (City College)
J.G Cao	Beijing University of Science and Technology
Gabriel Liu	Southwest University
Garry Li	Zhengzhou University
Aaron Ma	North China Electric Power University
Torry Yu	Shenyang Polytechnic University
Navy Hu	Qingdao University of Science and Technology
Jacob Shen	Hebei University of Engineering

Contents

Improved LBF Model Combined with Fisher Criterion

Xingming Yue, Mei Xie, and Liang Li

Image Processing and Information Security Lab, UESTC, Chengdu, China
mingming7318@126.com, xiemei@ee.uestc.edu.cn, liliangdx@yeah.net

Abstract. An improved local binary fitting (LBF) [1] model combined with Fisher criterion is introduced in this paper. We hope to improve collection degree within classification and separation degree between two classifications, pixels are distributed over local domain of two sides of the boundary. Based on this idea, a new energy term are constructed. The new model is more robust under global information of pixels in the local region of boundary. Furthermore, the narrow band implementation of new model is able to reduce the computational cost greatly.

Keywords: Image segmentation, Level set, LBF model, Fisher criterion.

1 Introduction

Level set method, which was first introduced by Osher and Sethian [2] in 1987, is one of the most popular methods for image segmentation today. Region-based level set methods [2] are particularly popular for it can be deal with topological deformation of curves and detection of weak boundaries. Most of region-based methods [3,4,5], however, are failure to segment images with intensity inhomogeneity. Recently, Li et al. [1] proposed a region-based level set method, called local binary fitting (LBF) model, which is able to deal with intensity inhomogeneities. However, the defect of this model is easily converge to the local minimum due to it has serious localization. Therefore, the result of curves evolution is often stop in the object or background if the initialization of contour is inappropriate.

In this paper, we propose an efficient and robust method that combines Fisher criterion and narrow band implementation [6] of LBF model. This method is effectively prevent the contour from being stuck in local minimum. Moreover, narrow band implementation be able to deal with interested objects and speed up the pace of evolution.

2 Narrow Band Implementation of LBF Model

2.1 Local Binary Fitting Model

The basic idea of LBF model is the intensities on the two sides of the object boundary can be locally approximated by two constants. Therefore, image segmentation can be formulated as a problem of seeking an optimal contour C and two spatially varying

D. Jin and S. Jin (Eds.): Advances in FCCS, Vol. 1, AISC 159, pp. 1–5.
springerlink.com

fitting functions f_1 and f_2 that locally approximate the intensities on the two sides of contour. This optimization problem can be formally defined to minimize the local binary fitting energy

$$\varepsilon_X^{Fit} = \lambda_1 \int_{inside(C)} K_\sigma(x-y)|I(y)-f_1(x)|^2 dy + \lambda_2 \int_{outside(C)} K_\sigma(x-y)|I(y)-f_2(x)|^2 dy, \quad (1)$$

where λ_1 and λ_2 are positive constants, $K_\sigma(x\text{-}y)$ is a weighting function that decreases to zero as y goes away from x. $f_1(x)$ and $f_2(x)$ are two fitting values that locally approximate image intensities in *outside(C)* and *inside(C)* respectively. This local binary fitting energy should be minimized for all x in the image domain Ω.

To smooth the contour C and preserve the regularity of the level set function φ, which is necessary to add the arc length term L(φ) and regularization term P(φ)

$$L(\varphi) = \int_\Omega |\nabla H(\varphi(x))| dx, P(\varphi) = \int_\Omega \frac{1}{2}((|\nabla \varphi(x) - 1)^2 |) dx. \quad (2)$$

Thus, the entire energy functional of LBF model [1] is defined as

$$\mathrm{E}(\varphi, f_1, f_2) = \varepsilon^{Fit}(\varphi, f_1, f_2) + \mu P(\varphi) + \nu L(\varphi), \quad (3)$$

where, μ and ν are positive constants.

Using the standard gradient descent method to minimize the energy functional Eq.(3), we derive the level set evolution equation as

$$\frac{\partial \varphi}{\partial t} = -\delta(\varphi)(\lambda_1 e_1 - \lambda_2 e_2) + \nu\delta(\varphi) div(\frac{\nabla\varphi}{|\nabla\varphi|}) + \mu(\nabla^2\varphi - div(\frac{\nabla\varphi}{|\nabla\varphi|})), \quad (4)$$

where, δ is Dirac Function, e_1 and e_2 are defined as

$$e_i(x) = \int K_\sigma(y-x)|I(x) - f_i(y)|^2 dy, i = 1,2, \quad (5)$$

2.2 Narrow Band Implementation

If either $\varphi_{i-1,j}$ and $\varphi_{i+1,j}$,or $\varphi_{i,j-1}$ and $\varphi_{i,j+1}$ are of opposite signs for pixel P=(i,j), we call P is a *zero crossing pixel* of a level set function φ. Given a set of zero crossing pixels Z of the function φ, we construct the corresponding narrow band of Z by $R = \cup_{x \in Z} N_r(x)$, where $N_r(x)$ is a (2r+1)×(2r+1) square block centered at the pixel x. This parameter r can be chosen as the minimum value r=1. The level set function is updated only at the points on the narrow band in each iteration. Readers can be refer to the description of narrow band algorithm in [6] for more details.

3 Improved LBF Model Combined with Fisher Criterion

3.1 Improved Fisher Criterion Function

To achieve classification effectively between two categories in pattern recognition theories, the classify principle can be described as improve collection degree of within and separation degree of between classification. The famous Fisher criterion function was proposed based on this idea [7].

To overcome the defect of LBF model which easily immerge into local minimum due to the localization property itself, we propose an improved Fisher criterion function combined with narrow band algorithm of LBF model and give the definition as follows:

$$J = \frac{\frac{1}{N_1}\int_{inside\ (C)\cap R}(I(x)-c_1)^2\,dx + \frac{1}{N_2}\int_{outside\ (C)\cap R}(I(x)-c_2)^2\,dx}{(c_1-c_2)^2}, \tag{6}$$

where, N_1 and N_2 are the number of pixels in the region of *inside(C)∩R* and *outside(C)∩R* respectively. c_1 and c_2 represent mean value of pixels in the region of *inside(C)∩R* and *outside(C)∩R* respectively.

We give an analysis for Eq.(6). There is always a certain difference between the local intensities on the two sides of real boundary, while the local intensities on each side of real boundary are basically the same. Therefore, we expect the difference between c_1 and c_2 as far as possible, while the variance of all pixels on each side of contour C as small as possible. That is to say, we hope the Eq.(6) can be minimized by the result of curves evolution.

We can rewrite the Eq.(6) by introducing Heaviside function:

$$J(\phi, c_1, c_2) = \frac{\frac{1}{N_1}\int_R (I(x)-c_1)^2 H(\phi(x))\,dx + \frac{1}{N_2}\int_R (I(x)-c_2)^2[1-H(\phi(x))]\,dx}{(c_1-c_2)^2}. \tag{7}$$

Add Eq.(7) to the LBF model and the result as followed:

$$E(\phi, f_1, f_2, c_1, c_2) = \varepsilon^{Fit}(\phi, f_1, f_2) + \mu P(\phi) + \nu L(\phi) + \kappa J(\phi, c_1, c_2). \tag{8}$$

where κ is a positive constant. Thus, global information is introduced into the improved LBF model due to the value of c_i (i=1,2) is the mean value of all pixels on the narrow band in the region of *inside(C)* or *outside(C)* and this will avoid the curves immerging in local minimum during the evolution.

3.2 Energy Minimization

We minimize the energy functional E in Eq.(8) with respect to φ using the standard gradient descent method by solving the gradient flow equation as follows

$$\begin{aligned}\frac{\partial\phi}{\partial t} &= -\delta(\phi)(\lambda_1 e_1 - \lambda_2 e_2) + \nu\delta(\phi)\,div\left(\frac{\nabla\phi}{|\nabla\phi|}\right) + \mu\left(\nabla^2\phi - div\left(\frac{\nabla\phi}{|\nabla\phi|}\right)\right) \\ &\quad - \kappa\frac{\delta(\varphi)\left[\frac{1}{N_1}(I(x)-c_1)^2 - \frac{1}{N_2}(I(x)-c_2)^2\right]}{(c_1-c_2)^2}.\end{aligned} \tag{9}$$

N_1, N_2, c_1, c_2 are updated after every iteration of φ according to

$$N_1 = \frac{1}{\int_R H(\phi(x,y))\,dxdy},\quad N_2 = \frac{1}{\int_R [1-H(\phi(x,y))]\,dxdy}, \tag{10}$$

$$c_1 = \frac{\int_R I(x,y)H(\phi(x,y))\,dxdy}{\int_R H(\phi(x,y))\,dxdy},\quad c_2 = \frac{\int_R I(x,y)[1-H(\phi(x,y))]\,dxdy}{\int_R [1-H(\phi(x,y))]\,dxdy}. \tag{11}$$

4 Experimental Results

4.1 Accuracy of Segmentation

We apply our improved LBF model for a group of images and compare it with the results of the LBF model. In Fig.1, the corresponding results of LBF and improved LBF models in Rows 1 and 2 respectively. Obviously, despite significant intensity inhomogeneities and different positon of initial contour, the results of improved LBF model are all in good agreement with the true object boundaries.

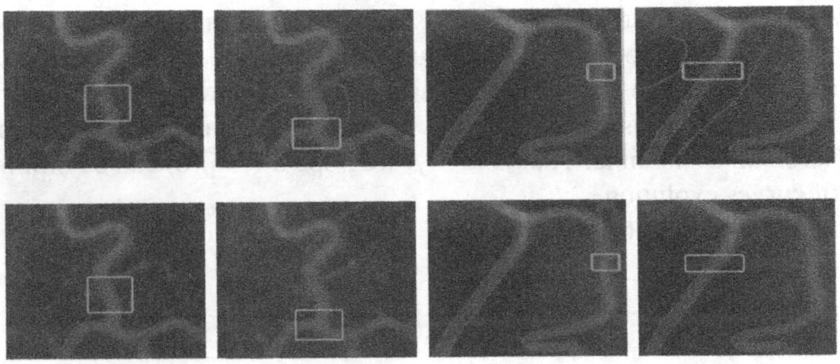

Fig. 1. Row 1: Segmentation results by LBF; Row 2: Segmentation results by improved LBF. (The green contours are initial contours, while the red contours are segmentation results.)

4.2 Segmentation of the Interested Objects

We also used our algorithm to segment a group of synthetic images to extract the interested object boundaries. Fig.2 shows these images as examples, with the corresponding segmentation results of the LBF and improved LBF models in Rows 1 and 2 respectively. We assume that the right side of images contains the interested objects that boundaries are expected to be extracted. Row 1 shows the uninterested region also be segmented as well due to level set function updated in full domain in LBF model. Row 2 shows only the interested objects be detected on account of level set function updated only on the narrow band in improved LBF model.

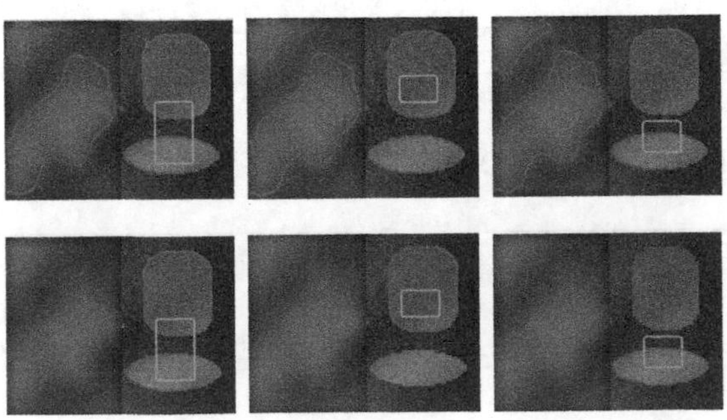

Fig. 2. Row 1: Segmentation results by LBF; Row 2: Segmentation results by improved LBF.

4.3 Computational Efficiency

We compare the computation time of our algorithm with that of the LBF algorithm for two images in Column 1 and Column 3 of Fig.1. The CPU times in these experiments are obtained by running Matlab programs on a Intel PC with E7400 2.80GHz, 2GB RAM, with Matlab7.8.0, on Windows XP.

The CPU times for the two images are listed in Table 1 which show that the cost time of the our improved algorithm reduce dramatically compared with the original LBF algorithm.

Table 1. Iteration number and CPU time (in seconds) of the Improved LBF and original LBF algorithms for the results in Column 1 and Column 3 of Fig. 1.

Image	Iteration number	CPU time (in second)	
		Original LBF	Improved LBF
1	150	5.797527	0.763261
2	220	9.042302	1.327420

5 Conclusion

We have proposed an improved LBF model for image segmentation. This algorithm is more robust to initialization due to introduced global information of pixels on the narrow band. Furthermore, the narrow band implementation can realize segment the interested objects only and greatly reduce the computational cost. Comparisons with original LBF model demonstrate the superior performance of our model in terms of accuracy, robustness, and efficiency.

References

1. Li, C., Kao, C., Gore, J.C., Ding, Z.: Minimization of region-scalable fitting energy for image segmentation. IEEE Trans. Imag. Proc. 17(10), 1940–1949 (2008)
2. Osher, S., Sethian, J.: Fronts propagating with curvature-dependent speed: Algorithms based on Hamilton-Jacobi formulations. J. Comput. Phys. 79(1), 12–49 (1988)
3. Chan, T., Vese, L.: Active contours without edge. IEEE Trans. Imag. Proc. 10(2), 266–277 (2001)
4. Paragios, N., Mellina-Gottardo, O., Ramesh, V.: Gradient vector flow fast geometric active contours. IEEE Trans. Patt. Anal. Mach. Intell. 26(3), 402–407 (2004)
5. Ronfard, R.: Region-based strategies for active contour models. Int'1 J. Comp. Vis. 13(2), 229–251 (1994)
6. Li, C., Xu, C., Gui, C., Fox, M.D.: Distance regularized level set evolution and its application to image segmentation. IEEE Trans. Imag. Proc. 19(12), 3243–3254 (2010)
7. Bian, Z.-Q., Zhang, X.-G.: Pattern Recognition, 2nd edn., pp. 87–89. Tsinghua University Press, Beijing (2000) (in Chinese)

A Background Model Combining Adapted Local Binary Pattern with Gauss Mixture Model

YuBo Jiang[1] and XinChun Wang[2]

[1] School of Computer Science and Technology, Shandong Jianzhu University, Shandong Jinan 250014
ybj2008@126.com
[2] Computer Department of Jinan Vocation College, Shandong Jinan 250014
wxcwjl@126.com

Abstract. Detecting moving objects using background subtraction is a critical component for many vision-based applications, but it is extremely sensitive to dynamic scene changes due to lighting and extraneous events, especially when the dynamic texture such as swaying trees and waving water surface is existed. In this paper, a reliable background modeling scheme has been proposed to combine adapted Local Binary Pattern (LBP) of local block with Gauss Mixture Model (GMM) of pixel level. A pixel in new frame is judged as foreground not only by comparing intensity with background image, but also by comparing its block histogram with relevant background block model. Experiment results show this algorithm is efficient than the classical GMM algorithm using only color intensities.

Keywords: Background subtraction, Gauss Mixture Model, Local Binary Pattern, Jaccard Coefficient.

1 Introduction

Detecting moving object using background subtraction provides a classification of the pixels in the video sequence into either foreground (moving objects) or background, which is a critical component for many vision-based applications such as remote sensing, video surveillance and traffic monitoring [1]. Background subtraction involves three steps: first, construct background reference image; secondly, subtract the background from the new frame; the last, get the foreground area (objects) from the image by threshold. The heart is background modeling and background adoption to both sudden and gradual changes in background, but natural scenes put many challenges on background modeling since they are usually dynamic in nature including illumination changes, swaying vegetation, rippling water, flickering monitors, fluttering flags and so on. According to information used, background modeling methods can be divided into two categories. The first kind of methods only extracts the color or intensity information of single pixel, which can be called pixel-based method, for example, W4 method [2], Single Gauss Model [3], and Staffer's Gauss Mixture Model (GMM) [4]. Approaches of this type have the advantage of extracting detailed shapes of moving objects, but may suffer from the drawback that their segmentation

D. Jin and S. Jin (Eds.): Advances in FCCS, Vol. 1, AISC 159, pp. 7–12.
springerlink.com

results are sensitive to dynamic texture (e.g. waving trees) in background, and will generate many false negatives [5]. Another kind of methods study block-based representations for background modeling. Compare with pixel, block can give more spatial distribution information, which makes block-based methods are more capable of dealing with non-stationary backgrounds. The Non-parametric model [6] not only matches the pixel intensity, but also compares the position of neighbor, which makes the computation is complex. Reference [7] introduces the temporal analysis of blocks into background subtraction, and gets preferable detect results. The algorithm proposed by Li and Leung [8] combines the texture of 5*5 blocks with color feature, which is insensitive to illumination changes as the texture feature less affected by light change. How to utilize effectively both the color intensity and texture features is still a difficult problem. The local binary pattern (LBP) technique is very effective to describe the image texture features, and has advantage of fast computation, so we improve the classical LBP from computation aspects, fusion this adapted LBP with Gauss Mixture Model to construct the background model and identify foreground. Experiment results show the detection of this algorithm is insensitive to illumination change, better than other algorithms in complicate scenes with swaying vegetation and rippling water.

2 Adapted LBP Feature

LBP presented by Ojala is a block-based texture descriptor [9]. The operator labels the pixels of an image region by threshold the neighborhood of each pixel with the center value and consider the result as a binary number (binary pattern). The outstanding of LBP is insensitive to light change and very fast to compute. Taking pixel position (x_c, y_c), 3×3 neighborhood as example, the basic LBP pattern at the pixel can be expressed as follow.

$$LBP(x_c, y_c) = \sum_{n=0}^{7} s(i_n - i_c) \times 2^n \tag{1}$$

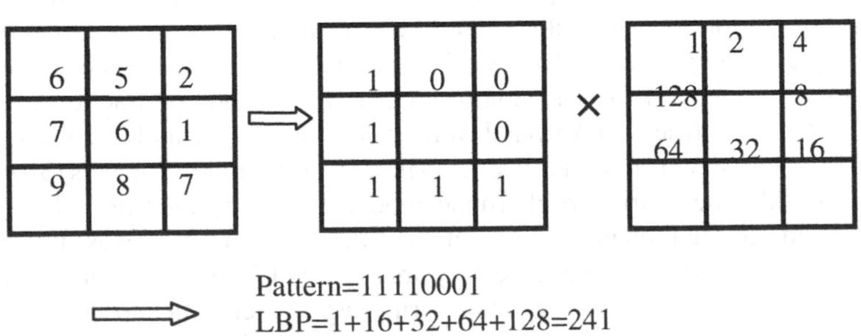

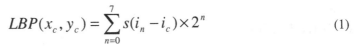

Fig. 1. An example of basic LBP

Where, i_c corresponds to the grey value of the central pixel (x_c, y_c), i_n are the grey values of the 8 surrounding pixels. If $i_n - i_c > 0$, $s(i_n - i_c) = 1$, otherwise $s(i_n - i_c) = 0$. Fig. 1 shows an example of how to compute the basic LBP value.

The calculation speed of Basic LBP is slow for it calculates each pixel's pattern code in certain neighborhood. In order to improve the computing speed, we consider improve classical LBP from computational aspects.

Figure 2 shows a fragment of one gray-scale image. Regarding the pixel of gray value 6 in Fig. 2(a) (referred to as pixel1 later), the LBP of this pixel is Pattern1=11110001, as far as the pixel of gray value 7 is concerned (referred to as pixel2 later), the LBP is Pattern2=00100101. The last code 1 of Pattern1 is get by comparing pixel2 with pixel1, and the first code 0 of Pattern2 is get by comparing pixel1 with pixel2 (drawn in Fig.2 (b) by red boxes), so pixel 1 and pixel 2 compare twice, and there is a repeat comparison. From further analysis one can get every code of Local Binary Pattern has this repeat comparison, so with regard to the whole image, half of LBP can describe local texture completely (drawn in Fig.2 (c) by red lines). Based on this we propose semi-LBP as following.

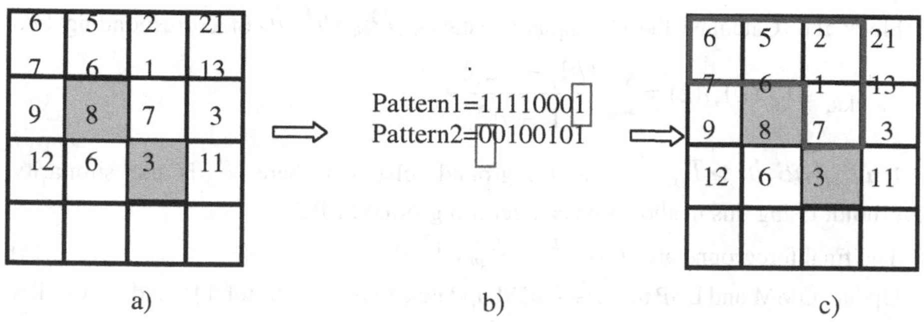

Fig. 2. Comparison of two neighbor pixels' LBP

$$SLBP(x_c, y_c) = \sum_{n=0}^{3} s(i_n - i_c) \times 2^n \tag{2}$$

The classical LBP does 8 comparisons to calculate the LBP of each pixel, while in semi-LBP only half comparison is enough, and the code range narrowed to 0~15 from 0~255, which can greatly reduce the computation in calculation the LBP histogram following. Although this adapted technique is not rotation invariance, it is applicable in background modeling with monocular camera, for there is small rotation in raw video get by single camera.

3 Background Model Combining Semi-LBP with GMM

Stauffer and Grimson in Ref. [4I model the background by a mixture of Gaussians:

$$p(x_t)=\{p_{i,t} \mid i=1,\cdots,K\} \quad p_{i,t}=[w_{i,t},m_{i,t},\sigma^2_{i,t}] \quad \sum_{i=1}^{K} w_{i,t}=1 \tag{3}$$

The background model combining semi-LBP with GMM extracts the adopted Local Binary Pattern (LBP) feature of blocks, and combines these features with GMM of pixels to construct the background reference image. The procedures are:

- Construct the background Gauss mixture model according to Ref. [4I, match each pixel $x^t_{i,j}$ in new frame with all distributions in gauss mixture background model, one can get foreground Fgmm.
- Block background image with 8*8 windows, calculate LBP histogram hB_{mn} of each block, here m, n is the block index of horizontal and vertical. As the new frame coming, block it with 8*8 windows; calculate semi-LBP histogram h_{mn} of each block too. Calculate the Chi-square distance $D_{chi}(hB,h)$ of corresponding two blocks, $$D_{chi}(h1,h2)=\sum_{i=1}^{l}\frac{(h1_i-h2_i)^2}{h1_i+h2_i} \tag{4}$$

  If $D_{chi}(hB,h)>T_D$, $x^t_{i,j}$ is foreground, else not, here T_D is the similarity threshold. Using this method, one can get foreground FLBP.
- The final foreground area F is: $F=F_{gmm}\bigcap F_{LBP}$ (5)
- Update GMM and LBP models. GMM update can be find in Ref. [4I, and we use IIR filter to update the LBP models: $hB^t_{mn}=(1-\gamma)\cdot h^{t-1}_{mn}+\gamma\cdot h^t_{mn}$ (6)

4 Experiments

Test this semi-LBP+GMM algorithm, GMM+LBP scheme and GMM methods with two sequences. Test one (Water Surface sequence): In this sequence, water surface in background is waving all the time and Fig.3 shows the results. From left to right, images are a) original frame; b) only GMM detection result; c) GMM+LBP result and d) semi-LBP+GMM result. In this test, number of model K in GMM is 3, $T_D=2.5$. Although the GMM detection result is better in getting contour and suppressing noise, a few of noise still exist, because the intensity diversification of the rippling water cannot represent by several gauss model completely. GMM+LBP detection result and semi-LBP+GMM detection result are better than the GMM result.

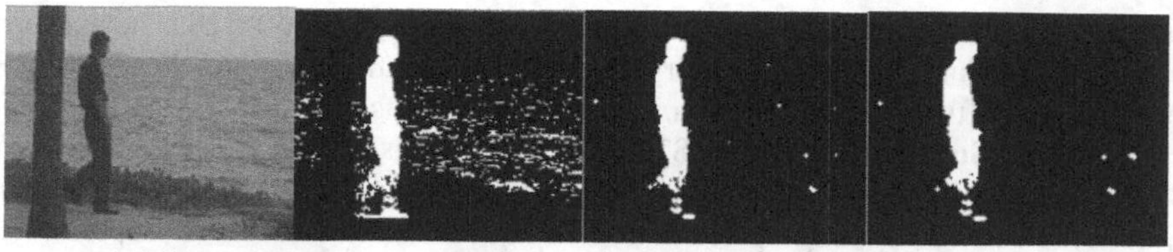

Fig. 3. Detection results of Water Surface sequence

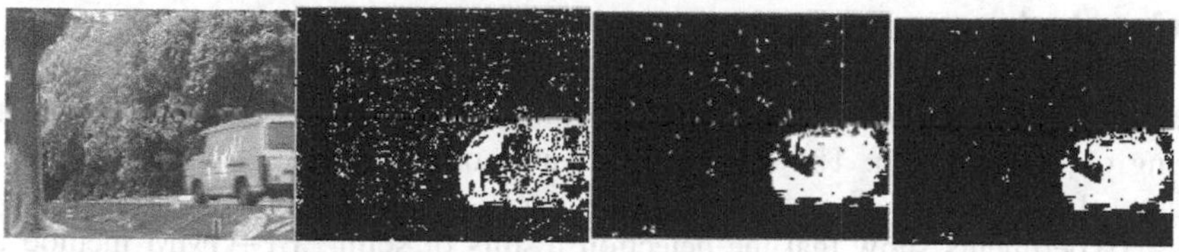

Fig. 4. Detection results of Campus sequence

Test two (Campus sequence): In this sequence, the tree branches are continuous swaying substantially because of strong wind, and the results are in Fig.4.The order of images is same as Fig. 3. In this test, $T_D = 2.8$ .GMM+LBP detection result and semi-LBP+GMM detection result are the best too.

In order to evaluate the testing results, we first generate the ground truth images, and then calculate the Jaccard Coefficient [10] of this algorithm and many classical schemes: $JC = \dfrac{TP}{TP + FP + FN}$ (7)

Where, TP (True Positives) is the number of foreground pixels correctly detected; FP (False Positives) is the number of background pixels incorrectly detected as foreground; FN (False Negatives) is the number of foreground pixels incorrectly detected as background; the Jaccard Coefficients of five methods show in Table 1, which demonstrate this semi-LBP+GMM algorithm is efficient than the classical algorithms.

Table 1. Jaccard coefficients of four algorithms in test one and test two

sequence	SoG	GMM	Temporal analysis	GMM+ LBP	Semi-LBP +GMM
Water Surface	0.4304	0.6703	0.7012	0.8348	0.8342
Campus	0.3517	0.6023	0.6856	0.7467	0.7468

In the two tests above, we also calculate the time costs of semi-LBP+GMM scheme and LBP+GMM scheme. 20 frames are using for background initialization, only one frame are tested for object detection, and resolution of image is 160*128. The platform is: CPU, Intel Core 2 3GHz; memory, 3.25 GB; operating system, Windows XP; programming software, matlab 2010a. The time costs of semi-LBP+GMM are 0.853 seconds (Water Surface sequence), 0.872 seconds (Campus sequence), the time costs of LBP+GMM are 2.461 seconds (Water Surface sequence), 2.624 seconds (Campus sequence). The computing performance of semi-LBP+GMM is superior to LBP+GMM.

5 Conclusions

This paper fuse block-based local texture feature with pixel-based color feature for moving object detection. The semi-LBP features of 8*8 blocks are used as texture features of blocks, and Gaussian mixture model of intensity is used as features of pixels. Experiments show that the detection results of semi-LBP+GMM method are better than results of traditional algorithms, and in terms of computing performance, semi-LBP superior to the traditional LBP.

Acknowledgements. We would like give the thanks to Liyuan Li for providing the test sequences, and thanks to Jinan Science and Technology Bureau (the contracts numbers is No.201004002).

References

1. Elhabian, S., El-Sayed, K., Ahmed, S.: Moving object detection in spatial domain using background removal techniques-state-of-art. Recent Patents on Computer Science 1, 32–54 (2008)
2. Ismail, H., David, H., Larry, S.D.: A fast background scene modeling and maintenance for outdoor surveillance. In: International Conference of Pattern Recognition, pp. 179–183 (2000)
3. Christopher, R.W., Ali, A., Trevor, D., Alex, P.P.: Pfinder: real-time tracking of human body. IEEE Transactions on Pattern Analysis and Machine Intelligence 19(7) (1997)
4. Stauffer, C., Grimson, W.E.L.: Adaptive background mixture models for real-time tracking. In: Proc. of CVPR 1999, pp. 246–252 (1999)
5. Yong-Zhong, W., Yan, L., Quan, P., et al.: Spatiotemporal background modeling based on adaptive mixture of guassians. Acta Automatica Sinica 35(4), 371–378 (2009)
6. Elgammal, A., Harwood, D., Davis, L.: Non-parametric model for background subtraction. In: IEEE Frame Rate Workshop (1999)
7. Spagnolo, P., Orazio, T.D., Leo, M., et al.: Moving object segmentation by background subtraction and temporal analysis. Image and Vision Computing 24, 411–423 (2006)
8. Liyuan, L., Leung, K.H.: Integrating intensity and texture differences for robust change detection. IEEE Trans. Image Process 11(2), 105–112 (2002)
9. Ojala, T., Pietikäinen, M., Harwood, D.: A comparative study of texture measures with classification based on feature distributions. Pattern Recognition 29, 51–59 (1996)
10. Rosin, P., Ioannidis, E.: Evaluation of global image thresholding for change detection. Pattern Recognition Letters 24(14), 2345–2356 (2003)

Driver Fatigue Detection Based on AdaBoost

Chiliang Xiong, Mei Xie, and Lili Wang

Image Processing and Information Security Lab, School of Electronic Engineering, UESTC, Chengdu, China

Abstract. Depending on CCD camera to detect driver fatigue has been proposed in past research. This paper proposes a rapid method of driver fatigue detection by applying AdaBoost algorithm,and PERCLOS to judge the degree of the driver fatigue. Compared to the original method, this method reduces redundant facial texture information, and accelerates the speed of eye state detection while improving detection accuracy rate of driver fatigue. The experiments results show that this method is fast , effective and easy which can be used in real-time detection of driver fatigue.

Keywords: CCD, fatigue detection, AdaBoost, PERCLOS, real-time.

1 Introduction

Driver fatigue of detection method can be divided into contact and non-contact detection . A camera-based method for fatigue detection is non-contact. It is less intrusive to the driver, as opposed to contact methods, which are more accurate, but intrusive. Isaam Saeed et al. present a camera-based driver fatigue detection system, which uses the AAM [6] as a means to track rigid and non-rigid motion of a face in an input video sequence and propose an Active Appearance Model specific to the ocular region of the face to place greater focus on the eyes, as fatigue cues based on ocular features [7]. However because Isaam Saeed et al. directly use AAM on face, their model contains too many facial textures which are useless for eyes and reduces the fitting speed and detection rate. This paper proposes a method of using Adaboost three times to locate and identify eyes.Viola et al. have introduced a rapid object detection scheme based on a boosted cascade of simple feature classifiers [1]. Boosting is a fast and powerful learning concept. It combines the performance of many“weak” classifiers to produce a powerful ‘committee’. Here, gentle adaboost is selected.

2 The Theory of AdaBoost

2.1 The Development of Boost

The Adaboosting algorithm,introduced in 1995 by Freund and Schapire,solved many of the practical difficulties of the earlier boosting algorithms,and is one of the key pointsof this paper.The algorithm takes as input a training set $(x_1, y_1),...(x_m, y_m)$ where each

D. Jin and S. Jin (Eds.): Advances in FCCS, Vol. 1, AISC 159, pp. 13–17.
springerlink.com

x_i belongs to some set X,and each label y_i is in some label set Y.for convenience,we assume Y={-1,+1},adaboost calls a given weak or base learning algorithm repeatedly in a series of rounds t = 1,...,T.One of the main ideas of the algorithm is to maintain a distribution of weights over the training set .The weight of this distribution on training example i on round t is denoted $D_t(i)$ .initially,all weights are set equally,but on each round ,the weights of wrong classified examples are increased so that the weak learner is set to focus on the hard examples in the training set.The weak learner's job is to find a weak hypothesis $h_t : X -> \{-1,+1\}$ right for distribution D_t .The goodness of a weak hypothesis is measured by its error

$$\varepsilon_t = \mathrm{P}_{r_i \sim D_t}[h_t(x_i) \neq y_i] = \sum_{i:h_t(x_i) \neq y_i} D_t(i) \tag{1}$$

Practically, the weak learner may be an algorithm that can use the weights D_t on the training examples. Alternaatively, when this is not possible,a subset of the training examples can be sampled according to D_t ,and these(unweihgted) resampled examples can be used to train the weak leaner. Given: $(x_1, y_1),...,(x_m, y_m)$ where $x_i \in X$, $y_i \in Y$ ={-1,+1}, initialize $D_1(i) = 1/m$ for t = 1,...,T: :

1.Train weak leaner using distribution D_t 2. Get weak hypothesis $h_t : X \rightarrow \{-1,+1\}$ with error

$$\varepsilon_t = P_{r_i \sim D_t}[h_t(x_i) \neq y_i] \tag{2}$$

1. choose

$$\alpha_t = \frac{1}{2}\ln(\frac{1-\varepsilon_t}{\varepsilon_t}) \tag{3}$$

2. Update:

$$D_{t+1}(i) = \frac{D_t(i)\exp(-\alpha_t y_i h_t(x_i))}{Z_t} \tag{4}$$

Where Z_t is a normalization factor(chosen so that D_{t+1} will be a distribution).Output the final hypothesis:

$$H(x) = \text{sign}(\sum_{t=1}^{T} a_t h_t(x)) \tag{5}$$

The above is the procedure of algorithm adaboost.

3 Localization of the Eyes

Here AdaBoost is used to detect the eyelids,detecting contains three steps as follows

(1) Use Gentle Adaboost to train a face classifier to detect face we can get a box that contains face

(2) Cut the box according to the geometric position of the eyes in the face , and the eyes are located roughly. Then we use Gentle Adaboost the second time to train an eye classifier to detect eye in the two small images cut out before。

(3) We use Adaboost the third time to train a more accurate eyes classifer to detect eye in step (2)'s images.

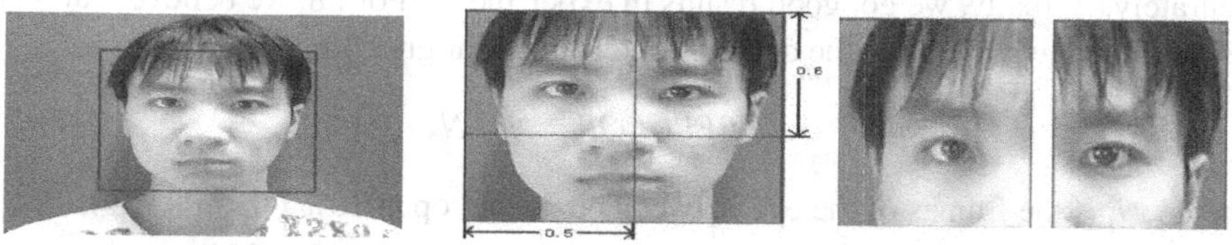

(a) (b) (c)

Fig. 1. (a)Using AdaBoost to detect the face. (b)Cut the image into four parts. (c) two eye images which have been cut

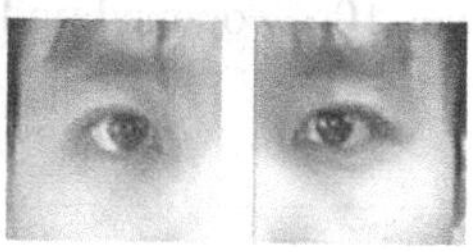

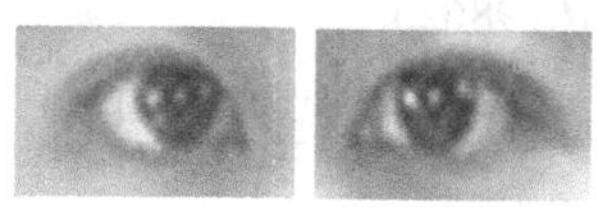

Fig. 2. The eyes detected by AdaBoost **Fig. 3.** The eyes detected by AdaBoost a second time

4 PERCLOS and Results

PERCLOS is the percentage of eyelid closure over the pupil over time and reflects slow eyelid closures rather than blinks [7].

PERCLOS had three drowsiness metrics:

(1) P70, the proportion of time the eyes were closed at least 70 percent.
(2) P80, the proportion of time the eyes were closed at least 80 percent.
(3) EYEMEAS (EM), the mean square percentage of the eyelid closure rating. Of the three, P80 has the best correlation with driver fatigue.

P80 has the best correlation with driver fatigue comparing with the two other metrics.

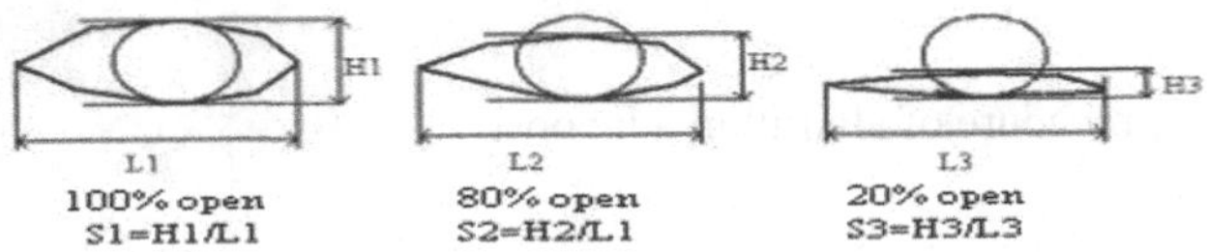

Fig. 4. The eye contour aspect ration is 100%, 80%, and 20% of the cases

In the former part, we have got the contours of the eyes. Here we propose to convert the eye contour aspect ratio of the real-time tracking into PERCLOS. Eye contour ration can not be affect by the distance between the eyes and camera. From Figure 5, we define the contour ration,

$$S = L / H$$

where, L is the length of the eye, H is the height of the eye. As long as the model has a good match with the eye, S can be used to reflect the eye open state simply and accurately. Actually, we got good results in experiments. For P8, we believe that when PERCLOS value f >0.15, the driver is determined fatigue. So

$$PERCLOS = N_1 / N_2$$

where, N_1 is the frames of the eye between close and open 20% ($0 < S < S3$), N_2 is the frames of the eye between open 20% and open 80% ($S3 < S < S2$).

Result were generated on an Intel pentium4 (3GHZ) processor with 2GB RAM, and input camera of 640×480pixels. We got 7 frames per second speed of eye state detection. Figure 6 show the difference of S under imitative normal and fatigue state in 30 frames and calculate PERCLOS separately. We can see in the lower figure eye blink slower. Here the aspect ratio S of the tester's eyes 100% opened under normal state was 0.16198625.

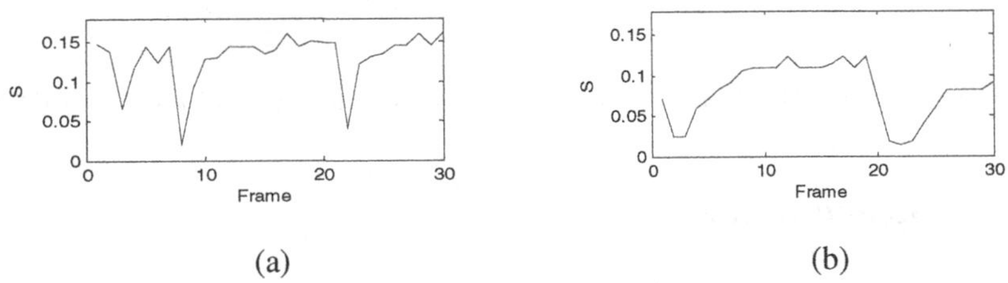

Fig. 5. (a)figure is under imitative normal state and $PERCLOS = 0.125 < 0.15$. (b) figure is under imitative fatigue state and $PERCLOS = 0.1667 > 0.15$.

Now, We have made the driver fatigue detection based on AdaBoost come true. Compared with the original methods,this paper used only one algorithm to detect the driver's fatigue.It is more convenient and more fast.We had lots of eye state data with a certain tine , which can make us get more accurate recognition in PERCLOS determine.

5 Conclusions

This paper abandon using AAM that originally was proposed by Isaam Saeed et al.[6] in driver fatigue detection, and reduces redundant facial information to accelerate the eye fitting speed. We also advise to convert the eye contour aspect ratio into PERCLOS to detect driver fatigue. Though the whole fatigue detection system have shown encouraging results in our experiments, further work is required in selecting more effective statistics time interval and threshold. In addition, one should remember that this method is for personal characteristics. The aspect ratio S of the driver's eyes 100% opened under normal state must be known in this paper ,we get a conclusion that we use AdaBoost three times to recognize whether a driver is in fatigue with PERCLOS.

Acknowledgments. This paper is supported by Image Processing and Information Security Lab.especially pay my thanks to Huazhi Dong.

References

1. Viola, P., Jones, M.J.: Rapid Object Detection using a Boosted Cascade of Simple Features. In: IEEE CVPR (2001)
2. Cootes, T.F., Edwards, G.J., Taylor, C.J.: Comparing active shape models with active appearance models. In: Proc. British Machine Vision Conf., pp. 173–182 (1999)
3. Bookstein, F.L.: Landmark methods for forms without landmarks: localizing group differences in outline shape. Medical Image Analysis 1(3), 225–244 (1997)
4. Stegmann, M.B.: Active Appearance Models. IMM-EKS-2000-25
5. Moriyama, T., Kanade, T.: Automated Individualization of Deformable Eye Region Model and Its Application to Eye Motion Analysis. In: IEEE CVPR (2007)
6. Saeed, I., Wang, A., Senaratne, R., Halgamuge, S.: Using the Active Appearance Model to Detect Driver Fatigue. In: ICIAFS (2007)
7. PERCLOS: A Valid Psychophysiological Measure of Alertness As Assessed by Psychomotor Vigilance, Washington, Office of Motor Carriers (1998)

Research of Cascaded Conditional Random Fields Model for Sentence Sentiment Analysis Based on Isotonic Constraints

Yu Zhao and Wandong Cai

College of Computer Science, Northwestern Polytechnical University, China
zhaoyu_mail@126.com, caiwd@nwpu.edu.cn

Abstract. The sentence sentiment analysis is a key task in sentiment analysis. Existing methods ignored the contextual information, the negative effect of the redundancy between labels, or the relationship from sentiment words to annotation labels. Aiming at these problems, this paper present a novel cascaded model based on isotonic constraints, which respectively classify sentiment polarities and strength in different layers. Different from traditional cascaded model, the proposed method incorporates a kind of domain knowledge about sentiment words through enforcing a set of monotonic constraints on the CRF parameters. Experimental results indicate that the proposed algorithm has strong discrimination ability between different labels, and thus validate the effectiveness of our model in sentence sentiment analysis for Chinese texts.

Keywords: Sentence sentiment analysis, conditional random fields, cascaded model, isotonic constraints, Information retrieval.

1 Introduction

Recently, the number of product reviews from ecommerce increases surprisingly. This huge amount of reviews makes it difficult to analysis them comprehensively. To overcome this problem, a variety of methods have been developed for analyzing subjective information. All of these methods belong to the field of sentiment analysis. Sentiment polarity and strength of sentences play a key role in sentiment analysis. In recent years, two kinds of approaches have been proposed to deal with sentence sentiment analysis: one is based on classification methods [1, 2]; the other one is based on sequence annotation methods [3-5]. However, the first kind of methods ignores the contextual information between sentences; moreover, the second kind of methods ignores domain knowledge or the negative effect of the redundancy between different annotation labels.

Aiming at these problems, we propose a cascaded CRF model based on isotonic constraints. The analysis procedure can be divided into two steps: Firstly, sentences sequence is classified by standard CRF, according to sentiment polarity. Secondly, a generalized isotonic CRF is adopted to annotate the result sequence from the first step with respect to sentiment strength labels. The domain knowledge about the relationship between sentiment words and annotation labels is used in the second step.

D. Jin and S. Jin (Eds.): Advances in FCCS, Vol. 1, AISC 159, pp. 19–24.
springerlink.com

The rest of this paper is organized as follows: Section 2 surveys related work. In Section 3, we introduce the proposed approach in detail. Section 4 presents and discusses the experimental results. Section 5 is conclusion and future work.

2 Related Work

Sentences sentiment analysis includes three subtasks: (1) objective\ subjective classification; (2) sentiment polarity classification; (3) sentiment strength annotation. Many methods adopted in these subtasks treat the data as categorical [1, 6]. These techniques are mainly formulated as supervised classification methods. Although classification methods report over 90% accuracy on objective information categorization, their performance degrades drastically when applied to sentence sentiment analysis.

Because context plays a vital role in determining sentence sentiment, another kind of methods, treating texts as local sentiment flow, could be more effective[3-5]. Liu et al. [3] pointed out that multiple-layer CRF is an effective method for sentence sentiment analysis. The advantage of [3] is considering redundancy between annotation labels. Mao et al. [4] incorporated domain knowledge about the relationship from sentiment words to annotation labels. McDonald et al. [5] reduced the joint sentence and document level analysis to a sequential classification problem. Since both methods perform sentiment polarity classification and sentiment strength annotation in single-layer CRF model, they can not avoid the negative effect of redundancy between labels in sentence sentiment annotation.

Different from those methods discussed above, our model simultaneously takes into account the relationship from sentiment words to annotation labels and adopts a multiple-layer annotation frame.

3 Cascaded CRF Model Based on Isotonic Constraints

3.1 CRF

Sequential annotation is the task of associating a sequence of labels $y=(y_1, \ldots, y_n)$, $y_i \in Y$ with a sequence of observed values $x=(x_1, \ldots, x_n)$, $x_i \in X$. CRF [7]are parametric families of conditional distributions $p_\theta(y|x)$. In the sentence sentiment analysis, CRF is a chain structure. The clique set is $C=\{(y_{i-1}, y_i), (y_i, x): i=1, \ldots, n\}$. The formula of $p_\theta(y|x)$ in our study is

$$p_\theta(y \mid x) = Z^{-1}(x,\theta) * \exp\Big(\sum_i \sum_{\sigma,\tau \in Y} (\lambda_{<\sigma,\tau>} * f_{<\sigma,\tau>}(y_{i-1}, y_i)) + \sum_i \sum_{\tau \in Y} \sum_k (\mu_{<\tau,A_k>} * g_{<\tau,Ak>}(y_i, x, i))\Big). \quad (1)$$

Where $\theta=(\lambda,\mu)$ is the parameter vector. The valuesσ, τ correspond to arbitrary labels in Y and A_k corresponds to binary functions of both observation x and some position i in the sequence. $Z(x,\theta)$ represents the conditional normalization term. The model training is carrying out by maximizing likelihood estimation. In our experiments, we adopt quasi-Newton implementation.

3.2 Isotonic Constraints

Despite the great popularity of CRF in sequence annotation, they are not appropriate for ordinal data such as sentiment strength. Mao et al. [4] found that the presence of sentiment words to increase (or decrease) the conditional probability of particular labels is consistent with the ordinal relation of sentiment. The parameter set of $\mu_{<\tau,Ak>}$, where A_k measures appearance of a sentiment word, is candidates for ordinal constraints. Assuming that A_w measures the presence of word w, if w belongs to the positive sentiment lexicon V_P or the negative sentiment lexicon V_N, it enforces

$$\tau \leq \tau' \quad \Rightarrow \quad \mu_{<\tau,A_w>} \leq \mu_{<\tau',A_w>} \quad \forall w \in V_P . \tag{2}$$

$$\tau \leq \tau' \quad \Rightarrow \quad \mu_{<\tau,A_w>} \geq \mu_{<\tau',A_w>} \quad \forall w \in V_N . \tag{3}$$

[4] presented a re-parameterization method to lead a simpler optimization scheme. The re-parameterization method has the benefit of converting the complex totally ordered relation in (2)-(3) to simple non-negative constrains.

3.3 Cascaded CRF Model Based on Isotonic Constraints

Cascaded CRF model compute sentiment polarity and sentiment strength of sentence respectively. Different from standard cascaded CRF [3], our model adopts standard CRF in the first layer and uses isotonic CRF in the second layer. We give the detailed process of our model for sentence sentiment analysis as follows.

Step 1. Get sentiment polarities of sentences.
Standard CRF only takes the sentiment polarities into account, including praised, objective, and criticized (the corresponding values is 1, 0, -1). The features, which are used in model training and inferring, can be divided into three categories.

The first kind of features is about the appearance of sentiment words, topic words, transitional words and advocating words. The vocabulary of sentiment words and advocating words is obtained from HOWNET. Those words, which don't appear in the experiment corpus, are removed from sentiment words list. The size of the sentiment vocabulary in the experiment is 617, including 358 positive words and 259 negative words. If the appearance number of a noun or a noun phrase is lager than 10 times in the test corpus, this word is identified as a topic word.

The second kind of features is about position information of sentences in a text. In our experiment, a text is divided into three sections. The first two sentences and the last two sentences may contain conclusion or evaluation. The rest of a text may be about description of evaluation objects. The third kind of features is about the relation between two adjacent sentences, such as f (y_{i-1}=-1, y_i=1).

Step 2. Get sentiment strength of sentences.
The objective sentences, get in the first layer, are removed from the test corpus. An isotonic CRF annotates the result data from the first-layer with respect to sentiment strength, including high praised, praised, objective, criticized, and high criticized (the corresponding values is 2,1, 0, -1,-2). The features, used in this layer, are divided into two types. The first kind is the same as the property of words in the first layer, excepting advocating words. The other is about the relation between two adjacent

sentences with respect to sentiment strength. The model training in step 2 is a constrained MLE. These constraints are $\mu_{<\tau, Aw>} \geq 0$, $w \in Vp$ and $\mu_{<\tau, Aw>} \leq 0$, $w \in V_N$, for all $\tau \neq -2$, where A_k is the feature measuring appearance of w.

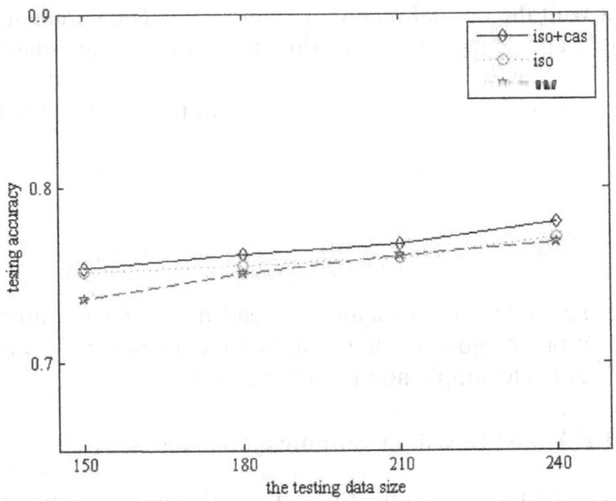

Fig. 1. Total results for the proposed method and baseline methods

Table 1. Annotation effects based on iso+cas, 80% of the data for training

label	manually labeled	model labeled	correct	Accuracy [%]	Recall [%]	F-value [%]
high praised	34	45	23	51.1	67.6	58.2
praised	122	131	54	41.2	44.3	42.7
objective	987	942	839	89.1	85	87
criticized	104	120	60	50	57.7	53.6
high criticized	49	58	36	62.1	73.5	67.3
total	1296	1296	1012	78.1	78.1	78.1

4 Experiment Result and Discussion

We collected 300 Chinese review texts about digital cameras. We use LTP v1.4 (http://ir.hit.edu.cn/demo/ltp/) to preprocess these reviews. All of sentences are manually labeled as high praised, praised, objective, criticized, and high criticized. The detailed information of the data sets is shown in Table 1. The code for experiment is a modified version of the Mallet toolkit. The ideas of implementation for isotonic CRF are from [4]. The experiments are performed in ten-fold cross-validation. We respectively use 80%, 70%, 60% and 50% of the testing corpus for training, the others as test data. The iterative times are 300. In the figure 1, let iso+cas, iso, cas denote respectively the proposed method, isotonic CRF, and cascaded method. From Fig. 1 and Table 1, one can see that:

(1) Among all the 4 experiments, the total accuracy of the proposed method is superior to comparison methods. It indicates that our model is an effective algorithm for sentence sentiment analysis in Chinese texts.

(2) In the fourth experiment, 80% data used for training, the proposed method gets satisfactory result for high sentiment strength annotation. The accuracy, recall, and F-value are all more than 50%. This data provide evidences that our model has a strong discrimination ability of sentiment strength between different labels.

5 Conclusions

In this paper, we propose a cascaded CRF model based on isotonic constraints, which use multiple variant of CRF models to classify sentences sentiment and sentiment strength in different layers. The cascaded frame incorporates a particular of domain knowledge through generalized isotonic constraints on the model parameters. We conduct experiments on four different sizes of training data sets. The experiment results show that our model is an effective algorithm for sentence sentiment analysis in Chinese texts.

In the future work, we will try other methods for feature selection to improve the performance of our algorithm. Furthermore, we experiment on relatively small-scaled corpus and we will verify our model on bigger scaled data set.

Acknowledgments. Supported by the National High-tech R&D Program of China (863 Program), under Grant No. 2009AA01Z424.

References

1. Pang, B., Lee, L.: A sentimental education: sentiment analysis using subjectivity summarization based on minimum cuts. In: Proceedings of the 42nd Annual Meeting on Association for Computational Linguistics, pp. 271–278. Association for Computational Linguistics, Barcelona (2004)
2. Pang, B., Lee, L., Vaithyanathan, S.: Thumbs up? Sentiment Classification using Machine Learning Techniques. In: Proceedings of EMNLP 2002, pp. 79–86. Association for Computational Linguistics, Philadelphia (2002)
3. Liu, K., Zhao, J.: Sentence Sentiment Analysis Based on Cascaded CRFs Model. Journal of Chinese Information Processing 22, 123–128 (2008)
4. Mao, Y., Lebanon, G.: Isotonic Conditional Random Fields and Local Sentiment Flow. In: the 21st Annual Conference on Neural Information Processing Systems, pp. 961–968. MIT Press, British Columbia (2007)

5. McDonald, R., et al.: Structured Models for Fine-to-Coarse Sentiment Analysis. In: Proceedings of the 45th Annual Meeting of the Association of Computational Linguistics, pp. 432–439. Association for Computational Linguistics, Stroudsburg (2007)
6. Kim, S.-M., Hovy, E.: Automatic Identification of Pro and Con Reasons in Online Reviews. In: The Proceedings of the Conference on Computational Linguistics/Association for Computational Linguistics, pp. 483–490. Association for Computational Linguistics, Sydney (2006)
7. Lafferty, J., McCallum, A., Pereira, F.: Conditional Random Fields: Probabilistic Models for Segmenting and Labeling Sequence Data. In: The International Conference on Machine Learning, pp. 282–289. Omni press, Williamstown (2001)

A Tree Routing Protocol Based on Ant Colony Algorithm for Mine Goaf Monitoring

Zheng Sun[1], Jili Lu[1], and Hui Li[2]

[1] College of Mechanical and Electrical Engineering, Zaozhuang University, Zaozhuang 277160, China
[2] College of Mechanical and Electrical Engineering, China University of Mining and Technology, Xuzhou 221116, China
cumt_sz@163.com

Abstract. This paper proposes a novel routing protocol called ZACATREE, which will prolong the Zigbee network lifetime, reduce the end-to-end delay, and improve the reliability of mine goaf monitoring. Using ant colony algorithm to search for the next hop would effectively reduce energy consumption while tree structure is helpful to reduce the end-to-end delay and packet loss rate. The simulation results show that the ZACATREE protocol can improve the network lifetime by 27.7% compared with the AODV protocol, while the end-to-end delay can be reduced by 16.92% and 46.04% abatement in the packet loss rate is obtained. Therefore, the performance of mine goaf monitoring network can well be improved by using the ZACATREE protocol.

Keywords: routing protocol, Zigbee, ant colony algorithm, mine goaf.

1 Introduction

At present, wireless sensor networks (WSN) are used in coal mine more and more. As a kind of new monitoring technology, WSN can overcome the shortcoming of tradition system with wire, such as poor reliability and wiring complexity. Reference [1] proposes a multi-path routing protocol based on energy for safety monitoring of coal mine. Compared with AOMDV, this protocol improves the reliability and reduces the time delay. Reference [2] proposes a wireless system called SASA for monitoring the underground environment, which can discover the change of network structure from subsidence damage, and locate the inanition of subsidence damage.

The safety accidents are happened easily in coal mine goaf because of the failure of rock mass, which affects the safety in coal production. The existed monitoring system can't be applied to the region, but the technology of WSN can solve this problem. Reference [3] proposes a clustering algorithm for reducing the energy consumption of monitoring system in the coal mine goaf. This paper proposes a routing protocol called ZACATREE, which will prolong the Zigbee network lifetime, reduce the end-to-end delay, and improve the reliability of mine goaf monitoring.

2 Protocol Description

The ZACATREE is similar with AODV, where the source node sends the information of discovering the path when there is not a path to the destination node, and the

D. Jin and S. Jin (Eds.): Advances in FCCS, Vol. 1, AISC 159, pp. 25–31.
springerlink.com

destination node sends the data packet of routing setup to the source node. The ZACATREE has the follows features.

(1) The protocol uses the tree structure for routing discovery, which can achieve the path with less number of hops, avoid the shortcoming of discovery redundant path with AODV, and reduce the end-to-end delay of network. As figure 2, when the source 1 sends data to the node 2, the algorithm base on broadcasting can discover the rout 2, but the algorithm with tree structure avoid the communication between the node1 and the node 2 in the same layer, and discover the rout 1 with less number of hops. The tree structure is used in the network with mass data for the sink node, where the sink node is the root node and builds the tree [4]. When multi-source and multi-destination exist, the layer number of node makes a mistake easily, and it is difficult to discover the route. In figure 1, source 1 reaches node 3 from node 1, source 2 reaches node 3 from node 1, the level number of node 1 is 2 in source 1, and 1 in source 2. To solve this problem, the protocol contains a virtual address table besides the routing table, storing the level number of node when each node is the source node. The routing table and the virtual address table are illustrated in figure 2.

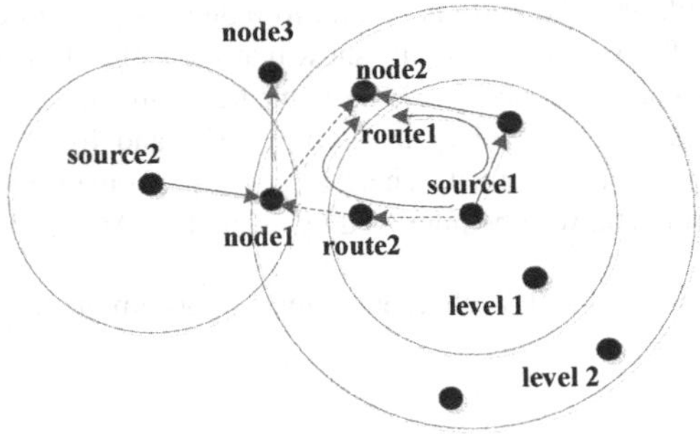

Fig. 1. Existing problems of route discovery with ZACATREE

destination	nexthop

CBR source ID	my level

Fig. 2. Items of the routing table and the virtual address table

(2) The protocol finds out the next hop with the ant colony algorithm. For balancing the energy consumption and prolonging the network lifetime, the algorithm considers the rest energy and the number of CBR data packets transmitted, calculates the probability that the node is selected as next hop with formula (1) [5], and selects the node with maximum probability as the next hop.

$$p = \frac{p_{\text{old}} + \Delta p}{1 + \Delta p} \tag{1}$$

Where $\Delta p = k\left(\frac{E_{\text{rem}}}{E_{\text{init}}}\right)^{\alpha}$, E_{rem} is the rest energy of node, E_{init} is the initial energy of node. Considering the transmitted data packet, the formula (1) is changed to the formula (2).

$$p = \frac{p_{\text{old}} + \Delta p}{1 + \Delta p}(1 - \Delta \tau) \qquad (2)$$

Where $\Delta \tau = \frac{q_{\text{f}}}{q_{\max}}$, q_{f} is the total number of CBR data packets which the node has transmitted and received, $q_{\max}$ is the maximum of the packet number in the network, $q_{\max} = \max\{q_{\text{f1}}, q_{\text{f2}}, q_{\text{f3}}, \cdots\cdots, q_{\text{fn}}\}$, n is the number of nodes.

2.1 Route Discovery

When the source node needs to transmit the monitoring data, it checks the route table whether containing the route from source node to destination node. If the route does not exist, the source node set its level number as 0, add the CBR data to the temporary storage queue of CBR data, and transmit the data packet of Findrouting. The source node ID, the destination node ID , the node ID and level are transmitted in the Findrouting packet, and the TTL is 1. The algorithm is illustrated in figure3.

1. I will send packet to node desCBR
2. if SearchRTable(desCBR) == 0
1. Lmylevel = 0
2. Add CBR Packet to CBRQueue
3. Broadcast Findrouting Packet

Fig. 3. Source node transmit Findrouting packet

When the neighbor nodes receive the packet, they check the route table whether containing the information of the node transmitting Findrouting or CBR packet. If the information does not exist, set the node transmitting Findrouting packet as destination node and next hop. Simultaneously, set the node transmitting CBR packet as the destination node, and the node transmitting Findrouting packet as the next hop. Then, they check the virtual address table whether containing the information of the nodes transmitting CBR packets. If the information does not exist, the level number increases 1 itself, and the node transmitting CBR packet and the level number are added to the virtual address table. Finally, if the node itself is the destination node of CBR, sent the Found packet to the source node of CBR, containing the node ID, destination node ID, destination node ID of CBR and the selecting probability P, as shown in figure 4.

The node receiving Found packet sets the destination node of CBR as the destination node and the node transmitting Found packet as the next hop, and the P to the routing table. When the source node of CBR receives the Found packet, according to the information of destination node of CBR, it transmits the relevant data packet in the temporary storage queue. If the node is not the destination node of Findrouting packet, it judges whether the level number greater than the node transmitting Findrouting packet. If it isn't greater, reject the data packet, otherwise broadcasted the data packet. The algorithm of setting up the route backward is shown in figure 5.

1. I receive Findrouting packet
2. if SearchRTable(des_{CBR}) == 0
1. Add des_{CBR} to routing table, next hop is $sou_{findrouting}$
3. if SearchRTable($sou_{findrouting}$) == 0
1. Add $sou_{findrouting}$ to routing table, next hop is $sou_{findrouting}$
4. if SearchVirtualTable(sou_{CBR}) == 0
1. $L_{mylevel} = L_{findrouting} + 1$
2. Add sou_{CBR} and $L_{mylevel}$ to virtual table
5. else $L_{mylevel}$ = SearchVirtualTable(sou_{CBR})->level
6. if I am the CBR destination
1. Send Found Packet to sou_{CBR}
7. else if $L_{mylevel} > L_{findrouting}$
1. Broadcast Findrouting Packet

Fig. 4. Algorithm of receiving the Findrouting packet

1. I receive Found Packe
2. if SearchRTable(des_{CBR}) == 0
1. Add des_{CBR} to routing table netxt hop is sou_{found}
3. if SearchRTable(sou_{found}) == 0
1. Add sou_{found} to routing table netxt hop is sou_{found}
4. if I am sou_{CBR}
1. Start resending timer(0.5)
5. else
1. Send Found Packet to sou_{CBR}
6. resend timer expire
1. SearchCBRQueue(des_{CBR}) and resend CBR Packet

Fig. 5. Algorithm of setting up the route backward

2.2 Information Update

When transmitting the CBR data packet, if the protocol always calculates the selecting probability P, the network lifetime would be reduced. ZACATREE updates the selecting probability P by broadcasting the update packet to neighbor nodes periodically. The coordinator node counts the number of nodes sending CBR data, selects the maximum and broadcasts it in the whole network. Each node calculates P with the formula (2), and broadcasts it to the neighbor nodes. After a node receiving update packet, it updates P of transmitting node as the next hop node.

3 Protocol Simulation

This paper utilizes the NS2 as simulation software, and the scene is the demo4 of wpan demo. Parameters of simulation scene are shown in table 1.

Table 1. Parameters of simulation scene

Parameter Name	Parameter Value
Node number	21
Scene size	50×50
Data stream	CBR
Time interval	2s
Distance between nodes	10m
Communication radius of the node	10m

3.1 Network Lifetime

In simulation, the initial energy of the node was 4J, the energy consumption was 0.5J. So a node with less 0.5J energy was regarded as a dead node. The network lifetime was defined as the time from starting to appearing the first dead node. The simulation result was shown in figure 6, where simulation time was 400s, and the number of dead nodes was counted during the simulation. The first dead node was appeared at 213 second with AODV, and at 272 second with ZACATREE. Compared with the AODV, ZACATREE can improve the network lifetime by 27.7%. ACA algorithm can efficiently balance the node energy consumption, avoiding premature death of a node.

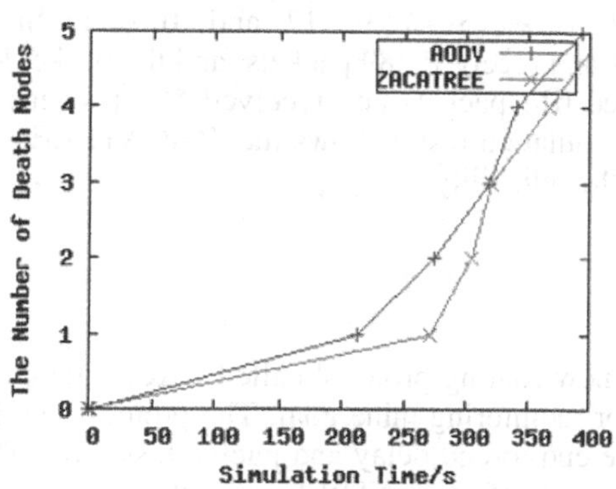

Fig. 6. Number of dead nodes during the simulation

3.2 End-to-End Delay

The time of transmitting CBR packet from source node to destination node was counted, and shown in figure 7. The simulation time was 200s, and each protocol transmitted and received 285 CBR packets successfully. By analyzing the trace file, the average delay of AODV was 0.022653s , and the ZACATREE was 0.018820s, reduced by 16.92%.

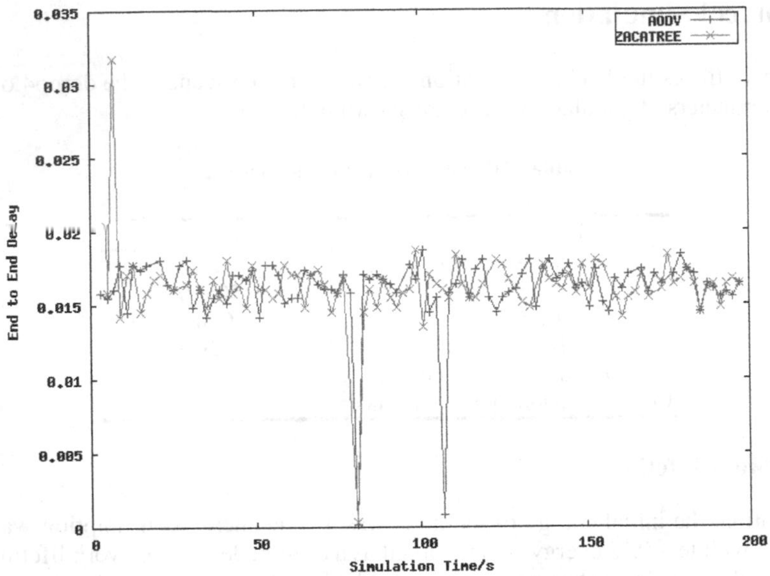

Fig. 7. End-to-end delay of CBR data packets

3.3 Packet Loss Ratio

The packet loss ratio is the rate between received packets with sent packets in the whole network, mainly reflecting the reliability of network. In simulation, we set four data streams of CBR, 0→3, 5→8, 9→17 and 10→11. In simulation, AODV transmitted 633 packets and received 589 packets, and the packet loss ratio was 6.95%. ZACATREE transmitted 614 packets and received 591 packets, and the packet loss ratio was 3.75%. The simulation result shows that ZACATREE can reduce the packet loss ratio and improve the reliability.

4 Conclusion

This paper proposes a new routing protocol called ZACATREE based on Zigbee and ant colony algorithm for monitoring mine goaf. The protocol can improve the network lifetime, and reduce the end-to-end delay and packet loss ratio. The simulation results show that the performance of ZACATREE is better than AODV in the Zigbee. Furthermore, the ZACATREE can combine with clustering structure to improve the network performance greater.

Acknowledgments. The authors would express their appreciation for the financial support of Shandong Natural Science Foundation, grant NO. ZR2011EL016. The authors also would express their thanks for Zaozhuang Science and Technology Development Project, grant NO. 201128, and Zaozhuang University Doctor Science Foundation.

References

1. Xiao, S., Wei, X.Y., Wang, Y.: A Multipath Routing Protocol for Wireless Sensor Network for Mine Security Monitoring. Mining Science and Technology 20, 148–151 (2010)
2. Li, M., Liu, Y.H.: Underground Coal Mine Monitoring with Wireless Sensor Networks. ACM Transactions on Sensor Networks 5, 10:1–10:10 (2009)
3. Wang, Q.F., Zhang, S., Yang, Y., Tang, L.: The Application of Wireless Sensor Networks in Coal Mine. In: 7th International Conference on Information, Communications and Signal Processing, pp. 1–4. IEEE Press, NJ (2009)
4. Ding, G., Sahinoglu, Z., Orlik, P., Zhang, J., Bhargava, B.: Tree-Based Data Broadcast in IEEE 802.15.4 and ZigBee Networks. IEEE Transactions on Mobile Computing 5, 1561–1574 (2006)
5. Duan, H.B.: Ant Colony Algorithms: Theory and Application. Science Press, Beijing (2005)

Coverage Optimization Algorithm of Wireless Sensor Network

Xuezheng Han[1], Shuai Li[1], and Xun Pang[2]

[1] College of Mechanical and Electrical Engineering, Zaozhuang University, Zaozhuang 277160, China
[2] Materials Department, Yanzhou Coal Mining Company Limited, Jining 273500, China
hxz6131@sina.com

Abstract. To the problem of randomly deployed nodes failing to guarantee the network coverage and the objects detection probability, this study puts forward a coverage optimization algorithm of wireless sensor network based on simplified virtual forces-oriented particle swarm with mutate mechanism (SVF-PS-MA), which would improve the update speed of particle swarm algorithm, guide the direction of particle evolution and speed up the algorithm convergence by dislodging attractive forces from the virtual forces. And this algorithm introduces the mutate mechanism, take the mutation for particles, improve the particles diversity and reduce the probability of particles falling into the local optimization. The result shows that the algorithm can improve the network coverage effectively, and guarantee less time-consumption.

Keywords: virtual force, particle swarm, wireless sensor network, coverage.

1 Introduction

Wireless sensor network has strong ability for information collection and co-processing, so it is widely used in target tracking, environmental monitoring, and so on. It can improve the network coverage and target monitoring probability, and reduce the network energy consumption that realizing the dynamic optimization of wireless sensor networks according to the application environment [1]. Therefore, dynamic optimization has become an important research field of WSN [2].

Reference [2] proposed a deployment optimization algorithm of wireless sensor network based on virtual forces. Reference [3] proposed a coverage optimization algorithm based on particle swarm optimization and voronoi diagram. Reference [4] proposed an algorithm based on particle swarm optimization and grid strategy. In reference [5] the problem of network coverage and energy efficiency was optimized by using ant colony algorithm with different pheromone. To the coverage problem of WSN, this paper proposed a coverage optimization algorithm based on simplified virtual forces-oriented particle swarm with mutate mechanism.

D. Jin and S. Jin (Eds.): Advances in FCCS, Vol. 1, AISC 159, pp. 33–39.
springerlink.com

2 Deployment Optimization Model of WSN

In the wireless sensor network, every node has the same measuring range, communicating range and initial energy. Assumed that the position of node s_i is (x_i, y_i), the position of monitoring point is (x, y), and the distance between s_i and the monitoring point is $\sqrt{(x_i - x)^2 + (y_i - y)^2}$.

In practical applications, considering the interference of environment and noise, the measurement probability model between a node and the target point is,

$$c_{xy}(s_i) = \begin{cases} 0 & r + r_e \le d(s_i, P) \\ e^{(-\alpha_1 \lambda_1^{\beta_1} / \lambda_2^{\beta_2} + \alpha_2)} & r - r_e \le d(s_i, P) \le r + r_e \\ 1 & d(s_i, P) \le r - r_e \end{cases}$$

Where $r_e\ (0 < r_e < r)$ is the parameter of measurement reliability. $\alpha_1, \alpha_2, \beta_1, \beta_2$ are the measuring parameters about the node features, λ_1, λ_2 are the input parameters. The joint measurement probability of the multi-nodes is,

$$c_{x,y}(S_{ov}) = 1 - \prod_{s_i \in S_{ov}} \left(1 - c_{x,y}(s_i)\right)$$

Where S_{ov} is the set of measuring nodes. Set c_{th} as the measurement probability threshold, and the condition of measuring the target effectively is,

$$\min_{x,y} \{ c_{x,y}(s_i, s_j) \} \le c_{th}$$

In this paper, we meshed the network as the grid with certain size, computed the measurement probability between a grid center and the nodes around. If the probability was bigger than the threshold, the grid would be regarded as full coverage.

3 Optimization Algorithm of WSN Based on Simplified Virtual Forces-Oriented Particle Swarm with Mutate Mechanism

3.1 Simplified Virtual Forces Algorithm

Virtual forces algorithm assumes that all various types of objects including wireless node, obstacle and hot-region can exert attraction or repulsion on the nodes of WSN. In this paper, the nodes were deployed uniformly with the virtual repulsion forces. Supposed that the total virtual repulsion force of node s_i is $\vec{F}_i$, the total virtual repulsion force from other nodes is $\vec{F}_{i-nodes}$, the virtual repulsion forces from hot-region and obstacle are $\vec{F}_{i-A}$ and $\vec{F}_{i-R}$. The total repulsion force of s_i is,

$$\vec{F}_i = \vec{F}_{i-nodes} + \vec{F}_{i-A} + \vec{F}_{i-R}$$
$$= \sum_{j=1, j\neq i}^{k} \vec{F}_{ij} + \vec{F}_{i-A} + \vec{F}_{i-R}$$

Where $\vec{F}_{ij}$ is the repulsion force between the nodes, using the distance threshold d like the virtual forces algorithm. When the distance between nodes is bigger than the threshold, the force will be ignored. The relation of virtual repulsion forces and the distance threshold is,

$$\vec{F}_{ij} = \begin{cases} 0 & d_{ij} \geq d_{th} \\ \left(w\left(\frac{1}{d_{ij}} - \frac{1}{d_{th}}\right), \alpha_{ij} + \pi \right) & d_{ij} < d_{th} \end{cases}$$

Where α_{ij} is the azimuth angle from s_i to s_j, w is the coefficient of virtual repulsion force for adjusting the level of the force with the distance. In this paper, the network is on the ideal region without hot-region and obstacle.

The virtual forces can be divided into two-components of x-axis and y-axis as follows,

$$\Delta x = \frac{F_x}{F_{xy}} \times MaxStep \times e^{\frac{-1}{F_{xy}}}$$

$$\Delta y = \frac{F_y}{F_{xy}} \times MaxStep \times e^{\frac{-1}{F_{xy}}}$$

Where F_{xy}, F_x, F_y are the total virtual force, x-axis virtual force and y-axis virtual force. $MaxStep$ is the maximum displacement distance.

3.2 Simplified Virtual Forces-Oriented Particle Swarm Optimization

Supposed that the initial position, speed and the best position of the node s_i is (x_i, y_i), (v_{xi}, v_{yi}) and P_i, the best history position of a node is P_g. The update speed and position of node is,

$$v_{ix}(t+1) = w(t) \times v_{ix}(t) + c_1 r_{1x}(t)\left(P_{ix}(t) - x_i(t)\right) + c_2 r_{2x}(t)\left(P_g(t) - x_i(t)\right)$$
$$v_{iy}(t+1) = w(t) \times v_{iy}(t) + c_1 r_{1y}(t)\left(P_{iy}(t) - y_i(t)\right) + c_2 r_{2y}(t)\left(P_g(t) - y_i(t)\right)$$
$$x_i(t+1) = x_i(t) + v_{ix}(t+1)$$
$$y_i(t+1) = y_i(t) + v_{iy}(t+1)$$

Where c_1 and c_2 are the accelerating coefficient for adjusting step length of local optimum and global optimum.

For speeding up the convergence of PSO, and guiding the global evolution of the particles, add the virtual forces to the PSO and the update speed is,

$$v_{ix}(t+1)=w(t)\times v_{ix}(t)+c_1r_{1x}(t)\left(P_{ix}(t)-x_i(t)\right)+c_2r_{2x}(t)\left(P_g(t)-x_i(t)\right)+c_3r_{3x}(t)\,g_{ix}(t)$$

$$v_{iy}(t+1)=w(t)\times v_{iy}(t)+c_1r_{1y}(t)\left(P_{iy}(t)-y_i(t)\right)+c_2r_{2y}(t)\left(P_g(t)-y_i(t)\right)+c_3r_{3y}(t)\,g_{iy}(t)$$

Where c_3 is the influence factor between virtual repulsion forces and node searching speed, r_{3x},r_{3y} are the random number from 0 to 1, $g_{ij}(t)$ is the displacement distance of node s_i by the virtual forces.

$$g_{ix}=\frac{F_x^i}{F_{xy}^i}\times MaxStep\times e^{\frac{-1}{F_{xy}^i}}$$

$$g_{iy}=\frac{F_y^i}{F_{xy}^i}\times MaxStep\times e^{\frac{-1}{F_{xy}^i}}$$

3.3 Algorithm Description

The global optimization capability of PSO is strong, but the initial process is very important for final optimization results, so if the initial process is bad, the particle swarm falls into the premature convergence problem easily. Therefore, this paper introduced a self-adaptive mutation to the optimization. If the network coverage is better with the mutate particle, the mutate particle would instead of the primary particle. Because the premature convergence problem happened in the later stage generally, the mutate probability need to change with the evaluation generation as follows,

$$P_m=0.01+\frac{t}{MaxG}\times 0.5$$

Where $MaxG$ is the maximum evaluation generation, t is the current generation, P_m is the mutate probability of the particle. In the algorithm, each particle generates a random number, and then if the random number is smaller than P_m, the mutate operation would happen. The mutate operator is,

$$Mut=1+0.5\times\eta$$

Where η is the random number from 0 to 1 obeying the standard normal distribution.

The algorithm process is shown as follows,

Step 1 Initialize the wireless sensor network, set the population size as m, and generate the initial position and speed of each particle randomly.

Step 2 Compute the fitness value of particle with the statistical method based on the grid.

Step 3 Update the particle position and the best global position, and optimize all particles with simplified virtual forces-oriented particle swarm.

Step 4 According to the fitness of the new position, update the position and speed.

Step 5 Judge whether mutation is need, and if yes, mutate the particle and update the particle based on the mutation position.

Step 6 Judge whether meeting coverage requirement, if no, back to Step2, otherwise, the algorithm is over, and returns the position of nodes and the network coverage.

4 Simulation and Analysis

Supposed that there are 30 wireless sensor nodes random deployed in the square region with 20m length, measuring radius r=2.5m, and communicating radius R=3r=7.5m. For improving the authenticity, the parameters of probability measuring model $\alpha_1 = 1, \alpha_2 = 0, \beta_1 = 1, \beta_2 = 0.5, c_{th} = 0.9$. The parameters of simplified virtual forces $w = 20$, $MaxStep = 0.5$, $r = 1.25$, impact factors for accelerating c_1, c_2, c_3 are set as 1, the maximum evolution generation $MaxG = 200$. The simulation platform is Matlab 2008 in the PC of dual-core with 1.8GHz frequency. Optimize the deployment of wireless sensor network with VFA, PSO, VF-PSO and SVF-PS-AM.

The results is shown that the initial coverage is 70% as fig. 1, the PSO falls into the local optimal easily as fig. 2, part of nodes gather in the lower right corner, and the coverage is 92%, not improved efficiently. In fig. 3, the VFA use the local nodes information as distance threshold, which impacts the network deployment, so the network coverage is only 86%. In fig. 4, VF-PSO has favorable global property and directivity, and the network coverage is up to 95%. In fig. 5, the SVF-PS-MA has better capability of global search, and the coverage is up to 98% after the optimization. In fig. 6, the absolute convergence rate of VFA is achieved fastest, where the network get the optimization coverage 86% in 35 generation. But considered the relative convergence rate, the coverage of the proposed algorithm is up to 93%, increased by 7% than VFA.

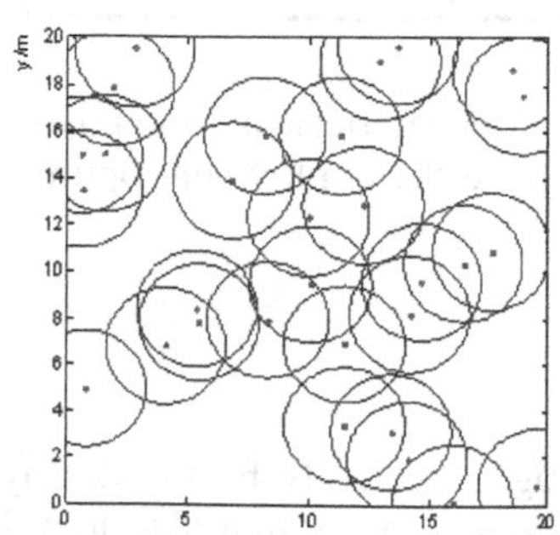

Fig. 1. Initial random deployment of WSN

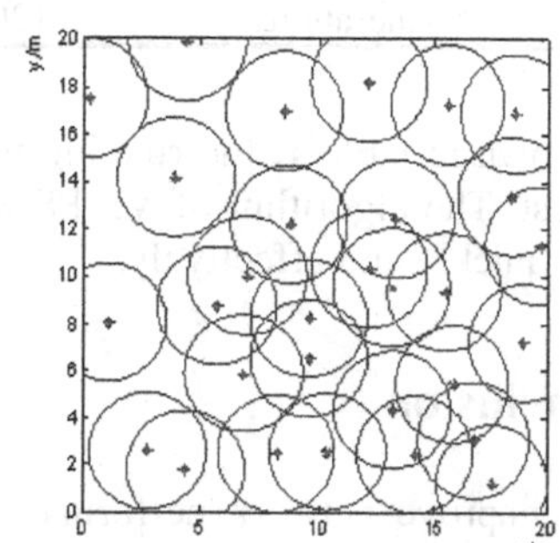

Fig. 2. Deployment of WSN using VFA

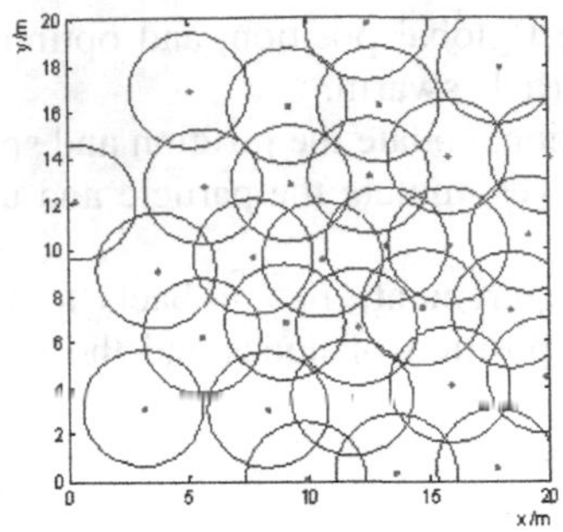

Fig. 3. Deployment of WSN using PSO

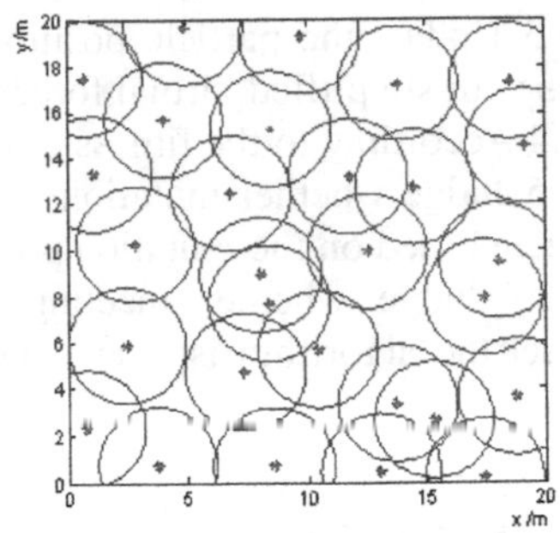

Fig. 4. Deployment of WSN using VF-PSO

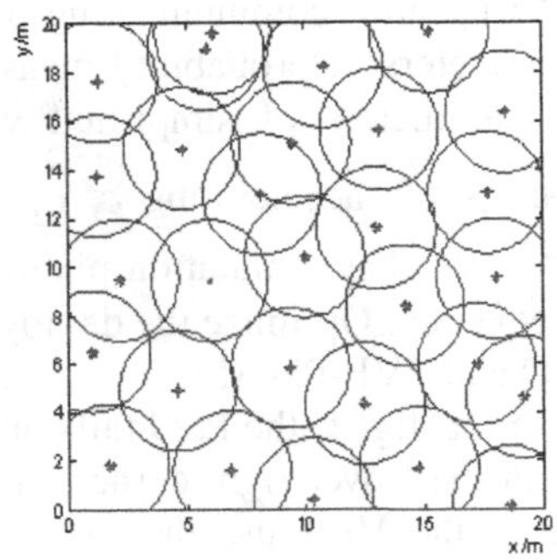

Fig. 5. Deployment of WSN using SVF-PS-MA

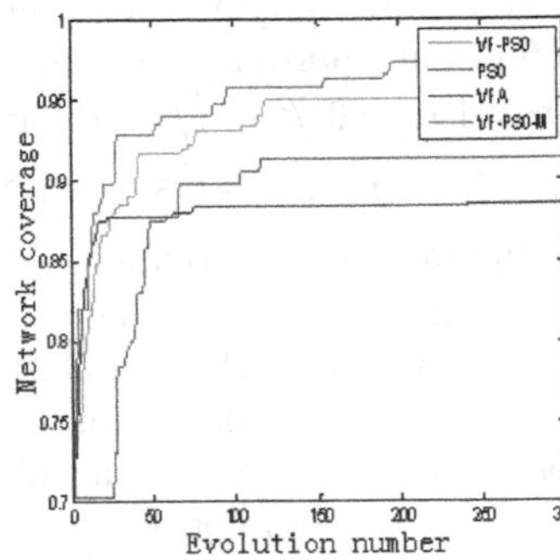

Fig. 6. Performance comparison of the optimization algorithms

Table 1. Average performance of optimization algorithms after running fifty times

	PSO	VFA	VF-PSO	VF-PS-MA
Coverage (%)	92.31	86.23	95.04	98.52
Time consumption (/s)	18.77	12.23	16.98	15.86
Generations	120	243	127	190

As shown in table 1, the running time of VFA is the shortest, but the coverage is the lowest. The algorithm of VF-PS-MA can complete the deployment optimization of WSN quickly and effectively.

5 Conclusion

It could improve network performance and coverage effectively by the deployment optimization of WSN. This paper proposed a coverage optimization algorithm based on simplified virtual forces-oriented particle swarm with mutate mechanism. This algorithm used the probability measurement model, maximized network coverage as the target of evolution fitness, introduced the simplified virtual forces as a component of particle evolution to the normal PSO, and guided the particle to move to better direct with a weight. Simultaneously, this algorithm used mutation operation for preventing the particle falling into local optimization position. The simulation result shows that the algorithm of SVF-PS-MA can improve the network coverage effectively, and guarantee less time-consumption.

Acknowledgments. The authors would express their appreciation for the financial support of Shandong Natural Science Foundation, grant NO. ZR2011EL016. The authors also would express their thanks for Zaozhuang Science and Technology Development Project, grant NO. 201128.

References

1. Wang, X., Jiang, A., Wang, S.: Mobile Agent Based Wireless Sensor Network for Intelligent Maintenance. In: Huang, D.-S., Zhang, X.-P., Huang, G.-B. (eds.) ICIC 2005. LNCS, vol. 3645, pp. 316–325. Springer, Heidelberg (2005)
2. Zou, Y., Chakrabarty, K.: Sensor Deployment and Target Localization based on Virtual Forces. In: 32nd Annual Joint Conference of the IEEE Computer and Communications, pp. 1293–1303. IEEE Press (2003)
3. Aziz, N.A.B.A., Mohemmed, A.W., Alias, M.Y.: A Wireless Sensor Network Coverage Optimization Algorithm based on Particle Swarm Optimization and Voronoi Diagram. In: International Conference on Networking, Sensing and Control, pp. 602–607. IEEE Press (2009)
4. Wan Ismail, W.Z., Manaf, S.A.: Study on Coverage in Wireless Sensor Network using Grid based Strategy and Particle Swarm Optimization. In: IEEE Asia Pacific Conference on Circuits and Systems, pp. 1175–1178. IEEE Press (2010)
5. Lee, J.W., Choi, B.S., Lee, J.J.: Energy-Efficient Coverage of Wireless Sensor Networks using Ant Colony Optimization with Three Types of Pheromones. IEEE Transactions on Industrial Informatics 7, 419–427 (2011)

Adaptive Frame Selection for Multi-frame Super Resolution

Cuihong Xue, Ming Yu, Chao Jia, Shuo Shi, and Yandong Zhai

School of Computer Science and Engineering,
Hebei University of Technology, Tianjin, 300401
xuecuihong@scse.hebut.edu.cn

Abstract. Image super-resolution (SR) is a process to reconstruct a high-resolution (HR) image by fusing multiple low-resolution (LR) images. A critical step in image SR is accurate registration of the LR images, however the larger inter-frame motion can significantly affect the sub-pixel image registration, then it can also affect the output of HR reconstruction. So a novel Adaptive Frame Selection method is proposed in this paper for the reconstruction of multi-frame SR. It devises a framework to resolve the image SR reconstruction problem into two steps. Firstly, using the Optical flow algorithm to calculate the inter-frame motion estimation, designing an adaptive frame selection method to discard some of the larger inter-frame motion frames, then the less inter-frame motion of successive frames is obtained. Secondly, using the maximum a posteriori (MAP) based SR algorithm for the SR reconstruction. The experimental results indicate that the proposed algorithm has considerable effectiveness in terms of both objective measurements and visual evaluation.

Keywords: Super-resolution, MAP, Adaptive frame selection, Optical flow, motion estimation.

1 Introduction

The main aim of multi-frame SR reconstruction is to reconstruct one or a set of HR images from a sequence of LR images of the same scene, by using image processing techniques. HR images are useful in many applications such as video surveillance, medical diagnostics, satellite imaging and military information gathering etc. However, because of the high cost and physical limitations of the high precision optics and image sensors, it is not easy to obtain the desired HR images in many cases. So, SR image reconstruction techniques have been widely researched in the last two decades.

The multi-frame SR reconstruction usually contains three main steps: registration[1-3] (motion estimation), interpolation, and reconstruction. It was originally proposed by Tsai and Huang[4]. They recovered a HR unaliased image from a sequence of LR images. Kim et al.[5] proposed a recursive least-square solution by extending Tsai and Huang's approach to include motion blurring and additive noise. Kim et al.'s approach is further refined by Kim and Su[6] to consider different blurrings for different LR multi-frame.

D. Jin and S. Jin (Eds.): Advances in FCCS, Vol. 1, AISC 159, pp. 41–46.
springerlink.com

Narayanan et al.[7] proposed a computationally efficient video SR. Farsiu et al.[8] propose an fast and robust multi-frame super resolution. Liangpei Zhang et al.[9] propose an edge-reserving maximum a posteriori estimation based SR algorithm. Zhang zhi et al.[10] proposed frame selection in multi-frame image SR reconstruction. Raghavender R et al.[11] proposed adaptive frame selection for improved face recognition in LR videos.

This paper focus on the above problem, it discard the larger inter-frame motion frame by adaptive frame selection method, and then using the MAP method for SR reconstruction. This method can dismiss the registration error, improves flexibility, and also demonstrates both effectiveness and robustness in visual and PSNR effect.

2 Image Observation Model

The image observation model is employed to relate the desired referenced HR image to all the observed LR images. Typically, the imaging process involves warping, followed by blurring and down-sampling to generate LR images from the HR image. Let the underlying HR image be denoted in the vector form by $X = [x_1, x_2, x_3, ... x_N]^T$, where $N = L_1 N_1 \times L_2 N_2$ is the HR image size. Letting L_1 and L_2 denote the down-sampling factors in the horizontal and vertical directions, respectively, each observed LR image has the size $N_1 \times N_2$. Thus, the LR image set can be represented as $F = \{f_1, f_2, f_3, ... f_k\}$, $k \in \{1,2,...n\}$, $k \leq n$, the LR image can be represented as $f_k = [f_{k,1}, f_{k,2}, ..., f_{k,n_1 \times n_2}]^T$, where $k = 1,2,...,n$, with n being the number of LR images. Assuming that each observed image is contaminated by additive noise, the observation model can be represented as:

$$f_k = DB_k M_k X + \eta_k \qquad k = 1,2,...,n . \tag{1}$$

Where $M_k \in R^{L_1 N_1 L_2 N_2 \times L_1 N_1 L_2 N_2}$ is the warp matrix, $B_k \in R^{L_1 N_1 L_2 N_2 \times L_1 N_1 L_2 N_2}$ represents the camera blur matrix, $D \in R^{N_1 N_2 \times L_1 N_1 L_2 N_2}$ is a down-sampling matrix, and $\eta_k \in R^{N_1 N_2 \times 1}$ represents the noise vector.

3 Adaptive Frame Selection

Motion estimation/registration plays a critical role in SR reconstruction, where it is necessary to select a frame from the sequence as the reference one. Gradient-based optical flow methods such as Lucas-Kanada's[12] achieve high accuracy for scenes with small displacements(<1~2 pixels/frame) but fail when the displacements are large, it attractive when applied to high frame rate sequences for the following reasons.

The assumption of brightness constancy, which states that the rate of change in intensity I along the motion trajectory is zero, so the optical flow constrain formula is as follow:

$$f_x u + f_y v + f_t = 0 . \tag{2}$$

In formula (2) $f_x = \frac{\partial f}{\partial x}$, $f_y = \frac{\partial f}{\partial y}$, $f_t = \frac{\partial f}{\partial t}$ stands for the gradient of one pixel point along motion trajectory x, y and t respectively, the motion field vector $u = \frac{\partial x}{\partial t}$ and $v = \frac{\partial y}{\partial t}$ is called optical flow of point x and y respectively.

It is becomes more valid as frame rate increases, while when the frame rate becomes the standard rate, it effect the motion estimation/registration significantly, so this paper proposed the adaptive frame selection (AFS) algorithm to discard the larger inter motion frame, it over come the registration errors caused by inter-frame motion in order to improve the performance of the SR reconstruction. The main features of this algorithm are the quantification of inter-frame motion and selection of frames for SR.

To quantify the motion between two consecutive frames based on the optical flow field, this paper define the Inter-Frame Motion Parameter(IFMP) β . The two consecutive frame f_k and f_{k+1}, the size is $M \times N$ pixels, the optical flow matrices $X_{k,k+1}$ and $Y_{k,k+1}$, which contain the pixel intensity displacements between the f_k and f_{k+1} in X and Y direction respectively , which is represented as u and v :

$$\begin{bmatrix} X_{k,k+1} \\ Y_{k,k+1} \end{bmatrix} = \begin{bmatrix} u \\ v \end{bmatrix} = \begin{bmatrix} \sum f_x f_x & \sum f_x f_y \\ \sum f_x f_y & \sum f_y f_y \end{bmatrix}^{-1} \begin{bmatrix} \sum f_x f_t \\ \sum f_y f_t \end{bmatrix} . \tag{3}$$

We than uses the Mean Absolute Deviation(MAD) to considering the IFMP. This can be denoted as :

$$\beta(f_k) = MAD(u,v) = \frac{1}{M*N} \sum_{i=0}^{M-1} \sum_{j=0}^{N-1} \left| f_k(i,j) - f_{k+1}(i+u_i, j+v_j) \right| . \tag{4}$$

Where u_i and v_j denote the pixel intensity displacement between the frame f_k and f_{k+1} at the pixel location (i, j) along the X an Y directions respectively.

The MAD algorithm only have the plus and absolute value calculation, have less computational time. After obtaining the inter-frame motion values $\beta(f_k)$ for individual frames in the video, a threshold Γ is used for the adaptive selection of frame. Selection the value for Γ is an important step, as it helps detect frames possessing large inter-frame motion values. The value of Γ in our experiments was decided for 4 empirically. Successive frames whose β values is less than Γ is registered through optical flow and used for SR reconstruction, whose β values is larger than Γ indicate the successive frame have larger inter-frame motion, can larger the motion/register error and degrade the SR processing, so we discard the frame. The AFS is as follows:

1) Input: Given a low resolution video Set F containing n frames $F = \{f_1, f_2, f_3, \ldots f_n\}$, a threshold Γ, null Set $\overline{F}$

2) When $\beta(f_i) \leq T, i \in (1,2,...,k)$, the first frame as the reference frame, calculate the motion estimation of $\overline{f_i}$ through the optical flow, $\overline{F} = \overline{F} \cup \{\overline{f_i}\}$; when $\beta(f_i) \geq T, i \in (1,2,...,k)$, we discard the frame.
3) Output: the less inter-frame motion registration image $\overline{F} = \{\overline{f_1}, \overline{f_2}, \overline{f_3}, ...\overline{f_l}\}, l \leq k$.

4 Experiment Results

In this section, we demonstrate the efficacy of the proposed method to multi-frame LR observation, and present the SR reconstruction results of using the AFS algorithm and MAP Bayes' rule reconstruction. For motion estimation, the Gradient-based optical flow methods was used to select the larger inter-frame motion frame to discard it, and then use the MAP SR image reconstruction algorithm for the left less inter-frame motion frame sequences. All the tests were run on a personal computer with Pentium D CPU 2.10GHz PC. The following peak signal-to-noise ratio (PSNR) was employed as a quantitative measure and defined as follows:

$$PSNR = 10\log_{10} \frac{255*255(L_1N_1 * L_2N_2)}{\sum_{m=0}^{M-1}\sum_{n=0}^{N-1}(f(m,n) - f(m_2,n_2))^2} \quad . \tag{5}$$

where $L_1N_1 * L_2N_2$ is the total number of pixels in the HR image, $f(m,n)$ and $f(m_2,n_2)$ represent the reconstructed HR image and the original image, respectively.We select the 512x512 pixel standard lena image as the test image, use the proposed model(1) to generate LR image sequences simulation. In the experiments, using the Gaussian blur model to generate the blur image, down-sampling factor set to 4,output image size set to 100x100 pixel, output frame Set set to 80. Through the AFS algorithm to select the less inter-frame motion frame as a reference frame, the frame set is 4,7,16,20 - 28,35,40-43,55-60,65,69-75 frame, sum to 31 frame. From the new frame Set of LR frame sequences we select the 4th frame as the reference frame, and register the 7th frame through the optical flow method, then use the MAP SR reconstruction algorithm for reconstruction of the 7th high-resolution image frame, and select the larger inter-frame motion frame of 8th to reconstruct for comparison.

From figure 1 c) we can see that in the 8th images the reconstruction effect is too bed, for it has a big movement from the reference frame, so using the optical flow algorithm for motion estimation and registration will cause larger registration error, and then have a serious distortion of SR reconstruction algorithm, figure d) used the proposed method, Figure b) used the bilinear interpolation algorithm, while Figure d) can keep much detail information than Figure b) from the above experiment results, that's because this paper use the AFS algorithm, so it makes the optical flow image registration error decreases, and then have a better visual effect than other method. This paper proposed the method also has a higher peak signal to noise ratio, for figure b), it's PSNR is 28.7 DB; for figure c), it's PSNR is 25.32 DB; while for figure d), it's PSNR is 30.12 DB, through the analysis we can see that the proposed algorithm has a higher peak signal to noise ratio.

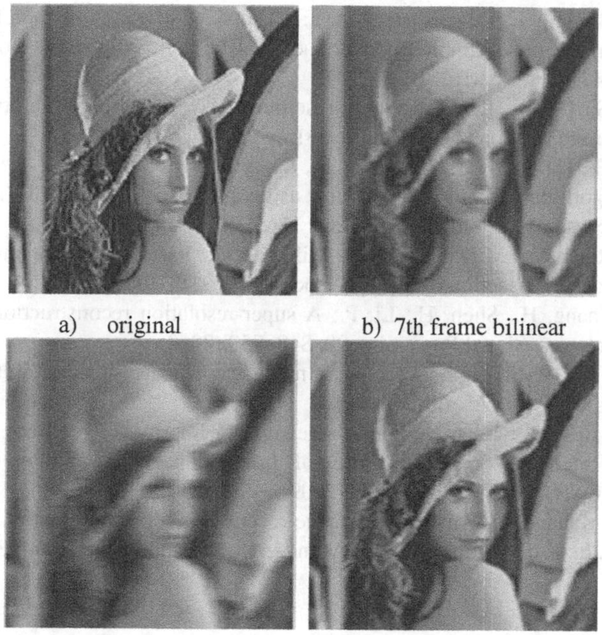

a) original b) 7th frame bilinear

c) 8th frame SRR d) 7th frame SRR

Fig. 1. SR reconstruction results

5 Conclusion

In this paper we present a novel adaptive frame selection algorithm for multi-frame super resolution. The proposed method based on the optical flow algorithm to adaptive select the less inter-frame motion frame sequences, and then joint the MAP for the SR frame sequences. Comparing to existing approaches, this method perform better visual effect, and higher peak signal to noise ratio.

References

1. Lucas, B.D., Kanade, T.: An iterative image registration technique with an application to stereo vision. In: International Joint Conference on Artificial Intelligence, pp. 674–679 (1981)
2. Irani, M., Peleg, S.: Improving resolution by image registration. CHIP: Graphical Models and Image Processing 53(3), 231–239 (1991)
3. Hardie, R.C., Barnard, K.J., Bognar, J.G., Armstrong, E.E., Watson, E.A.: High-resolution image reconstruction from a sequence of rotated and translated frames and its application to an infrared imaging system. Optical Engineering 37(1), 247–260 (1998)
4. Tsai, R.Y., Huang, T.S.: Multiframe image restoration and registration. In: Huang, T.S. (ed.) Advances in Computer Vision and Image Processing, vol. 1, pp. 317–339. JAI Press, Greenwich (1984)

5. Kim, S.P., Bose, N.K., Valenzuela, H.M.: Recursive reconstruction of high resolution image from noisy undersampled multiframes. IEEE Transactions on Acoustics, Speech, Signal Processing 38(6), 1013–1027 (1990)
6. Kim, S.P., Su, W.Y.: Recursive high-resolution reconstruction of blurred multiframe images. IEEE Transactions on Image Processing 2(4), 534–539 (1993)
7. Narayanan, B., Hardie, R.C., Barner, K.E., Shao, M.: A computationally efficient super-resolution algorithm for video processing using partition filters. IEEE Transactions on Circuits and Systems for Video Technology 17(5), 621–634 (2007)
8. Farsiu, S., Robinson, M.D., Elad, M., Milanfar, P.: Fast and robust multiframe super resolution. IEEE Transactions on Image Processing 13, 1327–1344 (2004)
9. Zhang, L., Zhang, H., Shen, H., Li, P.: A super-resolution reconstruction algorithm for surveillance images. Signal Processing 90, 848–859 (2010)
10. Zhang, Z., Wang, R.-S.: Frame selection in multi-frame image super-resolution restoration. Signal Processing 25, 1775–1780 (2009)
11. Jillela, R.R., Ross, A.: Adaptive Frame Selection for Improved Face Recognition in Low-Resolution Videos. Appeared in Proc. of International Joint Conference on Neural Networks (IJCNN), Atlanta, USA (June 2009)
12. Lucas, B.D., Kanade, T.: An iterative image registration technique with an applicatioin to stereo vision. In: Proceedings of DARPA Image Understanding, pp. 121–130 (1981)

The Development of Web Site Construction in E-commerce Based on MVC Structs

CaiQian Zhang* and Lei Ge

Software College of Kaifeng University, Kaifeng, Henan, 475004, China
zhangcaiqian2012@sina.com

Abstract. The paper first describes the development of relevant technical and business framework for the site of knowledge, then develop a retail Web site from the start, in the function of the system and databases to analyze and design based on the Struts MVC framework in order to concrete realization of the system, the final completion of an online shopping site. The site can browse products, inquiries, purchase features, and comments to the site customers can view information such as shopping help to achieve the basic requirements of the general shopping site. The experimental results indicate that this method has great promise.

Keywords: e-commerce, web site, MVC structs.

1 Introduction

E-commerce activities on the network, to expand the company's corporate influence, increase business, improve customer satisfaction, has become the focus of many companies to consider. Which are based on having a well-functioning shopping site on how to most effectively way to choose the best tools technology, rapid development, deployment, application shopping site as the focus of the current issue. By using this semantic web technique, the information necessary for resource discovery could be specified as computer-interpretable.

At present, the domestic sales of most goods business and not have their own online sales site, this is difficult to adapt to the future development of information technology. Although there are some small shopping site, but the general lack of development of these sites a good development model and framework to support, develop greater complexity, development time is longer, while the latter part of difficult to maintain. With a number of java related technologies as platform-independent, portable, open source, etc., has been widely recognized by the development community, and gradually formed a good model and framework to make development relatively simple and efficient. Struts framework is one such technology that allows the development of small and medium sized shopping sites only focus on business logic, and rarely involved site architecture, website development to reduce the complexity of [1]. The use of sophisticated models and frameworks, you can quickly build a well-structured, post-maintainable online sales system, in order to achieve convenient and fast online shopping.

---

* Zhang CaiQian (1979.3), male, Software College of Kaifeng University.

D. Jin and S. Jin (Eds.): Advances in FCCS, Vol. 1, AISC 159, pp. 47–51.
springerlink.com

Shopping site is a typical Web application, it is a distributed application. From the technical point of view, Web development techniques can be divided into client-side technology and services into two categories. Web client's main task is to show the information content, and show the HTML language is an effective carrier of information. CSS (Cascading Style Sheets) and DHTML (Dynamic HTML) technology is so rich Web pages more dynamic show. Dynamically generated HTML pages the first technology is the CGI technology, it is now mainstream server-side technology PHP, ASP, JSP have been produced and widely used.

Shopping site for current construction requirements, this article from a practical point of view, and the development of online shopping sites to explore. Paper first describes the development of relevant technical and business framework for the site of knowledge, then develop a retail Web site from the start, in the function of the system and databases to analyze and design based on the Struts MVC framework based on theoretical knowledge and develop technologies to applied to the concrete realization of the system, the final completion of an online shopping site. The site can browse products, inquiries, purchase features, and comments to the site customers can view information such as shopping help to achieve the basic requirements of the general shopping site.

2 J2EE Platform and the Struts Framework, MVC Architecture

SUN's J2EE as the development platform has been to promote a number of information technology vendors. MVC architecture is the platform of a development model, as an important separation technology, so that the coupling is greatly reduced, while improving the reusability of the system. Struts are an implementation of the MVC architecture is a popular Web development framework, has been widely used. This chapter introduces the J2EE platform, MVC architecture, Struts framework related content.

J2EE (Java 2 Platform Enterprise Edition) is Sun's proposed development of a distributed enterprise application technology architecture, which uses Java 2 Platform, Enterprise Solutions to simplify the development, deployment and management.

Its technology systems include: EJB, JSP, Servlet, JNDI, JMS, RMI and so on. The middle layer of J2EE integration framework allows for the case of existing systems into new components and applications makes it possible to retain the original premise of enterprise information assets, the expansion of system functionality [2]. At the same time, it is entirely based on Java 2 platform, so with the Java technology platform-independent, be able to "write once, run anywhere".

MVC (Model-View-Controller) model - view - controller is a classical architecture, which the application is divided into model, view, controller of three parts. A *generation* relation between a domain and its corresponding category means the “is-kind-of” relationship. The formula 1 which calculates the similarity of data acquisition is as follows.

$$\sum_{j=1}^{n+1} f_{i,j}T_{i,j} + \sum_{j=1}^{n} f_{j,i}Rx_i \le E_i \tag{1}$$

Model model is a major part of the application, it is generally said that business data and business logic to provide data for the view layer. Multiple views of a model can be used to improve the reusability of the system. View to interact with the user interface. Can be used to display information and accept user input. It generally does not carry out business processes, but can handle the resulting changes in view of the model changes to synchronized view.

The Controller to accept the user's request then calls the appropriate modules and views to complete the request. J2EE project in general to serve as this task by the Servlet. Based on the MVC architecture of J2EE application model is shown in Figure 1.

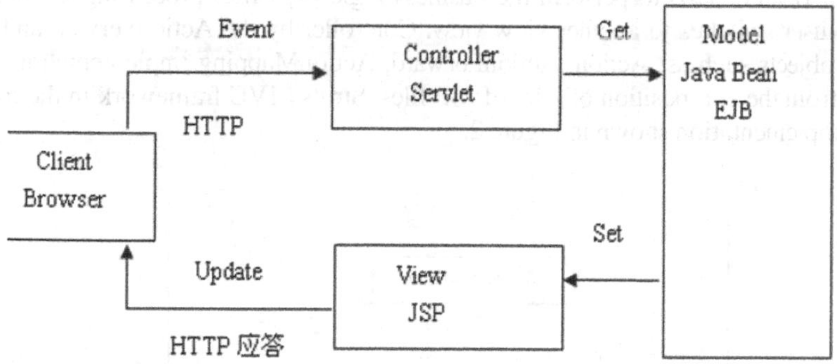

Fig. 1. Web Mining Ontology System Structure in E-Commerce based on FCA.

In Figure 1 the client browser makes an HTTP request, the Servlet as the controller receives the request, and call the appropriate model to deal with this request, the model completely, generate response pages, and then sent to the client browser device, thus completing a request-response process.

3 System Requirements and Design Description

Jakarta-Struts of the Apache Software Foundation provide an open-source project that provides a model for Web applications - View - Controller (MVC) architecture [3]. Analysis of the general system requirements including functional and performance requirements analysis requirements analysis, system discussed in this section two of this Web site needs analysis.

Implement the view of the JSP file does the business logic, and no model information, only the labels that can be standard or custom JSP tag labels (such as Struts comes with the label). Usually also put in the Struts framework is divided into the ActionForm Bean is also the view module. ActionForm Bean is actually a Java Bean, Java Bean addition to the conventional methods, it also contains some special way: as Vlidate () method (used for verification of the submitted value), reset () method (used for Form in the value of re-assignment). Struts framework with the ActionForm Bean is passed between the view and controller form data. Submit the form view, the controller automatically synchronized to update ActionForm Bean. The formula 2 which calculates the similarity of Agents is as follows.

$$l_i = \frac{R_i(0)}{E_{Tx}^i(T) + n_i(T) \times E_{Rx}} \tag{2}$$

Model represents the state of the application and business logic. Action by the ActionServlet class and controller classes. In a Struts framework using applications, ActionServlet object at run time to only one, and is automatically provided by the Struts framework. ActionServlet in the Struts framework controller to play a central role, it receives an HTTP request message, then the configuration files Struts-config.xml configuration information, the request is forwarded to the appropriate Action object. Action after receipt of the request object, call it's execute () method. This method is by using Java Bean or EJB to perform the business logic [4]. After processing the business logic, the user switches to another view view. Controller by the ActionServlet, and other auxiliary objects such as: Action, ActionForward, ActionMapping implementation, while the view from the composition of a set of JSP files. Struts MVC framework in the form of specific implementation shown in Figure 2.

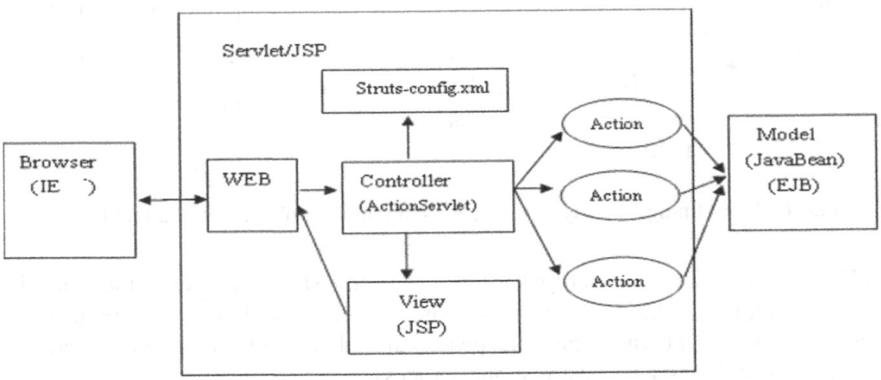

Fig. 2. Struts implementation of MVC framework map.

4 The Application of MVC-Based Struts Framework in E-commerce Web Site Construction

This paper first describes the site in the Struts framework to achieve the general idea, and then according to the MVC pattern is described in detail in the Struts framework model layer, View and Controller layers of concrete realization. Be developed using the Struts framework is introduced in the project development when the Struts framework model, and then follow the specifications of this model to achieve the system. The system implementation is to first create a Web project under Myeclipse myshop, then automatically generate the corresponding Myeclipse Web project files and libraries needed. After the wizard and then follow the introduction of Myeclipse Struts framework Struts 1.1 (version), but works automatically in the framework of the package files and components needed. JSP page is the most tedious part of the development, under the Myeclipse development tools can be completed on the part of the control of the drag, but most of the information displayed in the form of writing code to use. JSP pages on this site is mainly HTML tags, JSP tags, custom labels (mainly Struts tag), CSS and JavaScript. The system is primarily used HTML tags, Bean label, and Tiles label. When the page is to introduce a java Bean, you can use the JSP page tags import attribute, you can use jsp tag usebean properties. Fig.3 shows the struts-config.xml diagramthe in application of construction e-commerce web site based on MVC struts framework. The experimental results indicate that this method has great promise.

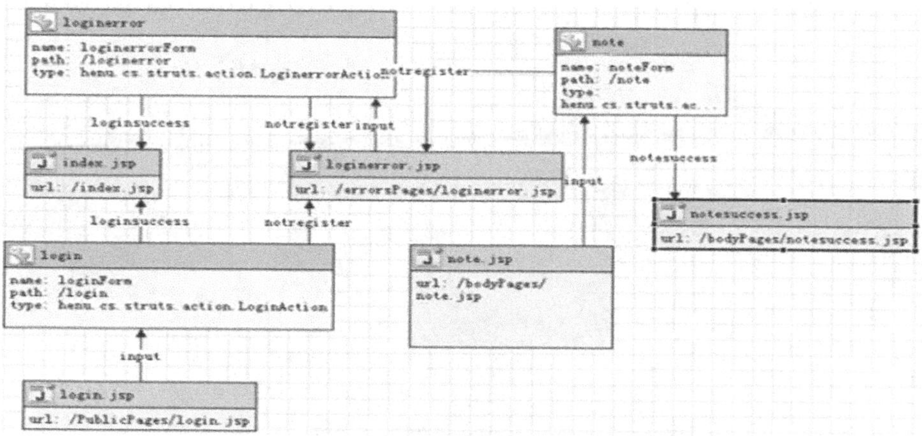

Fig. 3. Struts-config.xml diagram.

5 Summary

Development process in the shopping site, select the appropriate development framework will not only speed development and improve development quality, but also improve the maintainability, reusability. The performance of the system layer, model layer, business logic layers of their commitment to the task, reducing the inter-linkages between, the systems greatly reduces the complexity of the development. In the Struts framework, the development of the corresponding JSP, ActionForm, Action, the first design of the page to submit data, then Struts-config.xml for graphics development to generate ActionForm, while allowing the system to automatically generate the JSP page, with a little modification, you can use . Finally, generate the corresponding Action class, which is a more appropriate order.

References

1. Chen, L., Gao, Q.: Research on Framework Developing Technology based on MVC. AISS 3(3), 25–31 (2011)
2. Zeng, W., Zeng, W.: Exploration and Study of Chinese Thesaurus Automation Construction for Digital Libraries. JCIT 6(4), 109–117 (2011)
3. Fensel, D.: Ontologies: Silver Bullet for Knowledge Management and Electronic Commerce. Springer, Berlin (2000)
4. Liu, G., Lou, Z., Hou, Y., Shen, C.: Rural Emergency System Based on WebGIS. JCIT 6(2), 342–346 (2011)

Using XML Technology to Build E-commerce System

Qing Duan*

Office of Academic Affairs, The Central Institute for Correctional Police,
Baoding 071000, Hebei, China

Abstract. XML as a markup language used to describe the data, a description of its powerful features, scalability, and platform-independent structured semantic characteristics such as the Internet and distributed and heterogeneous environment as a major carrier of data transmission and exchange, in e-commerce has been widely used. Finally, the application of XML technology in E-Commerce is proposed based on XML. XML data exchange technology in the presence of a large number of e-commerce has been widely used, its increasingly urgent security needs. The experimental results indicate that this method has great effective promise.

Keywords: e-commerce, XML, DOM, DTD.

1 Introduction

Networks, especially Internet technology, the emergence and development of the flow of information has greatly enhanced the speed and efficiency, attracting more and more individuals, companies engaged in related activities through the network, network-based data exchange and business cooperation more and more frequent[1]. XML is a markup language used to describe the data, with a strong descriptive, extensible, cross-platform feature. These features of the Internet and fully meet the needs of distributed heterogeneous environment, is considered to be network data exchange and distribution of standard format. The HMDT platform serves as a transcending service broker, which uses ontology-based metadata and web services technologies.

Traditional information security technologies such as Secure Sockets (SSL), IP layer security standards, to some extent to meet the security needs of XML, but XML language for its structural characteristics, information security technology for data raised new demands. XML security in an increasingly strong demand for the premise, the international, W3C (World Wide Web Consortium), OASIS (Organization for the Advancement of Structured Information Standards), IETF (the Internet Engineering Task Force) and several other groups to participate XML security standards development work, a series of new services, XML security standards, to the exchange of XML as a data carrier applications to provide security protection. Ontology are formal, explicit specifications of shared conceptualizations of a given domain of discourse.

---

* DUAN Qing(1981.12-), Male, Han, Master of education technology of Northeast Normal University, Research area: XML and E-commerce.

D. Jin and S. Jin (Eds.): Advances in FCCS, Vol. 1, AISC 159, pp. 53–58.
springerlink.com

Currently, the area of safety-related standards development is the most important part of the XML Encryption (XML Encryption), XML Digital Signature (XML digital Signature), XML Key Management Specification XKMS (XML Key Management Specification), Security Assertion Markup Language SAML (Security Assertion Markup Language) and XML access Control markup Language XACML (XML Access Control Language). E-learning is an alternative concept to the traditional tutoring system. Some of these norms have become a recommended standard; some are still in the initial stage, the need for further improvement and development.

The XML Web-based e-commerce enterprise exchange order is the most frequent form of data, transmit it or store on the Internet may be subject to theft, altered, playback, camouflage and other attacks. How to ensure that the XML Web-based security e-commerce orders the successful completion of the project is a key issue. With the XML security specifications to develop and improve, XML security technology continues to expand. The security of XML-related areas of research has been supported by many large companies such as Microsoft, IBM and so on in their products by adding the XML security technology. Security tools developed by IBM to support IBM XML Security Suite XML signature and encryption, also joined the XML access control, only authorized users can access the corresponding document.

XML with its own characteristics, in e-commerce has been widely and in-depth applications. XML-based e-business needs for data security is more urgent. XML security technologies in e-commerce applications, in keeping with international standards, Chinese Academy of Sciences of the joint 8848 software, Lenovo, in cooperation with UF and other co-created a situation of China, with independent intellectual property rights of e-series standard cnXML. One of the e-message security guarantees, is based on XML security technology. At present, software development projects, has been extensive use of XML technology, XML security research in this area is in the initial stage of development.

2 The Research of XML Technology

XML is an extensible markup language; a description of its powerful, scalable, structured, platform-independent features such as data exchange in the presence of a large number of e-commerce has been widely used. This chapter describes the characteristics of XML documents, format, display, conversion and processing.

XML is Extensible Markup Language (extensible Markup Language) for short, is SGML (Standard Generalized Markup Language) is an optimal subset of the W3C in 1998 to become an official standard published by the organization [2]. XML is a markup language used to describe the syntax of the other, a powerful description ability and simplicity for network applications. Compared with HTML, which is different from HTML only a fixed set of tags, all fields can define their own special set of tags used in the field of information sharing and exchange, and highly scalable. And the separation of content and form of XML, you can use different style sheets to make the same data showing a different display appearance. In addition, XML is platform independent, strong interoperability. These advantages and features of XML, the markup language to design, communication between heterogeneous systems, data exchange, Web services, content management, Web integration, and many other aspects of the various configuration files have been widely used.

2.1 XML Language Specifications

XML documents can be divided into two main parts: the preamble and the document instance. Provide examples to explain the document preamble information, such as the XML version and its compliance with the document type. In this information-exploding era, the user expects to spend short time retrieving really useful information rather than spending plenty of time and ending up with lots of garbage information. Document type definition, abbreviated DTD, and XML Schema files are used to define the syntax rules of XML tags in two ways [3]. XML Schema is a standard XML file, and uses their own DTD is a special syntax. DTD is a very broad use of XML Schema; XML Schema has already become a formal W3C Recommendation, and the trend of alternative DTD. When in position s_i and s_j the distance between the sensor nodes and $K_\theta(\cdot)$ is associated with the model, the covariance function of the distance variable based on a decreasing non-negative functionof attributes defined as follows: Eqs1.

$$corr\ \{s_i, s_j\} = \rho_{i,j} = K_\theta(d_{i,j}) = \frac{E\lfloor s_i s_j \rfloor}{\sigma_s^2} \tag{1}$$

DTD is an XML markup language definition language built, defines the allowable element types, attributes and entities, and their combinations can make some limitations. If an XML document with its DTD consistent claimed that it is effective.

1. The document must begin with an XML declaration.
2. All elements must have start and end tags, empty element "/>" as the closing tag.

Managed by the W3C XML Schema specification of two components: XML Schema Part 1: Structure (http://www.w3.org/TR/xmlschema-1) and XML Schema Part 2: Datatypes (http://www.w3. org/TR/xmlschema-2). Compared with the DTD, XML Schema has the following advantages.

1. Consistency: Schema allows the definition of XML do not have to use a particular formal language, but to use the basic syntax rules of XML to define the structure of XML documents, making the XML from the inside out to achieve the perfect unity, but also for the further development of XML has laid a solid foundation.

2. Scalability: Schema of the DTD has been expanded, the introduction of data types, namespaces, so that it has strong scalability.

3. Interchangeability: the use of Schema, to write an XML document and verify the legality of the document. In addition, through a specific mapping mechanism can also be converted to different Schema to achieve a higher level of data exchange.

2.2 Apply XML Processing to Build E-commerce System

Application is provided by the XML parser API for XML document processing. W3C and XML_DEV proposed two standard application programming interfaces: DOM and SAX. DOM is short for Document Object Model, it is a structured document for the application program interfaces, rather than dedicated to XML documents. For example, DOM can use the same interface to the HTML document model. It is a widely used and accepted document for structured programming model.

DOM is to use the object to control the memory interface to structured data collection. Support for the DOM parser for XML document manipulation, it is constructed by an object in memory tree of the XML document to describe, DOM tree model used to access XML documents. Nodes from the document object tree main components. The object tree to tree through the XML document to read, search, modify, add, and delete operations. List 1-2 of the text structure of XML DOM tree shown in Figure 1.

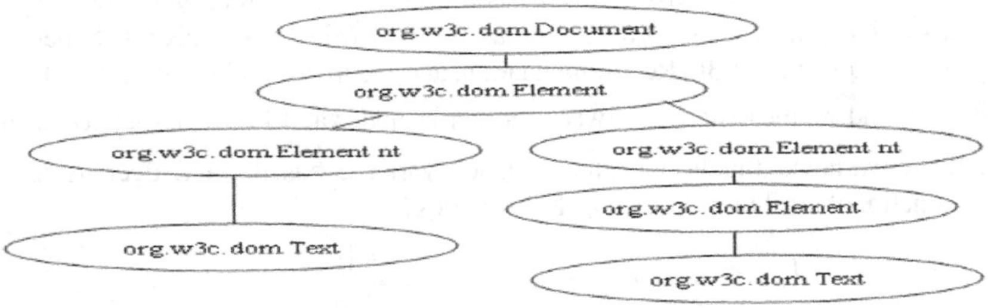

Fig. 1. The DOM Document system.

SAX (Simple API for XML) is another standard application programming interface, called a simple XML processing interface. SAX is completely different from the DOM object-based interface; it is the structure of the document as a stream of events, rather than build in memory to match the document's logical structure. It parses the XML document can trigger a series of events, as found in a given tag, it can activate a callback method, the method developed telling the label has been found. The XML document as a stream to deal with a performance advantage, because it does not generate any logical memory structure, the programmer can be quite flexible storage they need some. However, work with the SAX parser code would be more difficult, and difficult to simultaneously access the same document in multiple different data.

3 The Development of E-commerce System Based on XML Technology

To facilitate the PKI and digital certificates with XML applications and Web services using the integration of these programs, Microsoft, VerSign and WebMethods to develop an open specification that XML Key Management Specification (XML Key Management Specification, XKMS). Subsequently, the specification was submitted to the W3C, W3C XML Key formed a working group, in collaboration with other interested participants to further their development [4]. This work is still developed in the future there may be further improved.

The XPath language is a kind of location information within an XML document standard, it can be used to access the text data in XML documents, elements, attributes, and other information. XPath is generally not used independently, mainly embedded in the XSLT, XPointer, DOM and other host language applications.

For example, XPointer uses XPath to build part of the XML document referenced Web address. The equation 2 which calculates XML similarity of data acquisition is as follows.

$$D(M) = E\left[(S - \hat{S}(M))^2\right] \tag{2}$$

Location path is the XPath expression is the most important type. It selects a node associated with the context node set as its results. Location step has three parts: an axis, a node test and zero or more predicates. Axis positioning of the steps specified node and select the context node in the node tree, and there are children, ancestors, parents, and so on. Node test: Specify the type of the selected node location step, such as elements, attributes, and so on. Predicates using the built-in function to further filter unwanted nodes to achieve the desired node. Its syntax is: <Axis>:: <node test> [<predicate expression>], para [position () = 4], that the use of a predicate to select the context node of the first four para child nodes. Another model, the expression is id (string) function call, select the id element node with the same string. XPath is a W3C specification, the syntax can refer to the detailed documentation in the augmented matrix is the equation 3 as follows.

$$D(M) = \sigma_S^2 - \frac{\sigma_S^4}{M(\sigma_S^2 + \sigma_N^2)}(2\sum_{i=1}^{M}\rho(s,i) - 1) + \frac{\sigma_S^6}{M^2(\sigma_S^2 + \sigma_N^2)}\sum_{i=1}^{M}\sum_{j\neq i}^{M}\rho(i,j) \tag{3}$$

XSL has three main parts: XSL transformation (XSLT), XPath, XSL Formatting Objects (XSL-FO). XSL transformation is the conversion of an XML document, which the XML document into XSL described goals. XPath is used to refer to specified parts of an XML document. XSL-FO is the formatting objects and formatting properties of the vocabulary used to describe how content is displayed to the readers. XSL uses XSLT to transform XML documents that come in the form of a tree, then the tree generated by the information described in XSL result tree. Finally, be expressed by the XSL-FO.

4 Using XML Technology to Develop E-commerce System

Descriptive XML, scalability, and platform-independent structural characteristics of the Internet and fully meet the needs of distributed heterogeneous environment, as the main carrier of data exchange network, a strong impetus to the development of e-commerce and other network applications, the security has also been a lot of attention. XML data exchange technology in the presence of a large number of e-commerce has been widely used, its increasingly urgent security needs. XML as a markup language used to describe the data, a description of its powerful features, scalability, and platform-independent structured semantic characteristics such as the Internet and distributed and heterogeneous environment as a major carrier of data transmission and exchange, in e-commerce has been widely used. Fig.2 shows the detailed application of XML in e-commerce system. The experimental results indicate that this method has great promise.

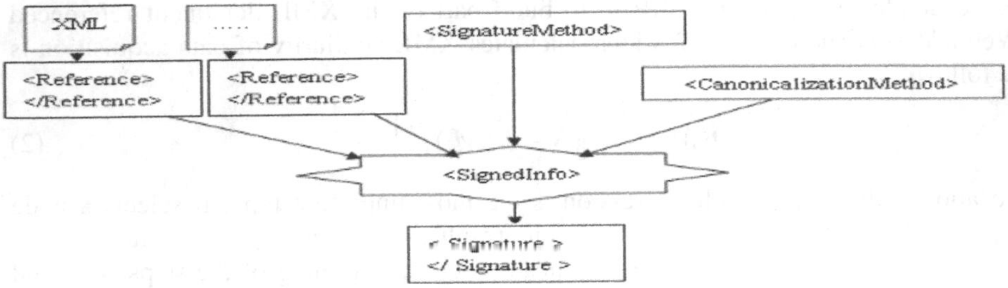

Fig. 2. The application results of XML technology in e-commerce system.

5 Summary

This paper analyzes the security of e-commerce data exchange in real applications, combined with the technical standards for XML security, designed for an integrated e-commerce data exchange XML security technology solutions, and to achieve some functionality. Solve the Web-based XML storage and security of e-commerce order exchange problem.

References

1. Boyer, J.M.: Bulletproof, Business Process Automation: Securing XML Forms with Document Subset Signatures. In: Proceedings of the 2003 ACM Workshop on XML Security, pp. 104–111. ACM Press (2003)
2. Wu, T.C., Huang, C.C., Guan, D.J.: Delegated Multi-signature Scheme with Document Decomposition. Journal of Systems and Software 55(3), 321–328 (2001)
3. Lim, C., Park, S., Son, S.H.: Access Control of XML Documents considering Update Operations. In: ACM Workshop on XML Security (2003)
4. Takase, T., Uramoto, N., Baba, K.: XML Digital Signature System Independent of Existing Applications. In: Proceedings Applications and the Internet (SAINT) Workshops, pp. 150–157 (2002)

The Application of .NET Technology in Remote Test System on Internet

Guo Li*

Network and Information Center, Qingyuan Polytechnic,
Qingyuan, 511510, Qingyuan, China
liguoliguo2011@sina.com

Abstract. Remote online test system is becoming one of the main areas of the design of educational software. This paper proposes the B / S structure of the remote examination system research, design and development, first of all system development platform and tools for a careful choice, and then to achieve the system objectives and functions of the system's development status, trends and demands for a more detailed analysis. Then the analysis based on the system's database and each module has been designed in conjunction with the system is given by the implementation of specific modules.

Keywords: Remote Test System, .NET technology, Database.

1 Introduction

With the development of computer technology in all areas of society, education Normalization has made a great improvement. Remote online examination system is becoming one of the main areas of the design of educational software. Meanwhile, the current overall education trends also put forward a very high request to this field. Traditional teaching methods, each organization must go through many steps in an examination, such as: paper organization, validation printing, sending the collection, registration distribution, archiving and evaluation, and teachers to the questions, printed papers, arranging examination, invigilation, collecting papers, marking papers, commenting on papers and analytical papers and so on[1]. With the increase of type examination and examination requirements continue to increase, the workload of teachers will be growing, but this work is very tedious and error-prone. At this time the traditional test method has been unable to meet the needs of modern education.

Based on Client / Server architecture of the examination system has changed this situation. Use of computer local area network, each client will be able to install the relevant software and server-side communication. The advantage is the use of a computer network, no longer requires a lot of manpower to achieve the consistency of questions and examination of information collection, automation and more efficient. According to the structural requirement, remote test system can be divided into two categories: C/S structure and B/S structure, that is to say, browser / server model and

* Guo Li, Female, Han, Network and Information Center, Qingyuan Polytechnic, Research area: Remote education, computer application and research.

D. Jin and S. Jin (Eds.): Advances in FCCS, Vol. 1, AISC 159, pp. 59–64.
springerlink.com

client/server model. This paper introduces the design method and Implementation of the remote test system, in which B/S structure and NET technology are employed. Remote test system can take examinations in any time and any place by using an accurate, efficient and networking information technology. Thus, it has the superiority which traditional examination will never exceed.

Based on actual demands of the remote test system, this paper discusses analysis of system requirements, development environment, and the design and implementation of the module function. A simple remote test process has been achieved. In the presented system, student users can login, register, test themselves and inquiry examination results; and administrators can manage test information, test results and so on. Remote examination system with accurate, efficient, network-based information technology tools to make the examination without the time and place constraints, better service for candidates and teachers.

Remote examination system uses B/S architecture, the language is. NET development environment, and using C #, database using Visual 2005 environment, the built-in SQL Server database. By the system to combine paper to explain to the remote examination system analysis, design and development process. To include the system requirements analysis, development environment, the choice of functional modules of the system design, database design, system implementation, etc. Achieve the basic functions of remote examination system to ensure the security of examination papers and management convenience. Different users have different permissions. Which administrators can conduct the relevant information for students to add, delete, query, etc., for maintenance of paper, add paper, questions, etc., students can check their examination results, select the papers for examinations.

2 System Development Environment

The system chooses development environment and NET platform, programming language in Visual Studio 2005 C #. NET. . NET Framework is a Microsoft development platform for the network. On this platform, you can use multiple languages to develop Windows applications, ASP.NET Web applications, mobile Web applications and XML WEB Service, etc.

As part of Microsoft's next-generation object-oriented language, C # from C and C + + evolved from a simple, object-oriented, type-safe new programming language. It maintains the familiar C + + syntax, but also contains a large number of efficient code and object-oriented features [2]. C # is almost all high-level language combines the advantages of the present, has many features not available in other languages, such as: the syntax is simpler, rapid application development (RAD) features, language, freedom, a strong-side components of Web services, to support cross-platform , and XML integration, the integration of C + +, etc.. Closely integrated with the Web, complete security and error handling techniques, easy-to-use version of the processing technology, good flexibility and compatibility, many advantages decided to develop a good C # development tool. Therefore, the development of this system is selected it as the main language.

Examination system in practice there will be a lot of data access operations, Visual 2005 environment, the built-in ADO.NET database access technology to promote access and manipulate data sets, to achieve greater scalability and flexibility. Application software running on the current network model there are two categories: C / S (Client / Server) mode and B / S (Browser / Server) mode. In the B / S structure, the user interface work is mainly achieved through the WWW browser, the main business logic on the server side implementation. This greatly simplifies the client computer's load, thereby reducing system maintenance and upgrade costs and workload.

This paper discusses the remote test system uses more advanced B / S mode. Based on C / S network exams, it is more stable, more suitable for the examination on the Internet. Meanwhile, the remote online exam is based on the operation of the exam, they are able to achievc intelligent automatic test paper, automatic scoring and automatic analysis, greatly reducing the test cycle, reducing the client's requirements.

SQL Server introduces SQL Server Management Studio, this is a new unified management tools. This tool set will include some new features to develop, configure the SQL Server database, fault find and repair them. The tool also features the former group also made some improvements, to improve data management efficiency, reduce operational complexity and maintenance costs, security, reliability, scalability, usability has a great economic advantage.

Application software running on the current network model there are two categories: C / S (Client / Server) mode and B / S (Browser / Server) mode. In the B / S structure, the user interface work is mainly achieved through the WWW browser, the main business logic on the server side implementation. This greatly simplifies the client computer's load, thereby reducing system maintenance and upgrade costs and workload. The C / S While the use of an open model, but only the openness of the system development level, in particular the application of either Client side or Server-side software to support specific needs, not available to the user really want an open environment.

In the remote test system development, using the traditional C / S structure of the main drawback is maintenance, upgrades more troublesome and difficult to achieve. Should test the content on remote servers, each exam before going to the machine installation, configuration; In addition, the test program on the client machine, security is also affected to some degree. Because the system is to achieve long-range test, but also to satisfy multiple users simultaneously, so the B / S structure will be the perfect choice.

3 Functional Analysis of Remote Examination System

System from the functional test can be roughly divided into modules and test candidate management module. Student test module candidates login authentication, test taking, examination time records and the selected candidates to receive the answers [3]. Test management module a student information database and test database management and maintenance. Required to achieve the initial user registration, examination time control, automatic graders, test input, modify exam, user management, account management, fractional management and other important functions, basically meet the requirements of paperless examinations. According to the system requirements analysis, system functional block diagram of the whole as shown in Figure1.

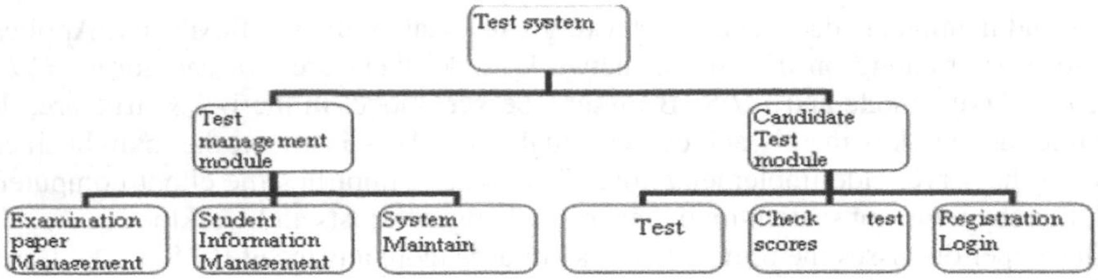

Fig. 1. Overall system block diagram

The system does not have exams to achieve the traditional features to meet any licensing examination candidates and quickly get results anywhere, truly reflects the advantages of remote examination system. In this paper, the following aspects of user need analysis.

(1) The user's information needs: Candidates can check their examination results to; test database administrator can query information, paper information, student information, etc.
(2) Processing needs of users: administrators can test database and test subjects to add, modify, and delete operations. Candidates must be able to carry out an assignment after the automatic generation of test scores; test time should be strictly limited.
(3) General system requirements: require the system to change the curriculum without limitation, can be applied to test a variety of subjects.
(4) System security requirements: All persons entering the system must be authenticated.

The distance education system is for the current status of the separation of teaching designed to achieve the students from examinations to take the exam location restrictions. Students only need to log in examination system, and then select to participate in the test subjects begin. System to be able to automatically calculate the student's answer time, the system immediately after assignment marking papers, and exam results are given.

Work to be done is the following data flow analysis. Data flow diagram can fully describe the process of examination system to achieve the specific flow data, integrated with a few symbols reflect the flow of data in the system, storage and handling conditions.

Remote examination marking system is a very important module. First look at the type of questions, and then compared with the standard answer questions number, if the answer is right, cumulative score, or starting the next question, and finally arrive at the total score of papers.

4 System Design and Implementation Based on .NET Technology

Remote examination system database management function is mainly reflected in the provision of a variety of information, save, update, query and other operations, so the database design to achieve ease of operation, and to ensure data security. System to achieve a few basic functions are: administrator papers, questions to add, modify, on the candidate's registration information, papers and information management, automatic grading feature, student registration log, enter the examination room test, to check their examination results, etc. System to achieve these functions, one can test a variety of dynamic management information, on the other test time and place flexibility, speed and scoring, test scores fair and reasonable. Functional modules from the system analysis, the remote interface of online examination system should be divided into the following sections: system login interface, user registration interface, students first interface test interface, the administrator first interface, the interface to add paper, add interface and other questions.

System administrators and users are mainly two categories of students; the system used in main data table has the following.

(1) Student Information Sheet. Record student registration information, including user ID, user name, password, password question, answer and other data items, based on registration information to prepare the user to check their examination results, and in case of lost password recover password.
(2) Student performance information table. Recording students' test scores, including the student's name, paper name and other data items.
(3) Paper information table. All existing paper records, including paper number, paper name, total score, and the examination time data entry.
(4) Test database information table. Record the questions, including questions number, item scores, item types and other data items.

First-line examination system to create a remote database, then create the required database tables and fields. To ensure the system has good scalability, and some business rules can be stored on the database server process on the way. Stored procedures provide a data-driven application many advantages; the use of stored procedures, database operations can be encapsulated in a single command, in order to obtain the best performance, optimization and additional security enhancements through the system security.

A complete system, to ensure information security, so users must be authenticated before logging. The first remote system interface test system to be able to achieve the old user's login, and can be linked to the new user registration page.

Login required entering a user name, password and verification code. When the user does not accidentally forget the password, you can click on the "Forgot Password" link to "retrieve password" page, according to the registration information related to retrieve the password input.

Add selected when the first questions the type of questions, there are multiple-choice, multiple-choice questions, and to determine other, then the provisions of the theme scores. Then begin to add questions, and then the standard answer questions. Click "Save" questions of the road into the corresponding papers.

Registered user interface, users need to enter a user name, password and password question and the answers to complete the registration information will be stored in the user information table. Because of the subjective question of judging ideas and technologies involved in a variety of algorithms to present learned and can not solve the problem, so this thesis is to achieve the objective questions such as: multiple choice examination to determine such questions. In this discourse, it is not involved in the organization on paper how to generate random content, it remains to be further expanded later to continue to learn and improve. The next step is to solve the major work of the details of the system and not yet implemented, to further improve the system functionality, truly paperless exam.

5 Summary

This paper describes the B / S structure of the remote examination system research, design and development, first of all system development platform and tools for a careful choice, and then to achieve the system objectives and functions of the system's development status, trends and demands for a more detailed analysis. Then the analysis based on the system's database and each module has been designed in conjunction with the system is given by the implementation of specific modules.

References

1. Chang, C.C., Hsiao, K.-C.: A SOA-Based e-Learning System for Teaching Fundamental Information Management Courses. JCIT 6(4), 298–305 (2011)
2. Hua, Z., Li, H., Li, Y.: Remote Sensing Image Fusion Algorithm Using Dyadic Contrast Contourlet Transform. IJACT 3(7), 132–140 (2011)
3. Nie, J.Y.: A general logical Approach to inferential information retrieval. Encyclopedia of Computer Science and Technology, 203–226 (2001)

The Research of Intelligent Recommendation System in E-commerce Based on Variable Precision Rough Set

XiaoYing Sun*

School of Computer and Information Engineering, Nanyang Institute of technology, Nanyang, 473000, Henan, China
sunxiaoying2012@sina.com

Abstract. Variable precision rough set model allows a certain degree of misclassification rate existing, so that, on one hand, it complements the concept of approximation space, on the other hand it benefit from the use of variable precision rough set of data to fine the relevant data in the thought irrelevant data. The paper puts forward the construction and application of intelligent recommendation system in e-commerce based on variable precision rough set, to suffice the needs of theory and application in E-commerce recommendation system. Therefore, the experimental results show that this method can effectively improve the performance of the recommendation system and recommended items.

Keywords: variable precision rough set, e-commerce, intelligent recommendation system.

1 Introduction

In 1993, Ziarok published a paper "Variable precision rough set model" in Journal of Computer and System Sciences, which marks the emergence of variable precision model. But the knowledge acquisition of information systems based on variable precision rough set also has no systematic research, such as classification of the variable precision rough set, attribute reduction, decision-making rules and other aspects of theoretical research and application is not mature enough, which need further research [1]. One of the most important reasons is to make decisions in a certain situation most reasonable by efficiently collecting relevant knowledge from heterogeneous domains. Navigation system to establish the main purpose of the search engine is to narrow the search, and search results in the association is established between them, so that users associated with the discovery of new knowledge of these.

With the deepening development of E-commerce, E-commerce recommendation system has been used more and more widely. E-commerce recommendation system is to provide users with information and advice recommend products to users they may

* SUN XIAO-ying, female, School of Computer and Information Engineering . Nanyang Institute of technology, Graduated from the Changsha Railway Institute, Mster degree,Research: Theory and Application of Computer.

D. Jin and S. Jin (Eds.): Advances in FCCS, Vol. 1, AISC 159, pp. 65–70.
springerlink.com

be interested in, help users to complete the process of purchase and provide personalized service based on the understanding and learning of user's needs and preferences. Navigation style search engine in the establishment of the database is based on extracts from the pages of the same or similar search terms on the organization's web pages. Construction of lattice conceptual clustering process is the process of building the concept lattice cell has a very important position. For the same data, the resulting grid is unique, that is, data or properties from the impact of the order, which is one of the advantages of concept lattice. The navigation of search results is an important way for users to gain the results they need and to narrow search scope effectively. Through the navigation of search results, users can gain the search results they need more quickly and the search scope is also narrowed effectively.

The expansion model established on variable precision rough set model which is available to deal with the mechanism of incomplete information and the noise, such as the variable precision rough set model based on similarity. Traditional search engines are difficult to solve using technology users "to find the information difficult" issue. Essence of this difficulty is the lack of knowledge of search engine capabilities and understanding of the information is only used to retrieve the keyword matching to achieve mechanical. Through the use of concept lattice model, the user's search results are in accordance with the different attributes of the document, taking the form of their background, and then generate a concept lattice, the concept of search results in different forms presented to the user.

Finally, this paper puts forward the construction intelligent recommendation system in e-commerce based on variable precision rough set. Variable precision rough set on the search results in using the second method of dealing with the classification. For the huge number of search results for the query words, allowing users to narrow the search scope, you can more accurately find the content they need. Therefore, in order to better handle the noise data, increase the necessary redundancy to the attribute reduction process, Zero, who proposed a variable precision rough set (Variable Precision Rough Set, VPRS) model, greatly improving the coverage and generalization ability of extraction of rules, to better reflect the data correlation of the data analysis, so as to laid the fundation for obtaining similar decision-making rules.

2 Variable Precision Rough Set Model

In order to solve these problems, Ziarko, a Professor variable precision rough set model (Variable Precision Rough Set Model, VPRS), it is in Pawlak rough set model based on the introduction of the threshold β ($0 \leq \beta < 0.5$), the model has some fault tolerance. On the one hand improve the approximation space concept, it also help with the rough set theory from the data that is not found in the relevant data. Variable precision rough set attribute the main task is to solve the uncertain relationship between the non-function or data classification problem [2]. A formal context K: = (G, M, I) consists of two sets G and M and G, M relationship between the composition $I \subseteq G \times M$, G in the form of the background elements are called objects, M elements are known as the formal context of the property, if gIm or (g, m) I, said, "the object g has attribute m".

Proposed a matrix based on variable precision to distinguish the rules for attribute reduction algorithm. Variable precision rough set theory in the approximate range of equivalence class assigned to a certain degree of error allowed when there is, distinguished by the nuclear matrix, and in accordance with the importance of various attributes, and properties dependent on the extent of attribute reduction, which makes the algorithm a certain degree of fault tolerance. Formalization is that the data is handled by formal mathematical entities do not have the concept of human thinking exactly the same, it also pointed out that the formal concept analysis and processing of basic data in the form is the form of background, the background is a form of background knowledge in human small part.

Approximation space form a partition of the domain; if an equivalence relation on to represent the equivalence classes that constitute the set of all equivalence classes, ie quotient set; of all equivalence classes form a partition into blocks and the corresponding equivalence class. Set $AS = (U, R)$ approximation space, where the domain of non-empty finite set, as the equivalence relations, or basic set of equivalence classes consisting of a collection $U/R = \{E_1, E_2, \cdots E_n\}$. For U, define the lower approximation of formula 1 and formula 2.

$$X \overset{\beta}{\subseteq} Y \Leftrightarrow C(X, Y) \leq \beta$$

$$\underline{R}_\beta X = \bigcup \{E \mid E \in U / \mathrm{R} \text{ 且 } X \overset{\beta}{\supseteq} E\} \tag{1}$$

In the concept lattice nodes can establish a partial order, given C1 = (X, X), C2 = (X2, X2), then C1 <C2 X2 <X2, we can understand this partial order for the sub-concept - the concept of super-relationship. Partial order can be generated based on the Hasse diagram grid, if C1 <C2, and there is no other element C3, making C1 <C3 <C2, then there is from C1 to C2 is an edge [3]. Both intelligent agent and semantic web service technologies are able to reach remarkable achievements and in some cases have overlapping functionalities.

$$\overline{R}_\beta X = \bigcup \{E \mid E \in U / R \text{ 且 } C(E, X) < 1 - \beta\} \tag{2}$$

Positive domain (or the lower approximation) may be understood as the object to the classification error of not more than points in the set. X is understood as the negative region corresponding to the object to the classification error of not more than points in the complement (ie, $\cap$) collection.

The node represents a concept (with the sensor nodes in order to distinguish the concept we call nodes), each concept is formed by the extension and intension, outside the node displays the node's attributes and objects. Can be quickly read from this linear map out all the connotation and extension, if an object belongs to the concept of looking for nodes along the route to find an increase in the rise of the concept of nodes on the line containing the Object is the same by the concept node along the fall line search, the concept of nodes on the line with this concept are contained in the node attributes.

3 The Intelligent Recommendation System of E-commerce

In the increasingly fierce competitive environment, e-commerce recommendation system can effectively retain customers, increase sales of e-commerce systems an effective means. Recommended introduction of e-commerce technology and systems research status, analysis of existing technologies and systems exist in the quality, timeliness and scale of the problem is proposed based on variable precision rough set intelligent recommendation system of e-commerce idea[4]. Order theory and lattice theory as a practical application combined with the product, concept lattice has been considered a powerful tool for data analysis, due to the mathematical concept lattice has a good nature and suitable for batch processing, etc.

With the development of electronic commerce, e-commerce website providing customers with more choice at the same time, its structure becomes more complex. On the one hand, customers face a lot of helpless product information, customers are often lost in the information space in a large number of commodities; the other hand, businesses have lost contact with consumers. Recommended technologies and systems modeling shop sales staff recommendation of goods to customers to help customers find merchandise. To the successful completion of the purchase process, so you can effectively retain customers, increase sales of e-commerce systems; businesses can also recommend the system to maintain contact with customers, rebuilding customer relationships. Currently, almost all large-scale e-commerce system, Amazon, CDNOW, eBay, online bookstore Dangdang varying degrees recommended using e-commerce technology and systems. A variety of Web sites to provide personalized service recommendation system also needs to support.

While e-commerce technology and systems are recommended by the relatively great progress, but how timely the vast amounts of information from the network needed to find the goods has become increasingly difficult. IEEE Internet Computing2001 year statistics also show that the Internet produces 2 × 10 bytes each year the amount of information, the famous search engine Google can only search to 1.3 × 10 pages. For U, define the lower approximation of formula 3.

$$\sup_h(X) = \frac{1}{k} \times \sum_{i_p \in X} h_p \times \sup(X) \tag{3}$$

Which sup (*X*) is the classic X support. This definition uses the right to take the average weighted value of the level of ideology seek support, not to a large extent the weight of larger projects highlight the. Current e-commerce recommendation system is based on a single application of e-commerce sites, recommended products and services are based on e-commerce sites. Customer base is specific, and therefore can not meet the conditions of large-scale web-based e-commerce recommendation applications, and there is real poor, the problem of high quality recommendations. Recommended based on knowledge of its customers the advantage of less demanding, so for customers of the temporary random browsing. It is recommended to make the majority of customers will be able to use. However, knowledge-based recommendation technology is the biggest difficulty is the acquisition of knowledge, and ontology technology can effectively achieve the acquisition of knowledge, aggregation and intelligent recommendations problems.

4 The Development of Intelligent Recommendation System of E-commerce Based on Variable Precision Rough Set

Knowledge-based recommendation is important and difficult is the need to acquire knowledge, it involves three kinds of knowledge: a catalog of knowledge, that knowledge of recommended products and their characteristics; b. Functional knowledge, the recommended system should be able to meet customer needs and possible characteristics of the demand for more goods; c. customer knowledge. Knowledge-based reasoning is recommended with the fundamental principles to guide the user through the user interface features clear requirements for the product in order to gain knowledge of customer needs, recommend the system according to user needs access to knowledge and knowledge of the product catalog knowledge, the use of functional knowledge reasoning to identify the products to meet user needs and recommend to the user. From the collection of literature, the present systematic study of such recommendation in sporadic research stage, the lack of systematic research. It is noteworthy that the body as a shared conceptual model of formal specification, are increasingly being used in recommendation systems, mainly used to describe user needs and products, and recommended to the user through the variable precision rough set products. Based on this background create the result of intelligent recommendation system of e-commerce based on variable precision rough set as shown in Figure 1.

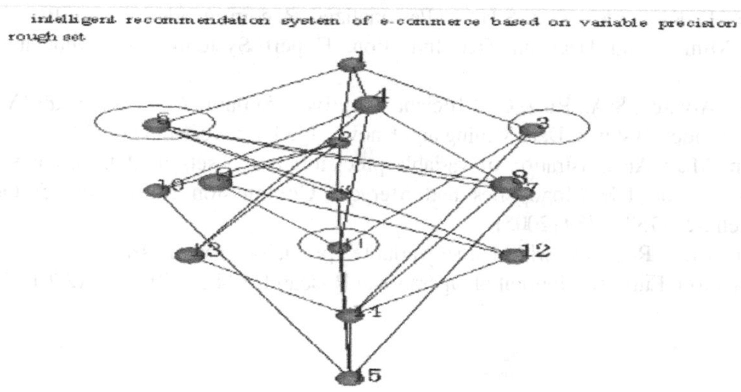

Fig. 1. The result of intelligent recommendation system of e-commerce based on variable precision rough set

VPRS-based intelligent recommendation system of e-commerce is in understanding and learning needs and preferences of customers, based on the product to provide users with information and advice, may be of interest to users recommend their products to help users complete the purchase process, providing personalized service. One collaborative filtering recommendation system of e-commerce technology in the most successful, largely determines the quality of recommendation system recommended.

Currently recommended in e-commerce in China is mainly used to find or search technology, which is based on the recommendation of the content to find the target. Relatively narrow recommendation technology, technology exists to find or retrieve a low degree of automation, poor motivation, low level of personalization and real-time shortcomings and poor.

5 Summary

Recommended and automatically recommend the personality of theoretical studies have just begun, then the variable precision rough set and e-commerce recommendation technologies, the product knowledge to meet network conditions, user needs and recommend functional knowledge of effective knowledge acquisition, aggregation and intelligent work recommended requirements, to meet large-scale e-commerce recommendation system, high quality and strong real-time requirements. With the OGSA and WSRF or P2P based grid technology continues to mature, and the continuous development of the semantic grid, variable precision rough set-based technology, e-commerce recommendation system is currently recommended as the systematic study of a new direction.

References

1. Cho, Y.H., Kim, J.K., Kim, S.H.: A Personalized Recom-mender System Based on Web Usage Mining and Decision Tree Induction. Expert Systems With Applications 23(3) (2002)
2. Lin, W., Alvarez, S.A., Ruiz, C.: Efficient Adaptive—Support Association Rule Mining for Recommender Systems. Data Mining and Knowledge Discovery 6(1) (2002)
3. Beynon, M.J.: An illustration of variable precision rough sets model: an analysis of the endings of the UK Monopolies and Mergers Commission. Computers & Operations Research 32, 1739–1759 (2005)
4. Beynon, M.: Reduces within the variable precision rough sets model: A further investigation. European Journal of Operational Research 134(3), 592–605 (2001)

The Application of Ontology Technology in E-commerce Recommendation System

TingZhong Wang*

College of Information Technology, Luoyang Normal University, Luoyang, 471022, China
wangtingzhong2@sina.cn

Abstract. Ontology is an explicit specification of a conceptual model. Semantic ontology and e-commerce recommendation technologies can the product knowledge to meet network conditions, user needs and recommend functional knowledge of effective knowledge acquisition; aggregation and intelligent recommendations work requirements, to meet large-scale e-commerce recommendation system, quality and strong real-time requirements. Finally, the recommendation system in E-Commerce is proposed based on ontology technology. Our experiments showed that we adopt ontology to develop the recommendation system in order to improve the efficiency of recommendation in e-commerce. The experimental results indicate that this method has great effective promise.

Keywords: ontology, recommendation system, e-commerce.

1 Introduction

At present, the current situation of ontology building can be analogy with the early period of the development of software engineering is still in individual (or the clique) manual mill stage [1]. The concept of the body from the field of philosophy, in the computer industry, changing the definition of the body. In 1993, Gruber gives the body of one of the most popular definition that "ontology is an explicit specification of a conceptual model. However, such approaches also leave room for improvements in several aspects such as interpretability, modularity and accuracy. Ontology allow web resources to be semantically enriched. Well-structured and simple problems can be solved with regular rules and principles. They have knowable and comprehensible solutions where the relationship between decision choices and all problem states is known or probabilistic.

With the deepening development of e-commerce, e-commerce recommendation system more widely. E-commerce recommendation system is in understanding and learning needs and preferences of customers, based on the product to provide users with information and advice, may be of interest to users recommend their products to help users complete the purchase process, providing personalized service. However, it is not easy for a tourist to search the information what exactly he really wants from a

* TingZhong Wang(1973.7-), Male, Han, Master of Henan University of Science and Technology, Research area: e-commerce, recommendation system, ontology.

D. Jin and S. Jin (Eds.): Advances in FCCS, Vol. 1, AISC 159, pp. 71–76.
springerlink.com

large amount of information available on the Internet. Information appliances also benefit from the specialization of function in that it allows customization in terms of operation, look, shape and feel. Ontology building and its representation in a formal language is usually carried out by knowledge engineers (KE), sometimes with the assistance of domain experts. Ontology is formal, explicit specifications of shared conceptualizations of a given domain of discourse.

However, the existing e-commerce recommendation system itself has shortcomings, mainly in two aspects: First, the data sparseness problem is the evaluation of the user when there are fewer goods to buy, recommended quality will be reduced. Second, the cold start problem, into the "new goods problem" and the "new user problem", the performance when adding a new product without any score data, as well as a new user does not have any goods to score on the resulting system can not be recommended to the user recommend.

Finally, this paper puts forward building recommendation system in E- Commerce based on ontology technology. First, create the user's personal information, which records the customer's basic data types and preferences. Then use the classification results and to construct the body of the product preferences of the group tree (Group Preference Tree, GPT), the final recommendation by the preferences of the module tree recommend appropriate products to users. Which provided the user's preferences when too few classes. Ontology techniques also play an important role in e-commerce and eservices, proving to be useful tools for understanding how ecommerce and e-service Web sites and services are used. The use of ontology-based GPT preferences to find a relationship type, category and increase the user's preferences and increase the number of recommended items.

2 The Research of Information Management System in E-commerce Based on Ontology

Ontology is the conceptualization of explicit description. He provided Practical knowledge of the formal semantic representation, knowledge management is the most promising technology Surgery, has been widely used in knowledge management, semantic Web and other areas. Development of ontology for a specific domain is not yet an engineering process, but it is clear that ontology must include descriptions of explicit concepts and their relationships of a specific domain.

To meet these requirements, ontology (Ontology) as a level of knowledge in semantic and describe the concept of information system modeling tools, has been proposed since it caused the concern of many researchers at home and abroad, and many fields in the computer a wide range of applications, such as knowledge engineering, digital libraries, software reuse, information retrieval and processing of heterogeneous information on the World Wide Web, semantic Web, etc. These formal ontological structures are concerned with the description of concept types and relations types and generally are not concerned with physical or process objects[2]. Ontology is one of the branches of philosophy, which deals with the nature and organization of reality.

The web services execution environment supports common B2B and B2C (Business to Consumer) scenarios, acting as an information system representing the central point of a hub-and-spoke architecture. Recommender systems are generally two methods to formulate recommendations both depending on the type of items to be recommended and the way that user models are constructed.

"Conceptualization" refers to the objective world of abstract concepts and a number of phenomena to be an overview of the model, the meaning of the concept of model performance independent of the specific state of the environment, for example: it is assumed that A, B on behalf of two boards, with ON (A , B) or ON (B, A), the meaning of the concept remains the same ON.

In short, the body's objectives are: to obtain, describe, and express knowledge in related fields, providing a common understanding of knowledge in this area, determine the areas of common recognition of the word, and from different levels of formal models of these terms is given relationship between vocabulary and a clear definition of[3]. The formula 1 which calculates the ontology recommendations nodes of data mining is as follows.

$$\hat{x}_{k,m} = \overline{u}_k + \frac{\sum_{u_a \in S_u(u_k)} s_u(u_k, u_a)(x_{a,m} - \overline{u}_a)}{\sum_{u_a \in S_u(u_k)} s_u(u_k, u_a)} \tag{1}$$

Ontology: a specific area of research and related words or terms, such as medicine, business simulation. Ontolingua is a language based on KIF, which uses a unified standard format to describe the body. Its characteristics are: the construction and maintenance of anthologies provides a unified, computer-readable form; its ontological structure can easily be converted to a variety of knowledge representation and reasoning systems, which use the body to maintain its target system with the separation.The formula 2 which calculates the ontology recommendations nodes of data mining is as follows.

$$\hat{x}_{k,m} = \frac{\sum_{i_b \in S_i(i_m)} s_i(i_m, i_b)(x_{k,b})}{\sum_{i_b \in S_i(i_m)} s_i(i_m, i_b)} \tag{2}$$

In user-based collaborative filtering approach, with the growing number of users in a large number of users within the class, "nearest neighbor search" will become the bottleneck of the algorithm. For the items concerned, the similarity between them to stabilize, and therefore the greatest amount of work can be done offline similarity calculation step, thereby greatly reducing the online computation and improve the efficiency of recommended.

3 Using Ontology Technology to Build Recommendation System

Recommended system (Recommender Systems) is the use of statistical and knowledge discovery techniques to solve the interaction with the target customers to provide products recommended problem. Is now widely cited recommendation system (Recommender System) is defined Resnick & Varian given in 1997: "It's e-commerce system to provide customers with product information and recommendations to help customers decide what products to buy, analog sales recommend products to customers to complete the purchase process[4]. "Recommended system recommended in the e-commerce site which is commodity purchases of merchandise, customer demographics or purchase history of customers on the analysis produced.

E-commerce site visitors into buyers: e-commerce system, visitors are often in the process, and no desire to buy, personalized recommendation system to recommend to the user they are interested in the goods, thus contributing to the purchase process.

In recent years, with the development of the Internet, a series of Web-based ontology language, also called ontology markup language, such as SHOE, XOL, RDF, RDF-S, Oil, Oil + DAML, OWL.

At this stage, the usual practice is for the industry with experts in the field to summarize and extract information in order to manually build the body, the body of the manual is a huge workload and build abnormal complex task, following the birth of these methods are described in the specific ontology development projects among the practice of service for specific projects. The equation 3 which calculates the ontology recommendation nodes of e-commerce is as follows.

$$S_{ui} = \frac{1}{\sqrt{(1/s_u(u_k, u_a))^2 + (1/s_i(i_m, i_b))^2}} \tag{3}$$

The λ and δ the value can be determined by the experimental method, λ determines the size of the value relevance of the importance of the user, if λ 1, our algorithms into a user-based collaborative filtering algorithm; if λ 0, we algorithm into a collaborative filtering algorithm based on the item. Parameters used to adjust the SUIR impact, if 0, then the integration framework based on the mere combination of two methods of users and items. Concepts and relationships are basic components in ontology. Web documents are the most important source for deriving concepts and relationships. Built entirely by hand in order to solve a problem caused by the body, where the second ontology construction method, which uses automated or semi-automatic method to build ontology. In this way, you can manually build a body of work to simplify and improve the quality of body.

Improve the ability to cross-sell e-commerce websites: personalized recommendation system in the user process to the user to purchase other valuable commodities recommended, the user can provide the recommended list from the system to purchase their own needs but does not think in the purchase process of goods to improve cross-selling e-commerce system. E-commerce sites to increase customer loyalty: Compared with the traditional business model, e-commerce system enables users to have more choices; the user is extremely easy to replace the business, just click the mouse once or twice can be in different electronic jump between business systems.

4 The Recommendation System in E-commerce Based on Ontology Technology

Recommended system recommended in the e-commerce site which is commodity purchases of merchandise, customer demographics or purchase history of customers on the analysis generated. Broadly speaking, these considerations make the e-commerce with a personalized color, but also for different customers, with the recommendation system of e-commerce website to show some self-adaptive. Body has two properties: static and dynamic static nature means that it reflects the conceptual model does not involve dynamic behavior; dynamic nature means that it's content and services targeted at changing for different areas, you can define and construct different body. Fig.1 shows the application of recommendation in e-commerce by ontology. The experimental results indicate that this method has great promise.

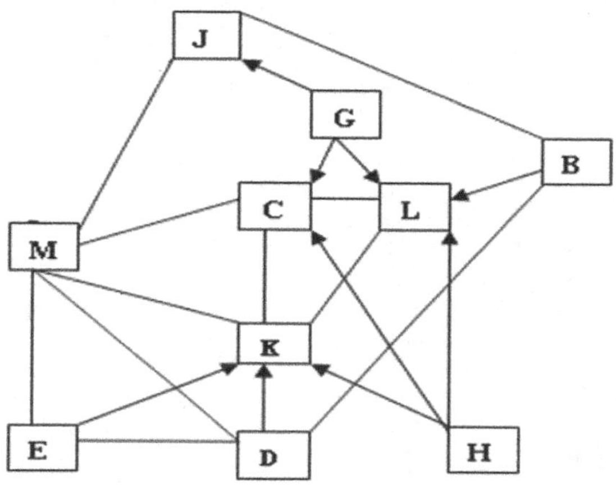

Fig. 1. The application results of e-commerce recommendation system by ontology.

5 Summary

In order to transform the present disorder Web into the orderly knowledge which computer can understand, semantic Web using multi-level showing frame model, ontology is located in the document description to the knowledge inference turning point position, so the construction of the ontology is the key link of implementing semantic Web. In this paper, we adopt ontology to develop the recommendation system in order to improve the efficiency of recommendation in e-commerce. In the increasingly competitive environment, the personalized recommendation system can effectively retain customers; increase the service capacity of e-commerce systems. Successful referral system will bring huge benefits.

References

1. Christakou, C., Stafylopatis, A.: A Hybrid Movie Recommender System Based on Neural Networks. In: Proceedings of 5th International Conference on Intelligent Systems Design and Applications (2005)
2. Green, P., Rosemann, M.: Integrated process modelling: an ontological analysis. Inf. Syst. 25(2), 73–87 (2000)
3. Yuan, S. T., Cheng, C.: Ontology-Based Personalized Couple Clustering for Heterogeneous Product Recommendation in Mobile Marketing. Expert Systems With Applications 26(4) (2004)
4. Kavalec, M., Maedche, A., Svátek, V.: Discovery of Lexical Entries for Non-taxonomic Relations in Ontology Learning. In: Van Emde Boas, P., Pokorný, J., Bieliková, M., Štuller, J. (eds.) SOFSEM 2004. LNCS, vol. 2932, pp. 249–256. Springer, Heidelberg (2004)

The Application of Clustering Algorithm in Intrusion Detection System

Lei Ge* and CaiQian Zhang

Software College of Kaifeng University, Kaifeng, Henan, 475004, China
gelei2012@sina.cn

Abstract. Intrusion is the use of unauthorized behavior, by scanning the loopholes in the system, access to user accounts, user file tampering. Clustering is a clustering process; the clustering technique applied to intrusion detection in a supervised learning algorithm to overcome the requirements of the training set data in pure question mark, and can detect unknown intrusions. Discuss how the clustering algorithm is applied to intrusion detection and analysis of intrusion detection algorithm based on clustering problems.

Keywords: intrusion detection, cluster algorithm, Information system.

1 Introduction

Two decades, Internet has changed the human way of life. However, as more and more people use the Internet, the computer system itself, gradually exposed the loopholes, so take advantage of malicious intrusion, such as computer viruses, data theft, hacking and so on. According to statistics, 99% of large companies have too much of the invasion occurred, such as the world's leading business website Yahoo, Buy, Amazon and so has been hacked, resulting in huge economic losses, and even specialized in network security of the RSA website also attacked by hackers. Russel and Gangemi (1991) proposed computer system security is based on the confidentiality, integrity and availability requirements of the above [1]. Confidentiality (Confidentiality) means that only authorized users can access information; Integrity (Integrity) refers to the system to maintain data consistency and accuracy, will not be accidentally or maliciously modified; availability (Availability) refers to the need for system resources when the authorized user , the system has been providing resources, and will not refuse access to authorized users.

Intrusion is the use of unauthorized behavior, by scanning the loopholes in the system, access to user accounts, user files tampered with, such behavior is the intrusion. According CIDF (Common Intrusion Detection Framework) standard, IDS (Intrusion Detection System) is a computer network or system automatically analyzes the data, and found the ultimate network or system security policy is a violation of the act or signs of attack network security technology. While the issues for exploring environmental sustainability are well rehearsed and known, the issues that should form the social dimension are less appreciated and addressed by stakeholders

* GeLei(1978.10), male, Software College of Kaifeng University.

D. Jin and S. Jin (Eds.): Advances in FCCS, Vol. 1, AISC 159, pp. 77–82.
springerlink.com

involved in the development process. As new services appear with high performance requirements, mechanisms to ensure quality of service and metrics to monitor this quality become necessary.

With more and more complex network environment, only rely on firewall technology can not prevent attacks from within, and there are all kinds of loopholes and firewall itself back door and can not provide real-time intrusion detection capabilities. In order to enhance the computer system or network security systems, need to adopt proactive strategies and more powerful program, which is an effective solution to intrusion detection. Since Denning proposed intrusion detection model, intrusion detection people had a great interest in the study, made a number of intrusion detection system prototype. This chapter of the existing intrusion detection model and its implementation technology classification, indicating the intrusion detection system's main features and implementation techniques of the pros and cons, from the grasp of the overall research and development of intrusion detection.

When the invasion patterns and network behavior changes, traditional intrusion detection systems can not do anything with the clustering algorithm can detect new and unknown intrusions. However, the clustering algorithm processing, describe the characteristics of the limitations of network behavior, once the attack is treated as normal data included in the training set, we can not detect such attacks and their variants. Therefore, the current intrusion detection algorithms can not accurately detect unknown attacks, false alarm rate and other issues. In order to improve intrusion detection algorithm to improve detection rate and reduce false alarm rate, this paper based on semi-supervised clustering algorithm for intrusion detection.

2 The Research of Intrusion Detection Systems

With more and more complex network environment, only rely on firewall technology can not prevent attacks from within, and the firewall itself has all kinds of loopholes and back doors, and can not provide real-time intrusion detection capabilities. In order to enhance the computer system or network security systems, need to adopt proactive strategies and more powerful program, which is an effective solution to intrusion detection. Since Denning proposed intrusion detection model [2], intrusion detection people had a great interest in the study, made a number of intrusion detection system prototype. This chapter of the existing intrusion detection model and its implementation technology classification, indicating the intrusion detection system's main features and implementation techniques of the pros and cons, from the grasp of the overall research and development of intrusion detection.

Intrusion detection system should have the following characteristics of six aspects:

(1) To monitor, analyze user and system activity, the illegal users and legitimate users find unauthorized operation.

(2) Testing the correctness of the system configuration and security holes, and prompts managers to fix vulnerabilities.

(3) To identify the type of attack known to the network administrator alarm.

(4) Analysis of abnormal behavior patterns.

(5) Of the operating system audit trail management, and identify users who violate security policy behavior.
(6) Evaluation of critical systems and data integrity of the document.

From Denning's statistical algorithms used to establish the normal behavior of network and user intrusion detection model so far, intrusion detection technology has gone through a decade of development. Misuse detection is the use of signatures in the invasion of well-defined pattern matching with the audit data to detect intrusions. Misuse detection systems intrusion marked the first mode of coding, the establishment of the invasion pattern library, and then analyze the network data to detect whether the intrusion match. Misuse detection systems are faced with the problem is how to describe the characteristics of a model of attack and its variants, and the model can not match the behavior of non-invasive.

Misuse intrusion detection model can accurately detect intrusions library already has a low false alarm rate. But the emergence of new attacks, new attack signatures need to be manually added to the invasion mode model library, which means it needs to constantly upgrade and update in order to ensure the completeness of the system detection capabilities. In addition, because the misuse detection system highly dependent on the target system, so the system portability is not good; inability to detect unknown intrusion, so the detection rate.

The original misuse detection system is rule-based expert system. It is known as the invasion of coding a rule set, which if-then rules with the structure, conditions for the invasion of some features, then part of the precautionary measures for the system [3]. When the rule condition part is met, then part of the implementation of the action. The establishment of expert system depends on the completeness of the knowledge base; knowledge base in turn depends on the completeness of data completeness and timeliness.

$$S_j(P_i) = 1/(1+\alpha \cdot D_{ij})^{\beta} \quad (1)$$

The formula 1 which is $S_j(P_i)$ Sensor node is the perception of accuracy in point is the distance between nodes and points, to reflect the physical characteristics of the sensor node device parameters. Expert system can use it as an autonomous black box; users do not need to interfere in the internal reasoning process of expert systems. Its disadvantages are: features difficult to extract the invasion, there is massive data processing efficiency, speed, real-time requirements difficult to meet; as to change the rules in the rule base must be considered when the dependence between different rules, so it is difficult to maintain the rule base.

$$C(P_i) = 1 - \prod (1 - S_j(P_i)) \quad (2)$$

State transition is misuse detection for the analysis of algorithms, it uses the system state and state transition expressions to describe and detect known intrusions. Invasion of the state transition to achieve two main models: the state transition analysis, and colored Petri-Net.

State transition analysis uses the state transition diagram to represent the known intrusion, the system state changes by the state diagram of the initial state, intermediate state and end state consisting of a sequence. System status or user rights through the system properties to be described by the system state transitions are event-driven, the engine maintains the state transition table of a state transition, and each event is to refresh this table.

3 Intrusion Detection Model

State transition algorithm has the advantage that the state transition rules easier to create and update, and transfer rules easier to understand; only triggered state transition analysis of the events, providing an independent intrusion and data description. The disadvantage is that the state declares a list of events and actions need to be manually coded, can not fully express the invasion of more complex models, the system is also difficult to detect intrusion simple variant of the low efficiency.

Another technology uses state transition model is colored Petri nets (Colored Petri Nets, CPNs), the model by Kumar and Spafford of Purdue University proposed to optimize the misuse detection system, the specific implementation is IDIOT system. It uses CPN to represent and detect intrusion patterns; each pattern is represented as an invasion of CPN[4]. CPN token in the event that the properties of color, movement of the token that the progress of the invasion process, as CPN tokens from the initial state to move to the end of the state, said the successful completion of the invasion process. The formula 3 which calculates the similarity of j and R is as follows.

$$\Delta(P_i) = C(P_i) - C^{'}(P_i) = S_n(P_n) \cdot \prod (1 - S_j(P_i)) \tag{3}$$

Denning intrusion detection model is a generic intrusion detection model, it is independent of the specific system, application environment and attack types, for the subsequent detection of the study provide a reference model and the system value. If IDES / NIDES are Denying model based on extended. Realization of these functions using a standard component technology. The system frame structure shown in Figure 1.

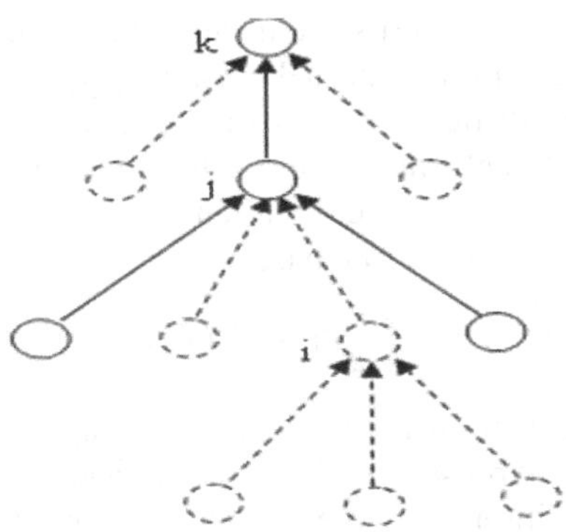

Fig. 1. System frame structure of Time-Based Inductive Machine.

Model prediction is based on anomaly detection intrusion detection algorithm, which assumes that the system sequence of events is not random but follow a recognizable pattern, the characteristics of the algorithm is to consider the relationship between events. Teng and Cheng gives time-based reasoning model prediction algorithm, the application time rules identify the characteristics of normal user behavior. Dynamically generated by inductive learning rules, and can adjust to make real-time online high predictability, accuracy and credibility. If the rules are mostly accurate, but can successfully predict the observed data, the higher the credibility of the rule (if rule 1 to rule 2 successfully predicted more events, the rule 1 to rule 2 more predictable).

Validity means that IDS has a high detection rate and low false alarm rate. Adaptability refers to the fast update by encoding intrusion patterns, to detect variations of known attacks and unknown attacks. Scalability refers to the network configuration, the system can be incorporated into the detection module, or other services tailored. Complex because of the current network environment, the audit records increasingly large, ever-changing attack, you need a more systematic, automated algorithms to construct intrusion detection models.

4 The Application of Clustering Algorithm in Intrusion Detection System

The basic idea of clustering is to define the distance between the data from the data on behalf of the similarity between the measure of the degree of similarity according to the size of the data are classified one by one, until the completion of all data gathered. Division of clustering is the first rough classification of the data, and then revised in accordance with certain principles until the classification is reasonable so far. The basic idea is: for a given data set of data, based on experience for the first set number of classes, in accordance with the provisions of generating a particular cluster center, followed by calculating the distance between each data center, select the distance minimum value of the center, the data included in this class, get a clustering. K-means algorithm based on the error sum of squares criterion. For optimum results, the first division of the initial data set, the sample mean as the center of the cluster, through an iterative, successive lower objective function value of an error, until the objective function does not change, to get classification results. K-means algorithm described below.

Algorithms: data based on class division of the mean clustering algorithm.

Input: data set $X = \{x_i\}_{i=1}^{n}$, $x_i \in \mathrm{R}^d$, k number of clusters.

Output: to meet the minimum squared error criterion disjoint clusters $\{X_h\}_{h=1}^{k}$.

1) Select k an initial cluster centers $\{\mu_h^{(0)}\}_{h=1}^{k}$;

2) Repeat to (3) to (5), the objective function iteration until convergence;

3) Calculate the mean of each cluster in the data k, the data will be re-assigned to the most similar to each class $h^*, h^* = \arg\min_h \left\| x - \mu_h^{(t)} \right\|^2$;

4) Recalculate the mean class $\mu_h^{(t+1)} \leftarrow \frac{1}{X_h^{(t+1)}} \sum_{x \in X_h^{(t+1)}} x$, assigned to each class.

5) Calculation O_{random} instead O_j of the total cost S;

6) If $S < 0$, then O_{random} replace O_j, the formation k of a new collection center

Although the K-means algorithm to handle large data sets, but the nature of the algorithm is iterative, the implementation process need to repeatedly scan the entire data set, changing the data belongs to the cluster center and cluster until all cluster centers no longer change. Time required for this clustering of data sets with increasing size and rapid expansion, the calculation is large, so large network data sets is very difficult to divide.

5 Summary

This paper systematically studies the basic theory of intrusion detection systems, intrusion detection model analysis of the research status and existing problems. Discuss how the clustering algorithm is applied to intrusion detection and analysis of intrusion detection algorithm based on clustering problems.

References

1. Barbara, D., Couto, J., Jajodia, S., Wu, N.: ADAM: A Testbed for Exploring the Use of Data Mining in Intrusion Detection. SIGMOD Record 30(4), 15–24 (2001)
2. Mukkamala, S., Janoski, G., Sung, A.: Intrusion detection using neural networks and support vector machines. In: Proceedings of the International Joint Conference on Neural Networks, pp. 1702–1707. IEEE Computer Society Press, New Jersey (2002)
3. Lee, W., Stolfo, S.: Data mining approaches for intrusion detection. In: Proceedings of the 7th USENIX Security Symposium, San Antonio (1998)
4. Basu, S., Banerjee, A., Mooney, R.J.: Semi-supervised clustering by seeding. In: Proceedings of 19th International Conference on Machine Learning (ICML 2002), Sydney, Australia, pp. 19–26 (2002)

The Modeling and Forecasting of Variable Geomagnetic Field Based on Artificial Neural Network

Yihong Li, Wenliang Guan, Chao Niu, and Daizhi Liu

Xi'an Research Inst. of Hi-Tech, Xi'an, China
gyihongli@126.com

Abstract. The time series of variable geomagnetic field can be considered as nonlinear and nonstationary signal. The view and methods of intelligent signal processing can be used in modeling and forecasting for variable geomagnetic field. This paper proposed a new point of view that Artificial neural network (ANN) with good self-organizing and self-learning could be used for modeling and forecasting variable geomagnetic field and tested this view with four ANN which are RBF, BP, GRNN and ELMAN. Lastly, the results of forecasting were compared and analyzed.

Keywords: variable geomagnetic field, Artificial neural networks (ANN), time series forecasting, modeling.

1 Introduction

Variable geomagnetic field of the earth is generated by the space current system outside the solid Earth, also known as external source field[1]. Although the Variable geomagnetic field of the earth is only 1% of the geomagnetic field of the Earth, it carries rich information about Earth's space environment. Many applications need to consider effects of Variable geomagnetic field , such as space weather monitoring and forecasting[1], earthquake forecasting based on geomagnetic method[2], geomagnetic navigation and geomagnetic homing technique[3]. Accurate modeling and forecasting of Variable geomagnetic field has become an important research topic. Variable geomagnetic field with complex physical properties changes rapidly over time. It is difficult to achieve high forecasting accuracy just from the physical model. The Variable geomagnetic field should has characteristics of high non-linear and non-stationary from the signal processing point based on observations of the the variable geomagnetic field data. Therefore, modeling and forecasting based on non-linear, non-stationary time series analysis methods is a good idea. ANN with strong self-organization and self-learning ability is now widely used in nonlinear model forecasting method.

ANN is new information processing systems which imitate the human brain cell structure and function, human brain structure and brain handling problem function[4]. The significant features of ANN are: a high degree of nonlinearity, good self-learning and self-adaptive, good fault tolerance and associative memory capacity and a high degree of parallelism[5]. It has achieved good application results in pattern recognition, artificial intelligence, nonlinear dynamic processing, automatic control, signal processing, forecasting, evaluation and other fields.

D. Jin and S. Jin (Eds.): Advances in FCCS, Vol. 1, AISC 159, pp. 83–88.
springerlink.com

Since we can get rich historical data by geomagnetic observation to train the network, so several neural network with learning instructors are choose to forecast the neural network, including RBF, BP, GRNN and Elman neural network. The neural network toolbox of Matlab is used to model and forecast the variable geomagnetic field, and the forecasting results were compared.

2 Forecasting Based on ARTIFICIAL Neural Networks

The artificial neural networks were used to model and forecast the variable geomagnetic field, the 10-minute average data of 1996 at Beijing geomagnetic declination D was choose as the experimental data. The data of the first 30 days was selected as training samples, the data of the later 2 days as test samples, so the length of training samples is 1430 and the length of test data is 288. Input vector length is 512 and the output vector length is 1. That is to say, the top 512 historical data is used to forecast one future data, the input vector is choose through a sliding window automatically. The length of the sliding window is the length of input vector. This is the forecasting process of single-step. Multi-steps forecasting can be forecasted by recursive step, the output of single-step forecasting was used as the last element of the input vector, the forecasting process of single-step was repeated to achieve multi-step forecasting.

Since the number of the data is too large, only the data of two days that between 0:00 on May 1, 1996 to 23:50 on May 3rd of 10-minute average data of 1996 at Beijing geomagnetic declination D which benchmark value has been subtracted as the test data forecasting map. The geomagnetic activity was in its low level in this periods, was quiet period. Test samples is shown in Fig.1.

Experimental data and experimental methods described above were used in all forecasting experiment in this paper, it will not be described later.

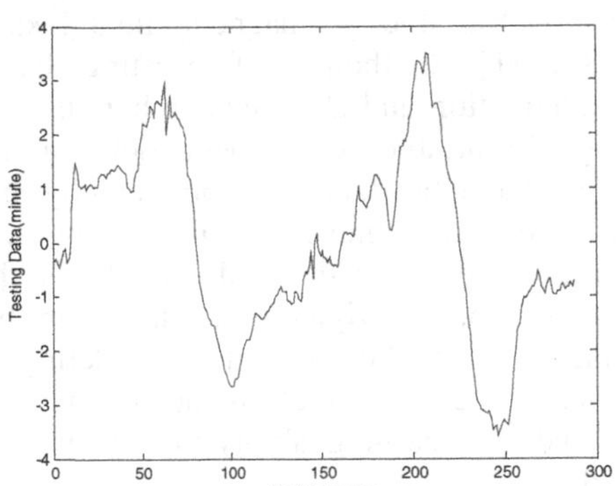

Fig. 1. Testing data of forecasting with neural networks.

2.1 Forecasting Based on RBF Neural Network

RBF neural network is a three-layer feed-forward network. The first layer is input layer, the second layer is the hidden layer, and the third layer is output layer. The characteristics of RBF neural network is simple structure, simple training, and fast learning convergence. The parameters RBF neural network using to forecasting is mainly the choice of spread function and fitting error. If the fitting error is too small, "over-fitting" will be leaded, and the generalization ability of the model will be reduced. Of course, the fitting error can not be set too large, otherwise, the neural network are not adequately trained. The greater the extension function, the smoother the fitting function. But the extension function can not be too large, or the individual differences of neurons will be weaken, leading to all neurons have similar performance.

Single-step forecasting and multi-step forecasting results are shown in Fig.2 and Fig.3. It can be seen that single-step forecasting for the geomagnetic declination has high accuracy based on RBF network. More than 96% of the errors can be controlled within 0.5 minute through error statistics, but the multi-step forecasting is poor, the basic forecasting curve can not reflect the true curve trends.

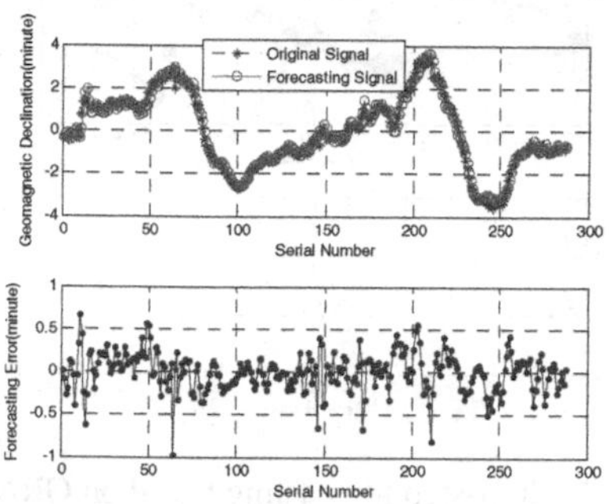

Fig. 2. Single-step forecasting based on RBF.

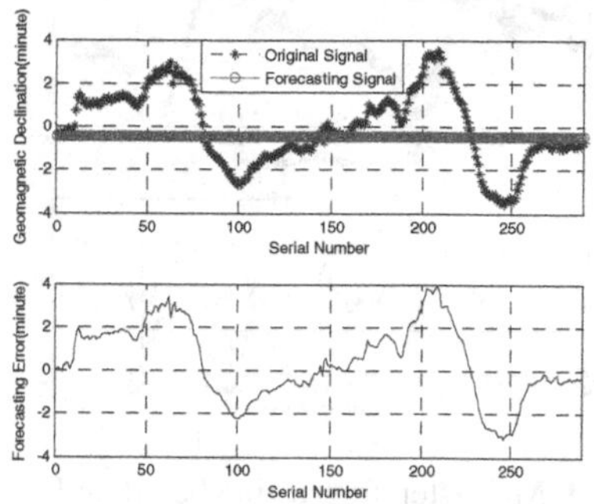

Fig. 3. Multi-step forecasting based on RBF.

2.2 Forecasting Based on BP Neural Network

BP neural network is a multi-layer forward network, usually contains one or more hidden layers, error back propagation algorithm is used to train the network. The structure is simple , the plasticity is good. Negative gradient descent method is used to adjust weight, this method of weight adjustment has its limitations, it may fall into local minima. Momentum method is used in learning function to reduce the sensitivity of the local details of the error surface, network is effectively inhibited into a local minimum.

The parameters that BP neural network used to forecast are much more, including network layers, hidden nodes, the network training function, learning function, performance function and so on. Multi-step forecasting geomagnetic declination results based on BP network is shown in Fig.4. It can be seen that the trends of real curve can be reflected by forecasting curve basically, but the error is larger where the changes is rapid.

2.3 Forecasting Based on GRNN Neural Network

General regression neural network(GRNN) is a radial basis neural network, GRNN has a strong nonlinear mapping ability for solving nonlinear problems. The approximation ability and learning speed of GRNN neural network is better than RBF neural network. The structure of GRNN contains four layers, including the input layer, pattern layer, summation layer and output layer.

Multi-step forecasting results of geomagnetic declination based on GRNN is shown in Fig.5. It can be seen that multi-step forecasting results of geomagnetic declination based on GRNN is better, the trends of real curve can be reflected by forecasting curve basically, the error is smaller where the changes is rapid , all the errors can be controlled within 1.5 minute through error statistics.

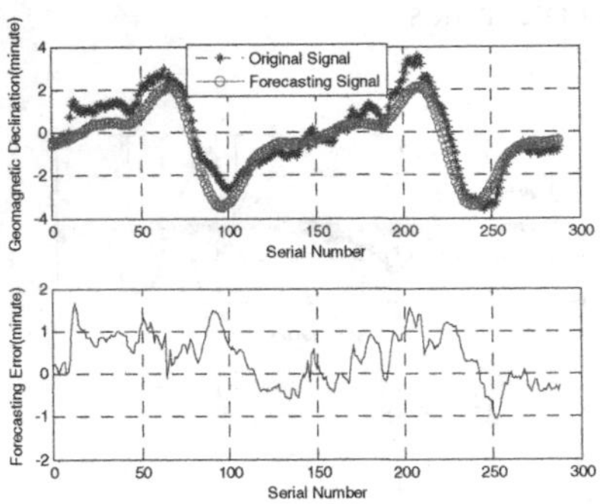

Fig. 4. Multi-step forecasting based on BP.

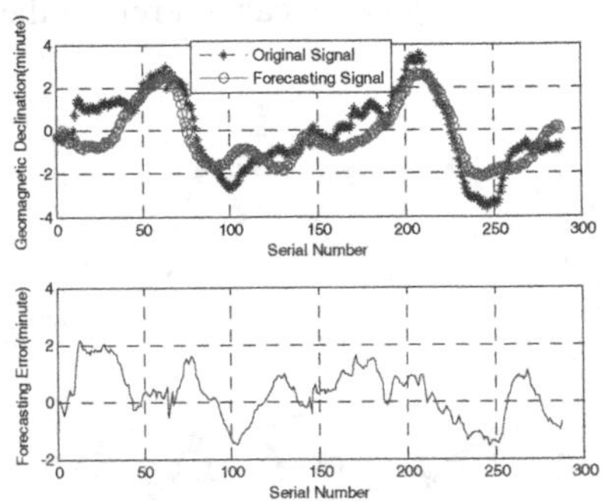

Fig. 5. Multi-step forecasting based on GRNN.

2.4 Forecasting Based on ELMAN Neural Network

Elman neural network is a feedback neural network, its structure contains four layers, including input layer, hidden layer, continued layer and output layer. The connection of input layer, hidden layer and output layer to the network is similar to the forward network, the continued layer is used to carry the memory unit of the the output value of hidden layer one time before and returns to the input of the hidden layer. Internal feedback network is used to increase the ability of network to handle dynamic information. BP algorithm is used to correct weight of Elman neural network.

Multi-step forecasting results of geomagnetic declination based on Elman network is shown in Fig.6. It can be seen that the trends of real curve can be reflected by forecasting curve basically Elman network on the geomagnetic declination of the multi-step forecasting curve basically reflects the trends in the real curve, but the error is larger where the changes is rapid.

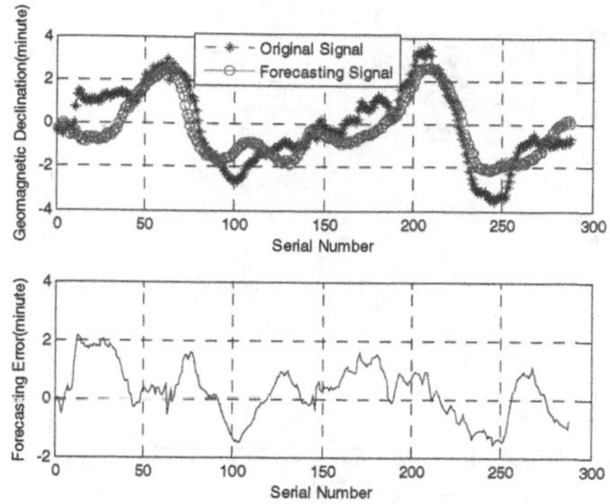

Fig. 6. Multi-step forecasting based on ELMAN.

3 Analysis of Experimental Results

Statistical analysis was done based on multi-step forecasting error of Variable geomagnetic field of the earth described in Section 2. The number that the absolute value of each method of forecasting error is less than the specified threshold was calculated, the proportion of the calculated number to the total number was shown in Fig.7. The results show that:

(1) Forecasting based on RBF is the worst, it is shown that the RBF network does not have the capability to forecast geomagnetic declination of Variable geomagnetic field of the earth with the long-term.
(2) Forecasting based on GRNN is the best, all the forecasting errors of geomagnetic declination can be controlled within 1.5 minute through error statistics. GRNN network is much better than RBF, poor capability of multi-step forecasting has been overcomed. Compared with BP network, the parameters need to be adjusted of GRNN neural network is fewer, only one spread parameter, the calculation is greatly reduced, the calculation speed is greatly improved. And multi-step forecasting accuracy is also higher than the BP network.
(3) BP network and the Elman network forecasting are quite effective, 97% of forecasting errors of geomagnetic declination based on BP network can be controlled within 2 minute through error statistics, 98% of forecasting errors of geomagnetic declination based on Elman network can be controlled within 2 minute.
(4) Compared with RBF and GRNN network, the computing amount of BP network and Elman network is larger, forecast time is longer.

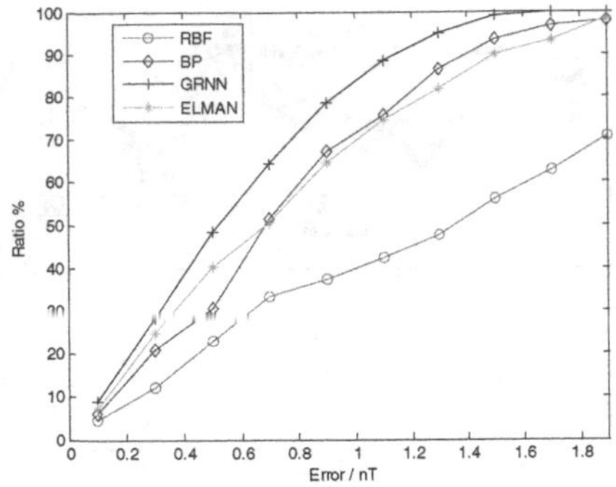

Fig. 7. Multi-step forecasting error of geomagnetic time-series

4 Conclusion

Neural network has its unique advantage on handling the non-linear issues, modeling and forecasting were done to the variable geomagnetic field with non-linear, non-stationary characteristics based on neural network. Modeling and forecasting, simulation analysis were done based on RBF, BP, GRNN and ELMAN. Simulation results show that, GRNN network forecasting accuracy is ideal, and the parameter set is simple, computing value is small; Forecasting effects of BP network and ELMAN network is general, and the parameter set is complex, computing value is large; Single-step forecasting based on RBF network is good, but multi-step forecasting is poor.

Modeling and forecasting based on artificial intelligence methods to the variable geomagnetic field were just done as an initial attempt in this paper. How to choose more suitable methods to model and forecast variable geomagnetic field with intelligent forecasting methods is the work we should do in the future.T

References

1. Xu, W.: Earth physics of electromagnetic phenomena. University of Science and Technology of China Press (2009)
2. Peng, C., Chen, X.: Applied Research on Short-term forecasting of Earthquakes Based on Geomagnetic Methods. Northeastern Seismological Research 23(3), 28–37 (2007)
3. Peng, F.: Geomagnetic model and geomagnetic navigation. Ocean Mapping 26(2), 73–75 (2006)
4. Ren, L.: Research on Medium-term Load Forecasting Model Based on Elman Neural Network. Master's Tesis of Lanzhou Technology University (2007)
5. Yang, J.: Practical Tutorial of Artificial Neural Networks. Zhejiang University Press (2001)

The Implementing of Tourism Resource Management Information System in E-commerce

ZhongXia Hu*

Department of Computer Application, Qingyuan Polytechnic College,
Qingyuan, 511510, Guangdong, China
huzhongxia2011@sina.com

Abstract. With the development of tourism, tourist information will also greatly affect the progress of society, but also on the tourism industry of information technology has had a profound impact on the management. This paper develops tourism resource management information system by ASP and Access database management system in e-commerce. The system uses Dreamweaver MX 2004 software as a tool for system development, using ASP technology to solve. The entire system is based on ASP technology to develop, create the database using Microsoft Access 2000, database query and call through the ASP built-in ADODB components to achieve.

Keywords: Tourism resource, ASP, Access, Management Information system.

1 Introduction

Ordinary users (usually tourists): can be based on area attractions information for simple queries, travel accommodation for tourists services and facilities; and with the growing prevalence of people travel, some related news should also be able to provide timely information to passengers, visitors can post suggestions for the site administrator can give back, etc. Administrator (business operator): the average user has all privileges, all data can be added, deleted and modified; database backup and restore system. While the system is based in Henan Province to develop, but should have broad applicability enter new data, we can build into a simple and practical field of tourism resources management system [1]. With the development of tourism, tourist information will also greatly affect the progress of society, while information technology industry, tourism management also had a profound impact. Since the 1990s, more and more areas of computer applications and information management play a very important role in various types of tourism enterprises in order to meet the needs of market competition have to use computer tools in this modern.

The system uses Dreamweaver MX 2004 software as a tool for system development, using ASP technology to solve. ASP is a server-side scripting in the operating environment, this environment, users can create and run dynamic, interactive Web server applications such as interactive dynamic web pages, including

* ZhongXia Hu (1970.6-),Male, Han, Master of computer department of JiNan University, Research area: computer application, e-commerce, web design.

D. Jin and S. Jin (Eds.): Advances in FCCS, Vol. 1, AISC 159, pp. 89–94.
springerlink.com

the use of HTML forms to collect and process information, upload and download, etc. and so on. Dreamweaver MX 2004 tools with its powerful, flexible data window, object-oriented development capabilities of its advantages in the field of database application development occupies a leading position, these features provide the comprehensive application development capabilities to create favorable conditions. The tourism resources system, the interface clear, easy to read, modify and easy, very easy to the customer. Therefore, the operation has a good feasibility.

In today's growing popularity of computer, network become indispensable part of life, use of holiday trips has become a way to ease the pressure. The development of tourism resources management system for two reasons: First, travel agents in order to better promote their own brand names, the use of network resources to promote his own company, to provide users with tourist routes, developing tourism projects, so as to create benefits for the company; two The user is in desperately work, it is also necessary to enjoy their lives, the new system by using computer networking to promote integration of various departments, improve efficiency, reduce manpower for the travel agents expenditure, improve operational efficiency, the implementation of the new system will be powerful . Tourism resources management system brings us a wealth of travel information, can not go out that the latest travel information, find a suitable own play, to enjoy the pleasure of travel.

With the development of tourism, tourist information will also greatly affect the progress of society, while information technology industry, tourism management also had a profound impact. Since the 1990s, more and more areas of computer applications and information management play a very important role in various types of tourism enterprises in order to meet the needs of market competition have to use computer tools in this modern.

2 Browser/Server Mode and ASP Technology

Browser / Serve: is called a browser / server mode, abbreviated as B / S. It is with the development of network technology and the rapidly growing up as a user to complete the service, users can be realized only through the server browser services provided and it is replacing old-fashioned client / server mode (Client / Serve: abbreviated C / S). And C / S mode compared, B / S method has the following advantages.

(1) For the back-end database is distributed structure of the system, with B / S mode can easily achieve the exchange of information; and C / S mode can only work in the two systems can not meet the multi-layer structure of the distribution system.

(2) B / S mode ultimately provide to the user's information is fully rendered in the browser, and the client operating system independent; and C / S mode in the user's software mostly direct effect on the server back-end database-specific, so whether it is server-side or a client's environment has changed, we should rewrite the source code, and B / S has obvious limitations compared to. Achieve B / S modes of Web technology are many.

ASP (Active server pages) is similar to HTML, Script and CGI (Common GAteway Interface Common Gateway Interface) combination, but its higher operating efficiency than CGI, HTML programming is more than convenient and more flexible, application security and confidentiality of better than Script.

HTML is a hypertext markup language, file format can be interconnected through the Internet browser, users will be using a web browser tool you can browse the files, now more commonly used tools include Microsoft Internet Explorer, Netscape Communicator, etc., as HTML files by tag (tag) of the composition, making it more suitable for static pages, in addition, due to inherent limitations on the HTML is not directly access the database, so access to the database work mostly rely on CGI to handle. ASP can not only contain HTML tags, you can also directly access the database and use the unlimited expansion of the ActiveX control, so the programming on the convenient and richer than the flexibility of HTML. HTML-based ASP, HTML code, or the main, only where required to join dynamically generated ASP ASP code. The final result is shown ASP HTML code, can accurately control the output of the ASP.

$$confidence(X \Rightarrow Y) = P(Y|X) = support(X \cup Y)/support(X) \qquad (1)$$

Script (script) is a group in the WEB server or client browser running on a combination of command, the current preparation of the more popular web scripting language, including VBScript, JavaScript. These scripts are mostly run on the client, so the client can clearly get the script content. So, in terms of security, these client-side scripting language is indeed dangerous [2]. ASP scripting language, while having the convenience, but because he is in the WEB server is running, running and then run the results in HTML format will be sent to the client browser. Therefore, ASP scripting language, compared with the general, to secure more.

ASP can be in HTML or other scripting languages (VBScript, JavaScript) nested within each other. ASP is a server running in the WEB scripting language, program code security. ASP object-based, so you can use ActiveX controls continue to expand its capabilities. ADO ASP built-in components, so you can easily access the various databases. ASP can run the results in HTML format sent to the client browser, which can be applied to a variety of browsers.

Active Server Page technology provides a script-based application developers an intuitive, fast and efficient application development, has significantly improved development results. ASP script is in clear text (plain text) way to write. ASP script is a series in a specific syntax (currently supports two scripting languages vbscript and jscript) written, standard HTML page with the mixed composition of the script text file. When the client's end users with a WEB browser to access through INTERNET ASP script based applications, WEB WEB server the browser will send the HTTP request. WEB server analysis, determine the request after the application of ASP script is automatically invoked by ASP script ISAPI interface to explain the engine running (ASP.DLL)[3]. ASP.DLL internal buffer from the file system or ASP script to get the specified file, and then they parse and interpreted. The results will form the final content in HTML format, through the WEB server "backtrack" to return to the WEB browser from WEB browser on the client form the final results show. This completes a full ASP script calls. Number of organic ASP script calls on the formation of a complete application of the ASP script.

3 The Analysis and Design of Tourism Resource Management Information System

The entire system is based on ASP technology to develop, create the database using Microsoft Access 2000, database query and call through the ASP built-in ADODB components to achieve. According to the previous design analysis, system development in accordance with the basic point of view of tourism information systems break down, from content to do the following division: travel dynamic, tourist attractions, specialty style, service facilities, visitors comments, contact us etc.

The system consists of system management, information browsing sites, add and delete features style management, facilities management services, visitor message management, such as video on demand feature six functional modules. Site administrators have the highest administrative privileges, it can add additional administrators, its permissions set. Attractions include updated information: Attractions to add, modify, and delete. Style features to add, delete, and modify. News and information to add, delete, and modify. The updated service information, including: Services and facilities to add, delete modify. The following figure 1 is a functional block diagram of the system.

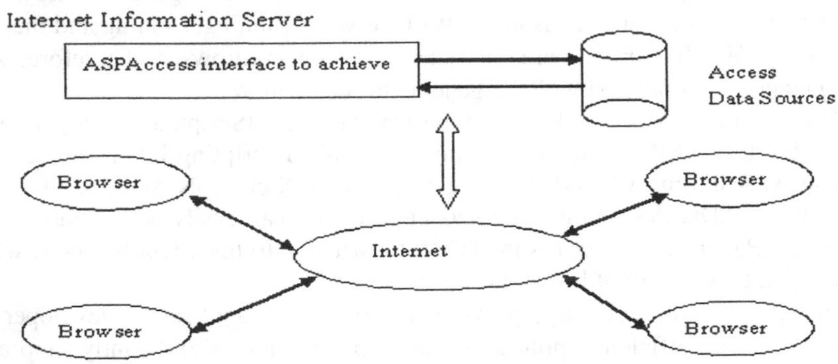

Fig. 1. Functional block diagram of the tourism resource management information system.

Database in the construction site play an important role in database design is good or bad will report directly to the application system efficiency and achieve the effect of an impact. The department opened the main page, "Attractions", "character style", "services", "Video Center" four operating sections, and in order to facilitate communication with visitors, but also set up a "Contact Us" section. Among them, the "Attractions" and "character style" two columns is the focus of web design, travel agents in the background by the system to control its output. "Contact us" compared to travel agents and tourists provide a simple means of communication.

Management is to manage all the information your web site front operations, information on attractions, special customs, services, news and information release, amended to delete the site front a variety of information dissemination to modify delete, etc., a good background, you can easily manage all information on the site.

4 The Implementing of Tourism Resource Management Information System by ASP

The system meets the dual aspects of tourists and travel requirements. System with its convenient, easy operation and beautiful interface for users to save time, the full realization of tourism resources and information to add, change management, etc., thus greatly reducing the workload of the staff travel, improve management efficiency and service quality. The entire system is based on ASP technology to develop, create the database using Microsoft Access 2000, database query and call through the ASP built-in ADODB components to achieve. The system consists of system management, information browsing sites, add and delete features style management, facilities management services, visitor message management, such as video on demand feature six functional modules. It can be seen from the figure, the system is based on B / S model developed. The E-R diagram entity relationship of the overall site as shown in Figure 2.

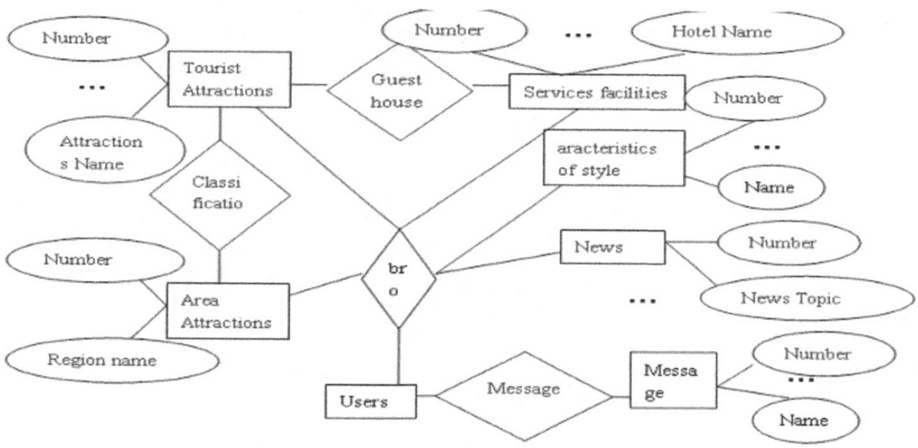

Fig. 2. E-R diagram entity relationship.

5 Summary

The system uses Dreamweaver MX 2004 software as a tool for system development, using ASP technology to solve. ASP is a server-side scripting in the operating environment, this environment, users can create and run dynamic, interactive Web server applications such as interactive dynamic web pages, including the use of HTML forms to collect and process information, upload and download, etc. and so on. Dreamweaver MX 2004 tools with its powerful, flexible data window, object-oriented development capabilities of its advantages in the field of database application development occupies a leading position, these features provide the comprehensive application development capabilities to create favorable conditions. So use it as development of tourism resources management system, is entirely feasible.

References

1. Min, W.: A Research on Statistical Information Applied to Tourist Traffic and Transport System Design Based on ASP. NET. JCIT 6(1), 147–156 (2011)
2. Yang, Q., Shi, J., Tang, B.: Distributed Dynamic Channel Access Scheduling Method for Wireless Ad Hoc Network. JDCTA 5(3), 310–319 (2011)
3. Torpelund-Bruin, C., Lee, I.: Geospatial Web Data Tessellation and Visualization. AISS 3(2), 87–93 (2011)

People Flow Control Using Cellular Automata and Computer Vision Technologies

Roberto Di Salvo, Alberto Faro, Daniela Giordano, and Concetto Spampinato

Department of Electrical, Electronics and Computer Engineering, University of Catania, viale A.Doria 6, Catania, Italy
{roberto.disalvo,afaro,dgiordan,cspampin}@dieei.unict.it

Abstract. The paper proposes an automatic fuzzy system to control in real time people flows in crowded scenes. An integration of cellular automata and computer vision technologies is envisaged to control the people flow in different conditions and in real time. An adaptive control paradigm has been adopted to manage real scenarios since they usually differ slightly from the theoretical ones. The proposed fuzzy control is not only simple, but it may incorporate measurements derived from both fast and accurate computer vision techniques and perceptions of people operating on the field.

Keywords: Cellular Automata, Fuzzy systems, People Behavior Understanding, Computer Vision, Computing with words.

1 Introduction

The paper proposes an automatic fuzzy system to control in real time people flow in crowded scenes. An integration of cellular automata and computer vision technologies is envisaged to control the people flow in different conditions. Although many papers present how modeling people flows, few feasible proposals are available to control such critical systems in real time, since this needs to implement an adaptive control where both simulation tool and sensing technologies should be integrated suitably.

Both the parameters that characterize the geometry of the context of a crowed scene, and the ones dealing with the people (e.g., density, speed, and flow) are measured by two effective computer vision (CV) technologies [1], [2] to set off-line the basic parameters of the model that controls the flows in typical scenes such as workers entering/exiting to/from their farm, flows of commuters at the rail station, flows of tourists at the airport, evacuation of employees from offices. Such CV technologies are also used on-line to fine tuning the control indications and to measure obstacles that are not present in the original scenarios, or objects traversing the scene. To this aim, when the control system verifies in real time that density and speed of the flow differ from the one foreseen by the model, it will modify the electronic signs until such difference is within a certain threshold.

Sect.2 illustrates the cellular automata based tool able to simulate the behavior of people in typical flowing scenarios. Sect.3 derives from specific simulation tests, the fuzzy rules that help finding the right indications to be given to the people. Sect.3 illustrates some policies for managing the people flow in some realistic scenarios following the adaptive control paradigm.

D. Jin and S. Jin (Eds.): Advances in FCCS, Vol. 1, AISC 159, pp. 95–104.
springerlink.com

2 Modeling People Flow by Cellular Automata

Ideally, pedestrians should proceed with low speed variations to allow other people to enter into the same path under the condition that each person has a suitable space around her/him. Since we are interested in monitoring and controlling large flows of people, in the paper we assume a macroscopic model of the people flow, where flow q, density k and velocity v obey to the classical continuity equation governing any traffic stream [3], i.e., $q = v\ k$. As for the car flow, also in the people flow $q < q_{max}$, since after q_{max} the density increases, but the speed decreases more rapidly until a complete saturation is achieved.

People flow depends also on the context. Indeed, boundary conditions greatly influence the flow and the speed along the path. Also, impediments contrasting the people flow may decrease suddenly the flow. Therefore finding the optimal flow profile requires solving a complex problem that depends on the flow sources insisting on the path, on street dimensions, on the impediments that may vary the street width and on climate conditions. To solve this problem we may adopt an analytical approach, e.g. [4], [5]. However, such models are not appropriated for real time control, as the ones of our interest, whose parameters may vary during the time. Thus, we adopt a set of fuzzy rules derived from a simulation tool able to inform the people by means of electronic signs on the walking regime to be adopted.

In particular, the simulation tool proposed in the paper is based on the cellular automata technique since it is known from the literature its effectiveness and simplicity in modeling complex pedestrian and car flows, e.g.,[6], [7], [8], [9]. Currently, this tool allows us to simulate the presence of obstacles of any dimension placed at either the right and left sides or in the center of the street interested by the people flow. It is possible to insert both check-in systems provided with one or more doors and traffic lights, which may influence the people flow dynamics.

Fig. 1 shows the cellular automaton modeling a pedestrian. It simulates her/his lateral and forward move in a crowd and in absence of any indication. In particular, given the proximity matrix **M** of fig.2 and a pedestrian at the position (i_p , j_p), the cellular automaton assumes that the preferred movement is to go forward unless on the left or on the right of the people in front to her/him there is the possibility to go faster. Although its simplicity, the pedestrian model is enough to simulate a realistic pedestrian behavior as pointed out in the literature, e.g., [5]. An extension of the proposed pedestrian model based on the notion of social force is for further study.

To demonstrate how the fuzzy control rules and related fuzzy sets may be derived from our tool, first we simulate the behavior of the people flow in absence of any indication to change the walking direction and speed. The chosen scenario (see fig.3) deals with the typical people flowing towards a place (e.g., a stadium) hosting a mass event. The main gates, located at distance from the event, may be used to regulate the flow. An internal stop signal avoids that the people flow interferes with other flows.

GIVEN	
the typical velocity $\nu = 72$ meters per minute of the pedestrian flow	
the cell area $A_c = 60x60cm^2$ of the cellular automaton	
the time interval $\Delta T = 1\ sec$ used in the simulation tool	
the pedestrian coordinates (ip, jp) at time t0	
the proximity matrix $M(i_p : i_p + 2, j_p - 1 : j_p + 1)$ around (ip, jp), since $v/\Delta T$ = two times the cell length	
the probability $p(i,j)$ that the pedestrian reaches during time $t_0 + \Delta T$ the cells (i,j) of M	
$f = min(e_{forward}, v/\Delta T)$, $l = min(e_{left}, v/\Delta T)$, $r = min(erigth, v/\Delta T)$, where $e_{forward}$, e_{left} and e_{rigth} are the empty cells respectively in front, to the left and to the right of the pedestrian	
WHEN	**THEN**[a]
$f = l = r = 0$	$p(i_p, j_p) = 1$
$f \neq 0, f \geq l, f \geq r$	$p(i_0 + s, j) = \frac{1}{f}$, where $s = 1 \ldots f$
$l \neq 0, l > f, l > r$	$p(i_0 + s, j_0 + 1) = \frac{1}{l}$, where $s = 1 \ldots l$
$r \neq 0, r > f, r > l$	$p(i_0 + s, j_0 - 1) = \frac{1}{r}$, where $s = 1 \ldots r$
$l = r \neq 0, l > f, r > f$	$p(i_0 + s, j_0 + 1) = \frac{0.5}{l}$ $p(i_0 + s, j_0 - 1) = \frac{0.5}{r}$, where $s = 1 \ldots l$

Fig. 1. Cellular automaton modeling a pedestrian

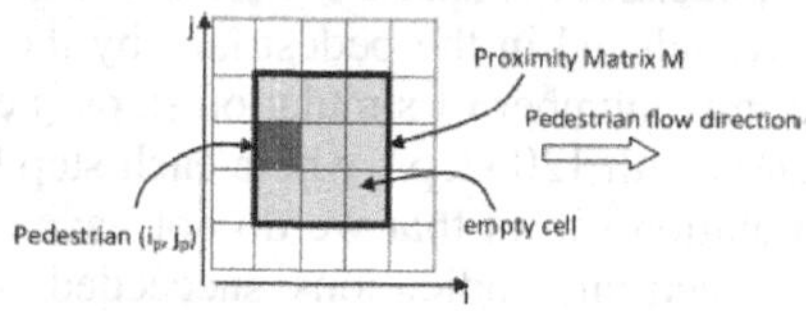

Fig. 2. Proximity Matrix for a pedestrian at position (i_p, j_p)

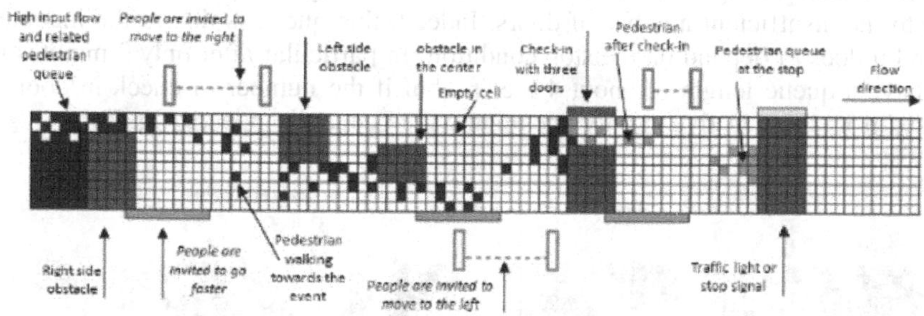

Fig. 3. General structure of the cellular automaton modeling the people flow possibly controlled by electronic signs in presence of various flow impediments.

In principle the queues outside the street, leading to the check-in, may not be so dangerous since they usually develop in a big square, whereas internal queuing should be avoided. Fuzzy rules concerning evacuation could be found by simulating a similar scenario with people moving in the opposite direction. The result of the simulation (fig. 4) shows that there is a clear tendency of the pedestrians to thicken themselves towards the obstacles which behave as a sort of stop condition.

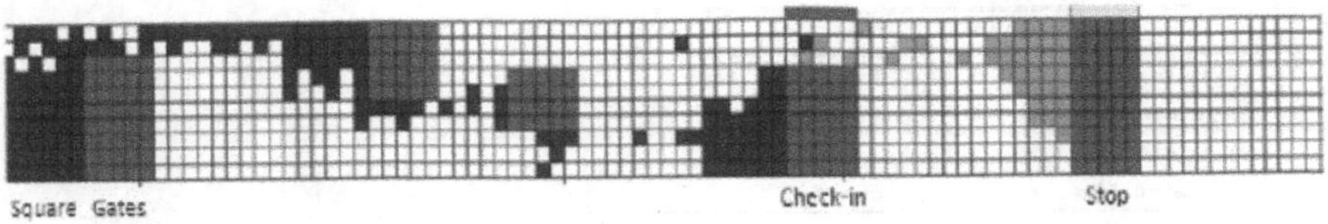

Fig. 4. Simulation of the people's flow going towards a stadium without any control system.

Thus, we introduce in the simulation some electronic signs to invite the people to go in a direction that allows them to avoid an incoming obstacle which is timely known by the control system thanks to the data collected by the webcams. In this case the pedestrian model described in fig. 1 should be extended to simulate the lateral move induced in the pedestrians by the electronic signs. Fig.5 shows the results after the same number of simulation steps (i.e. 60 sec.) during which the people finishes 72 meters with 120 steps, where each step length is assumed to be 60 cm long. Since the simulation shows that we do not have any congestion near to the obstacles, certainly the electronic indications succeeded to avoid accidents. However, the simulation shows also that two queues are increasing more and more at the check-in and at the internal traffic light. This implies that we have to increase the green cycle and, if possible, the number of check in doors. This rule is confirmed by fig 6, which shows that even if the stop condition is removed (e.g., by organizing that the pedestrian flows never intersecting with other flows) there is a dangerous situation at the check in due to the insufficient number of doors. Indeed, this queue is due to the incoming flow and it doesn't depend on the stop condition. In particular after only 2 minutes we could have a queue length of about 4 meters, but if the number of check-in doors is increased from three to five in only one minute the queue disappears (fig. 7).

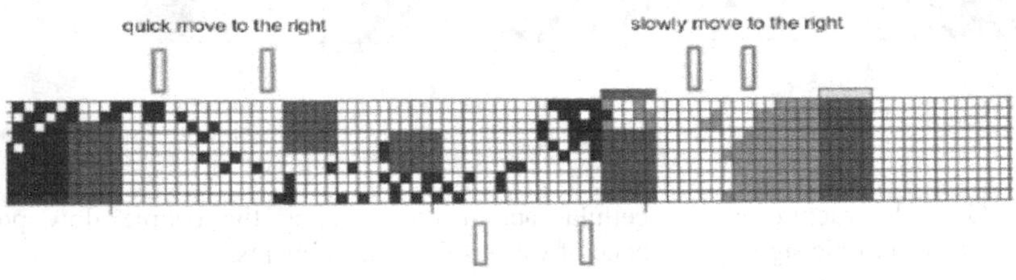

Fig. 5. Simulation of people's flow going towards the stadium guided by electronic signs.

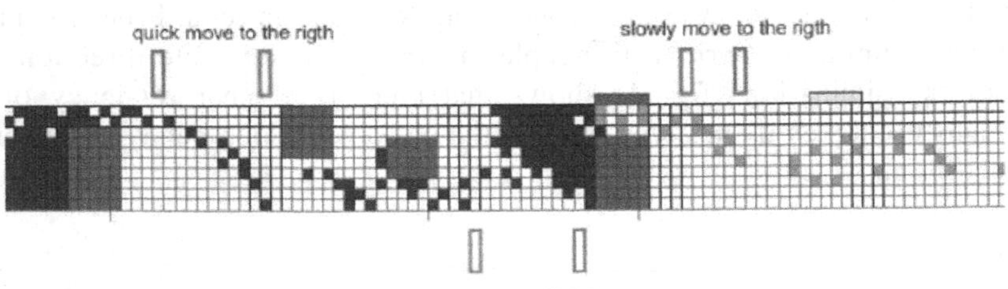

Fig. 6. Flow simulation when the stop condition is removed.

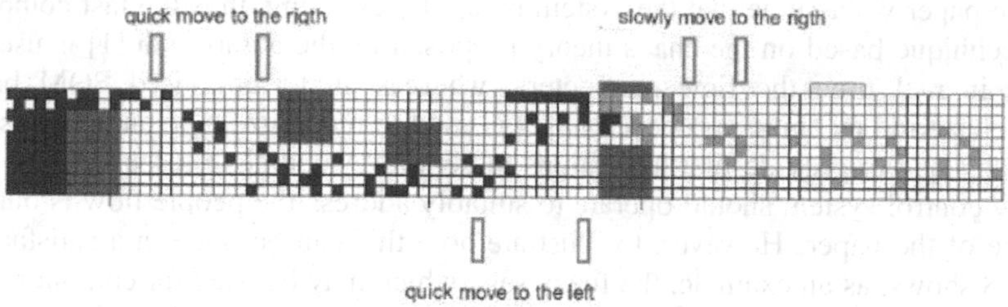

Fig. 7. Flow simulation while increasing the number of check-in doors.

3 Controlling the Flow by Fuzzy Rules

From the above discussion, we derive the following fuzzy rules:

- Achieving high throughput without causing unsustainable queues: a) if the density in a zone of a street slowly/greatly increases/decreases (e.g., due to obstacles, check in points or light traffic), then the input flow of the street should moderately/highly decrease/increase, alternatively the conditions which caused the congestion should be relaxed, e.g., by increasing the number of check-in doors or by increasing the green cycle of the traffic light and so on; b) The people that surpassed an obstacle or the check-in should be invited to go faster;
- Avoiding collisions with obstacles: if an obstacle moderately/highly reduces the street section for a small length/significant length, then the flow of the previous sections should directed to the opposite direction/s to the one in which the obstacle is located with moderate/high speed;
- Avoiding of being involved in some brawl: if a small/great brawl arises in the street, then the input flow of the street should be suddenly decreased, whereas the incoming flow should be directed to zones distant from the brawl with moderately/high speed, and the police intervention should be planned shortly/suddenly;
- Restoring the normal condition after the traffic problem has been removed: if an obstacle or a brawl, which reduced the trait section for a small length/its total length is removed, then the incoming flow should be distributed along all the front of the street with moderate/high speed and the input flow of the street should be moderately increased.

The domains of the fuzzy variables appearing at both the right and left sides of the above fuzzy sets are people density, people flow, people speed, obstacle dimension, and brawl dimension. Thus an automated monitoring system is needed to identify in real time these values. In particular, density, speed and flow should be monitored for a sufficient number of street sections, whereas the control system has to compute the dimensions of the flow impediments to suggest the most appropriate lateral move. Such computations can be done from the images taken by a camera either using analytical models or soft computing techniques depending on the available computing resources and on the speed of the flow under control.

In the paper we assume that the system is rapidly evolving, thus the fast computer vision technique based on the chaos theory proposed by the authors in [1] is used to measure in real time the flow parameters, whereas an unsupervised SOM based parallel clustering technique proposed in [2] is used to recognize type and dimensions of obstacles and to provide first indications to people. A complete discussion on how the fuzzy control system should operate to suitably address the people flow is outside the scope of the paper. However, to illustrate how this can be done in a satisfactory way, fig.8 shows, as an example, the fuzzy sets, which may be used for computing the fuzzy rules and related indications for the optimal flow control in presence of obstacles recognized from the sensed data.

In general, the length and the width of the detected obstacle would lead to issue contrasting invitations. For example in the situation depicted in fig. 9 the fuzzy rules address people respectively to go quick and very slowly to the right. However, due to the different weights of such indications, it would be easy for the fuzzy control system to average such indications, thus finding the best invitation to display to people, as shown in fig.8c, with respect to the type of the detected obstacle.

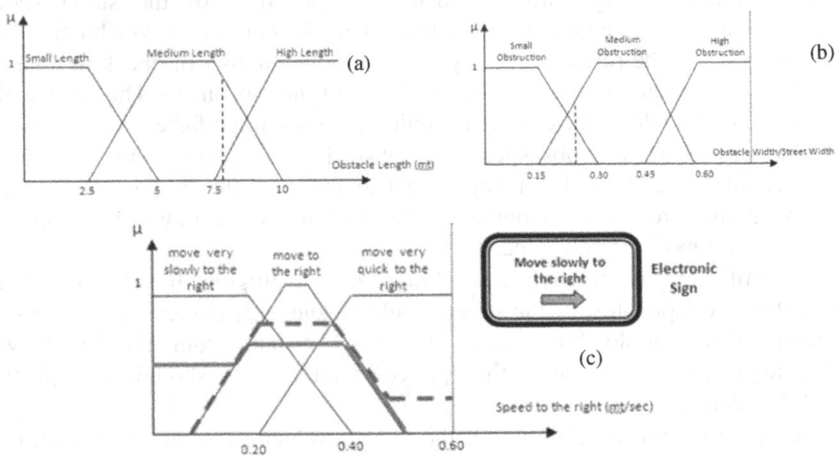

Fig. 8. An example of how the fuzzy control system operates: the sensed data indicates an obstacle 8 meters long a) and occupying 20% of the street section; b) the weight of the indications to go quick or slowly to the right due to the fuzzy rules are shown in fig.8c, where the dotted line depends on the obstacle length L, and the solid line on the obstacle width W. The indication resulting on the electronic sign derives from the average of such two indications.

4 Adaptive People Flow Control: An Example

To better understand how the overall proposed system works, let's consider the scenario shown in fig. 9 related to a train station. By processing a set of 600 frames (20 seconds) using the SOM like clustering [2] we detect, as shown in the same fig.9, a different behavior in proximity of the doors to get on the trains. For each part of this region we extract the size. Moreover we measure the mean speed v and the average flow q, by using the chaos theory based technique [1]. This allows us to discover if such zones the people does not move. Both the techniques have an accuracy of 90% and allow us to process the images in few tens of milliseconds.

Fig. 9. Detected event over a set of 600 frames.

The information about the configuration of the regions close to the train doors containing people which is not in motion is very important to define the policy to manage the people flow. In fact, the simulation results show that there are two main interventions to implement in this situation: a) to invite the people to walk on the center and b) to put some barriers near to the trains, e.g., two yellow lines, which address the flow to the center. Fig.10 shows the flow without any intervention, whereas fig. 11 and fig. 12 show the benefits of the above mentioned interventions in terms of absence of long queues near to the trains.

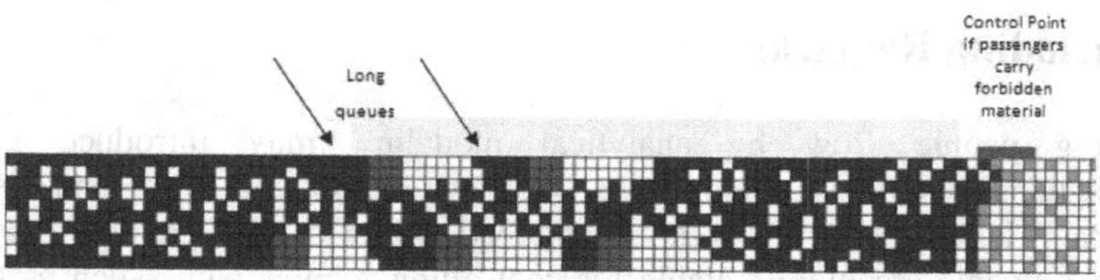

Fig. 10. The people flow is not controlled. There are long queues near to the trains. The throughput is equal 3.34 persons/sec..

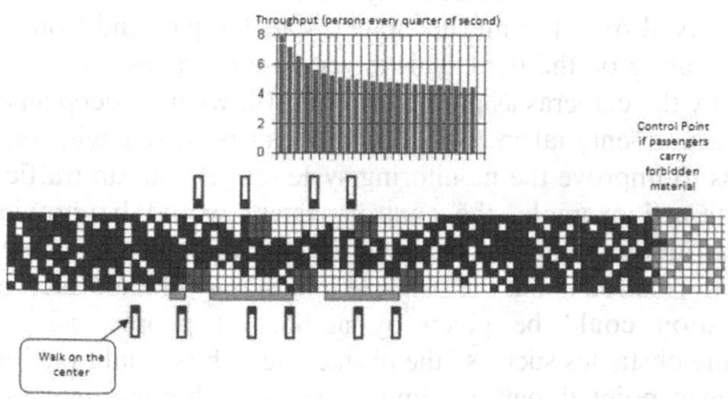

Fig. 11. The people flow is controlled by invitations to walk on the center. There are small queues near to walk on the center. The throughput is equal 4.4 persons/sec.

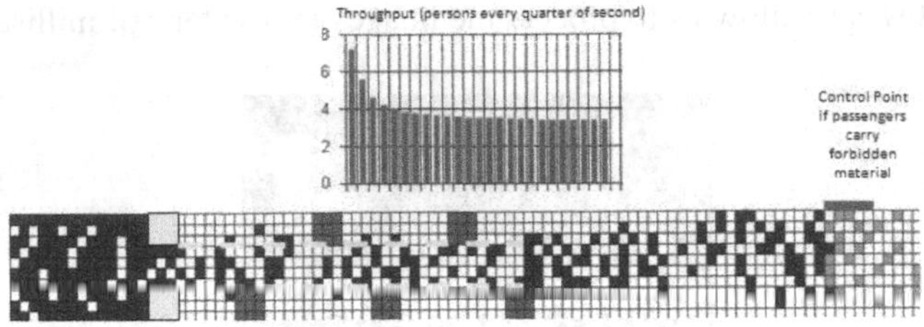

Fig. 12. The people flow is controlled by "not passable" yellow lines. There are no queues near to the trains. The throughput is 3.35 persons/sec.

Let us note that fig.12 does not present queues, but it has a throughput of about 30% less than the one depicted in fig.11. However, in case the throughput is not important, e.g., since there is a control at the end of the binary, the solution b) seems to be the preferable one. Also, solution b) is recommended, possibly integrated by invitations to walk on the center, at the increase of the distance between two successive exit doors from the trains. In fact in such cases, the simulation shows that inviting people by electronic signs to walk on center produces a few increase of the throughput (i.e., from 3, 4 to 3,7 persons/sec in our case) with respect to the situation controlled by only the yellow lines.

5 Concluding Remarks

Controlling people flow by analytical modeling may introduce excessive approximations to model outdoor paths, such as a street with trees and disordered parked cars, bad weather conditions, and bad status of the paving. Indeed, the analytical models are more suitable for controlling regular paths such as the flow through building or metro indoor corridors. On the contrary a simulation tool integrated by data coming from the field may be fast and accurate to identify the fuzzy rules that support effectively people flow control even in complex contexts.

The proposed fuzzy control methodology is not only simple, but it may incorporate measurements derived from fast and accurate CV techniques and from the perceptions of the people operating on the field. This is very useful in case some parts of a scene cannot be taken by the cameras as pointed out in [10], where a deep analysis is carried out on how measurements taken from cameras can be fused with the ones derived from perceptions to improve the monitoring system of the urban traffic. Indeed, if, in the above mentioned example, the control system would be provided with only information about the obstacle width, it will invite the people to move slowly to the right, whereas information about the obstacle length may change such indication. The missing information could be given by authorized people using simple fuzzy descriptions of the obstacles such as "the obstacle length is small, medium, or high".

Also, the paper pointed out the importance of adopting the adaptive control paradigm because the real scenarios always differ from the theoretical ones, e.g., the control of the flows at the rail stations is obtained by setting the parameters of the fuzzy control system with respect to typical situations such as the arrival of commuter trains, tourist trains, and so on, whereas, in practice, a train has a composition which differs from these typical compositions. In future we plan to define the terms most used in the people flow control applications so that the visual/audio information that address pedestrians towards safe zones may be given on their mobiles following standard format instead of proprietary codes, e.g., [11], [12], [13]; in addition the adaptive control will be facilitated by setting the starting parameters of the model equal to the parameters of the model retrieved from a design memory [14] that best fit the current people behavior.

Advanced graphical interfaces that alert people by proper electronic signs and by clear indications on their mobiles are also under development due to the importance of providing the mobile users with effective visual information [15]. Also, very effective computer vision technologies will be integrated in the application for denoising the images taken by the cameras, e.g., [16], [17]. so that it will be possible not only to control the overall people flow but also to recognize people, e.g. [18], responsible of illegal activities.

References

1. Spampinato, C., Faro, A., Palazzo, S.: Event Detection in Crowds of People by Integrating Chaos and Lagrangian Particle Dynamics. In: Proc. 4th International Conference on Advanced Computer Theory and Engineering (ICACTE 2011), pp. 779–784. ASME (2011)
2. Faro, A., Giordano, D., Maiorana, F.: Mining massive datasets by an unsupervised parallel clustering on a GRID: Novel algorithms and case study. Future Generation Computer Systems 27(6), 711–724 (2011)
3. Chowdhury, M., Sadeh, A.: Fundamentals of ITS Planning. Artech House (2003)
4. Chen, M., Barwolff, G., Schwandt, H.: A Derived Grid-based Model for Simulation of Pedestrian Flow. J. of Zhejiang University - Science A 10(2), 209–220 (2009)
5. Seyfried, A., Steffen, B., Lippert, T.: Basics of modeling the pedestrian flow, vol. 368, pp. 232–238 (2006)
6. Schadschneider, A.: Modelling of Transport and Traffic Problems. In: Umeo, H., Morishita, S., Nishinari, K., Komatsuzaki, T., Bandini, S. (eds.) ACRI 2008. LNCS, vol. 5191, pp. 22–31. Springer, Heidelberg (2008)
7. Wąs, J., Gudowski, B., Matuszyk, P.J.: New Cellular Automata Model of Pedestrian Representation. In: El Yacoubi, S., Chopard, B., Bandini, S. (eds.) ACRI 2006. LNCS, vol. 4173, pp. 724–727. Springer, Heidelberg (2006)
8. Tonguz, O.K., Viriyasitavat, W., Bai, F.: Modeling urban traffic: a cellular automata approach. Comm. Mag. 47(5), 142–150 (2009)
9. Sarmady, S., Haron, F., Talib, A.Z.H.: Modeling groups of pedestrians in least effort crowd movements using cellular automata. In: Asia International Conference on Modeling and Simulation, pp. 520–525 (2009)
10. Faro, A., Giordano, D., Spampinato, C.: Evaluation of the traffic parameters in a metropolitan area by fusing visual perceptions and CNN processing of webcam images. IEEE Transactions on Neural Networks 19(6), 1108–1129 (2008)
11. Faro, A., Giordano, D., Spampinato, C.: Integrating location tracking, traffic monitoring and semantics in a layered ITS architecture. Intelligent Transport Systems, IET 5(3), 197–206 (2011)

12. Stoffel, E.-P., Lorenz, B., Ohlbach, H.J.: Towards a Semantic Spatial Model for Pedestrian Indoor Navigation. In: Hainaut, J.-L., Rundensteiner, E.A., Kirchberg, M., Bertolotto, M., Brochhausen, M., Chen, Y.-P.P., Cherfi, S.S.-S., Doerr, M., Han, H., Hartmann, S., Parsons, J., Poels, G., Rolland, C., Trujillo, J., Yu, E., Zimányie, E. (eds.) ER Workshops 2007. LNCS, vol. 4802, pp. 328–337. Springer, Heidelberg (2007)
13. Faro, A., Giordano, D., Musarra, A.: Ontology based intelligent mobility systems. In: Proc. IEEE Conference on Systems, Man and Cybernetics, vol. 5, pp. 4288–4293. IEEE, Washington (2003)
14. Faro, A., Giordano, D.: Design memories as evolutionary systems: socio technical architecture and genetics. In: Proc. IEEE International Conference on Systems Man and Cybernetics, vol. 5, pp. 4334–4339 (2003)
15. Giordano, D.: Evolution of interactive graphical representations into a design language: a distributed cognition account. International Journal of Human-Computer Studies 57(4), 317–345 (2002)
16. Cannavò, F., Nunnari, G., Giordano, D., Spampinato, C.: Variational method for image denoising by distributed genetic algorithms on grid environment. In: Proceedings of the 15th IEEE International Workshops on Enabling Technologies: Infrastructure for Collaborative Enterprises, WETICE 2006, pp. 227–232. IEEE, Washington, DC (2006)
17. Crisafi, A., Giordano, D., Spampinato, C.: GRIPLAB 1.0: Grid Image Processing Laboratory for Distributed Machine Vision Applications. In: Proc. Int. Workshops on Enabling Technologies: Infrastructure for Collaborative Enterprises, WETICE 2008. IEEE (2008)
18. Faro, A., Giordano, D., Spampinato, C.: An automated tool for face recognition using visual attention and active shape models analysis. In: Proc IEEE Conference in Medicine and Biology Society, EMBS 2006, pp. 4848–4852 (2006)

System Dynamics Model on Military—Supply Control

Yi-Shan Wang

Dept. of Automobile Management, Auto Management Bengbu Automobile
N.C.O Academy, Bengbu China
ngcwyz@sina.com

Abstract. SMI (Supply Management Inventory) is divided into two types：SMI based on passive supply inventory control and SMI based on active distribute inventory control. The traditional supply chains, passive supply—SMI and active distribute—SMI are modeled and simulated with System Dynamics tools. And some traditional constraints on Distribute Centers are loosened, such as unlimited capacity. And inventory management methods of Distribute Centers based on SMI are led into the model. After the simulation, in terms of these comparable data, the active distribute—SMI out performs the other two is concluded.

Keywords: military supply control, Distribute Centers, System Dynamics.

1 Foreword

It requires massive supply that to do system battle base on information. Therefore, 《compendium of construct modern supply completely》 put forward： "construct gradually modern military supply system that integrate purchase, storage, transport, distribute base on the national supply system and social ensure resources". Transform supply mode form passive supply to active supply, reduce inventory and curtail reaction time. The article simulate the military supply system, open out rule of military supply system behavior, simulate characteristic of future supply mode behavior, offer reference for decision-making by building model with system dynamics.

System Dynamics commonly didn't aspire rigorous arith logic, didn't pursues optimal solution, didn't according as abstract presumption, but according present world as premise, face object, time or incidence, consider much immeasurable complexity, randomicity and dynamic state, for seeking strategy improving system behavior, capability [1]. This is an important orientation analyzing complex supply system.

When decision-making complex supply system, conventional supply chain decision-making issue and supply management inventory (SMI) decision-making issue is hotspot study issue. So far as, a obvious deficiency is immeasurable capability presume of distribute center [2] ; however in present, require to think over inventory capability and military-economic benefits, presume that distribute center capability is

D. Jin and S. Jin (Eds.): Advances in FCCS, Vol. 1, AISC 159, pp. 105–111.
springerlink.com

finite. To put forward that strategy or champion unit-SMI divide passive supply-SMI and active distribute-SMI, to discuss detailed supply chain function under finite capability. The article will compare with conventional supply chain model and two strategy or champion unit management inventory model.

2 Building Model

Building the model with simulation software-Ithink5.0, the model element includes mainly four sorts: stock, flow, converter and connector. The article use storage to standing for inventory, use conveyor to represent the transport process and production process [2]. Using array function of ithink, we simulate inventory operations of five independent combat troops.

Traditional supply is a structure that don't share information and coordinate among all tache. We hypothesis that strategy or champion unit have a supplier and a distribute center, there are five different combat troops in downstream; distribute center supply directly when can meeting down stream's require; the supplier need to organize production when the distribute center inventory is deficiency. Producing periods of war materials is n day, combat troops check inventory everyday. Combat troops require supply when quantity lows require point. Strategy or champion unit doesn't exchange information about war materials consuming with battle troops. Five combat troops of downstream are located in different geographic locations; it is teams need t days that shipping war materials from the distribute center to each combat troops.

For the sake of describing every tache of supplier receive, retain, perform and complete the supply request, we design two inventories: supply-backlog and unit-backlog, representing respectively request accumulation that strategy or champion unit and battle troops received; withal design request flow: ordering-at-unit, ordering-with-supply u-discharge and s-discharge, cooperating that. Combat requirement affect inflow valve of supply request of battle troops, the value of supply-backlog and unit-backlog control directly shipping-to-unit and shipping-to-battle. The quantity of supply request of every battle troops control directly supply request flow valve of strategy or champion unit namely ordering-with-supply. The strategy or champion units decide work order by comparing supply request of passive supply inventory control with s-effective-inventory for controlling F-production-process.

According to the control logic, raw and processed material flow F-production-process, Supply-inventory, in-transit-to-unit, Unit-inventory and combat in turn. Model simulate demand increase suddenly after two days war break out, show in 1.

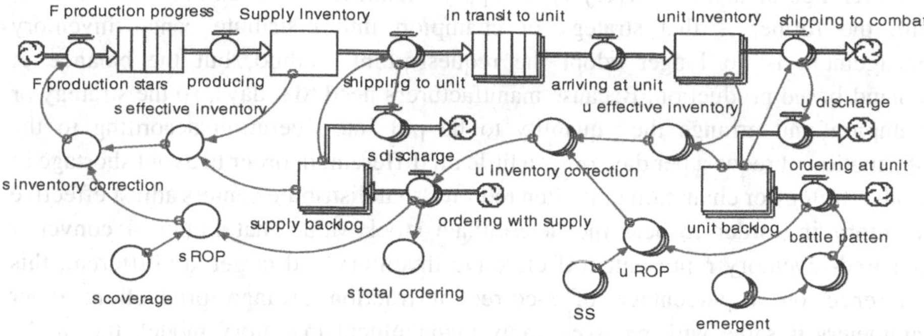

Fig. 1. Traditional supply inventory system model

Difference of Passive-SMI with traditional supply is that battle troops existed communication channels with strategy or champion unit; the consumed information each day must be sent to the strategy or champion unit distribute central library; strategy or champion unit entirely responsible for arranging purchase and quantity when asked to the battle troops' inventory. Shown in Figure 2, five different battle troops supply request information (flow valve array ordering-at-unit) were sent directly to the strategy or champion unit of the supply request processing system (valve ordering-with-supply) through a connection, strategy or champion unit decide the lates request point according to value; strategy or champion unit supply distribute center adopt passive-type inventory management strategy, and other parameters is same with the traditional supply model, detailed model of the building see Figure 2.

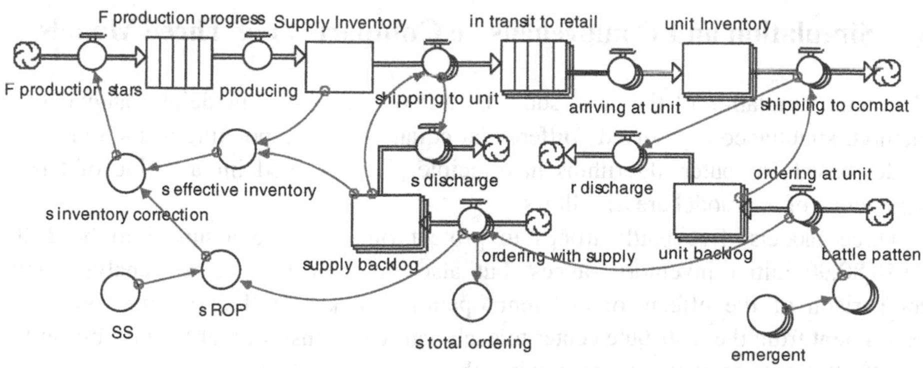

Fig. 2. passive supply inventory system model

Difference of active delivery type supply management inventory (Active-SMI) with the former is that strategy or champion unit distribute center inventory management is no longer adopt the request-point method, but the balance of demand-based production. Because manufacturers need to n days, so the strategy or champion unit arrange the quantity to be put into operation according to the information of request per day; other a little bit different, in order to avoid shortage of goods, strategy or champion unit often requires that distribute centers adjust effective inventory in order to near the determinate stock value that Figure 4 converter s-desired-inventory represented, if effective inventory and target are different, this difference on a percentage of s-correction-fraction arrange production; other parameters is same with passive supply management inventory model, the model shown in Figure 3.

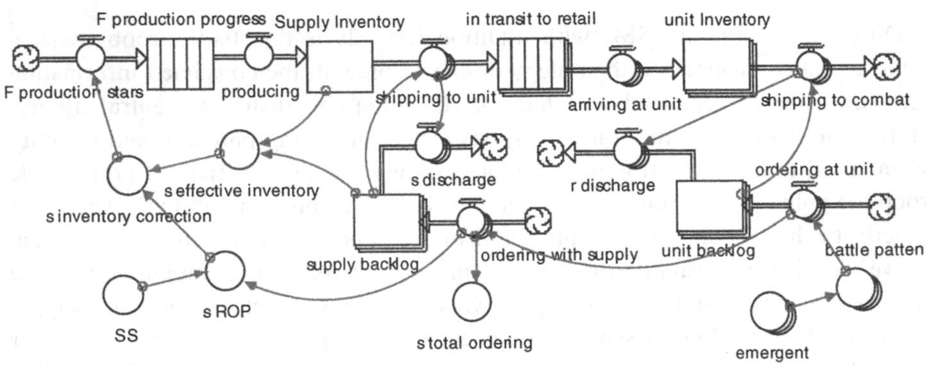

Fig. 3. Active supply inventory system models

3 Simulation and Comprehensive Comparison of Three Models

Computer simulation of the case, adopt DANIMO simulation modeling language in method, simultaneously solved differential equations corresponding to the dynamic model based computer algorithms in principle [3]. Assumed initial value of basic parameters of the model are as follows:

Three models, five battle troops to protect objects were assumed to be 120, 70,50,80,90 initial inventory ranges, but also in order to facilitate analysis and comparison of the effects of different opening stocks; production time one day experiment from the distribute center to each warhead transport (sub) teams are set to one day time; strategy or champion unit is the central link in supply security, upstream suppliers in the strategy or champion unit built around the inventory, ready to supply . After 25 days of simulation, the following analysis:

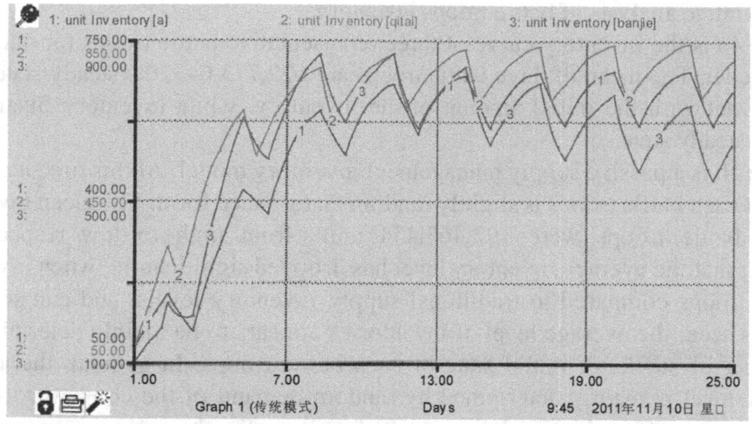

Fig. 4a. Traditional supply inventory

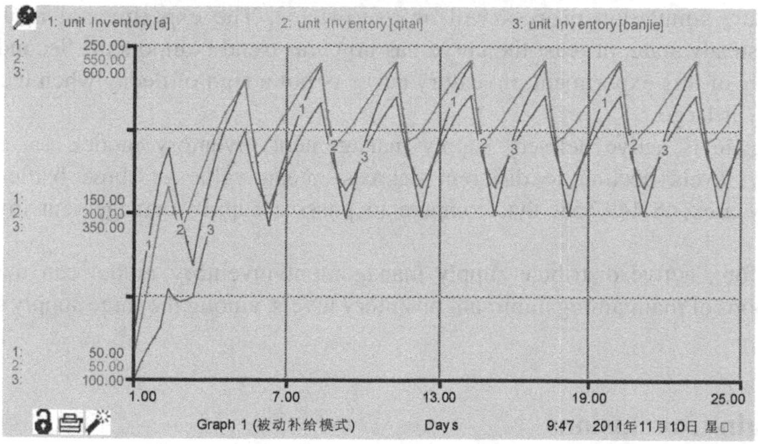

Fig. 4b. Passive supply inventory

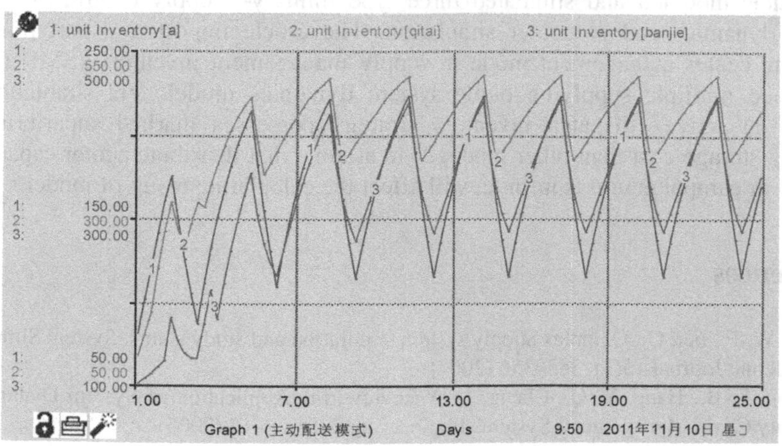

Fig. 4c. Active supply inventory

Comparative analysis of battle troops inventory:

Figure 4a is the inventory curve of three representative battle troops for traditional supply model. The mean of three battle troops are 622,773,643, the steady-state value isn't relevant with the initial amount of the inventory, while inventory fluctuations reaches a steady state.

Figure 4b is a passive supply management inventory model. At this time inventory change of each battle troops is slightly random disturbance form. The mean inventory of three battle troops were 192,465,454 units from high to low respectively, indicating that the average inventory level has dropped significantly, when inventory of battle troops compared to traditional supply inventory levels, and can see from Figure As seen, the average level of inventories appears to be mainly determined by the number of stocks of initial state of each battle troops. In general, the average inventory level is mainly determined by random demand of the combat troops, the initial inventory affect the simulation results for a long time to a certain extent, unless it is 1) the initial inventory than the actual average demand a lot, or 2) not reached steady-state simulation process will be over[4] [5]. The experimental results has reached steady state, at least the curve has no clear trend component; Second, from the results of this experiment, inventory curve is not a sign of decay when the initial inventory is large.

Figure 4c is active delivery supply management inventory model, the average inventory levels decline to different degrees, mean value of three battle troops inventory are 168,448,358, that compare to passive supply management inventory models.

Therefore, active distribute supply management inventory model can make the battle troops of maintaining minimum inventory levels, among the three supply support model.

4 Main Conclusions

The article modeled and simulated three type military—supply control system by system dynamics and computer simulation. The article import transport time and distribute center management mode in supply management inventory (SMI) model, researched multiple supplying battle system dynamics model. Via simulating we proved: 1) active dispatch inventory strategy possesses marked superiority at reducing storage cost than other mode, 2) to assume that distribute center capacity of strategy or campaign unit is infinite, will affect the calculating result of model.

References

1. Zhu, W.-F., Fei, Q.: Complex supply system simulation and study status. System Simulation Academic Journal 15(3), 353–356 (2003)
2. Zhang, L.-B., Hang, Y.-Q., Cheng, J.: A Review: the Application of System Dynamics in Supply Chain Management. Systems Engineering 31(6), 8–15 (2005)

3. Sun, B.-F., Lv, X.-W., Li, J.: Research on Order Point of Distribute Center Based on VMI. Application Research of Computers 23(2), 48–49 (2006)
4. Tang, Q.-S., Xie, R.-H.: Problems and countermeasures of sties park construction. Chinese Journal of Chongqing Jiaotong University 23(3), 107–110 (2004)
5. Gui, S.-P., Zhu, Q.: System dynamics research in the area of logistics system. Chinese Journal of Handling Equipment 20(12), 8–1 (2002)

Design of Management Information System for Online Police Exam Based J2EE

Yanhong Yu

School of Business, Shandong Polytechnic University, Jinan, 250353, China
yuyh13@163.com

Abstract. A management information system for online police exam is designed in this paper, which based on B/S structure and using JSP technique. MVC design pattern and Struts framework are used in the three-tier structure online exam system. This system adopts SQL Server database for data storage and realizes a series of functions such as paper database management, automatic test question setting, online exam, marking papers and score multi-analysis. The practice indicates that this system is stable and reliable, can adapt to the examination for a large number of students at the same time, and significantly reduces the workload of workers, which has good practical value as well.

Keywords: J2EE, MVC, Struts, B/S, SQL Server.

1 Introduction

The traditional exam mode is a system of written work. The cost of this mode is too high and the efficiency is too low. The work of going over examination papers is difficult with the increase of exam amount and types. The traditional inconsequential exam mode, as the bottleneck factor of restriction of teaching reform, has come out. Reforming the Examination form with network and automatic technology is sorely needed. Using the network exam, learners can break through the limitation of the traditional method, and the constraint of time and space[1]. At the same time, the new method can reduce the burden of the manual, improve efficiency and quality, strengthen the safety test process, analysis test data quickly. Therefore, the application of this research has important theoretical and practical significance.

In the process of promoting information technology, significant development and remarkable achievements are acquired with the rapid development of information technology[2]. Online examination system draw more and more attention because online examination system has high efficiency, low consumption, operability and objectivity. And it gradually become a platform for various types of examinations. Today, whether domestic or foreign major manufacturers are constantly introduced a series of online examinations, such as Microsoft's MCSE, CISCO's CCNA etc. Police system test is developing toward information technology[3]. In this paper a management information system for online police exam is designed. This management system is based on B/S structure, JSP technique, MVC design pattern, Struts framework. This system adopts SQL Server database for data storage and realizes a series of functions such as paper database management, automatic test

D. Jin and S. Jin (Eds.): Advances in FCCS, Vol. 1, AISC 159, pp. 113–117.
springerlink.com

question setting, online exam, marking papers and score analysis. The practice indicates that this system is stable and reliable, can adapt to the examination for a large number of students at the same time, and significantly reduces the workload of workers, which has good practical value as well.

2 Technology Structure

2.1 Java EE

Java Platform, J2EE or Java EE is a widely used platform for server programming in the Java programming language [4]. The Java platform (Enterprise Edition) differs from the Java Standard Edition Platform (Java SE). It adds libraries which provide functionality to deploy fault-tolerant, distributed, multi-tier Java software. Fig. 1 illustrates the Java EE architecture framework. As a framework for the development of many applications, it simplifies the task of developing applications for multi-tier architecture by providing containers. Containers provide certain complex function, and software developers can concentrate on writing the business logic. The platform was known as Java 2 Platform, Enterprise Edition or J2EE until the name was changed to Java EE in version 5. The new version is called Java EE 6. Java EE is defined by its specification. As with other Java Community Process specifications, providers must meet certain conformance requirements in order to declare their products as Java EE compliant.

Presentation and access	JSP/Servlets	Java foundation/ Swing	Web Services
Business logic	Session Enterprise JavaBeans	Entity Enterprise JavaBeans	Message Driven Beans
Connectivity	JCA JDBC JMS SOAP		
Runtime	Java Runtime Engine (JRE) (Java Byte Code)		

Fig. 1. Java EE architecture framework

2.2 Struts Frame

Struts is an open-source web application framework for developing Java EE web applications. It was at first created by Craig McClanahan and donated to the Apache Foundation in 2000. Formerly located under the Apache Jakarta Project and known as Jakarta Struts, it became a top-level Apache project in 2005. It extends the Java Servlet API to support developers to adopt a model-view-controller (MVC) architecture[5]. The client will typically submit information to the server via a web form in a standard Java EE web application. The information is then either handed over to a Java Servlet that processes it, interacts with a database and produces an HTML-formatted response, or it is given to a Java Server Pages (JSP) document that intermingles HTML and Java code to achieve the same result.

Along with the development of software technology, in the multilevel software project, reusable, extensional software module is certain to carry weight. Give an abstract of the same type of question, an application framework is formed. As a widely used framework, Struts is mainly used to application development in web layer. According to the Java EE standard, struts and Jsp are both exists on the web layer. Struts has two branch: struts action framework and struts shale framework. Struts has own controller. At the same time, it integrates other technique to realize Model and View layer. At Model layer, Struts can easily combine with date access technology like EJB, JDBC. At View layer, Struts can combine with JSP presentation layer assembly. Struts is composed of a group of classes, Servlet and Jsp TagLib. Web application programs based on Struts framework are conformed to the design standard of Jsp Model. Struts can be taken as a changed form of MVC.

3 System Design

As shown in Fig.2, Online police exam system is composed of four parts: Test database management, Test paper generation, Online exam and Correcting paper. The detailed function of every module is as following.
A. Background Management

1. login and exit of Background : Verify if the name and password are correct of logon user.

2. Examinee management module
The Examinee management
Add examinee : Add new Examinee information to the database. Modify/delete examinee: Modify or delete the information of the examinee in the database.
Regional management
Add region : Add new region information to the database. Examine whether the information accord with the format.
Modify/delete region : Modify or delete the information of the region in the database.

3. Test paper management module
Type Management
Add type : Add new type of test paper information to the database. Examine whether the information accord with the format.
Modify/delete type : Modify or delete the information of the type of the test paper in the database.
Test database management
Online exam : Add new test paper to the database.
Modify/delete test paper : Modify or delete the information of the test paper in the database.

4. Test paper generation module

a. Generate Test paper rules: Including paper title, the time limit, the type of test paper, the number of kinds of questions and various kinds of questions.
b. online composing test paper: According to the rules to generate new papers. Include Auto-generating test paper and Manual generating test paper.

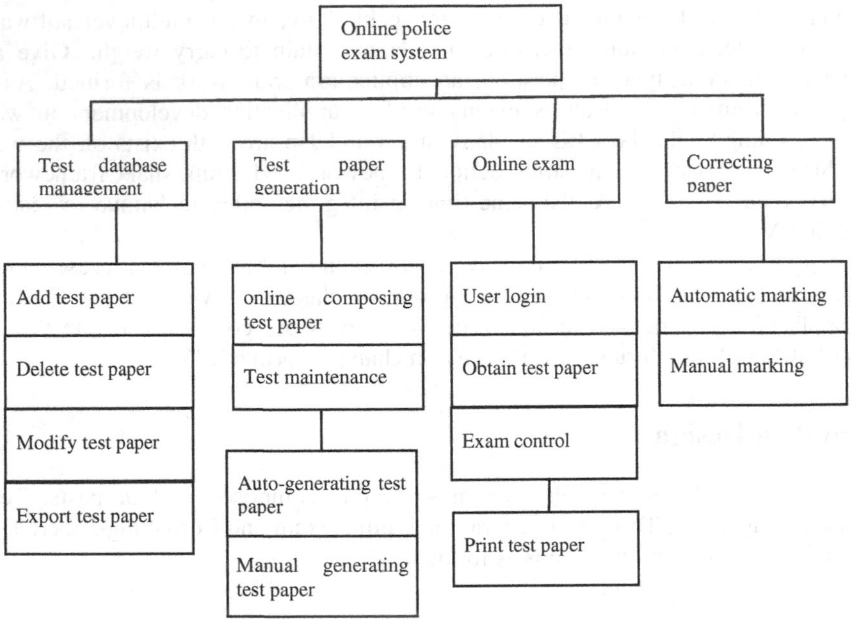

Fig. 2. Organizational chart of online police exam system

5. Correcting paper: Include Automatic marking and Manual marking.

B. Foreground Partition

a. Login of foreground : Verify the logon user name and password if are correct.

b. Display personal information : personal information will be displayed after the successful login, check the information with database.

c. Use the timer : display the time of the exam. Can not respond the test while remaining time is 0, and automatically submit the test papers.

4 Conclusion

In this paper, a management information system for online police exam is designed in this paper, which based on B/S structure and using JSP technique. MVC design pattern and Struts framework are used in the three-tier structure online exam system. This system adopts SQL Server database for data storage and realizes a series of functions such as paper database management, automatic test question setting, online exam, marking papers and score multi-analysis. The practice indicates that this system is stable and reliable can adapt to the examination for a large number of students at the same time, and significantly reduces the workload of workers, which has good practical value as well.

Acknowledgement. This work is supported by A Project of Shandong Province Higher Educational Science and Technology Program(J10LG20), China, and by Natural Science Foundation of Shandong Province (ZR2011FQ038), China.

References

1. Yue, J.B., Qu, Z.C., Kong, Q.K.: Design and Implementation of On-line Examination System Based on J2EE. Agriculture Network Information 9, 128–130 (2010)
2. Priestley, M.: Practical Object—Oriented Design with UML. Tsinghua University Press, Beijing (2004)
3. Wang, K.H.: E-commerce software technique tutorials. Tsinghua University Press, Beijing (2004)
4. Bass, L., Che, L.H.: Software Architecture in Practice. Tsinghua University Press, Beijing (2004)
5. Johnson, R.: J2EE Development Frameworks. IEEE Computer 38, 107–110 (2005)

An Enhanced Slotted ALOHA Algorithm for Anti-collision in RFID System

Jing Li, Meishan Jin, and Jing Liu

Aviation University of Air Force 130022, Changchun China
568026388@qq.com

Abstract. The anti-collision algorithm is one of the key problems in the RFID application system and the Time Division Multiple Access (TDMA) technique can be adopted to solve this problem; its relevant methods involve such anti-collision methods as ALOHA mehod, slotted ALOHA method, binary search, dynamic binary search, and so on. This paper mainly involves the study on the technology of multi-tag anti-collision that is key in RFID system, with a view to optimizing the anti-collision technology now available and enabling the identification of multi-tag technology more efficient to a certain extent.

Keywords: RFID, Anti-collision, Enhanced slotted ALOHA algorithm.

1 Introduction

RFID (Radio Frequency Identification) is a technology[1] in which the non-contact bi-directional data transmission is conducted between the reader and tag by means of radio frequency in order to achieve target identification and data exchange. The radio frequency identification technology is provided with many advantages and has been applied widely in some fields; however, there are still many restraints hindering the comprehensive promotion of RFID system, among which the integrity of RFID data transmission is one of the most important factors[2] to restrict the development of RFID system and has a direct bearing on the performance of the whole system. The problems affecting the integrity of data transmission mainly involve two aspects, one is the external interference, and the other is the interference of RFID system itself. The interference of RFID system itself consists of two collisions: one is called the collision of reader, and the other is called tag collision. As the reader has more functions than the tag, comparatively speaking, the collision of tags is more difficult to settle.

To solve the problem of tag collision is the main content studied in this paper. The so-called collision of tags is referred to as follows: during the operation of RFID system, there will be more than one tag being within the range simultaneously where the reader is operated; when these tags send data simultaneously to the reader, certain interference of signals will occur; in this way, the collision of tags is caused due to the reason that the tags fail to be identified correctly. This type of communication in which multiple tags transmit data to reader simultaneously is called multiple access. In radio technology, the multiple access has been well-known for a long time and there are four[3] methods to solve this problem: Space Division Multiple Access (SDMA),

D. Jin and S. Jin (Eds.): Advances in FCCS, Vol. 1, AISC 159, pp. 119–123.
springerlink.com

Frequency Division Multiple Access (FDMA), Code Division Multiple Access (CDMA) and Time Division Multiple Access (TDMA). As to the RFID system, the method of Time Division Multiple Access (TDMA) has become the largest portion of multiple access. At present, the study on anti-collision algorithm of TDMA applied targeting at RFID system can be summarized as two categories: one is the deterministic anti-collision algorithm and the other is nondeterministic (or probabilistic) anti-collision algorithm[4].

2 The Enhanced ALOHA Algorithm

The random time is used for the tags in nondeterministic anti-collision algorithm to respond to the order given by the reader; these kinds of methods are mostly based on ALOHA system, such as pure ALOHA algorithm, slotted ALOHA algorithm, frame slotted ALOHA algorithm. The main content studied in this paper is nondeterministic algorithm and the improvement of traditional nondeterministic algorithm is also referred to. The drawback of traditional ALOHA algorithm and frame slotted ALOHA algorithm is that the throughput rate of system is lower and the performance of system will drop dramatically during exchange of mass data. In this paper, an enhanced dynamic frame slotted ALOHA algorithm in RFID system is put forward by synthesizing several ALOHA algorithms now available and combining the information transmission characteristic of RFID system. The enhanced ALOHA algorithm is presented as follows:

The RFID reader sends request based on certain initial frame length (i.e. 8); the number of tags on site can be estimated by means of the number of tags and length of frame correctly identified and the number of tags that fails to be identified can be estimated according to the number of slots during collision as well as the collision efficiency of system; then the length of frame can be adjusted based on the number of unidentified tags. In case that the number of unidentified tags is less than the amount that enables the throughput rate of system to reach the maximum, the length of frame shall be reduced, otherwise be increased. In case that the number of unidentified tags exceeds the maximum allowed for the system, the modular arithmetic is used to classify the tags into several subsets and then the methods mentioned above can be followed for reading. In each cycle of reading, the reader can estimate the number of unidentified tags on site and adjust the length of frame in order to ensure that the system can operate in maximum efficiency (throughput rate).

2.1 Enhanced Algorithm Date

Several hypotheses are arranged:

(1) One frame includes 256 slots at most;

(2) In case that the reader can still receive the data of this tag without considering the capture effect, all the tags neighboring the reader can be regarded to comply with the same statistical law;

(3) In case that the tags in the same frame enjoy equal opportunity to send data in all the time slots, the probability of the tags (numbered r) occurring in certain given time slot complies with the binomial distribution; it can be expressed as follows[5]:

Defining the throughput of system $S = P_1/(P_0 + P_1 + P_{22})$, noting that

$$(P_0 + P_1 + P_{22}) = 1, \; S = P_1 = B_{n,1/N} = \binom{r}{n}(1/N)^r(1-1/N)^{n-r}$$

Defining the load capacity G = n, so the relation between S and G can be expressed as: $S = (G/N)(1-1/N)^{G-1}$. It can be calculated as follows:

$$N = G, \; S_{max} = (1-1/G)^{G-1}$$

The maximum throughput rate S_{max} can be calculated as follows:

$$\lim_{G\to\infty} S_{max} = \lim_{G\to\infty}(1-G)^{G-1} = e^{-1} \approx 0.368$$

The real number of tags n, estimated on site, and the minimum mean-square deviation

$$\delta^2_{min} = |C - C_1|^2$$

In case that the estimated number of tags n is less than the number of current time slot N, the number of time slots in one frame shall be reduced; in case that n is more than the number of current time slots N, the number of current time slots N shall be increased in order to enable the system can operate in the maximum throughput rate[6].

Classify the tags into M groups[7]: $M = n/N_{max}$

Design of program design for the estimation of tags to be identified:

In case that the tags (numbered n) are within the effective area of reader, the number of tags selecting the time slot (i) in certain frame can be distributed as follows:

P_x {tags (numbered x) select the time slot i} = $P_x = \dfrac{C_n^x (N-1)^{(n-x)}}{N^n}$

In case that collision occurs in time slot (i), it can be known that the number of tags selecting the time slot (i) is 2 at least (the number of tags being 0 means that no tag exists in time slot (i); the number of tags being 1 means only 1 tag exists in time slot (i)); the distribution function can be inferred to be:

P_r {tags (numbered x) select time slot (i) | the collision occurs} = P_X^0 =

$$\left\{ \begin{array}{l} 0 \ldots\ldots\ldots\ldots x=0, x=1 \\ \dfrac{p_x}{1-p_0-p_1} \ldots\ldots\ldots\ldots\ldots x>2 \end{array} \right\}$$

Then the number of tags in time slot (i) is expressed as follows on average:

$$e = \sum_{x=2}^{n} xP_x^0 = \frac{\sum_{x=2}^{n} xC_n^x (N-1)^{(n-x)}}{N^n - (N-1)^{(n-1)}(N-1+n)}$$

The system efficiency can be expressed as follows:

$$\text{System efficiency} = \frac{\text{Number of time slots when there is only one tag}}{\text{Length of frame}} = \frac{P_1}{N}$$

The formula[8] expressing the number of tags yet to be correctly read: $N_0 = e \cdot C2$

However, in fact the number of tags is unknown for the reader before it checks the tags out. In specific and real situations, we can realize the check of reader for multi-tags based on the formula mentioned above and by means of the follow procedures.

2.2 Main Procedure of Algorithm

Initialization parameter: N=N0: N0 is the initial value of the length of frame; generally, it can be set according to the specific application mode; as the length of frame is too small, which is not in conformity with the real application, suppose the power of index first, in which '0' is 8 and the variation law is 2.

Initialization within the reading cycle: $C=C_l=C_2=O$: Counting of initial collision and counting of correct receiving; slotCounter=N: Counting of time slot for initialization reader;

① Reader sends the check order in which N is set to be the parameter, waiting the tags for reply;

② The reply received is possibly judged as the three situations mentioned as follows:

a) Collision: counting of collision plus 1: C_2++;

b) Correct receiving: counting of correct receiving plus 1:C_1++

c) No signal: no action; under any circumstances, the counting of time slot for the reader minus 1: slotCounter--;

③slotCounter==0?a)Yes:C2=0? i)Yes: ending this check. ii)N. :call the length of frame, adjust the subprogram and reset, return to step (I); b) N. :send orders, enter the next time slot, return to step ③;

2.3 The Length of Frame is Used to Adjust the Subprogram

Initialization parameter: e=2;N_0= round(e*CZ)

①Searching e'=f(N_0, N) Curve (Look up table 3.2, using the calculated data): e=f(N_0, N), namely the formula (4-6);

②N'=round(e'*C_2);

③N_0=N'?

Yes:N=N-C_2:, the adjustment of subprogram by means of length of frame ends.

No:N_o =N', return to step 1.

2.4 Grouping the Subprograms

According to the analysis conducted above, in order to solve the problem where DFSA is not suitable for the collision of large amount of tags, the method listed below can be followed:

Through referring to the implementation of Binary Tree algorithm, divide all the tags into many groups at first, then read the DFSA algorithm of each group successively for the solution of this problem. The detailed implementation is listed as follows:

① In accordance with the formula $M=n/N_{max}$, group the tags into M groups, the value of N_{max} in the formula is 256.

② In case that the number of tags N increases sharply, the tag generates the number of groups, which represents the grouping and are the random number ranges from 1 to M; at the same time, the tag selects a random number in the group to answer the grouping order sent from the reader.

③The reader reads the tag from the first group; when the DFSA algorithm is used to read tags within the group, 1 is added to the number of groups after all the tags in this group finish being read; and by this analogy, this step is repeated.

References

1. Liu, Q.: An Application of RFID Technology in Logistics Management. Modern Enterprise (8), 27–28 (2007)
2. Chen, X.: Review on RFID Development. Information Technology and Standardization (7), 20–24 (2005)
3. Ma, Q., Zhang, G., Yu, J.: Promote RFID Industry of China. Information Technology and Standardization (2004)
4. Yu, S., Zhan, Y., Peng, W., Zhao, Z.: An Anti-collision Algorithm Based on Binary-tree Searching of Regressive Index and its Practice. Computer Application Engineering 16, 26–28 (2004)
5. Cheng, W., Zhao, M., Xu, J.: Enhanced Dynamic Slotted ALOHA Algorithm in RFID Systems. Journal of Huazhong University of Science and Technology (Natural Science)
6. Choi, H.S., Cha, J.R., Kim, H.H.: Fast Wireless Anti-collision Algorithm in Ubiquitous ID System. In: IEEE 60th Vehicular Technology Conference, VTC 2004-Fall, vol. 6, pp. 4589–4592 (September 2004)
7. Chen, X., Zhang, S., et al.: Study on RFID Anti-collision Technology. Cards World (9), 34–37 (2005)
8. Xie, X.: Computer Network. Dalian University of Technology Press (1989)

Study of Tight Bivariate Wavelet Frames with Multi-scale and Application in Information Science

Zhang Yu* and Lv Tingqin

Dept. of Inform. Technology, Zhengzhou Normal University, Zhengzhou 450044, P.R. China
sxxa11xauat@126.com

Abstract. Information science is an interdisciplinary science primarily concerned with the analysis, collection, classification, manipulation, storage, retrieval and dissemination of information. Frame theory has been the focus of active research for twenty years, both in theory and applications. In this paper, the notion of the bivariate generalized multiresolution structure of subspace $L^2(R^2)$, which is the generalization of frame multiresolution analysis, is proposed. The biorthogona-nality traits on wavelet wraps are researched by using time-frequency analysis approach and variable separation approach. The construction of a bivariate generalized multiresolution structure of Paley-Wiener subspace of $L^2(R^2)$ is studied. The pyramid decomposition sch- eme is obtained based on such a GMS and a sufficient condition for its existence is provided. A procedure for designing a class of orthogonal vector-valued finitely supported wavelet functions is proposed by virtue of filter bank theory and matrix theory.

Keywords: Mathematical modelling methods, Banach frames, bivariate wavelet frames, wavelet frame, Bessel sequence, the local Hardy space.

1 Introduction and Notations

Information science is an interdisciplinary science primarily concerned with the analysis, collection, classification, manipulation, storage, retrieval and dissemination of information. Practitioners within the field study the application and usage of knowledge in organizations, along with the interaction between people, organizations and any existing information systems, with the aim of creating, replacing, improving or understanding information systems. Information science is often (mistakenly) considered a branch of computer science. However, it is actually a broad, interdisciplinary field, incorporating not only aspects of computer science, but often diverse fields such as archival science,cognitive science, communications, law, library science, museology, management, mathematics, philosophy, public policy, and the social sciences.Information science focuses on understanding problems from the perspective of the stakeholders involved and then applying information and other technologies as needed. In other words, it tackles systemic problems first rather than individual pieces of technology within that system. In this respect, information science can be seen as a response to technological determinism, the belief that technology "develops by its

* Corresponding author.

D. Jin and S. Jin (Eds.): Advances in FCCS, Vol. 1, AISC 159, pp. 125–131.
springerlink.com

own laws, that it realizes its own potential, limited only by the material resources available, and must therefore be regarded as an autonomous system controlling and ultimately permeating all other subsystems of society. The main advantage of wavelet function is their time-frequency localization property. There exist a great many kinds of scalar scaling functions and scalar wavelet functions. Although the Fourier transform has been a major tool in analysis for over a century, it has a serious laking for signal analysis in that it hides in its phases information concerning the moment of emission and duration of a signal. The frame theory has been one of powerful tools for researching into wavelets. Duffin and Schaeffer introduced the notion of frames for a separable Hilbert space in 1952. Later, Daubechies, Grossmann, Meyer, Benedetto, Ron revived the study of frames in[1,2],and since then, frames have become the focus of active research, both in theory and in applications, such as signal processing, image processing and sampling theory. The rise of frame theory in applied mathematics is due to the flexibility and redundancy ofis due to the flexibility and redundancy of frames, where robust-ness, error toleranceand noise suppression play a vital role [3,4]. The concept of frame multiresolution analysis (FMRA) as described in [2] generalizes the notion of MRA by allowing non-exact affine frames. However, subspaces at different resolutions in a FMRA are still generated by a frame formed by translates and dilates of a single function. This paper is motivated from the observation that standard methods in sampling theory provide examples of multiresolution structure which are not FMRAs. Inspired by [2] and [5], we introduce the notion of a bivariate generalized multiresolution structure(BGMS) of $L^2(R^2)$, which has a pyramid decomposition scheme. It also lead to new constructions of affine frames of $L^2(R^2)$.

In the following, we introduce some notations. Z and Z_+ denote all integers and all nonnegative integers, respectively. R denotes all real numbers. R^2 denotes the 2-dimentional *Euclidean* space. $L^2(R^2)$ denotes the square integrable function space. Let $x=(x_1,x_2)\in R^2$, $\omega=(\omega_1,\omega_2)\in R^2$, $k=(k_1,k_2)\in Z^2$, $z_1=e^{-i\omega/2}$, $z_2=e^{-i\omega_2/2}$. The inner product for any functions $\hbar(x)$ and $\lambda(x)$ ($\hbar(x),\lambda(x)\in L^2(R^2)$) and the Fourier transform of $\hbar(x)$ are defined, respectively, by

$$\langle \hbar, g\rangle = \int_{R^2}\hbar(x)\overline{g(x)}\,dx, \quad \hat{\hbar}(\eta)=\int_{R^2}\hbar(x)e^{-i\eta\cdot x}dx,$$

where $\omega\cdot x=\omega_1x_1+\omega_2x_2$ and $\overline{g(x)}$ denotes the complex conjugate of $g(x)$. Let R and C be all real and all complex numbers, respectively. Z and N denote, respectively, all integers and all positive integers. Set $Z_+=\{0\}\bigcup N, a,s\in N$ as well as $a\geq 2$ By algebra theory, it is obviously follows that there are a^2 elements $d_0,d_1,\cdots,d_{a^2-1}$ in $Z_+^2=\{(n_1,n_2):n_1,n_2\in Z_+\}$ such that $Z^2=\bigcup_{d\in\Omega_0}(d+aZ^2)$; $(d_1+aZ^2)\bigcap(d_2+aZ^2)=\phi$, where $\Omega_0=\{d_0,d_1,\cdots,d_{a^2-1}\}$ denotes the aggregate of all the different representative elements in the quotient group $Z^2/(aZ^2)$ and order $d_0=\{\underline{0}\}$ where $\{\underline{0}\}$ is the null element of Z_+^2 and d_1,d_2 denote two arbitrary distinct elements in Ω_0 .Let $\Omega=\Omega_0-\{\underline{0}\}$ and Ω,Ω_0 to be two index sets.

Definition 1. A sequence $\{\hbar_n(x)_{n\in Z^2} \subset L^2(R^2, C^s)\}$ is called an orthogonal set, if

$$\langle \hbar_n, \hbar_v \rangle = \delta_{n,v} I_s, \quad n, v \in Z^2, \tag{1}$$

where I_s denotes the $s \times s$ identity matrix and $\delta_{n,v}$, is generalized Kronecker symbol.

Let H be a separable Hilbert space and Λ is an index set. We recall that a sequence $\{\xi_v : v \in Z\} \subseteq H$ is a frame for H if there exist positive real numbers C, D such that

$$\forall \eta \in W, \quad C\|f\|^2 \le \sum_{v\in\Lambda} |\langle f, \xi_v \rangle|^2 \le D\|f\|^2, \tag{2}$$

A sequence $\{\xi_v : v \in Z\} \subseteq W$ is a Bessel sequence if (only) the upper inequality of (2) holds. If only for all $\hbar \in \Gamma \subset W$, the upper inequality of (2) holds, the sequence $\{\xi_v\} \subseteq W$ is a Bessel sequence with respect to (w.r.t.) Γ. If $\{f_v\}$ is a frame, there exists a dual frame $\{f_v^*\}$ such that

$$\forall\ \Upsilon \in W, \quad \Upsilon = \sum_{v\in\Lambda} \langle \Upsilon, f_v \rangle f_v^* = \sum_{v\in\Lambda} \langle \Upsilon, f_v^* \rangle f_v. \tag{3}$$

For a sequence $c = \{c(v)\} \in \ell^2(Z)$, we define its discrete-time Fourier transform as the function in $L^2(0,1)^2$ by

$$Fc(\omega) = C(\omega) = \sum_{v\in Z^2} c(v) e^{-2\pi i x\omega} dx, \tag{4}$$

Note that the discrete-time Fourier transform is 1-periodic. Let $T_v\phi(x)$ stand for integer translates of a function $\phi(x) \in L^2(R^2)$, i.e., $(T_{va}\phi)(x) = \phi(x - va)$, and $\phi_{n,va} = 4^n \phi(4^n x - va)$, where a is a positive real constant number. Let $\hbar(x) \in L^2(R^2)$ and let $V_0 = \overline{span\{T_v\hbar : v \in Z^2\}}$ be a closed subspace of $L^2(R^2)$. Assume that $\text{H}(\omega) := \sum_v |\hat{\hbar}(\omega + v)|^2 \in L^\infty[0,1]^2$. In $[5]$, the sequence $\{T_v\hbar(x)\}_v$ is a frame for V_0 if and only if there exist positive constants L_1 and L_2 such that

$$L_1 \le \text{H}(\omega) \le L_2, \quad \text{a.e.}, \omega \in [0,1]^2 \setminus N = \{\omega \in [0,1]^2 : \text{H}(\omega) = 0\}. \tag{5}$$

2 The Characterization of Tight Bivariate Wavelet Frames

We begin with introducing the concept of pseudoframes of translates.

Definition 2. Let $\{T_{va} f,\ v \in Z^2\}$ and $\{T_{va} \tilde{f},\ v \in Z^2\}$ be two sequences in $L^2(R^2)$. Let Ω be a closed subspace of $L^2(R^2)$. We say $\{T_{va} f,\ v \in Z^2\}$ forms an affine pseudoframe for U with respect to $\{T_{va} \tilde{f},\ v \in Z^2\}$ if

$$\forall\, \Gamma(x) \in \Omega, \ \Gamma(x) = \sum_{v\in Z^2} \langle \Gamma, T_{va} \tilde{f} \rangle T_{va} f(x) \tag{6}$$

Define an operator $K:\Omega\to\ell^2(Z^2)$ by

$$\forall\,\Gamma(x)\in\Omega,\quad K\Gamma(x)=\{\langle\Gamma,T_{va}f\rangle\},\tag{7}$$

and define another operator $F:\ell^2(Z^2)\to W$ such that

$$\forall c=\{c(k)\}\in\ell^2(Z^2),\ Fc=\sum_{v\in Z^2}c(v)T_{va}\widetilde{\Upsilon}.\tag{8}$$

Theorem 1. Let $\{T_{va}\Upsilon\}_{v\in Z^2}\subset L^2(R^2)$ be a Bessel sequence with respect to the subspace $U\subset L^2(R^2)$, and $\{T_{va}\widetilde{\Upsilon}\}_{v\in Z^2}$ is a Bessel sequence in $L^2(R^2)$. Assume that K be defined by (7), and S be defined by(8). Assume P is a projection from $L^2(R^2)$ onto U. Then $\{T_{va}\Upsilon\}_{v\in Z^2}$ is pseudoframes of translates for U with respect to $\{T_{va}\widetilde{\Upsilon}\}_{v\in Z^2}$ if and only if

$$KSP=P.\tag{9}$$

Proof. The convergence of all summations of (7) and (8) follows from the assumptions that the family $\{T_{va}\Upsilon\}_{v\in Z^2}$ is a Bessel sequence with respect to the subspace Ω, and he family $\{T_{va}\widetilde{\Upsilon}\}_{v\in Z^2}$ is a Bessel sequence in $L^2(R^2)$ with which the proof of the theorem is direct forward.

We say that a bivariate generalized multiresolution structure (BGMS) $\{V_n, f(x), \widetilde{f}(x)\}$ of $L^2(R^2)$ is a sequence of closed linear subspaces $\{V_n\}_{n\in Z}$ of $L^2(R^2)$ and two elements $f(x)$, $\widetilde{f}(x)\in L^2(R^2)$ such that (i) $V_n\subset V_{n+1}$, $n\in Z$; (ii) $\bigcap_{n\in Z}V_n=\{0\}$; $\bigcup_{n\in Z}V_n$ is dense in $L^2(R^2)$; (iii) $h(x)\in V_n$ if and only if $h(4x)\in V_{n+1}$ $\forall n\in Z$. (iv) $g(x)\in V_0$ implies $T_{va}g(x)\in V_0$, for $v\in Z$; (v) $\{T_{va}f\}_{v\in Z^2}$ forms pseud-oframes of translates for with respect to $\{T_{va}\widetilde{f}, v\in Z^2\}$.

Proposition 1[6]. Let $f\in L^2(R^2)$ satisfy $|\widehat{f}|$ a.e. on a connected neighbourhood of 0 in $[-\frac{1}{2},\frac{1}{2})^2$, and $|\widehat{f}|=0$ a.e. otherwise. Define $\Lambda\equiv\{\omega\in R^2:|\widehat{f}(\omega)|\ge C>0\}$, and $V_0=PW_\Lambda=\{\phi\in L^2(R^2):\mathrm{supp}(\phi)\subseteq\Lambda\}$.

Then for $\widetilde{f}\in L^2(R^2)$, $\{T_v f: v\in Z^2\}$ is pseudoframes of tran-anslates for V_0 with respect to $\{T_v\widetilde{f}: v\in Z^2\}$ if and only if

$$\overline{\hat{h}(\omega)\hat{\widetilde{f}}}(\omega)\chi_\Lambda(\omega)=\chi_\Lambda(\omega)\quad\text{a. e.},\tag{11}$$

where χ_Λ is the characteristic function on Λ. Moreover, if $\widetilde{f}(\omega)$ is the above conditions then $\{T_v f : v \in Z^2\}$ and $\{T_v \widetilde{f} : v \in Z^2\}$ are a pair of commutative pseudo-frames of translates for V_0, i. .e.,

$$\forall \Gamma(x) \in V_0, \quad \Gamma(x) = \sum_{k \in Z^2} \langle \Gamma, T_k \widetilde{f} \rangle T_k f(x) = \sum_{k \in Z^2} \langle \Gamma, T_k f \rangle T_k \widetilde{f}(x). \tag{12}$$

Proposition 2[5]. Let $\{T_{va} f\}_{v \in Z^2}$ be pseudoframes of translates for V_0 with respect to $\{T_{va} \widetilde{f}\}_{v \in Z^2}$. Define V_n by

$$V_n \equiv \{\Upsilon(x) \in L^2(R^2) : \Upsilon(x/4^n) \in V\}, \quad n \in Z, \tag{13}$$

Then, $\{f_{n,va}\}_{v \in Z^2}$ is an affine pseudoframe for V_n with respect to $\{\widetilde{f}_{n,va}\}_{v \in Z^2}$.

The filter functions associated with a BGMS are presented as follows. Define filter functions $D_0(\omega)$ and $\widetilde{D}_0(\omega)$ by $D_0(\omega) = \sum_{s \in Z^2} d_0(s) e^{-2\pi i \omega}$ and $\widetilde{B}_0(\omega) = \sum_{s \in Z^2} \widetilde{b}_0(s) e^{-2\pi i \omega}$ of the sequences $d_0 = \{d_0(s)\}$ and $\widetilde{d}_0 = \{\widetilde{d}_0(s)\}$, resp-ectively, wherever the sum is defined. Let $\{b_0(v)\}$ be such that $D_0(0) = 2$ and $B_0(\omega) \neq 0$ in a neighborhoood of 0. Assume also that $|D_0(\omega)| \leq 2$. Then there exists $f(x) \in L^2(R^2)$ (see ref.[3]) such that

$$f(x) = 2 \sum_{s \in Z^2} d_0(s) f(4x - sa). \tag{14}$$

There exists a scaling relationship for $\widetilde{f}(x)$ under the same conditions as that of d_0 for a seq. $\widetilde{d}_0$, i.e.,

$$\widetilde{f}(x) = 2 \sum_{s \in Z^2} \widetilde{d}_0(v) \widetilde{f}(4x - sa). \tag{15}$$

3 The Traits of Nonseparable Bivariate Wavelet Packs

To construct wavelet packs, we introduce the following notation: $a = 3$, $\varphi_0(x) = f(x)$, $\varphi_v(x) = g_v(x), d^{(0)}(n) = b(n)$, $d^{(v)}(n) = q^{(v)}(n)$, where $v \in \Delta$ We are now in a position of introducing orthogonal bivariate nonseparable wavelet wraps.

Definition 3. A family of functions $\{\varphi_{an+v}(x) : n = 0, 1, 2, 3, \cdots, v \in \Delta\}$ is called a nonseparable bivariate wavelet packs with respect to an orthogonal scaling function $\varphi_0(x)$, where

$$\varphi_{an+v}(x) = \sum_{k \in Z^2} d^{(v)}(k) \varphi_n(ax - k), \tag{13}$$

where $v = 0, 1, 2, 3.$ By taaking the Fourier transform for the both sides of (12), we have

$$\hat{\varphi}_{an+v}(\omega) = D^{(v)}(z_1, z_2) \cdot \hat{\varphi}_n(\omega/2). \tag{14}$$

$$D^{(v)}(z_1, z_2) = D^{(v)}(\omega/2) = \sum_{u \in Z^2} d^{(v)}(u) z_1^{u_1} z_2^{u_2}. \tag{15}$$

Lemma 1[6]. Let $\phi(x) \in L^2(R^2)$. Then $\phi(x)$ is an orthogonal one if and only if

$$\sum_{k \in Z^2} |\hat{\phi}(\omega + 2k\pi)|^2 = 1. \tag{16}$$

Lemma 2[6]. Assuming that $f(x)$ is an orthogonal scaling function. $D(z_1, z_2)$ is the symbol of the sequence $\{d(k)\}$ defined in (3). Then we have

$$\Pi = |D(z_1, z_2)|^2 + |D(-z_1, z_2)|^2 + |D(z_1, -z_2)|^2 + |D(-z_1, -z_2)|^2 \tag{17}$$

Lemma 3[6]. Let $n \in Z_+$ and n be expanded as (17). Then we have

$$\hat{\varphi}_n(\omega) = \prod_{j=1}^{\infty} D^{(v_j)}(e^{-i\omega_1/2^j}, e^{-i\omega_2/2^j})\hat{\varphi}_0(0).$$

Lemma 4 can be inductively proved from formulas (14) and (18).

Theorem 1[6]. For $n \in Z_+$, $k \in Z^3$, we have

$$\langle \varphi_n(\cdot), \varphi_n(\cdot - k) \rangle = \delta_{0,k}. \tag{18}$$

Theorem 2[7]. For every $k \in Z^2$ and $m, n \in Z_+$, we have

$$\langle \varphi_m(\cdot), \varphi_n(\cdot - k) \rangle = \delta_{m,n}\delta_{0,k}. \tag{19}$$

Theorem 3[8]. Let $\phi(x)$, $\tilde{\phi}(x)$, $\psi_\iota(x)$ and $\tilde{\psi}_\iota(x)$, $\iota \in \Lambda$ be functions in $L^2(R^2)$. Assume conditions in Theorem 1 are satisfied. Then, for any $\hbar \in L^2(R^2)$, and $n \in Z$,

$$\sum_{k \in Z^4} \langle \hbar, \tilde{\phi}_{n,k} \rangle \phi_{n,k}(x) = \sum_{\iota=1}^{15} \sum_{v=-\infty}^{n-1} \sum_{k \in Z^4} \langle \hbar, \tilde{\psi}_{\iota:v,k} \rangle \psi_{\iota:v,k}(x). \tag{20}$$

$$\hbar(x) = \sum_{\iota=1}^{15} \sum_{v=-\infty}^{\infty} \sum_{k \in Z^4} \langle \hbar, \tilde{\psi}_{\iota:v,k} \rangle \psi_{\iota:v,k}(x). \quad \forall \hbar(x) \in L^2(R^2) \tag{21}$$

Consequently, if $\{\psi_{\iota:v,k}\}$ and $\{\tilde{\psi}_{\iota:v,k}\}$, $(\iota \in \Lambda, v \in Z, \ k \in Z^2)$ are also Bessel sequences, they are a pair of affine frames for $L^2(R^2)$.

References

1. Telesca, L., et al.: Multiresolution wavelet analysis of earthquakes. Chaos, Solitons & Fractals 22(3), 741–748 (2004)
2. Iovane, G., Giordano, P.: Wavelet and multiresolution analysis: Nature of ε^{∞} Cantorian space-time. Chaos, Solitons & Fractals 32(4), 896–910 (2007)
3. Li, S., et al.: A theory of generalized multiresolution structure and pseudoframes of translates. Fourier Anal. Appl. 7(1), 23–40 (2001)
4. Chen, Q., Huo, A.: The research of a class of biorthogonal compactly supported vector-valued wavelets. Chaos, Solitons & Fractals 41(2), 951–961 (2009)
5. Shen, Z.: Nontensor product wavelet packets in $L_2(R^s)$. SIAM Math. Anal. 26(4), 1061–1074 (1995)
6. Chen, Q., Qu, X.: Characteristics of a class of vector-valued nonseparable higher dimensional wavelet packet bases. Chaos, Solitons & Fractals 41(4), 1676–1683 (2009)
7. Chen, Q., Wei, Z.: The characteristics of orthogonal trivariate wavelet packets. Information Technology Journal 8(8), 1275–1280 (2009)
8. Yang, S., Cheng, Z., Wang, H.: Construction of biorthogonal multiwavelets J. Math. Anal. Appl. 276(1), 1–12 (2002)

The Features of Poly-scaled Non-orthogonal Bivariate Wavelet Packages and Applications in Finance Science

Yonggan Li *

Office of Financial Affairs, Henan Quality Polytechnic, Pingdingshan 467000, P.R. China
zas123qwe@126.com

Abstract. In this paper, a sort of bivariate wavelet packages with poly-scale are introduced, which are generalizations of univariant wavelet packets. Finitely Supported wavelet bases for Sobolev spaces is researched. Steming from a pair of finitely supported refinale functions with poly-scaled dilation factor in space $L^2(R^2)$ satsfying a very mild condition, we provide a novel method for designing wavelet bases, which is the generalization of univariate wavelets in Hilbert space. The definition of biorthogonal nonseparable bivariate wavelet wraps is provided and a procedure for designting them is proposed. The biorthogonality trait of binary wavelet wraps is studied by virtue of time-frequency analysis method and iterative method.Three biorthogonality formulas regarding these wavelet wraps are erected. Moreover, it is shown how to get new Riesz bases of $L^2(R^2)$ from the wavelet wraps.

Keywords: Short support, Fourier coefficeients, poly-scaled bivariate wavelet packages, Bessel sequence, oversampling theorem, time-frequency analysis approach, iterative method, homomorphism.

1 Introduction and Notations

Finance management is a branch of finance that refers to the management of financial resources of a company. The main objective of corporate financing is to maximize the company value by making proper allocation of financial resources, along with taking care of the financial risks. Finance management focuses on analyzing the financial problems and devising the universal solutions, which are applicable to all kind of companies. Finance management is an absolute necessity for all types of business organizations. Earlier it used to be the part of overall finance management of a firm. But, over the last one decade, it has emerges as a separate discipline altogether. Today, in both large and medium sizes corporations, there is a dedicated department involved in taking care of the corporate finance management of the company. The main advantage of wavelet function is their time-frequency localization property. Construction of wavelet functions is an important aspect of wavelet analysis, and multiresolution analysis approach is one of importment ways of designing all sorts of wavelet fun-ctions. There exist a great many kinds of scalar scaling functions and scalar wavelet functions. As we know, the selection of method and algorthm

* Corresponding author.

D. Jin and S. Jin (Eds.): Advances in FCCS, Vol. 1, AISC 159, pp. 133–139.
springerlink.com

determine the accuracy of fault information. In the signal denosing method, wavelet analysis is a new analy-tical tool which develops on the basis of the Fourier analysis. The advantages of wavelet packets and their promising features in various application have attracted a lot of interest and effort in recent years. In addition, wavelet packets provide better frequency localization than wavelets while time-domain localization is not lost. Wavelet wraps[1],due to their good characterisics, have attracted considerable attention. They can be widely app- lied in science and engineering [2,4]. Coifman R. R. and Meyer Y. firstly introduced the notion for orthogonal wavelet wraps which were used to decompose wavelet components. Chui C K. and Li C. [3] generalized the concept of orthogonal wavelet packets to the case of non-orthogonal wavelet w- wraps so that wavelet wraps can be employed in the case of the spline wavelets and so on. Tensor pr- oduct multivariate wavelet wraps has been constructed by Coifman and Meyer. The introduction for the notion on non-tensor product wavelet wraps attributes to Shen Z [5]. Since the majority of infor- mation is multidimensional information, many researchers interest themselves in the investigation into multivariate wavelet theory. Nowadays, since there is little literature on biorthogonal wavelet wraps. It is necessary to investigate biorthogonal wavelet wraps. The aim of this paper is to gene-ralize the concept for univariate orthogonal wavelet packets to biorthogonal bivariate wavelet wraps. The definition for nonseparable biorthogonal bivariate wavelet wraps is given and a procedure for constructing them is described. Next, the biorthogonality properties of nonseparable bivariate wavelet packages are investigated.

In the benow, we introduce some notations. Z and Z_+ denote all integers and all nonnegative integers, respectively. R denotes all real numbers. R^2 denotes the 2-dimentional *Euclidean* space. $L^2(R^2)$ denotes the square integrable function space. Let $x=(x_1,x_2)\in R^2$, $\omega=(\omega_1,\omega_2)\in R^2$, $u=(u_1,u_2)\in Z^2$, $z_v=e^{-i\omega_v/a}$, $v=1,2$. The inner product for any two functions $\hbar(x)$ and $\phi(x)$ ($\hbar(x),\phi(x)\in L^2(R^2)$) and the Fourier transform of $\phi(x)$ are defined, respectively, by

$$\langle \hbar,\phi\rangle=\int_{R^2}\hbar(x)\overline{\phi(x)}\,dx,\quad \hat{\phi}(\omega)=\int_{R^2}\phi(x)e^{-2\pi i\omega\cdot x}dx,$$

where $\omega\cdot x=\omega_1x_1+\omega_2x_2$ and $\overline{\phi(x)}$ denotes the complex conjugate of $\phi(x)$. Let R and C be all real and all complex numbers, respectively. Z and N denote, respectively, all integers and all positive integers. Set $Z_+=\{0\}\bigcup N, a,s\in N$ as well as $m\geq 2$ By algebra theory, it is obviously follows that there are a^2 elements $\lambda_0,\lambda_1,\cdots,\lambda_{a^2-1}$ in $Z_+^2=\{(n_1,n_2):n_1,n_2\in Z_+\}$ such that $Z^2=\bigcup_{\lambda\in\Delta_0}(\lambda+aZ^2)$; $(\lambda_1+aZ^2)\bigcap(\lambda_2+aZ^2)=\phi$, where $\Delta_0=\{\lambda_0,\lambda_1,\cdots,\lambda_{a^2-1}\}$ denotes the aggregate of all the different representative elements in the quotient group $Z^2/(mZ^2)$ and order $\lambda_0=\{\underline{0}\}$ where $\{\underline{0}\}$ is the null element of Z_+^2 and λ_1,λ_2 denote two arbitrary distinct elements in Λ_0 .Let $\Delta=\Delta_0-\{\underline{0}\}$ and Λ, Λ_0 to be two index aggregates. By $L^2(R^2,C^s)$, we denote the set of vector-valued functions $L^2(R^2,C^s):=\{\hbar(x)=(h_1(x), h_2(x),\cdots,h_u(x))^T: h_l(x)\in L^2(R^2), l=1,2,\cdots,s\}$,where T means the transpose of a vector.

Definition 1. A sequence $\{\hbar_n(y)_{n\in Z^2} \subset L^2(R^2,C^s)\}$ is called an orthogonal set, if

$$\langle \hbar_n, \hbar_v \rangle = \delta_{n,v} I_s, \quad n,v \in Z^2, \tag{1}$$

where I_s denotes the $s\times s$ identity matrix and $\delta_{n,v}$, is generalized Kronecker symbol.

2 The Bivariate Multiresolution Analysis

Let $h(x)\in L^2(R^2)$ satisfy the following refinement equation:

$$f(x) = a^2 \cdot \sum\nolimits_{u\in Z^2} b_k f(ax-u), \tag{2}$$

where $\{b(n)\}_{n\in Z^2}$ is real number sequence which has only finite terms and $f(x)$ is called scaling function. Formula (1) is said to be two-scale refinement equation. The frequency form of formula (1) can be written as

$$\hat{f}(\omega) = B(z_1,z_2)\hat{f}(\omega/a), \tag{3}$$

$$B(z_1,z_2) = \sum_{(n_1,n_2)\in Z^2} b(n_1,n_2)\cdot z_1^{n_1}\cdot z_2^{n_2}. \tag{4}$$

Define a subspace $U_j \subset L^2(R^2)$ $(j\in Z)$ by

$$U_j = clos_{L^2(R^2)}\left\langle a^j f(a^j x-u):\ u\in Z^2\right\rangle. \tag{5}$$

Definition 2. We say that $f(x)$ in (2) generate a multiresolution analysis $\{U_j\}_{j\in Z}$ of $L^2(R^2)$, if the sequence $\{U_j\}_{j\in Z}$ defined in (4) satisfy the following properties: (i) $U_j \subset U_{j+1}$, $\forall j\in Z$; (ii) $\bigcap_{j\in Z} U_j = \{0\}$; $\bigcup_{j\in Z} U_j$ is dense in $L^2(R^2)$; (iii) $\psi(x)\in U \Leftrightarrow \psi(ax)\in U_{k+1}, \forall k\in Z$ (iv) the family $\{f(a^j x-k): k\in Z^2\}$ forms a *Riesz* basis for the spaces U_j.

Let $W_k (k\in Z)$ denote the complementary subspace of U_j in U_{j+1}, and assume that there exist a vector-valued function $H(x)=\{\hbar_1(x),\hbar_2(x),\cdots,\hbar_{a^2-1}(x)\}$ constitutes a *Riesz* basis for W_k, i.e.,

$$W_j = clos_{L^2(R^2)}\left\langle \hbar_{\lambda:j,k} : \lambda = 1,2,\cdots,a^2-1;\ k\in Z^2\right\rangle, \tag{6}$$

where $j\in Z$, and $\hbar_{\lambda:j,k}(x) = a^{j/2}\hbar_\lambda(a^j x-k)$, $\lambda=1,2,\cdots,a^2-1$; $k\in Z^2$. Form condition (5), it is obvious that $\hbar_1(x),\hbar_2(x),\ \cdots,\hbar_{a^2-1}(x)$ are in $W_0 \subset U_1$. Hence there exist three real number sequences $\{q_u^{(\lambda)}\}(\lambda\in\Delta,\ u\in Z^2)$ such that

$$\hbar_\lambda(x) = a^2\cdot\sum_{u\in Z^2} q_u^{(\lambda)} f(ax-u), \tag{7}$$

Formula (7) in frequency domain can be written as

$$\hbar_{\lambda}(\omega)=Q^{(\lambda)}(z_1,z_2)\widehat{f}(\omega/a),\quad \lambda=1,2,\cdots,a^2-1. \tag{8}$$

where the signal of sequence $\{q_u^{(\lambda)}\}(\lambda=1,2,\cdots,a^2-1,\ u\in Z^2)$ is

$$Q^{(\lambda)}(z_1,z_2)=\sum_{(u_1,u_2)\in Z^2} q^{(\lambda)}_{(u_1,u_2)}\cdot z_1^{u_1}\cdot z_2^{u_2}. \tag{9}$$

A bivariate function $f(x),\ \widetilde{f}(x)\in L^2(R^2)$ is called biorthogonal ones, if

$$\left\langle f(\cdot),\widetilde{f}(\cdot-u)\right\rangle=\delta_{0,u},\quad u\in Z^2. \tag{10}$$

We say $H(x)=\{\hbar_1(x),\hbar_2(x),\cdots,\hbar_{a^2-1}(x)\}$ is an orthogonal bivariate vector-valued wavelets associated with the scaling function $f(x)$, if they satisfy:

$$\left\langle f(\cdot),\hbar_{\nu}(\cdot-u)\right\rangle=0,\quad \nu\in\Delta,\ u\in Z^2, \tag{11}$$

$$\left\langle \hbar_{\lambda}(\cdot),\hbar_{\nu}(\cdot-u)\right\rangle=\delta_{\lambda,\nu}\delta_{0,u},\ \lambda,\nu\in\Delta,\ u\in Z^2 \tag{12}$$

3 The Characters of Nonseparable Bivariate Wavelet Packages

To construct wavelet packs, we introduce the following notation: $a=2$, $\psi_0(x)=f(x)$, $\psi_\nu(x)=\hbar_\nu(x), b^{(0)}(n)=b(n),\ b^{(\nu)}(n)=q^{(\nu)}(n)$, where $\nu\in\Delta$ We are now in a position of introducing orthogonal bivariate nonseparable wavelet wraps.

Definition 3. A family of functions $\{\psi_{ak+\nu}(x):n=0,1,2,3,\cdots,\ \nu\in\Delta\}$ is called a nonseparable bivariate wavelet packages with respect to an orthogonal scaling function $\psi_0(x)$, where

$$\psi_{ak+\nu}(x)=\sum\nolimits_{u\in Z^2} b^{(\nu)}(n)\psi_k(ax-u), \tag{13}$$

where $\nu=0,1,2,3.$ By taaking the Fourier transform for the both sides of (12), we have

$$\widehat{\psi}_{nk+\nu}(\omega)=B^{(\nu)}(z_1,z_2)\cdot\widehat{\psi}_k(\omega/2). \tag{14}$$

$$B^{(\nu)}(z_1,z_2)=B^{(\nu)}(\omega/2)=\sum_{k\in Z^2} b^{(\nu)}(k)\,z_1^{k_1}z_2^{k_2} \tag{15}$$

Lemma 1. Assuming that $f(x)$ is an semiorthogonal scaling function. $B(z_1,z_2)$ is the symbol of the sequence $\{b(k)\}$ defined in (3). Then we have

$$\Pi=|B(z_1,z_2)|^2+|B(-z_1,z_2)|^2+|B(z_1,-z_2)|^2+|B(-z_1,-z_2)|^2 \tag{16}$$

Proof. If $f(x)$ is an orthogonal bivariate function, then $\sum_{k\in Z^2}\left|\widehat{f}(\omega+2k\pi)\right|^2=1$. Therefore, by Lemma 1 and formula (2), we obtain that

$$1=\sum_{k\in Z^2}|B(e^{-i(\omega_1/2+k_1\pi)},e^{-i(\omega_2/2+k_2\pi)})\cdot\widehat{f}((\omega_1,\omega_2)/2+(k_1,k_2)\pi)|^2$$

$$=|B(z_1,z_2)\sum_{k\in Z^2}\widehat{f}(\omega+2k\pi)|^2+|B(-z_1,z_2)\cdot\sum_{k\in Z^2}\widehat{f}(\omega+2k\pi+(1,0)\pi)|^2$$

$$+|B(z_1,-z_2)\cdot\sum_{k\in Z^2}\widehat{f}(\omega+2k\pi+(0,1)\pi)|^2+|B(-z_1,-z_2)\cdot\sum_{k\in Z^2}\widehat{f}(\omega+2k\pi+(1,1)\pi)|^2$$

$$=\left|B(z_1,z_2)\right|^2+\left|B(-z_1,z_2)\right|^2+\left|B(z_1,-z_2)\right|^2+\left|B(-z_1,-z_2)\right|^2$$

This complete the proof of Lemma 2. Similarly, we can obtain Lemma 1 from (3), (8), (13).

Lemma 2[6]. If $\psi_\nu(x)$ ($\nu=0,1,2,3$) are orthogonal wavelet functions associated with $h(x)$. Then we have

$$\sum_{j=0}^{1}\{B^{(\lambda)}((-1)^j z_1,(-1)^j z_2)\overline{B^{(\nu)}((-1)^j z_1,(-1)^j z_2)}+B^{(\lambda)}((-1)^{j+1}z_1,(-1)^j z_2)$$

$$\cdot\overline{B^{(\nu)}((-1)^{j+1}z_1,(-1)^j z_2)}\}:=\Xi_{\lambda,\mu}=\delta_{\lambda,\nu},\quad \lambda,\nu\in\{0,1,2,3\}. \tag{17}$$

For an arbitrary positive integer $n\in Z_+$, expand it by

$$n=\sum_{j=1}^{\infty}\nu_j 4^{j-1},\quad \nu_j\in\Delta=\{0,1,2,3\}\ . \tag{18}$$

Theorem 1. For any $n\in Z_+$, $u\in Z^2$, we have

$$\langle\psi_n(\cdot),\psi_n(\cdot-u)\rangle=\delta_{0,u}. \tag{19}$$

Proof. Formula (20) follows from (10) as n=0. Assume formula (20) holds for the case of $0\le n<4^{r_0}$ (r_0 is a positive integer). Consider the case of $4^{r_0}\le n<4^{r_0+1}$. For $\nu\in\Delta$, by induction assumption and Lemma 1, Lemma 2, we have

$$(2\pi)^2\langle\psi_n(\cdot),\psi_n(\cdot-u)\rangle=\int_{R^2}\left|\widehat{\psi}_n(\omega)\right|^2\cdot\exp\{iu\omega\}d\omega$$

$$=\sum_{j\in Z^2}\int_{4\pi j_1}^{4\pi(j_1+1)}\int_{4\pi j_2}^{4\pi(j_2+1)}\left|B^{(\nu)}(z_1,z_2)\cdot\widehat{\psi}_{[n/4]}(\omega/2)\right|^2\cdot e^{ik\omega}\,d\omega$$

$$=\int_0^{4\pi}\int_0^{4\pi}\left|B^{(\nu)}(z_1,z_2)\right|^2\sum_{j\in Z^2}|\widehat{\psi}_{[n/8]}(\omega/2+2\pi j)|^2\cdot e^{ik\omega}d\omega$$

$$=\int_0^{4\pi}\int_0^{4\pi}\left|B^{(\nu)}(z_1,z_2)\right|^2\cdot e^{ik\omega}d\omega=\int_0^{2\pi}\int_0^{2\pi}\Pi\cdot e^{ik\omega}d\omega=\delta_{o,k}$$

Thus , we complete the proof of theorem 1.

Theorem 2. For every $u \in Z^2$ and $m, n \in Z_+$, we have

$$\langle \psi_m(\cdot), \psi_n(\cdot - u)\rangle = \delta_{m,n}\delta_{0,u}. \tag{20}$$

Proof. For the case of $m = n$, (20) follows from Theorem 1. As $m \neq n$ and $m, n \in \Omega_0$, the result (20) can be established from Theorem 2, where $\Omega_0 = \{0, 1, 2, 3\}$. In what follows, assuming that m is not equal to n and at least one of $\{m, n\}$ doesn't belong to Ω_0, rewrite m, n as $m = 4m_1 + \lambda_1$, $n = 4n_1 + \mu_1$, where $m_1, n_1 \in Z_+$, and $\lambda_1, \mu_1 \in \Omega_0$.

Case 1. If $m_1 = n_1$, then $\lambda_1 \neq \mu_1$. By (17), formulas (21) follows, since

$$(2\pi)^2 \langle \psi_m(\cdot), \psi_n(\cdot - k)\rangle = \int_{R^2} \widehat{\psi}_{4m_1+\lambda_1}(\omega)\overline{\widehat{\psi}_{4n_1+\mu_1}(\omega)} \cdot \exp\{ik\omega\}d\omega$$

$$= \int_{[0,4\pi]^2} B^{(\lambda_1)}(z_1, z_2) \sum_{s\in Z^2} \hat{h}_{m_1}(\omega/2 + 2s\pi) \cdot \overline{\hat{h}_{m_1}(\omega/2 + 2s\pi) B^{(\mu_1)}(z_1, z_2)} \cdot e^{ik\omega} d\omega$$

$$= \frac{1}{(2\pi)^2} \int_{[0,2\pi]^2} \Xi_{\lambda_1,\mu_1} \cdot \exp\{ik\omega\}\, d\omega = O.$$

Case 2. If $m_1 \neq n_1$ we order $m_1 = 4m_2 + \lambda_2$, $n_1 = 4n_2 + \mu_2$, where $m_2, n_2 \in Z_+$,
and $\lambda_2, \mu_2 \in \Omega_0$. If $m_2 = n_2$, then $\lambda_2 \neq \mu_2$. Similar to Case 1, we have (21) follows.
That is to say, the proposition follows in such case. As $m_2 \neq n_2$, we order $m_2 = 2m_3 + \lambda_3$, $n_2 = 2n_3 + \mu_3$, once more, where $m_3, n_3 \in Z_+$, and $\lambda_3, \mu_3 \in \Omega_0$. Thus, after taking finite steps (denoted by r), we obtain $m_r, n_r \in \Omega_0$, and $\lambda_r, \mu_r \in \Omega_0$. If $\alpha_r = \beta_r$, then $\lambda_r \neq \mu_r$. Similar to Case 1, (21) holds. If $\alpha_r \neq \beta_r$, Similar to Lemma 1, we conclude that

$$\langle \psi_m(\cdot), \psi_n(\cdot - k)\rangle = \frac{1}{(2\pi)^2} \int_{R^2} \widehat{\psi}_{4m_1+\lambda_1}(\omega)\overline{\widehat{\psi}_{4n_1+\mu_1}(\omega)} \cdot e^{ik\omega} d\omega$$

$$= \frac{1}{(2\pi)^2} \int_{[0,2^{r+1}\pi]^2} \{\prod_{t=1}^{r} B^{(\lambda_t)}(\omega/2^t)\} \cdot O \cdot \{\prod_{t=1}^{r} B^{(\mu_t)}(\omega/2^t)\} \cdot e^{ik\omega} d\omega = O.$$

Theorem 3[7]. If $\{\Psi_\beta(x), \beta \in Z_+^2\}$ and $\{\Psi_\beta(x), \beta \in Z_+^2\}$ are vector-valued wavelet packs with respect to a pair of biorthogonal vector-valued scaling functions $\Psi_0(x)$ and $\widetilde{\Psi}_0(x)$, then for any $\alpha, \sigma \in Z_+^2$, we have

$$\langle \Psi_\alpha(\cdot), \widetilde{\Psi}_\sigma(\cdot - k)\rangle = \delta_{\alpha,\sigma}\delta_{0,k} I_s, \quad k \in Z^2. \tag{21}$$

References

1. Telesca, L., et al.: Multiresolution wavelet analysis of earthquakes. Chaos, Solitons & Fractals 22(3), 741–748 (2004)
2. Iovane, G., Giordano, P.: Wavelet and multiresolution analysis:Nature of ε^{∞} Cantorian space-time. Chaos, Solitons & Fractals 32(4), 896–910 (2007)
3. Zhang, N., Wu, X.: Lossless Compression of Color Mosaic Images. IEEE Trans. Image Processing 15(16), 1379–1388 (2006)
4. Shen, Z.: Nontensor product wavelet packets in $L_2(R^s)$. SIAM Math. Anal. 26(4), 1061–1074 (1995)
5. Chen, Q., Wei, Z.: The characteristics of orthogonal trivariate wavelet packets. Information Technology Journal 8(8), 1275–1280 (2009)
6. Chen, Q., Qu, X.: Characteristics of a class of vector-valued nonseparable higher-dimensional wavelet packet bases. Chaos, Solitons & Fractals 41(4), 1676–1683 (2009)
7. Chen, Q., Huo, A.: The research of a class of biorthogonal compactly supported vector-valued wavelets. Chaos, Solitons & Fractals 41(2), 951–961 (2009)
8. Chen, Q., Shi, Z.: Biorthogonal multiple vector-valued multivariate wavelet packets associated with a dilation matrix. Chaos, Solitons & Fractals 35(2), 323–332 (2008, 2009)

The Stability of Bivariate Gabor Frames and Wavelet Frames with Compact Support*

Qingjiang Chen** and Zhenni Wei

School of Science, Xi'an University of Architecture & Technology, Xi'an 710055, China
chen66xauat@126.com

Abstract. Information science is an interdisciplinary science primarily concerned with the analysis, collection,classification, manipulation, storage, retrieval and dissemination of information. Wavelet analysis has been developed into a new branch for over twenty years. The window functions and bivariate Gabor frames are introduced. The existence of bivariate Gabor frames with compact support is discussed. Sufficient conditions for irregular bivariate Gabor system to be frames are pres-ented by means of frame multiresolution analysis and paraunitary vector filter bank theory. We show how to construct bivariate Gabor frames with compact support.

Keywords: Sobolev space, iterative method, bracket product, vector subdivision scheme, bivariate Gabor frames, time-frequency analysis.

1 Introduction

Information science is an interdisciplinary science primarily concerned with the analysis, collection, classification, manipulation, storage, retrieval and dissemination of information. Practitioners within the field study the application and usage of knowledge in organizations, along with the interaction between people, organizations and any existing information systems, with the aim of creating, replacing, improving or under-tanding information systems. Information science is often (mistakenly) considered a branch of computer science. A major advantage of the application of information engineering is that it virtually forces the organization to address the entire spectrum of its information systems requirements, resulting in a functionally integrated set of enter-prise systems. In contrast, ad hoc requirements may result in a fragmented set of systems (islands of automation), which at their worst may be incompatible, contain duplicate (perhaps inconsistent) information, and omit critical elements of information. The concept of wavelet frames have received much research attention and evidence has shown that they can be used to improve the localization of the frequency field of wavelet bases. Although the Fourier transform has been a major tool in analysis for over a century, it has a serious laking for signal analysis in that it hides in its phases information concerning the moment of emission and duration of a

* Foundation item: This work was supported by the Science Research Foundation of Education Department of Shaanxi Provincial Government (Grant No:11JK0468).

** Corresponding author.

D. Jin and S. Jin (Eds.): Advances in FCCS, Vol. 1, AISC 159, pp. 141–147.
springerlink.com

signal. In image analysis and processing, there is a classical choice between spatial or frequency domain representations. The former, consisting of a two-dimensional (2D) array of pixels, is the standard way to represent discrete images. This is the typical format used for acquisition and display, but it is also common for storage and processing. Space representations appear in a natural way, and they areimportant for shape analy-sis, object localization and description (either photometric or morphologic) of the sce-ne. The frame theory plays an important role in the modern time-frequency analysis. It has been developed very fast over the last twenty years, especially in the context of wavelets and Gabor systems. In her celebrated paper[1], Daubechies constructed a family of finitely supported univariate orthogonal scaling functions and their corres-ponding orthogonal wavelets with the dilation factor 2. Since then wavelets with compact support have been widely and successfully used in various applications such as image compression and signal processing. The frame theory has been one of pow-erful tools for researching into wavelets. To study some deep problems in nonharmon-ic Fourier series, Duffin and Schaeffer[2] introduce the notion of frames for a separ-able Hilbert space in 1952. Basically, Duffin and Schaeffer abstracted the funda-mental notion of Gabor for studying signal processing [3]. These ideas did not seem to generate much general interest outside of nonharmonic Fourier series however (see Young's [4]) until the landmark paper of Daubechies, Grossmann, and Meyer [5] in 1986. After this ground breaking work, the theory of frames began to be more widely studied both in theory and in applications [6,7], such as signal processing, image processing, data compression and sampling theory. The notion of Frame Multiresolution Analysis (FMRA) as described by [6] generalizes the notion of MRA by allowing non-exact affine frames. However, subspaccs at different resolutions in a FMRA are still generated by a frame formed by translares and dilates of a single bivariate function. Inspired by [5] and [7], we introduce the norion of a Generalized Bivariate Multireslution Structure (GBMS) of $L^2(R^2)$ generated by several functions of integer translates $L^2(R^2)$. We demonstrate that the GBMS has a pyramid decomposition scheme and obiain a frame-like decomposition based on such a GBMS.It also lead to new constructions of affine of $L^2(R^2)$. Since the majority of information is multi-dimensional information, many researchers interest themselves in the investigation into multi-variate wavelet theory. The classical method for constructing multivariate wavelets is that separable multivariate wavelets may be obtained by means of the tensor product of some univariate wavelet frames. It is significant to investigate nonseparable multi-variate wavelet frames and Gabor frames.

2 Notations and Fundamentals

Let Ω be a separable Hilbert space .We recall that a sequence $\{\eta_v\}_{v\in\Lambda} \subseteq \Omega$ is a frame for Ω, if there exist two positive real numbers L, M such that

$$\forall\, \wp \in \Omega, \quad L\|\wp\|^2 \le \sum_{v\in\Lambda} |\langle \wp, \eta_v \rangle|^2 \le M\,\|\wp\|^2 \,, \tag{1}$$

where Λ is an index set. A sequence $\{\eta_v\}_{v\in\Lambda}\subseteq\Omega$ is called a Bessel sequence if only the upper inequality of (1) follows. If only for all element $g\in Q\subseteq\Omega$, the upper inequality of (1) holds, the sequence $\{\eta_v\}_{v\in\Lambda}\subseteq\Omega$ is a Bessel sequence with respect to (w.r.t.) Q. If $\{\eta_v\}$ is a frame, there exist a dual frame $\{\tilde{\eta}_v\}$ such that

$$\forall\varphi\in\Omega,\ \ \varphi=\sum_{v\in\Lambda}\langle\varphi,\eta_v\rangle\tilde{\eta}_v=\sum_{v\in\Lambda}\langle\varphi,\tilde{\eta}_v\rangle\eta_v. \tag{2}$$

The mathematical theory for Gabor analysis in $L^2(R^2)$ is based on two classes of operators on $L^2(R^2)$, that is

Translation by $a\in R$, $T_{na}:L^2(R^2)\to L^2(R^2)$, $(T_{na}f)(x)=f(x-na)$, $n\in Z^2$,

Modulation by $b\in R$, $E_b:L^2(R)\to L^2(R)$, $(E_bf)(x)=e^{2\pi ibx}f(x)$, $x\in R^2$.

Gabor analysis aims at representing functions $\phi(x)\in L^2(R^2)$ as superpositions of translated and modulated versions of a fixed function $\phi(x)\in L^2(R^2)$.

Definition 1. A bivariate Gabor frame is a frame for space $L^2(R^2)$ of the form $\{E_{mb}T_{na}\phi(x)\}_{m\in Z,n\in Z^2}$, where $a,b>0$ and $\phi(x)\in L^2(R^2)$ is a fix function.

Frames of the above type are caled Weyl-Heisenberg frames. The function $\phi(x)\in L^2(R^2)$ is called the window function or generator. The Fourier transform of an integrable function $h(x)\in L^2(R^2)$ is defined by

$$Fh(\omega)=\hat{h}(\omega)=\int_{R^2}h(x)e^{-2\pi ix\omega}dx,\ \ \omega\in R^2. \tag{3}$$

For a sequence $k=\{k(v)\}\in\ell^2(Z^2)$, we define the discrete-time Fourier tramsform as the function in $L^2(0,1)^2$ given by

$$Fk(\omega)=K(\omega)=\sum_{v\in Z^2}k(v)\cdot e^{-2\pi iv\omega}. \tag{4}$$

For $s>0$, we denote by $H^s(R^2)$ the Sobolev space of all quarternary functions $h(x)\in H^s(R^2)$ such that

$$\int_{R^2}|\hat{h}(\omega)|(1+\|\omega\|^{2s})\,d\omega<+\infty.$$

The space $H^\lambda(R^2)$ is a Hilbert space equipped with the inner product given by

$$\langle h,g\rangle_{H^\lambda(R^2)}:=\frac{1}{(2\pi)^2}\int_{R^2}\hat{h}(\gamma)\,\overline{\hat{g}(\gamma)}(1+|\gamma|^{2\lambda})\,d\gamma,\ \ h,g\in L^2(R^2). \tag{5}$$

We are interested in wavelet bases for the Sobolev space $H^\lambda(R^2)$, where λ is a positive integer. In this case, we obtain

$$\langle h,g\rangle_{H^\lambda(R^2)}=\langle h,g\rangle+\langle h^{(\lambda)},g^{(\lambda)}\rangle,\ \ h,g\in L^2(R^2). \tag{6}$$

Suppose that $h(x)$ is a function in the Sbolev space $H^{\lambda}(R^2)$. For $j \in Z$, $k \in Z^2$, setting $\phi_{j,u}(x) = 2^{j/2}\phi(2^j x - u)$, we have

$$\| h_{j,k} \|_{H^{\lambda}(R^2)} = \| h_{j,k}^{(s)} \|_{L^2(R^2)} = 2^{j\lambda/2} \| h^{(s)} \|_{L^2(R^2)} .$$

Note that the discrere-time Fourier transform is Z^2-periodic.Let r be a fixed positive integer, and J be a finite index set, i.e., $J = \{1, 2, \cdots, r\}$. We consider the case of multiple generators, which yield multiple pseudoframes for subspaces.

Definition 2. Let $\{T_k \hbar_j(x)\}$ and $\{T_k \tilde{\hbar}_j(x)\}$ $(j \in J, u \in Z^2)$ be two sequences in subspace $M \subset L^2(R^2)$, We say that $\{T_k \hbar_j(x)\}$ forms a multiple pseudoframe for V_0 with respect to (w.r.t.) $\{T_k \tilde{\hbar}_j(x)\}$ $(j \in J, k \in Z^2)$, if

$$\forall \Upsilon(x) \in M, \quad \Upsilon(x) = \sum_{j \in J} \sum_{v \in Z^2} \langle \Upsilon, T_v \hbar_j \rangle T_v \tilde{\hbar}_j(x) , \tag{7}$$

where we define a translate operator, $(T_u \lambda)(x) = \lambda(x - u)$, $u \in Z^2$, for a function $\lambda(x) \in L^2(R^2)$. It is important to note that $\{T_k \hbar_j\}$ and $\{T_k \tilde{\hbar}_j\}$ $(j \in J, k \in Z^2)$ need not be contained in M. The above example is such case. Consequently, the position of $\{T_k \hbar_j\}$ and $\{T_k \tilde{\hbar}_j\}$ are not generally commutable, i.e., there exists $\Upsilon \in M$ such that

$$\sum_{j \in J} \sum_{v \in Z^2} \langle \Upsilon, T_v \tilde{\hbar}_j \rangle T_v h_j(x) \neq \sum_{j \in J} \sum_{v \in Z^2} \langle \Upsilon, T_v \hbar_j \rangle T_v \tilde{\hbar}_j(x) = \Upsilon(x) .$$

Definition 3. A generalized multiresolution structure (GBMS) $\{V_j, \hbar_j, \tilde{\hbar}_j\}$ is a sequence of closed linear subspaces $\{V_j\}_{j \in Z}$ of $L^2(R^2)$ and $2r$ elements $h_j, \tilde{\hbar}_j \in L^2(R^2)$, $j \in J$ such that (a) $V_j \subset V_{j+1}$ $\forall\, j \in Z$; (b) $\bigcap_{j \in Z} V_j = \{0\}$; $\bigcup_{j \in Z} V_j$ is dense in $L^2(R^2)$; (c) $h(x) \in V_j$ if and only if $Dh(x) \in V_{j+1}$, $\forall j \in Z^2$, where $D\phi(x) = \phi(2x)$, for $\forall \phi(x) \in L^2(R^2)$; (d) $g(x) \in V_0$ implies $T_v g(x) \in V_0$, for all $v \in Z^2$; (e) $\{T_v h_j(x), j \in J,\ v \in Z^2\}$ forms a multiple pseudoframe for V_0 with respect to $\{T_v \tilde{\hbar}_j(x), j \in J, v \in Z^2\}$.

3 Characterization of GBMS of Paley-Wiener Subspace

A necessary and sufficient condition for the construction of multiple pseudoframe for Paley-Wiener subspaces of $L^2(R^2)$ is presented as follow.

Theorem 1. Let $h_j \in L^2(R^2)$ $(j \in J)$ be such that $|\hat{\hbar}_j| > 0$ a.e. on a connected neighbourhood of 0 in $[-\frac{1}{2}, \frac{1}{2})^2$, and $|\hat{\hbar}_j| = 0$ a.e. otherwise. Define $\Lambda \equiv \bigcap_{j \in J} \{\gamma \in R :$ $|\hat{\hbar}_j| \geq c > 0,\ j \in J\}$ and

$$V_0 = PW_{\Lambda} = \{f \in L^2(R^2) : \ \text{supp}(\hat{f}) \subseteq \Lambda\}. \tag{8}$$

$$V_n \equiv \{\Upsilon(x) \in L^2(R^2) : \ \Upsilon(x/2^n) \in V\}, \quad n \in Z , \tag{9}$$

Then, for $\tilde{h}_j \in L^2(R^2), \{T_v h_j : j\in J,\ k\in Z\}$ is a multiple pseudoframe for V_0 with respect to $\{T_v \tilde{h}_j, j\in J, v\in Z^2\}$ if and only if

$$\sum_{j\in J} \overline{\hat{h}_j(\gamma)}\, \hat{\tilde{h}}(\gamma).\chi_\Lambda(\gamma) = \chi_\Lambda(\gamma) \quad a.e., \tag{10}$$

where χ_Λ is the characteristic function on Λ.

Proof. For all $\Upsilon(x)\in PW_\Lambda$ consider

$$F(\sum_{j\in J}\sum_{v\in Z^2}\langle \Upsilon, T_v\hbar_j\rangle T_v\tilde{\hbar}_j = (\sum_{j\in J}\sum_{v\in Z^2}\langle \Upsilon, T_v\tilde{\hbar}_j\rangle F(T_v\hbar_j)$$

$$=(\sum_{j\in J}\sum_{v\in Z^2}\int_{R^2}\hat{\Upsilon}(\mu)\overline{\widehat{\tilde{\hbar}}_j(\mu)}.e^{2\pi i v\mu}d\mu\,\hat{\hbar}_j(\omega)e^{-2\pi jv\omega}$$

$$=\sum_{j\in J}\sum_{v\in Z^2}\int_0^1\sum_{n\in Z^2}\hat{\Upsilon}(\mu+n)\overline{\hat{\tilde{\hbar}}_j(\mu+n)}.e^{2\pi i v\mu}d\mu\,\hat{\hbar}_j(\omega)e^{-2\pi jv\omega}$$

$$=\sum_{j\in J}\widehat{\hbar}_j(\omega)\sum_{n\in Z^2}\hat{\Upsilon}(\omega+n)\overline{\hat{\tilde{\hbar}}_j(\omega+n)} = \sum_{j\in J}\hat{\Upsilon}(\omega)\widehat{\hbar}_j(\omega)\overline{\hat{\tilde{\hbar}}_j(\omega)}.$$

where we have used the fact that $|\widehat{\hbar}|\neq 0$ only on $[-\frac{1}{2},\frac{1}{2})^2$ and that $\sum_{n\in Z^2}\hat{\Upsilon}(\omega+n)\,\overline{\hat{\tilde{\hbar}}(\omega+n)},\ j\in J$ is 1-periodic function. Therefore

$$\sum_{j\in J}\hat{\hbar}_j\overline{\hat{\tilde{\hbar}}_j}\cdot\chi_\Lambda = \chi_\Lambda, \qquad \text{a.e.,}$$

is a neccssary and sufficient condition for $\{T_k\hbar_j, j\in J,\ k\in Z^2\}$ to be a multiple pseudoframe for V_0 with respect to $\{T_k\tilde{\hbar}_j, j\in J, k\in Z^2\}$.

Theorem 2. Let $\hbar_j, \tilde{\hbar}_j \in L^2(R^2)$ have the properties specified in Theorem 1 such that (6) is satisfied. Assume V_ℓ is defined by (9). Then $\{V_\ell, \hbar_j, \tilde{\hbar}_j\}$ forms a GBMS.

Proof. We need to prove four axioms in Definition 2. The inclusion $V_e \subseteq V_{e+1}$ follows from the fact that V_e defined by (9) is equivalent to PW_{2A}. Condition (b) is satisficd bccause the set of all band- limited signals is dense in $L^2(R^2)$. On the other hand ,the intersection of all band-limited signals is the trivial function. Condition (c) is an immediate consequence of (9). For condition (d) to be prooved, if $f\in V_0$, then $f=\sum_{j\in J}\sum_{k\in Z^2}\langle f, T_k\tilde{\hbar}_j\rangle T_k\hbar_j$. By taking variable substitution, for $\forall n\in Z^2$, $f(t-n)=\sum_{j\in J}\sum_{v\in Z^2}\langle f(\cdot),\tilde{\hbar}_j(\cdot-v)\rangle\hbar_j(t-v-n)=\sum_{j\in J}\sum_{v\in Z^2}\langle f(\cdot-n),\tilde{\hbar}_j(\cdot-v)\rangle\hbar_j(x-v)$ That is, $T_n f = \sum_{j\in J}\sum_{v\in Z^2}\langle T_n f\cdot T_v\tilde{\hbar}_j\rangle T_v\hbar_j$. Or, it is a fact $f(T_v\hbar_j)$ has support in Ω for arbitrary $k\in Z$. Therefore, $T_n f\in V_0$.

Example 1. Let $\hbar_j \in L^2(R^2)$ be such that

$$\tilde{\hbar}_j(\gamma)=\begin{cases}1/r & a.e., \quad \|\omega\|\le \frac{1}{2}, \\ (3-4\|\omega\|)1/r, & a.c., \quad \frac{1}{2}<\|\gamma\|<\frac{3}{4}, \\ 0, & otherwise.\end{cases} \qquad \bar{\tilde{\hbar}}(\omega)=\begin{cases}1, & a.e., \quad \|\gamma\|\le \frac{1}{2}, \\ 3-4\|\omega\|, & a.e., \quad \frac{1}{2}<\|\gamma\|<\frac{3}{4}, \\ 0. & otherwise.\end{cases}$$

$j\in J$. Choose $\Lambda=\{\omega\in R^2 : |\hbar(\omega)|\ge \frac{1}{r}\}=[-\frac{1}{2},\frac{1}{2}]^2$, and define $V_0=PW_\Lambda$, select $\tilde{\hbar}_j \in L^2(R^2)$ such that Then, since $\sum_{j\in J}\tilde{\hbar}_j(\omega)\overline{h_j(\omega)}=1$ a.e. on Λ, by Theorem 1, $\{T_k\hbar_j\}$ and $\{T_k\tilde{\hbar}_j\}$ form a pair of pseudoframes for $V_0=PW_A$. We begin with introducing the concept of pseudoframes of translates.

Define an operator $K: U\to \ell^2(Z^2)$ by

$$\forall\, \Upsilon(x)\in U, \quad K\Upsilon=\{\langle \Upsilon, T_v f\rangle\}, \tag{11}$$

and define another operator $S:\ell^2(Z^2)\to W$ such that

$$\forall c=\{c(u)\}\in \ell^2(Z^2), \quad S\{c(k)\}=\sum_{u\in Z^2} c(u) T_u \tilde{f}. \tag{12}$$

Theorem 3[8]. Let $\phi(x)$, $\tilde{\phi}(x)$, $\psi_\iota(x)$ and $\tilde{\psi}_\iota(x)$, $\iota\in\Lambda$ be functions in $L^2(R^2)$. Assume conditions in Theorem 1 are satisfied. Then, for any $\hbar\in L^2(R^2)$, and $n\in Z$,

$$\sum_{k\in Z^2}\langle \hbar, \tilde{\phi}_{n,k}\rangle \phi_{n,k}(x)=\sum_{\iota=1}^{3}\sum_{v=-\infty}^{n-1}\sum_{k\in Z^4}\langle \hbar, \tilde{\psi}_{\iota:v,k}\rangle \psi_{\iota:v,k}(x). \tag{13}$$

$$\hbar(x)=\sum_{\iota=1}^{3}\sum_{v=-\infty}^{\infty}\sum_{k\in Z^2}\langle \hbar, \tilde{\psi}_{\iota:v,k}\rangle \psi_{\iota:v,k}(x). \quad \forall\, \hbar(s)\in L^2(R^2) \tag{14}$$

Consequently, if $\{\psi_{\iota:v,k}\}$ and $\{\tilde{\psi}_{\iota:v,k}\}$, $(\iota\in\Lambda, v\in Z, k\in Z^2)$ are also Bessel sequences, they are a pair of affine frames for $L^2(R^2)$.

References

[1] Daubechies, I.: The wavelet transform, time-frequency localization and signal analysis. IEEE Trans. Inform. Theory 39, 961–1005 (1990)

[2] Iovane, G., Giordano, P.: Wavelet and multiresolution analysis:Nature of ε^∞ Cantorian space-time. Chaos, Solitons & Fractals 32(4), 896–910 (2007)

[3] Chen, Q., Wei, Z.: The characteristics of orthogonal trivariate wavelet packets. Information Technology Journal 8(8), 1275–1280 (2009)

[4] Shen, Z.: Nontensor product wavelet packets in $L_2(R^s)$. SIAM Math. Anal. 26(4), 1061–1074 (1995)

[5] Chen, Q., Qu, X.: Characteristics of a class of vector-valued nonseparable higher dimensional wavelet packet bases. Chaos, Solitons & Fractals 41(4), 1676–1683 (2009)

[6] Chen, Q., Huo, A.: The research of a class of biorthogonal compactly supported vector valued wavelets. Chaos, Solitons & Fractals 41(2), 951–961 (2009)
[7] Charina, M., Chui, C.K., He, W.: Tight frames of compactly supported multivariate multiwavelets. J. Comput. Appl. Math. 233, 2044–2061 (2010)
[8] The characterization of a class of subspace pseudoframes with arbitrary real number translations. Chaos, Solitons & Fractals 42(5), 2696–2706 (2009)

[6] Chen, Q., Huo, [illegible] compactly supported vector-valued [illegible] Chaos, Solitons [illegible] (2009)
[7] [illegible] M., Chen, Q.: [illegible] frames of compactly supported multivariate [illegible] wavelets [illegible] Math. [illegible] (2010)
[8] The [illegible] subspace [illegible] with arbitrary real number [illegible] 24 [illegible] (2009)

The Features of a Sort of Seven-Variant Wavelet Wraps Concerning a Seven-Variant Scaling Functions

Baozhen Wang and Huimin Wang

Department of Mathematics Science, Huanghuai University, Zhumadian 463000, China
{fghjkp147,txxpds}@126.com

Abstract. Information science is an interdisciplinary science primarily concerned with the analysis, collection,classification, manipulation, storage, retrieval and dissemination of information.Wavelet wraps have been the focus of active research for twenty years, both in theory and applications. In this work, the notion of orthogonal nonseparable seven-variant wavelet wraps is introduced. A new approach for designing them is presented by iteration method. We proved that the seven-variant wavelet wraps are of the orthogonality trait. We give three orthogonality for-mulas regarding the wavelet packets. We show how to construct nonseparable seven-variant wavelet packet bases. The orthogonal seven-dimensional wavelet wraps may have arbitrayily high regularities.

Keywords: Sobolev space, seven-variant wavelet wraps, iterative method, bracket product, vector subdivision scheme, biorthogonality formulas.

1 Introduction

A major advantage of the application of information engineering is that it virtually forces the organization to address the entire spectrum of its information systems requirements, resulting in a functionally integrated set of enterprise systems. In contrast, ad hoc requirements may result in a fragmented set of systems (islands of automation), which at their worst may be incompatible, contain duplicate (perhaps inconsistent) information, and omit critical elements of informationThe concept of wavelet wraps have received much research attention and evidence has shown that they can be used to improve the localization of the frequency field of wavelet bases. Although the Fourier transform has been a major tool in analysis for over a century, it has a serious laking for signal analysis in that it hides in its phases information concerning the moment of emission and duration of a signal. The main feature of the wavelet transform is to hierarchically decompose general functions, as a signal or a process, into a set of approximation functions with different scales. Wavelet packets[1], owing to their good properties, have attracted considerable atten-tion. They can be widely applied in science and engineering [2,3]. Coifman R. R. and Meyer Y. firstly introduced the notion for orthogonal wavelet packets. Chui C K. and Li Chun [4] generalized the concept of orthogonal wavelet packets to the case of non-orthogonal wavelet packets so that wavelet packets can be employed in the case of the spline wavelets and so on. The introduction for biorthogonal wavelet packs attributes to Cohen and Daubechies [5]. The introduction for the notion of nontensor product

D. Jin and S. Jin (Eds.): Advances in FCCS, Vol. 1, AISC 159, pp. 149–155.
springerlink.com

wavelet packets attributes to Shen [6]. Since the majority of information is multidimensional information, many researchers interest themselves in the investigation into multivariate wavelet theory. The classical method for constructing multivariate wavelets is obtained by means of the tensor product of some univariate wavelets. But, there exist a lot of obvious defects in this method, such as, scarcity of designing freedom. The objective of this paper is to generalize the concept of univariate orthogonal wavelet packets to orthogonal seven-variant wavelet wraps.

2 Notations and Fundamentals

We beigin with the following notations. Z and N stand for integers and nonnegative integers, respectively. Let R be the set of all real numbers. R^s denotes the n-dimensional Euclidean space, where s is an integer and $2 \le s \in N$. By $L^2(\mathrm{R}^s)$, we denote the square integrable function space on R^s. Set $x=(x_1, x_2, \cdots, x_n) \in \mathrm{R}^s$, $k=(k_1, k_2, \cdots, k_s)$, $\eta=(\eta_1, \eta_2, \cdots, \eta_s)$, $z_\iota = e^{-i\eta_\iota/2}$, where $\iota = 1, 2, \cdots, s$ and $n \in N, n \ge 2$. The inner product for arbitrary $f(x), \varphi(x) \in L^2(\mathrm{R}^s)$, and the Fourier transform of $\varphi(x)$ are defined as, respectively

$$\langle f, \varphi \rangle := \int_{R^s} f(x)\overline{\varphi(x)}\,dx, \quad \hat{\varphi}(\omega) := \int_{R^s} \varphi(x) e^{-ix\cdot\omega} dx,$$

where $\omega \cdot x = \omega_1 x_1 + \omega_2 x_2 + \cdots + \omega_s x_s$ and $\overline{\varphi(x)}$ denotes the conjugate of $\varphi(x)$.

The space $H^\lambda(R^n)$ is a Hilbert space equipped with the inner product given by

$$\langle h, \varphi \rangle_{H^\lambda(R^s)} := \frac{1}{(2\pi)^s} \int_{R^s} \hat{h}(\eta)\,\overline{\hat{\varphi}(\eta)}\,(1+|\eta|^{2\lambda})d\eta, \quad h, \varphi \in L^2(R^s). \tag{1}$$

The bracket product of compactly supported functions $g(x), \varphi(x) \in L^2(\mathrm{R}^n)$ is given by

$$[f, \varphi](\omega) := \sum_{u \in Z^s} \langle f, \varphi(\cdot - u) \rangle e^{-iu\omega} = \sum_{u \in Z^s} \hat{f}(\omega + 2u\pi)\overline{\hat{\varphi}(\omega + 2u\pi)}. \tag{2}$$

We are interested in wavelet bases for the Sobolev space $H^\lambda(R^s)$, where λ is a positive integer. In this case, we obtain

$$\langle h, \varphi \rangle_{H^\lambda(R^s)} = \langle h, \varphi \rangle + \langle h^{(\lambda)}, \varphi^{(\lambda)} \rangle, \quad h, \varphi \in L^2(R^s). \tag{3}$$

By algebra theory, it is obvious that there are 2^s elements $\mu_0, \mu_1, \cdots, \mu_{2^s-1}$ in space $Z_+^s = \{(k_1, k_2, \cdots, k_s) : k_\iota \in Z_+, \iota = 1, 2, \cdots, s-1, s\}$ such that $Z^s = \bigcup_{d \in \Omega_0} (d + 2Z^s)$; $(d_1 + 2Z^s) \bigcap (d_2 + 2Z^s) = \phi$, where $\Gamma_0 = \{d_0, d_1, \cdots, d_{2^s-1}\}$ denotes the aggregate of all the different representative elements in the quotient group $Z^s/(2Z^s)$ and order $d_0 = \{\underline{0}\}$ where $\{\underline{0}\}$ is the null element of Z_+^s and d_1, d_2 denote two arbitrary distinct elements in Γ_0.Let $\Gamma = \Gamma_0 - \{\underline{0}\}$ and Γ, Γ_0 to be two index sets.

Definition 1. A sequence $\{\varphi_k(x)\}_{k\in Z^s} \subset L^2(R^s)$ is called an orthonormal set, if

$$\langle \varphi_u, \varphi_v \rangle = \delta_{n,v}, \quad u, v \in Z^s, \tag{4}$$

where I_s stands for the $s \times s$ identity matrix and $\delta_{n,v}$, is generalized Kronecker symbol, i.e., $\delta_{n,v} = 1$ as $n = v$ and $\delta_{n,v} = 0$, otherwise.

Definition 2. A sequence $\{\xi_v : v \in Z\} \subseteq W$ is a frame for H if there exist two positive real numbers C, D such that

$$\forall\, \eta \in W, \quad C\|\eta\|^2 \le \sum_{v\in\Gamma} |\langle \eta, \xi_v \rangle|^2 \le D\|\eta\|^2 , \tag{5}$$

where W be a separable Hilbert space and Γ is an index set. A sequence $\{\xi_v : v \in Z\} \subseteq W$ is a Bessel sequence if (only) the upper inequality of (6) holds. If only for all $\hbar \in \Omega \subset W$, the upper inequality of (6) holds, the sequence $\{\xi_v\} \subseteq W$ is a Bessel sequence with respect to (w.r.t.) Ω. For a sequence $c = \{c(v)\} \in \ell^2(Z^s)$, we define its discrete-time Fourier transform as the function in $L^2(0,1)^s$ by

$$Fc(\omega) = C(\omega) = \sum_{v\in Z^s} c(v) e^{-2\pi i x\omega} dx , \tag{6}$$

Note that the discrete-time Fourier transform is Z^n-periodic. Let $T_v\phi(x)$ stand for integer translations of a function $\phi(x) \in L^2(R^s)$, i.e., $(T_v\phi)(x) = \phi(x-v)$, and $\phi_{j,v} = 2^{nj/2}\phi(2^j x - v)$, where a is a positive real constant number. Let $\hbar(x) \in L^2(R^s)$ and let $V_0 = \overline{span}\{T_v\hbar : v \in Z^s\}$ be a closed subspace of $L^2(R^s)$.

3 Seven-Dimensional Multiresolution Analysis

The multiresolution analysis method is an important approach to obtaining wavelets and wavelet packs. We introduce the notion of multiresolution analysis of $L^2(R^7)$. Set $n = 7$, $\hbar(x) \in L^2(R^7)$ satisfy the refinement equation:

$$\hbar(x) = 128 \cdot \sum_{u\in Z^7} d(u)\hbar(2x-u), \tag{7}$$

where $\{b(u)\}_{u\in Z^7}$ is a real number sequence and $\hbar(x)$ is called a scaling function. Taking the Fourier transform for the both sides of refinement equation (1), we have

$$\hat{\hbar}(\omega) = B(z_1, z_2, z_3, z_4, z_5, z_6, z_7)\,\hat{\hbar}(\omega/2). \tag{8}$$

$$D(z_1, z_2, z_3, z_4, z_5, z_6, z_7) = \sum_{v\in Z^7} b(v) \cdot z_1^{v_1} z_2^{v_2} \cdots z_7^{v_7} . \tag{9}$$

Define a subspace $U_l \subset L^2(R^7)$ ($l \in Z$) by

$$U_l = clos_{L^2(R^7)} \left\langle 2^l \hbar(2^l \cdot - u) : u \in Z^7 \right\rangle, \tag{10}$$

We say that $\hbar(x)$ in (8) generates a multiresolution analysis $\{U_j\}_{j\in Z}$ of $L^2(R^7)$, if the sequence $\{U_j\}_{j\in Z}$, defined in (4) satisfies the below: (a) $U_l \subset U_{l+1}, \forall\, l\in Z$; (b) $\bigcap_{l\in Z} U_l = \{0\}$; $\bigcup_{l\in Z} U_l$ is dense in $L^2(R^7)$; (c) $\hbar(x)\in U_l \Leftrightarrow \hbar(2x)\in U_{l+1}$, $\forall\, l\in Z$; (d) the family $\{2^l \hbar(2^l\cdot - n) : n\in Z^5\}$ is a Riesz basis for U_l ($l\in Z$).

Let W_l ($l\in Z$) denote the orthogonal complementary subspace of U_l in U_{l+1} and assume that there exists a vector valued function $\Psi(x)=(\psi_1(x),\psi_2(x),\cdots,\psi_{127}(x))^T$ (see [7]) forms a Riesz basis for W_l, i.e.,

$$W_l = clos_{L^2(R^7)}\langle \psi_v(2^l\cdot - u) : v=1,2,\cdots,127;\, u\in Z^7\rangle,\quad l\in Z. \tag{11}$$

By (5), it is clear that $\psi_1(x),\psi_2(x),\cdots,\psi_{127}(x)\in W_0 \subset U_1$. Therefore, there exist fifteen real sequences $\{b^{(\iota)}(v)\}$ ($\iota=1,2,\cdots 127$, $v\in Z^7$) such that

$$\psi_\iota(x) = 128\cdot\sum_{u\in Z^7} b^{(\iota)}(u)\hbar(2x-u),\ \iota\in\Gamma, \tag{12}$$

where $\Gamma=\{1,2,3\cdots,127\}$, $\Gamma_0=\Gamma\bigcup\{0\}$. Formula(12) can be written in frequency domain as follows:

$$\widehat{\psi}_\iota(\omega) = B^{(\iota)}(z_1,z_2,\cdots,z_7)\hat{\hbar}(\omega/2),\ \iota\in\Gamma, \tag{13}$$

where the symbol of the real sequence $\{b^{(\iota)}(u)\}$ ($\iota\in\Gamma$, $u\in Z^7$) is

$$B^{(\iota)}(z_1,z_2,\cdots,z_7) = \sum_{n\in Z^7} d^{(\iota)}(n)\, z_1^{n_1} z_2^{n_2}\cdots\cdots z_7^{n_7}. \tag{14}$$

A five-variant scaling function $\hbar(x)\in L^2(R^7)\in$ is orthogonal one, if

$$\langle \hbar(\cdot), \hbar(\cdot - u)\rangle = \delta_{0,u},\quad u\in Z^7. \tag{15}$$

We say that $\psi_1(x),\psi_2(x),\cdots,\psi_{127}(x)$ are orthogonal five-dimensional wavelets associated with the scaling function $\hbar(x)$, if they satisfy:

$$\langle \hbar(\cdot), \psi_\iota(\cdot - v)\rangle = 0,\quad \iota\in\Gamma,\ v\in Z^7, \tag{16}$$

$$\langle \psi_\iota(\cdot),\psi_\tau(\cdot - u)\rangle = \delta_{\iota,\tau}\delta_{0,v},\ \iota,\tau\in\Gamma,\ v\in Z^7 \tag{17}$$

4 The Features of Nonseparable Seven-Variant Wavelet Wraps

We are now ready to introduce the below notations: $f_0(x)=\hbar(x), f_\iota(x)=\psi_\iota(x)$, $d^{(0)}(v)=d(v), d^{(\iota)}(v)=b^{(\iota)}(v), \iota\in\Gamma \quad u\in Z^7$. We are now in a position discuss the traits of orthogon-al nonseparable five-variant wavelet wraps.

Definition 3. A family of seven-variant functions $\{ f_{128n+\nu}(x): n=0,1,2,\cdots, \nu=0, 1,2,\cdots,127 \}$ is called a nonseparable seven-dimensional wavelet packs with respect to the orthogonal scaling function $\hbar(x)$, where $\Gamma_0=\{0,1,2,\cdots,127\}$ and

$$f_{128n+\iota}(x)=\sum_{v\in Z^7} d^{(\iota)}(v)\, f_n(2x-v),\ \iota\in\Gamma_0, \tag{18}$$

Taking the Fourier transform for the both sides of (19) yields

$$\widehat{f}_{128n+\iota}(\omega)=D^{(\iota)}(z_1,z_2,\cdots,z_7)\,\widehat{f}_n(\omega/2). \tag{19}$$

where $\iota\in\Gamma_0$, $D^{(\iota)}(z_1,z_2,\cdots,z_7)=D^{(\iota)}(\omega/2)=\sum_{v\in Z^7} d^{(\iota)}(v)\, z_1^{v_1} z_2^{v_2}\cdots z_7^{v_7}$.

Lemma 1[6]. Assuming that $\hbar(x)$ is an orthogonal five-variant scaling function and $P(z_1,z_2,z_3,z_4,z_5)$ is the symbol of the sequence $\{b(v)\}$. Then we have

$$\begin{aligned}
&|P(z_1,z_2,z_3,z_4,z_5)|^2+|P(-z_1,z_2,z_3,z_4,z_5)|^2+|P(z_1,-z_2,z_3,z_4,z_5)|^2\\
&+|P(z_1,z_2,-z_3,z_4,z_5)|^2+|P(z_1,z_2,z_3,-z_4,z_5)|^2+|P(z_1,z_2,z_3,z_4,-z_5)|^2\\
&+|P(-z_1,-z_2,z_3,z_4,z_5)|^2+|P(-z_1,z_2,-z_3,z_4,z_5)|^2+\cdots\cdots+\\
&+|P(-z_1,z_2,-z_3,-z_4,-z_5)|^2+|P(z_1,-z_2,-z_3,-z_4,-z_5)|^2+|P(-z_1,-z_2,-z_3,-z_4,-z_5)|^2=1
\end{aligned}$$

Lemma 2[6]. If $\psi_\nu(x)$ ($\nu=0,1,\cdots,31$) are orthogonal wavelet functions associated with $\hbar(x)$. Denoting by $y_\iota=(-1)^j z_\iota$, $\iota=1,2,3,4,5,$ then we have

$$\begin{aligned}
\Xi_{\lambda,\mu}=\sum_{j=0}^{1}\{&D^{(\lambda)}(y_1,y_2,y_3,y_4,y_5)\overline{D^{(\nu)}(y_1,y_2,(-1)^j z_3,(-1)^j z_4,(-1)^j z_5)}\\
&+D^{(\lambda)}((-1)^{j+1}z_1,y_2,y_3,y_4,y_5)\cdot\overline{D^{(\nu)}((-1)^{j+1}z_1,(-1)^j z_2,y_3,y_5,(-1)^j z_5)}\\
&+D^{(\lambda)}((-1)^j z_1,(-1)^{j+1}z_2,y_3,y_4,(-1)^j z_5)\cdot\overline{D^{(\nu)}(y_1,-y_2,y_3,y_4,y_5)}\\
&+D^{(\lambda)}((-1)^j z_1,(-1)^j z_2,(-1)^{j+1}z_3,y_4,(-1)^j z_5)\cdot\overline{D^{(\nu)}(y_1,y_2,-y_3,y_4,y_5)}\\
&+D^{(\lambda)}(y_1,y_2,y_3,-y_4,y_5)\cdot\overline{D^{(\nu)}((-1)^j z_1,(-1)^j z_2,y_3,-y_4,(-1)^j z_5)}
\end{aligned}$$

$$+D^{(\lambda)}(y_1,y_2,y_3,y_4,-y_5)\cdot\overline{D^{(\nu)}((-1)^j z_1,(-1)^j z_2,(-1)^j z_3,(-1)^j z_4,(-1)^{j+1} z_5)}$$

$$+D^{(\lambda)}(-y_1,-y_2,y_3,y_4,y_5)\cdot\overline{D^{(\nu)}(-y_1,-y_2,(-1)^j z_3,(-1)^j z_4,(-1)^j z_5)}$$

$$+D^{(\lambda)}(-y_1,y_2,-y_3,y_4,y_5)\cdot\overline{D^{(\nu)}(-y_1,y_2,-y_3,(-1)^j z_4,(-1)^{j+1} z_5)}$$

$$+\cdots\cdots-\delta_{\lambda,\nu},\quad \lambda,\nu\in\{0,1,2,\cdots,31\}. \tag{20}$$

Theorem 1[6]. For every $u\in Z^5$ and $n\in Z_+$, $\iota\in\{0,1,2,\cdots,127\}$, we have

$$\langle f_{128n}(\cdot), f_{128n+\nu}(\cdot-u)\rangle=\delta_{0,\nu}\delta_{0,u}. \tag{21}$$

Theorem 2. For every $u\in Z^7$ and $m, n\in Z_+$, we have

$$\langle f_m(\cdot), f_n(\cdot-k)\rangle=\delta_{m,n}\delta_{0,k}. \tag{22}$$

Proof. For the case of $m=n$, (22) follows from Theorem 1. As $m\neq n$ and $m, n\in\Gamma_0$, (22) can be established from Theorem 1. Assuming that m is not equal to n and at least one of $\{m, n\}$ doesn't belong to Γ_0, rewrite m, n as $m=128m_1+\lambda_1$, $n=128n_1+\mu_1$, where $m_1, n_1\in Z_+$, and $\lambda_1,\mu_1\in\Gamma_0$.

Case 1. If $m_1=n_1$, then $\lambda_1\neq\mu_1$. By (14), (16) and (18), (22) holds, since

$$(2\pi)^7\langle f_m(\cdot), f_n(\cdot-k)\rangle=\int_{R^7}\widehat{f}_{128m_1+\lambda_1}(\omega)\overline{\widehat{f}_{128n_1+\mu_1}(\omega)}\cdot\exp\{ik\omega\}d\omega$$
$$=\int_{R^7}D^{(\lambda_1)}(z_1,z_2,\cdots,z_7)\widehat{f}_{m_1}(\omega/2)\cdot\overline{\widehat{f}_{n_1}(\omega/2)}\,\overline{D^{(\mu_1)}(z_1,z_2,\cdots,z_7)}\cdot e^{ik\omega}\,d\omega$$
$$=\int_{[0,2\pi]^7}\Xi_{\lambda_1,\mu_1}\cdot\exp\{ik\omega\}\,d\omega=O.$$

Case 2. If $m_1\neq n_1$, we set $m_1=128m_2+\lambda_2$, $n_1=128n_2+\mu_2$, where $m_2, n_2\in Z_+$, and λ_2, $\mu_2\in\Gamma_0$. If $m_2=n_2$, then $\lambda_2\neq\mu_2$. Similar to Case 1, we have $\langle\Lambda_m(\cdot),\Lambda_n(\cdot-k)\rangle=O$. That is to say, the proposition follows in such case. As $m_2\neq n_2$, we order $m_2=128m_3+\lambda_3$, $n_2=128n_3+\mu_3$, once more, where $m_3, n_3\in Z_+$, and $\lambda_3,\mu_3\in\Gamma_0$. Thus, after taking finite steps (denoted by r),we obtain that $m_r, n_r\in\Gamma_0$, and $\lambda_r,\mu_r\in\Omega_0$. If $\alpha_r=\beta_r$, then $\lambda_r\neq\mu_r$. Similar to Case 1, (22) follows. If $\alpha_r\neq\beta_r$, Similar to Lemma 1, we conclude that

$$\langle f_m(\cdot), f_n(\cdot-k)\rangle=\frac{1}{((2\pi)^7)}\int_{[0,2^{r+1}\pi]^7}\{\prod_{\iota=1}^{r}D^{(\lambda_\iota)}(\frac{\omega}{2^\iota})\}\cdot O\cdot\{\prod_{\iota=1}^{r}D^{(\mu_\iota)}(\frac{\omega}{2^\iota})\}\cdot e^{ik\omega}d\omega=O.$$

Corollary 1. For $n\in Z_+$, $\nu\in Z^7$, we have $\langle f_n(\cdot), f_n(\cdot-u)\rangle=\delta_{0,u}$.

References

1. Telesca, L., et al.: Multiresolution wavelet analysis of earthquakes. Chaos, Solitons & Fractals 22(3), 741–748 (2004)
2. Chen, Q., Huo, A.: The research of a class of biorthogonal compactly supported vector-valued wavelets. Chaos, Solitons & Fractals 41(2), 951–961 (2009)
3. Li, S., et al.: A theory of generalized multiresolution structure and pseudoframes of translates. Fourier Anal. Appl. 7(1), 23–40 (2001)
4. Chen, Q., et al.: Existence and characterization of orthogonal multiple vector-valued wavelets with three scale. Chaos, Solitons & Fractals 42(4), 2484–2493 (2009)
5. Shen, Z.: Nontensor product wavelet packets in $L_2(R^s)$. SIAM Math. Anal. 26(4), 1061–1074 (1995)
6. Chen, Q., Qu, X.: Characteristics of a class of vector valued nonseparable higher-dimensional wavelet packet bases. Chaos, Solitons & Fractals 41(4), 1676–1683 (2009)
7. Chen, Q., Wei, Z.: The characteristics of orthogonal trivariate wavelet packets. Information Technology Journal 8(8), 1275–1280 (2009)

Investigation on Traffic Signs Recognition Based on BP Neural Network and Invariant Moments

Hongwei Gao, Xuanxuan Liu, Zhe Liu, and Kun Hong

School of Information Science & Engineering, Shenyang Ligong University, Shenyang, 110159 China
ghw1978@sohu.com

Abstract. According to the traffic sign recognition process converges slowly, easy to fall into local minimum of defects in general BP algorithm, a BP algorithm is discussed in the paper, which is based on the improved conjugate gradient method on the basis of Fletcher-Reeves linear search method. At the beginning, the basic principle of the algorithm is introduced; meanwhile, an in-depth analysis is discussed from the theoretical aspects in the paper, and then the trained BP neural network is applied into the function approximation. Experimental results show that the algorithm has many advantages, such as it has a good effect in solving the rotation and scale invariance with the image, the extracted features with translation, scale and rotation invariance. The fast convergence speed and less iteration of this algorithm can meet the requirements of unmanned vehicle autonomous navigation.

Keywords: Traffic Signs Recognition, BP Neural Network, Invariant moments, Conjugate grads.

1 Introduction

Since the 1980s, unmanned technology began to study abroad. For example, Carnegie Mellon University developed Navlab series of prototype vehicles covered wagon, train, utility vehicles, large buses and other vehicles. In 1995, Navlab-5 prototype acrossed the United States, the total distance 4587 km, of which accounted for 98.2% of independent driving, the longest continuous autonomous driving distance of 111 km, average speed throughout the 102 kilometers per hour. In July 2010, the unmanned vehicle developed by Palma Italy University to across two continents, travelled 13000 km, and at the end of October arrived in Shanghai World Expo.

A number of units start tracking and researching the unmanned technique in a very early time in China, such as the National University of Defense Technology, Nanjing University of Science and Technology, Zhejiang University and Tsinghua University. In 2008, a major research plan-"audio-visual information cognitive computing" was established by the National Natural Science Fund Committee, which is related to the intelligent vehicle technology. And two sessions of China "intelligent car challenge in the future" contest were initiated and hosted by the fund commission in 2009 and 2010. Each race has more than 10 units sent people or car to participate in this competitive game at least.

D. Jin and S. Jin (Eds.): Advances in FCCS, Vol. 1, AISC 159, pp. 157–162.
springerlink.com

In today's rapid development of unmanned vehicles, image recognition has become a hot topic today. The most important events of unmanned vehicle on the road are reliability and security, and in the adaptability of complex environments, unmanned vehicle image recognition system reliability is essential. The image recognition guide method is a recently developed, in order to achieve proper self-guided vehicle navigation, this method uses camera to shoot the road image, the use of computer image recognition technology to recognize the path and the various signs. Facing a variety of traffic signs set up by the traffic department, how to transfer the traffic signs information to the driver accurately and usefully is crucial now. In the intelligent transportation system (ITS), traffic sign recognition requires a correct classification, fast operation speed, and also requires not sensitive to translation, rotation and the changes of the size, so how to realize the automatic recognition of traffic signs is very important to automatic driving or driving assistance system. To address the above issues, a method based on moment invariants and improved BP neural network in traffic sign recognition system is presented in the paper.

2 Moment Invariant Theory

In the traffic sign recognition process, there are many different ways to extract the image feature, such as texture features, geometric features, etc. All of these are susceptible to image rotation, size changes, translation and other conditions. So selecting an appropriate feature extraction, reducing the feature space dimension and stability is the foundation of solving the traffic sign classification and identification. In 1961, Hu was the first to put forward the seven geometric moment invariants for image recognition.

Hu moment invariant feature is one of the shape features of the image, Hu first proposes the concept of geometric moment with the algebraic invariant theory, and define as follows:

Given a two-dimensional image, so (p+q) order geometric moments are defined as follows:

$$M_{pq}=\sum_{x}\sum_{y}f(x,y)x^{p}y^{q} \qquad (p,q=0,1,2,\Lambda) \qquad (1)$$

In order to make the invariant moment in image translation, scaling, rotation remained unchanged, *Mpq* had to turn to make the following three transform.

(1) the definition of image center moment is:

$$\mu_{pq}=\sum_{x}\sum_{y}(x-x_0)^{p}(y-y_0)^{q}f(x,y) \qquad (p,q=0,1,2,\Lambda) \qquad (2)$$

There (x_0, y_0) is the center of mass of the image, defined as:

$x_0= m_{10}/m_{00}$, $y_0= m_{01}/m_{00}$.Obviously, the center distance u_{pq} is the image translation invariant.

(2) For the above center distance of standardization, we can get the form of the moment as follow (it is image translation and scaling invariant):

$$V_{pq}=\mu_{pq}\Big/\mu_{00}^{1+(p+w)/2} \qquad (p+q)\geq 2 \tag{3}$$

(**3**) Applying the algebraic invariant moment's principle, we can get seven invariant moments as follows; they are the image translation, scaling, and rotation invariant. This is the famous moment invariant formula:

$$\varphi_1=\nu_{20}+\nu_{02} \tag{4}$$

$$\varphi_2=(\nu_{20}-\nu_{02})^2+4\nu_{11}{}^2 \tag{5}$$

$$\varphi_3=(\nu_{30}-3\nu_{12})^2+(3\nu_{21}-\nu_{03})^2 \tag{6}$$

$$\varphi_4=(\nu_{30}+\nu_{12})^2+(\nu_{21}+\nu_{03})^2 \tag{7}$$

$$\begin{aligned}\varphi_5&=(\nu_{30}-3\nu_{12})(\nu_{30}+\nu_{12})\cdot\left[(\nu_{30}+\nu_{12})^2-3(\nu_{21}+\nu_{03})^2\right]\\&+(3\nu_{21}-\nu_{03})(\nu_{21}+\nu_{03})\cdot\left[3(\nu_{30}+\nu_{12})^2-(\nu_{21}-\nu_{03})^2\right]\end{aligned} \tag{8}$$

$$\varphi_6=(\nu_{20}-\nu_{02})\left[(\nu_{30}+\nu_{12})^2-(\nu_{21}+\nu_{03})^2\right]+4(\nu_{30}+\nu_{12})(\nu_{03}+\nu_{21}) \tag{9}$$

In practical application, because of the value φ_1 to φ_6 is very small and also appear positive and negative questions, the calculation using the logarithm of the absolute value of moment invariants form to avoid the accuracy limits, that is $\log|\varphi_i|$,i=1,2,3,4,5,6,as the features of image. The Hu invariant moments is applied widely in the computer image processing and pattern recognition fields relies on these features ,such as :scale, translation and rotation invariance, simply calculation etc.

3 Conjugate Gradient Method BP Neural Network

3.1 BP Neural Network Theory

BP neural network algorithm process is composed by the forward propagation and reverse transmission. In the forward propagation process, the input information transmit from the input layer to the output layer through the hidden units processing layer by layer, each layer state of neurons is only influence the next layer of neurons in the state. If you cannot get in the output layer of the desired output, then transferred to the reverse transmission, the error signal to return along the original connection path, and then get the smallest error signal by modifying the weights of each layer of neurons. BP model changes a set of sample input-output problem into a nonlinear optimization problem, the most common use of the optimized gradient descent method, with iteration solving the problem by adding hidden nodes is to optimize the adjustable parameters increase, thereby to be more fine-cut solution. This mapping is a highly non-linear mapping, if the neural network as a mapping from input to output. Although BP model from the aspects of all has the vital significance, it has some problems. Judging from the mathematics, it is a nonlinear optimization problem, which inevitably exist the local minima problem and the network operation is a one-way transmission, no feedback. The current this model is not a nonlinear dynamic system, but a nonlinear mapping.

3.2 Conjugate Gradient Method

Conjugate gradient method is used in each iteration step at the current point the direction of steepest descent to generate convex quadratic function f on the Hesee array of G conjugate direction, and seeking to establish the function f in $\Re^2$ very small point on the way. This method was first proposed by Hesteness and Stiefel in 1952 for solving positive definite linear equations, then studied by Fletcher etc and applied to unconstrained optimization problems, and obtained a wealth of results, therefore the conjugate gradient method is the currently important method to solve unconstrained algorithm. Conjugate gradient algorithm Fletcher-Reeves's formula is as follow:

$$\vec{z}(k+1) = \vec{z}(k) + \vec{a}(k)\vec{s}(k) \tag{10}$$

Search direction $\vec{s}(k)$ is a set of conjugate vectors, $\vec{a}(k)$ is the step size and:

$$\vec{s}(k+1) = -\vec{g}(k) + \beta(k)\vec{s}(k) \tag{11}$$

$$\beta(k) = \frac{[\vec{g}(k+1)]^T \vec{g}(k+1)}{[\vec{g}(k)]^T \vec{g}(k)} \tag{12}$$

Therefore we can say conjugate gradient algorithm is the utilization of the gradient of the past and gradient information at a point of the present, use its linear combination to construct a better search direction. The formula $\vec{g}(k) = \nabla f[\vec{z}(k)]$, Here are the conjugate gradient method for solving unconstrained problems , the main steps of the algorithm 1-6 are as follow:

Step1: Give accuracy of iteration $0 \le \varepsilon \le 1$ and initial point x0, and calculate $g0 = \nabla f(x0)$. Make K=0

Step2: If $\|g_k\| \le \varepsilon$, stop calculating, then output $x^+ \approx x_k$

Step3: Calculate the direction of search operators d_k :

$$d_k = \begin{cases} -g_k & k=0 \\ -g_k + \beta_{k-1} d_{k-1} & k \ge 1 \end{cases}$$

Step4: Use the method of the exact line searching to determine the search step size a_k

Step5: $x_{k+1} = x_k + a_k d_k$,then calculate $g_{k+1} = \nabla f(x_{k+1})$

Step6: Make k=k+1 ; then return step1.

4 Pretreatment of the Traffic Signs and the Extraction of the Traffic Signs

The traffic sign is circle, square and triangle, etc. The background color of the traffic signs is blue, red, and yellow. For example (blue, white), (yellow, white). This paper firstly confirm the traffic sign's geometric shape, then set up a few pairs of colors. The experiment shows that this method used to extract traffic signs can achieve above 95%. As shown in figure 1:

The pretreatment of the traffic signs directimpact on the efficiency of recognition, so a good recognition system of traffic signs must have a good image preprocessing function. This article uses a series of algorithm to deal with noise, deformation, and defect treatment of pictures, which works efficiently.

Fig. 1. The extraction of traffic signs

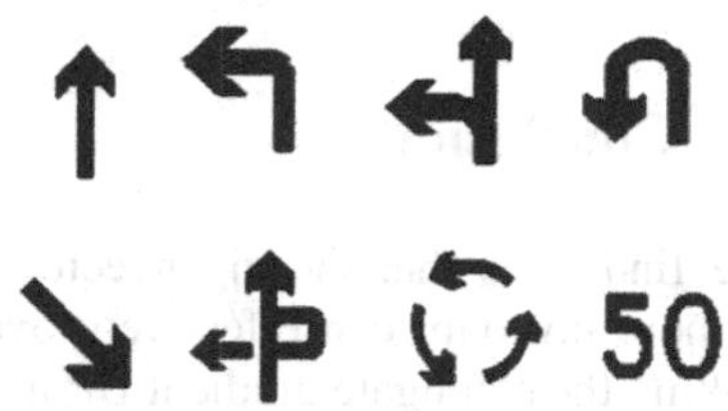

Fig. 2. Traffic signs sample figure

5 Analysis and Research of Experimental Results

This paper uses twelve color images as sample pictures, we make UGV collect a series of pictures in highway, select six pictures of them at will as test picture. The pictures must be preprocessed before we extract features form them. We do our best to deal with the pictures to the simplest state, which makes the feature we extract clearly separated from each other. The classification using BP neural network is much better. The characteristic value of these twelve pictures is shown in table 1.

Compare the improved BP algorithm with the traditional BP algorithm in the same picture recognition of traffic sign, we can see that the improved BP algorithm's iteration times and full learning time is shortest. With the improvement of accuracy requirement, the difference of the two algorithms is more and more obvious; the superiority of the improved BP algorithm is more identifiable. This article uses twelve traffic signs as sample, makes the picture collected by UGV rotated, magnified and deformed. The finding is that the identification accuracy of traditional BP algorithm is 92.3%, and the identification accuracy of traditional BP algorithm is 94.6%, we can see the latter is more superior.

Table 1. The training sample and expected output

Invariant moment by logarithm							Expecting value
1	1.3672	9.6723	16.4745	14.1304	28.6533	16.9599	10010000
2	1.3982	8.5637	17.0913	10.0197	28.0204	19.3872	10001000
3	1.2923	9.3822	16.2430	11.2538	29.5302	20.9534	10000010
4	1.3793	10.2103	14.1533	10.8715	29.7234	19.8743	01001000
5	1.3145	8.1208	14.8098	11.9024	24.0209	20.8092	01000010
6	1.3824	9.4722	14.7238	12.9234	28.5327	19.2503	00101000
7	1.5602	10.4721	15.3218	14.4538	29.8754	20.2533	00100100
8	1.3103	9.9801	15.2714	11.0703	28.1233	16.1530	00100010

6 Conclusion

The finding is that the eigenvector of traffic signs image which is extracted from moment invariant can effectively overcome the impact of the external environment. Making the conjugate gradient BP neural network as classifier works efficiently to the characteristics classification of the invariant moments. Experiments show that the method in this article has a good classification and recognition ability, conjugate gradient BP neural network is effective to reduce the number of iteration, to increase the convergent speed, and can satisfy the requirement of real-time autonomous navigation system.

Acknowledgement. This work is supported by China Liaoning Provice Educational Office Fund (No.L2011038).

References

1. Takashi, O., Kiyokazu, T.: Lane Recognition LIDAR,Intelligent VehiclesSymposium. Tokyo,Japan(2006).
2. Turchetto, R., Manduchi, R.: Visual Curb Localization for Navigation,Proceedings. The 2003 IEEURSJ Intl.Conference on Intelligent Robots and Systems Vegas, Nevada (2003).
3. Michael, S., Axel, G., Uwe, F., Uwe-Philipp, K., Uwe, F. et al.: Detecting Reflection Posts-Lane Recognition on Country Roads. IEEE Intelligent Vehicles Symposium University of Parma,Italy (2004).
4. Cyganek, B.: Colour Image Segmentation with Support Vector Machines for Road Sign Recognition.Technical Report,AGH-University of Science and Technology,(2007).
5. Cyganek, B.: Real-time Detection of the Triangular and Rectangular Shape Road Signs. Lecture Notes in Computer Science 4687, pp.744—755, Germany (2006).

Applying Soft Computing in Material Molding

Juan Wang[1], Siyu Lai[2], and Yang He[3]

[1] Department of Computer Science, China West Normal University, 637002 NanChong, China
[2] Department of Medical Image, North Sichuan Medical College, 637000 NanChong, China
[3] Winer Sanitary Ware Corporation, 511490 GuangZhou, China
{wjuan0712,lsy_791211}@126.com

Abstract. Soft computing is a collection of methodologies including fuzzy logic, neural network and evolutionary computing. They have become important manner for complex systems modeling in varied fields. The paper mainly discusses the application of soft computing in material research from material performance prediction modeling.

Keywords: soft computing, modeling, preparation, process, material science.

1 Introduction

Soft computing is the general name of several computing technology with the development of information technology and computer intelligence, which mainly include fuzzy logic, neural networks, simulated annealing algorithm, evolutionary computation and probability reasoning [1]. It differ from the traditional "hard computing", since soft computing try not to find the exact solution of the problem entirely, imprecision and uncertainty can be tolerated sometimes, the results obtained are accurate or inaccurate approximate solution of the problem. In general, these techniques are complementary and not competitive. They have their own advantages and strengths, both the neural networks and genetic algorithms simulate the biological principles. But genetic algorithm is based on mechanism of biological evolution and is often serve as a searching and optimization algorithm while neural network is the representation of typical human brains' performance and is mainly used in learning and curve fitting. In recent years, due to the tireless efforts of national research workers, soft computing got a rapid development and has been successfully applied to many complex, non-linear modeling fields, such as system administration [2], pattern recognition [3] and medical diagnosis [4].

Material properties is determined by many factors, such as component, organizational structure, material preparation or processing, which are of typical non-linear relationship and are difficult to describe accurately using mathematical model. The evaluation of material properties tend to have many forms, take the mechanical property as an example, the often used evaluation indexes are fracture toughness (K_{IC}, J_{IC}), conventional tensile yield strength σ_s, tensile strength σ_b, elastic modulus E, fatigue crack growth rate da/dN and so on. Although part of the datum of conventional materials can be found in the manual, there is always a certain amount

D. Jin and S. Jin (Eds.): Advances in FCCS, Vol. 1, AISC 159, pp. 163–168.
springerlink.com

of dispersion degree arising from the heterogeneity of the material itself and the differences in refining, rolling, heat treatment, aging, and mechanical molding process. In practice, there exists a problem that how to identify and evaluate specified material.

Soft computing provides a new way to predict material properties and is a highly efficient method in optimizing and controlling the preparation process, processing technique or heat treatment. This article mainly focuses upon predictive modeling of materials properties and introducing neural networks, fuzzy systems, genetic algorithms, simulated annealing method and the combination of them in application of materials science.

2 Neural Networks Based Molding

Artificial neural network is, built on the basis of the results of modern neuroscience research, an abstract mathematical model. Human brains are the directly model foundation and come into being parallel architecture through the simulation and simplification to structure and function of human brain, which reflect a number of basic characteristics of human brain function. Artificial neural network research began in the 1940s, experienced the fluctuation of climax-hard time-climax period [5]. And after 1980s, the artificial neural network researches and applications have been greatly promoted due to Hopfield's work.

Neural network technology is a leading material research method in soft computing modeling, which involves almost all aspects of materials research; the neural network technology is the most widely used method in material properties forecast, especially the combination of neural networks, genetic or simulated annealing algorithms make it possible to achieve global optimal solution, which is the effective and accurate system modeling manner.

Mechanical properties of materials and their application primarily depend upon the microstructure which is formed during the thermal processing; namely, "hot processing parameters" determine the "microstructure" and "microstructure" determine "mechanical properties". As a result, establish the relationship between thermal and mechanical processing parameters is necessary. Utilizing the infinite approximated non-linear mapping feature of neural networks, you can very easily set up such a relationship model. For example, as far as titanium and its alloy are concerned, such excellent properties as high strength-weight ratio, good corrosion and temperature resistance performance mainly depend on microstructure of materials in the absence of sufficient phase transition studies. S. Malinov et al. [6] built the relationship between titanium alloy heat treatment process parameters and their mechanical performance using neural network technology and the system architecture is shown in Figure 1. The input parameters are components for the alloy (including Al, Mo, V et al.), heat treatment parameters (including annealing, solution treatment, aging et al.) and working (or test) temperature (-100°C~600°C). The output parameters are nine key performance indicators, including tensile strength, yield strength, elongation, area reduction rate, impact strength, hardness, elastic modulus, fatigue strength and fracture strength. The established network model can be used to predict the performance of titanium alloy in varied temperatures and can also guide the parameters optimization in thermal processing.

P. Korczak set chemical composition, rolling process and microstructure et al. 14 parameters and ferrite grain size, yield strength, hardness, tensile strength as input and output nodes respectively. And he established the BP neural network model which fit the non-linear relationship between steel composition, microstructure, cooling rate and the final mechanical properties preferably as shown in Figure 2.

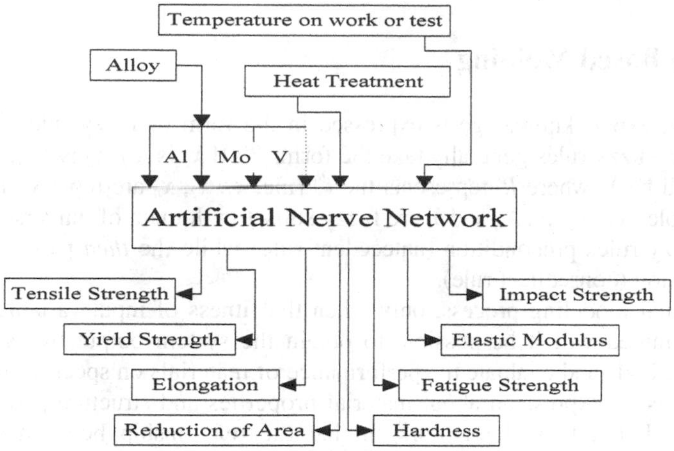

Fig. 1. Architecture of titanium alloy performance

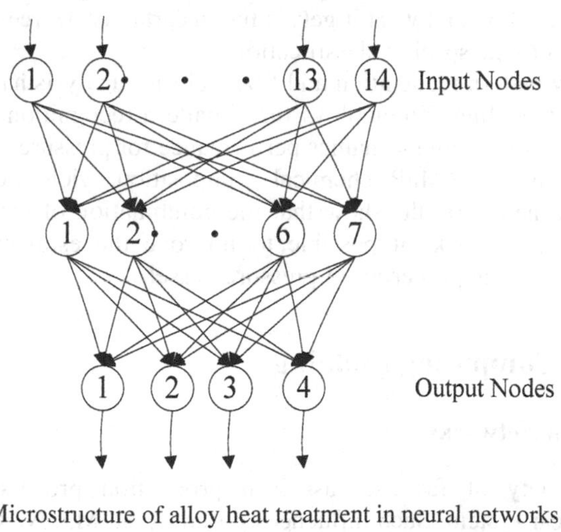

Fig. 2. Microstructure of alloy heat treatment in neural networks modeling

Soft computing, in particular, the neural networks describe the relationship between input and output data by learning and facilitate the nonlinear modeling in such a complex system [7]. It is a trend to acquire the properties of a new substance by forecasting before experiment. C. L. Philip Chen et al. [8]. using self-organizing neural network established the model between the structure and performance of substance to predict the new materials (unknown compound) performance.

They classify compounds by using K-medoid method (unauthorized learning algorithm) at first to guaranty the output not affected by external factors. And then train the OFBNN (Orthogonal Functional Basis Neural Network) to learn the performance of the same sort of compounds utilizing the OFBFM (Orthogonal Functional Basis Functional Mapping) and the trained network can be used in performance prediction later.

3 Fuzzy Based Molding

As we know, expert knowledge is expressed in the form of fuzzy rules (If-then) in fuzzy system, fuzzy rules generally take the form: R^i: If x_1 is X^i_1, x_2 is X^i_2,..., x_n is X^i_n, then y^i would be Y^i where R^i represents the ith rule, x_1, x_2, x_n are input variables, y^i is output variable, X_1, X_2, ..., X_n and Y are fuzzy subset in domain of universe; *If (•)* part is called fuzzy rules precondition (antecedent rule) while the *then* part be the fuzzy rules conclusion (consequent rule).

In the actual modeling process, only when the fitness of input variables for each rule is determined will it be possible to obtain the system output by using certain reasoning methods and evaluate the performance of materials on specific condition. A lot of relationship expression about material properties and structure parameters has been proposed, but they don't work in that the relationship between the two is complex combine with the exploratory stage of micro-scale mechanical processing. The most commonly used method of fuzzy pattern recognition are high-valve membership and nearest selection principles; Fuzzy clustering is soft style compared with the traditional method and what it gets is the uncertainty degree of the sample to be identified belongs to the specific classification.

Nowadays, fuzzy pattern recognition and fuzzy cluster analysis have been put into material properties modeling. Yu employs fuzzy pattern recognition in identification and characterization of fracture mechanics performance for pressure vessel and put up membership function on 16MnR chemical composition, yield point and tensile strength. The experimental results show that the combination of pattern recognition and fuzzy clustering [9] make it possible to improve the estimation accuracy in determining the mechanical properties of pressure vessel.

4 Other Soft Computing Molding

4.1 Fuzzy Neural Networks

Affected by a variety of factors, cast iron production process is a complex multivariate nonlinear system. Such influences posed by Si, Mn, Ni, Cu, etc. are non-linear and the experimental data are vulnerable to serious noise pollution. The conventional correlation analysis and statistical pattern recognition methods are not so good to predict material strength. Xia et al. proposed the fuzzy modeling methods and established the fuzzy neural network prediction model of gray cast iron strength, achieved a more automatic fuzzy prediction updating model.

ADI (Austempered Ductile Iron) enjoys similar chemical component to traditional cast iron and contain Ni, Mo or Cu in addition. The ADI relationship between skill and performance is multivariate non-linearly, interactions also exist among various factors, and as a result, conventional mathematical methods are hard to model the relationship. L.Arafeh et al. [10] set up four impact-strength prediction model based on fuzzy and neural fuzzy theory: (1) fuzzy system model; (2) Neural-fuzzy model based on BP algorithm; (3) Model based clustering algorithm; (4) Neural-fuzzy model based on BP-clustering algorithm. The experiments show that the four forecasts are accurate to the impact strength, the correlation between predictive value and expectation of model (1) is 0.88; model (2) have five samples with correlation of 0.97 except two samples with too large difference in predictive value and expectation; model (3) have the best correlation while model (4) have the worst one, in that the BP Neural-fuzzy model based algorithm requires more samples for the study.

4.2 Genetic and Simulated Annealing Algorithm

The genetic algorithm is a sort of biological evolution mechanism like stochastic global optimization manner developed in the past 20 years, which was first proposed by Professor J.H. Holland. At present, it is an important component of soft computing, which is characterized by structural and gradient information about the problem is not needed and insensitive to dimensions of the problem.

Simulated annealing method is another optimization algorithm inspired by the natural phenomena, the core idea of which is quite similar with the principle of thermodynamics, and in particular, it is similar like the liquid flow and crystallization and the metal cooling and annealing.

Simulated annealing algorithm uses a group of parameters called cooling schedule to control the process, the algorithm would finally achieve the relative global optimal solution about the optimization problem when cooling parameter t that trends to zero slowly. One solution of the optimization problem i and the objective function $f(i)$ are relative to microstate of the solid i and its energy E_i respectively. We refer to control parameter t that decreasing during the process as the role that temperature parameter T plays in solid annealing process. As for each t value, the algorithm takes Metropolis criteria, conduct iterative process continuously consists of new solution generation, judging, acceptance or abandonment to achieve the "balance point" under each "temperature", and thc final "balance point" is just the solution to the question.

As global optimization algorithm, genetic and simulated annealing algorithms are currently applied in materials research and developments combine with the neural networks. Achieve the materials design, equipment or process optimization through establishing the prediction model of material properties.

5 Conclusions

The system is composed of material properties or the preparation process and their impact factors are generally non-linearly. Establish the forecast model of material properties by using soft computing to bring about the optimization and administration of material preparation and processing.

The mainly used soft computing methods is currently the fuzzy system and neural network technology in materials field while the genetic and simulated annealing algorithms are seldom used as single method. And there are remarkable disparities between material and other fields for the genetic and simulated annealing algorithms are generally used to model material properties, optimize the design and process of materials combine with neural network technology.

References

1. Whitley, D.: Genetic Algorithms and Neural Networks. Genetic Algorithm in Engineering and Computer Science 3, 203–216 (1995)
2. Kim, J., Moon, Y., Zeigler, B.P.: Designing Fuzzy Net Controllers Using Genetic Algorithms. IEEE Contr. Syst. Mag. 15(3), 66–72 (1995)
3. Baraldi, A., Blonda, P.: A Survey of Fuzzy Clustering Algorithms for Pattern Recognition. IEEE Trans. Syst. Man. Cybern. B Cybern. 29(6), 786–801 (2000)
4. Eberhart, R.C., Dobbins, R.W., Hutton, L.V.: Neural Network Paradigm Comparisons for Appendicitis Diagnoses. In: Proceedings of the Fourth Annual IEEE Symposium on Computer Based Medical Systems, pp. 298–304 (1991)
5. Lu, J.J., Chen, H.: Researching Development on BP Neural Networks. Control Engineering of China 32(5), 449–451 (2006) (in Chinese)
6. Malinov, S., Sha, W., McKeown, J.J.: Modelling the Correlation between Processing Parameters and Properties in Titanium Alloys Using Artificial Neural Network. Computational Materials Science 21, 375–394 (2001)
7. Liau, L.C.K., Huang, C.J., Chen, C.C., et al.: Process Modeling and Optimization of PECVD Silicon Nitride Coated on Silicon Solar Cell Using Neural Networks. Solar Energy Materials & Solar Cells 71, 169–179 (2002)
8. Chai, D.L., Liu, J.H., Shen, P.Y.: The Application of Fuzzy Analysis Method to the Design forthe Structures and Properties of the Complex Materials. Journal of Xi'an Jiao Tong University 13(5), 54–60 (2003)
9. Yang, Z., Gu, Q.S., Liang, X.Z., Li, H.J., et al.: H J Li: Evaluation of High-Temperature Mechanical Properties of a Composite Based on Fuzzy Clustering-Fuzzy Pattern Recognition. Journal of East China University of Science and Technology (Natural Science Edition) 6, 39–45 (2001) (in Chinese)
10. Arafeh, L., Singh, H., Putatunda, S.K.: A Neuro Fuzzy Logic Approach to Material Processing. IEEE Trans. Syst. Man. Cybern. C Cybern. 29(3), 362–370 (1999)

Research on Computer Application in Material Science

Yi Liu[1], Siyu Lai[2], and Yang He[3]

[1] Office of Academic Affairs, China West Normal University, 637002, NanChong, China
[2] Department of Medical Image, North Sichuan Medical College, 637000 NanChong, China
[3] Winer Sanitary Ware Corporation, 511490 GuangZhou, China
{liu4fire,lsy_791211}@126.com

Abstract. The application fields of computer technology in material science study are introduced. The specific applications in material science with computer assisted are discussed. And the research, development and application of material are bound to have a prosperous boost.

Keywords: computer simulation, material science, metal forming, plastic forming, heat treatment.

1 Introduction

Computer simulation technology has been widely used in varied material forming process which is composed of liquid forming, plastic forming, connection forming, polymer material forming, powder metallurgical forming, composite and other material forming. The computer aided simulation that make material forming goes from qualitative description to quantitative prediction, provide theoretical basis and optimization project for material process and new technology development. And the simulation advanced to the knowledge-based computing inspection stage from the traditional error experience, which is extremely significant for the realization of small volume, high quality, low cost, short delivery, flexible production and friendly environment manufacturing model. Computer simulation is the way that future material preparation process must go and it trends to multi-scale simulation and integration.

2 Applying Computing in Material Science

2.1 New Material Development

Material design refer to prediction of new materials of composition, structure and properties by using theoretical and calculation or produce the new specific material with the best process and preparation method according to the requirements. Material design is primarily employs artificial intelligence, pattern recognition, computing simulation, knowledge base and database technology to integrate physics, chemistry theory and a large number of experimental data together. And make decisions for new material design with induction and deduction and provide effective techniques and methods for implementation of material development [1] [2].

D. Jin and S. Jin (Eds.): Advances in FCCS, Vol. 1, AISC 159, pp. 169–174.
springerlink.com

2.2 Computer Simulation in Material Science

To simulate real system, provide simulation results and guide the new materials utilizing computer is one of the effective manner for material design. The simulated objects cover material development and use and include synthesis, structure, performance and preparation.

Computer simulation is a computer based simulated experiment in the light of the actual architecture. By doing so, we can test the accuracy of the model, if the simplification of analytical theory derived from the model is success and provide solution for prediction of real model and laboratory model that is hard to achieve by comparing simulated results with real experimental data.

2.3 Automation and Optimization of Materials Processing

The development of materials processing is mainly reflected in the rapid improvement of control technology while the application of microcomputer and programmable controller (PLC) in materials processing nearly embody such a trend. The uses of computer not only reduce labor intensity but also better the quality and precision of product and improve output.

Computers can be used to optimize the materials processing. For example, it would be possible to control the nitriding and carburizing process after the mathematical model about the process has been made by using computer. The material preparation can be controlled accurately, such as the surface treatment (heat treatment) in the furnace temperature administration.

2.4 Image and Data Processing

The properties of material have an inseparable relationship with its condensed matter and one of the research means for them both is optical and electron microscope which interpret condensed matter in two-dimensional image model.

It is feasible to realize the structure of materials by using image processing and analysis functions of computer. To get useful such structural information as crystal size, distribution, aggregation methods from images and establish correlation between the information and materials properties to guide the study on structure.

3 Concrete Applications of Computer

3.1 Computer Simulation of Liquid Metal Forming

(1) Computer Simulation in Liquid Metal Filling Process

The major method of numerical simulation used in liquid metal filling process is based upon SO-LA-VOF (Solution Algorithm) approach method combine with the volume function in free surface processing and improve the heat transfer and flow correction.

Although algorithms on the field have been conducted so far, such as parallel algorithm, 3-D finite element method, 3-D finite difference method, hybrid numerical and analytical method, the best algorithm has not been found until now and various algorithms have different advantages and disadvantages and focus on varied aspects.

(2) Numerical Simulation of Solidification Process

It will be possible to determine the internal temperature field within casting, draw isothermal line distribution of casting in any section and isochron distribution by numerically simulating solidification process. And display the casting solidification process dynamically in three dimensions to determine thc final location and analyze the size and place of shrinkage cavity and contraction [3] [4]. Shrinkage and contraction quantitative prediction methods are currently applied in foundry and achieved good economic benefits.

(3) Numerical Simulation of Solidification Forming

The main numerical simulation of solidification models are Monte Carlo, phase field and interface stability based crystal growing model and each specializes in the specific fields. Microstructure simulation has made great progress in crystal form, grain growth, collision, redistribution of solute in liquid after ten years of research. At the moment, there are a lot of practical liquid forming simulation software has been developed, which is shown in Table 1.

3.2 Computer Simulation Applied in Heat Treatment

Computer simulation rely on the development of computational heat transfer, transformation kinetics, thermo-elastic mechanics, numerical calculation and other disciplines, which not only offer solid foundation for research and application of heat treatment simulation, but also promote the rapid development of heat treatment simulation. And success has been achieved in the following aspects including hardenability control, induction heating, vacuum heating, atmosphere control, phase-change kinetics calculation and temperature control. The heat management simulation technology is currently widely used and great attention has been paid.

The numerical computing simulation of steel quenching has been put into use in recent years and thermo-elastic model also has been put up. The ions carburizing mathematical model has been established in constant-value gas carburizing process and the digital model was recorded in satellite computer memory by applying computer simulation and then input parameters to calculate the carbon distribution curve that consistent with the actual data.

Heat treatment computer simulation is also applied in calculating the kinetics curve of carbon eutectoid steel A→P transformation, which build mathematical model that depict the transformation between DTA and kinetics curves to work out isothermal transformation kinetics curve steel A→P using differential thermal analysis and simulation technology combined method.

Table 1. The mainstream of liquid forming software and application

Software	Developer	Algorithm	Focus
FLOW-3D	US/ Liquid Science Company	FVM FDM	Flow and Heat Transfer
MAGMAsoft	Germany/Aachen University	FDM PEM	Flow and Heat Transfer, Stress and Microstructure
PROCAST	US/UES Company	FEM	Flow and Heat Transfer, Stress and Electromagnetic Casting
SOLSTAR	Britain/FOSECO Company	FVM A/D	Heat Transfer Analysis
NOVACAST	Sweden/NovacastAB Company	FDM A/D	Heat Transfer Analysis
CALCOMP	Swiss	FEM	Flow and Heat Transfer
RAPID/CAST	US/Parallel Tech Company	FDM	Flow and Heat Transfer, Microstructure
PHOENICS	Britain/Imperial College	FVM	Flow and Heat Transfer
FIDAP7.0	US	FEM	Flow, Heat and Mass Transfer
SOLIDA	Japan/Komatru Lab	DFDM	Flow and Heat Transfer
AFsolid	US/Foundry Association	FDM	Flow and Heat Transfer
LS-DYNA3D	Britain/Com&Een association	FEM	Flow and Heat Transfer, Stress and Strain
CAPS-3D	Britain/Argonne Lab	FVM	Flow and Heat Transfer
SIMULOR	France/Pechiney	FVM	Flow and Heat Transfer

3.3 Computer Simulation of Plastic Forming

Computer simulation of plastic forming is trace metal forming process on computer real-timely and demonstrates the whole process by using graphics system which is shown in Figure 1. Numerical simulation of plastic forming research concentrated in universities and research institutions initially and mainly used 2-dimensional simulation software. And with the rapid development of computer technology and improvement of mathematical algorithms in recent years, three-dimensional simulation technology has been getting feasible and extended to be widely used in industry to solve the forming problems of complex forging.

The existing simulation methods for plastic forming are partition method, slip-line method, upper bound method, finite element method and boundary element method, where the finite element method and especially the copy for rigid is widely used. The commercial software are DEFORM, AU TO2FORGE, SUPERFORGE for volume plastic molding and DYNA3D, PAM2STAMP, ANSYS, etc. for sheet plastic forming simulation.

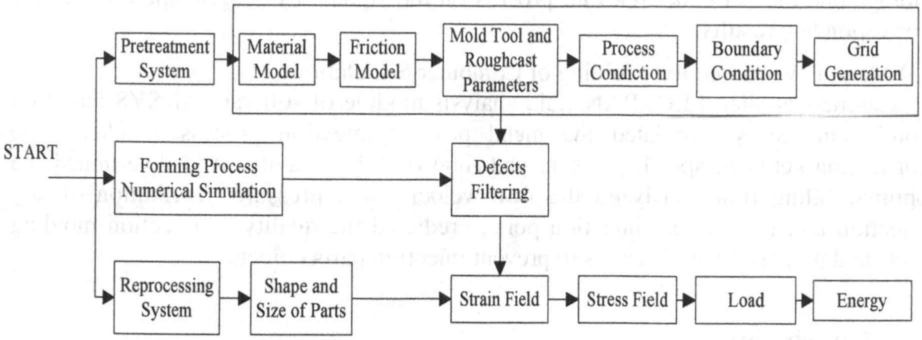

Fig. 1. Plastic forming of computer simulation process

3.4 Computer Simulation of Plastic Forming

There are four main means of computer microstructure simulations for welding process, that is, empirical method, statistical method, direct formation simulation and internal variable method [5] [6]. The direct formation method is appropriate for macroscopic change and incapable of predicting the microstructure and transition process of grain size and is the most commonly used manner. The internal variable method is ideal for establishment of organizational model for non-isothermal transformation and has been widely used in forecasting non-isothermal transition behavior of steel, cast and aluminum alloy and also achieved the microstructure formation simulation, which is a very promising approach.

Weld microstructure model tends to couple thermal field, stress and microstructure field together in the future and establish the model truly reflects the welding process changing and is able to predict the performance of welding joints to guide the welding process directly.

3.5 Computer Simulation of Plastic Forming

(1) Application of Computer Simulation in Corrosion and Protection Field. Computer simulation is presently applied for current density distribution, electric field strength analysis, conductivity, etc. in material corrosion and protection and correctly reflected the variation in corrosion process toward accurately prediction and rigorous quantization.

It is confirmed effective by taking the finite element analysis, finite difference and boundary element methods to solve the Laplace equation (potential distribution equation control of battery in electrochemical). The equation can be applied in varied complex situation by computer simulation.

(2) Application of Material Detection
Computer applications in material detection are currently focus on material composition, structure, phase, physical performance detection and non-destructive mechanical parts testing etc. The basic principle of which is to transform the analog signal detected with specific detector into digital one and transfer it into computer storage and then conduct relevant process on the signals by programmers to get the corresponding results.

(3) Metal Power Injection Molding of Computer Simulation
Researchers applied FLOTRAN fluid analysis module of software ANSYS based on continuum theory simulated the metal powder injection process, disclosed the correlation between speed, pressure and time of filling-cavity. And determined the optimal filling time, analyzed the flow velocity and pressure distribution of key injection unit in different injection ports, predicted the quality of injection molding parts and proposed the measures to prevent injection parts defects.

4 Conclusions

Material science is a newly emerging multi-disciplinary and immature subject, which is largely depend on the accumulation of facts and experiences, and there is a long way to go to study on this field systematically. Computer, as a modern tool in all areas of the world, plays an increasingly significant role and penetrated into different disciplines and daily life. Computer is becoming a very important tool in material field and in which the application applied is just the important promotion to boost the rapid development of material science.

References

1. Liu, B.C.: Material Processing Technology in the 21st Century. Aeronautical Manufacturing Technology 6, 17–69 (2003) (in Chinese)
2. Wen, S., Farrugia, D.C.J.: Conf. on Advances in Materials and Processing Technologies, pp. 123–127. Ireland University of Tehran (1999)
3. Zhu, Y.X., Zhang, X.F., Yang, B., et al.: The Numeric Simulation of Weld Residual Stress of Several Weld-Repaired Based on Finite Element. Transactions of The China Welding Institution 29(1), 68–72 (2002) (in Chinese)
4. Zhao, H.T., Mi, G.F., Wang, K.F.: The Simulation of Casting and solidification process. Aerospace Manufacturing Technology 38(1), 105–108 (2007)
5. Zhao, X.R., Zhu, Y.X., Sun, Q.M.: Analysis of Finite Element of Residual Stresses in Butt Welds. Welding Technology 34(5), 14–15 (2003) (in Chinese)
6. Yurioka, Y., Koseki, T.: Modeling Activities in Japan. Theoretical Prediction in Joining and Welding, 39–58 (1996)

Applying Data Mining in Score Analysis

Yi Liu

Office of Academic Affairs, China West Normal University, 637002, NanChong, China
liu4fire@126.com

Abstract. Data mining is one of the important technologies in data warehouse and one of the fields with the fasted development in computer industry. The skill boosts database technology into a more advanced stage and quickly forming a separate area. The rough set theory and data mining techniques are briefly discussed on the basis of analyzing the existent problem in colleges and universities currently. Take the college student score as an example, the solution to the score analysis is proposed based on rough set theory and data mining techniques.

Keywords: data mining, score analysis, rough set theory, data warehouse.

1 Introduction

It is needed to conduct a comprehensive analysis on student score management system in universities for the scores are affected by many factors. Traditional analysis are nothing more than mean, variance, reliability and validity etc., which are often still based on teaching itself. In fact, there are some factors inperceptible in teaching process or factors come from other channels affecting student achievement, which require further analysis and get conclusion for managers to make appropriate decisions by using data mining techniques [1].

2 Rough Set Based Data Mining

Data mining is the process that extracting latent, unknown and potentially available information from a large number of incomplete, noised, fuzzy datum [2]. Analyze and reason complex, incomplete data to find latent relationship and extract useful model, simplify information processing, determine inaccurate, incomplete knowledge representation, which is the crucial task of data mining.

Polish scientist Paw lak proposed the rough set theory in 1982 which drawing on the imprecise, fuzz definitions in logic and philosophy fields, and formed a complete theoretical system, the rough set theory according to the vague knowledge base. Rough set theory is established on the basis of classification mechanism and interprets it as indiscernible relation in a particular space and the relation constitute the space division.

D. Jin and S. Jin (Eds.): Advances in FCCS, Vol. 1, AISC 159, pp. 175–180.
springerlink.com

3 Solutions

Here, we take classification algorithms of data mining to transform a large number of data into classification rules in order to better analyze the data and the steps are shown in Figure 1.

Step 1: mining objects determination. It is an important part to clearly define the problem and identify the purpose of data mining. The result of data mining is unpredictable but the issue itself should be foreseen.

Step 2: data acquisition. This is a step occupy much time and work, it is needed for teachers to collect information in a variety of ways in the past teaching practice.

Step 3: data pretreatment. Integrate data of different type and transform them into a data analysis model that based on different algorithms, different algorithms support different models.

Step 4: data classification mining. The object of classification mining is to establish a classification model. You need first select the appropriate mining algorithm and use the right software to implement and then conduct mining process on these converted data.

Step 5: classification rules analysis. This step is mainly to explain and evaluate the classification results.

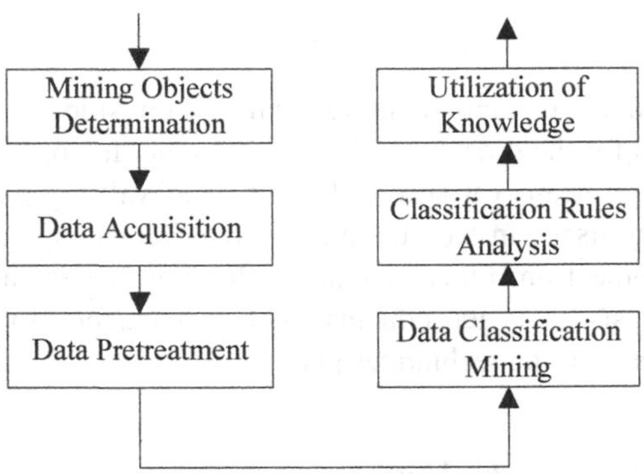

Fig. 1. Classification mining process

4 Instance of Program Implementation

4.1 Mining Objects Determination

We carry out our research on students of computer department and hope to learn about what factor is possibly influence the scores in the light of the student's basic routine learning including interests, preview and study time on computer after class etc. and to guide future teaching with the results obtained.

4.2 Data Acquisition

1) Students basic information. Data structure lists below: ID, name, gender, birthplace, department, specialty, class, which can be accessed in SIMS (Student Information Management System).
2) Student information survey. The item consists of ID, interest, preview, learning effect on class and review on computer after class etc. The acquisition of information is mainly relying on questionnaire fill by students. In order to reduce the big amount of statistical data, we plan to produce an online learning survey system considering the high error rate arising from artificial statistic.
3) Score. The score database consists of usually score, examination and final grade, which has been built during teaching process by teachers.

4.3 Data Pretreatment

1) Data integration. Integrate data from multiple data sources together. In this study, we generate the basic student score database by integrating different database files, which is shown in Figure 2.

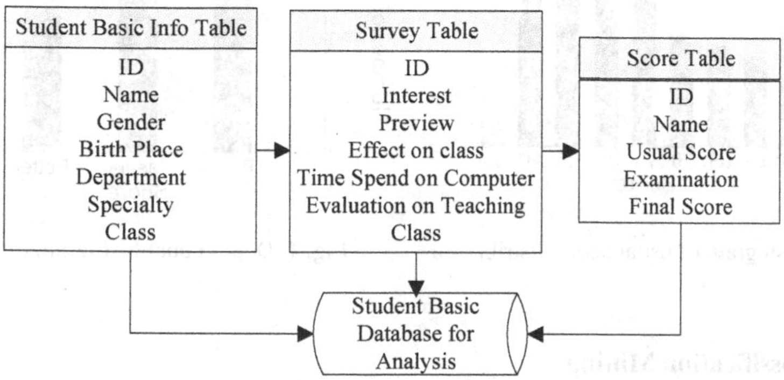

Fig. 2. Basic structure of student score database

2) Data cleaning. The main task of data cleaning is to fill the missing data value. Being lack of some attribute values we are interested in basic student score database, data cleaning could be used to fill the null. There are many ways to fill null value for property: ① Tuple ignore. The method is usually adopted when class label or more than one attributes of tuple is missing; ② Manual fill. Generally speaking, the method is a time-consuming work and it can not work when the data set is large or many values are lost; ③ Fill the null with a global constant value. Replace the null value with such a same constant as "Unkown". But if you do so, mining program may possibly take it for granted that these values come into being an interesting concept for they all have the same value "Unkown". Hence, this method would not be recommended despite of simplicity; ④ Fill the null with the average; ⑤ Use the average of all samples that belong to the same class with a given tuple; ⑥ The most likely value. The regression, Bayesian or Decision tree methods could be used to fill these null values.

In this case, we choose item ① to delete records with large number of null and item ② to fill other individual null for fewer amounts according to the property that used as screening condition.
3) Data conversion [3]. Data conversion aim to normalize the data. Most properties are of discrete while only individual continuous attributes need discretization operation such as usual score, examination and final score. Histogram analysis is a relatively simple discrete method and it can be divided into width-equal and depth-equal categories. The former partitions property values with equal intervals while the latter divides values into different parts that are possibly contain the same number of samples. Figure 3 illustrates the histogram of usual score distribution where depth-equal method is used to discrete the data into three categories. Here, scores from 0 to 65 belong to "Poor", 65 ~ 85 are "Media" and 85 ~ 100 are labelled "Better" and the depth-equal histogram of usual score are shown in Figure 4.

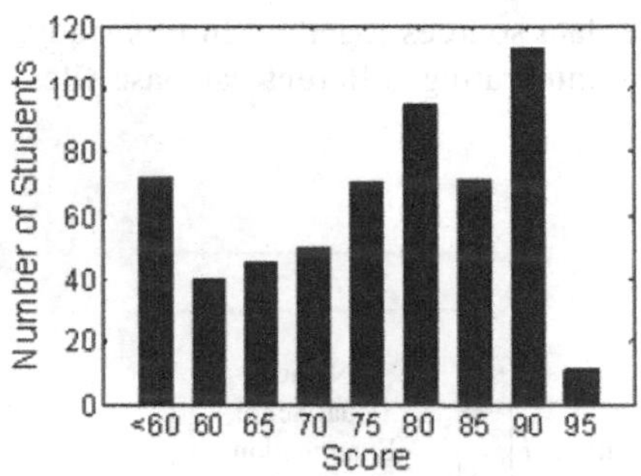

Fig. 3. Histogram of usual score distribution

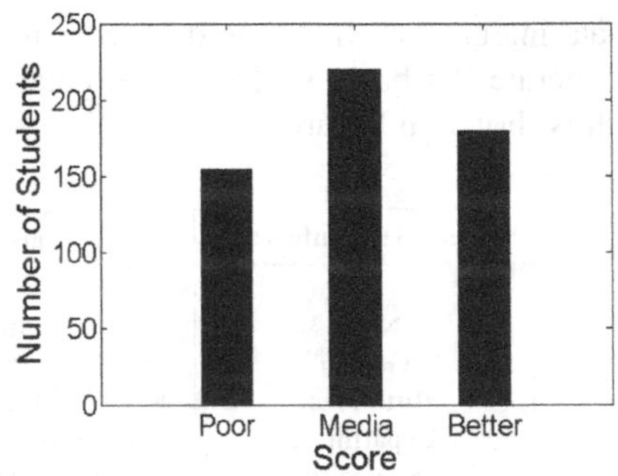

Fig. 4. Depth-equal histogram of score

4.4 Classification Mining

We first simplify the attributes of basic student score Table S and simplify the values of Table S' got from Table S and then generate the classification rules.

1) Attribute simplification [4]. Let M be the differential matrix of Table S in which elements M_{ij} is called items. Suppose there are K categories M_{ij} in M and we put each category Mij into array DA, $da(t)$ denote element in DA. $p(t)$ represent the number of M_{ij} of each category, $t = 1, 2,\ldots,K$.

$C = \{C1, C2, \cdots, Cn\}$ is the set of all attributes in S and in which any $C_r \in C$.

$f(Cr)$ is the attribute frequency function of any C_r in DA, then $f(C_r) = f(C_r) + |C| / |da(t)| \times p(t)$ in which sign | | means the number of elements in set.

Algorithm is described as follows:

Step 1: cone = $U \{| da(t) | = 1\}$, R = core;
Step 2: $Q = \{da(t) \mid da(t) \cap R = \varphi, t = 1, 2,\ldots, k\}$, $B = C - R$;
Step 3: if $Q = \varphi$, goto step 6, otherwise goto step4;
Step 4: compute $f(C_r)$ for all $C_r \in B$ and let $f(C_v) = \max \{F(C_r)\}$.

If the $f(C_r)$ value of several attributes are the same largest we select attribute C_v with the least number of simplified combination to current number q({attribute combination}); And if q of several attributes are the same smallest we select C_v that placed at the front of the table.

Step 5: $R = R \cup \{C_v\}$ goto step2;

Step 6: perform the following redundancy elimination operation on simplified attribute to get more optimal one.

Retain the simplified attributes for each $da(t)$ in array DA and find out the kernal, that is, the optimal simplified Table S'.

2) Value simplification [5]. We set attributes kernal as the initial candidate set and append the highest important attributes in it before deciding whether a current candidate set is a simplified value repeatedly until a simplified valuse has been found.

Input: the attributes simplified decision table S' $=< U, C \cup \{d\}, V, f >$;

n condition attributes expressed with C_i (i = 1, 2, …, n);

m records, the C_{ik} (i = 1, 2, …, n; k = 1, 2, …, m) represents the ith condition attribute of the kth record valued as V_{ik}.

Output: the decision rule set RULE after value simplified.

Step 1: RULE = Φ;

Step 2: inspect all records that are not deleted in decision table S';

① Judge by column: whether conflict would happen if the column is deleted and if it occurs, retains its original attribute or label it as "?" otherwise. At this point the unmarked record is attribute kernal that correspond to the attribute value set for the P_k (k = 1, 2,…, m).

② Only by judging attribute kernal the decision is made or decision can not be made but the number of mark "?" is 1, then append the condition attribute to the candidate set directly. And extract rule r_k, RULE = $r_k \cup$RULE, delete the records in decision table that make it possible to make the right decision by using the rule.

Step 3: inspect records that are not deleted until all records in decision table are deleted.

① $R_k = P_k$;

② Find an attribute $C_{ik} = V_{ik}$(i= 1, 2, …, p) exclude R_k where p refers to the number of condition attribute making SIG ($C_{ik} = V_{ik}$) achieve maximum value;

③ Add $C_{ik} = V_{ik}$ to rear end of R_k, $R_k = P_k \cup \{ C_{ik} = V_{ik} \}$, determine whether the decision can be reached derive from R_k, if not goto ②;

④ Inspect each attribute $C_{ik} = V_{ik}$ start from the end of R_k in inverted sequence: if $\{ C_{ik} = V_{ik} \} \in P_k$, attributes ahead of $C_{ik} = V_{ik}$ are belong to the kernal then jump out of ④; Otherwise, judge one by one if removing the attribute would not affect the decision-making and then find the minimum set, jump out ④;

⑤ Extract rule r_k, RULE = $r_k \cup$RULE, remove records that can make the right decision in decision table using the rule r_k, then goto step 3.

4.5 Classification Rules Generation

The most obvious advantage by using rough set theory to mining data is the direct classification rules extraction property [6]. As we wish to learn about those factors that infect students' score in this case, therefore the extracted rules are mainly considering the rules that classified as "Yes". The resulting classification rules are as follows:

IF usual score = better AND class effect = fully master THEN final grades for better ratio = 100%

IF usual score = better AND class effect = basic master THEN final grades for better ratio = 75%

IF usual score = better AND class effect = not so good AND time on computer ≥ 4 THEN final grades for better ratio = 65%

IF usual score = better AND class effect = not so good AND time on computer = 2 to 4 THEN final grades for better ratio = 62%

… …

5 Conclusion

In short, with the rapid growth of the amount of information and higher requirement for information extraction, now we can hardly find basis of decision-making in huge amounts of data according to the traditional methods. And we should explore the hidden data or model with the help of data mining to provide more effective support. Data mining, as a tool, can never replace teachers, but it offer scientific proof. Data mining technology is crystallization of huge amount of practice, which provides a shortcut to establish a traditional model that it is difficult to obtain. The practice shows that data mining provide educational reform effective theory and practice foundation, which play an important role in analyzing the factors that influence the score. And if we apply the technology in fields of colleges and universities, it may bring a big quantity of meaningful results so as to formulate appropriate measures to achieve our ultimate goal — improve education quality.

References

1. Peng, Y.Q., Zhang, H.M., He, H., et al.: Data Mining Technology and Its Application in Teaching. Journal of Hebei University of Science and Technology 4, 33–36 (2001) (in Chinese)
2. Nicolas, P., Yves, B., Rafik, T., et al.: Efficient Mining of Association Rules Using Closed Itemset Lattices. Information Systems (1999)
3. Zhang, L., Lu, X.Y., Wu, H.Y.: An Improved Heuristic Algorithm Used in Attribute Reduction of Rough Set. In: Proceedings of the First International Symposium on Data, Privacy and E-Commerce, pp. 192–199. Science (2007)
4. Ding, Z.B., Yuan, F., Dong, H.W.: Application of Data Mining to Analysis of University Students' Grades. Computer Engineering and Design 4, 590–592 (2006) (in Chinese)
5. Agrawal, R., Imielienski, T., Swami, A.: Mining Association Rules between Sets of Items in Large Databases. In: Proceedings of the 1993 ACM SIGMOD International Conference on Management of Data, pp. 145–152 (1993)
6. Gao, J.S., Guo, J.: Research and Application of Data Mining in Information Technology. China Education Info. 17, 75–76 (2007) (in Chinese)

Cross Message Authentication Code Based on Multi-core Computing Technology

Feng Yang[1], Cheng Zhong[1], and Danhua Lu[2,*]

[1] School of Computer and Electronics and Information, Guangxi University, Nanning, 530004, Guangxi, P.R. China

[2] The College of International Education, Guangxi University for Nationalities, Nanning, 530006, Guangxi, P.R. China

{Yf,chzhong}@gxu.edu.cn, luanua@sina.com

Abstract. The high speed block cipher-based message authentication code algorithms are required for different networking applications. This paper presents the Cross Message Authentication Code based on Multi-core computing technology (CMMAC), which is a parallel block cipher working mode on multi-core computing environment and secure for messages with any bit length. In CMMAC mode, the large file is divided into a number of blocks that are encrypted based on a symmetric key block cipher, keeping the relation among blocks by message cross transfer among processing cores. The algorithm analysis and experiment result indicates this method is expandable and obtains linear speedup.

Keywords: Block cipher, Message Authentication Code (MAC), Advanced Encryption Standard, multi-core machines.

1 Introduction

In cryptography, Message Authentication Code (MAC) can be constructed from a block cipher [1]. There were many method used to construct MAC, such as Cipher Block Chaining Message Authentication Code (CBC-MAC) [2]. In CBC-MAC, the message is encrypted with some block cipher algorithm in Cipher Block Chaining (CBC) mode to create a chain of blocks such that each block depends on the proper encryption of the previous block [3]. On the other way, the requirement for speed of authentication code generating to high-capacity file is higher with the development of high speed network, which limits the use of the cryptographic service [4-6]. The Chip Multi-Processor (CMP) technology integrates many cores into one processor, because it has the advantages of high performance and low consumer, The CMP technology has become mainstream CPU chip technology [7-9].

To improve the speed of generating MAC, the Cross Message Authentication Code based on Multi-core computing technology (CMMAC) is presented in this paper, which is a parallel block cipher working mode applying to multi-core environment.

The rest of this paper is organized as follows. Section 2 presents the CMMAC mode and detailed analysis of the mode. Section 3 provides the performance results. We conclude the paper in section 4.

* Corresponding author.

D. Jin and S. Jin (Eds.): Advances in FCCS, Vol. 1, AISC 159, pp. 181–186.
springerlink.com

2 CMMAC Mode

In CMMAC mode, many block cipher algorithms (such as DES, AES) can be used. We analyze the CMMAC mode by AES. All notations are listed in Table 1.

Table 1. The Notations

parameter	description
P	The message will be authenticated
P_{siz}	Sizes of P, $P_{siz} = \lvert P \rvert$, (bit)
K	The key of AES
n	Number of CPU core
T_{cap}	Block sizes of AES, such as 128, 192, or 256 (bit)
N	Number of the message block, $N = \lceil P_{siz} / T_{cap} \rceil$
P_i	The i block, $1 \le i \le N$
O_i	The ciphertext of P_i, $1 \le i \le N$
$core_k$	The k core, $1 \le \kappa \le n$
$thread_k$	The k thread, , $1 \le \kappa \le n$
L	Sizes of MAC, such as 64,128, 192, or 256 (bit)
T	The MAC which is constructed by sender
T'	The MAC which is received by receiver

2.1 Message Dividing

To guarantee this mode is secure for messages of any bit length, P is divided into P_1, P_2, ..., P_N. $N = \lceil P_{siz} / T_{cap} \rceil$, which is show in Fig 1. In particular, if the last block size is less than T_{cap}, it need to be filled by the rule filling "1" to the first position.

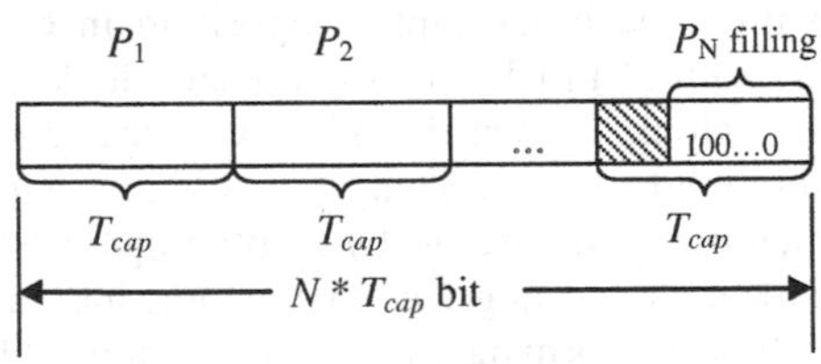

Fig. 1. Message P dividing

2.2 Processing of CMMAC Mode

In CMMAC mode, after dividing, the $thread_k$ must be created for the $core_k$, then each core process some blocks. To keep relation among blocks through AES, the multiple value equation (1) is used to control the relationship between round i and round i-1, thus any difference of message will affect the final MAC.

$$O_{k+in} = \begin{cases} = E_K(P_{k+in}), & i=0 \cap 1 \le k \le n \\ = E_K(P_{k+in} \oplus O_{k+in-1}), & 0 < i < \lceil N/n \rceil \cap k = 1 \\ = E_K(P_{k+in} \oplus O_{k+in-n-1}), & 0 < i < \lceil N/n \rceil \cap k \ne 1 \end{cases} \quad (1)$$

In equation (1), when i=0, the block P_{k+in} will be encrypted by $O_{k+in} = E_K(P_{k+in})$. When $0 < i < \lceil N/n \rceil \cap k = 1$, in the other rounds except round 0, and in $core_1$, the block P_{k+in} will be encrypted by $O_{k+in} = E_K(P_{k+in} \oplus O_{k+in-1})$. When $0 < i < \lceil N/n \rceil \cap k \ne 1$, in the other rounds except round 0, and from $core_2$ to $core_n$, the block P_{k+in} will be encrypted by $O_{k+in} = E_K(P_{k+in} \oplus O_{k+in-n-1})$.

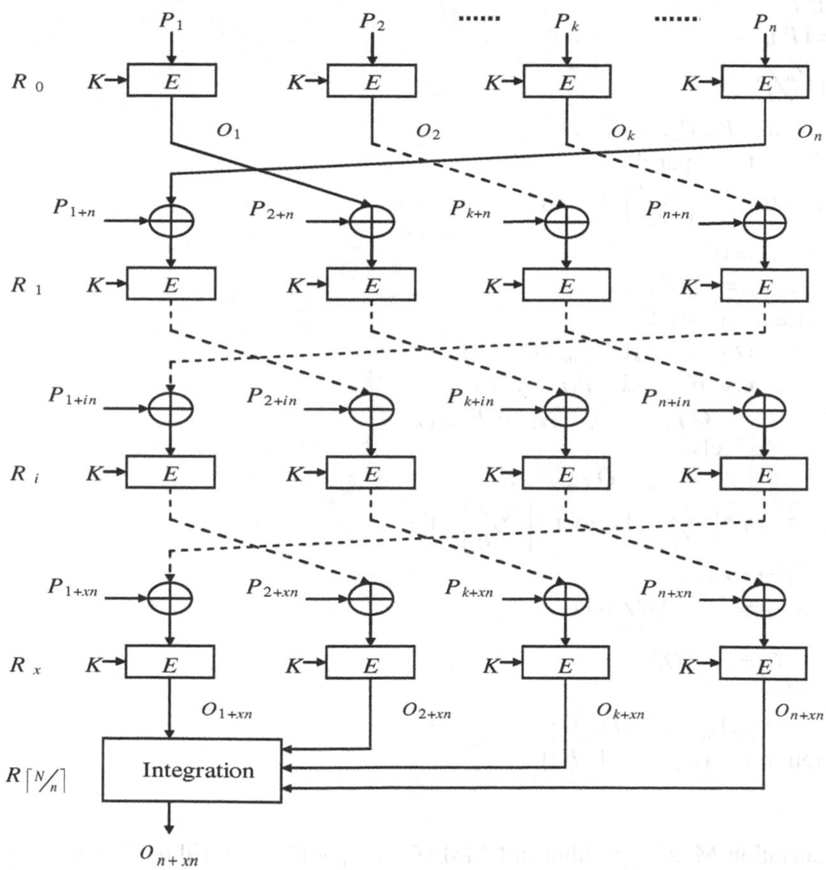

Fig. 2. The key-schedule in CMMAC mode

Figure2 shows that the relation among blocks is achieved by message cross transfer realized through the shared cache among cores. In addition, in the round x ($x = \lceil N/n \rceil - 1$), if $(k + in) > N$, then $O_{k+in} = O_{k+in-1-n}$ because the block P_{k+in} that appears in $O_{k+in} = E_K(P_{k+in} \oplus O_{k+in-1-n})$ does not exist. Otherwise, the MAC would loss the relationship with block $P_{k+in-1-n}$, which can bring security weaknesses. In the last round, the output of the all processing cores in round x should be integrated in $core_1$. At last, the MAC of P can be O_{n+xn}, or some part of it.

2.3 Generation MAC Algorithm in CMMAC

The pseudo code of generation MAC algorithm in CMMAC is described as follows.

Input: P, K, T_{cap}, L and n
Output: T
{ $P_{siz} = |P|$;
  $N = \lceil P_{siz} / T_{cap} \rceil$;
  divide P into $P_1, P_2, \ldots, P_N$;
   for $k = 1$ to n par-do
   for $i = 0$ to $(\lceil N/n \rceil - 1)$ do
     {if $i == 0$
      $O_{k+in} = E_K(P_{k+in})$
     else if $(k == 1)$
        $O_{k+in} = E_K(P_{k+in} \oplus O_{k+in-1})$
        else if $((k + in) =< N)$
          $O_{k+in} = E_K(P_{k+in} \oplus O_{k+in-1-n})$
          else
          $O_{k+in} = O_{k+in-1-n}$;}
  for $j = (n * \lceil N/n \rceil - 1)$ to $n * \lceil N/n \rceil$ do
     {$O_j \rightarrow core_1$;
      if $j == (n * \lceil N/n \rceil - 1)$
       $O_j = E_K(O_j)$
      else
       $O_j = E_K(O_j \oplus O_{j-1})$;
     return $T = O_{n * \lceil N/n \rceil} \mid L$;}
 }

The generation MAC algorithm in CMMAC is a parallel algorithm. The speedup of this algorithm is a primary parameter in CMMAC mode. Let b represent the time-consuming of encrypting a block using AES, b is a constant. Scale of the problem here can be represented by N. Let c represent the time-consuming of creating a thread, c is a constant. Let e represent the time-consuming of cross transferring one ciphertext block, e is a constant too. We then have equation (2).

$$s_p = t_s / t_p = (b * N) / (b * N / n + c * n + e * N) = \frac{b}{b/n + (c*n)/N + e} \quad (2)$$

In equation (2), if $N \to \infty$ and $e \to 0$, then $s_p = n$. Because the cross transferring of ciphertext is achieved through the shared cache among cores, the time-consuming e is very small. That is to say, when the size of the message authenticated is bigger, the speedup of generation MAC algorithm in CMMAC is closer linear.

3 Implementation and Performance Analysis

For the performance analysis of CMMAC mode in the Multi-core system, we use DELL PowerEdge 2950 with Intel Xeon E5405-2.0 GHZ, 2 GB DDR RAM and 12MB cache as a hardware device. The Microsoft VC+ + and Openssl package are used as development tools.

The CMMAC mode speedup is show in Figure 3. For example, when the size of message is 50MB, the speedup with 2 cores is 1.62, the speedup with 3 cores is 2.65 and the speedup with 4 cores is 3.58. In addition, when the size of message is larger, the accelerate effect of CMMAC mode is more obvious. For example, when the size of message is 10MB, the speedup with 4 cores is 3.58. When the size of message is 90MB, the speedup with 4 cores is 3.78.

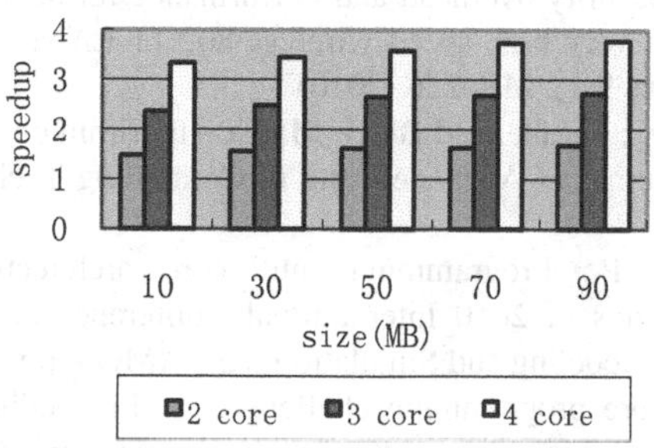

Fig. 3. The speedup of CMMAC mode

4 Conclusion

In this paper, the CMMAC mode is proposed to improve the speed of generating MAC. In this mode, the message is split into a number of blocks first. Then, each core process some blocks parallel. In order to guarantee security and basically invariable memory space, all blocks keep relation by message cross transfer which can be realized through the shared cache among cores. The experiments show that the mode in multi-core environment basically achieves a linear speedup. This paper mainly studies how to improve the speed of MAC generation through the parallel working mode. In order to achieve better acceleration effect, we should make a study of block cipher algorithm itself parallel. Our future work also includes studying the authentication encryption mode based on block cipher in order to meet the applications not only needing secret service but also authenticate service.

Acknowledgment. This paper is supported by Guangxi Natural Science Foundation under Grant NO. 2011GXNSFA018152 and Project of Outstanding Innovation Teams Construction Plans at Guangxi University.

References

[1] Jutla, C.S.: Encryption Modes with Almost Free Message Integrity. In: Pfitzmann, B. (ed.) EUROCRYPT 2001. LNCS, vol. 2045, pp. 529–544. Springer, Heidelberg (2001)

[2] Bellare, M., Kilian, J., Rogaway, P.: The Security of the Cipher Bloek Chaining Message Authentication Code. Journal of Computer and System Sciences 61(3), 362–399 (2000)

[3] Black, J.A., Rogaway, P.: CBC MACs for Arbitrary-Length Messages:The Three-Key Constructions. In: Bellare, M. (ed.) CRYPTO 2000. LNCS, vol. 1880, pp. 197–215. Springer, Heidelberg (2000)

[4] Liberatori, M., Otero, F., Bonadero, J.C., Castifieira, J.: AES-128 cipher: hi speed, low cost FPGA implementation. In: Proceedings of Third Southern Conference on Programmable Logic SPL 2007, Mar Del Plata, Argentina, pp. 195–198 (2007)

[5] Krawczyk, H.: The Order of Encryption and Authentication for Protecting Communications (or: How Secure Is SSL?). In: Kilian, J. (ed.) CRYPTO 2001. LNCS, vol. 2139, pp. 310–331. Springer, Heidelberg (2001)

[6] Olteanu, A., Xiao, Y.: Security overhead and performance for aggregation with fragment retransmission (AFR) in very high-speed wireless 802.11 LANs. IEEE Transactions on Wireless Communications 9(1), 218–226 (2010)

[7] Jiao, L., Zhou, J., Shu, X.: Parallelized Block Match Algorithm on Multi-core Processors. IJACT: International Journal of Advancements in Computing Technology 3(6), 252–259 (2011)

[8] Arandi, S., Evripidou, P.: Programming multi-core architectures using data-flow techniques. In: Proceedings of 2010 International Conference on Embedded Computer Systems: Architectures, Modeling and Simulation, IC-SAMOS, pp. 152–161 (2010)

[9] Cooke, D.: The multi-core programming challenge. In: Proceedings 22nd International Conference on Software Engineering & Knowledge Engineering (SEKE 2010), pp. 3–4 (2010)

Video-Based Face Detection Using New Standard Deviation

Jianfeng Wang

Chongqing Aerospace Vocational College, Jiangbei District, 400021

Abstract. While traditional face recognition is typically based on still images, face recognition from video sequences has become popular recently. To achieve the high accuracy of the face detection, we introduce a visual perception model that aims at quantifying the local tolerance to noise for arbitrary imagery. Based on this model, extract the features of video frame, the two-dimensional statistical parameters and new standard deviation from DCT coefficients without its inverse transform was computed. Results are reported for a database comprising 940 face images of 34 video clips under a variety of challenging circumstances. These results indicate significant performance improvements over previous methods and demonstrate the usefulness of the confidence data.

Keywords: face detection, compressed domain, standard deviation, statistical parameters.

1 Introduction

For decades human face detection has been an active topic in the field of object detection. Donner et al. [11] proposed a fast AAM using the canonical correlation analysis (CCA) that modeled the relation between the image difference and the model parameter difference to improve the convergence speed of fitting algorithm. Matthews and Baker [12] proposed an efficient fitting algorithm that did not require a linear relationship between the image difference and the model parameter difference. This model has faster convergence and better fitting accuracy than the original. Sung and Kim [13] proposed a new fitting algorithm of 2D + 3D for a multiple-calibrated camera system, called stereo (STAAM), to increase the stability of the fitting of 2D + 3D. STAAM reduces the number of model parameters, and resultantly improves the fitting stability. A general statement of this problem can be formulated as follows: Given still or video images of a scene, identify one or more persons in the scene using a stored database of faces [1]. A lot of algorithms have been proposed to deal with the image to image, or image-based, detection where both the training and test set consist of still face images. Some examples are Principal Component Analysis (PCA) [2], Linear Discriminate Analysis (LDA) [3], and Elastic Graphic Matching [4]. However, with existing approaches, the performance of face detection is affected by different kinds of variations, for example, expression, illumination and pose. Thus, the researchers start to look at the video-to-video, or video-based detection [5][6][7][8][9][10], where both the training and test set are video sequences containing the face.

D. Jin and S. Jin (Eds.): Advances in FCCS, Vol. 1, AISC 159, pp. 187–192.
springerlink.com

In this paper, we take an opposite direction in investigating the similarity measurement by introducing a correlation test into an existing distance matching algorithm to improve its performance on face detection. The rest of the paper is structured into three sections, where Section 2 describes the proposed algorithm, section 3 report experimental results in evaluating the proposed algorithm based on existing benchmark technique, and finally, section 4 provides concluding remarks.

2 Proposed Method

A. Neutral Face Model and Feature Points

The FAPs are a set of parameters defined in the MPEG-4 visual standard for the animation of synthetic face models. There are 68 FAPs including 2 high-level FAPs used for visual phoneme and expression, and 66 low-level FAPs used to characterize the facial feature movements over jaw, lips, eyes, mouth, nose, cheek, ears, etc.

*B. Transform the 8*8 Block to the 6*6 Block*

In fact, the frame of the MPEG video as the JPEG image was divided into the 8*8 block before transformed into the frequency domain . in the other words, after transformed ,the coefficients was made of the 8*8 block in the DCT domain,, there are many coefficients approach to zero and less influence the quality of the image, so we extract the 36 coefficients based on the sequence and construct the histogram.

C. Definition of New Standard Deviation

In this correspondence, we implement the proposed human visual model via the following filters.

$$E(k,r) = -\sum_{i=1}^{256} q(m_i, u_i)\lg[q(m_i, u_i)] \tag{1}$$

Let X=x(i,j) (i,j =0,1, N-1) be gray values in a N*N block of an image or frame(in next steps, we take the N=6). If X is regarded as a random variable inside the block, the kth moment of X in this block is defined to be the expectation of X^k, which can detailed as follows:

$$m_k = E(X^k) = \frac{1}{N^2}\sum_{i=0}^{N-1}\sum_{j=0}^{N-1} x^k(i,j) \tag{2}$$

Similarly, its kth central moment, μ_k of X, is by definition $E[(X-E(X))^k]$, or the kth moment of (X-E(X)):

$$\mu_k = E[(X - E(X))^k] = \frac{1}{N^2}\sum_{i=0}^{N-1}\sum_{j=0}^{N-1}[x(i,j) - E(X)]^k \quad (3)$$

The 1st order moment m_1 and the 2nd order central moment μ_2 (also called variance σ^2) satisfy the following relationship:

$$\sigma^2 = m_2 - m_1^2 \quad (4)$$

D. The Standard Deviation Computation with DCT Coefficient

Inverse DCT transform is:

$$f(i,j) = (\frac{2}{N})\sum_{i=0}^{N-1}\sum_{j=0}^{M-1} A(i)A(j)\cos[\frac{\pi \cdot u}{2N}(2i+1)]\cos[\frac{\pi \cdot v}{2N}(2j+1)]F(u,v) \quad (5)$$

By considering the mean definition, and replace equation (5) for $x(i, j)$ in the definition of m_1, we have:

$$m_1 = \frac{1}{N^2}\sum_{i=0}^{N-1}\sum_{j=0}^{N-1} f(i,j) = \frac{2}{N^3}\sum_{u=0}^{N-1}\sum_{v=0}^{N-1} A(u)A(v)F(u,v)\sum_{i=0}^{N-1}\cos\left(\frac{(2i+1)u\pi}{2N}\right)\sum_{j=0}^{N-1}\cos\left(\frac{(2j+1)v\pi}{2N}\right)$$

From:

$$\sum_{i=0}^{N-1}\cos\left(\frac{(2i+1)u\pi}{2N}\right) = \begin{cases} N & u = 0 \\ 0 & otherwise \end{cases} \quad (6)$$

we have:

$$m_1 = \frac{1}{N}F(0,0) \quad (7)$$

For the 2nd moment, since

we have:

$$E(k,r) = -\sum_{i=1}^{256} q(\frac{1}{N^2}\sum_{i=0}^{N-1}\sum_{j=0}^{N-1} f(i,j), u_i)\lg[q(\frac{1}{N^2}\sum_{i=0}^{N-1}\sum_{j=0}^{N-1} f(i,j), u_i)]$$

By examining the nature of the basis function, it is revealed that:

$$\sum_{i=0}^{N-1}\left[\cos\left(\frac{(2i+1)u\pi}{2N}\right)\cos\left(\frac{(2i+1)s\pi}{2N}\right)\right] = \begin{cases} N & s = u = 0 \\ N/2 & s = u \neq 0 \\ 0 & s \neq u \end{cases} \quad (8)$$

Thus:

$$m_2 = \frac{1}{N^2}\sum_{u=0}^{N-1}\sum_{v=0}^{N-1} F^2(u,v) \quad (9)$$

Considering the relationship between m_1, m_2 and σ^2 in equation (3), Lastly, The standard deviation $S(m,\sigma)$ is defined as

$$S(m,\sigma) = \left| \frac{1}{\sigma^2 - 1} \sum_{i=1} (i - \frac{1}{\sigma^2} \prod m)^2 \right| \tag{10}$$

replace equation (10) for (8)、(9)in the definition, we have:

$$S(m,\sigma) = \frac{N^2}{\sum_{u=0}^{N-1}\sum_{v=0}^{N-1} F^2(u,v) - N^2} \sum_{i=1} i - \frac{N^2 \prod m}{\sum_{u=0}^{N-1}\sum_{v=0}^{N-1} F^2(u,v)} \tag{11}$$

The flow-chart of our proposed algorithm

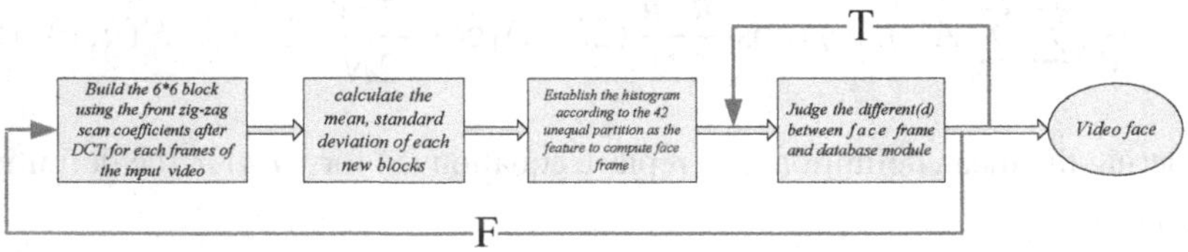

3 Algorithm Evaluations via Experiments

We test the baseline algorithm on the Task database using 940 face images of 34 video clips . Generally speaking, the more states used to train one video, the better modeling we have, while we also have more parameters to estimate. For the Task database, we found 11 states is a good compromise between modeling and estimation. Each state uses one Gaussian distribution to model the observation probability. Eventually we obtain 4.1% recognition error rate, which is better than the best result we can get from the baseline algorithm. We also apply the new standard deviation on the same test set and obtain 3.5% recognition error rate. Also, we do the same comparison with the newly captured test set. As shown in Table 1, although the overall recognition rate is a lot higher than the first test set because of the tradition standard deviation, we still see that our proposed methods work better than the baseline algorithm.

Table 1. Experimental Results for face detection

	baseline	New standard deviation	Tradition standard deviation
Set-1	0.43	0.65	0.56
Set-2	0.19	0.76	0.52
Set-3	0.64	0.88	0.59
Set-4	0.71	0.95	0.67

Fig. 1. Test result (discontinuous video)

Fig. 2. Test result (continuous video)

4 Conclusions

The paper shows that video-based face detection is one promising way to enhance the performance of current image-based retrieval. To achieve the high accuracy of the face detection, we introduce a visual perception model that aims at quantifying the local tolerance to noise for arbitrary imagery. Based on this model, extract the features of video frame, the two-dimensional statistical parameters and new standard deviation from DCT coefficients without its inverse transform was computed.

Acknowledgements. This paper Supported by 2010 youth science and technology project of China Aerospace Science and Technology Corporation.

Supported by 2011 science and technology project of ChongQing education committee(project number:KJ113201).

References

1. Chellappa, R., Wilson, C.L., Sirohey, S.: Human and machine recognition of faces: a survey. Proceedings of the IEEE 83(5), 705–741 (1995)
2. Turk, M., Pentland, A.: Eigenfaces for Recognition. Journal of Cognitive Neuroscience 3(1), 71–86 (1991)
3. Belhumeur, P.N., Hespanha, J.P., Kriegman, D.J.: Eiegnfaces vs. Fisherfaces: Recognition Using Class Specific Linear Projection. IEEE Transaction on Pattern Analysis and Machine Intelligence 19(7), 711–720 (1997)
4. Lades, M., Vorbruggen, J.C., Buhmann, J., Lange, J., von der Malsburg, C., Wurtz, R.P., Konen, W.: Distortion Invariant Object Recognition in the Dynamic Link Architecture. IEEE Transactions on Computers 42(3), 300–311 (1992)
5. Li, Y.: Dynamic face models: construction and applications. PhD Thesis, Queen Mary, University of London (2001)
6. Edwards, G.J., Taylor, C.J., Cootes, T.F.: Improving Identification Performance by Integrating Evidence from Sequences. In: Proc. of 1999 IEEE Conference on Computer Vision and Pattern Recognition, Fort Collins, Colorado, June 23-25, pp. 486–491 (1999)
7. Zhou, S., Krueger, V., Chellappa, R.: Face Recognition from Video: A CONDENSATION Approach. In: Proc. of Fifth IEEE International Conference on Automatic Face and Gesture Recognition, Washington D.C., May 20-21, pp. 221–228 (2002)

8. Liu, X., Chen, T., Thornton, S.M.: Eigenspace Updating for Non-Stationary Process and Its Application to Face Recognition. To appear in Pattern Recognition, Special issue on Kernel and Subspace Methods for Computer Vision (September 2002)
9. Roy Chowdhury, A., Chellappa, R., Krishnamurthy, R., Vo, T.: 3D Face Recostruction from Video Using A Generic Model. In: Proc. of Int. Conf. on Multimedia and Expo, Lausanne, Switzerland, August 26-29 (2002)
10. Baker, S., Kanade, T.: Limits on Super-Resolution and How to Break Them. IEEE Transactions on Pattern Analysis and Machine Intelligence 24(9), 1167–1183 (2002)
11. Donner, R., Reiter, M., Langs, G., Peloschek, P., Bischof, H.: Fast Active Appearance Model Search Using Canonical Correlation Analysis. IEEE Trans. Pattern Analysis and Machine Intelligence 28(10), 1690–1694 (2006)
12. Matthews, I., Baker, S.: Active Appearance Models Revisited. Int'l. J. Computer Vision 60(2), 135–164 (2004)

A New Approach to Recognize Online Handwritten NǔShu Characters

Yunchao Li and Jiangqing Wang

School of Computer Science, South-Central University for Nationalities, Wuhan, China
liyunchao0826@126.com

Abstract. In this paper, we propose a new stroke-based approach to recognize online handwritten NǔShu characters. The proposed approach extracts 14 kinds of basic strokes, applies an improved secondary search method to extract feature points, uses the 8-directional feature to establish stroke feature dictionary in order to recognize strokes. In the procedure of the recognition, strokes and stroke order feature are used for the first classification, then the position relationships of key strokes for secondary classification to recognize the similar characters. As a result of testing, the recognizing rate of a system based on this approach can achieve 95.7%.

Keywords: online handwritten NǔShu characters, stroke-based, preprocessing, feature extraction, pattern recognition.

1 Introduction

Character recognition is an important task in many applications such as license plate recognition. The corresponding recognition algorithm can be divided into two major categories, namely online (stroke trajectory-based) or off-line (image-based). As a real-time recognition method, online handwritten recognition has high recognition efficiency and plays an important role in protecting this precious nation civilization gene. Online recognition systems are available commercially for symbolic languages like English, Chinese etc. But there has been a corresponding recognition system for NǔShu, which is a "female script" as the only system of ancient characters, used solely by women in Jiangyong Country, Hunan province [1].

Wang Jiangqing and Zhu Rongbo [2] have attempted recognition of NǔShu characters based on hidden Markov model, with recognition accuracy of 93.7%. Wang Jiangqing and Wan Chen [3] utilized peripheral direction contribution to recognize handwritten off-line NǔShu characters, but their training data is taken from only 442 individuals. So far, the online handwritten recognition results have not appeared yet.

The work in this paper aims at realizing a new stroke-based approach to recognize online handwritten NǔShu characters. We extract 14 kinds of basic strokes, apply an improved secondary search method to extract feature points, use the 8-directional feature to establish stroke feature dictionary in order to recognize strokes. In the

D. Jin and S. Jin (Eds.): Advances in FCCS, Vol. 1, AISC 159, pp. 193–199.
springerlink.com

procedure of the recognition, strokes and stroke order feature are used for the first classification, then the position relationships of key strokes for secondary classification to recognize the similar characters.

Section 2 presents a short introduction about NǚShu characters and the flow chats of online handwritten NǚShu character recognition. Sample collecting and preprocessing are explained in Section 3. The proposed feature extraction method is introduced in Section 4. Recognition results are shown in Section 5 and concluding remarks are given in Section 6 respectively.

2 About NǚShu Characters and Recognition Process

2.1 NǚShu Characters

NǚShu is not only the unique gender literature, rare syllable-marking literature, but also a world ancient literature which has been going on for now. According to studies by some famous experts, there are more than 2,000 NǚShu characters, almost all the words present as a long rhombus, write from right to left, top to down, shown in Fig. 1. In our system, we extract 14 kinds of basic strokes, named point, short horizontal, short vertical, left bias, right bias, left arc, right arc, upper arc, lower arc, left fold, right fold, left vertical fold, right vertical fold and singular. Fig. 2 shows the first 13 strokes, if a stroke does not belong to the above 13 strokes, we call it singular.

2.2 Recognition Process

Online handwritten NǚShu Characters recognition system is composed of sample collection, feature extraction and pattern recognition, shown in Fig. 3. Raw data was obtained from the stylus (pen) and written board connected to the computer. The pressure sensor array in the digitizer can capture pen position, pen pressure and writing time. We save them into defined ink file and establish NǚShu dictionary after preprocessing and extracting features. When users finish writing a word, preprocess it and extract features code again, match the feature code with our dictionary, output the word with the highest match ratio.

3 Preprocessing

The raw data acquired from written board is ill-conditioned for feature extraction and the artifacts observed in data are noises, jitters and overlapping of samples. To remove these issues from the data, de-noising and smoothing are necessary.

3.1 De-noising

Because of the written board has a high inductive level and exists hardware noise, there are so much noise and overlapping points in our raw data. Moreover, many fractured strokes occur while writing.

So for these reasons, we must deal with the noise first:

1) If the coordinates of two adjacent points are same, wipe off the latter one;

2) If the length of a stroke less than the threshold we set, we consider it as an interferential stroke.

3.2 Smoothing

Roughness of the pen tip or writing surface may cause jitters in the pen trajectory. We apply weighted mean filter algorithm to avoid jitters. The new coordinates after smoothing are:

$$X_i' = \frac{(X_{i-1} + 2X_i + X_{i+1})}{4} \qquad Y_i' = \frac{(Y_{i-1} + 2Y_i + Y_{i+1})}{4} \tag{1}$$

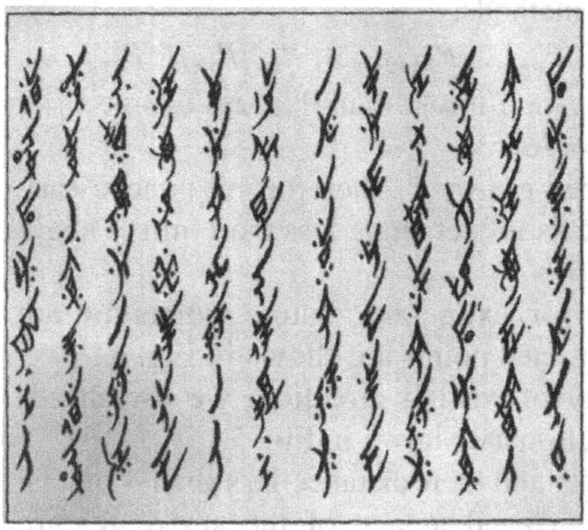

Fig. 1. NǔShu Character Set. [4]

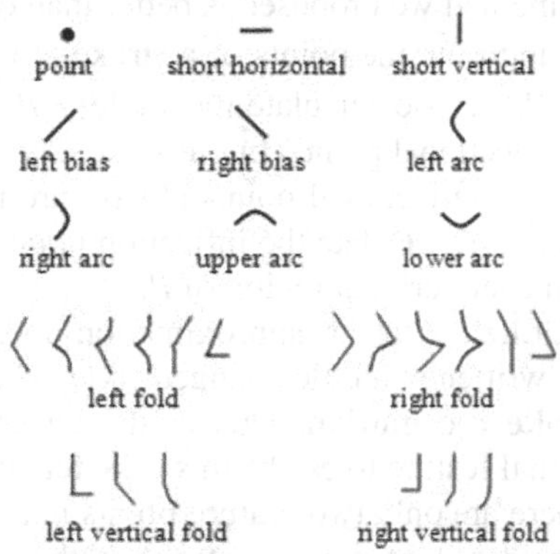

Fig. 2. 13 Kinds of NǔShu Strokes

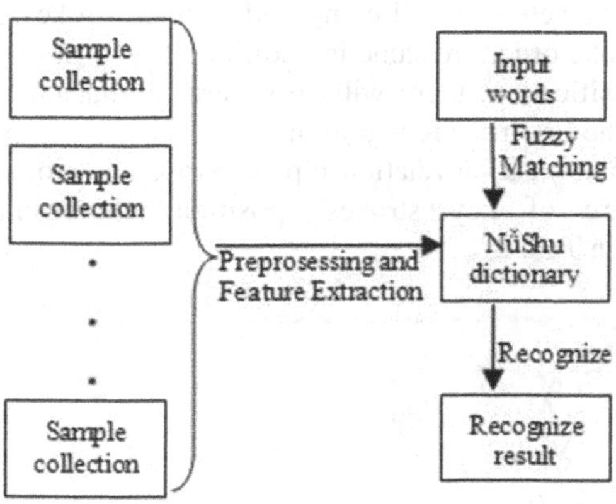

Fig. 3. Flow chats of online handwritten NǔShu characters recognition

4 Feature Extraction

4.1 Feature Types

Features extracted from the raw data, which will be used in classification are the most important. We design a secondary classifier, use the strokes and stroke order for the fist classification, and the position relationships of key strokes for the secondary classification.

4.2 Feature Extraction Steps

1) Stroke separation. While writing a new character, there is always a time-lag or distance between two adjacent strokes, maybe 200ms, 8 pixels, or even longer. So that, combining time-lag with distance of two sample points is the best principle for separating strokes.

2) Fature points extraction. Experimental results show that the improved secondary search method we proposed is better than other methods.

a) Supposing the points of a stroke are P_{i-k}, P_{i-k+1}, P_{i-k+2}, $\ldots P_i$, P_{i+1}, P_{i+2}, $\ldots$ P_{i+k-1}, P_{i+k}, we calculate the angle $\angle P_{j-k}P_jP_{j+}$k , if it less than θ_{max} ,recognize P_i as a new inflectional point. (In our system, $k=2$, $\theta= 160°$)

b) In *a)*, inflectional points always are ignored nearby P_i, therefore, it is necessary to search twice. Delete the inflection point P_i' if its inflect angle less than inflect angle of P_i in every critical region of P_i.

c) Utilizing linear approximation method for extracting feature points of arc strokes written in a little changed style. The extracted points are shown in Fig. 4.

3) Stroke recognition. Due to the uncertainty of writing direction, we use the 8-directional feature to establish stroke feature dictionary, shown in Fig. 5.

If there are only two feature points in a stroke, and their distance less than 4 pixels, we consider it as a point stroke. Otherwise, take advantage of the stroke feature dictionary (shown in Tab. 1) to recognize another 13 kinds of strokes.

4) Position relationships of key strokes. After separating and recognizing the whole strokes, feature code for the first classification is obtained. However, as Fig. 6 showed, these three characters are all composed of three strokes, left bias, right bias and left bias, their stroke order are same in addition. If we just use strokes and stroke order feature for classification, there will be several similar characters which could not be recognized in the first-time recognition.

Aim at this question, position relationships of strokes dictionary is established to recognize the feature of key strokes' position relationships for secondary classification, shown in Tab. 2.

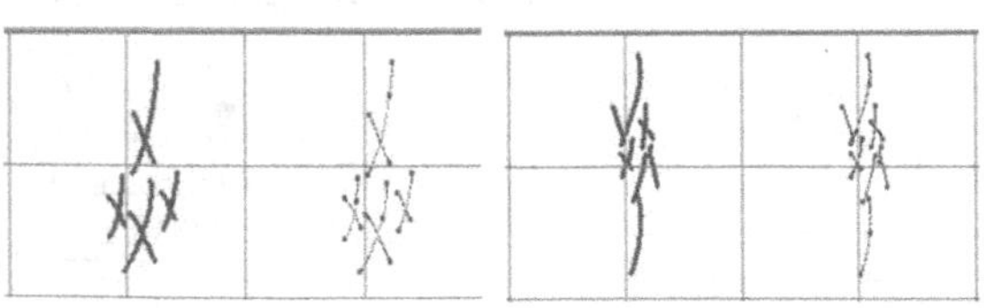

Fig. 4. Results of fature points extraction

Table 1. Stroke Feature Dictionary

Stroke Name	Stroke Code	Segments Direction
Point	0	Random
Short horizontal	1	D1/D5
Short vertical	2	D3/D7
Left bias	3	D2/D6
Right bias	4	D4/D8
Left arc	5	D6D8
Right arc	6	D8D6
Upper arc	7	D2D8
Lower arc	8	D8D2
Left fold	9	D6D8/D6D8D7/D7D6D8 /D7D6D8D7/D6D7/D6D1
Right fold	a	D8D6/D8D6D7 /D7D8D6/D8D7/D8D5
Left vertical fold	b	D7D1/D7D8
Right vertical fold	c	D7D5/D7D6
Singular	d	Otherwise

Table 2. Stroke Position Relationships

Position Relationships	Description	Feature Code	Position Relationships	Description	Feature Code
Left-Right	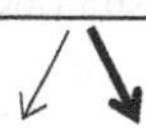	e	Tail-Head Link		k
Right-Left	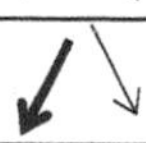	f	Head-Middle Link		l
Up-Down		g	Middle-Head Link		m
Down-Up		h	Tail-Middle Link		n
Cross		i	Middle-Tail Link		o
Head-Tail Link		j	Singular	Otherwise	p

(In the description row, the arrow describes writing direction, the thin line describes the first stroke and the rough line describes the second stroke)

Table 3. Stroke similarity matrix

	0	1	2	3	4	5	6	7	8	9	a	b	c	d
0	0.0	0.3	0.3	0.3	0.3	0.5	0.5	0.5	0.5	0.7	0.7	0.7	0.7	1.0
1	0.3	0.0	1.0	0.6	0.6	0.9	0.9	0.9	0.9	1.0	1.0	1.0	1.0	1.0
2	0.3	1.0	0.0	0.6	0.6	0.9	0.9	0.9	0.9	0.9	0.9	0.6	0.6	1.0
3	0.3	0.6	0.6	0.0	1.0	0.6	0.9	0.9	0.9	0.3	0.9	0.7	0.7	1.0
4	0.3	0.6	0.6	1.0	0.0	0.9	0.6	0.9	0.6	0.9	0.3	0.7	0.7	1.0
5	0.5	0.9	0.9	0.6	0.9	0.0	1.0	0.7	0.7	0.3	0.9	0.4	0.9	1.0
6	0.5	0.9	0.9	0.9	0.6	1.0	0.0	0.7	0.7	0.9	0.3	0.9	0.4	1.0
7	0.5	0.6	0.9	0.9	0.9	0.7	0.7	0.0	0.9	0.9	0.9	0.9	0.9	1.0
8	0.5	0.6	0.9	0.9	0.6	0.7	0.7	0.9	0.0	0.9	0.9	0.9	0.9	1.0
9	0.7	1.0	0.9	0.3	0.9	0.3	0.9	0.9	0.9	0.0	1.0	0.6	0.9	1.0
a	0.7	1.0	0.9	0.9	0.3	0.9	0.3	0.9	0.9	1.0	0.0	0.9	0.6	1.0
b	0.7	1.0	0.6	0.7	0.7	0.4	0.9	0.9	0.9	0.6	0.9	0.0	1.0	1.0
c	0.7	1.0	0.6	0.7	0.7	0.9	0.4	0.9	0.9	0.9	0.6	1.0	0.0	1.0
d	1.0	1.0	1.0	1.0	1.0	1.0	1.0	1.0	1.0	1.0	1.0	1.0	1.0	0.0

(Characters '0-d' represent of the strokes code, 0.0 means two strokes are same, 1.0 means completely different, numbers between 0.0 and 1.0 represent of the similarity of two strokes)

5 Pattern Recognition

Patter recognition methodology is discussed in the current section. Distance-based recognition is widely used in online character recognition, we build a stroke similarity matrix (shown in Tab. 3) in order to finish matching strokes and calculating distance between feature code extracted from input word and feature code in NǔShu dictionary. Finally, output the character with nearest distance as the first time recognize result, and list characters with the first ten nearest distance for convenience.

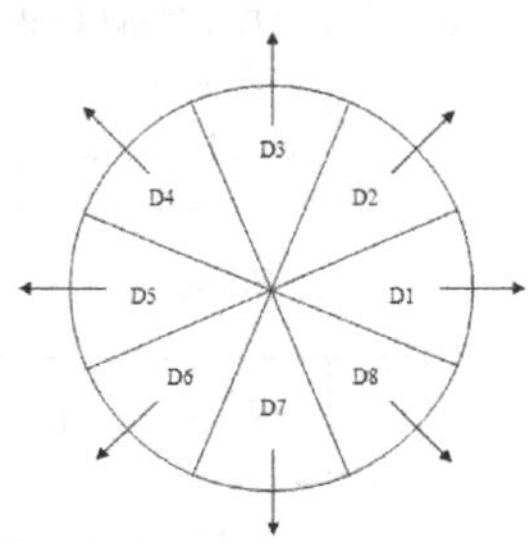

Fig. 5. 8-directional feature extraction method

Fig. 6. Three characters with same strokes and stroke orders

Table 4. Recognition result

Method	First-time Recognition Accuracy	Ten-time Recognition Accuracy
Strokes and Stroke Order Based Method	75.70%	94.50%
Our Approach	91.10%	95.70%

6 Conclusion

This paper presents a new approach to recognize NǔShu characters. 14 kinds of strokes are extracted and two classifiers are designed, strokes and stroke order feature for the first one, the position relationships of key strokes for the second. As Tab. 4 shows, it has been found that the proposed method gives higher recognition accuracy than the general method for handwritten NǔShu characters, which will be a powerful contribution to protect this endangered precious language.

Acknowledgments. This work is supported by the National Natural Science Foundation of China (60975021).

References

1. He, Q.: Only script used solely by women. Women 3, 42–43 (2007)
2. Wang, J., Zhu, R.: Handwritten nushu character recognition based on hidden Marko model. Journal of Computers 5(5), 663–670 (2010)
3. Wang, J., Wan, C.: Application of Peripheral Direction Contribution in Feature Extraction for Handwritten NǔShu Characters. Journal of South-Central University for Nationalities (Nat. Sci. Edition) 29(3), 65–67 (2010)
4. Xie, Z., Zou, J.: NǔShu, Women's Secret Script. Women of China 4, 31–33 (2001)

Research and Application of Maintenance Scheme Evaluating Model in Hot-Blast Stove

Yan Xie, Yane Liao, and Zhe Mao

School of Electrical and Electronic Engineering, Wuhan Polytechnic University, Wuhan, Hubei, China
limimimm@163.com

Abstract. With the increasing demand for steel production in recent years, the smelting with high air temperature is an effective measure to increase production, energy efficiency and saving. Therefore, it is very important strengthen the maintenance of the hot stove and to increase the transmission capacity of the hot air. In hot-blast stove maintenance scheme evaluating, due to the uncertainty of the selection of evaluation indexes and its weight, the extent and scope of sustainable use and so on, it caused the individual to evaluate a number of useful information is often missing, thus affecting the reliability and accuracy of evaluation results. In this paper, the grey relational analysis is introduced to use to hot-blast stove maintenance scheme evaluating. The result indicated that this method is credible, and can provide a new method and a possible new way for hot-blast stove maintenance scheme evaluating.

Keywords: hot-blast stove maintenance, scheme, grey relational analysis, evaluation.

1 Introduction

Hot-blast stove is a heat exchange unit which heats the cold air sent from the confining tube blower and sent hot air into the blast hot metal smelting furnace. With the increasing demand for steel production in recent years, the smelting with high air temperature is an effective measure to increase production, energy efficiency and saving. Therefore, it is needed to strengthen the maintenance of the hot stove and to increase the transmission capacity of the hot air. How to maintain hot stove safe, reliable, economic operations, which need to strengthen the operation and maintenance work.

When the device is in an abnormal state, according to the device status assessment a variety of optional maintenance scheme can be developed. It is difficult for the operators to evaluate the pros and cons of each program and to choose the optimal scheme.

Grey relational analysis, which is an important part of grey system theory, can grasp principal contradiction and found the main features or main relationship of complicated things. So it has more complete, objective and comprehensive reflection of the quality level of things.

D. Jin and S. Jin (Eds.): Advances in FCCS, Vol. 1, AISC 159, pp. 201–206.
springerlink.com

In this paper, the grey relational analysis is introduced to use to hot-blast stove maintenance scheme evaluating. This method can provide a new method and a possible new way for hot-blast stove maintenance scheme evaluating.

2 Grey Relational Analysis Method

Grey system is a system which some of its information is clear and some of its information is not clear. Grey correlation, which knows as grey relation, is the uncertainty associated between things, or uncertainty associated between system factors and the main behavioral factors. Grey relational analysis, which is one of the important contents of grey system theory, is based on the degree of similarity or differences between the main development trends of factors to measure factors the extent of nearly level.

Assumed that the reference sequence is

$X_i=(x_i(1),\ x_i(2),\ \ldots,\ x_i(n))$

And sequence being evaluated is

$X_j=(x_j(1),\ xj(2),\ \ldots,\ x_j(n))$

To eliminate the impact of evaluating indicator's dimension, the sequence should be standardized.

If the greater factor is the better, it is standardized according to equation 1.

$$a_{ij}^{(k)} = \frac{v_{ij}^{(k)}}{[\max(v_{ij}^{(k)}) + \min(v_{ij}^{(k)})]} \quad (1)$$

If the smaller factor is the better, it is standardized according to equation 2.

$$a_{ij}^{(k)} = 1 - \frac{v_{ij}^{(k)}}{[max(v_{ij}^{(k)}) + min(v_{ij}^{(k)})]} \quad (2)$$

If Y_i is standardized value of X_i, and Y_j is standardized value of X_j .So the correlation coefficient between sequence being evaluated and reference sequence is defined as follows.

The absolute difference between Y_i and Y_j is

$$\Delta_{ij}(k) = |Y_i(k) - Y_j(k)| \quad (3)$$

The minimum absolute difference is

$$\Delta min = \min_j \min_k |Y_i(k) - Y_j(k)| \quad (4)$$

The largest absolute difference is

$$\Delta max = \max_j \max_k |Y_i(k) - Y_j(k)| \quad (5)$$

So the correlation coefficient is

$$\xi_{ij}(k)=\frac{\Delta min+\rho\Delta max}{\Delta_{ij}(k)+\rho\Delta max} \tag{6}$$

Where ρis the resolution coefficient, generally the value takes 0.5.

Associated degree,γ, is associated area ratio of between sequence being evaluated X_j and reference sequence X_i, and is defined as

$$\gamma_{ij}=\frac{S_{ij}}{S_{ii}} \tag{7}$$

Normally, we directly use the average correlation coefficient $\xi_{ij}(k)$ as the correlation, namely equal weight correlation:

$$\gamma_{ij}=\frac{1}{n}\sum_{k=1}^{n}\xi_{ij}(k) \tag{8}$$

For some sequence, when the various factors have the different importance, according to the role of the various indicators were given different weights, the weighted correlation between the being evaluated X_j and the reference sequence X_i is

$$\gamma_{ij}=\sum_{k=1}^{n}\omega(k)\xi_{ij}(k) \tag{9}$$

Accoding to equation 9, sequence being evaluated can be sorted, and the best sequence can be found out.

3 Hot-Blast Stove Maintenance Scheme

For the hot-blast stove maintenance, to achieve the same goal is often to have a lot of schemes. It is an important research topic of hot-blast stove maintenance how to finally make the right decision-making through a variety of schemes evaluation and comparison choice.

Effects and consequences of hot stove failures can be divided into the security, reliability and economy. The purpose of the maintenance is to eliminate the faults and to minimize the impact of security, reliability and economy. Also, it is taken into account such as the maintenance costs of each scheme, time to repair, maintenance, job satisfaction, and resource constraints.

According to the hot stove maintenance, four schemes are put forward.So six main indexs is used to assess these four schemes, which are security, reliability, loss of production process, maintenance costs, maintenance job satisfaction, repair time and resource constraints.These six indexs were defined as the weight of 0.282, 0.282, 0.201, 0.007, 0.110, and 0.118. It's rating level as shown in table 1.

Table 1. The schemes scoring are as follows.

Evaluation index	Scheme 1	Scheme 2	Scheme 3	Scheme 4
Security/%	94	81	64	58
Reliability/%	61	76	89	92
Loss of production process (ten thousand yuan)	5	80	200	1000
Maintenance costs (ten thousand yuan)	1.5	2	12	60
Maintenance job satisfaction	9	8	8	6
Repair time and resource constraints	1	3	6	8

4 Maintenance Scheme Evaluating Model

Based on grey relational analysis method, we do the optimal selection of hot-blast stove maintenance scheme, and sort four schemes and find the best scheme.

Firstly, all indexes are standardized according to equation 1 and equation 2. The result of standardization is given in table 2.

Table 2. Standardized indexes are as follows.

Evaluation index	Scheme 1	Scheme 2	Scheme 3	Scheme 4
Security/%	0.618	0.533	0.421	0.382
Reliability/%	0.399	0.497	0.582	0.601
Loss of production process (ten thousand yuan)	0.995	0.920	0.801	0.005
Maintenance costs (ten thousand yuan)	0.976	0.967	0.805	0.024
Maintenance job satisfaction	0.600	0.533	0.533	0.400
Repair time and resource constraints	0.889	0.667	0.333	0.111

From table 3, the reference sequence X_0 can be written:

X_0= {0.618, 0.601, 0.995, 0.976, 0.6, 0.889}

According to equation 3, the absolute difference between proposal being evaluated scheme 1(or 2, 3, 4) and reference proposal X_0 can be given in table 4.

Table 3. The absolute difference indexes are as follows.

Evaluation index	Scheme 1	Scheme 2	Scheme 3	Scheme 4
Security/%	0.000	0.086	0.197	0.237
Reliability/%	0.203	0.105	0.020	0.000
Loss of production process (ten thousand yuan)	0.000	0.075	0.194	0.990
Maintenance costs (ten thousand yuan)	0.000	0.008	0.171	0.951
Maintenance job satisfaction	0.000	0.067	0.067	0.200
Repair time and resource constraints	0.000	0.222	0.556	0.778

From table 4, it can be seen that the minimum absolute difference is 0, and the maximum absolute difference is 0.333.So according to equation 6, the correlation coefficient of the proposals to reference proposal can be written in table 5.

Table 4. The correlation coefficients of the schemes to reference scheme difference indexes are as follows.

Evaluation index	Scheme 1	Scheme 2	Scheme 3	Scheme 4
Security/%	1.000	0.581	0.375	0.333
Reliability/%	0.333	0.492	0.838	1.000
Loss of production process (ten thousand yuan)	1.000	0.869	0.718	0.333
Maintenance costs (ten thousand yuan)	1.000	0.983	0.736	0.333
Maintenance job satisfaction	1.000	0.600	0.600	0.333
Repair time and resource constraints	1.000	0.636	0.412	0.333

Considering the weight of each index, according to equation 9 the correlation of three proposals can be concluded as follows:

$\gamma_1 = 0.812$, $\gamma_2 = 0.625$, $\gamma_3 = 0.606$, and $\gamma_4 = 0.521$.

Four schemes are sorted according to its correlation, and the result is $\gamma_1 > \gamma_2 > \gamma_3 > \gamma_4$.

If the correlation is the greater, the scheme is the better. So the best is the scheme 1, and the better is scheme 2.

5 Conclusion

In this paper, grey relational analysis method is introduced and reasonably applied to the hot-blast stove maintenance scheme evaluating. It has a small amount of calculation and credible conclusion result. The result showed that it is scientific and applicable method in hot-blast stove maintenance scheme evaluating, and can provide for hot-blast stove maintenance scheme evaluating a new way of thinking and methods.

References

1. Sun, J., Yang, C., Ai, Z., Zhou, J.: Application of fuzzy comprehensive evaluation nethod in bridge type multi-scheme selection. Journal of Hebei University of Technology 38(4), 99–102 (2009)
2. Zheng, X.-H., Fu, K.-P., Tian, Q.-C., et al.: Application of grey relation analysis to evaluation sustainable of urban land. Resource Development & Market. Papers 24(10), 876–877 (2008)
3. Mingfeiluo, Kubnell, B.T.: Fault detection using grey relational grade analysis. The Journal of Grey System, Papers 16(4), 35–38 (1993)
4. Zhang, M.: The grey analysis of air quality in shenzhen city. The Journal of Grey System, Papers 21(3), 14–17 (1995)

5. Deng, J.-L.: Gray System Theory Course, pp. 55–62. Huazhong University of Science and Technology Publishing House (1990)
6. He, Z.-P., Li, Y.-P., Liu, Y.-F.: Comprehensive evaluation method of grey interaction degree and its application to R & D project investment decision. Industrial Engineering Journal, Papers 8(3), 89–91 (2005)
7. Bai, W.: Performance evaluation for commercial gas cooking oven based on grey relation analysis. HeNan Science, Papers 26(12), 1539–1541 (2008)
8. Du, R., Cai, Y.M.: Gray Connection Analysis of the Relationship between Informatization and Newly Leading Industry. Commercial Research, Papers 378, 108–111 (2008)

Triggering Collaborating in Converged Networks

Rongheng Lin and Hua Zou

State Key Lab of Networking and Switching Technology, Beijing University of Posts and Telecommunications, Beijing, China, 100876
{rhlin,zouhua}@bupt.edu.cn

Abstract. Context aware is a trend of converged network and service. In order to support context aware, network capability would be extended to collect and provide environment or user related information. At the same time, context based triggering schema would be introduced for service triggering which changes the way that people use service. New service could be triggered in a more feasible way. However, that would also introduce some other problems. One of the problems is how to collaborate different service together, as service would happen anytime which will lead to conflict. In this study, we introduce a gaming theory method, which models the procedure of service triggering and interaction. A service collaboration example is provided in the experiment. And result of this study would be combined with our previous study to become a framework to support human centric service.

Keywords: service triggering, network service, game theory, converged networks.

1 Introduction

As it is known, service is main motivation of converged networks development. And service is in fact driven by human's requirement. So how to fulfill human's requirement becomes high priority in the service related research.

Doing thing according to user's requirement is the word 'human centric' means. In order to do so, it is necessary to understand user's requirement. In a network environment, it becomes how to understand the status of user and how to predict user intention. In other word, it becomes aware of user status. Context aware in network can be discussed in different layers, such as core network transport layer, control layer, and service layer. In this study, we focus on the control layer and the service layer.

Current converged networks provide different kinds of capabilities. However, when services use those capability through gateway/server in control layer, it is hard for them to aware the environment. It is because current capabilities don't integrate with sensing utility or context information.

Targeting to provide context information or sensing capability, control layer needs to track or data mine environment information. Furthermore, those environment information need to be delivered to possible client as needed. In our previous works, an infrastructure that can sense human relationship and status is proposed. In that infrastructure, information are shared and disseminated in community, which helps

D. Jin and S. Jin (Eds.): Advances in FCCS, Vol. 1, AISC 159, pp. 207–212.
springerlink.com

'potential' nodes to know about some related information. From a goal perspective, our previous work provides a possible way to cluster related information together and deliver proper information to some proper nodes. And those nodes can trigger service as needed. Instead of Detected point schema, new style service triggering schema is based on information with policy on the nodes. Changes on the triggering help service to trigger in a more customized way.

However, the feasibility of service triggering will introduce new problems. Telecom services always need more than one party to participate. If service triggering actions in a pure distributed way without supervisors, there would be lots of unexpected conflict among different nodes. For instance, some services have the same user set or feature sets; they should be triggered at the same time.

In this study, we propose a gaming theory model for collaborating service triggering. And based on the schema, it is possible to regular the interaction between services. This paper is organized as follows. Section 2 introduces some related work; Main motivation of the new framework is proposed in Section 3; while Section 4 shows the architecture; some of the challenges are discussed in Section 5; after that, Section 6 use a service to validate the framework.

2 Related Work

Current trigger schema in IMS needs to be improved [1], as it is known ubiquitous service become a trend [2, 3]. Social cooperation triggering can be a new style[4, 5].

In early day of telecom industry, client has no intelligent at all. So that is why IN is called IN, because network provides all the intelligence.IN introduces detection point (DP), detection point schema models service in several different states. Each state is associated with some events. DP model will check whether the corresponding event comes. If comes, service would be triggered. This model helped people easy to develop service. And IMS still uses similar model. However, as people demand for customized service, this principle becomes a problem. Because customized service would be triggered in a dynamic way, which will require flexibility in triggering model.

Today network is more powerful, and so does clients' capability. It is no reason that the network service just involves network side. Some researchers began to focus on providing client intelligence as trigger method. For example, Client intelligence is discussed in [6,7]. And as an extension of network, client side application becomes one part of service.

Besides, lots of successful commercial service/application is client based. For instance, the Apple app store provides hundreds of application for people to use. The problem of app store is most of the application is a standalone program, which is not related to the network service. But even that, those applications are still widely used by people. Client application has become one important part in the telecom service domain. As client is near users, some context aware features can be implemented in client. So client would be a good entry point of service.

3 Triggering Coordinating

Current service triggering in converged networks is based on DP (detection Point) which was introduced in the IN (Intelligent Network). This kind of trigger schema helped developers easier to develop the service. However, as all the detection points are preset, all the triggering moments are preset, which means it hard to support dynamic triggering.

As the community structure proposed above, the new system is different from current network. All the nodes in community can trigger service, so it is necessary to find out some principles for nodes to follow. Otherwise, the triggering will be chaos.

Looking into triggering models, there might be two kinds of styles, one is competition, and the other is cooperation. Competition means two nodes that have two or more different conflicts in triggering targeting service. Cooperation means that two or more nodes join together to determine the triggering criteria. In order to fix this problem, some gaming theory models are introduced in here.

Table 1. Concept Mapping between game theory and problem domain.

GameTheory Domain	Network Concept
Player	Node
Strategy space	Solution space of node's action set
Payoff function	The "value" under certain action set
Game	Interactive model between nodes, including the rules

There are two kinds of models in our problems, which can be mapped to non-cooperative game and cooperative game situation. However there are still lots of models in non-cooperative game. Sex-war gaming model is selected for the problem.

The reason that we choose the sex-war model is because one of the players needs to give up his chose in our non-cooperative situation, as only one service can be triggered. But the real situation is this not simple. It is not complete information gaming. In the model, each player has two strategies: trigger or not trigger. So we assume each player will choose their decision based on some probability. q and r are the probability of player's decision. As the Fig shows, player 2 uses a possibility q to choose trigger, while player 1 use r to.

Assuming a set of players: $N = \{1, 2, 3...n\}$ and assuming the strategy space is $S_1, S_{2,} S_3 ... S_n$; Strategy payoff functions are $u_1, u_{2,} u_3 ... u_n$

So Payoff functions can be represent as follows $u_i : S_1 \times S_2 \times S_3 \times ... \times S_n \rightarrow R$

Considering the situation that player 1, 2 use probability of r and q to trigger the service.

	s_{21} (q)	s_{22} (1-q)
s_{11} (r)	$u_1(s_{11},s_{21})$, $u_2(s_{11},s_{21})$	$u_1(s_{11},s_{22})$, $u_2(s_{11},s_{22})$
s_{12} (1-r)	$u_1(s_{12},s_{21})$, $u_2(s_{12},s_{21})$	$u_1(s_{12},s_{22})$, $u_2(s_{12},s_{22})$

$$EU_2(s_{21},(r,1-r)) = r\times u_2(s_{11},s_{21}) + (1-r)\times u_2(s_{12},s_{21}) \quad (1)$$

$$EU_2(s_{22},(r,1-r)) = r\times u_2(s_{11},s_{22}) + (1-r)\times u_2(s_{12},s_{22}) \quad (2)$$

$$EU_1(s_{11},(q,1-q)) = q\times u_1(s_{11},s_{21}) + (1-q)\times u_1(s_{11},s_{22}) \quad (3)$$

$$EU_1(s_{12},(q,1-q)) = q\times u_1(s_{12},s_{21}) + (1-q)\times u_1(s_{12},s_{22}) \quad (4)$$

Formula (1), (2) show the payoff value when player 1 chooses trigger or not.

Formula (3) (4) show the payoff value when player 1 chooses trigger or not trigger

As the formulas (1, 2, 3, 4) above, problem is changed to choose the bigger EU. Obviously, EU is decided by q, r and u. So problem becomes how to choose the q, r and u.

u is function of s. As s is about the strategy, u represents the payoff of the strategy.

$$u_k(s_i,s_j) = w_k(h(s_i)) - w_k(h(s_j)) + w_k(h(s_i)\cap h(s_j)) \quad (5)$$

$$w_k(X) = \sum_{x\in(X\cap J_k)} (z(x)\times n(x)) \quad (6)$$

As formula (5), u is determined by w, h and s. $h(s)$ represents the set of service features under current strategy. As it is known, service is composed by the service features. So the value of service feature will affect real service value. While (6) $w(X)$ is the contribution of current service features, which is decided by the unit feature value and the unit number. Unit value of the service feature is determined by the network provider. In our experiment, we try some value based on the complexity of the features.

We divide the service feature into two levels, which is determined by independent use of the feature. Level1 features are those that can be used independent in the service. Level 2 features are some supplement features.

Table 2. Level 1 Service Feature Value

Service Feature	Label	Suggested Value
User Status	*SI*	2
Location	*LI*	2
User Pofile	*UP*	2
Third party information service(weather..)	*TI*	2

The service feature values above are determined by the complexity and the fee of the service. So the value can be changed according to different network provides.

Another issue that will affect service value is the probability q and r. The probability of q and r is determined by the history of success as (7).

$$p_i = p_{i-1} + \frac{1}{2} x e^{\frac{-count(x)}{10}}; x = -1, 1; p_0 = 0.5; \max(p) = 1 \quad (7)$$

In this model (7), all the probability begins with 0.5 and based on success history to generate new value. In a long term, if half success and half fail, the value will be around 0.5. Based on the value, service can be triggered as needed.

4 Scenario Evalution

Player A, is at a meeting, with his presence "not interrupt", and status "in a meeting", Player B needs to contact with A when he finishes the meeting. So he sets the rule, when A is out of his meeting, connecting two together with a video call. Player C is one of the friends of A. And he is just nearby, and tried to find some friend to hang out. And as our system can combine the social relationship with rich presence information, so the system will choose A as C's choice.

When Player A finishes his meeting, he changes the status to "free to talk". And currently, there are two services can be triggered at this time. (Besides, A have some customized on his video and audio call). The system now will now begin a gaming processing. We can treat the two services that would be AB and AC.AB would be a video call with some A's customized function.AC would be a simple audio call.

Based on the gaming equation, we can compute u value. In this case, we can easy to know uab > uac. However, the final decision should be associated with the probability of history.

If the AC's probability is much higher than AB, it would means that A and C are more familiar with each other, so A will more likely to talk with C. Otherwise, the system will choose AB to trigger. So network will get more profits than AC.

For the scenario AB, after service is triggered, video called is being processing. Also as the preference of the player A, a SMS will send out the caller's business card to A. Figure 1 simulates A's screen before setting up the video call and the business card that is received.

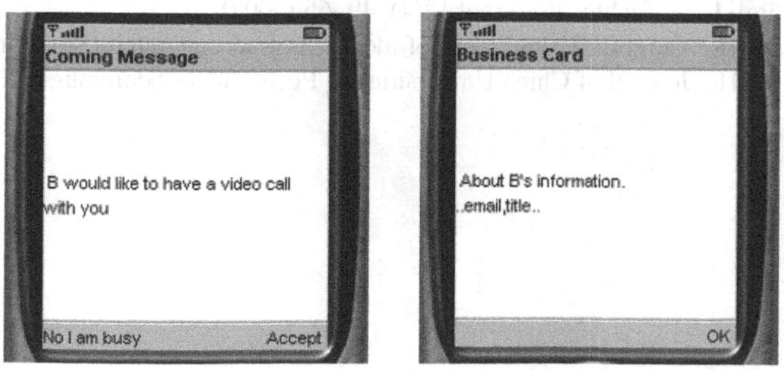

Fig. 1. Snapshot of the J2ME simulator

5 Conclusion

In this paper, we focus on the service triggering model, which is one of the most important problems to support the context aware service. And some example scenario is discussed using this framework.

In future work, we will continue to work on context aware/human-centric service [8]. And as discussed in this paper, there are some relationships between human, network and service. We will focus on the human and service in the future. Not just provide the framework for developer to create new service, but also provide methodology of creating service related with human factor.

Acknowledgments. This work was supported by the National Natural Science Fund China under Grant No. 2009CB320406, the National 863 High-tech Project of China under Grant No. 2008AA01A317, the Important National Science Technology Specific Projects under Grant No. 2009ZX01039-001-002, Funds for Creative Research Groups of China (60821001).

References

1. Tarkoma, S., Bhushan, B., Kovacs, E., van Kranenburg, H., Postmann, E., Seidl, R., Zhdanova, A.: Spice: A Service Platform for Future Mobile IMS Services. In: IEEE International Symposium on a World of Wireless, Mobile and Multimedia Networks, WoWMoM 2007, pp. 1–8 (2007)
2. Zhang, J., Helal, S.: UbiNet: A Generic and Ubiquitous Service Provider Framework. In: International Conference on Software Engineering Advances, pp. 11–11 (2006)
3. Bouguettaya: Infrastructure for ubiquitous services. In: Proceedings of the Nineteenth Conference on Australasian Database, vol. 75, pp. 9–9
4. Sarin, A.: The Future of Convergence in the Communications Industry. IEEE Communications Magazine 45(9), 12–14 (2007)
5. Eagle, N., Pentland, A., Lazer, D.: Inferring Social Network Structure using Mobile Phone Data. PNAS (2007)
6. Bachmann, A., Motanga, A., Magedanz, T.: Requirements for an extendible IMS client framework. In: Proceedings of the 1st international conference on MOBILe Wireless MiddleWARE, Operating Systems, and Applications (2008)
7. Isukapalli, R., Benno, S., Park, C., Feder, P.: Advanced IMS client supporting secure signaling. Bell Labs Technical Journal 12(4), 49–65 (2008)
8. Lin, R., Zou, H., Yang, F.: Using 'word of mouth' to find community structure in an online IMS system. The Journal of China Universities of Posts and Telecommunications 15, 88–93 (2008)

The Platform of Enterprise Third-Party Data Service Based on Distributed Mobile Communication

Hui Han[1], Congyi Sun[1], Baohua Bai[2], Jianping Xing[1], and Xiaojing Yao[3]

[1] School of Information Science and Engineering,
Shandong University, 250100 Jinan, China
[2] Navigation Communication Co-system Engineering
Center of Shandong Province, Jinan, China
deepbbh@hotmail.com
[3] China National Institute of Standardization, 100000 Beijing, China

Abstract. At present, China's enterprise data services face the problems of service processes unclear, un-standardized, non-realtime, and lack mobility. In this paper a design of platform based on Step-Task work-flow process model is presented. Experimental results show that this design can makes the data services process more secure, clear, and through this design, distributed mobile platform of data services can be realized in order to meet the service providers' and enterprises' mobile office needs.

Keywords: distributed Mobile communication, third-party data service, workflow.

1 Introduction

With the development of computer information systems, enterprise user's core business is increasingly dependent on reliable operation of information systems, and business-critical data has become user's most important asset. Human error, hard drive damage, natural disaster and so may cause data loss and bring incalculable damage to the development of country and enterprises [1]. In order to ensure data security, building a local or remote disaster recovery system will minimize disaster losses. However, a self-built disaster recovery center needs huge investment. Not any organization willing or has the ability to bear alone, especially for small and medium-sized organizations. Therefore, the choice of third-party data service has become an important development trend in the area of data disaster recovery[2].

Currently, data disaster recovery technologies include snapshots and data replication technologies. Snapshot technology is an available copy set of specified data, the copy contains the corresponding data image. Data replication technology include: disk-based system, technologies which is based on virtual storage, based on host storage, base on area networks, based on the database and applications [3].

China's third-party data storage services processes now is confusion, duplication, and lack of standard; the relevant person offer the services in a un-realtime, and poor mobile way, which greatly reduce the work efficiency and brought a lot of inconvenience.

D. Jin and S. Jin (Eds.): Advances in FCCS, Vol. 1, AISC 159, pp. 213–219.
springerlink.com

To solve the problems of our country's third-party data services, this paper presents a platform based on Step-Task model, trying to standardized data backup and recovery process; giving an implementation of distributed mobile platforms which is operate-easily and interface-friendly. Experimental results show that this platform can make the information services process more secure, clear, and mobile platform can be realized to meet the service providers' and enterprises' mobile office needs.

2 Architecture of Mobile Data Services Based on C /S Model

The platform architecture of Enterprise third-party disaster data recovery services consist of five parts: users, service process, disaster recovery data, workflow-process service engine, distributed mobile terminals.

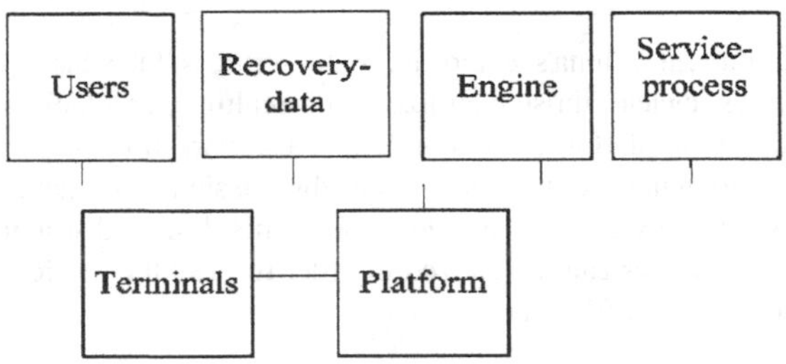

Fig. 1. This shows the architecture of enterprise third-party data service platform.

Among them, user is the main part of the mission. Service process is composed of activities, and task is the instance of activities. The users of platform are divided into two categories: service side and demand-side personnel. Each personnel is mapped to the virtual department of platform in accordance with their actual department functions, and is assigned appropriate permissions in accordance with the actual authority for the department.

Among them, the service process is the main line of the entire platform. Running service process instances constitute the entire platform area. Each service flow contains several tasks. Each task is to serve the specific process instance nodes. The context of the entire platform is composed of service processes and personnel.

Among them, the disaster recovery data is the information carrier. Data backup center uploads the customer data in the form of attachment storage, and describes them with XML language, in order to systematize the stored data and standardized the management. Documents are delivered in the service process constantly, and their trajectory is recorded. Data is also the process-driven material of entire service.

Among them, the process service engine is the core parts of the platform. Engine is responsible for promoting the service process of each task, it's the driving force behind the entire system. Engine is a service program, providing services for clients, while monitoring the process.

Among them, distributed mobile terminal supports the platform. Users and service providers get together with distributed mobile terminals.

Analytically this platform uses the technology of Step-Task workflow, and the workflow engine supports the operation of it [4].

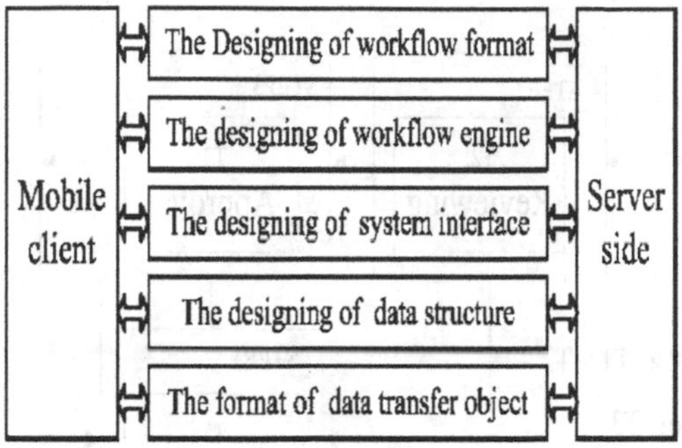

Fig. 2. The system is divided into three modules, client, server and core,This shows the Framework of the data service platform.

According to the business conditions of different units, the service sets different process templates, to make the service process customizable. The system interfaces define information transfer standard between modules. The designing of data structure defines the format of data storage. The format of the data transfer object defines the delivery format of data and description.

3 Step-Task Workflow Process Model

The Step-Task workflow process model model consists of step layer and task layer. Step layer define the process steps which the logical structure of operation, task layer describe the specific implementation of the business [5]. Modeling workflow process model with layers can separate the logical structure and specific tasks, allowing them to achieve loosely coupling by defining them in the relationship maps[6,7].

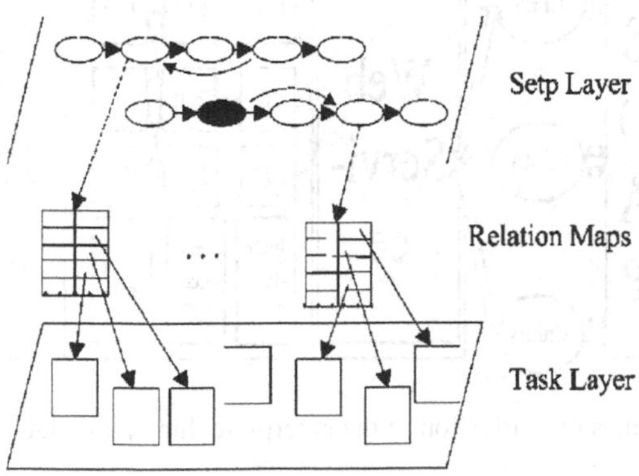

Fig. 3. This figure shows Step-Task workflow process model structure

Service providers, for example, a typical service process including four steps: acception , reviewing, approval and execution, each step including a specific task. Data recovery process shown in Figure 4.

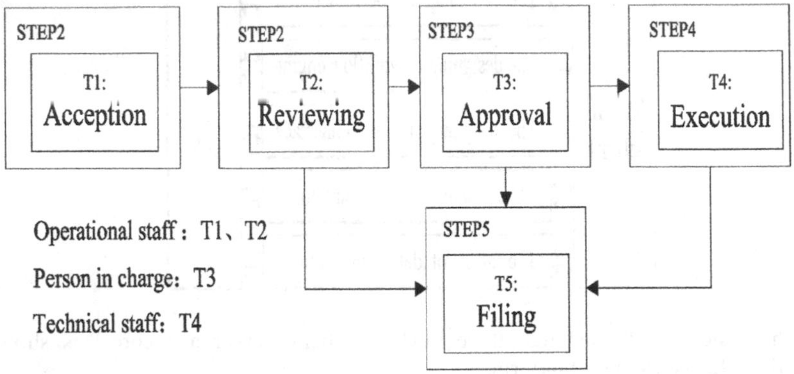

Fig. 4. This shows the implementation of the workflows for mobile platforms.

4 Realization of System

By analyzing the function of the platform, the system is divided into three layers, namely the mobile client, the application server and the data server (shown in figure 5). The routing service engine of this system uses the Step-Task workflow process model of Double-layers structure. The business logic of space information service is packaged by EJB, achieves the portability of the application layer. The client is based on the Android/Windows Mobile platform, with friendly interface and powerful functions. The application server uses EJB as the principal component and JBOSS as the operating environment. The data server uses Oracle.

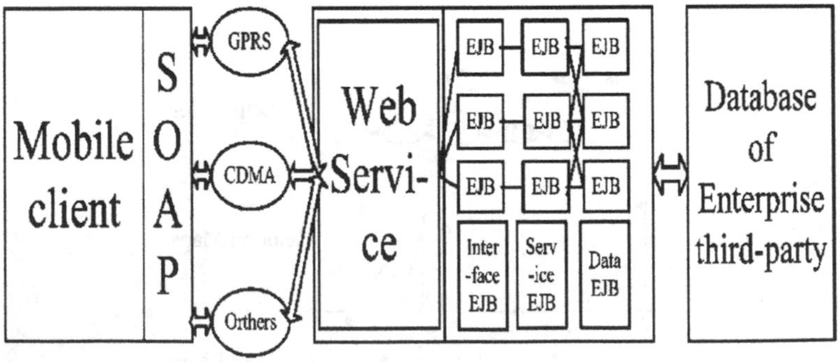

Fig. 5. This describes level division of the enterprise third-party data service system.

4.1 Mobile Terminal

The characteristics of the platform is to develop a moving embedded official terminal, realizing the target of the office environment anywhere at anytime . Mobile client is based on the distributed moving communication target of GPRS/CDMA/3G network. All business applications, data entry and other operation tasks are done here. When the user needs exchanging data, it is not allowed to directly get access to the database server, but through the service interfaces provided by the business logic layer. Thus it ensures the security of the background data, and achieves a true sense of the thin client.

The server-side program and the client-side program use the SOAP interface protocol . The protocol packages the XML data between applications through the network.

It is scattered in different mobile devices. Mobile database has a small kernel, high operating efficiency and strong handling capability[8,9]. The mobile data management system is developed by the .Net Compact Framework .

The main function modules for the client are the business management module, the user management module, the document browsing module, the Workflow browsing module and the document storage module.

Fig. 6. The running system interface of the client-side is shown in figure 6.

The functions can be accomplished by the mobile phone after login as follows: Business management function is the main module to the client .It is responsible for sending, receiving and processing tasks and the querying of relevant information. User management function is responsible for inputting, browsing and deleting accounts. Document browsing function is responsible for browsing and viewing the document. Workflow browsing function is responsible for browsing and checking the progress of the workflow. Document storage function is used to store the received data files.

4.2 Application Server

The application server has three components, namely "interface EJB", "service EJB" and "data EJB", dealing with three different functions respectively. Web Service is based on XML and HTTPS. Its communication protocol is based on SOAP. The system takes full advantage of EJB's three forms, so that it can easily get these functions [10].

4.3 Database Server

The database layer provides the necessary data services for the application service. For distributed spatial information service, the spatial data which is requested by the user on the client at one moment is certain, and it is also local relatively to the whole massive spatial data. So it is needed to establish the index for Massive spatial data, which is the core of the data services layer. Therefore, the database server consists of two layers, namely the storage layer and the index layer. When the user sends application, the required data is found by the search engine, and then sent to the application server.

5 Conclusions

After studying the enterprise third-party data service deeply, this paper puts forward the enterprise third-party communication service platform model based on mobile. The model makes the process clear and simple, strengthens the specification of the service platform and improves the system mobility. With a routing service engine as a driving force, this paper founds an automatic and intelligent service platform. Its operation is simple, its interface is friendly, and its function is strong .Finally it achieves the third-party routing service functions of data storage.

References

1. Yao, W., Wu, C.: Development of Standards and Industry of Disaster Backup and Recovery in China. ZTE Comunication 16(5) (2010)
2. Wang, D., Wang, L.: Research of Disaster Tolerance System. Computer Engineering 31(6) (2005)
3. Yang, Y.-X., Yao, W.-B., Chen, Z.: Review of Disaster Backup and Recovery Technology of Information System. Journal of Beijing University of Posts and Telecomunication 33(2), 1–6 (2010)
4. Luo, H.B., Fan, Y.S., Wu, C.: Analysis of event balance on the verification of workflow soundness. Journal of Software, 1686–1691 (2002)
5. Xing, J., Zhao, L., Meng, L.: The research of a new Workflow model with step-task layers based on XML documents. In: Proceedings - 2006 IEEE/WIC/ACM International Conference on Web Intelligence (WI 2006 Main Conference Proceedings), WI 2006, pp. 970–973 (2007)

6. Workflow Management Coalition. Process Definition Interface - XML Process Definition Language, WFMC-TC-1025, V 2.0 (2005)
7. Workflow Management Coalition. Workflow Standard - Interoperability Wf-XML Binding, WfMC-TC-1023, V 1.0 (2000)
8. Min, Y., Xiong, Q.: Research of mobile database technology. Journal of Science and Technology of Wuhan University 32(1), 156–159 (2008)
9. Lin, Y., Chiang, K., Lai, I.J.: The design and implementation of high performance data synchronization of high performance data synchronization server for mobile applications on education. Journal of Internet Technology 8(1), 75–87 (2007)
10. Wang, X., Xiong, Q.: Data synchronization model Based on Mobile Agent and Web Service. The Computer and Digital Project 38(4) (2010)

Study and Implementation of Virtualization Technology in Embedded Domain Using OKL4

HongYan Kang

Department of Computer and Information Engineering, Heze University, Shandong, China
khyky@sina.com

Abstract. With virtual machines have become mainstream in enterprise computing, virtualization is being an attractive technology in the embedded community. However, current popular virtualization technologies do not suit for embedded system, which are characterized by small memory footprint, power constraint, small trusted computing base, and fine-grained security. In this paper, we selected a well-known embedded microkernel as the virtualization layer and then proposed a virtualization architecture based on OKL4, and analyzed the characteristics of virtualization used in embedded domain. Finally, we have ported the OKL4 and OKLinux to PXA270 and test it on hardware board. Experimental results show that the design scheme is feasible and correct.

Keywords: Embedded microkernel, OKL4, Hypervisor, PXA270, OKLinux.

1 Introduction

Virtualization means software running on a virtual platform rather than on the real hardware. Virtualization technology can expand the capacity of the hardware and simplify software re-configuration process. Virtual machines have become mainstream in enterprise computing in recent years due to consolidation of servers and lower energy and maintenance costs. Virtualization allows a hardware platform to run multiple operating systems and applications can run in their own space, thus can significantly improving the efficiency and security of the computer. Data indicates that effect of virtualization brought to the enterprise is remarkable. Traditional virtualization technology is mainly used in the enterprise domain. As embedded systems are usually designed to deal with one or a few dedicated functions, it is used to be relatively simple and constrained by memory footprint, energy consumption and so on. So that enterprise-style virtualization technology is ill-matched to the requirements of the embedded domain, which are characterized by low-overhead communication, real-time capability, small memory footprint, small trusted computing base, and fine-grained control over security[1]. When it comes to embedded environments, the virtualization technique has to make some adjustment according to characters of embedded system, such as severe hardware resources and power management requirements, etc [2]. Embedded virtualization technology has brought an unprecedented new opportunity for the embedded system designer. In this paper we attempt to discuss performance and prospect of embedded system based on the introduction of virtualization technology.

D. Jin and S. Jin (Eds.): Advances in FCCS, Vol. 1, AISC 159, pp. 221–225.
springerlink.com

2 Background

Virtualization technology was first applied to operating system as IBM/370 in 1970 which through VM-CP provided a software emulation of the IBM 370 processor and channel I/O architecture so that OSes such DOS/360,TSS,OS/360 could be run to maintain legacy applications. Such technology supports services consolidation, load-balancing, power management, and running different operating systems.

Embedded virtualization design is categorized into two ways. The first one is to use a microkernel approach in which it provides a modular system construction fashion and another way is to use a traditional VMM system. Using the microkernel system makes guest systems suffer from lots of code modifications because the microkernel offers high level abstraction such as process and memory regions, and guest OSes need to tame the abstraction to execute them on the abstraction. Microkernel features a minimum of functionality, typically scheduling, memory management, process synchronization, and IPC. Performance was gained by moving memory management policy out of the kernel, reducing IPC cost, and reducing the cache footprint. The L4 microkernel has been the basis for most subsequent systems [3]. Traditional VMM system virtualizes most system resources and guest OSes can run upon it without any significant changes to the systems. But VMM itself is too complicated in regarding to embedded system design if there is no special hardware to help performing some operations, like the page table virtualization, and would therefore cause considerable power consumption [4].

Because many properties, such as efficient IPC, real-time capability, small memory footprint and so on, which make Xen and other hypervisors unsuitable for embedded system[1], we choose OKL4 as our research object.

3 Virtualization Technology in Embedded System

Virtual machine environment based on microkernel has many characteristics such as high reliability, high flexibility, and real-time support etc. With the development of virtualization technology and the introduction of new architecture, the performance obstacles of the embedded system virtualization also are being overcome gradually, so it has extensive application prospect in embedded system.

We can support multiple isolated operating systems to run on the same embedded platform by virtualization, such as the e-commerce and other applications which require high security packaged in a high degree of isolated guest operating system, and another operating system as a normal application environment, so can realize high safety and reliability of the operating mode.

With virtual machine based on microkernel architecture, we can convert hardware resources to various real-time system services, and deliver to client operating systems which run on virtual machine by mode of virtual devices. In this way, it can support real-time and non-real-time applications to run simultaneously, and provide a universal and transparent interactive interface between non-real-time applications and real-time system functions.

By using virtual machine, we can get a higher compatible software running environment on embedded platform. It makes the specificity of hardware of embedded system transparent to the maximum extent through software virtualization, and can realize the coordination between different platforms of embedded systems. And through virtualization, we can also achieve other complex functions, such as the migration process between different platforms, distributed operation, fault recovery and so on.

By introducing mechanisms of network interconnection and remote access, operation system can access to remote resources transparently through virtualization and localization technology, and expanded the embedded system function greatly.

4 Embedded Virtual Machine Model Based on OKL4

4.1 OKL4

OKL4[5] is not only a commercial, microkernel-based hypervisor targeted for use in embedded systems, but also a real-time operating system with simple memory protection. It from the OK-Labs company is one of the most complete solution at this time. It provides a micro-kernel derived from the L4 project with the ability of running guest OSes using the paravirtualization technique. The OKL4 hypervisor[6] is a member of the L4 microkernel family which originated with Liedtke's original L4. It has been successfully deployed in CE devices, including an estimated 250 million mobile phones. It supports para-virtualized high-level OSes as well as RTOSes. OKLinux, the para-virtualized version of Linux on OKL4 is in fact a port to an "OKL4 architecture", meaning that a new architecture subtree is introduced into the Linux source tree.

4.2 Embedded Virtual Machine Model

In this paper we propose a high performance virtualization architecture, showed in Figure 1, based on OKL4 hypervisor which has the following characteristics:

High Flexibility. The system supports both coarse-grained virtual machine, and fine-grained lightweight components. Guest OSes which run on virtual machine can be embedded non-real-time OSes such as Windows CE, and real-time OSes like RTLinux. Drivers could be run directly in the lightweight execution environment, and other subsystems could share them, so the reusability of the code was improved.

High Performance. OKL4 has been designed to meet the requirements of virtualization for embedded system. Specifically, it features high-performance IPC of less than 200 cycles for a one-way message-passing operation and provides efficient mechanisms for setting up shared memory regions. It has low interrupt latencies, able to deliver an interrupt to a driver running in a virtual machine within a few microseconds. Therefore, high-performance of OKL4 ensures that the system has high performance.

High Security. Because model adopts micro-kernel based structure, security can be guaranteed from the structure. The complete code size of OKL4 embedded hypervisor is less than 12 kLOC, so that can ensure the system with low error rate.OKL4 is based on an underlying fine-grained access-control model using capabilities. This mechanism is flexible and efficient, so can ensure the security of the system more finely.

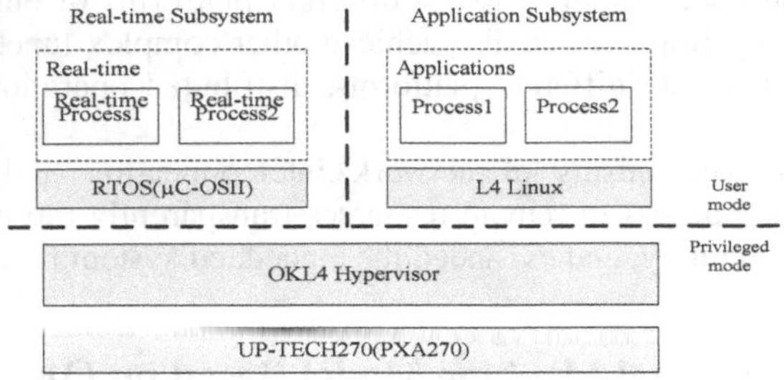

Fig. 1. Virtualization architecture

5 Experimental Work

To evaluate the technical feasibility and to demonstrate the use case, we selected OKL4 3.0 as the virtualization layer.

The experimental work consisted of generating, cross-compiling, and testing a paravirtualized version of OKLinux on PAX270.To get up and running there are basic prerequisites:

Toolchain, Phyton, OKL4 and OKLinux.

Step by step for install the toolchain:

Download the okl4_3.0.tar.gz and oklinux_2.6.24.9-patch.4.tar.gz from the Open Kernel Labs Website;

Cross compiler arm-linux-gnueabi-4.2.4.tar.gz and arm-linux-3.4.4.tar.gz should be download also;

Expand the tarball and add correct directory to your PATH;

Make, and img file generated;

Finally you can use TFTP to transfer img file to your hardware board. The resulting image successfully runs on hardware board as shown in Figure 2.

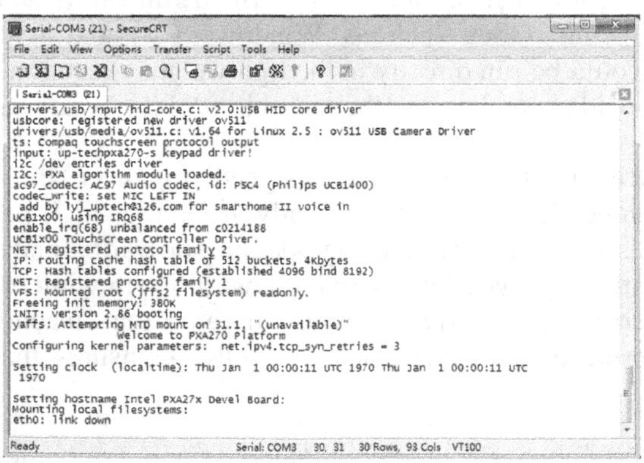

Fig. 2. Result of the system running

6 Conclusion

In this paper, we described design and implementation of OKL4 and OKLinux on ARM, which is a secure system virtualization of ARM architecture. We described related knowledge of the embedded virtualization and reviewed recent work on OKL4 for virtualization of embedded systems, and selected OKL4 for a proof-of-concept demonstration on hardware board up-tech270 which equipped with PXA270 processor. The results indicate that OKL4 and OKLinux can match the requirements of virtualization for embedded system. Further we are beginning to port some applications to virtualization system to enhance the performance of the system.

References

1. Heiser, G.: Hypervisors for Consumer Electronics. In: 6th IEEE Consumer Communications and Networking Conference, pp. 1–5. IEEE Press, New York (2009)
2. Heiser, G.: The role of virtualization in embedded systems. In: 1st Workshop on Isolation and Integration in Embedded Systems, pp. 11–16. ACM, Glasgow (2008)
3. Tsung-Han, L., Kinebuchi, Y., Shimada, H.: Hardware-assisted Reliability Enhancement for Embedded Multi-core Virtualization Design. In: 17 IEEE International Conference on Embedded and Real-Time Computing Systems and Applications, New Delhi, pp. 101–105 (2011)
4. Acharya, A., Buford, J., Krishnaswamy, V.: Phone Virtualization Using a Microkernel Hypervisor. In: 2009 IEEE International Conference on Digital Object Identifier, pp. 1–6. IEEE Press, New York (2009)
5. OKL4, `http://www.ok-labs.com/products/okl4-microvisor`
6. Heiser, G., Leslie, B.: The OKL4 Microvisor: Conver-gence point of microkernels and Hypervisors. In: 1st Asia-Pacific Workshop on Systems, New Delhi, pp. 19–24 (2010)

Parametric Reverse Modeling and Redesign Method of Prismatic Shapes

Jianhua Yang, Siyuan Cheng, Xuerong Yang, Xiangwei Zhang, and Shaoming Luo

Faculty of Electromechanical Engineering, Guangdong University of Technology, 510006 Panyu, Guangzhou, Guangdong, China
happyyangjianhua@163.com, imdesign@gdut.edu.cn

Abstract. The objective of this study is to propose a method of parametric reverse modeling and redesign for prismatic shapes. The main structural feature levels for the parametric feature extraction were divided firstly to distinguish core issues from the operation process. Three basic steps towards successful reverse engineering prismatic shapes were put forward, which include strategy analysis of parametric reverse modeling, extraction of sectional curves and reconstruction of feature models, and parametric redesign of prismatic shapes. In this study, a sample part was reverse modeled and redesigned to demonstrate the proposed method. This process proved an efficient way towards parametric reverse modeling and innovative redesign method of prismatic shapes.

Keywords: Reverse engineering, Prismatic shapes, Parametric modeling, Redesign.

1 Introduction

As one of the most important methods of modern advanced design, reverse engineering (RE) is playing an increasingly important role in product development and innovative design. With the rapid development of computer graphics, computer aided geometric design, high performance computers, 3D data acquisition devices and other relevant technologies, RE has being deeply researched, widely gained acceptance and extensively used in the design community [1-3]. Reverse engineering software extracts geometrical and topological information from the digitized point cloud and describes it to the user. And so, the design intent understanding and representation of geometric features in the physical model are important issues in product improvement, reproduction and quality control [4-5]. Due to these factors, the method of feature-based reverse modeling has become the new promising development direction and solution of RE, which combines with feature-based technologies of forward design and the flexibility of reverse design. The model allows for redesign modification and iteration and is well suited for downstream analysis and rapid prototyping.

At present with the last few decades development, the method of feature-based reverse modeling has become the main research direction of RE. There mainly have been existing three representative reverse modeling approaches, namely: the feature template matching approach, the feature primitive elements extracting approach and the directly point cloud feature processing approach [6-7]. The first approach could

D. Jin and S. Jin (Eds.): Advances in FCCS, Vol. 1, AISC 159, pp. 227–233.
springerlink.com

match feature templates with the point cloud for some particular case, and then make the point cloud analysis and process followed matching results to reconstruct surface models. The second approach could directly process the point cloud and extract different feature primitive elements of the surface model, then fit each primitive element to surface patch to achieve CAD models. The last approach evolves from the approximation algorithm brought up by Zhejiang University, which mainly include the feature extraction and region segmentation directly based on differential geometry estimates of the point cloud. Lots of attentions and researches are focused on this algorithm for speeding up the maturity of correlation design techniques, and then it will become the chief approach of the feature-based reverse modeling.

In the process of feature-based reverse modeling, it is considered not only to use different extracted and parameterized methods for different kinds of features, but also to draw up different modeling methods and workflows for different kinds of products. Prismatic shapes is this class product which have many obvious and easy distinguished structural features, and the reverse modeling of these products mainly could be processes by dividing structural feature levels and then implemented different feature extraction methods. This study intends to introduce an efficient way of parametric reverse modeling and redesign process for prismatic shapes. The goal is to create a feature-based parametric reverse model from the digitized point cloud for the relational design and change management. And a bearing seat included typical prismatic shape structural features was used as the case study to illustrate the process we presented.

2 The Feature Analysis and Modeling Strategy of Prismatic Shapes

Before reverse modeling, it is necessary to conduct preliminary analysis and understanding, which involved the functional features of products, the structural types of features, the modeling strategy, etc., finally to make the definite modeling purpose and draw up the strategy plan in the process. Corresponding to different user case scenarios, there are three major RE modeling strategies, namely: automatic free-form surface modeling, feature and parametric-based solid modeling, and curve-based surface modeling [8]. We should employ different RE modeling strategies for different user case scenarios. The choice of an appropriate modeling strategy is based on sufficient analysis of model's characteristics and features.

The most important step of the feature-based reverse modeling are the identification and extraction of features, the identification all kinds of parameters and constraints information to accord with the original design intent. On the background of the subsistent sectional feature extraction technique and the implementation behaviors of the feature-based reverse modeling, we divide the main structural feature levels of prismatic shapes as shown in Fig. 1.

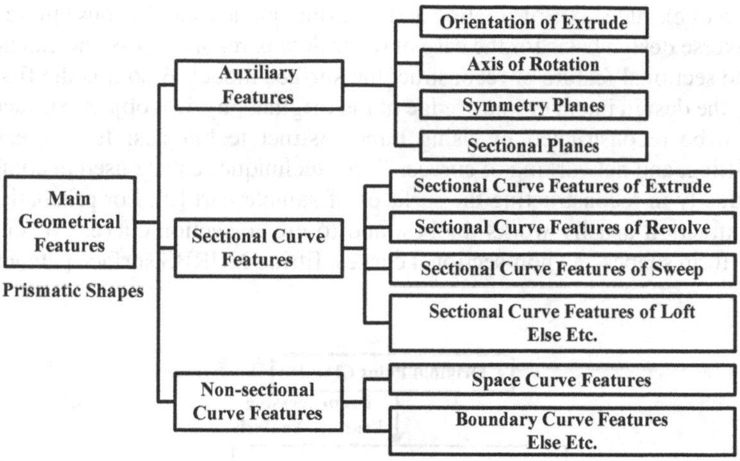

Fig. 1. The main structural feature levels

According to the main structural feature levels of prismatic shapes, these products are mainly build from regular features such as the straight line of extrude, the fixed angle of revolve, etc. In the process of their reverse modeling, regular features can be built from 2-dimensional (2D) profile sketches of sectional features by different construction operations. So the key step for prismatic shapes is the parametric extraction of sectional features in the reverse modeling.

Following the feature analysis of prismatic shapes, the modeling strategy for different features must be the drawn up, which includes constructing the global coordinate system, setting the structural method of every feature, understanding the parameters relationship of each feature depended on. So it could help us to make the specific plan for rebuilding different features. The first step is to construct auxiliary elements and extracting reference features. It is the basis for the follow-up step of extracting sectional and non-sectional curve features. In the step of extracting sectional curve features, sectional planes have been established and sectional curves can be extracted by using the microtomy method. Else, the different approximation algorithms could be used to fit curves in the extracted non-sectional curve features to achieve parametric curves.

3 The Parametric Reverse Modeling Based on Feature Extraction

3.1 The Workflow of the Reverse Modeling Based on Feature Extraction

According to the above section about the division of the main structural levels and feature analysis for prismatic shapes, it has formulated the feature-based modeling strategy for the whole model. And the overall workflow of the reverse modeling based on feature extraction for prismatic shapes can be seen in Fig. 2.

First of all, the preceding task is the extraction of the auxiliary and referential features, and the most important issues are the extraction of symmetry planes, orientation of extrude, axis of rotation and sectional planes, etc. Because these features are the reverse design basis for the ease of other downstream process and can help us to extract the sectional feature or reconstruct the surface model. Also it is the first step of capturing the design intent or knowledge of the original physical object. Surfaces of the model can be reconstructed by using three distinct techniques: feature extraction, surface fitting, and networking of curves. These techniques can be used in combination or individually in reconstructing the surfaces of sample part [9]. For prismatic shapes, feature extraction usually works the best and to create section curves. When dealing with free-form shapes and non-sectional curves, fitting NURBS surface patches is used primarily.

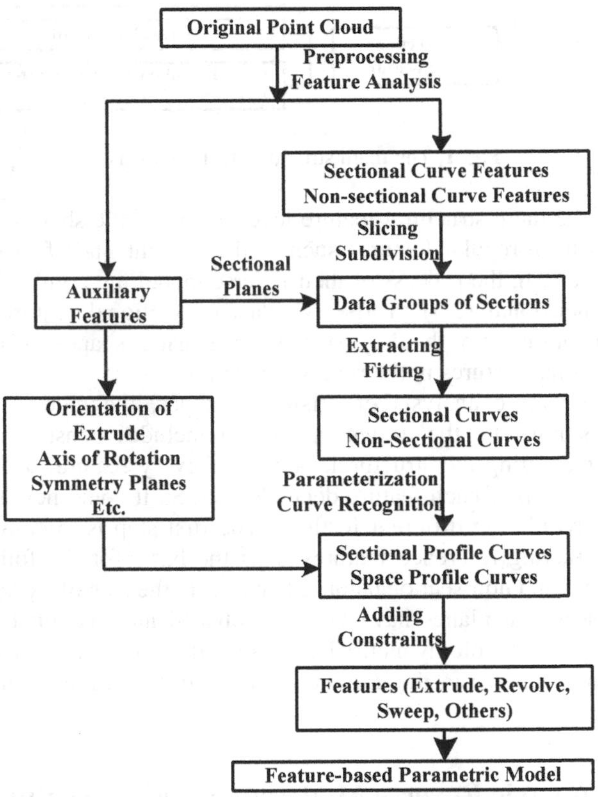

Fig. 2. The workflow of reverse modeling based on feature extraction

3.2 The Extraction and Reconstruction of Sectional Curve Features

This phase of the reverse-engineering process is suitable for prismatic shapes where the sectional curve feature can be extracted and parameterized. For the feature-based parametric reverse modeling, all the extracted sectional curves are constrained geometrically and dimensionally to be used as profile sketches for parametric sectional feature creation, such as extrude, revolve, sweep, loft, etc., and perform Boolean operations as well. In this process, a history of construction can be built simultaneously and used for modifications [9]. Then various complex parametric surfaces can be reconstructed according to the sectional curve feature and some required auxiliary elements through the construction operation. Every extracted curve represents a profile for feature creation, and the feature-based parametric solid model can be reconstructed after the extraction of feature curves. So the extraction of sectional curve features is the key factor for the parametric model reconstruction.

The general workflow of extracting the sectional curve feature is shown as Fig. 2, and an example is given below. As seen in Fig. 3, this is a part of surface of the extrude feature. The first step is to acquire the orientation of extrude and get the sectional sweep line by using the method of extracting the auxiliary features for the point cloud. In the sectional profile of this feature, there included the simple and obvious curves such as lines, circular arcs, ellipse. The following work is to extract these curves, which include ascertaining the join-points and the types of curves, judging the location of points through the fitted error. After fitting sectional curves, the constraints between curves should be added manually, as shown in Fig.4, and it could add varied dimension information in the section curve feature.

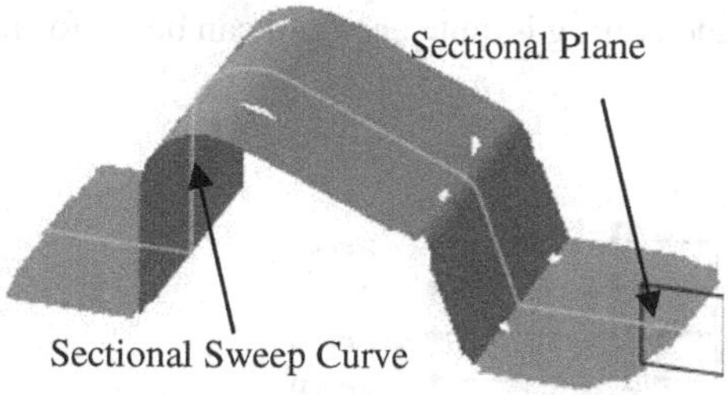

Fig. 3. The sectional curve

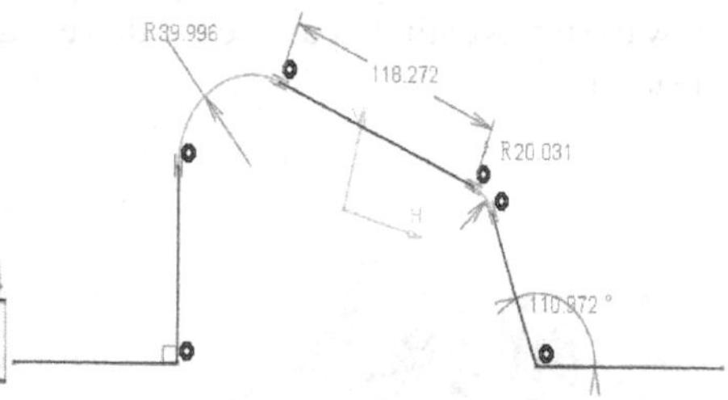

Fig. 4. The parameters and constraints

After extracting sectional curve features, the acquired sectional profile curves should be parameterized and adjustable. With these curve features, three-dimensional (3D) parametric surface feature could be constructed according to different surfacing functions in CAD software. Once the surface reconstruction is accomplished, modeling operations including surface extension, trimming, sewing, and thickening could be performed to form solid model for downstream operations such as CAE analysis and RP. Since achieving the parametric model, rapid redesign can be performed easily by changing the shape parameters of sectional curve feature.

4 The Parametric Reverse Redesign

The final objective of RE is to redesign or innovative design of products, and it is the soul of RE. In this section, we propose to use these extracted shape parameters from the scanned data to perform reverse redesign. The extracted features and construction history store the parametric information of CAD models acquired from the feature-based reverse modeling. In the process of redesign CAD models through different features, we can directly change and adjust the parametric information to acquire the new digital model satisfied our requirements.

As an example to illustrate the application of the proposed method, we performed the reverse engineering of a bearing seat shown in Fig. 5. By following the forward processes, we can achieve the feature-based parametric reverse model. The achieved model has various sectional features, and it is very important to ensure that different features were connected reasonably. Especially, the edges of all the adjacent features must be kept consistent with their geometric continuities.

However, during the product optimization, where the design changes are mostly iterative and based on engineering analysis data, it is essential to have a history-based parametric model. The accompanying parametric history tree shows the complete feature-based parameterization of the bearing seat, as represented in Fig. 6, and the finished parametric model was shown in Fig. 7 with parametric information. The profile shape and structure of the bearing seat can be easily redesigned by modifying the stored parameters of the extracted features. For instance, an example of making design variation is shown by editing the parameters of the bearing seat. With the consideration of the endurance and the stress, it must increase the thickness of the fillet size between the belly plate and the bottom plate, and the corresponding modification was made according to the design requirement as shown in Fig. 8. Notice that here the new design is entirely parametric driven, and editing is quite easy and can be performed instantly.

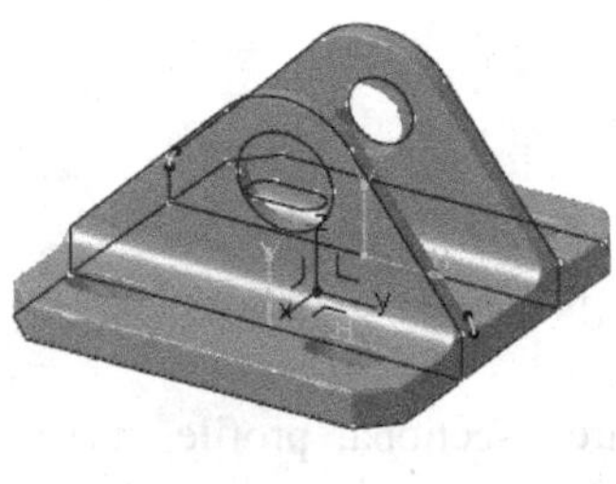

a) Extracted features

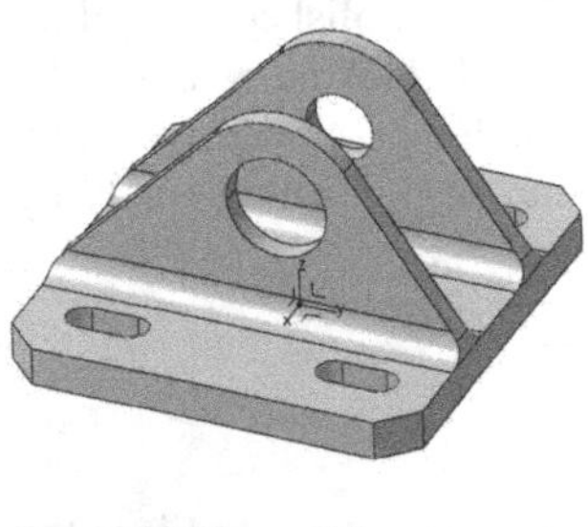

b) Created features

Fig. 5. Parametric feature model

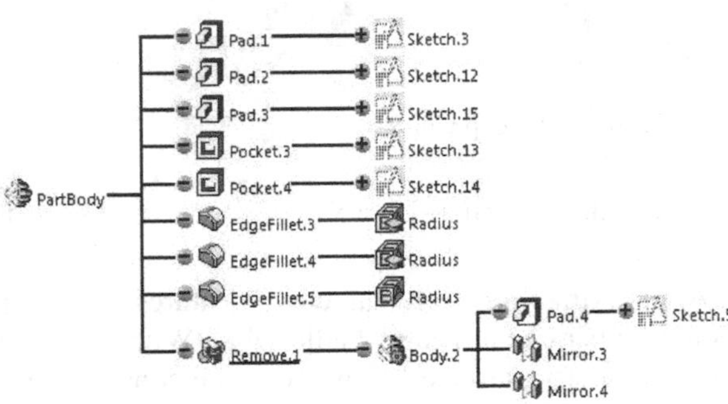

Fig. 6. Parametric history tree of features

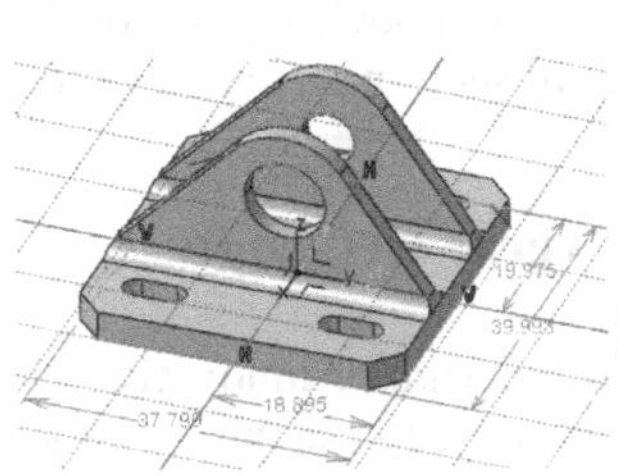

Fig. 7. Parametric information of the model

Fig. 8. Modified parameters

5 Conclusions

In this paper what we focus on is the method of reverse engineering for prismatic shapes. Furthermore, a method to create a parametric reverse modeling and redesign of prismatic shapes has been presented. This approach also has important advantages such as a relational design with associative parameters that would respond to changes quickly, and the profile sketches can be portable to any parametric CAD package. The core of the proposed method is to reconstruct a feature-based parametric shape from the scanned data. The parameter of the digital model can be used for driving the changes of the product model to form new designs. At last, the application example of a bearing seat has been presented to testify the feasibility of this method.

Fundamentally, it is merely enough to reconstruct the prismatic shape based on the feature extraction, especially for the model having the obvious structure levels. Therefore, further researches should be exploited in the parametric reconstruction of free-form shapes. And to improve the accuracy and quality of the shape, the distance analysis and smoothness evaluation can be performed during the process. Feedback from these analysis and evaluations will be used to improve the digital model through model parameters adjustment or local shape deformation.

Acknowledgements. This work was supported by the National Natural Science Foundation of China (No. 51105078, 50805025), the Guangdong province and Ministry of Education Industry- University- Research integration project (2011A091000040), and the higher education and research institutions science and technology project of Dongguan (201010810205).

References

1. Li, B.: The Research of CAD Modeling and Redesign Technology in Reverse Engineering. Guangdong University of Technology (2011) (in Chinese)
2. Fu, P.: Revisiting reverse engineering. Manuf. Eng. 134(4), 16–17 (2005)
3. Fu, P.: RE in the auto industry. Time-Compress Technol. 12(2), 2–4 (2004)
4. Tamas, V., Ralph, R.M., Jordan, C.: Reverse engineering of geometric models: an introduction. Computer-Aided Design 29(4), 255–268 (1997)
5. Zhu, G.-S.: The Research of Some Key Technologies in Reverse Engineering System Based on the Feature Modeling. Nanchang University (2009) (in Chinese)
6. Liu, Y.-F., Ke, Y.-L., Wang, Q.-C., Hu, X.-D., Peng, W.: Research on Reverse Engineering Technology Based on Features. Computer Integrated Manufacturing Systems 12(01), 32–37 (2006) (in Chinese)
7. Shan, D.-R.: Research on Parametric Modeling in Reverse Engineering Based on Features and Constraints. Mechanical Science and Technology 24(5), 522–525 (2005) (in Chinese)
8. Yea, X., Liua, H., et al.: Reverse innovative design — an integrated product design methodology. Computer-Aided Design 40, 812–827 (2008)
9. Soni, K., Chen, D., Lerch, T.: Parameterization of prismatic shapes and reconstruction of free-form shapes in reverse engineering. Int. J. Adv. Manuf. Technol. 41, 948–959 (2009)

5 Conclusions

In this paper, we [illegible] is the method of reverse engineering for prismatic shapes [illegible] parametric [illegible] modeling and redesign of prismatic shapes [illegible]. The approach has important advantages such as [illegible] design [illegible] that would [illegible] to changes [illegible] and [illegible] to a parametric CAD package. The [illegible]

[illegible] and the [illegible]
[illegible] model [illegible]
[illegible] presented [illegible]

Fundamentally, [illegible]
[illegible]
[illegible] explored [illegible]
[illegible] shape [illegible]
[illegible] evaluation can [illegible] process. Feedback
[illegible] and evaluations will be used to [illegible] the digital model through
[illegible] operations.

Acknowledgement. [illegible] National Natural Science Foundation [illegible] Guangdong province and [illegible] integration project (2011 [illegible]) and [illegible] institutions [illegible] and [illegible]

References

1. [illegible]
[illegible]
2. [illegible] 2005)
3. [illegible] 4(2 [illegible]
4. [illegible]
[illegible]
5. [illegible]
[illegible]
6. [illegible]
[illegible]
[illegible]
7. [illegible] based on [illegible]
and [illegible] 22 [illegible]
8. [illegible] product design
[illegible]
9. Sen[illegible] and [illegible]

Values Analysis of New Modular Multi-level Converter Components

Yan Xie[1,2], Hong Xie[1], and Zhe Mao[1]

[1] School of Electrical and Electronic Engineering, Wuhan Polytechnic University, Wuhan, Hubei, China
[2] School of Electrical Engineering, Wuhan University, Wuhan, Hubei, China
limimimm@163.com

Abstract. This paper introduced a new modular multi-level converter (MMC), which could enhance the voltage and power level by sub-converter modules in series and was easy to extend to any level of output. Its structure and working mechanism were described. The values of four main components were analyzed which were DC power supply, reactor L, sub-module's power switch and capacitor C, and corresponding affecting factors were discussed. Their value is critical to the reliable operation of the MMC system. The conclusions of the analysis given have some practical significance for studying on MMC system.

Keywords: modular multi-level converter, value analysis, current limit, power switch, parameter design.

1 Introduction

With the development of power electronics technology, a variety of high-power switching devices was used widely. In the high-voltage high-power application fields, the demand for advanced power electronic devices has become increasingly urgent, such as in the field of power transmission and distribution, voltage source converter(VSC) has be widely used. VSC can have various forms of topology, and now commonly used the two-level VSC and three-level VSC in practice. There are the main problems for two-level VSC , which are static voltage balance, dynamic voltage balance and electromagnetic interference (EMI) caused by the insulated gate bipolar transistor (IGBT) series, and excessive switching losses caused by the high switching frequency. For multi-level voltage source converter, common topologies include diode clamped, capacitor clamped and cascaded H-bridge type.

From the point of industrial production and demand for services, it is essential that the main inverter circuit is strictly modular structure. It requires that VSC main circuit can have a number of the same structure sub-modules, no additional such as positive and negative DC bus and other components, the VSC voltage and power requirements of different can be used to meet by varying the number of sub-modules, and VSC sub-module should use the same hardware and mechanical construction and can be widely used. Diode-clamped and capacitor-clamped in practice is few to get multi-level VSC level because that they are difficult to modular production.

D. Jin and S. Jin (Eds.): Advances in FCCS, Vol. 1, AISC 159, pp. 235–240.
springerlink.com

In 2002, a new topology of modular multi-level VSC is proposed by Germany Bundeswehr Munich University. The new modular multi-level VSC (MMC) satisfies the modular functional requirements, and it will capacitance and switching device as a whole to build a sub-module, to enhance the voltage and power level by sub-converter modules in series, easy to extend to any level of output, and it can be used to reduce switching loss and improve transmission efficiency with low harmonic distortion and lower switching frequency.

With all the above its advantages, the study on MMC is become a hot research. This paper describes the structure and working mechanism of the MMC, analyze affecting factors and the values of its components, and give the corresponding analysis results.

2 Structure and Working Principle of MMC

Three-phase MMC converter structure is shown in Figure 1. It has a modular structure. Each phase includes upper and lower bridge, and each bridge consists of n sub-modules and current limiting reactor L in series. U_{dc} is the DC side voltage. In order to maintain a constant DC output voltage, at any time into each phase of MMC is equal to the total number of sub-modules. So each phase of the MMC sub-module has 2n, the number of output level is n+1.

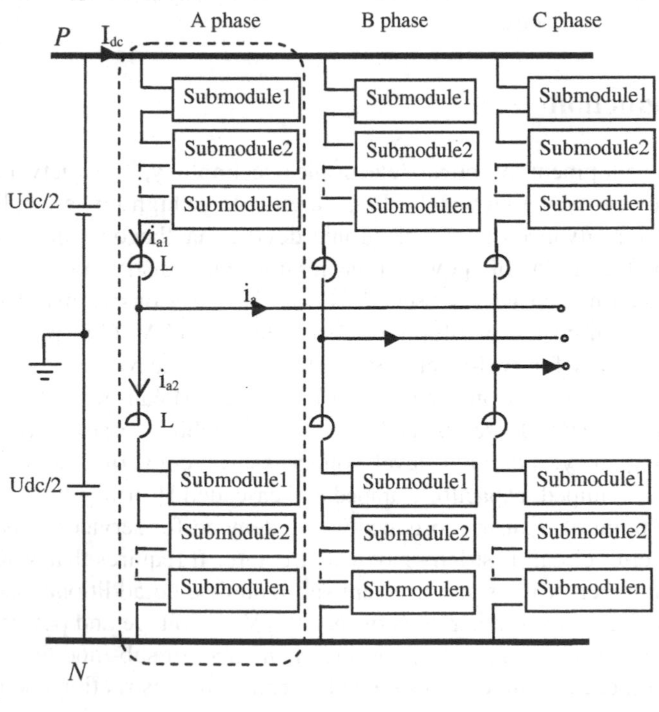

Fig. 1. It is the circuit structure of MMC system.

The sub-module structure is shown in Figure 2. Each sub-module structure is the same. It includes: T1 and T2 (insulated gate bipolar transistors, IGBT), D1 and D2 (diodes for the continued flow) and C (the sub-module capacitor). u_C is the voltage of the sub-module capacitor C.

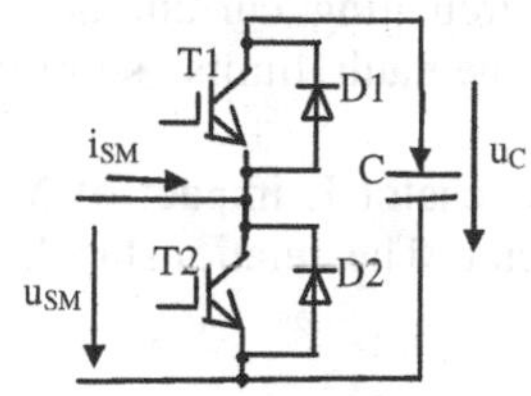

Fig. 2. It is the structure of MMC sub-module.

By controlling the switch IGBT T1 and T2, it can make the capacitor C into or removed from the bridge, and the corresponding output voltage u_{SM} is equal to u_c or 0. The work state of new modular multi-level converter (MMC) is changed by controlling its sub-modules (Fig.1). According to current direction of the bridge arm, switches T1 and T2 of sub-module are controlled on or off, and the capacitor C is achieve to be charged, discharged, or bypass, so the sub-module output voltage u_{sm} varies between zero and u_c. When the operating switches T1 and T2, it is equivalent to a DC power input or removal from the bridge. Through balancing the sub-module's capacitor voltage, it is considered the capacitor voltage as the voltage source u_c.

At different T1, T2, D1 and D2 switch state, the corresponding sub-module output voltage u_{sm} and the corresponding capacitance C state is shown in Table 1.

Table 1. Sub-module's work modality is changed according to power unit switch TI, T2, D1, and D2 state.

Model	i_a	T1	D1	T2	D2	u_{sm}	Capacitor C
1	>0	Off	On	Off	Off	u_c	Charging
2	>0	Off	Off	On	Off	0	-
3	<0	On	Off	Off	Off	u_c	Discharging
4	<0	Off	Off	Off	On	0	-

3 Values Analysis of MMC Components

MMC main components include DC power supply, reactor L, sub-module's power switch (T1, T2, D1, D2), and capacitor C (Fig.1 and Fig.2). To ensure MMC system reliable operation, the values of its components is affected by some conditions. The following is analysis and discussion of their values.

3.1 DC Power Supply

The value of DC power supply of MMC is selected according to system operating parameters working in rectifier or inverter state. DC bus voltage determines the total voltage of working sub-module capacitors in series.

3.2 Reactor L

Reactor L is in series in the MMC phase between the upper and lower arm, which is significantly different with the two-level topology. Its value is an important effective factor for MMC system. Its main functions are: output power limits on normal working state, two frequency circulating current between units constraints, short-circuit current of DC pole-to-pole fault limits, short-circuit current of AC output three-phase fault limits.

We knows that the values of reactor L impact of MMC system power output by analyzing MMC equivalent circuit. The relationship between them can be given in Equation 1.

$$P_f = \frac{3k^2 U_{dc}^2 R_f}{8[R_f^2 + \omega^2 (L_f + L/2)^2]} \tag{1}$$

Where P_f is output power of MMC system, R_f and L_f is resistance and inductance of the load, k is voltage modulation ratio. Equation 1 shows that when k and the load is constant, if the series reactor L increases, the phase output current decreases and P_f is also reduced, that is, L decreased will lead to increased output power P_f .

Equation 2 shows the values of reactor L can limit two frequency circulating current between units.

$$I_{2f} = \frac{S}{3}\left(\frac{1}{8\omega^2 CL\overline{U}_C - U_{dc}}\right) \tag{2}$$

Where I_{2f} is two frequency circulating current between units, S is total power of MMC system. Increasing the values of reactor L, the current I_{2f} will be limited.

Equation 3 shows the values of reactor L can limit short-circuit current of DC pole-to-pole fault.

$$i_1 = \left[\sqrt{\frac{C}{nL}} U_{dc} \sin(\omega_1 t) + I_1 \cos(\omega_1 t)\right] e^{-\frac{t}{\tau}} \tag{3}$$

Where i_1 is short-circuit current of DC pole-to-pole fault. Increasing the values of reactor L, the current i_1 will be limited.

Equation 4 shows the values of reactor L can limit short-circuit current peak value of AC output three-phase fault.

$$I_{peak} = \frac{kU_{dc}}{\sqrt{R_s^2 + (\omega L)^2}} + I_1' \tag{4}$$

Where I_{peak} is short-circuit current peak value of AC output three-phase fault. Increasing the values of reactor L, the current I_{peak} will be limited.

As can be seen from the above that the values of reactor L should be selected based on limiting the selection of three minimum conditions, and be checked based on the power transfer characteristic.

3.3 Power Switch (T1, T2, D1, D2)

By controlling the power switch T1, T2, D1 and D2, MMC system is in reliable work state. Their operating voltage is determined by the DC bus voltage and MMC number of level, and the current is determined from the output power. Another important parameter for the power switches is the switching frequency, which determines the power loss of the MMC and is related to the control style of the MMC system.

3.4 Capacitor C

The capacitor C is one of important factors to determine the total cost and area size of the MMC. Capacitor C parameter design will directly affect the economy of MMC system.

From the operation mechanism of the MMC we can see that the voltage of the capacitor C is determined by the DC bus voltage and MMC number of level, the current of the capacitor C is same as the flow capacity of power switches. The capacitance value of the capacitor C can be obtained according to the principle of energy pulsating as follows:

$$C = \frac{\Delta W_{SM}}{2n\varepsilon \overline{U}_C^2} \tag{5}$$

Where ΔW_{SM} is the pulse energy of sub-module, ε is the voltage ripple coefficient of sub-module, $\overline{U}_C$ is the rated voltage of the capacitor C.

4 Conclusion

In this paper, thc structure and working mechanism of MMC system is introduced. Its four main components are DC power supply, reactor L, sub-module's power switch and capacitor C. Their value is critical to the reliable operation of the MMC system. It is analyzed based on discussing the corresponding affecting factors.

The conclusions of the analysis given have some practical significance for studying on MMC system.

References

1. Lin, P., Wang, L., Li, J., et al.: Research on cascade multi-level converters with sample time staggered SVM and its application to APF. Proceedings of the CSEE 25(8), 70–74 (2005) (in Chinese)
2. Zheng, C., Zhou, X.: Small signal dynamic modeling and damping controller designing for VSC based HVDC. Proceedings of the CSEE 26(2), 7–12 (2006) (in Chinese)

3. Zheng, C., Zhou, X., Li, R., et al.: Study on the steady characteristic and algorithm of power flow for VSC-HVDC. Proceedings of the CSEE 25(6), 1–5 (2005) (in Chinese)
4. Lai, J.S., Peng, F.: Multilevel converter: a new breed of power converters. IEEE Trans. on Industrial Applications 32(3), 509–517 (1996)
5. Rodriguez, J., Lai, J.S., Peng, F.: Multilevel inverters: a survey of topologies, control, and applications. IEEE Trans. on Industrial Electronics 49(4), 724–738 (2002)
6. Wang, G.: Mechanism of DC bus voltage unbalance in diode-clamped multilevel inverters. Proceedings of the CSEE 22(12), 111–117 (2002) (in Chinese)
7. Lin, L., Zou, Y., Zhong, H., et al.: Study of control system of diode-clamped three-level inverter. Proceedings of the CSEE 25(15), 33–39 (2005) (in Chinese)
8. Song, Q., Liu, W., Yan, G., et al.: A neutral-point potential balancing algorithm for three-level NPC inverters by using analytically injected zero-sequence voltage. Proceedings of the CSEE 24(5), 57–62 (2004)
9. Wu, H., He, X.: Research on PWM control of cascade multilevel converter. Proceedings of the CSEE 21(8), 42–46 (2001) (in Chinese)
10. Ding, G.-J., Ding, M., Tang, G.-F., He, Z.-Y.: Submodule capacitance parameter and voltage balancing scheme ofa new multilevel VSC modular. Proceedings of the CSEE 29(30), 1–6 (2009) (in Chinese)
11. Pan, W.-L., Xu, Z., Zhang, J., Wang, C.: Dissipation analysis of VSC-HVDC converter. Proceedings of the CSEE 28(21), 7–14 (2008) (in Chinese)

A Slope Stability Analysis Method Combined with Limit Equilibrium and Finite Element Simulation

Jiawen Zhou[1,*], Junye Deng[2], and Fugang Xu[1]

[1] State Key Laboratory of Hydraulics and Mountain River Engineering, Sichuan University, Chengdu 610065, PR China
[2] Armed Police Hydropower Troops, Xinyu 338029, PR China
jwzhou@scu.edu.cn

Abstract. This paper presents a slope stability analysis method combined with the limit equilibrium and finite element simulation. Based on the simulated stress of a slope, the safety factory of one element can computed with Mohr-Coulomb criterion, and introduces the Dijkstra algorithm of graph theory to search the dangerous slide surface combined with limit equilibrium method. For an examine slope, the simulated results of safety factor and location of dangerous sliding surface is close to the strength reduction method. Sensitivity analysis results show that, the safety factor of slope is decreased with the increasing height and inclination of slope, but decreased with the decreasing cohesion and friction angle.

Keywords: slope stability, safety factor, limit equilibrium, finite element method, Dijkstra algorithm.

1 Introduction

Landslide is a typical geological disaster in the worldwide, and causes a huge threat to the human life and property [1-2]. The landslide will happen to a slope because of earthquake, rainfall and manmade excavation [3]. Therefore, it is very important to determine the dangerous sliding surface and safety factor of slope. Currently, the slope stability analysis methods include: Geo-mechanical method (such as stereographic projection), the limit equilibrium method, finite element method, discrete element method, Lagrange differential method [4-7]. These slope stability analysis methods based on mechanics theory cannot consider the internal stress-strain relationship of soil or rock mass, the failure process of the slope can not be determined [8]. But the numerical simulation methods can consider the stress and strain status of slope, so that combined with the limit equilibrium method can solve the slope stability problem more effectively [9].

In this paper, based on the simulated stress of slope through finite element method, the safety factor of one element is computed base on Mohr-Coulomb criterion, and then introduced the shortest path problem in Dijkstra algorithm of the graph theory search the dangerous sliding surface, combined with the limit equilibrium method, the safety factor of slope can be determined.

* Corresponding author.

D. Jin and S. Jin (Eds.): Advances in FCCS, Vol. 1, AISC 159, pp. 241–247.
springerlink.com

2 Slope Stability Analysis Based on Finite Element Method

In this section, two slope stability analysis methods base on finite element analysis are introduced, strength reduction and improved Dijkstra method. The improved Dijkstra method is combined with the simulated stress of slope and limits equilibrium analysis, and introduce Dijkstra algorithm to search the dangerous sliding surface.

2.1 Strength Reduction Method

Strength reduction is a slope stability method based on finite element analysis, with the decreasing of shear strength for slope materials (rock or soil), the safety factor and dangerous sliding surface can be determined.

Safety Factor of One Element. In the numerical simulation process of a slope, the principal stress of one element can determined, and the resistance shear strength can computed base on Mohr-Coulomb criterion,

$$\tau_f = c + \sigma \tan\phi \tag{1}$$

where σ is the normal stress; c is the cohesion; ϕ is the friction angle.

Fig. 1 shows the stress state of one element based on Mohr-Coulomb criterion.

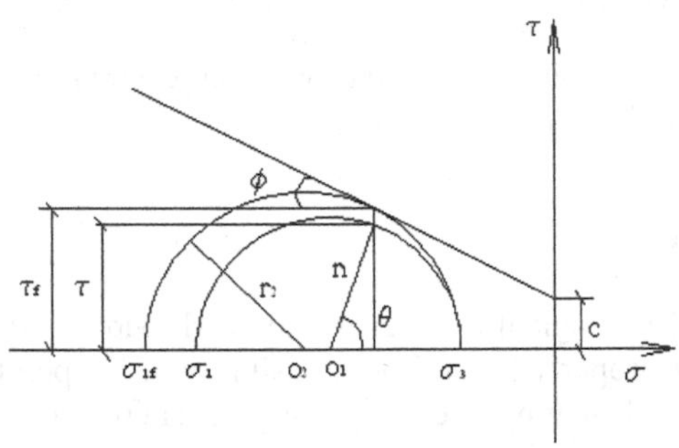

Fig. 1. Stress state of one element based on Mohr-Coulomb criterion.

As shown in Fig. 1, the normal stress σ and shear strength τ can computed as follow,

$$\sigma = p - q\cos\theta, \quad \tau = -q\sin\theta \tag{2}$$

where $p=(\sigma_1+\sigma_3)/2$; $q=(\sigma_1-\sigma_3)/2$; σ_1 is the first principal stress; σ_3 is the third principal stress.

Then the safety factor of element can computed base on limit equilibrium method,

$$F_s = \tau_f / \tau = [-(p - q\cos\theta)\tan\phi + c]/(-q\sin\theta) \tag{3}$$

Aimed to solve the minimum safety factor of slope, derivation is carried out for the Eq. (3), when

$$\theta = \arccos[(-q \tan \phi)/(c - p \tan \phi)] \tag{4}$$

The safety factor F_s can get minimum value.

Slope Failure Criterion. Strength reduction method is fulfilled through decreasing the shear strength of slope materials step by step, the relationship of shear strength and safety factor of slope is as follow,

$$c' = c/F_s\,, \quad \phi' = \arctan(\tan \phi / F_s) \tag{5}$$

In the numerical simulation process, the initial value of safety factor is 1.0, through a series of reduction for shear strength until the slope failure is happened, and the reduction factor is the minimum safety factor of slope. In the strength reduction process, the mechanical parameters used for numerical simulation are renewed by the Eq. (5). Several tests are carried out to find out the reasonable reduction factor for the shear strength, a slope failure criterion should be certain in the iterative calculation process. In this paper, an error of unbalanced force ε is used to make certain the slope is failure or not, if the calculation unbalanced force μ is less then ε, the compute results is stable, if not, the calculation should be carried out until the $\mu<\varepsilon$.

2.2 Improved Dijkstra Method Based on Simulated Results

Improved Dijkstra method is combined with the simulated stress results and limits equilibrium method, and introduce the Dijkstra algorithm to search the dangerous sliding surface. The Dijkstra algorithm is a method to solve the short path problem in the graph theory, and in this paper we use it to solve the dangerous sliding surface problem in slope stability analysis process [10].

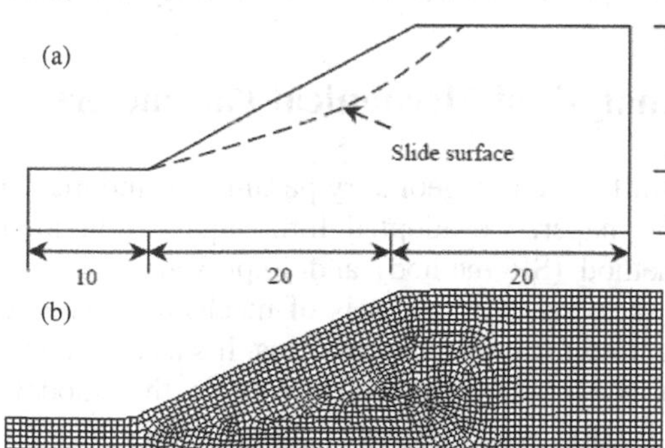

Fig. 2. Examine slope for stability analysis method: (a) geometry and (b) simulated grid.

In this paper, the thought, theory and calculation process are not described in this paper, reference [10] shows this method in details.

In the calculation process, the grid information, element information should be determined before the search of dangerous sliding surface. The grid information includes: grid number, coordinate of grid, stress value of each grid and shear strength parameters. The element information includes: simulation element, sliding boundary.

Here we take a typical examine slope to verification the improved Dijkstra method. The examine slope is designed by Donald and Giam [7], which is used for develop the ACADS slope stability analysis procedure. Fig. 2(a) shows the geometry of this two-dimensional soil slope, and Fig. 2(b) shows the simulated grid of this slope. The mechanical parameters are: density is 20.0 kN/m^3, elastic modulus is 10 MPa, Poisson's ratio is 0.25, cohesion is 3.0 kPa, and friction angle is 19.6°. The theory analysis results of safety factor for this slope is 1.000. Then we adopted improved Dijkstra method to solve this slope stability problem. Fig. 3 shows the simulated result of dangerous sliding surface and its safety factor.

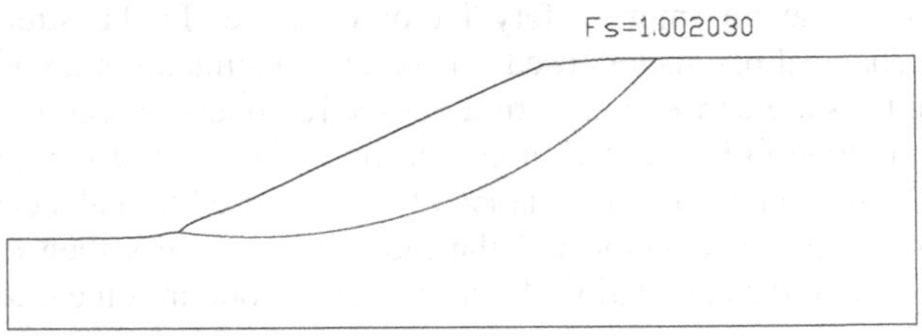

Fig. 3. Dangerous sliding surface and its safety factor of examine slope which is solved by improved Dijkstra method.

As shown in Fig. 3, the dangerous sliding surface solved by the improved Dijkstra method is close to the theory analysis result, and the minimum safety factor is 1.002, the recommend result is 1.000, the simulated error is very little, so that the improved Dijkstra method can be applied to solve the slope stability problem.

3 Sensitivity Analysis of Mechanical Parameters

The slope stability is influenced by geometry parameters and mechanical parameter of slope material. In this paper, we adopted limit equilibrium method (LE method), strength reduction method (SR method) and improved Dijkstra method (presented method) to carry out the sensitivity analysis of mechanical parameters. Fig. 4 shows the geometry of slope used for sensitivity analysis. It's a soil slope, only one layer, the seepage and external loading are not considered in this model. Four mechanical parameters are used for the sensitivity analysis of safety factor of slope, includes: slope height, slope angle, cohesion and friction angle of soil. According to one mechanical parameter, other parameters are fixed.

Table 1 show the mechanical parameters set of the soil slope. The slope height is range from 5 m to 20 m, slope angle is range from 25° to 55°, cohesion of soil is range from 10 kPa to 40 kPa, and friction angle of soil is range from 10° to 40°. And then the safety factor of this slope under different conditions can be computed.

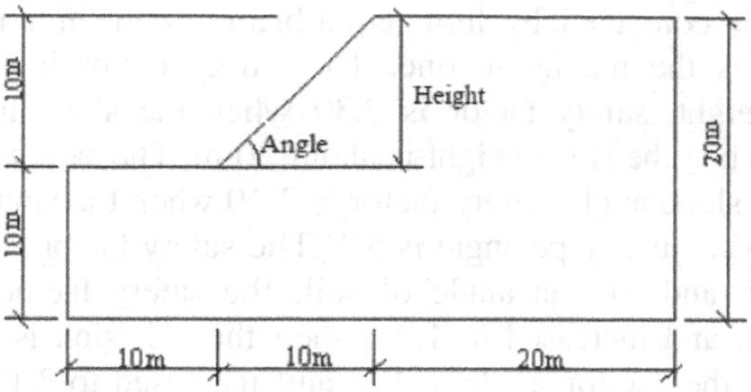

Fig. 4. Geometry of the slope used for sensitivity analysis.

Table 1. Mechanical parameters set of the soil slope.

Set	1	2	3	4	5
Slope Height(m)	5~20	10	10	10	10
Slope angle(°)	45	25~55	45	45	45
Cohesion (kPa)	25	25	10~40	25	25
Friction angle	20	20	20	10~40	20

Fig. 5 shows the sensitivity analysis results of mechanical parameters.

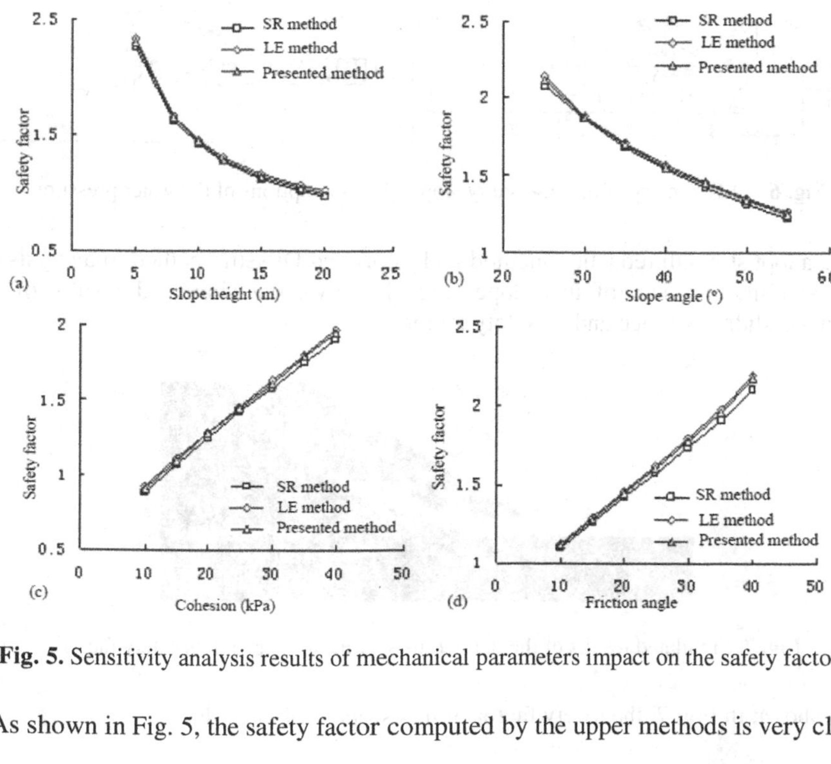

Fig. 5. Sensitivity analysis results of mechanical parameters impact on the safety factor.

As shown in Fig. 5, the safety factor computed by the upper methods is very close.

The safety factor computed by limit equilibrium is the minimum one, and the strength reduction is the maximum one. The safety factor is decreased with the increasing slope height, safety factor is 2.30 when the slope height is 5 m, and decreased to 1.00 when the slope height is about 20 m. The safety factor is decreased with the increasing slope angle, safety factor is 2.10 when the slope angle is 25°, and decreased to 1.24 when the slope angle is 55°. The safety factor is increased with the increasing cohesion and friction angle of soil, the safety factor is 0.90 when the cohesion is 10 kPa, and increased to 1.91 when the cohesion is 40 kPa; the safety factor is 1.11 when the friction angle is 10°, and increased to 2.12 when the friction angle is 40°.

4 Case Study

The case study slope is a two layers soil slope, Fig. 6(a) shows the geometry of this slope and Fig. 6(b) shows the assumptions of the water pressure. For the soil at the upper layer, the natural density is 14.7 kN/m^3, the saturated density is 17.64 kN/m^3, the cohesion is 10 kPa, and the friction angle is 20°. For the soil at the upper layer, the natural density is 19.6 kN/m^3, the saturated density is 22.54 kN/m^3, the cohesion is 10 kPa, and the friction angle is 25°.

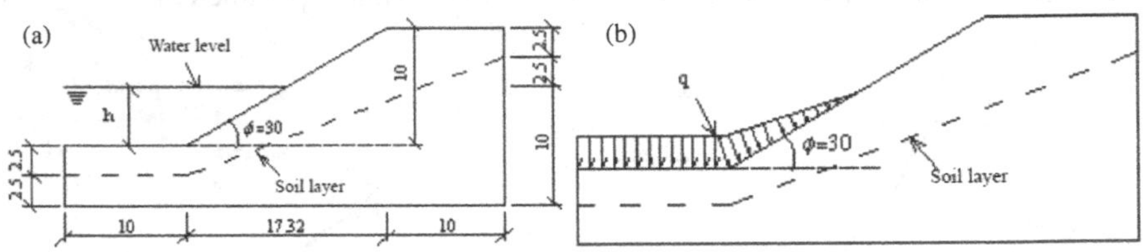

Fig. 6. (a) Geometry of the case study slope; (b) assumptions of the water pressure.

We adopt strength reduction method and improved Dijkstra method to analysis the slope stability problem of this slope. Fig. 7 shows the simulated results of the dangerous sliding surface and its safety factor.

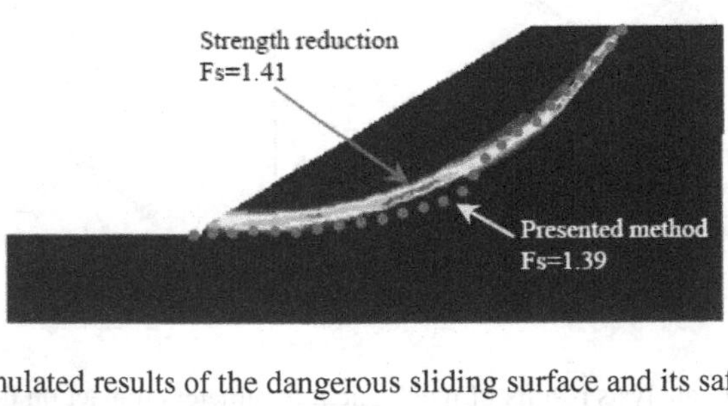

Fig. 7. Simulated results of the dangerous sliding surface and its safety factor.

As shown in Fig. 7, the safety factor of this slope is about 1.40.

The dangerous sliding surface computed by the improved Dijkstra method is very close to the strength reduction method, and the error of safety factor is also very small.

5 Conclusions

In this paper, a slope stability analysis method is presented base on the simulated stress of slope and the Dijkstra algorithm in graph theory. Through a verification of examine slope, the dangerous sliding surface and its safety factor is very close to the theory solution. The sensitivity analysis results show that, the safety factor of slope is decreased with the increasing slope height and angle, and increased with the increasing cohesion and friction angle of soil. Case study results show that the dangerous sliding surface and its safety factor is very close to the strength reduction method. The stress state of slope can be simulated by the finite element method, combined with the limit equilibrium method and introduce some optimization algorithm can solve the slope stability problem.

Acknowledgements. The support of Chinese National Natural Science Foundation (No. 41030742, 41102194) and China Postdoctoral Science Foundation (No. 20110491741) are gratefully acknowledged.

References

1. Chen, Z.Y.: Stability analysis of soil slope—Theory, Method and Programs. Chinese Water Press, Beijing (2003)
2. Kim, J.Y., Lee, S.R.: An improved search strategy for the critical slip surface using finite element stress fields. Computers and Geotechnics 21, 295–313 (1997)
3. Kentli, B., Topal, T.: Assessment of rock slope stability for a segment of the Ankara-Ppzantl motorway, Turkey. Engineering Geology 74, 73–90 (2004)
4. Jones, D.R.V., Dixon, N.: Landfill lining stability and integrity: the role of waste settlement. Geotextiles and Geomembranes 23, 27–53 (2005)
5. Bondy, J.A., Murty, U.S.R.: Graph Theory with Applications. The Macmillan Press Ltd., New York (1976)
6. Cherkassky, B,K., Goldberg, A.V., Radzik, T.: Shortest Paths Algorithms: Theory and Experimental Evaluation. Technical Report, pp. 93–1480, Computer Science Department, Stanford University (1993)
7. Bandini, P.: Numerical limit analysis for slope stability and bearing capacity calculations. Purdue University, America (2003)
8. Chen, Z.Y.: A generalized solution for tetrahedral rock wedge stability analysis. International Journal of Rock Mechanics and Mining Sciences 41, 613–628 (2004)
9. Zhou, J.W., Xu, W.Y., Yang, X.G., Shi, C., Yang, Z.H.: The landslide and analysis of the stability of the current Huashiban slope at the Liangjiaren Hydropower Station, Southwest China. Engineering Geology 114, 45–56 (2010)
10. Xu, W.Y., Zhou, J.W., Deng, J.Y., Shi, C., Zhang, Z.L.: Slope stability analysis of limit equilibrium finite element method based on the Dijkstra algorithm. Chinese Journal of Geotechnical Engineering 29, 1159–1172 (2007)

An Empirical Study of Correlation between Computer-Assisted College English Autonomous Learning and Learning Styles

Lin Yu and Yongjun Huang

Hubei University of Technology, 430068 Wuhan, Hubei, People Republic of China
Catherine_yulin@126.com

Abstract. The empirical correlation study shows that almost all types of learning styles have, more or less, positive or negative correlation with the academic achievement of computer-assisted English autonomous Learning. Only tactile learning style has no any correlations with language achievement at the significant level. This result implies that the tactile learning potential of college students is seriously ignored in computer-assisted English autonomous Learning. Based on the research findings, this essay discusses the possible reasons and gives some suggestions to English learners, English teachers, the university and technical workers.

Keywords: CALL, learning style, autonomous learning, learning effect, correlation.

1 Introduction

With the constant deepening of the teaching reform, College English Teaching Curriculum Requirements (For Trial Implementation in 2004) clearly points out that the new teaching mode should be based on modern information technology, and advocates autonomous learning and individual-characterized learning.

According to Wikipedia, computer-assisted language learning (CALL), a form of computer-based accelerated learning, originating in 1960s, just meets such requirements, which carries two important features: bidirectional learning and individualized learning. It is not a method but a student-centered accelerated learning material, which promotes self-paced accelerated learning. On the other hand, the concept of "autonomous learning" has been long a hot issue in the field of linguistics both at home and abroad, ever since its birth in the 1980s. The integrating of CALL and autonomy is universally applied in the individual learning of College English, and its learning effect depends on individual difference, in which learning style is one of the affecting factors, the individual's stable learning preference made gradually with many affecting factors during a long period. Few people have been engaged into the correlation study between learning style and computer-assisted English autonomous Learning, a new learning way and environment. This study aims to find out the possible correlation between these two elements, to enhance the effectiveness of computer-assisted English autonomous learning and increase learners' ESL language learning proficiency.

D. Jin and S. Jin (Eds.): Advances in FCCS, Vol. 1, AISC 159, pp. 249–256.
springerlink.com

2 Frameworks

2.1 CALL (Computer-Assisted Language Learning)

CALL was first conceived in the1950. In this 50 year period of development, CALL has experienced , according to Warschauer(1996) , three phases : Behaviorist CALL, Communicative CALL and Integrative CALL(Multimedia and the Internet). Each phase reflects a certain level of technology and corresponds to certain pedagogical theories. Until now, CALL activities has no longer been limited to interaction with the computer and with other students in the class, but included communication with learners in other parts of the world——either learners from specific classes chosen by instructors or self-selected participants who choose to spend time in computer-mediated communication for language learning. (Paramskas, 1993).

2.2 Autonomous Learning

For a definition of autonomous learning or autonomy, the most widely quoted and highly honored is Holec's. Holec defines autonomy as "the ability to take charge of one's own learning which is to have and to hold the responsibility for all the decisions concerning all aspects of learning: i.e. determining the objectives; defining the contents and progressions; selecting methods and techniques to be used; monitoring the procedure of acquisition properly speaking; and evaluating what has been acquired." He also explains that this ability has "a potential ability to act in a given situation—in our case learning—and not the actual behavior of an individual in that situation". So, for him leaner autonomy is not an action but a capacity that potentially exists in a person.

2.3 Learning Styles

The term Learning Style was first coined by Herbert Thelen in 1954(Liu Runqing and Dai Manchun, 2003). On the basis of previous researches, Joy Reid (1987:87-103) distinguishes six kinds of learning styles in perceptual way: visual, auditory, tactile, kinesthetic, group and individual. She also provides the"*Perceptual Learning Style Preference Survey*".Here we define learning style in this article as an individual's natural, habitual, and preferred perceptual ways of learning new information and skills. According to Reid(2002), every learner has all of these learning styles, but they function differently. Those learning styles function best for a learner are called major learning styles.

3 Study

3.1 Research Questions

As far as EFL learning styles of Chinese learners are concerned, several important investigations have been conducted by Reid (1987), Melton (1990), and Yu Xinle

(1997),but most of the participants are relatively advanced English learners. Therefore, the results of these investigations have limitations to some extent. In this essay, the author wants to find out the answers to the following different questions:

1) What is the general learning style for college students in the computer-assisted English autonomous learning?

2) Is there correlation between college students' learning styles and the effect of computer-assisted English autonomous Learning?

3) If so, how do different types of learning styles influence the effect of computer-assisted English autonomous Learning?

3.2 Participants

About 78 freshmen in Hubei University of Technology are chosen as the participants of this research, who come from 2 typical academic majors: 1) Mechanical Designing and Automation, 2) Chinese as a Foreign Language. They have got basic knowledge of English language and have never been exposed to computer-assisted English autonomous Learning before. They all have taken the College Entrance Examination (full mark: 150), which shows their English level is similar: mean(106), and Std. Deviation(4.6). When the effect test of the computer-assisted English autonomous Learning is administered, all the participants will have studied English in the school's Autonomous Learning Center for half an academic year.

3.3 Instruments

There are four instruments in this study. The first one is The Perceptual Learning Style Preference Questionnaire (Reid, 1984) is specially designed to identify adult ESL students' perceptual style preferences, which covers four perceptual(visual, auditory, tactile, and kinesthetic) and two social (group and individual) learning style preferences. The second instrument is the test paper of the Final-term Examination, which is designed by a group of College English teachers in HUT. The test paper has already been tested secretly in 10 classes and it has turned out that the paper has comparatively high reliability and validity. The third instrument is Statistical Package for the Social Sciences (SPSS 10.0), which is used to make an analysis of the data. The fourth instrument is the face-to-face interview between the author and those participants.

3.4 Data Collection and Analysis

The Perceptual Learning Style Preference Questionnaire (Chinese version), is administered at the beginning of the first semester of 2011-2012 academic years. Altogether 78 questionnaires are administered, the same number returned, and all of them are valid. The Final-term Examination (in the form of CET-4), is conducted at the end of the first semester of 2011-2012 academic years.

The questionnaire data collected are input and processed by the Statistical Package for the Social Sciences (SPSS 10.0) to do analysis work. Firstly, descriptive statistic, including means and standard deviations, are computed to summarize the students' responses to each style category and to determine the style preference which is used at high frequency. Secondly, Pearson correlation coefficient is used to indicate the correlation between different types of students and each item of their English achievement in computer-assisted English autonomous Learning.Thirdly, based on the results of this study, the face-to-face interview is used to find out why such results are formed.

4 Results and Discussions

4.1 The General Situation of All Participants' Learning Styles

The fact showed that in general, students seldom tend to the one extreme or the other in terms of their style preferences, which may result in the fact that the differences obtained are not so apparent as it has been expected. However, the variance could still be found.

Table 1. Descriptive Statistics of All Participants' Learning Styles

Learning Styles	Sample Number	Minimum Score	Maximum Score	Mean Score	Std. Deviation	Ranks
Visual	78	22	46	41.67	5.28	3
Auditory	78	16	44	34.40	5.83	5
Kinestheti c	78	20	48	41.68	5.89	2
Tactile	78	20	48	42.13	5.25	1
Group	78	14	48	32.67	6.39	4
Individual	78	18	50	30.33	6.92	6

The result in Table 1 shows that the students in the sample scored mostly above the average score 25. Among those variables, tactile learning style is the most preferred (mean=42.13) and individual style is least preferred (mean=30.33). These 6 types of learners are distributed widely among the subjects. But comparatively speaking, these participants are more varied and more heterogeneous among 2 modes of learning: individual (std. Deviation: 6.92) and group (std. Deviation: 6.39).

These findings partly corroborate the results of earlier research. Reid (1987), Melton and many other researchers found that kinesthetic style and visual style are both the students' preferred ones. Chinese researchers like Hu Xiaoqiong (1997), used the same questionnaire of Reid's PLSP (Perceptual Learning Style Preference Questionnaire) and both found that tactile style of learning was students' most favored one, and group style was the less preferred one.

The similarities and differences between this research and earlier researches indicate that relatively majority of the students in this college prefer hands-on learning, as in building models; writing notes or instructions. These findings may explain that traditional Chinese EFL teaching is reading and writing-oriented. Over the early past decades, the school system for English education in China totally neglected speaking and listening abilities. Another indication is that majority of the students in this college tend to learn best by experience, by being involved physically in classroom experiences. They remember information well when they actively participate in activities, field trips, and role-playing in the classroom. A combination of stimuli, for example, an audio tape combined with an activity will help them understand new material. Although it is not easy to make much progress in a short period of time, the "student-centered" principle contributes to students' being keen on participating in different class activities.Moreover,these findings also prove that people's learning style is possible to extend or change.

4.2 Final-Term Exam Scores among Participants

The test paper consists of 5 main kinds of question items including listening, reading, cloze, vocabulary and writing. In order to compare and discuss the score contribution of each test item of the examination, the present author changed the original scores into the ration scores, according to the different percentage of correctness of each test item. The mean of correctness percentage of each test item here is the most important evident to make analysis of the effect of computer-assisted English autonomous learning. Table 2 demonstrates that students score highest in reading (mean=0.706) and lowest in listening (mean=0.530).

Table 2. Descriptive Statistics of Ration Scores of Examination

Test Items	Sample Number	Mean Score	Std. Deviation	Minimum	Maximum
Listening	78	0.530	0.142	0.30	0.90
Reading	78	0.706	0.138	0.47	0.93
Cloze	78	0.651	0.130	0.40	0.90
Vocabulary& Grammar	78	0.698	0.173	0.35	1.00
Writing	78	0.684	0.104	0.40	0.90

This finding confirms some part of results of learning style preferences study. Chinese students are comparatively weak in listening abilities for the auditory learning style is the second or third least preferred style for Chinese students. By contrast, their reading abilities are the most contribution to the total score for the majority of students showed their preferences for visual and kinesthetic style. Traditional Chinese teaching is reading and writing-oriented. Most teachers in China emphasize learning through reading. With reading and writing priority, students like to read and obtain great deal of visual stimulation. The failure of students' listening achievement might result more or less from their dislike of auditory style. As shown in Table 1, auditory style is the second least preferred style. Therefore, listening remains one of comparatively poor skills which need to be strengthened in the future EFL teaching and more efforts should be made to improve students' auditory preference.

4.3 The Correlation Study between Participants' Academic Achievement in Computer-Assisted English Autonomous Learning and Their Learning Styles

The present study tried to find out the possible correlation between them. Pearson correlation coefficient was used to do analysis and the results were presented in the following table.

Table 3. Correlation between Participants' Academic Achievement in Computer Assisted English Autonomous Learning and Their Learning Styles

Styles		Listening	Reading	Cloze	Vocabulary	Writing	Total
Visual	Pearson Correlation	0.207	0.673**	0.261*	0.312**	0.107	0.520**
	Sig. (2 tailed)	0.069	0.000	0.021	0.005	0.352	0.000
Auditory	Pearson Correlation	0.821**	0.207	0.260*	0.292**	0.152	0.503**
	Sig. (2 tailed)	0.000	0.069	0.021	0.009	0.185	0.000
Kinesthetic	Pearson Correlation	0.197	0.366**	0.201	0.137	0.162	0.326**
	Sig. (2 tailed)	0.083	0.001	0.078	0.230	0.156	0.004
Tactile	Pearson Correlation	0.163	0.059	0.078	0.081	0.148	0.145
	Sig. (2 tailed)	0.154	0.606	0.496	0.483	0.195	0.206
Group	Pearson Correlation	0.095	0.110	-0.031	-0.247*	-0.006	-0.019
	Sig. (2 tailed)	0.407	0.340	0.788	0.029	0.959	0.867
Individual	Pearson Correlation	0.278*	0.310	0.230	0.800**	0.257*	0.588**
	Sig. (2 tailed)	0.014	0.006	0.043	0.000	0.023	0.000

**. Correlation is significant at the 0.01 level (2-tailed).

*. Correlation is significant at the 0.05 level (2-tailed).

The results in Table 3 indicates that in general, almost all types of learning styles have , more or less, correlations with some test item or total score. Only one type of learning styles has no any correlations with language achievement at the significant level. It is tactile style of learning. This result is just the same as other Chinese researchers' findings. As mentioned above, Hu Xiaoqiong (1997) and Wang Chuming

(1992) used the same questionnaire of Reid's PLSP(Perceptual Learning Style Preference Questionnaire) and all found that tactile style of learning was students' most favored one, but no significant difference was found between high language achievers and low language achievers, and no relationship was found between language achievement and tactile learning style by Yu Xinle (1997). Specifically speaking, only one type has significant negative correlation with language achievement at the level 0f 0.05—group style with coefficient: -0.247. In conclusion, the first 3 pairs with extremely significant correlation are individual and vocabulary, auditory and listening, visual and reading.

These results imply that Chinese college students have talent or potential abilities in tactile activities but these advantages are not made fully use of in learning. This is mainly due to the Chinese traditional "teacher-centered" teaching approach in which teachers spoon-feed learners and learners perceive the knowledge passively. In the past several decades, Chinese English teachers gave lessons with abundant information by means of speaking and writing what they had prepared for learners while learners memorized what they had heard and what they had taken down on the notebook. All of the learning tasks in class were controlled by teachers. Teachers took charge of learning content and even learning method. Students just followed teachers all the way. When they had questions, they would wait for the teachers' explanation rather than settle them by themselves. Their learning effect was determined by teachers to some degree. As a result, students' activity and creativity was ignored seriously. They had no opportunity to take active participation in hands-on activities and other physical movements, such as hands-on training, field trips, discussion, role-playing, using tools and working in laboratory settings, using communicative skills in the real life. Even though they were good at learning through physical activities, teachers unconsciously restrict them to take advantage of their potential. Further more, in the new computer-assisted autonomous learning environment, there are fewer chances for students to conduct physical learning activities. What a pity that their tactile ability was greatly wasted! That's the possible reason why most college students prefer tactile learning style but this style has no significant correlation with their effect in computer-assisted English autonomous Learning.

In contrast, this "teacher-centered" teaching approach enhanced learners' visual ability, auditory ability and individual ability at the same time. Students learned well from seeing words in books and on the chalkboard. They remembered and understood information and instructions better if they read them. They learned from hearing words spoken and from oral explanations. They might remember information by reading aloud or moving their lips as they read, especially when they were learning new material. Besides, owing to so many examinations they had to take alone in the stage of Middle School, those Chinese students learned best when they worked alone, especially when they were memorizing the vocabulary alone. These are the possible reasons why Chinese college students of individual learning style get higher score in vocabulary, students of auditory learning style get higher score in listening, and students of visual learning style have better learning effect in reading.

5 Suggestions

Based on the research findings, some suggestions should be given to English learners, English teachers, universities and technical workers. On the one hand, learners should raise the awareness of learning styles; self-diagnose their learning styles, stretch and balance their learning styles, and acquire learning strategies and techniques. On the other hand, teachers should devise the syllabus and curriculum, implement target-oriented teaching, provide facilities and resource, monitor and assess web-based autonomous learning, and foster teacher autonomy. Besides, the school and technical workers can offer the necessary hardware and software facilities.

References

1. Paramskas, D.M.: Computer-assisted Language Learning(CALL): Increasing Integrated into an Even More Electronic World. The Canadian Modern Language Review 50(1), 124–143 (1993)
2. Liu, R., Di, M.: Chinese college foreign language teaching reform present situation and the development strategy research. Foreign language teaching and research press, Beijing (2003)
3. Reid, J.M.: The Learning Style Preference of ESL Students. TESOL Quarterly 21, 87–111 (1987)
4. Reid, J.M.: Learning Styles in the ESL/EFL Classroom. Foreign Language Teaching and Research Press, Beijing (2002)
5. Melton, C.D.: Bridging the Culture Gap: A Study of Chinese Students Learning Style Preferences. RELC Journal 21(1) (1990)
6. Yu, X.: A Study About English learning style of Chinese undergraduate students. Foreign Language Teaching and Research (1) (1997)
7. Oxford, R.L., Burry-Stock, J.A.: Assessing the Use of Language Learning Strategies worldwide with ESL/EFL Version of the Strategy Inventory for Language Learning. System 23(2) (1995)
8. Hu, X.: An Investigation of Learning Way of English majors in Chinese Universities. Foreign Language World (2) (1997)

The Redesign Method Integrated Reverse Modeling with Deformation

Lufang Zhao, Siyuan Cheng, Xuerong Yang, and Xiangwei Zhang

Guangdong University of Technology, 510006 Guangzhou, China
lfzhao@gdut.edu.cn

Abstract. As one of advanced design method, reverse engineering is being widely used in industry. A redesign method integrated reverse modeling with deformation is proposed in this paper. Since their "natural" parameters are hard to be defined and extracted, in redesign with deformation we propose dealing with freeform product models by extracting global and local product definition parameters. Designers can produce new design variations by editing the product definition parameters. Use a car model as sample, and with the mature CAD commercial software of CATIA and Geomagic Qualify as the platform, this paper proposed the method of model reconstruction with extraction of feature, and use the deformation redesign theory and methods for such products to meet the requirement of re-design. The redesign method integrated Reverse Modeling with Deformation is proved to be effective and feasible, and have great significance in engineering practice.

Keywords: Reverse engineering, Reverse modeling, Extraction of feature, Redesign with deformation.

1 Introduction

Today's product designer is being asked to develop high quality, innovative products at an ever increasing pace. Reverse engineering(RE) starts in the physical environment with the goal of conversion from the physical models to digital models that can be further used by CAD/ CAE/CAM applications, and its application ranged from the simply copy to reverse modeling supporting rapidly innovative design, which is playing an important role in product design [1].

Reverse engineering (RE), as one of the advanced methods of design, has being widely used in product development, especially on auto industry [2].RE is an important tool with which to generate CAD models [3]. Freeform product design has been the main focus of conventional RE. Since their "natural" parameters are hard to be defined and extracted, in redesign with deformation we propose dealing with freeform product models by extracting global and local product definition parameters that are defined by international, domestic or industrial de-facto standards, or byuser-defined key parameters. Designers can produce new design variations by editing the product definition parameters [4].

The remainder of the paper is organized as follows: Section 2 gives an introduction about RE and presents the two methodologies of classical surfacing and rapid

D. Jin and S. Jin (Eds.): Advances in FCCS, Vol. 1, AISC 159, pp. 257–263.
springerlink.com

surfacing for reverse modeling. In Section 3 combined with case study the characteristics of these two methodologies are analyzed and compared. Section 4 conveys our conclusions.

2 Redesign Method with Reverse Modeling and Deformation

Reverse modeling is redesign and re-creation by surface data processing and model reconstruction of original sample, at the same time with absorption and digestion, and then can develop a similar type but more excellent products. Being high starting point and fast is the character of inverse modeling, but usually with limitations of form, it can't fully meet the requirement of product redesign. To facilitate in product redesign, it should firstly construct the digital models which are in line with the original design intent and easily modified, and there are variety way for different models to facilitate in redesign, restoring the design intent of physical sample and enhancing the redesign ability of re-model are the current bottleneck of inverse modeling, at the same time, it are key technology of reverse engineering.

For a large number of product shape which composited by the free-form surface, the concept of characteristic is very vague, so the technology of features or parameters can't provide very effective way to redesign process. However, the deformation of curves and surfaces can play a powerful role in for product redesign. Deformation technology of curve and surface focused on the modification of existing CAD model, which can quickly generate the new mode based on original model .

Deformation as a class of surface design method which the field of geometric shapes is root for be widely used in geometric modeling, editing and smoothing operations such as CAD. Deformation technique is mainly for B-spline, NURBS and other curves and surfaces with parametric sketch. In 1984, Barr [5] firstly proposed global and local free-form deformation method, and then the deformation technology is becoming more and more popular and got more attention and research. Currently, according to correlation degree with expression of model, the model can be divided into two types; for one thing, deformation technology associated with expression of model, for another, deformation techniques is independent to expression of model. The former achieve the B-spline or NURBS surface deformation by moving the control points, The latter can be divided into deformation with tool and directly act on the object space deformation per whether use deformation tool.

For surface product, parametric tool can't provide effective means of redesign, and the detail cause as following: (1) free curves and surface of product can't be fully express by parameters, (2) surface CAD model of inverse modeling contains a lot of fitting curves and surfaces, and it has no external adjustable parameters information, (3) parent-child relationship can't be fully established in the model tree of inverse modeling, the adjustment of parameters can't be inherited to the model results.

However, with the theory and tools of surface deformation, it is more effective in redesign for this surface CAD model, because the surface deformation do not care about the modeling process of the original CAD model, and it has a greater fault tolerance to the original model. Only needed to know design goals, the model can be quickly and intuitively changed to expected profile. Surface deformation has been in-depth studied in theory and practice, and it has been implanted into some commercial software, as an effective redesign way, it has become a powerful tool for product innovation. In this paper, a car model, as example, using the theory and methods of reverse modeling and redesign, is used to show actual process of reverse modeling and deformation redesign.

3 Reverse Modeling

3.1 Digitalization and Preprocess

Digitalization model as the first part of Reverse engineering, and its integrity and accuracy of data is key for reverse modeling. The surface of car model is digitally scanned with Optical scanner, and then removed the excess noise, point cloud data reduction, triangular grid, and so on, by these preprocess, the mesh surface model is obtained.

3.2 Feature Extraction and CAD Model Reconstruction

In reverse engineering, extraction based on symmetry plane can not only accurately reconstruct the symmetrical features, but also greatly reduce the workload of inverse modeling, so the extraction based on symmetry plane is important part of auxiliary feature in reverse modeling, and it is one of key steps in reverse modeling too. For the symmetry plane extraction, some scholars have done some research, such as Professor Ke[1] in Zhejiang University, proposed the definition of bilateral symmetry of point cloud based on the basic principles of symmetry, at the same time, construct a new symmetric surface extraction algorithm with using of more sophisticated computer vision data matching technology, Minho Chang[6] also presented optimize the alignment algorithm at the symmetry plane extraction process.

Obviously, surface of this car model is symmetry, so only one half of the model need to be created. And by setting up coordinate system based on the symmetry plane, it can greatly simplify the difficulty of the modeling. So made an obvious feature in theoretic symmetry plane of grid surface with CATIA software, and then create an arbitrary initial symmetry plane, and mirror the grid surface based on initial symmetry plane, then align these two mesh surface by alignment algorithm in Geomagic Qualify software. The created symmetry plane is shown in Figure 1, and by coincide the XZ plane and the symmetry plane, the created coordinate system is shown in Fig. 1.

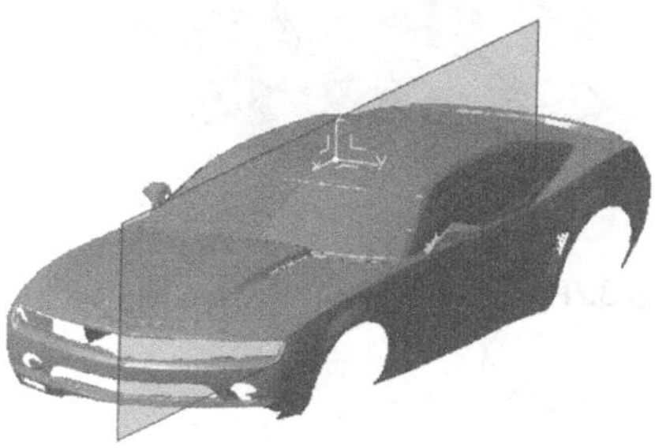

Fig. 1. Extract the symmetric plane and create the coordinate system

After set up the working coordinate system, each surface feature can be extracted separately. Feature lines must be extracted from point cloud manually according to user's judgment first, to segment the point cloud into several partitions, and each partition belongs to one surface primitive respectively. Then for each partition the best-suited surface fitting function must be found to construct the surface patch according to its characteristics. Finally the surface patches will be extended, trimmed, matched and merged to create a CAD model.

For the surface reconstruction of the engine hood sample, the first step is extracting the feature of cross-section, the resulting profile is parameterized plane curve, then we can construct other different three-dimensional parametric feature with these cross-sectional characteristics[7-8].The next step is to extract the transition feature above engine hood. First, curvature analysis and identify the position of contact curve, and then fit the free curve, then we can put projection to the multi-section feature, and reconstruct the surface with the software command of transitional feature. The overall result of surface reconstruction is shown in Fig. 2.

Fig. 2. The result of surface reconstruction

Mirror the reconstruction surface according to XZ plane, then get the whole surface of the car model, as shown in Fig. 3.

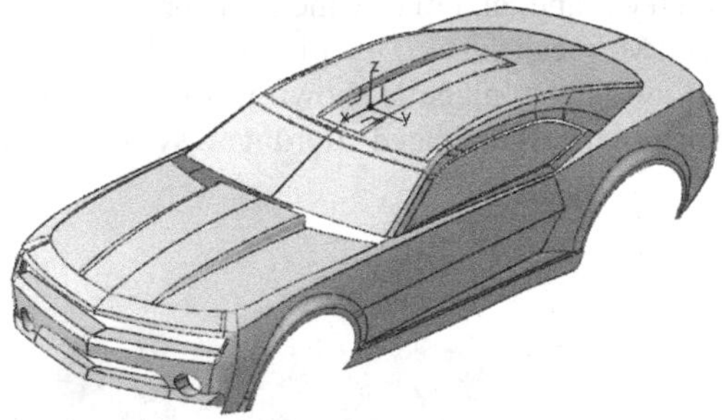

Fig. 3. The completed surface model of the car

4 Deformation Redesign

After completing the surface reconstruction process, the details and some partial modifications needed to be added. But if the completed surface or surface groups need to adjusted, it is a very painful thing for operational users, since each completion surface patch have determined topology and continuity relationship with adjacent surface patches, that is to say need to many of duplication effort to constantly repair the damaged relationships for adjustment.

As shown in Figure 4, for one certain consideration, now some adjustments need to be done in the car model. For deformation modification, at the beginning, it need to specify the target curve and initial curve to confirm the purpose of deformation, next is setting up the admissible region to restrict the deformation arca.The scope of the following limiting curve unchanged and upgrade the parts of the above limiting curve, the upgrade value be determined by the distance between initial curve and target curve, in the process of upgrading, the relationship between the various surface patches can't be changed. For such requirements, with using of GSM advanced tools of ThinkDesign in technical methods of surface deformation it can be easily achieve, the mutual position of the initial curve, the target curve and limiting curve shown in Fig. 4.

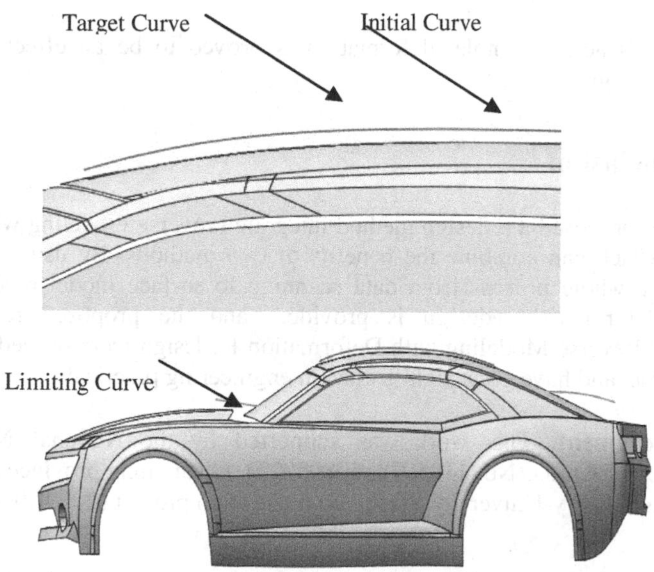

Fig. 4. The model with not deformation

Carrying out the overall deformation for the model, as shown in Fig. 5, where the pink area is the deformation zone, task of redesign can be easily completed and the original surface quality did not reduce.

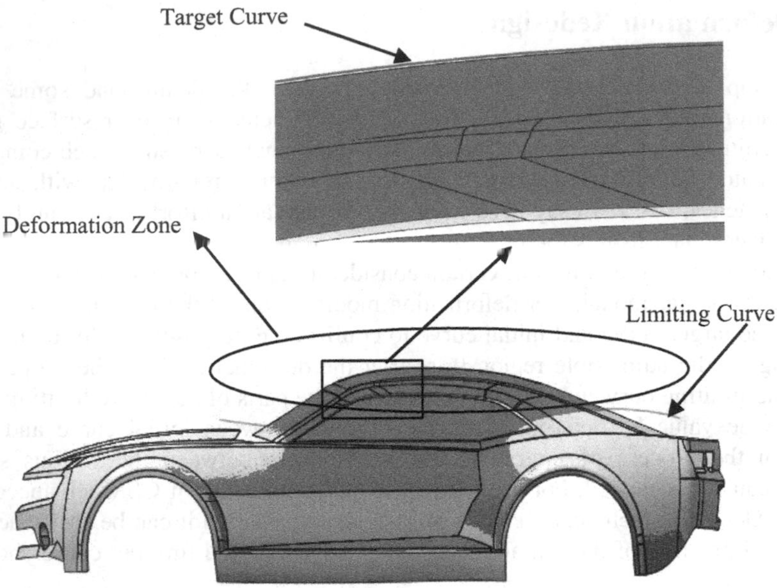

Fig. 5. The model with deformation.

From this actual sample, deformation is proved to be an effective method for surface redesign.

5 Conclusion

This paper proposed a redesign method integrated reverse modeling with deformation method, which can combine the benefits of two methods. By using a car model as example, a whole process from data scanning to surface model reconstruction and surface deformation redesign is provided, and the proposed redesign method integrated Reverse Modeling with Deformation Redesign were proved to be effective and feasible, and have great significance in engineering practical.

Acknowledgment. This work was supported by the National Natural Science Foundation of China (No. 1105078, 50805025), Guangdong province and Ministry of Education Industry-University- Research integration project (2009B090300044).

References

1. Ke, Y., Fan, S., Zhu, W., Li, A.: Feature-based reverse modeling strategies. Computer-Aided Design 38, 485–506 (2006)
2. Ye, X., Liu, H., Chen, L., Chen, Z., Pan, X.: Reverse innovative design- an integrated product design methodology. Computer-Aided Design 40, 812–827 (2008)

3. Barbero, B.R.: The recovery of design intent in reverse engineering problems. Computers & Industrial Engineering (2008), doi: 10.1016/j.cie.2008.07.023
4. Soni, K., Chen, D., Lerch, T.: Parameterization of prismatic shapes and reconstruction of free-form shapes in reverse engineering. The International Journal of Advanced Manufacturing Technology 41, 948–959 (2009)
5. Barr, A.H.: Global and local deformation of solid primitives. Computer Graphics 18(3), 21–23 (1984)
6. Chang, M., Park, S.C.: Reverse engineering of a symmetric object. Computer & Industrial Engineering 55, 311–320 (2008)
7. Bo, L., Cheng, S., Zhang, X., Yang, X.: Feature-based parametric reverse modeling method. In: 2010 WASE International Conference on Information Engineering, vol. 3, pp. 189–192 (2010)
8. Shan, D.-R.: Research on parametric modeling in reverse engineering based on features and constraints. Mechanical Science and Technology 24(5), 522–525 (2005)

Prediction Model of Tongguan Elevation Changes in Non-flood Seasons Based on T-S Fuzzy Reasoning

Xiaoping Du[1,2], Xufan Du[1], Xian Luo[3], and Fei Yang[1]

[1] College of Software, Beihang University, Beijing, China
[2] State Key Laboratory of Hydraulics and Mountain River Engineering Sichuan University, Chengdu, China
[3] Chongqing Jiaotong University, Chongqing, China
xpdu@buaa.edu.cn, tctcdtc@163.com, lxlh6085@sina.com, fyang.nicole@gmail.com

Abstract. Accurate prediction of Tongguan Elevation has important realistic significance in flood control of lower Wei River. This paper builds a prediction model of Tongguan Elevation changes in Non-flood seasons combined with T-S fuzzy reasoning and association rule learning, based on historical data of Tongguan of the Yellow River. Considering the "curse of dimensionality" during fuzzy reasoning, the model uses association analysis learning to prune the amount of fuzzy rules, ensuring the simplicity and effectiveness of the final reasoning rules. The results of prediction based on historical data indicate that this model has high efficiency and accuracy; the accuracy is better than the existing BP prediction model.

Keywords: prediction model, association analysis, T-S fuzzy reasoning, Tongguan Elevation.

1 Introduction

The height of Tongguan Elevation is closely related to the sediment scouring of Xiaobei Main Stream of the Yellow River and lower Wei River, having a great influence on flood control and waterlogging control in this area[1]. The existing researches about the changes of Tongguan Elevation are mainly comprised of qualitative analysis and quantitative prediction. Qualitative analysis is used much more often while quantitative prediction is comparatively rarely. From the perspective of the qualitative analysis, the [2] concludes that main factors of the changes of Tongguan Elevation include natural river conditions, runoff-sediment conditions and operation mode of Sanmenxia Reservoir. Reference [3] is one of the most effectual prediction models, aiming at runoff-sediment conditions and operation mode of Sanmenxia, used an improved BP network model to make effective predictions on change values of Tongguan Elevation during non-characteristic flood seasons and non-flood seasons.

The calculation of the absolute height of Tongguan Elevation does not reflect the impact of various factors with Tongguan Elevation and the calculation error is

D. Jin and S. Jin (Eds.): Advances in FCCS, Vol. 1, AISC 159, pp. 265–272.
springerlink.com

relatively large, existing researches on Tongguan Elevation focus on the analyses or predictions on the change value of Tongguan Elevation [3]. General rule of the change of Tongguan Elevation is that the siltation increases in non-flood seasons while it washes away and lows down in flood seasons. The factors of Tongguan Elevation in flood seasons and non-flood seasons are not the same [2][4], the models to compute Tongguan Elevation in flood seasons and in non-flood seasons are quite different.

This paper builds a non-flood-season Tongguan Elevation changes prediction model based on T-S fuzzy reasoning as well as the historical data of the natural river conditions in non-flood seasons, runoff-sediment conditions and operation mode of Sanmenxia Reservoir. The established model combines cluster analysis and association rule learning to complete data discretization and pruning the amout of fuzzy rules, which promises a good prediction effect.

2 Train of Thought for Modeling

2.1 Fuzzy Reasoning Methods

In recent years, fuzzy system modeling has attracted much attention[5][6]. Takagi and Sugeno (T-S) fuzzy models [7] can locally represent high nonlinear systems with convenience [8]. A nonlinear model will be divided into several fuzzy linear models; it used linear input output relationship to represent each rule. Rules in T-S fuzzy model [7] are constituted as:

$$R^i: if\ x_1\ is\ A_i^1\ and\ x_2\ is\ A_i^2\ and \cdots and\ x_n\ is\ A_i^n\ then\ y_i = g_i(x_1, x_2, \cdots, x_n) \quad (1)$$

Where $(x_1, x_2, \ldots, x_n)$ are variables of the antecedent part, $A_i^1, A_i^2, \ldots, A_i^n$ fuzzy sets, and $g_i(x_1, x_2, \ldots, x_n)$ functions of consequent part. The final output y^0 for inputs $x_1^0, x_2^0, \ldots, x_n^0$ is determined as a weighted mean value over all rules according to

$$y^0 = \sum_{i=1}^{M} \omega(i) f_i(x_1^0, x_2^0, \ldots, x_n^0) / \sum_{i=1}^{M} \omega(i), \qquad \omega(i) = \prod_{j=1}^{n} A_i^j(x_j^0) \quad (2)$$

Where $\omega(i)$ is the membership degree to i-th rule and M is the total number of the fuzzy rules.

2.2 Modeling Approaches

In hydrology, it is hard to describe the complex hydrological physical quantities and the various aspects of characteristics in the process of changes; the fuzzy reasoning can do a more scientific description of the characteristics in both its generation process and other aspects. Obtaining fuzzy inference rules and establishing membership functions are the key segment of constructing fuzzy systems [9].

T-S fuzzy model is selected in this paper to build the prediction model of Tongguan Elevation changes. There are 6 main steps to build the model:

1) data acquisition;
2) data discretization with cluster analysis;
3) generation and pruning of fuzzy rule sets with association analysis;
4) calculating membership degree;
5) calculating weighting parameters of rules;
6) establish the model.

3 Modeling

3.1 Data Acquisition

This paper picks data from Table 32 in [10] based on the historical data of the river regime and each section characteristic during non-flood seasons," among which sediment load in non-flood seasons", "time of different water levels" and " changes of Tongguan Elevation during non-flood seasons" are picked from 1973.11 to 2000.6. With regard to the value of "the proportion of sedimentation amount of each part in whole reservoir sections" of Huangyu (Huangyu means the Yellow River siltation), "Huangyu31~41" is selected to build the model because it is one of the main factors of non-flood seasons for Tongguan Elevation change [2]. The selected data are shown in Table1 after transferring; data in each year presents a non-flood reason from November last year to June this year, like the data of 1974 presenting the period of 1973.11-1974.6. All selected data are divided into the training set and the test set, the training set are data from 1974 to 1990 and the rest are for the test set. The part of training set is shown in table 1.

Table 1. Original Data of Training Set

year	days greater than some water level(d)				SL(sediment load in non-flood seasons)(10^8 t)	HY(Huangyu31-41)(10^8 m^3)	TG(Tongguan Elevation change)(m)
	>315m	>320m	>322m	>324m			
1974	140	121	77	14	2.02	0.6407	0.55
1975	129	72	61	0	2.08	0.5127	0.53
1976	149	73	31	17	2.15	0.4826	0.67

3.2 Data Discretization

Data discretization is to convert the collected continuous hydrological data into available discrete data for T-S reasoning model. The k-means clustering algorithm is used to discretize the input data into three clusters. Since all the original data are continuous data, each attribute can obtain three clusters and the clustering center of each cluster after discretization. Three clusters are signed to L (low), M (middle) and H (high) according to clustering values. The value of clustering center of each clusters is shown in table 2.

Table 2. Clustering Center of Each Attribute

Attribute		L	M	H
days greater than certain water level	>315	135.8	152.01	175.0
	>320	37.0	78.111	112.167
	>322	8.0	49.667	91.5
	>324	5.133	44.0	78.0
SL		1.168	1.386	2.015
HY		0.295	0.517	0.831
TG		0.214	0.545	1.25

Work out the absolute difference value of corresponding attributes of each precise data and all of the three clustering centers value. Discretize each precise data into L/M/H if the absolute difference value of this precise data and the corresponding L/M/H clustering center is the smaller than others. The result of discretization of the input data in table 1 is shown in table 3.

Table 3. Result of Data Discretization

record	>315	>320	>322	>324	SL	HY	TG
1974	L	H	H	L	H	M	M
1975	L	M	M	L	H	M	M
1976	M	M	M	L	H	M	M

3.3 Generation and Pruning of Fuzzy Rule Sets

The number of fuzzy rules increases with the number of input attributes and fuzzy subsets describe each attribute exponentially. In this paper, the association analysis [11] is used to prune the generated fuzzy rules, weed out the weak related or unrelated fuzzy rules and simplify the fuzzy rule base, to avoid the problem of "curse of dimensionality". After pruning the fuzzy rules by association analysis with the minimum support is 15% and the minimum confidence is 60%, 11 fuzzy rules are produced, part of the rules are shown in table 4.

Table 4. Fuzzy Rules

rules	>315	>320	>322	>324	SL	HY	TG
1		H		L		M	M
2	L			L	H	L	M
3	H	H					M
4		H	H		M		M
5	L	M	M	L		L	M
6	L	M	M	L	H	M	M

Since the "THEN" of T-S fuzzy model is a precise function, each rule can be expressed as a partial linear model according to (1). All the fuzzy rule sets can get an output R_M by (3).

$$R_M : y = f_M(x_1, x_2, \ldots, x_m) = b_{M0} + b_{M1}x_{1h} + \ldots + b_{Mm}x_{mh} \tag{3}$$

3.4 Calculating Membership Degree of Fuzzy Rules

For each historical record T_i, ω_{ir} is defined as membership degree of record T_i to rule r, r=1, 2, … , M. M is the amount of the fuzzy rules. ω_{ir} is the minimum of the membership degree of each attribute of record T_i towards rule r, as (4).

$$\omega_{ir} = \min\left\{\omega_{ir}^{1}, \omega_{ir}^{2}, \ldots, \omega_{ir}^{N}\right\} \tag{4}$$

ω_{ir}^{N} represents the membership degree of the N-th attribute of record T_i towards the N-th attribute of the rule r. The formula to get membership degree is shown as (5). The part of result of membership degree are shown in table 5.

$$\omega_L(x) = \begin{cases} 1, & (0 < x < k_l) \\ \dfrac{k_m - x}{k_m - k_l}, & (k_l \le x < k_m) \\ 0, & (x \ge k_m) \end{cases};\quad \omega_M(x) = \begin{cases} 0, & (0 < x < k_l) \\ \dfrac{x - k_l}{k_m - k_l}, & (k_l \le x < k_m) \\ \dfrac{k_h - x}{k_h - k_m}, & (k_m \le x < k_h) \\ 0, & (x \ge k_h) \end{cases};\quad \omega_H(x) = \begin{cases} 0, & (0 < x < k_m) \\ \dfrac{x - k_m}{k_h - k_m}, & (k_m \le x < k_h) \\ 1, & (x \ge k_h) \end{cases} \tag{5}$$

Where x is the original data, $k_l/k_m/k_h$ means the each attribute's clustering center of the subset of L/M/H. $\omega_L/\omega_M/\omega_H$ is the membership degree of subset L/M/H of the discretized original data.

Table 5. Membership degree of records to rules

record	rlues										
	1	2	3	4	5	6	7	8	9	10	11
1974	0.6061	0	0	0	0	0	0	0	0	0.2591	0
1975	0	0.0194	0	0	0	0.7291	0.0194	0	0	0	0
1976	0	0.1550	0	0	0.1857	0.1857	0.1550	0	0	0.5520	0

3.5 Calculating Weighting Parameters of Rules

$$A = \min\sum_{i=1}^{N}\omega_{ir}^{2}[Y_i - b_{r0} - b_{r1}X_{i1} - b_{r2}X_{i2} - \ldots - b_{rm}X_{im}]^2 \tag{6}$$

According to 11 fuzzy rules, build optimization models to each rule; calculate weighting parameters of attributes in fuzzy rules. To deal with rule r, we build a constrained optimization model based on sum of squares of minimum distance, bring each record into (6):

N is the amount of records in the training set, m is number of attribute of each record.

By calculating the value of b_{r0}, b_{r1}, … , b_{rm} when A gets the minimum value, the weighting parameters of rule r can be worked out. The part of result is shown in table 6.

Table 6. Parameters of fuzzy rules

rule	b0	b1	b2	b3	b4	b5	b6
1	-0.0047	0.0008	0.0027	0.0013	0.0027	0.0003	-0.0025
2	0.2207	0.0002	-0.0061	0.0051	0.0185	0.1083	0.3174
3	0	0.0173	-0.0351	0.0161	0.0097	0.0109	-0.0182
4	0.0747	0.0196	-0.0383	0.0160	0.0091	-0.0724	0.1035
5	0.0013	0.0032	0.0027	0	0.0007	-0.0028	-0.0004
6	-0.0010	0.0025	0.0008	0.0012	0.0005	-0.0018	-0.0004

3.6 Models Established

Bring the known weighting parameters into (4) to build 11 functions of fuzzy rules. For each record, the output model for Tongguan Elevation changes in Non-flood seasons prediction model is determined as a weighted mean value over all rules' functions according to (7), y is the final output of the model, which is the prediction result of Tongguan Elevation changes in Non-flood seasons.

$$y=\sum_{r=1}^{11}\omega_r f_r(x_1,x_2,\ldots,x_n)/\sum_{r=1}^{11}\omega_r \qquad (7)$$

4 Model Verification and Result Analysis

The first three columns in table 7 show the actual value of each year in non-flood seasons and fuzzy reasoning result of Tongguan Elevation changes in non-flood seasons. In order to evaluate the model further, the calculated errors of the prediction model for Tongguan Elevation in non-flood seasons based on T-S fuzzy reasoning and BP-network model [3] are shown in last two columns in table 7.

As the results shown in table 7, in our prediction model, there are 20 years that the prediction error are less than 20%, 14 years are less than 10% and 10 years are less than 5%, which are much more better than the prediction results of BP-Network model. Considering the great complexity of hydrological changes, the result is relatively more satisfactory than the existing prediction models. The fitting and predictions of Tongguan Elevation changes in non-flood seasons to the actual records in model provided by this paper is relatively accurate except in few years like in 1987 and in 1992-1994.

The reason why the error of output in 1987 and in 1992-1994 are relatively large is that there were continuous dry months with less sediment in these years, the low flow and sediment caused river channel shrinkage in both the Yellow River in Tongguan section and the lower Wei River, scattered changes of the river conditions and finally greatly impacted the Sanmenxia Reservoir area, which making the prediction more difficult.

Table 7. Predictive value and verification

record	actual value	calculated vale	Error of T-S model	Error of BP model
1973.11-1974.6	0.55	0.6048	9.970%	27.900%
1974.11-1975.6	0.53	0.5308	0.150%	27.150%
1975.11-1976.6	0.67	0.571	14.780%	0.830%
1976.11-1977.6	1.25	1.2467	0.270%	3.030%
1977.11-1978.6	0.51	0.5088	0.240%	4.600%
1978.11-1979.6	0.67	0.6814	1.700%	15.270%
1979.11-1980.6	0.2	0.2672	33.620%	21.130%
1980.11-1981.6	0.57	0.4707	17.430%	88.200%
1981.11-1982.6	0.5	0.4308	13.840%	20.090%
1982.11-1983.6	0.33	0.3418	3.590%	33.410%
1983.11-1984.6	0.62	0.6279	2.930%	9.090%
1984.11-1985.6	0.21	0.232	10.480%	12.160%
1985.11-1986.6	0.44	0.4426	0.600%	13.480%
1986.11-1987.6	0.12	0.1673	39.420%	17.300%
1987.11-1988.6	0.21	0.1938	7.730%	24.070%
1988.11-1989.6	0.54	0.4848	10.230%	15.330%
1989.11-1990.6	0.39	0.4147	6.330%	25.820%
1990.11-1991.6	0.42	0.4494	7.010%	24.390%
1991.11-1992.6	0.5	0.3533	29.330%	8.130%
1992.11-1993.6	0.48	0.3117	35.070%	23.100%
1993.11-1994.6	0.17	0.27	58.840%	42.430%
1994.11-1995.6	0.43	0.411	4.430%	27.130%
1995.11-1996.6	0.14	0.1453	3.760%	31.840%
1996.11-1997.6	0.33	0.4138	25.41%	8.580%
1997.11-1998.6	0.35	0.3586	2.450%	42.560%
1998.11-1999.6	0.15	0.1178	21.470%	0.230%
1999.11-2000.6	0.36	0.3022	16.060%	

5 Conclusion

In this paper, a prediction model for Tongguan Elevation changes in non-flood seasons based on T-S fuzzy reasoning, association analysis and cluster analysis is provided. The changes of Tongguan Elevation are a complex nonlinear process due to lots of factors; what's more, the processing of boundary conditions is difficult because of the special geographical location. The T-S fuzzy model combined with association analysis and cluster analysis can deal with the complex nonlinear hydrological progress, suit to deal with the changes of Tongguan Elevation. Comparing with the existing model, this model has relatively smaller computational complexity and the error is relatively less; the results of prediction of Tongguan Elevation changes in non-flood seasons are relatively accurate. The model can obviously reflect the trend of Tongguan Elevation changes in different water levels or different sediment load conditions, so it can provide a reliable method to deal with complex nonlinear hydrological records.

References

1. Zhou, J., Lin, B.: Awareness of Tongguan Elevation in the Yellow River. China Water (6), 47–49 (2003)
2. Li, W., Wu, B.: Analysis of Tongguan Elevation and factors. Yellow River 32(7) (2010)
3. Chen, J., Liu, Y., Hu, M., Zhang, J.: Prediction of Tongguan Elevation based on improved BP Neural Networks. Journal of Hydraulic Engineering (8), 96–100 (2003)
4. Zhang, J., Wang, Y.: Analysis of Tongguan Elevation in non-flood seasons and operation mode of Sanmenxia Reservoirr. Water Resources and Hydropower Engineering 6, 54–58 (2002)
5. Guo, F.: Development of fuzzy reasoning. Shanxi RTVU Journal (4), 71–74 (2007)
6. Jia, W., Jia, W.: Introduction of researches and applications on fuzzy reasoning. Science & Technology Information (6) (2008)
7. Taksgi, T., Sugeno, M.: Fuzzy indentification of systems and its application to modeling and control. IEEE Trans. on SMC 15(1), 36–39 (1985)
8. Zhong, Q., Yu, Y., Xu, S.: Impulsive control for T-S fuzzy model based Chaotic Systems with Adaptive Feedback. In: International Conference on Communications, Circuits and Systems, ICCCAS 2009, pp. 872–875 (2009)
9. Zhang, Z., Zhang, P.: Neural Fuzzy and Soft Computing. Xi'an Jiaotong University Press (2000)
10. Yellow River Water Resources Research Institute, The effect of operation of Sanmenxia Reservoirr to Tongguan Elevation, Yellow River Water Resources Research Institute (March 2004)
11. Agrawal, R., Srikant, R.: Fast algorithms for mining association rules in large databases. In: Proceedings of the 20th International Conference on Very Large Data Bases, VLDB, Santiago, Chile, pp. 487–499 (September 1994)

Impact on Foreign Direct Investment of RMB Real Effective Exchange Rate—Empirical Analysis Based on Quarterly Data of China

XiaoJun Lu and Lu Chen

Department of Public Economics, Xiamen University, China
axiaojunlu@yeah.net, b912708733@qq.com

Abstract. This paper using co-integration theory and error correction model to check the influence of RMB real effective exchange rate on Chinese FDI. The results show that in the long term, there is a co-integration relationship between these two variables and the RMB appreciation will promote the growth of FDI; in the short term, RMB real effective exchange rate doesn't have a significant effect on FDI. In this regard, this paper proposes the corresponding policy recommendations.

Keywords: Foreign Direct Investment, Co-integration Theory, Error Correction Model, RMB Real Effective Exchange Rate.

1 Introduction

Since 1978, the economy of China has been keeping a high growing trend. It is benefit from a large amount of foreign fund in a great degree, especially FDI. As we known before the foreign merchant decide to invest a market, he has to exchange his home currency. And then exchange his profits into home currency. The exchange rate has been involved into this process twice successively. Obviously, the exchange rate will affect the decision-making of multinational companies, furthermore the scale of FDI. Therefore, it is significant to study the relationship between RMB exchange rate and FDI. Based on the previous researchs, this paper is arranged as follows: the second part will do a theoretical analysis about the relationship between the RMB exchange rate and FDI; and then combined with the data of china, we used the co-integration theory and error correction model to check the influence of RMB real effective exchange rate on Chinese FDI; at last, we proposed some corresponding policy recommendations.

2 Theoretical Analysis of RMB Exchange Rate and FDI

The variation of RMB exchange rate has an influence on FDI in many ways, and different factors have different results. These results can be even approximately canceled out to some extent. Firstly, the variation of RMB exchange rate will change the relative price of raw material and labor, and then change the investment cost. As the cost of the investment has been changed, the expected return of the foreign merchant has changed too. It ends up with

D. Jin and S. Jin (Eds.): Advances in FCCS, Vol. 1, AISC 159, pp. 273–277.
springerlink.com

the change of the inflow amounts of FDI. Secondly, the exchange rate will affect the international competitiveness of the export product. RMB appreciation will raise the relative price of the export product and lower the relative price of the import product. As a result, it increases the inflow amounts of the market-oriented FDI and decreases the export leading ones. Finally, the movement of RMB exchange rate will lead to the change of China's economic structure, thereby affect the inflow of FDI.

3 Empirical Analysis of RMB Exchange Rate Variation on FDI

3.1 Variables and Data Selection

Based on previous studies, combined with China's specific conditions, this paper will introduce some relevant variables to explain the impact of RMB real effective exchange rate on China's FDI.

1) GDP. The variable mainly reflects China's market size and measure the impact of China's economic development on FDI. Many studies have shown that the larger size of the market, the higher country's overall level of economic development and then it is more attractive to foreign capital. Figure 1 shows the relationship between China's FDI and GDP of each quarter since 1997. It can be seen from Figure 1, FDI and GDP have a clear consistent trend and positive relationship. Therefore, we can expect the variable coefficient of GDP is positive in the empirical model.

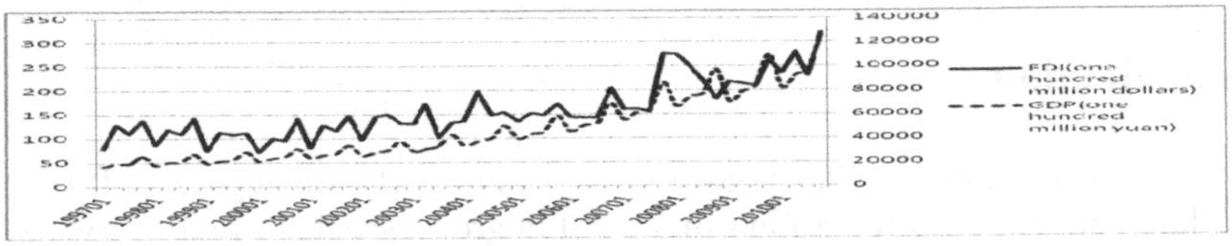

Fig. 1. The relationship between FDI and GDP

2) RT. The variable is mainly used to denote the degree of trade openness. As we all known, the higher the degree of trade openness, the FDI inflows will inevitably increase, and vice versa. Thus, this paper defines RT as the ratio of the total tariff and gross import and export. That is RT = total tariff/gross import and export. Figure 2 reflects the relationship between RT and every quarter of China's FDI since 1997. We can see from Figure 2, the relationship between RT and FDI is difficult to observe directly from the graphics. It is hard to expect the coefficient signs of RT, because the changes of them are in the same direction in some quarters, others are opposite. So the relationship needs the empirical model to be tested.

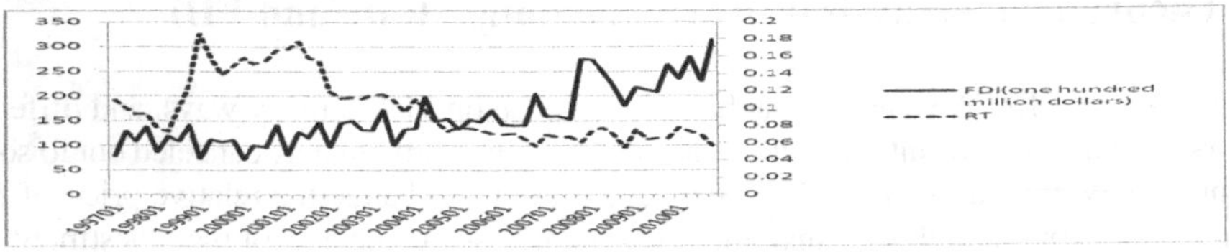

Fig. 2. The relationship between FDI and RT

3) REER. It is called real effective exchange rate, which is a variable to reflect a country's real value of goods relative to the foreign. It is used to measure a country's export competitiveness and the actual import cost in the international market. Only the real effective exchange rate can reject the deviation from the empirical results which is caused by the currency price changes. The value of REER rising means the appreciation of RMB.

4) μ. The variable mainly reflects other factors which have impact on FDI inflows, such as policy factors, social factors, political factors, and other economic factors. These factors are difficult to find suitable indicators to quantify, and some studies have shown that many factors have little influences on the level of FDI, so here we define these other factors as other factors.

Now we can specify the RMB real effective exchange rate and FDI Empirical model as follows:

$$FDI=\beta 0+\beta 1GDP+\beta 2RT+\beta 3REER+\mu. \quad (1)$$

This paper's sample space is from Q1 1997 to Q4 2010, a total of 56 samples. Meanwhile in order to weaken the degree of co-linearity and heteroscedasticity between these indicators in the empirical process and smooth the data, this paper took the logarithm of these four indicators. Time series of quarterly data are required seasonal adjustment before empirical process to reflect the objective laws of economic events' motion. So, in this paper, we use Eviews5.0 software and Census X12 seasonal adjustment method to adjust the quarterly data.

3.2 The Establish of Co-integration Model

Before using regression methods to do research of time series, we must first examine whether the original sequence is stationary, otherwise there will be "spurious regression". In this paper, ADF unit root test method is used to test the stationary of the variables. The ADF unit root test results of variables' original sequence and first-order differential sequence are as follows:

Table 1. ADF unit root test results of variables

[Variable]	[Method (C,T,K)]	[ADF Statistic]	[1% Critical value]	[Conclusion]
[FDI]	[c, 0, 4]	[0.522167]	[3.565430]	[nonstationary]
[GDP]	[c, t, 1]	[-2.406235]	[-4.137279]	[nonstationary]
[RT]	[0, 0, 4]	[0.141768]	[2.611094]	[nonstationary]
[REER]	[c, 0, 1]	[-1.481056]	[-3.557472]	[nonstationary]
[ΔFDI]	[c, 0, 3]	[-6.035471]	[-3.565430]	[stationary]
[ΔGDP]	[c, 0, 0]	[-5.127234]	[-3.557472]	[stationary]
[ΔRT]	[c, 0, 3]	[-6.264863]	[-3.565430]	[stationary]
[ΔREER]	[0, 0, 4]	[-2.647640]	[-2.612033]	[stationary]

Note: Test method (c, t, k) expresses if there was Constant C, Time trend term T and Lag order k in the ADF test. Δ means first difference of variables. Lag order selection following the AIC and SC criteria.

The unit root test results can be seen from the table. Under 1% significance level, the absolute values of ADF test of the original sequence are less than the absolute critical value. And this indicates that the original sequences of all variables exist unit root, they are nonstationary. But the absolute ADF test values of their first difference sequences are all greater than the absolute critical value of 1%.And this indicates that the first difference sequences of all variables doesn't exist unit root, they are stationary. So we can judge that the indicators all meet the premise for co-integration test.

Therefore, we can get the estimated regression equation as follows:

FDI=-5.15806+0.473249GDP+10.61831RT+0.029001REER

(2) t=(-54.29346)(195.1365) (106.1878) (2.377188)

R2=0.999256 F=24616.51 DW=1.741757

Given a significant level of 1%, we can get the critical value of t: t0.05(52)=2.01067. Obviously, the estimators are all significant. The values of F and R2 show that the model fits very well.And then, we do unit root test for the residuals $\hat{\mu}$ of the equation (2). The estimated residuals can be obtained by the regression equation:

$\hat{\mu}$ =FDI+5.15806-0.473249GDP-10.61831RT-0.029001REER

(3) The ADF test value of $\hat{\mu}$ is -3.263120. so we can reject the null hypothesis at the 1% significance level and ensure that $\hat{\mu}$ is a stationary series. And this can verify that the co-integration relationship is correct and reflect that there exist a long-term equilibrium relationship between the FDI and the measurement indicators.

3.3 The Establish of Error Correction Model

The traditional economic model usually expresses a "long-run equilibrium" relationship between variables, while the actual economic data generates in "non-equilibrium process". Therefore, it needs data's dynamic process of non-equilibrium to approximate long-run equilibrium process of economic theory when do the modeling. In Section 3.2, we have tested the co-integration between FDI and the real effective exchange rate. In order to examine the dynamic relationship between these two variables, we use the ECM model (error-correction model) to analyze.

First step, we can create the following error correction model:

ΔFDI=α0+α1ΔGDP+α2ΔRT+α3ΔREER+α4 $\hat{\mu}_{t-1}$ +μt

(4) And we can get the equation after estimation:

ΔFDI= -0.00188+0.538364ΔGDP+10.58629ΔRT-0.015542ΔREER-0.879232 $\hat{\mu}_{t-1}$

(5) t=(-0.726271) (7.363053) (120.4287) (-0.255004) (-6.157068)

R2=0.996452 DW=1.992027

In the formula (5), the coefficients of variables are significant in addition to the constant term and the coefficient of ΔREER. Differential term reflects the short-term fluctuations. Short-term fluctuations of FDI can be divided into two parts: one is the short-term effect of GDP, RT and other factors' fluctuations; the other is the variables' deviation from the long-run equilibrium. The coefficient of error correction term reflects the intensity of the adjustment for its deviation from the long-run equilibrium.

From the coefficient estimation, we can see it will be -0.879232 to pull the non-equilibrium back to equilibrium state, when a short-term fluctuation deviates from the long-run equilibrium.

4 Conclusions

From empirical analysis, we can conclude: In the long run, the RMB exchange rate and FDI have a stable equilibrium relationship, every 1% appreciation of RMB will increase 0.029001% of FDI inflows; in the short term, the RMB exchange rate changes have little effect on FDI. So this paper makes the following recommendations, which is from two aspects: exchange rate reform and FDI stable development respectively.

References

1. Wang, Z.: Research on the impact of exchange rate variation on foreign direct investment. China Economic Publishing House, China (2009)
2. Li, D., He, J., Gong, X., Wang, S.: Chinese foreign direct investment forecasting, performance analysis and international comparative study. Hunan University Press, China (2008)
3. Xie, L., Wang, S.: Submitted to Journal of Beihua University (Social Sciences) (2007)
4. Gao, T.-M.: Econometric Analysis and Modeling-Application and examples of Eviews. Tsinghua University Press, China (2009)
5. Zhou, J., Zhang, Z., Zhu, X.: Exchange rate and capital market. China Financial Publishing House, China (2008)

Design of Rotational Speed Measurement System of Engine on UAV Based DSP

JunQin Wang

Department of Mechanical and Electronic Engineering, Xi'an University of Arts and Science, Xi'an 710065, China
doctorzxj@yahoo.com.cn

Abstract. Rotational speed is one of the most important parameters for closed loop control on engine of UAV. Working principle of rotational speed measurement and futures of Enhanced Capture Unit on DSP TMS320F28335 were introduced. The design of rotational speed measurement system based on DSP was put forward. The software and hardware methods of the system were described. System can measure rotational speed real-time with high precision. Evidences show that the system works with stability and reliability and generalizes with practicability and value of popularize.

Keywords: Rotational speed measurement, Engine, UAV, DSP.

1 Introduction

Rotational speed is one of the most important parameters for controlling on engine of UAV, the accuracy of speed acquisition will have a major impact on results. Currently mainly use magnetic to measure rotational speed, that is generated by the magnetic effects associated with the rotational speed of the pulse, using the optical encoder as measuring element to measure, this approach is of simple structure, high anti-electromagnetic interference capability, high accuracy at low field. However, in the field of high speed, the real-time of rotational speed acquisition and accuracy etc are difficult to guarantee.

U.S. production of TI's high-performance DSP processor TMS320F28335 ‘s capture unit can respond to the input pin level changes in a timely manner, and capture the time when level changes occurred, using an edge detection unit to measure the time difference of the external signal which is used to determine signal frequency [1] [2]. This paper using the event manager general-purpose timers module and capture unit, designed a new DSP-based rotational speed measurement system, which can measure engine rotational speed in real-time, both in the field of high-speed and low-speed with high accuracy and reliability.

2 TMS320F28335 and Capture Unit Working Principle

2.1 About TMS320F28335

TMS320F28335 is a 32-bit floating point DSP processor introduced by TI company, which is a Harvard bus architecture, can clock up to 150MHz, with high system

D. Jin and S. Jin (Eds.): Advances in FCCS, Vol. 1, AISC 159, pp. 279–284.
springerlink.com

processing ability; core voltage and I / O voltage of 3.3V, with low power consume; integrated chip Flash memory, read-only memory, random single-port memory, Boot ROM, and external memory interface for external memory expansion; chip integrates a high performance ADC module, eCAN module, McBSP module, SPI module, SCI modules and event management module, with a powerful controlling and signal processing ability [1] [2].

In TMS320F28335 contains EVA and EVB two event managers, each event manager contains general-purpose timer, comparator, PWM unit, capture unit, and orthogonal encoder pulse circuit. Among them, the capture unit is specifically designed to measure the time of the external hardware input signal [1] [2].

2.2 Capture Unit Working Principle

Event Manager, EVA and EVB have three capture units, respectively, timer 1, 2 or 3 and 4 as the time base timer to count independently. When the input pin detected changes in a particular state, the capture unit will capture the current value of the selected general-purpose timer, and lock it in the corresponding two deep FIFO stack. When there are one or more of the effective capture value is latched to the FIFO, the corresponding interrupt flag bit is seted, if not mask the interrupt flag, there will generate a peripheral interrupt request. Every time after a new count value captured and stored in the FIFO, FIFO status register corresponding bits will adjust in real time.

From the capture unit input pin transitions to the corresponding general-purpose timer the count value is latched, it takes two CPU clock delay, and each input of capture unit is synchronized with the internal CPU clock, so, in order to accurately capture the input signal transition change the input signal must be maintained at the current level of two CPU clock rising / falling. Capture unit works is shown in Fig.1.

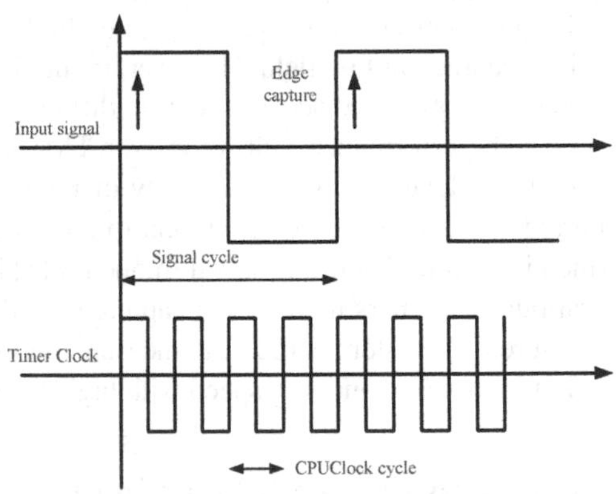

Fig. 1. Capture unit works

3 System Design

3.1 Rotational Speed Measurement Principle

In UAVs, often coaxially installed three-phase alternator with the engine, namely three-phase alternator mounted on the engine shaft, the engine rotates a circle, the engine will drive the three-phase alternator shaft rotating a circle, generator speed is the same with the engine rotational speed. With the engine rotating, three-phase AC generator and their speed is proportional to the rotational speed、 phase separated by 120 °、 the form of three-phase sine wave (A phase, B phase, C phase) alternating current. Therefore, we can use the AC signal that the generator outputs to measure the generator rotational speed, enabling the measurement of engine rotational speed.

Between AC sin signal frequency that the generator outputs and the generator rotational speed generally have the following relationship [3].

$$n=f*60/P \tag{1}$$

Where n is the generator rotational speed, measured in r / min;

f is the frequency of sine wave;

P is the generator rotor poles, in this paper P = 6.

Therefore, only shaping the sine wave generator, can get the corresponding pulse signal, for the count, measuring the sine wave frequency of the generator, according to (Eq. 1) to calculate the generator rotational speed.

3.2 System Hardware Design

In the three-phase AC generator outputs, selecting any one of the phases, as the system input signal, to male cut and filter processing, converting the AC sine wave into DC voltage, filter out high frequency signal crosstalk spikes spark and the ground line noise; by amplification shaping circuit, make the signal amplification process to get the stability of the signal amplitude of the whole formation of the waveform; by voltage comparator circuit to reshape the signal to obtain the square wave signal that has the same frequency with the three-phase AC generator base signal ; in order to prevent external signal interference on the DSP processor, make the square wave signal through the optical isolator for isolating and shaping and by Schmitt inverter output to the capture unit of TMS320F28335, the wave signal's rising or falling edge captured by each capture unit, edge detection unit is used to measure the time difference of the adjacent square wave (ie, general-purpose timer counts) to get square-wave frequency [3] [4].The hardware block diagram is shown in Fig.2.

In addition to the first cycle after, starting the engine, square wave frequency and the adjacent square wave time difference (ie, general-purpose timer counts) generally have the following relationship.

$$f=k*150*10^6/N \tag{2}$$

Where, f is the frequency of the input signal;

K is the input signal frequency coefficient;

N is the time difference between adjacent square wave.

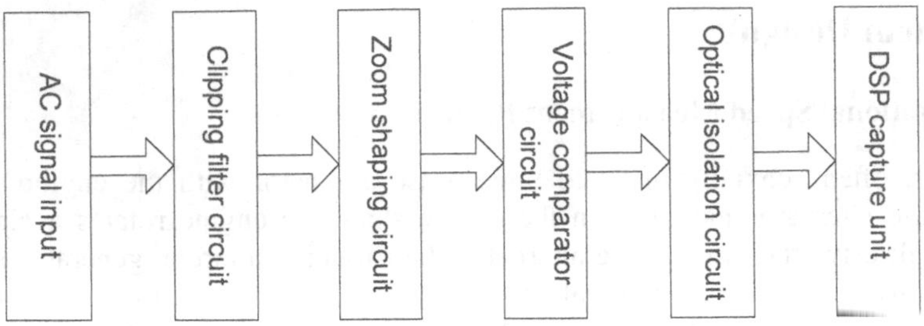

Fig. 2. The hardware block diagram

The input square wave can make 2 to 62 even-frequency by prescaler, can also choose direct way. Using frequency mode, the value will increase as meter, so the frequency measurement will improve too, but it will reduce the measurement in real time.

Before starting the engine, AC generator does not work, AC sine wave frequency is 0; generator maximum speed is not more than 30,000 r / min, AC sine wave frequency does not exceed 3,000 Hz, the frequency measurement range is 0 ~ 3,000 Hz.

When the engine is working at the maximum speed, the use of a direct counting method, the count is about 50,000, the absolute error caused by the timer count is approximately 0.002%, at this moment, rotational speed measurement error is less than 1r/min, much higher than the requirements of that the system on the engine speed error is less than 1%, therefore, the system takes direct way without the divider.

By (type 1), (type 2) shows that between the engine speed n and general-purpose timer counts N have the following relationship.

$$N=1.5*10^9/N \quad (3)$$

4 Software Design

System software includes DSP system initialization, capture counts, interrupt response, calculation of frequency and speed etc. The basic system software flow is shown in Fig. 3.

Starting the engine, the timer does not overflow, during the software designing, each interval is 20s, the timer is cleared, the timer will start counting from the beginning. DSP system initialization mainly includes: initialize the system controller, PLL, Watchdog and clock; set GPIO; set the PIE interrupt vector table; setting general-purpose timer and capture unit etc.

After initialization, when the capture pin input level specified changes occur, the capture unit will write the specified general purpose timer counts two FIFO, when the number of FIFO data is greater than or equal to 2, the capture interrupt flag is setted and generates an interrupt request.

In response to the interrupt, DSP reads general-purpose timer count value in the FIFO, the FIFO is cleared, and according to (Eq. 3), calculating the frequency of square wave and then getting the engine speed.

As TMS320F28335 timer is a 32-bit timer, when the DSP works in the frequency of 150MHz, the timer overflow time is about 28.63s, nearly half a minute, so when the AC sine wave frequencies above 3Hz, the timer will not overflow, less than 3Hz, the timer will overflow.

In order to ensure that before in response to the interrupt, DSP reads general-purpose timer count value in the FIFO, the FIFO is cleared, and according to (Eq.3),calculating the frequency of square wave and then getting the engine speed.

Full use of assembly language programming software for the hardware interface control and high-level language (C + +) programming is simple, readable advantages, the choice of mixing assembly language and high-level language programming style, by certain steps, the original used for control and PC-algorithm program directly into the C language running on the DSP,not only shorten DSP products development cycle, but also improve the readability and portability of the program [5] [6].

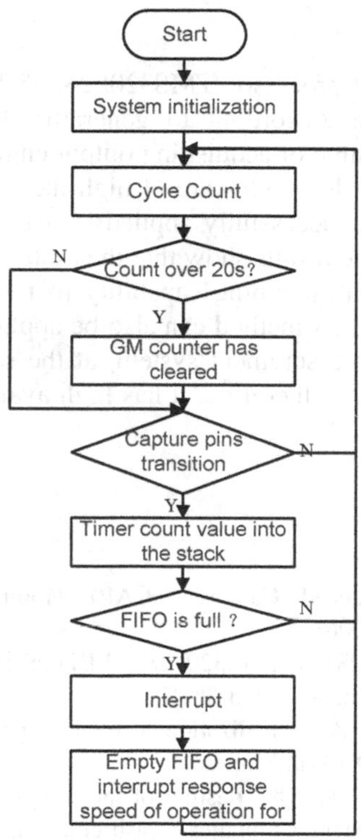

Fig. 3. The basic process of system software

5 Speed Measurements Results

Rotational speed in accordance with the above method, in the lab, within the engine's working area, 6 speed points were selected, with the analog sine wave signal generator for testing, the test results shown in Table 1.

Table 1. Rotational speed acquisition comparison

Frequency for a given value	*Speed set point*	*Speed measurements*	**Error**
500	5,000	4999	-0.02%
1,000	10,000	10,002	+0.02%
1,500	15,000	14,996	-0.027%
2,000	20,000	19,992	-0.04%
2,500	25,000	25,007	+0.028%-
3,000	30,000	30,010	+0.033%

6 Conclusions

In this paper, the core DSP processor TMS320F28335 build the engine rotational speed measurement system. According to generator three-phase AC sine wave frequency, the system is capable of acquisiting output engine rotational speed in real-time, both in high fields and low fields are of high measurement accuracy and real-time. This method has been successfully applied to a UAV engine rotational speed measurement and control. The results show that the system is stable and reliable, with high precision and strong anti-jamming capability to meet the requirements of the engine speed measurement. This method can also be applied to other rotational speed measurement or frequency measurement system, at the same time, as the system is designed simple, with low cost, therefore, it has high availability and promotion of a certain value.

References

1. TMS320x28xx,28xxx Enhanced Capture (eCAP) Module Reference Guide, Texas Instruments Incorporated (2006)
2. TMS320F28335, TMS320F28334, TMS320F28332 Digital Signal Controllers (DSCs) Data Manual, Texas Instruments Incorporated (2007)
3. An, Z., Li, Y., Xu, Y., Su, Z.: A realization of engine speed measurement. Sensors and Microsystems 25, 36–38 (2006) (in Chinese)
4. Hu, Q., Wang, Y., Ren, D.: The design of air engine rotational speed intelligent measurement and control system. Computer Measurement and Control 13, 49–51 (2005) (in Chinese)
5. Su, K., Lv, Q., Chang, T., Zhang, Y.: TMS320X281x DSP working principle and C program development. Beijing University of Aeronautics and Astronautics Press, Beijing (2008) (in Chinese)
6. Wang, N., et al.: DSP-based and application system design. Beijing University of Aeronautics and Astronautics Press, Beijing (2001) (in Chinese)

Accelerate x86 System Emulation with Protection Cache

Weiyu Dong, Liehui Jiang, Yun Zeng, and Yudong Guo

National Digital Switching System Engineering and Tenological Research Center
450002 Zhengzhou, Henan, China
xinbaoer.dong@gmail.com

Abstract. x86 ISA has implemented a very complex protection mechanism, and to fully emulate which means a great labor for system emulators. The paper provides a technic, named protection cache, that can accelerate x86 protection mechanism emulation, by caching and reusing recently accessed descriptors and recently generated protection checking results. Evaluation demonstrates that, when protection cache is enabled, emulation performance of x86 instructions or operations related to protection checking gets a speedup of 11%~24.5%, and performance of system call, page fault handling and I/O intensive applications that frequently using aforementioned instructions or operations are all steadily improved.

Keywords: System Emulation, Protection Cache, System Virtual Machine.

1 Introduction

System emulator emulates the ISA (Instruction Set Architecture) of a specified computer on another computer, as accurately as possible, to construct an equivalent VM (Virtual machine). x86, although a extremely complicated ISA, is widely used and get supported by most system emulators, and the generated x86 system VM is widely used in fields such as cross-platform OS transparent migration[1][2] and dynamic binary analysis[3][4].

Emulated x86 system VM usually suffers from performance degradation compared with the hardware machine, so performance optimization has always been one of the key problems of system emulator. There are already works using technology like dynamic binary translation and optimization[5][6] to address this problem, but due to the complexity of x86 ISA, performance gap between virtual and physical machine is still remarkable, and it is still worthy to continuously reduce the gap.

x86 ISA has very complex protection mechanism. When working in protected mode, many x86 instructions can't be executed unless processor's protection checking is passed, or a protection exception will be raised. Some x86 instructions or operations, such as INT and dispatching of interrupt or exception, are associated with tedious (if not unnecessary) protection checking. Emulating protection mechanism precisely may introduce heavy overhead to emulator.

The paper provides a mechanism, named ***Protection cache (P-cache)***, that can decrease system emulation overhead without sacrificing x86 ISA compatibility. The idea of P-cache is based on two observations. First, x86 processor apply complex

D. Jin and S. Jin (Eds.): Advances in FCCS, Vol. 1, AISC 159, pp. 285–292.
springerlink.com

protection checking operation for some instruction and operation, and it's hard to translate those operations directly into host instruction, and in fact, we have to emulate them by C functions, which further degrade the performance. Second, some x86 instructions or operations related to protection checking may be executed frequently, and it isn't necessary to repeat protection checking if environment has not changed, and we can cache the result of protection checking of afore-executed instructions or operations for their later execution.

2 Emulation Overhead Analysis

Let's take Linux system call as example to illustrate the complexity of x86 processor's protection mechanism and to spot the sources of emulation overhead.

In Linux, system call may be initiated by **INT 0x80**, and the main actions of x86 processor when executing **INT 0x80** instruction are listed as follows: (1) Taking 0x80 as index, get the gate descriptor (GateDesc) from IDT, and check whether the index goes beyond the boundary of IDT, whether the fields in GateDesc is valid. (2) Taking the segment selector in GateDesc as index, get code segment descriptor (CodeDesc) from GDT (Global Descriptor Table), and check whether the descriptor referred by the segment selector goes beyond the boundary of GDT, whether the fields in CodeDesc is valid. (3) Get stack pointer of privilege level 0 from the current TSS (Task State Segment), get stack segment descriptor (StackDesc) based on stack segment selector, and do various validation works. (4) At last, push the return address and EFLAGS register onto stack, load SS and ESP register to complete stack switch, and load CS and EIP register to complete control transfer.

From above case study, we can see that there are mainly two sources of overhead for x86 protection mechanism emulation. The first comes from memory accessing operation, which is very expensive in emulated system VM, when getting various descriptors from descriptor tables. System emulators usually use a region of process address space to simulate the physical memory of virtual machine, and use software to simulate the action of x86 MMU and TLB, So a memory accessing may involve querying virtual TLB to transfer GVA (Guest Virtual Address) into GPA (Guest Physical Address) and further into HVA (Host Virtual Address), and in worse case that a virtual TLB miss happens, may also involve walking guest OS page table to fill virtual TLB. The second comes from various checking on descriptors, and emulating those operations means using a lot of branch statements for protection rules violation detecting, which may affect the host processor pipeline performance.

3 Protection Cache

We argue that, like memory accessing, there is also some kind of ***locality*** attribute on x86 protection checking operation. For OS running on x86 platforms, the configuration of descriptor table is usually correct and seldom changes. Once an instruction or operation passes protection checking, it will pass checking on their later execution with great possibility. So, it makes sense to maintain a cache for an x86 processor to save latest accessed descriptors and the result of protection checking applied for instructions or operations that use these descriptors. Later, when these instructions or operations are executed again, we can obtain the descriptor and protection checking result from the cache, avoiding expensive VM memory accessing and tedious checking. We name the mechanism as ***Protection cache (P-cache)***. Since most OS on x86 rarely use LDT, we only cache descriptors from GDT and IDT in P-cache.

By analyzing x86 instruction set, we recognized two kinds of complex protection checking. First, before entering an interrupt handler, CPU makes checking on the gate descriptor indexed by interrupt type number. Instructions and operations associated with such checking include INT n, INTO, INT 3, hardware interrupt, and exception. Second, on loading segment registers, CPU validates the segment descriptor in GDT indexed by segment selector. Related Instructions include MOV or POP with segment register as destination operand, IRET, LCALL, LRET, LJMP, LSS, LDS, LES, LFS, and LGS. Instructions or operations such as INT and exception, which may conduct inter-segment control transferring or stack switching, also need load segment register, so the second kind of protection checking is also applicable for them.

According to the sources of descriptors, we divide P-cache into two parts, named ***INTR P-cache*** (P-cache for INTeRrupt) and ***SRL P-cache*** (P-cache for Segment Register Loading), and each is used to cache descriptors from IDT or GDT, and corresponding protection checking result, respectively. INTR and SRL P-Cache have the same number of entries as the number of descriptors in IDT and GDT, respectively, and the entries are one-to-one mapped to the descriptors, so given an interrupt type number or segment selector, we may easily locate an entry in P-Cache, which eliminate the needs for cache replacement policy.

As shown in figure 1a, INTR and SRL P-cache have similar logic structures. Each cache entry has two components. The first component, marked as ***Descriptor***, is used to cache descriptor from descriptor table, and the second component, marked as ***Checking Result***, is used to cache the protection checking result.

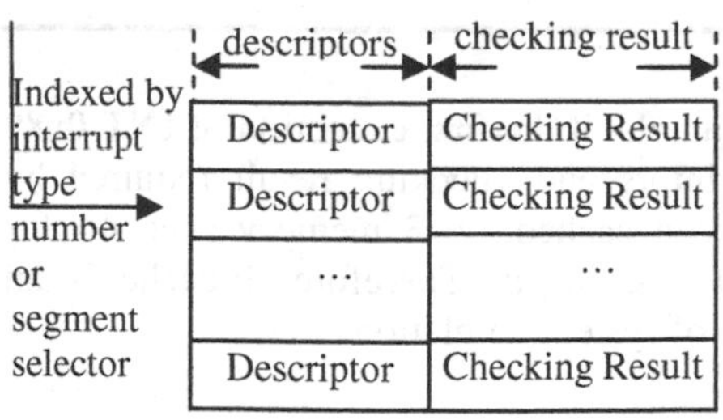

Fig. 1a. structure of P-cache.

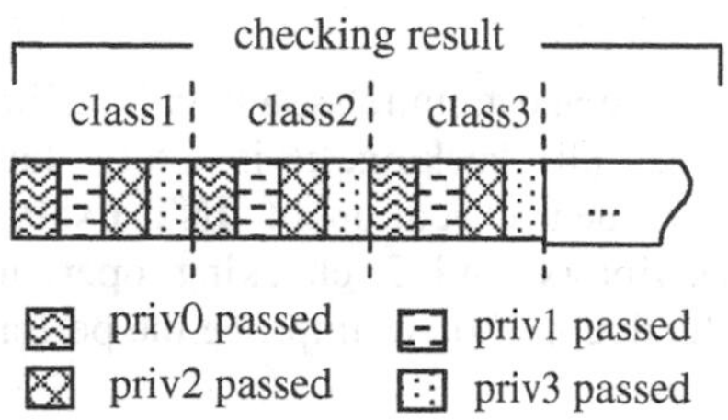

Fig. 1b. layout of protection checking result

Different instructions or operations that refer to the same descriptor may have different protection rule. Meanwhile, an instruction or operation may get different checking result under different privilege level. For example, supposing a selector SEL indexes a read-only data segment descriptor with DPL (Descriptor Priviledge Level) set to 0 in GDT, when CPL (Current Priviledge Level) is 0, instruction "**MOV DS, SEL**" will load descriptor in DS, but "**MOV SS, SEL**" will fail on protection checking (stack segment must be writable). To distinguish various protection rules, we classify aforementioned instructions and operations into 7 categories, as shown in table 1. Accordingly, the Checking Result field in cache entry is organized into a bitmap, as shown in Figure 1b. Each kind of instruction or operation takes 4 bits in the bitmap, and the nth (n=0,1,2,3) bit indicates if the instruction or operation has passed checking under privilege level n, in its last execution.

Given an instruction or operation and its arguments (interrupt type number, selector, CPL, etc.), we call the situation as ***P-cache hit*** if corresponding bit in the bitmap of the located cache entry is set, else we call it as ***P-cache miss***. A P-cache hit means that the emulator can directly extract descriptor from P-cache and no checking is needed for the instruction or operation being simulated. A P-cache miss may occur due to that P-cache has been flushed, or that the instruction or operation is being executed for the first time. On P-cache miss, emulator will read descriptor from the VM memory and do checking operation. If checking can be passed, it save descriptor in P-cache and set corresponding bit in the bitmap, and we call this action as ***P-cache fill***. Some time, when VM is initializing, or when changing to IDT or GDT is detected, system emulator will perform ***P-cache flush*** action, which clear all cache entries.

Table 1. Categories of instructions or operations with different segment level protection checking rules

category	instructions or operations
1	INT n, INTO, INT 3, hardware interrupt, and exception
2	IRET
3	MOV/POP that takes a segment register (excluding SS) as destination operand, and LDS, LES, LFS and LGS
4	MOV/POP that takes SS as destination operand , and LSS
5	LCALL
6	LRET
7	LJMP

Take above linux system call again for example, if the first execution of **INT 0x80** passes all checking, it is likely that descriptors and checking result required by subsequent execution of **INT 0x80** have been cached, so 3 memory accesses for descriptors, and 3 checking operations will be saved. Therefore, P-cache is an effective method to improve the performance of system emulation.

4 Implementation

We give an implementation of P-cache based on the QEMU[7], to which two main augments have been done. The first, which is straightforward, is to add P-cache related fields in the virtual CPU state structure, and modify the simulation methods of instructions and operations listed in table 1, which makes QEMU inquire P-cache before accessing descriptor or doing protection checking. The second is related with P-cache consistency maintenance, i.e., to add codes to QEMU to monitor the changes of the descriptor table to determine if the P-cache contents are stale.

In the remaining of the section, we take INTR P-cache as example to describe our approach used to maintain P-cache consistency. The approach is also applicable for SRL P-cache. The INTR P-cache can be deemed as stale when one of following feathers of IDT has been changed: linear address (Guest Virtual Address, GVA), size, physical address (Guest Physical Address, GPA) and the descriptor contents.

The first two changing to IDT is relatively easy to detect. Since the GVA and size of IDT are kept in the IDTR, which is set by LIDT instruction, changing to them can be detected by instrumenting LIDT. When changes are found, we just flush INTR P-cache.

The latter two changing to IDT is some how complicated to detect. We detect changing to the content of IDT by write-protect it, so GVA of IDT must be obtained at first, and before GVA of IDT is determined, we don't think the contents of INTR P-cache as valid.

For x86, when paging is not enabled, linear address is equal to physical address, so the GPA of IDT can be obtained directly from IDTR. But, when VM enables paging, in theory, although unlikely, IDT may be located at different physical address for different process context. The GPA of IDT can't be determined by querying the page tables of VM, for that IDT may not be in physical memory at all. We locate the GPA of IDT by intercepting virtual CPU's TLB fill events, and detailed method is described as follow. Cache entries in the INTR P-cache are grouped, called ***PCEG*** (P-Cache Entry Group), and cache entries belonging to the same PCEG come from the same physical page. We use data structure ***pceg_desc_t*** describing the relationship among PCEG, descriptor table, virtual page and the physical page, as shown in Figure 2. The ***cache*** field of ***pceg_desc_t*** points to P-cache entries, and the ***va*** and ***pa*** fields record the GVA and GPA of the IDT page that is associated with the PCEG. When virtual CPU accesses IDT, for example to read the associated descriptor for protection checking for some instruction or operation, if there is no mapping from virtual address to physical address in TLB, a TLB-FILL event will be generated. We intercept the event, and get the physical address of the page that is associated with the PCEG, and record the address in the ***pa*** field of ***pceg_desc_t***. Actually, ***pceg_desc_t*** can be seen as an extension of the TLB, which records the mappings between GVA and GPA of IDT only. The mappings in the TLB extension are not affected by TLB replacement policy, but when VM reloads CR3 or executes INVLPG instruction for IDT, we have to invalidate them.

Once GPAs of IDT are obtained, the changing to IDT can be detected by intercepting QEMU's physical memory writing methods, such as stl_kernel, __stl_mmu, etc. Each time QEMU writes VM physical memory, we determine whether the physical address falls into the physical address ranges of IDT, and flush INTR P-cache if it dose.

For P-cache consistency maintenance, following instructions or operations' emulation codes in QEMU are modified: LIDT and LGDT instruction (to detect changing to GVA or size of IDT and GDT), writing to CR0 (to detect protected mode and paging mode enabling or disabling), writing to CR3 and INVLPG (to detect changes of GPA of IDT or GDT), operations of virtual TLB (to obtain the GPA of GDT or IDT), physical memory write operation (to detect changing to the content of GDT or IDT).

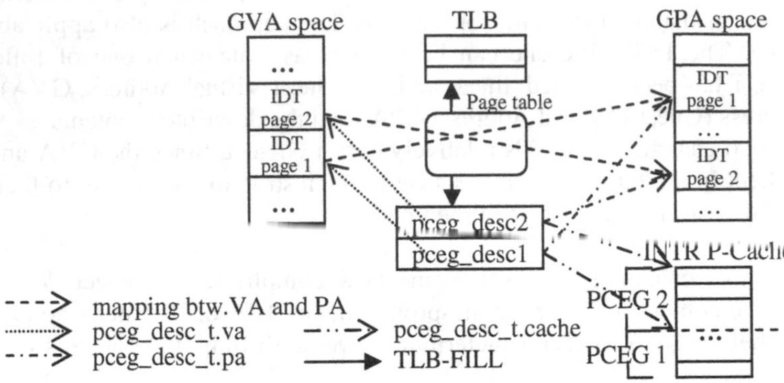

Fig. 2. Relation between associative data structures

5 Evaluation

We evaluate the performance of VM with and without P-cache. The host machine environment is an x86 PC running Fedora linux, and the emulated VM is an x86 PC running Debian linux.

We first evaluate the effect of P-cache by comparing average clock cycles spent by instructions or operations with or without P-cache. The gauge of clock cycle is determined by reading x86 host's time stamp counter register. Result given in Figure 3 shows that, when P-cache is enabled, performance of x86 instructions or operations related with protection checking gets a speedup of 11%~24.5%.

P-cache mainly promotes efficiency of interrupt handling and segment related instructions, so applications that frequently do system calls, that have poor memory locality and frequently generate page exception, or that are I/O-intensive, will benefit.

To test the impact of P-cache on system call and page exception, we construct two micro benchmarks: syscall and mtouch. The former repeatedly calls ***gettimeofday***, and the latter accesses a large range of memory randomly. Figure 4 illustrates the time spent by 2000,000 ~ 10,000,000 times of ***gettimeofday*** call and memory access with or without P-cache. The result shows that, when P-cache is enabled, the efficiency of the system call increases from 9% to 27%, and the efficiency of memory access increases from 7% to 14%. We select tiobench[8], which frequently does system calls (***read*** and ***write***) and triggers much interrupt, to evaluate the impact of P-cache on I/O-intensive applications. Figure 5a and 5b show the average I/O bandwidth and latency when using tiobench to read and write 1GB disk 5 times with or without P-cache. The result shows that, when P-cache is enabled, the bandwidth of write and read is increased by 12% and 13% at most, and read and write delays is decreased by about 9% and 12% at most.

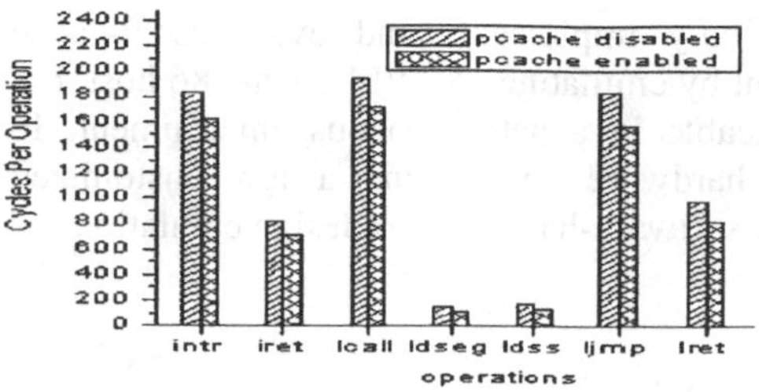

Fig. 3. Clock cycles of instructions or operations

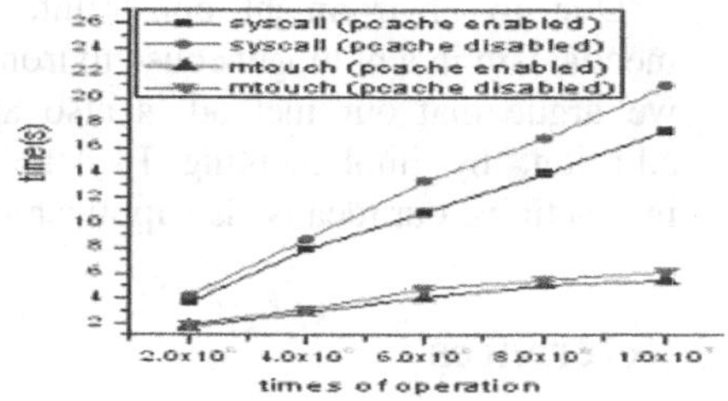

Fig. 4. Performance of syscall and mtouch

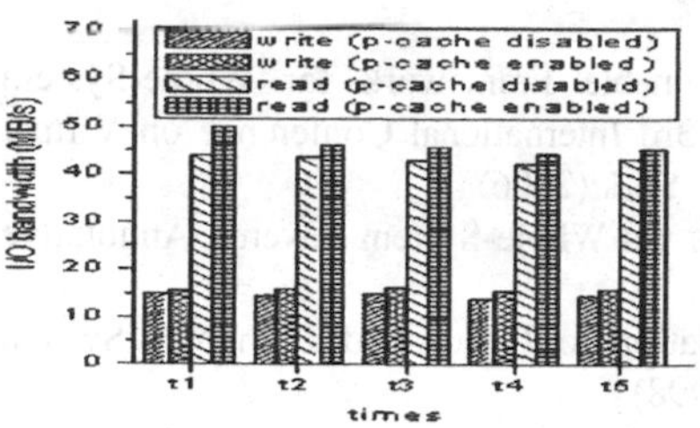

Fig. 5a. Tiobench I/O bandwidth

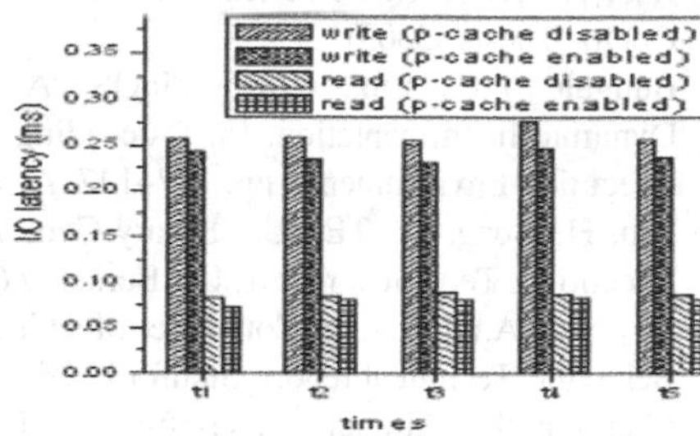

Fig. 5b. Tiobench I/O latency

The overhead of P-cache comes mainly from three aspects, i.e., detecting the GPA of descriptor table in the handler of TLB fill event, detecting changes of descriptor table when writing physical memory of VM, and flushing cache entries caused by the modification of descriptor table. The first two sources of the overhead are negligible since they only add few branch statements to the code base. As to the third overhead, it's observed that OS rarely modify descriptor table. Taking linux for example, IDT is only modified during system boot or driver loading, and GDT is modified only on thread switching since the 6~8th descriptors of GDT are used for thread local storage. We run nbench[9], in which most programs are CPU-intensive, to test impact of P-cache on normal applications, and find no negative effect, as shown in Table 2.

Table 2. Nbench performance with or without P-cache

	P-cache disabled (iteration/s)	P-cache enabled (iteration/s)
Numberic Sort	132.64	132
String Sort	7.0644	7.0644
Bit-Field	3.82E+07	4.31E+07
FP-Emulation	9.4084	9.46
Fourier	4574	4494
ASSI-OPNMENT	2.0355	2.031
IDEA	409.68	408
Huffman	103.48	103.56
Neural Net	1.1891	1.19
LU-Decomposition	48.483	49.12

Due to environment constraint, we only implements and evaluates P-cache mechanism in a homogeneous environment by emulating x86 VM on an x86 host, but we argue that our method is also applicable in a heterogeneous environment. In addition, by implementing P-cache in hardware and adding a few customized instructions, our idea is also applicable for software-hardware co-design emulation.

References

1. Hu, W., Jian, W., Xiang, G., et al.: GODSON-3: A Scalable Multicore RISC Process with x86 Emulation. IEEE Micro 29(2), 17–29 (2009)
2. Cao, H.: Research on Micro-Processor Oriented Dynamic Binary Translation. Philosophy Doctor Thesis (2005)
3. Bungale, P.P., Luk, C.-K.: PinOS: A Programmable Framework for Whole-System Dynamic Instrumentation. In: Proceedings of the 3rd International Conference on Virtual Execution Environments, pp. 137–147. ACM, New York (2006)
4. Yin, H., Song, D.: TEMU - Binary Code Analysis via Whole-System Layered Annotative Execution. Technical report, UC Berkeley (2010)
5. Stephen, A.H.: Using Complete Machine Simulation to Understand Computer System Behavior. Technical report, Stanford University (1998)
6. Ebcioglu, K., Altman, E., Gschwind, M.: Dynamic binary translation and optimization. IEEE Transaction on Computer 50(6), 529–548 (2001)
7. Bellard, F.: QEMU: a Fast and Portable Dynamic Translator. In: Proceedings of the Annual Conference on USENIX Annual Technical Conference, pp. 41–46. USENIX Association, Berkeley (2005)
8. Manning, J., Kuoppala, M.: Threaded I/O Tester, http://sourceforge.net/projects/tiobench/
9. Nbench, http://en.wikipedia.org/wiki/NBench

Overland Flow Pattern of a Computer Simulation Study

Yuqiang Wang[1,*] and Lu He[2]

[1] Zhejiang Water Conservancy and Hydropower College, Zhejiang Hangzhou 10018
[2] Huzhou Water Conservancy Bureau of Zhejiang, Zhejiang Hu zhou 313000
wangyq@zjwchc.com, 329236068@qq.com

Abstract. As a result of overland flow pattern is affected by many factors, which makes the flow simulation is tedious and complicated, the overland flow pattern calculation equations are simplified, and the introduction of the computer carries on the simulation, the simulation process, according to the actual situation of adjustment parameters, realized the need for flow simulation.

Keywords: slope flow, flow, computer simulation.

0 Introduction

Flow pattern is characterized by the slope sheet flow dynamics of the basic arameters, calculated with runoff and sediment transport calculations is directly related to flow pattern of overland sheet flow has important scientific significance. However, overland flow is sheet flow areas, soil type, vegetation cover, land use types, rainfall characteristics and disturbance, sediment, flow, rill density and small changes in geometry [1,2,3], will lead to water accumulation, evacuation, and thus lead to changes in flow pattern, which makes thin-slope flow pattern is very complex, so far overland sheet flow exactly what kind of flow pattern is inconclusive. Overland flow is formed in the rain or melting snow, and gravity along the slope movement in the shallow water, it is more than the rainfall infiltration and build capacity resulting from low-lying ground, through the import channel surface is formed the main part of the river flow, sometimes referred to as sheet flow or cross flow.

As the water depth shallower slope, by land surface conditions (slope boundary layer) is significantly affected, measurement is difficult, making the overland flow of many difficulties. Moreover, the slope susceptible to water depth slope boundary layer, so the resistance is different from the overland flow and out flow pipe flow in the roughness, generally referred to as "effective roughness" [4,5]. In different media for different bed slope of overland flow erosion test data analysis process, as more factors, making the analysis process is very cumbersome and complicated. This paper proposes a mathematical model based on the use of computer simulation of their movement, making the study of soil water movement tends simple.

---

* Yuqiang Wang (1978 -), male, Shaanxi Fuping person. Lecturer, master, mainly engaged in teaching and research work of water conservancy project.

D. Jin and S. Jin (Eds.): Advances in FCCS, Vol. 1, AISC 159, pp. 293–298.
springerlink.com

1 Simplified Mathematical Model of Water Flow

As the effects slope flow pattern of many factors, such as: surface of the plant, slope, soil structure, etc., in general, for large areas of open water, the water caused by its own gravity is the main driver of water movement, so, that water was static pressure distribution along the depth, plus assumed far greater than the depth of water area, the trend of the vertical acceleration due to gravity is small compared to even the former may just be the latter's 10-6 orders of magnitude [6], can be negligible, so it can be simplified Navier Stokes equations in a shallow water flow equations The equations, written in component form[7]:

$$\frac{\partial v}{\partial t}+G\frac{\partial v}{\partial x}+H\frac{\partial v}{\partial y}=b' \tag{1}$$

In which

$$G=\begin{Bmatrix} u & h & 0 \\ g & u & 0 \\ 0 & 0 & u \end{Bmatrix},\ H=\begin{Bmatrix} v & 0 & h \\ 0 & v & o \\ g & 0 & v \end{Bmatrix}$$

Vector $V=\begin{Bmatrix} h \\ u \\ v \end{Bmatrix}$ is the equation for the unknown quantity, u,v are the horizontal and vertical direction of the flow rate; g is the gravitational acceleration, b' is the source term, and the surface atmospheric pressure, the bottom elevation, the surface wind stress, bottom friction related to the above equation (1) assuming u,v the changes in space velocity is not very severe, and ignore its bottom friction, it is possible to linear form:

$$\left.\begin{aligned} \frac{\partial u}{\partial t}+g\frac{\partial h}{\partial x}=0 \\ \frac{\partial v}{\partial t}+g\frac{\partial h}{\partial y}=0 \\ \frac{\partial h}{\partial t}+d\left(\frac{\partial u}{\partial x}+\frac{\partial v}{\partial y}\right)=0 \end{aligned}\right\} \tag{2}$$

Here, d is the water depth, so the above equation can be simplified equation on a high degree of surface of the water:

$$\frac{\partial^2 h}{\partial t^2} = gd(\frac{\partial^2 h}{\partial x^2} + \frac{\partial^2 h}{\partial y^2}) \tag{3}$$

For the listing, when g, d is constant, the equation a non-dissipative equations form a thin layer of water, before joining the appropriate amount of dissipation, the equationbecomes

$$\frac{\partial^2 h}{\partial t^2} - k\frac{\partial h}{\partial t} = gd(\frac{\partial^2 h}{\partial x^2} + \frac{\partial^2 h}{\partial y^2}) \tag{4}$$

On the type, k for dissipation coefficient, g, d for velocity.

We also will be three-dimensional shallow water flow equations, into two-dimensional problem to deal with. Depth of points along the way, and then along the average depth, average depth obtained along the two-dimensional shallow water flow equations, h is the depth setting, z_h for the River at the end of the height, and write $H = z_h + h$, can be can be[6,8]:

$$\frac{\partial h}{\partial t} + \frac{\partial hu}{\partial x} + \frac{\partial hv}{\partial y} = 0 \tag{5}$$

$$\frac{\partial hu}{\partial t} + \frac{\partial huu}{\partial x} + \frac{\partial huv}{\partial y} - fvh = -gh\frac{\partial (h + z_h)}{\partial x} + \frac{1}{\rho}(\tau_x{}^s - \tau_x{}^h) +$$

$$\frac{1}{\rho}(2\frac{\partial}{\partial x}(h\mu\frac{\partial u}{\partial x}) + \frac{\partial}{\partial y}(h\mu(\frac{\partial u}{\partial y} + \frac{\partial v}{\partial x}))) \tag{6}$$

$$\frac{\partial hv}{\partial t} + \frac{\partial hvu}{\partial x} + \frac{\partial hvv}{\partial y} + fvh = -gh\frac{\partial (h + z_h)}{\partial y} + \frac{1}{\rho}(\tau_y{}^s - \tau_y{}^h) +$$

$$\frac{1}{\rho}(2\frac{\partial}{\partial y}(h\mu\frac{\partial u}{\partial y}) + \frac{\partial}{\partial x}(h\mu(\frac{\partial u}{\partial y} + \frac{\partial v}{\partial x}))) \tag{7}$$

Among them, $\tau_x{}^s$ 、 $\tau_y{}^s$ the surface wind stress, $\tau_y{}^h$ $\tau_y{}^h$ the resistance for the bottom entry, which can be expressed as, respectively:

$$\tau_x{}^s = \rho_a C_w u_w{}^2 \cos\beta$$

$$\tau_y^{\ s} = \rho_a C_w u_w^{\ 2} \sin\beta$$

$$\tau_y^{\ h} = C_f u$$

$$\tau_y^{\ h} = C_f v$$

$$C_f = \rho g \sqrt{u^2 + v^2} / C^2$$

Here ρ is the atmospheric density, C_w is wind stress coefficient, u_w is the surface wind speed at 10 meters on the value of. β for the wind direction and the x-axis angle, $C = (1/n)H^{(1/6)}$ for the Chezy coefficient, n is the bottom roughness coefficient.

By the above equation continues to simplify the calculation, the result greatly simplifies the general problem of overland sheet flow numerical simulation of flow pattern work.

2 Finite Volume Method for Solving

2.1 The Basic Idea

The calculation area is divided into a control volume are not repeated, and to each grid point is surrounded by a control volume; will be the solution of differential equations for each control volume integral, which results in a set of discrete equations. One unknown is the dependent variable on the grid point values. To find the volume control points must be assumed that the value of the variation between grid points, assuming the distribution of the value of the distribution of sub-section.

2.2 From the Integral Method of Selecting the Region Seems

Finite volume method is the weighted residual method in the sub-region method; unknown solution from the approximate method seems finite volume method is using a local approximation of the discrete method. In short, sub-regional development of the finite volume method is the basic method.

The basic idea of the finite volume method is easy to understand, and can draw a direct physical interpretation. The physical meaning of the discrete equation is the dependent variable in the limited size of the control volume in the conservation principle, as the dependent variable in the equations that infinitesimal control volume as in the conservation principle. Finite volume method to get the discrete equations, integral conservation requirements of the dependent variable for any set of control volumes are met, the entire computational domain, naturally, are met. This is the finite volume method attractive advantages. There are a number of discrete methods such as finite difference method, only when the grid is extremely fine, the only discrete equations satisfy integral conservation; and the finite volume method even in the case of coarse grid also shows the exact integral conservation.

2.3 For Methods in Terms of Discrete

Finite volume method may be regarded as finite element and finite difference method in the middle of things. Finite element method must assume that the value of the variation between the grid points (both interpolation function), and as an approximate solution. Only consider the finite difference method on a numerical grid point value regardless of changes between the grid points. Only a finite volume method for the node values, which wears a similar finite difference; but finite volume method for controlling the volume integral, we must assume that the value of the distribution between the grid points, which in turn wears with the finite element similar. In the finite volume method, the interpolation function is only used to calculate the control volume integral of the discrete equations obtained after the interpolation function can be forgotten; if necessary, can the different items on the differential equations take a different interpolation function.

3 Conclusion

Visible, finite volume method can be for any such irregular grid, both on the control volume approach, with a nature conservation. But also has the characteristics (rather than flow)-based wind resistance. And, with the excellent performance at the same time, in dealing with similar efficiency and finite difference method, finite element method which is much higher than can be said that the finite volume method reflects the finite element geometric properties, characteristics method and finite difference method the accuracy of the efficiency and conservation of it the disadvantage is that in the irregular grid computing a viscous term, than on a rectangular grid finite difference method trouble, nor did the use of the weak as the finite element solution and the adjoint operator of the second derivative of the concept of the reduced-order.

Interpolation function can be forgotten; if necessary, can the different items on the differential equations take a different interpolation function[8].

Strike through the thin layer of finite volume water flow numerical solution of the equation of state to get at each time step the solution domain in the control of body water depth and velocity triangles, the film formed by the surfaces of these triangles drawn out, get the water surface the shape over time, change the shape of the surface, resulting in a computer simulation.

Acknowledgement. Foundation item: Zhejiang Province Department of education scientific research fund project (Y201016745).

References

[1] Pan, C., Shanguan, Z.: Grass slope erosion, dynamic parameters of the impact. Journal of Hydraulic Engineering 36(3), 371–377 (2006)

[2] Guo, Y., Zhao, T., Sun, B., et al.: Grass slope water dynamics anddelay runoff. mechanism. Soil and Water Conservation Research 13(4), 264–267 (2006)

[3] Sha, J., Bai, Q.: Clay slope rill hydraulic characteristic test of the flow. Sediment Research (6), 39–41 (2001)

[4] Shen, B.: Slope Controlled flooding during the effective roughness of the experimental study. Journal of Hydraulic Engineering (10), 61–68 (1994)

[5] Engman, E.T.: Roughness coefficient for routing surface runoff. J. I. D. Eng., ASCE 112(1), 39–54 (1986)

[6] Zhang, M.: Numerical methods for shallow water wave problem research and application of. China University of Science and Technology Doctoral Dissertation (1998)

[7] Wu, X., Dong, L., Lu, D.: Based on neighborhood propagation wave simulation method. Journal of University of Science & Technology China 3, 278–282 (2010)

[8] Zhu, L.: Based on physical model of the flow animation computer simulation. Northwestern Polytechnical University, a master's degree thesis

Mapping H.264/AVC Fractional Motion Estimation Algorithm onto Coarse-Grained Reconfigurable Computing System

KiemHung Nguyen, Peng Cao, and Xuexiang Wang

National ASIC System Engineering Research Center, Southeast University, Nanjing, China

Abstract. REMUS is a coarse-grained reconfigurable computing system for multimedia and communication baseband processing. Fractional motion estimation (FME) is one of several tools which contribute to H.264/AVC's excellent coding efficiency. ASIC-based hardware accelerators for FME are indispensable in real-time video encoding applications in the past. The paper introduces the mapping of H.264/AVC FME algorithm onto platform of REMUS system in order to demonstrate that complex applications can be mapped with competitive performance onto REMUS platform. Experimental results show that the REMUS can perform FME at real-time speed for CIF/SDTV@30fps video sequences with one reference frame. The implementation method therefore can apply for H.264/AVC encoder in mobile multimedia applications. REMUS system is designed and synthesized by using TSMC 65nm low power technology. The die size of REMUS is 23.7 mm^2. REMUS consumes about 194mW while working at 200MHz.

Keywords: H264/AVC, Fractional Motion Estimation, Reconfigurable Computing, REMUS, Dynamically Coarse-grained Reconfigurable Architecture.

1 Introduction

Many multimedia systems are hand-held systems, such as MP3 players, PDAs and 3G phones, which require capability of handling real-time functions such as communication, camera, video, audio, touch screen, TV, and GPS in time. Besides, because of aiming at mobile multimedia applications, they also have constraints of performance, cost, power, and size.

H.264/AVC is the latest video coding standard developed jointly by ITU-T Video Coding Experts Group and ISO/IEC Moving Picture Experts Group [1]. It is designed to replace all the past video standards in almost all kind of next-generation multimedia applications. So, it defines different profiles and tools to meet the various requirements in different applications. Fractional motion estimation (FME) is one of several tools which contribute to H.264/AVC's excellent coding efficiency, but it also consumes approximately 40% of the total encoding time (for the case of 5 reference frames, +/- 16 search range) [4]. Therefore, an efficient hardware accelerator for FME is indispensable in real-time video encoding applications. Besides ASIC [2-4], FPGA [6], or ASIP [5] approaches which were used for implementing FME in the past, a very promising solution is to map these kernel modules onto CGRAs (Coarse-Grained Reconfigurable

D. Jin and S. Jin (Eds.): Advances in FCCS, Vol. 1, AISC 159, pp. 299–309.
springerlink.com

Architectures) which can achieve high performance approximately ASIC while maintaining a degree of flexibility close to DSP processors. By dynamically reconfiguring hardware, CGRAs allow many hardware tasks to be mapped onto the same hardware platform, therefore also result in reduction in area and power of the design. In the last decade, several CGRAs have been proposed, such as MorphoSys [8], PACT XPP-III [9], ADRES [10], and so on. However, to my knowledge, it has not yet any previous works which map FME algorithm onto CGRAs.

In this paper, we first descript our architecture of REMUS reconfigurable computing system, and then we will present optimizing and mapping FME algorithm, which aims at mobile multimedia applications, onto the platform of REMUS. The experimental results show that implementation meets requirements for high performance, flexibility, and energy efficiency by mobile applications.

The rest of the paper is organized as follows. In section 2, we are introducing overview of the REMUS architecture. FME algorithm is presented in section 3. The section 4 gives implementing method of FME algorithm on REMUS reconfigurable computing system. The experimental results and conclusion is given in the section 5 and 6, respectively.

2 Architecture of REMUS

2.1 Overview of the REMUS Architecture

REMUS, stands for REconfigurable MUltimedia System, is a coarse-grained reconfigurable computing system for multimedia and communication baseband processing, the overall architecture of the REMUS is shown in Fig.1. The REMUS processor consists of an ARM7TEMI, two RPUs, and several assistant function modules, including an interrupt controller (IntCtrl), a Direct Memory Access (DMA) unit, an External Memory Interface (EMI) and a Micro Processor Unit (μPU), etc. All modules are connected with each other by ARM AMBA bus. The ARM7TEMI processor functions as a host processor which is used for controlling application and scheduling reconfigurable hardware tasks. The RPU is a powerful dynamic reconfigurable system, which consists of four Reconfigurable Computing Arrays (RCAs), each of RCA is an array of 8x8 RCs (Reconfigurable Cells). Each RCA can run independently to accelerate computing performance.

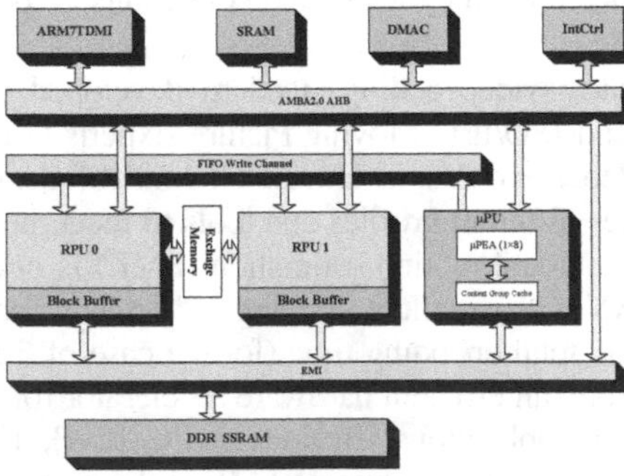

Fig. 1. The overall architecture of REMUS

In comparison with the previous version [11] of REMUS, this version has been improved by adding the μPU and optimizing Data-flow. The μPU is an array of 8 RISC processors (μPEA). The main functions of μPU are: generating configuration words at run-time, selecting configuration context, and executing other tasks which are not suited for RPUs. By mapping computation-intensive kernel loops onto the RPU, control-intensive tasks or bit-level-, float-point-operations based tasks onto μPEA, REMUS can achieve high performance as ASIC while maintaining a degree of flexibility close to DSP processors. To satisfy high data bandwidth requirement of multimedia applications, data flow of REMUS have been optimized to supports a three-level hierarchical memory, including off-chip memory (DDR SSRAM), on-chip memory (SRAM), and RPU internal memory (RIM). In order to accelerate the data flow, a block buffer is designed to cache data for each RPU, and an Exchange Memory (EM) for swapping data between two RPUs.

The function of REMUS is reconfigured dynamically at run-time according to required hardware tasks. Configuration of REMUS includes two aspects: functional configuration of RCA and configuration of data flow in each of RPU. To support such configuration, REMUS is equipped with a Configuration Interface (CI) for each RPU. The CI is responsible for receiving and buffering configuration words, or context, sent from the μPU through FIFO write channel. Function of one RCA can be reconfigured as a whole by the CI. Contexts can be dynamically pre-fetched and cached by a hierarchical configuration memory (CM) which includes off-chip CM, on-chip CM, and configuration register files.

2.2 Architecture of RCA Core

The 8x8 RCA core is composed of an Input FIFO, an Output FIFO, an array of 8x8 RCs, a PRE_INPUT unit, an OUTPUT_SELECT unit, two Constan_REG registers, RIM memory and a controller (Fig. 2(a)). The input and output FIFO is the I/O buffer between external data flow and RCA. RCs array consists of 64 RCs distributed into 8 rows and 8 columns. Each RC can get data from input FIFO or/and Constant_REG, and store back data to output FIFO. Interconnection among two neighboring rows of RCs is implemented by router. Through router, a RC can get results came from an arbitrary RC in the immediately previous row. The Controller generates control signal which maintain execution of RCA accurately and automatically according to configuration information in the Context Registers.

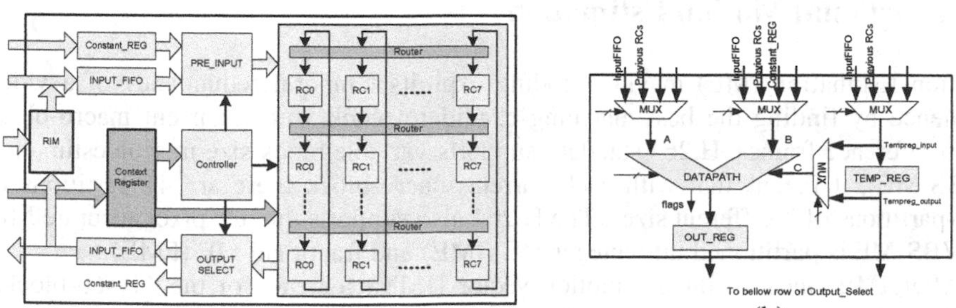

(a) (b)

Fig. 2. The REMUS's CGRA architecture (a) RCA architecture, and (b) structure of one RC

RC is basic processing unit of RCA. Each RC includes a data-path which has capability to execute signed/unsigned fixed-point 8/16-bit operations with two/three source operands, such as arithmetic and logical operations, multiplier, and multimedia application-specific operations (e.g. barrel shift, shift and round, absolute differences, etc.). Each RC also includes a register, called TEMP_REG. The register can be used either to adjust operating cycles of pipeline when a DFG is mapped onto the RCA, or to store coefficients during executing a RCA core loop.

2.3 Execution Model of RCA Core

Executing model of the RCA core is pipelined multi-instruction-multi-data (MIMD) model. In this model, each RC can be configured separately to process its own instructions, and each row of RCs corresponds to a stage of pipeline. The REMUS's RCA core architecture is basically loop-oriented architecture. A computation-intensive loop is first analyzed and represented by data-flow graphs (DFGs) [7]. After that, DFGs are mapped onto RCA by generating information which relate to assigning operation nodes to RCs and edges to interconnection. Finally, these DFGs are scheduled to execute automatically on RCA. Once configured for a certain loop, RCA is operating similar to a this-loop-specific hardware. When all iterations of loop have completed, the loop is removed from the RCA, and other loops are mapped onto the RCA.

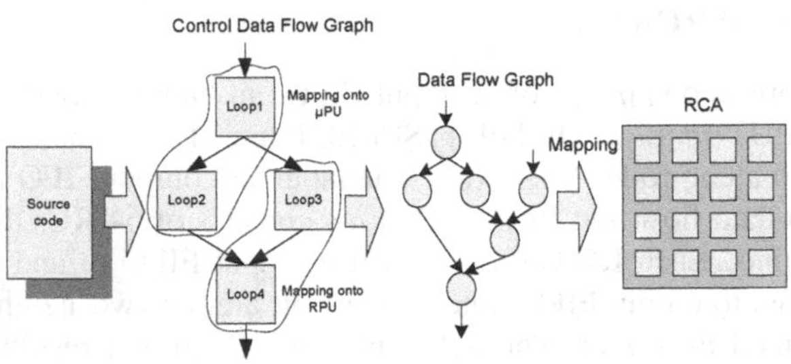

Fig. 3. Mapping of algorithm onto RCA

3 Fractional Motion Estimation

Motion Estimation (ME) in video coding exploits temporal redundancy of a video sequence by finding the best matching candidate block with a current macro-block from reference frames. H.264 standard supports variable block size motion estimation (VBS-ME), it means that with each current macro-block there are 41 partitions or sub-partitions of 7 different sizes. The H.264 also supports quarter-pixel accurate ME, so VBS-ME is partitioned into integer ME (IME) and fractional ME (FME).

After IME finds an integer motion vector (IMV) for each of the 41 sub-blocks, FME performs motion search around the center pointed by IMV and further refines 41 IMVs into fractional MVs (FMVs) of half- or quarter-pixels precision (Fig. 4(a)). FME interpolates half-pixels using a six-tap filter and then quarter-pixels by a two-tap one [1]. In the hardware implementation, FME is usually divided into two stages:

fractional pixels are interpolated from integer pixels, and then residual pixels between current block pixels and interpolated pixels are used to compute cost function for each search position. The cost function is given by:

$$J = SATD + \lambda \times MV\_bit\_rate \cdot \tag{1}$$

where, SATD is the sum of absolute values of Hadamard transformed residual pixels, MV_bit_rate is estimated by using a lookup table defined in the reference software by JVT, and λ is Lagranggian parameter that is derived from quantization parameter to make trade-off between distortion and bit-rate. The Hadamard transform is based on 4x4-block. Computation of SATD is as follows:

$$Diff(\mathrm{i,j}) = \mathrm{Original(i,j)} - \mathrm{Prediction(i,j)}. \tag{2}$$

$$DiffT = H * Diff * H = \begin{bmatrix} 1 & 1 & 1 & 1 \\ 1 & 1 & -1 & -1 \\ 1 & -1 & -1 & 1 \\ 1 & -1 & 1 & -1 \end{bmatrix} * Diff * \begin{bmatrix} 1 & 1 & 1 & 1 \\ 1 & 1 & -1 & -1 \\ 1 & -1 & -1 & 1 \\ 1 & -1 & 1 & -1 \end{bmatrix} \cdot \tag{3}$$

$$SATD = (\sum_{i,j} |DiffT(i,j)|)/2 \cdot \tag{4}$$

4 Implementation

The main challenge of FME implementation is to design efficiently SATD generators and interpolators and to schedule them for high utilizing ratio. In this section, we will present method for mapping FME algorithm, which may be applied to mobile multimedia applications, onto REMUS system. We first focus on analyzing loops of FME, some loop transformations are used to parallelize computation-intensive parts, and then mapping them onto RPU to increase total computing throughput and solve high computational complexity. Besides, some data-reuse schemes are also used to increase data-reuse ratio and, hence, reduce required data bandwidth.

4.1 Algorithm Analysis and HW/SW Partition

The two-stage-refinement FME algorithm of JM reference software [12] is presented by the flow chart as shown in Fig. 4(b). The algorithm is efficient in terms of video compressing quality as well as computational complexity, so it is also commonly adopted in VLSI implementation of FME. However, in order to fit on REMUS system as well as to meet the real-time constraint, loops of the algorithm need to be transformed and some computation-intensive loops must be mapped onto reconfigurable hardware for parallel processing.

As shown in Fig. 4(b), the input of the flow chart is 41 IMVs. In the worst case, these 41 IMVs may point to 41 different positions in the reference frame, therefore, each IMV need to be refined independently into sub-pixel precision. The FME process is implemented sequentially in two steps: half refinement and quarter refinement. There are nine positions searched in both half-pixel refinement and quarter-pixel refinement as shown in Fig. 4(a). In each refinement level, the computing process, in turn, is divided into four parts: sub-pixel interpolation, SATD computation, MVcost generation, the best candidate decision. Because computation of MVCost includes bit-level operations, it is more efficient to map this part onto the µPEA array. The other parts are mapped onto the reconfigurable hardware as shown in the next section.

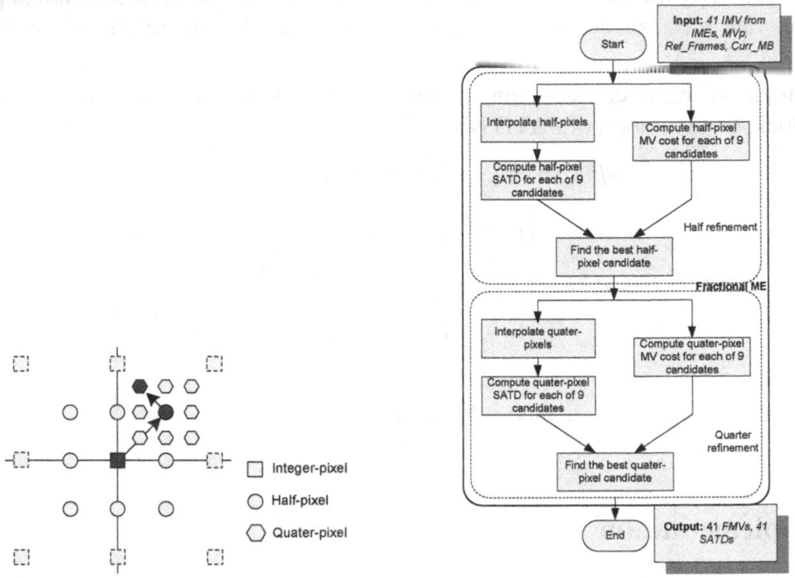

(a) (b)

Fig. 4. (a) Search positions of fractional samples, and (b) Flow chart of FME

4.2 Partitioning and Mapping Hardware Task onto RCA

In this sub-section, we will present mapping interpolation of fractional pixels and computation of SATD, two the most time-consuming tasks of FME, onto the RPU. Because computation of SATD values is based on 4x4-block, this is also the smallest size supported by the H.264. Therefore, we first decompose larger partitions into many 4x4-blocks, and then we can focus on mapping computation of 4x4-block and reuse it in all types of block size.

Sub-pixel Interpolation. Loops of half-pixel interpolating procedure are being unrolled for parallel processing according to available resource of each RCA. Fig. 5(a) shows the DFG of a half-pixel interpolator, which consists of five 6-tap filters. By such unrolling, ten input pixels can be applied to the interpolator so that it can generate 5 output pixels simultaneously. The process of half-pixel interpolation will first generate horizontal half-pixels, and then generate vertical half-pixels. Thanks to the transpose model of RIM (as shown in Fig. 5(b)), we implement interpolation of horizontal pixels as well as vertical pixels with the same DFG of interpolator. After generating half-pixel completely, quarter-pixels are generated by using the DFG shown in Fig. 6. Through the inputFiFo, a row of eleven horizontal adjacent integer-pixels and half-pixels are fed into the ten bilinear filters (denoted as BF) in the 1st row for interpolating ten horizontal pixels. The ten first pixels of inpuFiFo are also shifted down in the TEMP_REG of the RCA. After two cycles, twenty bilinear in the 2nd row generate vertical and diagonal pixels by filtering the corresponding pixels from TEMP_REG and the inputFiFo.

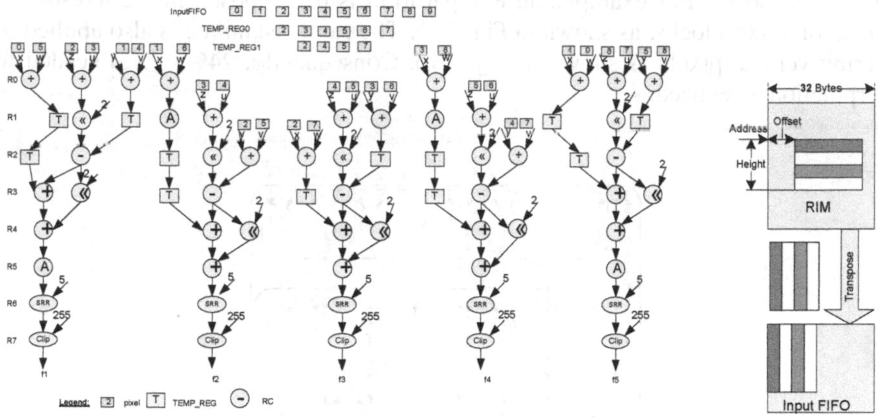

(a) (b)

Fig. 5. DFG for 1/2-pel interpolation (a) and transpose mode of RIM (b)

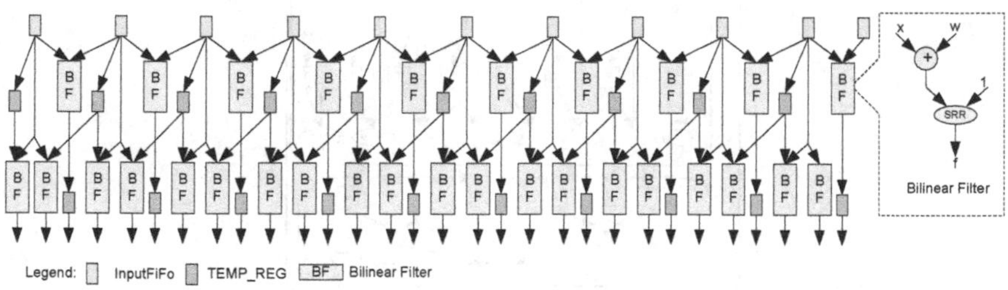

Fig. 6. DFG of Quarter-pixel Interpolator

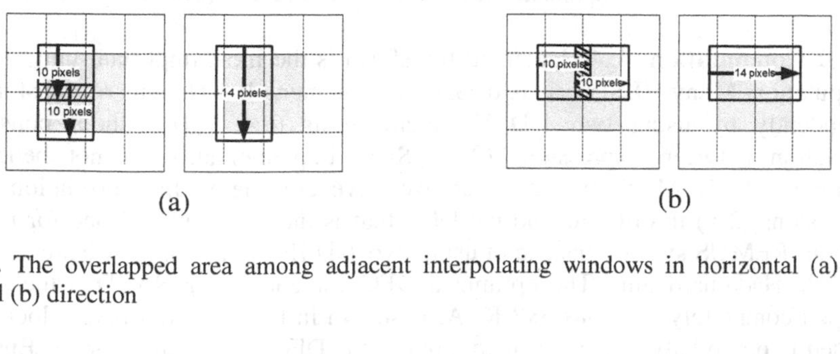

(a) (b)

Fig. 7. The overlapped area among adjacent interpolating windows in horizontal (a) and vertical (b) direction

So far, we have described how to map half-pixel and quarter-pixel interpolation of a 4x4-block onto the REMUS. In order to apply this method to blocks that is larger than 4x4-block, we can decompose them into some 4x4-blocks. However, such decomposing results in large redundant amount of interpolation due to the overlapped area among adjacent interpolating windows as shown in Fig. 7. To overcome this problem, when filtering horizontal pixels, we decompose a large partition into some 4-pixel-wide, 4/8/16-pixel-high blocks, instead of decomposing it into some 4x4-pixel-fixed blocks. For example, an 8x8-partition is decomposed into 2 4x8-blocks, instead of 4 4x4-blocks, as shown in Fig. 7(a). The same technique is also applied for filtering vertical pixels as shown in Fig. 7(b). Consequently, 24% cycles needed for interpolation is reduced.

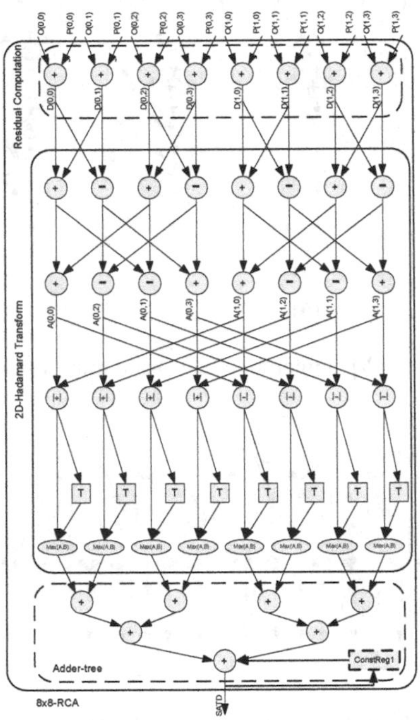

Fig. 8. The optimized DFG for implementing expressions (2-4)

SATD Computation. 4x4-Hadamard transform is the most important unit of SATD computation. Many VLSI designs of SATD implement 4x4-Hadamard transform unit dependently by using two 1-D Hadamard units (e.g. [2]) without considering correlation between expressions (2-4). Such implementation is not benefit for mapping onto REMUS. In the paper, we have considered the correlation among expressions (2-4) in order to find the DFG that is the most optimal one for mapping onto our REMUS system. Instead of using two 1-D Hadamard units, we are also using one 2-D Hadamard unit. The optimized DFG for computing SATD, which can be mapped completely onto one 8x8-RCA, is shown in Fig 8. A 4x4-pixel block is first divided into two halves of 8 pixels to input to the DFG in sequential order. Eight RCs in the 1st row generate eight residues in parallel and transmit them to 2-D Hadamard transform unit. The transformed residues of the 1st half is stored in TEMP_REGs waiting for transformed residues of the 2nd half. Once residues of the 2nd half finish transforming, they are compared with the transformed residues of the 1st half to find maximum values. The maximum values then are transmitted to the adder-tree in order compute SATD value, finish computing SATD of a 4x4-block. The process is fully pipelined with eight latency cycles. No intermediate data is buffered when computing SATD of a 4x4-block, therefore, do not require additional internal memory.

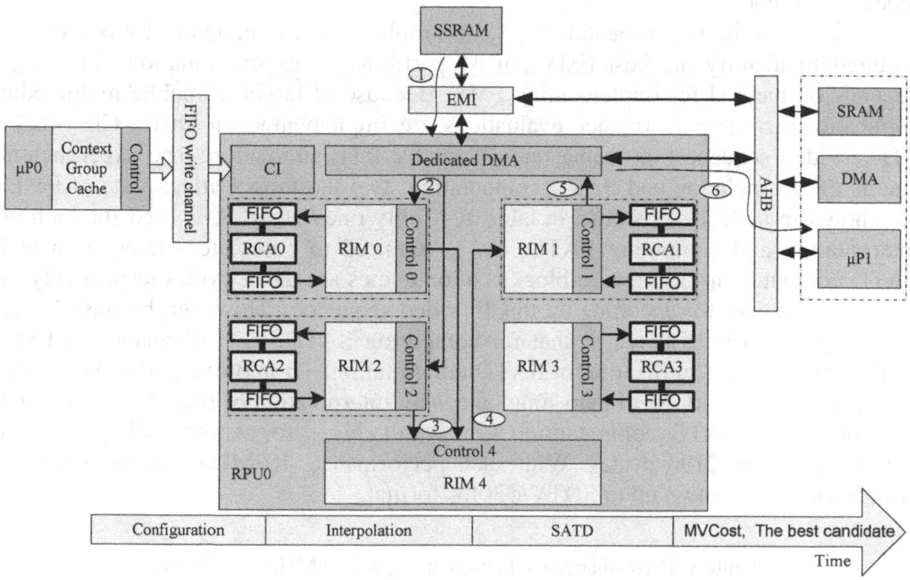

Fig. 9. Mapping of data flow and scheduling for tasks

4.3 Schedule of Tasks on the REMUS

Fig. 9 shows the simplified scheme of mapping and scheduling of tasks onto the hardware source of the REMUS platform. Execution of tasks is scheduled to implement in a four-stage pipeline: configuration, execution of interpolation, execution of SATD computation, and execution of selecting the best candidates. Two tasks, interpolation and SATD computation, are mapped onto {RCA0, RCA2} and RCA1 of the RPU0, respectively. Execution and Data-flow of RPU0 is reconfigured dynamically under controlling of μPO of the μPU. Meanwhile, tasks including MVCost generation, the cost function computation, and selecting the best candidate are assigned to the μP1 of the μPU.

5 Experimental Results

To evaluate the performance of REMUS, a functional RTL model is firstly designed in Verilog, and then synthesized by Synopsys DesignCompiler using TSMC 65nm low power technology. The die size of REMUS is 23.7 mm^2. REMUS consumes about 194mW while working at 200MHz (the maximum frequency is 400MHz).

Let Max_N_{cycles} be the maximum number of clock cycles allowed to identify the best FMVs of 41 partitons in the FME. Define R is the frame rate, is also the minimum rate at which video frames is processed per second. Relation between R and Max_N_{cycles} is given by expression:

$$Max\_N_{cycles} = F_{clk}/(R*N_{MBs}). \tag{5}$$

where, F_{clk} is clock frequency of hardware system, N_{MBs} is the number of macro-blocks per frame.

Also, define the experimental N_{cycles}, or simply N_{cycles}, is number of clock cycles required to identify the best FMVs of 41 partitons by experimentation. So, N_{cycles} depends on method for implementing FME. Because of target at mobile multimedia applications, some performance evaluations are implemented on QCIF, CIF, 4CIF, STDV video sequences at frame rate 30 fps by RTL simulation. The experimental results of interpolation and SATD computation, two the most critical parts of FME, are shown in table 1. As shown in table 1, if only one 8x8-RCA is used for each of interpolation and computing SATD, N_{cycles} required to complete interpolation and SATD computation for a macro-block is 4368 cycles and 2072 cycles respectively, it completely satisfy for encoding up to CIF video sequences. However, because N_{cycles} for interpolation is as twice as that for computing SATD, so performance of FME suffers from the low utilization of SATD computation. To improve performance, we have used two 8x8-RCAs to map simultaneously interpolation of two 4x4-blocks, and pipeline it with SATD computation. As a result, N_{cycles} for whole FME process is reduced to about 2196 cycles. With such performance, REMUS has capability to encode video sequences up to SDTV@30fps format.

Table 1. Performance evaluation at F_{clk} = 200 MHz, R = 30 fps

Frame size		QCIF (176x144)	CIF (352x288)	4CIF (704x576)	SDTV (720x480)
Max_N_{cycles}		67340	16835	4208	4938
Experimental N_{cycles} for interpolation	1 RCA	4368			
	2 RCA	2196			
Experimental N_{cycles} for computing SATD	1 RCA	2072			

6 Conclusion

FME is one of tools which have the highest computational complexity and the largest memory bandwidth in H.264/AVC encoder. Experiments in mapping FME algorithm onto REMUS demonstrate that complex applications can be mapped with competitive performance on REMUS platform. Performance evaluation shows that REMUS can perform FME at real-time speed for CIF/SDTV@30fps video sequences with one reference frame. The implementation method therefore can apply for H.264/AVC encoder in mobile multimedia applications.

Acknowledgments. This work was supported by the National High Technology Research and Development Program of China (863 Program) (grant no.2009AA011700). The authors would like to thank to M.Zhu, C.MEI, B.LIU, JJ.YANG, J.XIAO, and GG.GAO for their helpful discussions and technical support.

References

1. Richardson, I.E.: The H.264 advanced video compression standard, 2nd edn. John Wiley & Sons, Ltd. (2010)
2. Chen, T.-C., Huang, Y.-W., Chen, L.-G.: Fully utilized and reusable architecture for fractional motion estimation of H.264/AVC. In: Proceedings of IEEE International Conference on Acoustics, Speech, and Signal Processing, Montreal, Canada, pp. 9–12 (May 2004)
3. Yang, C., Goto, S., Ikenaga, T.: High performance VLSI architecture of fractional motion estimation in H.264 for HDTV. In: Proceedings of IEEE International Symposium on Circuits and Systems, Island of Kos, Greece, pp. 2605–2608 (May 2006)
4. Wu, C.-L., Kao, C.-Y., Lin, Y.L.: A high performance three-engine architecture for H.264/AVC fractional motion estimation. In: Proceedings of IEEE International Conference on Multimedia and Expo., Hanover, Germany, pp. 133–136 (June 2008)
5. Dang, P.P.: Architecture of an application-specific processor for real-time implementation of H.264/AVC sub-pixel interpolation. In: Real-Time Image Proc. (2009), doi:10.1007/s11554-008-0094-9
6. Ndili, O., Ogunfunmi, T.: Efficient fast algorithm and FPSoC for integer and fractional motion estimation in H.264/AVC. In: 2011 IEEE International Conference on Consumer Electronics, ICCE (2011)
7. Gajski, D., Dutt, N., Wu, A., Lin, S.: High-Level Synthesis, Introduction to Chip and System Design. Kluwer Academic Pub. (1992)
8. Singh, H., Lee, M.H., Lu, G., et al.: MorphoSys: an integrated reconfigurable system for data-parallel and computation-intensive applications. IEEE Transactions on Computers 49, 465–481 (2000)
9. X. Technologies, XPP-III Processor Overview, White Paper (July 13, 2006)
10. Mei, B., Vernalde, S., Verkest, D., et al.: ADRES: An architecture with tightly coupled VLIW processor and coarse-grained reconfigurable matrix, pp. 61–70 (2003)
11. Zhu, M., Liu, L., Yin, S., et al.: A Cycle-Accurate Simulator for a Reconfigurable Multi-Media System. IEICE Transactions on Information and Systems 93, 3202–3210 (2010)
12. JM reference software, `http://iphome.hhi.de/suering/tml/`

A Fault Detection Algorithm Based on Wavelet Denoising and KPCA

Xiaoqiang Zhao[1] and Xinming Wang[2]

[1] College of Electrical and Information Engineering, Lanzhou University of Technology, Lanzhou, China 730050
[2] Gansu Province Medical Science Institute, Lanzhou, China 730050
xqzhao@lut.cn, hpuwww@126.com

Abstract. Data of nonlinear chemical industry process have characterics of containing noises and random disturbances. An improved fault detection method based on wavelet denoising and kernel principal component analysis (KPCA) method is developed, it can not only denoise and anti-disturb, but also can transform nonlinear problems in the input space into linear problems in the feature space. So this can solves the poor performances of principal component analysis (PCA) method in nonlinear problems. The proposed method is applied to Tennessee Eastman (TE) process. The simulation results verify that the proposed method is superior to PCA method obviously in fault detection.

Keywords: Fault detection, KPCA, Wavelet denoising, TE processes.

1 Introduction

Chemical process has characterics of complex nonlinearity. Its actual production data inevitably contain noises and random disturbances. Due to complex production process, inflammable and explosive and toxic raw materials and products, large-scale and continuous production units, there are underlying risk factors in production processes. These risk factors will transform faults even accidents under certain conditions. The faults can damage regular production and endanger operators' safety. So the research of fault diagnosis of chemical process is very vital.

Principal component analysis (PCA) is a kind of data-driven method which can widely apply in process monitoring [1]. But PCA is a kind of method of linear transformation, its performance would greatly degrade for nonlinearity process. Schölkopf proposed kernel principal component analysis (KPCA) method which was an effective nonlinear method [2]. This method has obtained good effects for nonlinear process fault diagnosis. Its basic idea is to map an original input space to a high-dimensional feature space by nonlinear mapping firstly, then process principal component analysis in high-dimensional feature space, thus transform nonlinear problem of input space into linear problem of feature space.

By calculating partial derivative of kernel function, this method can get statistic T^2 and SPE contribution rates for each original variable and is applied wavelet denoising to preprocess data. Aim at characteristics of chemical process, a proposed

D. Jin and S. Jin (Eds.): Advances in FCCS, Vol. 1, AISC 159, pp. 311–317.
springerlink.com

fault detection method based on wavelet denoising and KPCA can not only denoise and anti-disturb, but also can solve fault detection problems like TE process under nonlinear condition.

Tennessee Eastman (TE) process is a practical chemical process model of Tennessee Eastman Chemical Corporation, proposed by Downs and Vogel[3]. Its data are broadly applied to research in many fields such as control, optimization, process monitor and fault diagnose. The proposed method is applied to TE process to illustrate its effectiveness.

2 Wavelet Denoising

Due to chemical process data with noises and disturbances, Probability of detecting and false alarm rate would increase greatly by directly using these data for fault diagnosis. So these data need to preprocess to achieve denoise and anti-disturbance.

Wavelet has characteristics of well time frequency localization, special denoising ability and convenient extracting weak signals for signal processing [4]. Owing to nonlinear wavelet transform threshold value method has merits of nearly complete restraining noises, comprehensive suitability and fast calculating speed, this method is used to this paper. The key step of nonlinear wavelet transform threshold value method for denoising is how to choose threshold value and treat threshold value[5]. When handling threshold value for wavelet coefficient, hard threshold value is coarser than soft threshold value. Soft threshold value is applied to compare signal absolute value with specified threshold value. It becomes 0 when less than or equal to the specified threshold and it becomes difference value of this value and threshold value when greater than the specified threshold:

$$\hat{\omega} = \operatorname{sgn}(\omega) \times \max(0, |\omega| - \lambda) \tag{1}$$

The denoising process is as following. First, actual signals are decomposed by wavelet, and decomposed level N is determined. So noises are usually contained in high frequency. Then high coefficient of wavelet decomposition is disposed by quantifying threshold value. Finally, wavelet reconfiguration is obtained by No.N level low frequency of wavelet decomposition and quantifying 1-N level high frequency coefficients. So this can eliminate noises.

3 Fault Detection Strategies Based on Wavelet Denoising and KPCA

3.1 KPCA Algorithm

PCA is a linear transform method, but KPCA is a nonlinear transform method which maps original input space ($x_1, x_2, \cdots, x_N \in \boldsymbol{R}^m$, here N is number of samples, m is dimension number of measured variables) into high-dimensional eigenspace F by nonlinear mapping $\boldsymbol{\phi}$ (namely $\boldsymbol{\phi} : \boldsymbol{R}^m \rightarrow \boldsymbol{F}$), then principal component analysis is processed in the high-dimensional eigenspace F. Consequently, the nonlinear problem in input space is transformed a linear problem in eigenspace [6]. Assumed $\phi(x_i) = \phi_i$ as x_i mapping, the covariance matrix of eigenspace F can expressed as:

$$C^F = \frac{1}{N}\sum_{i=1}^{N}\phi_i\phi_i^T \tag{2}$$

Assumed eigenvalue of matrix C^F is λ, eigenvector is as v, a coefficient α_i ($i = 1,\cdots,N$) can use ϕ_i to express v. Eigenvector v can be mapped by samples of eigenspace:

$$v = \sum_{i=1}^{N}\alpha_i\phi_i \tag{3}$$

$$\lambda \mathbf{v} = \mathbf{C}^F \mathbf{v} \tag{4}$$

By equation (4), maximum λ corresponding v is first principal component and minimum λ corresponding v is last principal component, so

$$\lambda\langle\phi_k, v\rangle = \langle\phi_k, C^F v\rangle,\ \phi_k = \phi(x_k),\ k = 1,\cdots,N \tag{5}$$

Combined equation (4) with equation (5), then:

$$\lambda\sum_{i=1}^{N}\alpha_i\langle\phi_k,\phi_i\rangle = \frac{1}{N}\sum_{i=1}^{N}\alpha_i\langle\phi_k,\sum_{j=1}^{N}\phi_j\rangle\langle\phi_j,\phi_i\rangle \tag{6}$$

Defining matrix $K \in R^{N\times N}$, let$[K]_{ij} = K_{ij} = \langle\phi_i,\phi_j\rangle$.

$$\lambda N\boldsymbol{\alpha} = K\boldsymbol{\alpha},\ \boldsymbol{\alpha} = [\alpha_1,\cdots,\alpha_N]^T \tag{7}$$

Standardization is done before processed principal component in eigenspace F. K could be replaced:

$$\hat{K} = K - I_N K - K I_N + I_N K I_N \tag{8}$$

Where I_N equals to multiplication of $1/N$ and $N\times N$ unit matrix $E \in R^{N\times N}$. So, this is equivalent to solve eigenvalue problem of equation (7) and to process principal component in eigenspace. Eigenvector v of matrix C^F can be worked out by eigenvector $\boldsymbol{\alpha}$ of matrix K and satisfied

$$\langle v_k, v_k\rangle = 1,\ k = 1,\cdots,p \tag{9}$$

Where p is number of principal components. So principal components can be calculated by calculating projection of mapping data on the eigenvector v_k.

$$t_k = \langle v_k, \phi(x)\rangle = \sum_{i=1}^{N}\alpha_i^k\langle\phi(x_i),\phi(x)\rangle = \sum_{i=1}^{N}\alpha_i^k k(x_i,x) \tag{10}$$

In order to solve eigenvalue problem of equation (7), principal component vectors of eigenspace can directly calculated by using equation (10) from input space and introducing kernel function $k(x, y) = \langle \phi(x), \phi(y) \rangle$. This can avoid calculating nonlinear mapping directly.

3.2 Fault Detection Strategies

Fault detection algorithm based on wavelet denoising and KPCA is similar to PCA. It uses statistics T^2 and SPE to detect faults in eigenspace. T^2 represents a kind of measure of model internal variation. It is standard quadratic sun of principal element vectors and is defined as:

$$T^2 = [t_1, \cdots, t_p]\Lambda^{-1}[t_1, \cdots, t_p]^T \tag{11}$$

Where t_i can get through equation (10), Λ^{-1} is an inverse matrix of diagonal matrix which is formed by eigenvalues of principal elements. The control limit of T^2 can get by F distribution:

$$T^2_{\lim} = \frac{p(N-1)}{N(N-p)} F_\alpha(p, N-p) \tag{12}$$

Where p is number of principal elements, N is number of samples.

Square prediction error (SPE) is called Q statistics. It represents error between variation trend and statistical model for every sample. It is a kind of measure of model external data variation and is defined as:

$$SPE = \| \phi(x) - \phi_p(x) \|^2 = \sum_{i=1}^{N} t_i^2 - \sum_{i=1}^{p} t_i^2 \tag{13}$$

The control limit of SPE statistic is:

$$SPE_{\lim} = g\chi_h^2 \tag{14}$$

The procedures of fault detection are:

(1) The previous nonlinear wavelet transform soft threshold method is applied to preprocess data for denoising.

(2) The model is established.

① Normal sample data after denoising are standardized by variable amplitude and variance.

② Kernel matrix K is calculated and deals with mean value centralization by using equation (8).

③ Principal element t_k is calculated for normal data in eigenspace.

④ T^2 and SPE statistics of normal data are calculated.

⑤ The control limits of T^2 and SPE are determined.

(3) On-line fault detection:

① New measuring data are denoised and standardized.

② Kernel matrix K is calculated and deals with mean value centralization by using equation (8).

③ T^2 and SPE statistics of test data are calculated.

④ If T^2 and SPE exceed control limits of T^2 and SPE for modeling under normal condition, this indicates some faults occurred.

4 Simulation Research

4.1 Tennessee Eastman (TE) Process

TE process is a large-sample-size complex nonlinear system, and data could be downloaded from http:// brahms. scs. uiuc. edu. Flow chart of Tennessee Eastman process is as Fig. 1. TE process includes 12 control variables, 41 measurement variables (including 22 continuous measurement variables and 19 components measure values), and 21 preset faults (including step types, random variation types, slow drift types, sticking types and constant position type and so on), all of the processes contain Gauss noises.

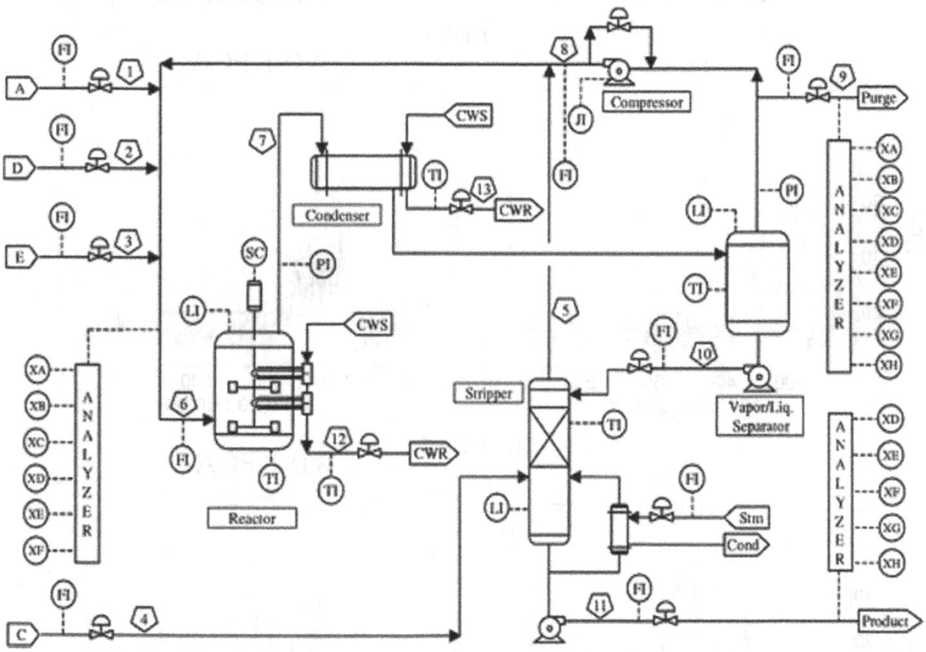

Fig. 1. Flow chart of Tennessee Eastman process

4.2 Fault Detection Based on Wavelet Denoising and KPCA

The fault detection method based on wavelet denoising and KPCA is applied to TE process to compare performances with PCA. These methods are used to detect and analyze 21 faults of TE process respectively. In the simulation experiment, data need to preprocess to denoise by nonlinear wavelet transform soft threshold. denoising, kernel function uses Gaussian kernel function: $k(x, y) = \exp(-\|x - y\|^2 / \sigma)$. Kernel parameter σ is chosen 400 and SPE statistic is used to monitor.

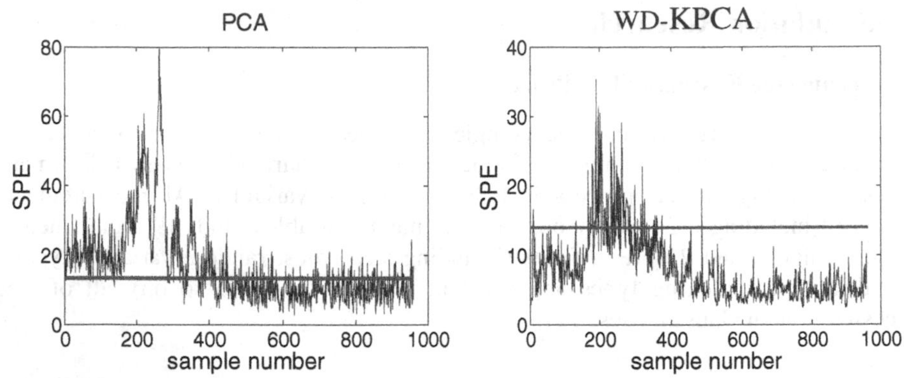

(a) Fault 1

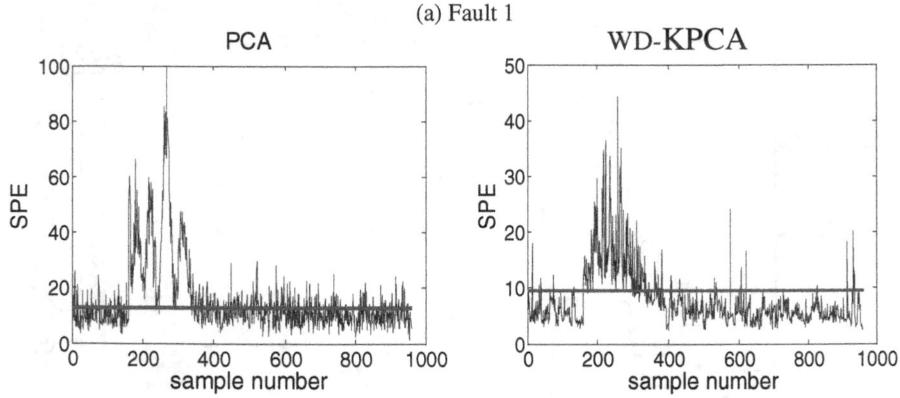

(b) Fault 5

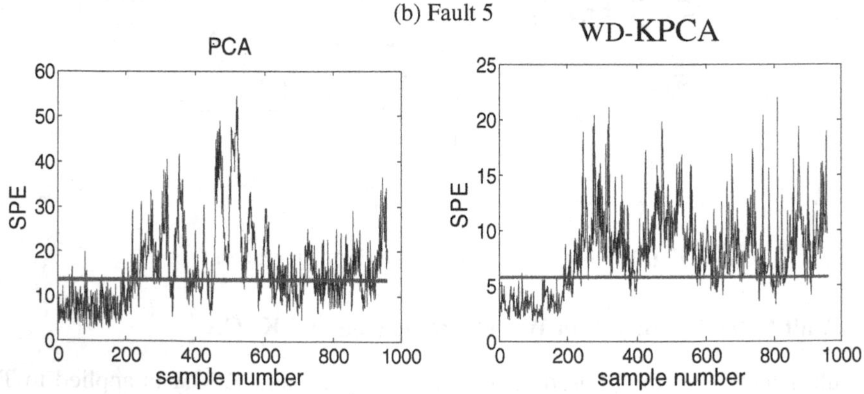

(c) Fault 8

Fig. 2. Detection results of Fault 1, 5, 8

Fig.2 is fault detection SPE charts of fault 1 and fault 5 and fault 8 respectively. Left charts are SPE graphs of fault detection by using PCA and right charts are SPE graphs of fault detection by using wavelet denoising and KPCA. From Fig.2, the fault detection method based on wavelet denoising and KPCA is obviously superior to PCA method. The former can accurately and timely detect occurred faults. Its fault detecting and false alarm rates are inferior to the latter. So fault detection can achieve good effects for chemical process using the fault detection method based on wavelet denoising and KPCA.

5 Conclusions

Aiming at data with noises and random disturbances of nonlinear chemical process, the fault detection method based on wavelet denoising and KPCA is proposed. First, data need to preprocess to denoise and anti-disturb by using wavelet denoising, then fault detection is processed by using KPCA. So this can solve shortcoming of poor performance using PCA to nonlinear system. The simulation results show that the proposed method is effective to detect fault for TE process.

Acknowledgments. This works is supported by Master Tutor Project of Education Department of Gansu Province (No. 1003ZTC085).

References

1. Wang, X., Kruger, U., Irwin, G.W.: Process monitoring approach using fast moving window PCA. Industrial & Engineering Chemistry Research 44(15), 5691–5702 (2005)
2. Schölkopf, B., Smola, A., Müller, K.R.: Nonlinear component analysis as a kernel eigenvalue problem. Neural Computation 10(5), 1299–1319 (1998)
3. Downs, J.J., Vogel, E.F.: A plant-wide industrial process control problem. Computers and Chemical Engineering 17(3), 245–255 (1993)
4. Monsef, H., Lotfifard, S.: Internal fault current identification based on wavelet transform in power transformers. Electric Power Systems Research 77, 1637–1645 (2007)
5. Singh, G.K., AlKazzaz, S.A.S.: Isolation and Identification of Dry Bearing Faults in Induction Machine Using Wavelet Transform 42, 849–861 (2009)
6. Cho, H.W.: Nonlinear feature extraction and classification of multivariate process data in kernel feature space. Expert Systems with Applications 32(2), 534–542 (2007)

Fig. [illegible] [illegible] charts of fault 1 and fault 5 and fault [illegible] respectively. Left charts [illegible] fault detection by using PCA and right charts are SPE graphs of fault detection [illegible] wavelet denoising and KPCA. From Fig. 2 the fault detection method based on wavelet denoising and KPCA is obviously superior to PCA [illegible] [illegible] its fault [illegible] [illegible] fault detection [illegible] [illegible] method based on wavelet [illegible] KPCA.

[illegible] Conclusions

[illegible] data with noises and [illegible] [illegible] process, the fault detection method [illegible] wavelet denoising and KPCA is proposed. First [illegible] [illegible] by using wavelet denoising, then fault detection [illegible] processed by [illegible] KPCA [illegible] [illegible] shortcoming of nonlinear [illegible] the PCA [illegible] [illegible] results show that the proposed method is effective [illegible] detect fault [illegible]

Acknowledgements. This [illegible] supported by [illegible] [illegible] of Education [illegible] [illegible]

References

[illegible] PCA [illegible] [illegible] 51, [illegible]
[illegible] component [illegible] a kernel eigenvalue problem [illegible] [illegible]
[illegible] [illegible] principal components and Neural Computation [illegible]
[illegible] wavelet [illegible] [illegible] Electric Power [illegible]
[illegible] [illegible] [illegible]
[illegible] [illegible]

Research on Application Integration in Digital Campus Based on JBoss ESB Platform

Haicun Yu[1], Jiang Ma[2], and Xuemin Yang[2]

[1] Modern Education Technology Center, TangShan Vocational & Technical College, TangShan, China
[2] Network & Education Center, TangShan College, TangShan, China

Abstract. ESB (Enterprise Service Bus) is a middleware technology which is a key technology to implement SOA. This paper discusses the relation among the Web Services、 SOA and ESB, and also proposes an implementation scheme of application integration based on JBoss ESB platform.

Keywords: Enterprise Service Bus, Service-Oriented Architecture, application integration, JBoss ESB platform.

1 Introduction

At present, the major shortcomings of information construction are lack of unified planning, construction level uneven and no unified standard construction in the university's information construction. As no account of the global university departments of general in the original design and development, application system of each college or department do their own way and in them own field, then has formed a lot of information isolated islands. These Colleges and Universities departments urgently need to integrate these independent application systems and their databases in order to realize information sharing.

Application Integration is the key point in the digital campus construction. By using different hierarchical integration technology the platform of application integration links all kinds of information isolated islands to share information, complete internal process integration, reduce resource consumption, strengthen the cooperation between various departments, thus to create more value for colleges and universities.

2 The Existing Technical Route on Information Integration

There are two kinds of solutions in information integration. One kind is based on data integration technology as the core to construct the new uniform application software platform by the global share data center. This plan would migrate all kinds of application systems to a new unified application management system. Combined with information portal, the uniform application platform solution is easy to achieve business integration, unified identity authentication and single point login, but it can't protect or make full use

D. Jin and S. Jin (Eds.): Advances in FCCS, Vol. 1, AISC 159, pp. 319–324.
springerlink.com

of the existing investment, and it takes a lot of financial and material, and system 'lack of flexibility and expansibility. In addition, data analysis and data process are needed after the data integration work and analyzed with the work, this integration mode does not apply to some simple applications.

Another solution is based on SOA integrated technology. In this mode, the function of software on the network is in the form of services which can be distributed in different physical locations. These services can discover and transfer each other. Because of the loose coupling and platform-independent, they can recombine according to the frequent change of the business demand. This solution not only integrates the existing application system and isolated legacy system according to the change of business need, but also accommodates future increasing and the dynamic changes of the information needs.

3 The Service Integration Technology Based on Web Services

Web Services is very suitable for realization of the SOA technology. In essence, Web Services is a self-description Modular Components based on the network and the Distributed. SOA is an architectural style for building software applications that use services available in a network such as the web. It promotes loose coupling between software components so that they can be reused. Applications in SOA are built based on services. In addition, SOA still needs safety, strategy management, reliable information and the support of the accounting system, so as to effectively work. Web Services is not the only technology to realize SOA, but one of the best choices.

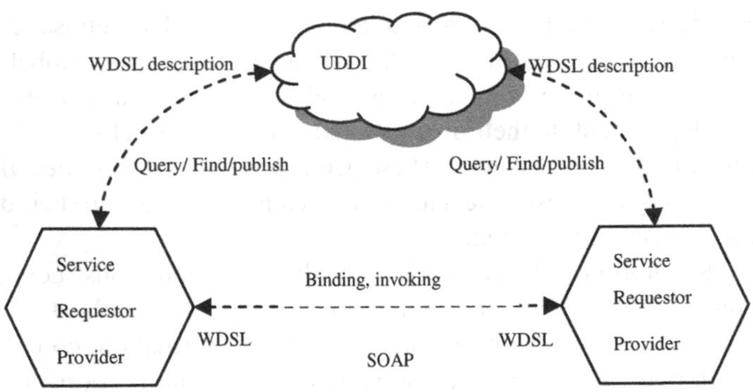

Fig. 1. The process of services transfer in Web services technique

3.1 The Process and Operating Mechanism of Web Service

Web services describes a standardized way of integrating Web-based applications using the XML, SOAP, WSDL and UDDI open standards over an Internet protocol backbone. XML is used to tag the data, SOAP is used to transfer the data, WSDL is used for describing the services available and UDDI is used for listing what services are available.

3.2 An Overview of the ESB (Enterprise Service Bus)

An enterprise service bus (ESB) is a software architecture model used for designing and implementing the interaction and communication between mutually interacting software applications in SOA. By use of SOAP, XML, WSDL and UDDI it is actually based on message queue to support SOA for distributed computing. Its core functions include message exchange, service mapping, context routing communication protocol conversion and so on. ESB provides a basic facilities to go further in decoupling between service provider and consumers, thus make application system has better openness, cooperation and expansibility.

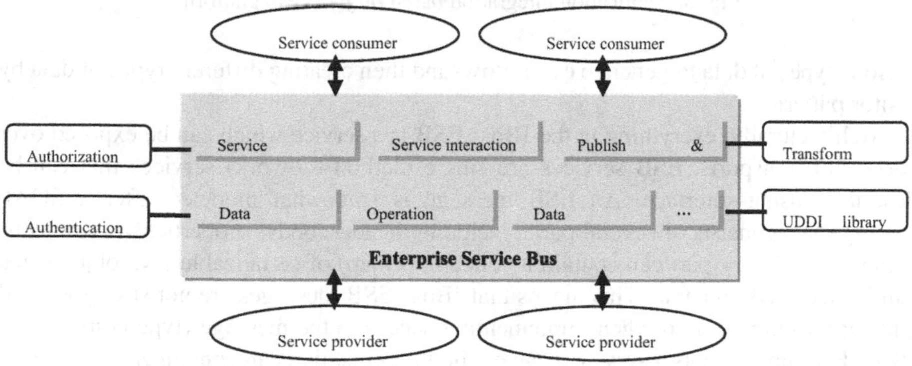

Fig. 2. The ESB Architecture

4 Digital Campus Application Integration Based on JBoss ESB Platform

4.1 The Overview of JBoss ESB

The JBoss Enterprise SOA Platform is free software/open-source Java EE-based Service Oriented Architecture (SOA) software. The JBoss Enterprise Service Bus software is part of the JBoss Enterprise SOA Platform. The software is middleware used to connect systems together, especially non-interoperable systems.The JBoss ESB is based on Rosetta which is producted by the JBoss company. In JBoss ESB, JBossMQ is the message layer, the JBoss rules provides routing Function, and the JBPM implements Web Services Choreography.

4.2 The Building of Application Integration Platform Based on JBoss ESB

At present, there are two ways to call web service in JBossESB. One is through the Gateway listener, another is directly through the Service Invoker API to call.. JBoss ESB use Smooks to realize data conversion. The basic principle of Smooks is using

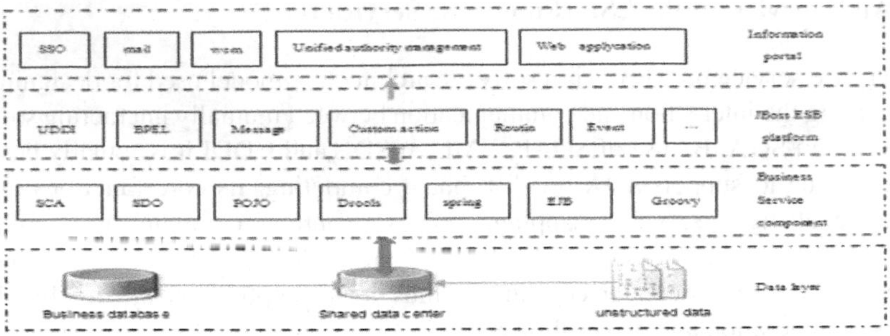

Fig. 3. Application integration based on jboss esb platform

various types of data to generate event flow, and then creating different types of data by visitor pattern.

Architecturally, everything in the JBoss ESB is a service which can be exposed over variety of transports. ESB services are single method (dowork) services that can be described using interface. An ESB message is somewhat modeled after a SOAP message and consists of several parts, including header, body, properties, attachments, context, etc. Every part can contain a collection (map) of serializable java objects that can be accessed by name. This means that JBoss ESB messages are not strongly typed and care should be used when implementing access to the message (type conversion). JBoss ESB implements a service as an explicit pipeline that can contain arbitrary set of actions, each of which implements either service business functionality or infrastructure support, for example storing service message in the message store.

JBoss ESB uses JUDDI to keep the service registered and the service metadata saved. Among them the service request is a through the network addressing entity, it accepts and execute user request. It will be his own service and interface contracts issued to service registration center, so that users can find and service access to the service. Any application system may be the role of the service request. By calling the ESB exposed the Web service to the public to existing service module calls.

In the construction of digital campus, data center is the core, it uses the perfect data synchronization mechanism to ensure that the data uniqueness and effectiveness. The JBoss ESB is the infrastructure framework to realize data share and data exchange between the web service providers and the web service consumers, and then realize the business process integration among applications. Finally, information portal integrates application system to presents information from diverse sources in a unified way and provide unified access control for different users though the standard application interface.

4.3 The Development Environment and Deployed

JBoss ESB 4.4 is used and deployed on the JBoss application server. Business process engine is implemented by the ActiveBPEL5.0 which is deployed on Tomcat 6.0. Business process modeling adopts the ActiveVOS Designer. Because JBoss esb 4.4 Series is based on JBoss AS, so there is a kind of expansion class - JBoss4ESB Deployer which is responsible to monitor analysis the JBoss esb bag. In $JBossesb.sar/META-INF/JBoss-service, the service of JBoss4ESBDeployer is definited. JBoss-esb. XML configuration is divided into two parts. Providers and services

```
<mbean code="org.JBoss.soa.esb.listeners.config.JBoss4ESBDeployer"
name="JBoss.esb:service=ESBDeployer">
<depends>JBoss.esb:service=ESBRegistry</depends>
<depcnds>JBoss.esb:service=JuddiRMI</depends>
</mbean>
```

JBoss AS supports hot deployment, that is to say the war files can be directly copied to the directory of $JBoss \ server \ default \ deploy. Data sources are configured by the means of the XML files. The following is the mysql-ds.xml profiles to set MySql database for data source:

```
<?xml version="1.0" encoding="UTF-8"?>
<datasources>
<connection-url>jdbc:mysql://127.0.0.1:3306/test</connection-url>
<driver-class>com.mysql.jdbc.Driver</driver-class>
</datasources>
```

The action of static message routing provided static and dynamic routing. The route is assigned by the order of the service name and catalogs. One sample of static routing is as follows:

```
<action  class="org.jboss.soa.esb.actions.StaticRouter" name="routeAction">
<property name="destinations">
<route-to service-category="routerToDisplay"
service-name="DisplayRouterListcner" />
<route-to service-category="routerToFile" service-name="FileRouterListener" />
</property>
</action>
```

5 Conclusion

In the ESB architecture, all sorts of system application services are connected to the bus to avoid the various system application service direct connection, furtherly realize decoupling of service. Open source ESB can well realize WebServices service integration, message routing, data transformation, and some other functions. Using JBoss ESB as bus type agent can flexible configurate different service from heterogeneous system and realize application integration accessible.

References

1. Krafzig, D., Banke, K., Slama, D.: Enterprise SOA:Service-Oriented Architecture. Best Practices. Tsinghua University Press (July 2006)
2. Polar lake White paper. Understanding the ESB, http://www.polarlake.com/files/esb.pdf
3. IBM White paper. IBM SOA Foundation:An architectural introduction and overview (December 2005), http://www.ibm.com/developerworks/webservices/library/ws-soa-whitepaper/
4. JBoss ESB Development, http://community.jboss.org/wiki/JBossESBDevelopment
5. Soo, H.C., Hyun, et al.: Design of Dynamic Composition. Handler for ESB-based Services. In: IEEE International Conference on e-Business Engineering, vol. 09, pp. 287–294 (2007)

The Research on DDoS Attack Based on Botnet

Ling Jia

The Chinese People's Armed Police Force Academy, Lang Fang, China
jialing123321@163.com

Abstract. In view of the current DDoS attack turning frequently, this paper mainly analysis characteristics of the DDoS attack based on botnet, and then provides a method of DDoS protection based on the push-back. This method is based on the classic push-back idea, ,and each router sends push-back request to its uplink routers when DDoS attack occurs; on receiving the push-back request, the router will match the feature of attack flow, and implement the flow control.

Keywords: Botnet, DDoS, Push-back, Router.

1 Introduction

With the spread of the Internet and maturity of network attack technology, the security of network service is suffering from a great threat. From 2007 Estonia DDoS information warfare to more than 30 Internet cafes in Guangxi Nanning DDoS extortion in 2010, and to sina.com not providing service more than 500 minutes because of suffering DDoS attack, it is very serious situation that DDoS attack is increasingly fierce, and the number of attacks has increased significantly, and the attack flow is becoming huge increasingly. Especially, at present, the attack flow more than 1G appears frequently, according to data CNCERT/CC released, the highest flow can arrive up to 12 G; even if the professional computer room can't stand such flow. Now, DDos attack is no longer the means of hackers and computer masters showing off their technologies, but the way of getting interests and extorting, stealing privacy information, and it has formed a complete industry chain! At the same time, the attacker's cost is very low, and just only the script and tools online, even the attacker has no advanced attacking technologies. In contrast, DDoS protection equipment is very expensive, but tracing to attack source is extremely hard; the protection cost is more than attack.

2 The Summary of Botnet and DDoS Attack

Thanks to the invasion technology grown mature increasingly and terrible low security awareness of ordinary Internet users, botnet is becoming more and more popular in the Internet, and that bring attackers great economic benefits and technical achievement, but botnet has made serious damage against Internet and people's life and social economy suffered from the actual loss at the same time. As well as, botnet

D. Jin and S. Jin (Eds.): Advances in FCCS, Vol. 1, AISC 159, pp. 325–330.
springerlink.com

provides an effective and convenient way to carry out DDoS attack against large commercial web site, and reduces the cost and risks.

2.1 Botnet

Botnet is a network where a host, which injects control programs into a large number of victim hosts by one or more spread ways, can control a lot of computers at the same time. It is not the physical network which has the topological structure, but just the logic network having control structure[1].

As shown in figure 1, botnet has three main parts: the attacker, bots and command and control channel[2].

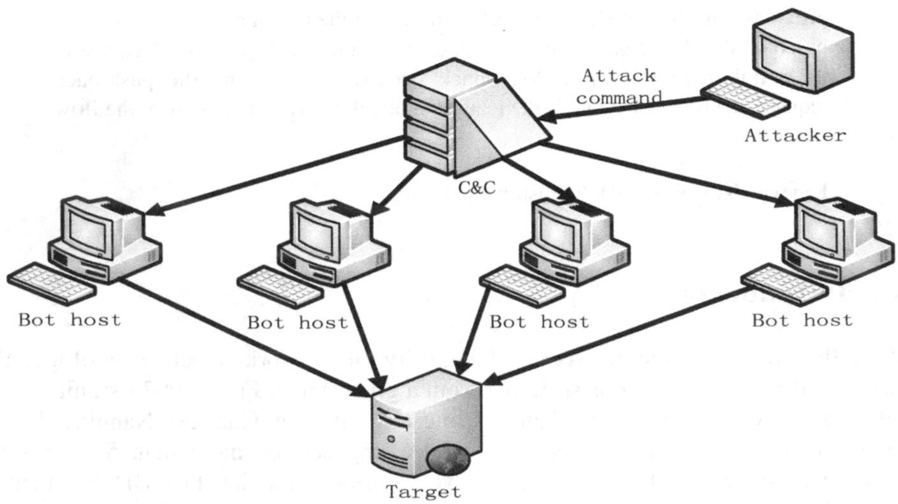

Fig. 1. Composition of Botnet

As shown in figure 1, the attacker in botnet is the host which can send all kinds of control commands to botnet, and carry out attack to specific IP or IP section by ways of IRC, P2P, AOL, etc; the attacks contains types such as spam mail flood, DDOS attack. In most cases, to hide himself and the channel of command and control, the attacker could make one or more bots as its springboard, and comprehensively use communication encryption and anti-tracking technology.

Generally, bots are the PCs Internet users use, which are infused bot program by an attacker in various means. Bot program has functions of network communication, command control, information stealing, hiding and anti-tracking; it is normal for users when the host runs daily, but it will open command and control port, and carry out a series of operations after receiving attack commands.

Command and control channel between the attacker and the bot host, is the center of botnet and the pivot of command and control distribution. In establishment and operation periods of the botnet, the command and control channels is generally stable, and command channel is fixed in the process of the attacker sending control command when the botnet is completed.

2.2 DDoS Attack

DDoS (Distributed Denial of Serviees) is that mangy attackers in a different position (local or network address) attack one or several network targets simultaneously, or one or more attackers control computers (bot computers) in different positions and use these computers to carry out attacks to targets at the same time on the network[3].

The principle of DDoS attack is as shown in figure 2, and the process of DDoS attack contains four roles: the attacker, the master, the puppet computer, and the target network[4]. The attacker is the source of the DDoS attack, which can be any computer on Internet; Thc master is the host which provide the attacker to do illegal invasion and manipulation control, and the attacker can control more puppet computers with the master; the puppet computer which is illegally controlled by thc master, runs all kinds of attack programs, is the ultimate attack performer; Target network is the goal the attacker will attack [5].

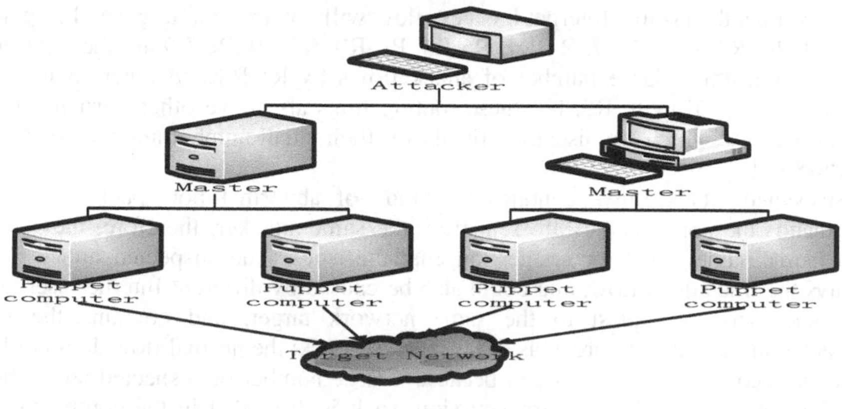

Fig. 2. The schematic diagram of DDoS attack

Firstly, the attacker attack the master(probably the computer with fault configuration and vulnerable system, or the server with poor protection ability), and get the super administrator privilege, and then install DDoS attack programs on the master; next, the attacker will make the master as the springboard to invade and control puppet computers, and install the DDoS attack program. DDoS attack program can coordinate puppet computers on the Internet to attack a fixed target, and cause the network target to not provide Internet services finally. The attacker can control multiple masters at the same time, and also each master can control multiple puppet computers at the same time, so the attacker can send commands to puppet computers in a specific time, and make them attack a specific networks target, and finally arrive to the purpose of denial of service. During the process of DDoS attack, the attacker can also control puppet computers' the number of DDoS attack processes, attack repeat number, attack packet size to adjust the intensity and frequency of attacks. After puppet computers receive attack commands sent by the master, each puppet computer will attack the target according to parameters of the number of DDoS attack processes, attack repeat number, attack packet size which the command set. And because the network protocol the attacker used is very common, protection programs are difficult to distinguish the normal flow and abnormal packets effectively. So, with the pass of time, even the server with strong computing ability and big network communication bandwidth will exhaust its resources.

3 DDoS Attack Restrain Method Based on Botnet

3.1 DDoS Attack Characteristics Based on Botnet

As shown in figure 3, in order to explain the DDoS attack based on botnet more directly, we present a simple DDoS attack network diagram through botnet. the networks target S is under attack, and all of the attack flows pass by R1-R8 routers, finally arrive to S with R9 router directly. In figure 3, the bold black line presents suspected attack flow, and the fine black line presents normal Internet business flow. When DDoS attack occurs, attack flow increases uo to the limit of each router instantaneously, if the router does not have relevant protection means and treatment strategy, then the normal Internet business flow will not arrive to target S. In figure 3, link R1-R6, R2-R6, R3-R7, R4-R7, R5-R8, R6-R9, R7-R9, R8-R9 are the bold lines; the bots transmit a large number of attack flows by R1-R8 and other routers, and finally send the flow to R9, but these routing links also have other normal Internet business flows, which are discarded finally for their small number and the congestion on R9-S link.

Suspected attack flows contains all kinds of abnormal bots packets, because commands the bots receives are sent from the same attacker, therefore, they should have some same attack or congestion characteristics. The suspected attack is not always the true attack flow, and could also be caused by different Internet users send the same service request to the same network target, and contains the same characteristics, so it is regarded as the suspected attack; the normal flow does not have attack and congestion features, but because a large number of suspected attack flows use the same routing link for transmission, so it is discarded in the communication process.

In figure 3, for each attack source is the bot host, so it will have three features at least as follows:

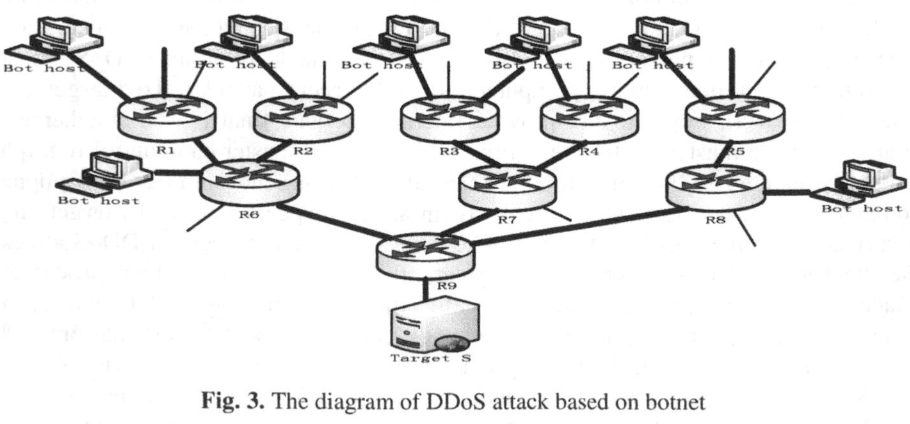

Fig. 3. The diagram of DDoS attack based on botnet

- When DDoS attack occurs, the flows between the routers in R1-R8 which bots are connected with will increase instantly.
- The increased flows arrive to the same network target S.
- By tracking communication of each bots, we can find bots will communicate with the same IP and port, and the flow is small; the IP is the command and control channel of botnet.

3.2 DDoS Attack Restrain Based on Push-Back

The basic thought of DDoS attack restrain based on push-back is that the router can find abnormal flows as much as possible by push-back mechanism, and discard them, when the router can't completely make sure whether flows belong to the normal flow or the attack flow; finally, we could provide a good service for the normal business.

When congestion on R9 happens, R9 will send push-back request to R6, R7, R8; even if their up-link and downlink don't have congestion, they must also be analysis the flow passing by them, and match the attack flow according to the characteristics preset, and find out the suspected attack flow, and limit its speed, and then the router R9 will discard part of or even all suspected flow R6, R7, R8 find out according to certain strategy. The same as R9, R6, R7 and R8 can also send push-back requests to R1, R2, R3, R4 and R5 in turn, and notify them to match and find out t suspected flow, and limit their speeds; R1, R2, R3, R4 and R5 will send push-back requests to their uplink routers until they find out the user-end access routers. If each router's bandwidth meet all normal flows, we can also allow part of suspected flows to pass according to certain strategy.

Figure 4 shows the work flow of suspected flow controlling in the router's push-back mechanism. In the figure, the input queue is the router's input link, and the output queue is the router's output link. When a router detect congestion or receive the push-back request from down-link router send, the packet in the input queue will be matched according to records(for example, the three features listed in 2.1) stored in the DDoS attack feature database, if don't match, then the packet will be put into the output queue without processing; If match, then the packet will enter the flow controller, and be decided to discard or not according to the current congestion situation and filtering strategy. This core of this part is DDoS feature database and the flow controller, where DDoS feature database can be constructed according to the DDoS characteristics based on botnet in 2.1, and the flow controller is implemented with Qos of the router.

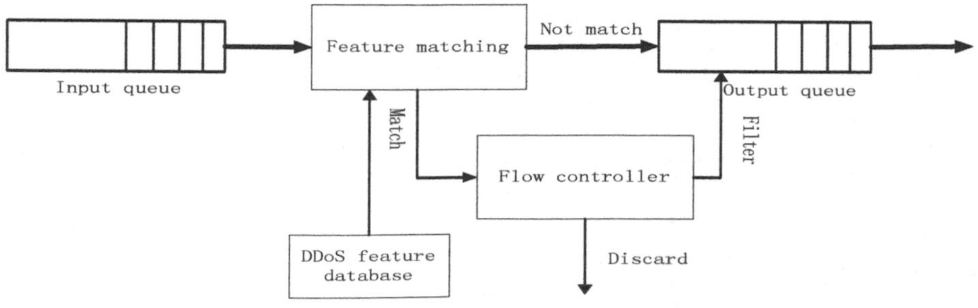

Fig. 4. The work flow of push-back mechanism in the router

4 Conclusion

As electronic commerce is booming increasingly, a large number of service requests cause great pressure to various Internet servers, and information system itself is already struggling to cope with all kinds of normal requests, such as the book half price activity of Jingdong mall before, its home page cannot be accessed for its servers which are stopped suddenly. If DDoS attack also attack the server, the result would be catastrophic, so how to eliminate DDoS attack on the router will be the research direction of DDoS protection in the future.

References

1. Dagon, D.: Modeling botnet propagation using time zones. In: 13th Annual Network and Distributed System Secu-rity Symposium, San Diego. pp. 235–249 (2006)
2. Provos, N.: A virtual honeypot framework. In: Proceedings of 13th USENIX Security Symposium, San Diego, CA (2004)
3. Xue, L.: DDoS Attack Detection and Protection. University of Electronic Science and Technology (June 2003)
4. Li, J.: Comprehensive Analysis of DDoS Attack. Technology and Applications of Network Security, 7–9 (2007)
5. Youy, Z., Haque, A.: Dctecting flooding-based DDos attaeks. In: Proe. of the IEEE International Conference on Communications, pp. 1229–1234 (2007)

The Application of Artificial Intelligence in Prison

ShuFang Wu[1], Jing Wang[2], and QingChao Jiang[1]

[1] Information Engineering Department,
HeBei Software Institute Baoding, China
[2] Department of Information Management,
The Central Institute for Correctional Police Baoding, China

Abstract. The cell allocation problem has long been a obstacle in the modern prison management, the way only depend on the experience of manager is not compatible to the development of modern science. In this paper, two intelligent algorithms are proposed to solve this problem. Bayesian algorithm and the algorithm based on optimal distance both can solve it effectively.

Keywords: Prison management, Cell allocation problem, Bayesian algorithm, optimal distance.

1 Introduction

As the development of society, progress of the technology and appearance of fresh achievements in nature science and social science, especially the growth of information technology by leaps and bounds have offered vast development space for the prison work. The informational management of prison is the gateway to the prison reform, the prison intelligent is also a hot topic. To be the mechanism of country Punishment of Criminals, it should be absorb the achievements of the civilization more positively and utilize many High-tech Means to keep the performance of functions.

The intelligence technology has been used in some areas of prison. Informational management system, multimedia digital safety monitor and control system [1]. More and more intelligence algorithms are used to guide the prison control such as pattern recognition and machine learning.

This paper has the following organization. Section 2 briefly reviews the concept of Artificial Intelligence. Section 3 proposes an intelligent algorithm for prison cell allocation problem based on Bayesian and optimal distance and Section 4 is a conclusion.

2 Introduction of Artificial Intelligence

Artificial intelligence is a very challenging science; people working in this area should know computer knowledge, psychology and philosophy. It is a comprehensive knowledge, and is composed of different research area, such as machine learning,

D. Jin and S. Jin (Eds.): Advances in FCCS, Vol. 1, AISC 159, pp. 331–335.
springerlink.com

computer vision and so on. All in all, One of most important researching goal of Artificial intelligence is making the machines to do some complicated job which needs human intelligence.

The development history of AI is related with the development history of computer science and technology. There are many areas which AI has referred to such as information theory, cybernetics, automation, Bionics, etc.

3 Prison Cell Allocation Problem Based on Bayesian and Optimal Distance

The living condition is very important to the reform of prisoners; the most important condition is the cell. Every prisoner in the same cell has his own character, age, family background, forte, native place, crime, etc. How to consider this factors comprehensively, and take the consideration into the cell allocation process to make the allocation result scientific and intelligent to make a better reform performance is a key point.

There are three problems exist in the cell allocation process.

The one-sided of cell allocation

(1)The administrators of the prison allocate the prison only depend on some characters of the prisoners, for example, usually we don't hope the prisoners in one cell come from the same area, because they are the same in life style and many other aspects, it is hard for them to promote each other. Also we don't want them have similar crime history, if all the prisoners in one cell are all thieves; it's bad for the reform. It is hard to offer help to the reform only depend some of the characters of the prisoner during the cell allocation process.

(2) Lack of scientific guidance

The cell allocation based on human feeling has long been the only way, it is obviously that the way based on experience is not scientific enough. In many conditions, it is difficult to allocate the cell from the human analysis. For example, A,B,C three cells are not full, the characters of them are there are many prisoner from HeBei Province, many thieves live in cell B and majority of the prisoners in cell C are violent temper, whenever a new one whose hometown is HeBei, he is thief and also a violent man come in, how to allocate his cell? Which is the optimal answer?

(3) Influence of the uncertainties of allocation result

The environment influence to a person is great; to be the smallest unit of prison work, the effect of prison is usually neglected. If the help of guard is the extraneous force, the inner Force is the small family cell. Both of this two force work together can make the reform success. Can a man live in the new condition? To what extent can he join in? What is the influence to the cell after his come? The problems are powerful enough to make level of this problem be raised.

The join of a new prisoner can or can't cause positive effect is the key to determinethe allocate result.

In this paper two algorithms are proposed to solve this problem

First the naive Bayesian algorithm is described as follow:

Naive Bayes is one of the most effective and efficient classifcation algorithms.Bayesian Theorem is a theorem of probability theory originally stated by theReverend Thomas Bayesian[2]. The theorem assumes that the probability of a hypothesis is a function of new evidence and previous knowledge. It can be seen as a way of understanding how the probability that a theory is true is affected by a new piece of evidence. Naive classifiers have several desirable features: First, they are simple to construct, requiring very little domain background knowledge, as opposed to general Bayesian networks which can require numerous intensive sessions with experts to produce the true dependence structure between features. Second, naive networks have very constrained space and time complexity [2].

Bayesian theorem provides a way to calculate the probability of a hypothesis, here the event Y, given the observed training data, here represented as X

$$p(Y \mid X) = \frac{p(X \mid Y)p(Y)}{p(X)} \tag{1}$$

Using the assumption for independence, according to (1), the joint probability for all n features can be obtained as a product of the total individual probabilities

$$p(x / y) = \prod_{i=1}^{n} p(x_i / y) \tag{2}$$

Finally the equation can be descried as follow

$$p(Y \mid X) = \frac{\prod_{i=1}^{n} p(x_i / y)p(Y)}{\sum_{k=1}^{m} p(y_k)\prod_{i=1}^{n} p(x_i / y_k)} \tag{3}$$

In the cell allocation problem here every prisoner is represented by a n dimensional vector $d = (d_1, d_2, \ldots, d_n)$ to describe the n attributes $A_1, A_2, \ldots, A_n$.

(1) $c_1, c_2, \ldots\ldots c_k$ are the k cells can absorb the new prison. The probability C_j of this new one belongs to one cell is

$$p(c_j|d) = \frac{p(c_j)p(d|c_j)}{p(d)} \tag{4}$$

The equation above means under the condition of given a prisoner d, the probability d belongs to c_j. So the problem is converted into calculate the $p(c_j|d)$, the cell which make the maximal $p(c_j|d)$ is the final cell.

Using the assumption for independence, there are no relationship between every attribute,

$$p(d|C_j) = \prod_{i=1}^{n} p(d_i|C_j) \tag{5}$$

Finding the $\max p(d|C_j)$, $j \in \{1, 2, \ldots\ldots, k\}$, and allocate him into the cell j.

This algorithm considers all the attributes of prisoners and solves the problem from probability angle. Making us have something to go by.

The second conception is the optimal distance. Measuring the performance based on the average distance between each two prisoners in the same cell. Here we suppose the average distance is optimal in a cell with good reform effect.

The distance between each two can be represented as:

$$d_{ij} = \sqrt{\sum_{k=1}^{n} (x_{ik} - x_{jk})^2} \tag{6}$$

n is the number of attributes, because of the evaluate criterion, first process is the discretization process. It is easy while the attributes are dispersion data; the discretization is mainly to preprocess the continuous features.

Many machine learning algorithms focus on learning in nominal feature space, but many real-world classification tasks exist that involve continuous features where such algorithm could not be applied unless the continuous features are first discretized, another reason for variable discretization aside from algorithm requirement is that the discretization can increase the speed of induction algorithms [3,4]

All the prisoners should be characterized [5]; here we try to describe a prisoner with the following characters, characteristics, age, family condition, strong point, native place, crime, term of penalty, performance, scholarships and prizes and psychological state. The algorithm is described as follow:

Step1: Train the optimal distance L from the best cell.

Step2: Given the new prisoner $d=(d_1,d_2,\ldots,d_n)$ m candidate houses $D_1,D_2,\ldots\ldots D_m$.

Step3: Calculate the average distance $L_1,L_2,\ldots\ldots L_m$ after d join the cell D_x, $x\in\{1,2,\ldots\ldots,m\}$.

Step4: Comparing $L_1,L_2,\ldots\ldots L_m$ with L, and then find the approximate cell to allocate.

The reason we allocate the new one into the cell with approximate L is based on the suppose that similar to the best one is better.

4 Conclusion

With the wide application of computer and information technology, combine the complicated management work with the modern technology to promote the continuous progress of prison work it the historic necessity. The two algorithm introduced apply machine learning algorithm to the cell allocation problem to overcome the old style of experience based model, becomes the new way for the intelligence and imformatioization of the prison management.

References

1. Chen, L.: The application of intelligent security defence system in prison. Guang Dong technology, vol. (9) (2007)
2. Dash, D., Cooper, G.F.: Exact model averaging with naive Bayesian classifiers. In: Proceedings of the Nineteenth International Conference on Machine Learning, pp. 91–98 (2002)
3. Dougherty, J., Kohavi, R., Sahami, M.: Supervised and unsupervised discretization of continuous features. In: Proc. Twelfth International Conference on Machine Learning, pp. 194–202 (1995)
4. Fayyad, U., Irani, K.: Multi-Interval Discretization of Continuous-Valued Attributes for Classification Learning. In: Proceedings of 13th Internatl. Joint Conference on Artificial Intelligence
5. Fan, L., Yu, L.M., Chang, L.Y.: Research about new methods of text feature extraction. Journal of Tsinghua University 41(7), 98–101 (2001)

Step [illegible] optimal distance [illegible] from the best cell.

Step [illegible] $d = (d_1, d_2, \ldots, d_n)$ [illegible] of candidate houses [illegible]

Step [illegible] Calculate the [illegible] [illegible] [illegible] the cell [illegible]

[illegible]

[illegible] that similar [illegible]

[illegible] Conclusion

W[illegible] combine the [illegible] continuous [illegible]

References

[illegible]

Weather Forecasting Using Naïve Bayesian

ShuFang Wu[1], Jie Zhu[2], and Yan Wang[1]

[1] Information Engineering Department,
HeBei Software Institute Baoding, China
[2] Department of Information Management,
The Central Institute for Correctional Police Baoding, China

Abstract. Various statistical methods are used to process operational Numerical Weather Prediction. In this paper we present an application of Bayesian in meteorology from a machine learning point of view. Due to the characteristic of attribute of continuous value, data discretization are done during the data preprocessing, then the naïve Bayesian are used to forecast the weather. Experiments results show that the proposed algorithm in this paper is feasible and effective.

Keywords: weather forecast, Naïve Bayesian, discretization.

1 Introduction

Because of the influence elements of weather is very complicated, Until now the weather forecasting result especially the weather forecasting in a long time is not good enough, it is based on the calculation and prediction, and also the experiences of reporter is playing an important role in it. The traditional weather forecasting is forming the math model based on the current files. The artificial Neural Networks has already been introduced in dealing with the problem because of its advantages of self learning, self adjusting and some like that, the forecast system has shown its effectiveness. [1].Data Ming knowledge are also used in the weather forecasting problem [2], probabilistic graphical models (Bayesian networks) in Meteorology as a data mining technique. Bayesian networks automatically capture probabilistic information from data using directed acyclic graphs and factorized probability functions. [3]

In recent years numerical models have become the focus of weather forecasting efforts. Various statistical methods are used to process operational Numerical Weather Prediction products with the aim of reducing forecast errors and they often require sufficiently large training data sets. One drawback associated with the use of numerical models is that they are still relatively constrained by computation feasibility. [4].

The size of the sample of forecast-observation pairs is crucial for the application of these algorithms, if there are many more examples of one class than another, some algorithms will tend to correctly classify the class with the larger number of examples. The data collected from history has one obvious character that the sunny day data is far more than other weather conditions. One advantage of SVM's is that they are

D. Jin and S. Jin (Eds.): Advances in FCCS, Vol. 1, AISC 159, pp. 337–341.
springerlink.com

remarkably intolerant of the relative sizes of the number of training examples of the two classes. So some papers try to solve this prediction problem based on multi-class SVM. There are still some factors to affect the performance, first, the time complexity of SVM is too high to train all the samples, even though the performance will be good. Second, the result given from the SVM is a single class, the weather condition between two classes such as sunny day with a few minutes rain is hard to be dscribed. Clustering algorithm such as K-means[5] classify the current weather condition data into the nearest cluster, but the performance is not good enough, because the clustering algorithm based on distance always believes that the importance of features are equal, the second shortcoming is that the choice of initial cluster centers, that is reason why clustering result are always different, it means that in some clusters, data may similar but not the same weather condition, noise points is also a big problem, the exist of noise data may affect the result greatly.

This paper has the following organization. Section 2 briefly reviews the concept of naïve Bayesian. Section 3 introduces the improved algorithm of naïve Bayesian algorithm based on discretization data. Section 4 contains description and discussion of the experiments, and Section 5 is a conclusion.

2 Bayesian Algorithm

Naive Bayesian is one of the most effective and efficient classification algorithms. Bayesian Theorem is a theorem of probability theory originally stated by the Reverend Thomas Bayesian. The theorem assumes that the probability of a hypothesis is a function of new evidence and previous knowledge. It can be seen as a way of understanding how the probability that a theory is true is affected by a new piece of evidence. It has been used in a wide variety of contexts, ranging from marine biology to the development of "Bayesian" spam blockers for email systems. Naive classifiers have several desirable features: First, they are simple to construct, requiring very little domain background knowledge, as opposed to general Bayesian networks which can require numerous intensive sessions with experts to produce the true dependence structure between features. Second, naive networks have very constrained space and time complexity[6]. Bayesian theorem provides a way to calculate the probability of a hypothesis, here the event Y, given the observed training data, here represented as X

$$p(Y \mid X) = \frac{p(X \mid Y)p(Y)}{p(X)} \tag{1}$$

This simple formula has enormous practical importance in many applications. It is often easier to calculate the probabilities, $P(X \mid Y)$, $P(Y)$, $P(X)$ when it is the probability $P(Y \mid X)$ that is required. This theorem is central to Bayesian statistics, which calculates the probability of a new event on the basis of earlier probability estimates derived from empirical data. The following section explains the different ways of performing statistical analyses using the classical and the Bayesian statistics [1,2].

Using the assumption for independence, according to (1), the joint probability for all n features can be obtained as a product of the total individual probabilities

$$p(x/y) = \prod_{i=1}^{n} p(x_i / y) \tag{2}$$

Inserting (2) into (1) yields

$$p(Y \mid X) = \frac{\prod_{i=1}^{n} p(x_i / y) p(Y)}{p(X)} \tag{3}$$

The denominator P(x) can be expressed as

$$p(x) = \sum_{k=1}^{m} p(y_k) \prod_{i=1}^{n} p(x_i / y_k) \tag{4}$$

Inserting (4) into (3) the formula used by the Naive Bayesian Classifier is obtained

$$p(Y \mid X) = \frac{\prod_{i=1}^{n} p(x_i / y) p(Y)}{\sum_{k=1}^{m} p(y_k) \prod_{i=1}^{n} p(x_i / y_k)} \tag{5}$$

One of the problems is that conditional independence assumption is rarely true in real-world applications. The data should be dispersion data is another problem restrained its application.

3 Weather Forecasting Algorithm Based on Bayesian

Many machine learning algorithms focus on learning in nominal feature space, but many real-world classification tasks exist that involve continuous features where such algorithm could not be applied unless the continuous features are first discretized, another reason for variable discretization aside from algorithm requirement is that the discretization can increase the speed of induction algorithms [7].

There are lots of discretization algorithms. There are mainly two kinds of discretization algorithms supervised and unsupervised algorithms, because the supervised algorithm fully use of the class attribute information, the classify precision is higher. Class-dependent discretization algorithm seeks to maximize the mutual dependence as measured by the interdependence redundancy between the discrete interval and the class label , some algorithms such as genetic algorithm and clustering are also applied to the discretization of continuous-valued.

In this paper EBD algorithm is choose to be the discretization algorithm, the algorithm is described as follow.

Given a sample set S, the number of class is m, suppose s is a sample in S, s_A is the value of attribute A of sample s. Because the attribute A is an attribute with continuous value, the value set of s_A is $SA = \{s_A \mid s \in S\} = \{a_1, a_2, ..., a_n\}$, the discretization process based on entropy is

1. To sort all the value appeared of attribute A according to the ascending order of values. Suppose the array is $a_1, a_2, ..., a_n$.

2. Every $T_i = \frac{a_i + a_{i+1}}{2}$ $(i = 1, 2, ..., n-1)$ is a possible boundary interval, the sample set is separated into two subsets $S_{1i} = \{s \in S \mid s_A \leq T_i\}$ and $S_{2i} = \{s \in S \mid s_A \succ T_i\}$ by T_i. Choosing the T_i, which will get the minimal entropy after recognized as partition point. The entropy is calculated as the follow eqution

$$E(S, T_i) = \frac{|S_{1i}|}{|S|} E(S_{1i}) + \frac{|S_{2i}|}{|S|} E(S_{2i}) \tag{6}$$

The $E(S_{2i})$ can be calculated as

$$E(S_{ki}) = -\sum_{l=1}^{m} p_{kl} \log_2 p_{kl} \quad k = 1 or 2 \tag{7}$$

p_{kl} is the probability of class l in S_k. The two subsets after partition is S_1 and S_2.

3. If the entropy $E(S, T)$ is larger than a given δ, recursive the partition process to S_1 and S_2.

Nearly every attribute used to describe the weather condition is continuous value such as Temperature and Humidity, so the performance of discretization process affects the prediction result greatly.

Some of the attributes are selected to be the feature of weather, atmospheric circulation, temperature, humidity, water vapor flux and some other features based on the experience of experts. The forecasting process based on naïve Bayesian is easily based on the history weather condition dataset after discretization. The final classification is sunny, cloudy, windy, raining, snow and so on, the result form is given as 70% possibility sunny, 20% possibility cloudy and 10% possibility rain. The advantages of this kind of expression is first it can show all the possible weather condition is one day, so it is a scientific way. Secondly, this way can conclude some special condition, for example the thunderstorm is common is summer, so the expression such as rainy is not good enough.

4 Conclusion

Motivated by designing a new algorithm for weather forecasting problem in the machine learning view, this paper proposed a Bayesian based algorithm to solve the problem. The algorithm is effective and efficient.

References

1. Feng, L.H.: Application of neural network in weather forecast. Information and Control 30(4), 335–337 (2001)
2. Chen, B.X., Yu, J.S.: Feasibility study of applications of data mining to weather forecast. Applied Science and Technology 31(3), 48–50 (2004)
3. Cano, R., Sordo, C., Gutierrez, J.M.: Applications of Bayesian Networks in Meteorology, pp. 309–327. Springer (2004)
4. de Kock, M., Haingura, P., Kent, M.: Weather Forecasting Using Dynamic Bayesian Networks
5. Huang, J.: The effect of K-means in meteorology data analysis. Computer Engineering and Application, 98–100 (2009)
6. Dash, D., Cooper, G.F.: Exact model averaging with naive Bayesian classifiers. In: Proceedings of the Nineteenth International Conference on Machine Learning, pp. 91–98 (2002)
7. Dougherty, J., Kohavi, R., Sahami, M.: Supervised and unsupervised discretization of continuous features. In: Proc. Twelfth International Conference on Machine Learning, pp. 194–202 (1995)

Calculation of Coverage Relationship between Camera Sensor Node and Scalar Sensor Node in Two-Tier Wireless Multimedia Sensor Networks

Qin Lu, Wusheng Luo, and Jingjing Xiao

Department of Instrument Science and Technology,
College of Mechatronics Engineering and Automation,
National University of Defense Technology, Changsha 410073, China
freda0126@gmail.com

Abstract. For energy-constrained wireless multimedia sensor networks (WMSNs), how to minimize the energy consumption while still capturing the events images with low latency becomes a crucial question in the design. A good solution for this question is using two-tier network, in which the camera sensor nodes should only be woken up when an event is detected by the scalar sensor nodes within its sensing field. In this solution, a key step is to know the coverage relationship between the camera sensor nodes and the scalar sensor nodes. Therefore we study the sensing models of these two kinds of sensor nodes and define the coverage relationship between them in this paper. An algorithm to calculate the coverage relationship between these two kinds of sensor nodes is presented. Thus, during the working process of the network, by saving the coverage relationship table in its memory, each camera sensor node can easy to know whether they cove the events or not.

Keywords: wireless multimedia sensor networks, two-tier network, camera sensor node, scalar sensor node, coverage relationship.

1 Introduction

Recently the availability of low-cost, small-scale CMOS cameras has fostered the development of wireless multimedia sensor networks (WMSNs) which promise a wide range of potential applications in both civilian and military areas [1]. A typical application of WMSNs is video surveillance for event detection. If we use a camera capturing 20 frames per second using 25Kpixels per frame with each pixel represented by 8 bits, 14GB information should be gathered and processed per hour. Obviously, this traditional strategy is not suitable for energy-constrained camera sensor nodes in WMSNs. How to minimize the energy consumption while still capturing the events images with low latency becomes a crucial question in the design.

A good solution for this question is using two-tier network [2]. The lower tier is formed by low power scalar sensor nodes which measure scalar physical quantities. The higher tier is formed by camera sensor nodes which have the responsibility for

D. Jin and S. Jin (Eds.): Advances in FCCS, Vol. 1, AISC 159, pp. 343–348.
springerlink.com

event image acquisition, computation and transmission. Then, a camera sensor node can only be woken up when an event is detected by the scalar sensor nodes within its sensing field.

In this solution, to wake up the proper camera sensor nodes, a key step is to determine the coverage relationship between the camera sensor nodes and the scalar sensor nodes which cover the events. In this paper, we study the sensing models of these two kinds of sensor nodes and define the coverage relationship between them. Then, an algorithm to calculate the coverage relationship between these two kinds of sensor nodes is presented.

2 Coverage Relationship Definition

The sensing area of a scalar sensor node s is represented as a disk sensing model $<P_s, R_s>$, where $P_s(m,n)$ is the position of s and R_s is the sensing range of s [3]. The sensing area of a camera sensor node c is represented as a directional sensing model $<P_c, R_c, \vec{V}, \alpha>$ [4], where $P_c(x,y)$ is the position of c, R_c is its sensing range, $\vec{V}$ is the vector representing the line of sight of the camera's field of view (FoV) which determines the sensing direction, and α is the offset angle of the FoV on both sides of $\vec{V}$, $0 < \alpha \leq \pi$. Fig. 1 illustrates the directional sensing model.

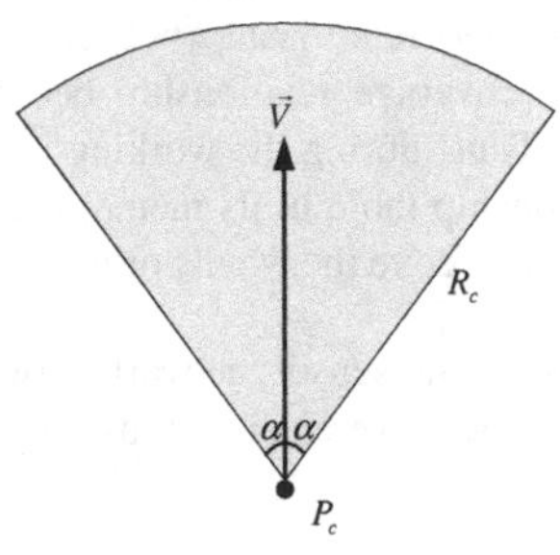

Fig. 1. Directional sensing model

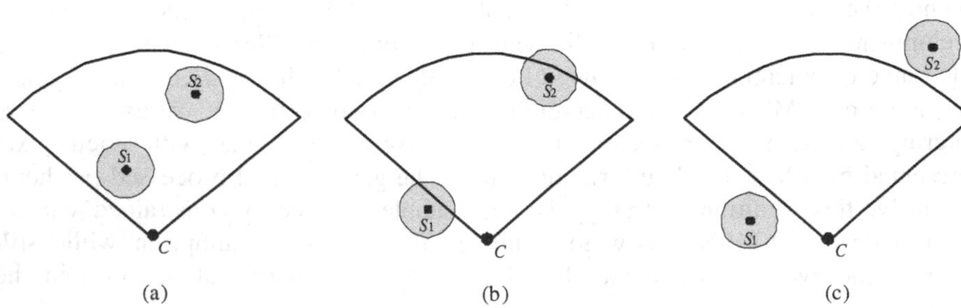

Fig. 2. Coverage relationships

Therefore, there are three types of coverage relationships: if the disk sensing area of a scalar sensor node s is totally covered by the directional sensing area of a camera sensor node c as shown in Fig. 2(a), s is called the inner-node of c; if the disk sensing area of a scalar sensor node s is partly covered by the directional sensing area of a camera sensor node c as shown in Fig. 2(b), s is called the fringe-node of c; if the disk sensing area of a scalar sensor node s is not covered by the directional sensing area of a camera sensor node c as shown in Fig. 2(c), s is called the outer-node of c.

From this definition, we can see that if s is the inner-node of c, it is sure that the events detected by s is covered by c. If s is the outer-node of c, no matter how close between s and c, we also can make sure that the events detected by s is not covered by c. If s is only the fringe-node of some camera sensor nodes, more than one camera sensor node should wake up because we cannot precisely localize the events detected by s.

3 Calculation of Coverage Relationship

Lemma 1. A point P is covered by a sensor node c with the directional sensing model $<P_c, R_c, \vec{V}, \alpha>$ if the following two conditions are true: (1) $\left\|\overrightarrow{P_cP}\right\| \le R_c$, and (2) $\overrightarrow{P_cP} \cdot \vec{V} \ge \left\|\overrightarrow{P_cP}\right\| \cdot \cos\alpha$, where $\overrightarrow{P_cP}$ is the distance vector from the sensor node c to the point P [4].

3.1 Calculation of Inner-Node

Theorem 1. Considering a scalar sensor node s with the disk sensing model $<P_s, R_s>$ and a camera sensor node c with the directional sensing model $<P_c, R_c, \vec{V}, \alpha>$, for a given point $P_c'(x + R_s \cdot V_x/\sin\alpha, y + R_s \cdot V_y/\sin\alpha)$ and $\beta = \alpha - \arcsin(R_s/(R_c - R_s))$, we have the following conditions:

(1) $\left\|\overrightarrow{P_cP_s}\right\| \le (R_c - R_s)$, and $\overrightarrow{P_cP_s} \cdot \vec{V} \ge \left\|\overrightarrow{P_cP_s}\right\| \cdot \cos\beta$;

(2) $\left\|\overrightarrow{P_c'P_s}\right\| \le \sin\beta(R_c - R_s)/\sin\alpha$, and $\overrightarrow{P_c'P_s} \cdot \vec{V} \ge \left\|\overrightarrow{P_c'P_s}\right\| \cdot \cos\alpha$;

(3) $R_s \le \left\|\overrightarrow{P_cP_s}\right\| \le (R_c - R_s)$, and $\overrightarrow{P_cP_s} \cdot \vec{V} \ge \left\|\overrightarrow{P_cP_s}\right\| \cdot \cos(\alpha - \pi/2)$.

For $0 < \alpha \le \pi/2$, s is the inner-node of c if the condition (1) and (2) are true;

For $\pi/2 < \alpha \le \pi$, s is the inner-node of c if the condition (1) and (2) are true, or the condition (3) is true.

Proof: By using different camera lens, we consider two situations: $0 < \alpha \le \pi/2$ and $\pi/2 < \alpha \le \pi$, as shown in Fig. 3.

For $0 < \alpha \le \pi/2$, the shadow is the overlap between the sector $<P_c, (R_c - R_s), \vec{V}, \beta>$ and the sector $<P_c', \sin\beta(R_c - R_s)/\sin\alpha, \vec{V}, \alpha>$. It is easy to find that when the point

P_s is covered by the shadow, s is the inner-node of c. Then according to Lemma 1, for $0<\alpha\le\pi/2$, s is the inner-node of c if the condition (1) and (2) are true.

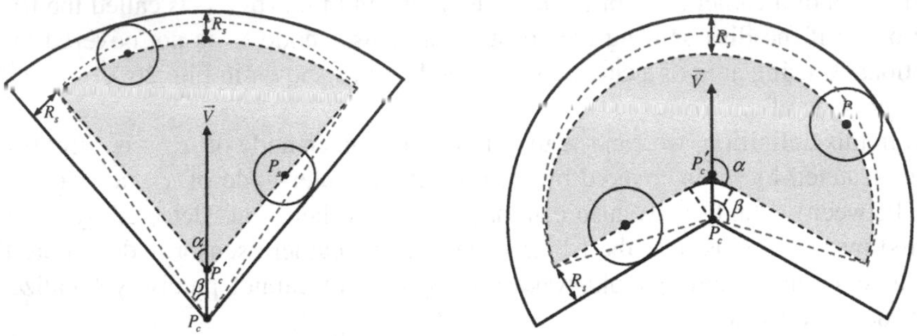

(a) $0<\alpha\le\pi/2$ (b) $\pi/2<\alpha\le\pi$

Fig. 3. Covering area when s is the inner-node of c

For $\pi/2<\alpha\le\pi$, the shadow includes the overlap between the sector $<P_c,(R_c-R_s),\vec{V},\beta>$ and the sector $<P_c',\sin\beta(R_c-R_s)/\sin\alpha,\vec{V},\alpha>$ and the space between the sector $<P_c,R_s,\vec{V},(\alpha-\pi/2)>$ and the sector $<P_c',\sin\beta(R_c-R_s)/\sin\alpha,\vec{V},\alpha>$. It is easy to find that when the point P_s is covered by the shadow, s is the inner-node of c. Then according to Lemma 1, for $\pi/2<\alpha\le\pi$, s is the inner-node of c if the condition (1) and (2) are true, or the condition (3) is true.

3.2 Calculation of Fringe-Node

Theorem 2. Considering a scalar sensor node s with the disk sensing model $<P_s,R_s>$ and a camera sensor node c with the directional sensing model $<P_c,R_c,\vec{V},\alpha>$, s is the fringe-node of c if one of the following conditions is true: (1) $(R_c-R_s)<\left\|\overrightarrow{P_cP_s}\right\|<(R_c+R_s)$, and $\overrightarrow{P_cP_s}\cdot\vec{V}\ge\left\|\overrightarrow{P_cP_s}\right\|\cdot\cos\alpha$; (2) For the point $A_1(x+R_c(V_x\cos\alpha+V_y\sin\alpha), y+R_c(V_y\cos\alpha-V_x\sin\alpha))$: $\left\|\overrightarrow{P_sA_1}\right\|<R_s$; (3) For the point $A_2(x+R_c(V_x\cos\alpha-V_y\sin\alpha), y+R_c(V_y\cos\alpha+V_x\sin\alpha))$: $\left\|\overrightarrow{P_sA_2}\right\|<R_s$; (4) For the point $B_1(m+R_s(V_x\cos(\pi/2-\alpha)-V_y\sin(\pi/2-\alpha)), n+R_s(V_y\cos(\pi/2-\alpha)+V_x\sin(\pi/2-\alpha)))$: $\left\|\overrightarrow{P_cB_1}\right\|<R_c$, and $\overrightarrow{P_cB_1}\cdot\vec{V}>\left\|\overrightarrow{P_cB_1}\right\|\cdot\cos\alpha$; (5) For the point $B_2(m+R_s(V_x\cos(\pi/2-\alpha)+V_y\sin(\pi/2-\alpha)), n+R_s(V_y\cos(\pi/2-\alpha)-V_x\sin(\pi/2-\alpha)))$: $\left\|\overrightarrow{P_cB_2}\right\|<R_c$, and $\overrightarrow{P_cB_2}\cdot\vec{V}>\left\|\overrightarrow{P_cB_2}\right\|\cdot\cos\alpha$; (6) $\left\|\overrightarrow{P_sP_c}\right\|<R_s$.

Proof: As shown in Fig. 4, no matter $0<\alpha\le\pi/2$, or $\pi/2<\alpha\le\pi$, we have six situations of s to be the fringe-node of c:

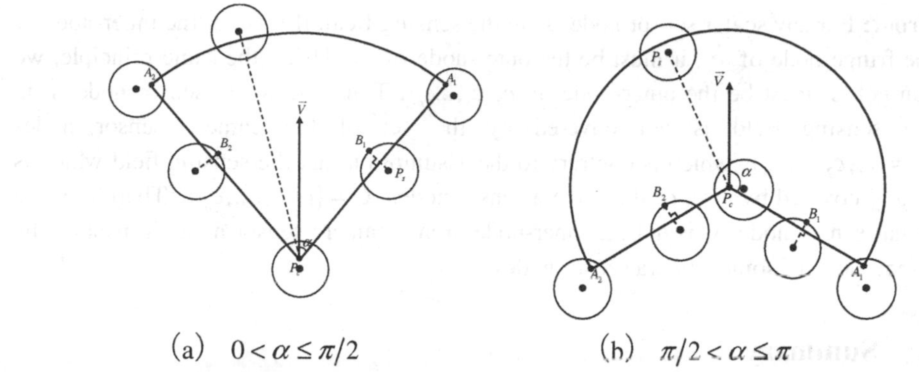

Fig. 4. Six situations of s to be the fringe-node of c

(1) the disk sensing area of s passes across the arc $\overset{\frown}{A_1A_2}$ of the directional sensing area of c. That is to say $(R_c - R_s) < \left\|\overrightarrow{P_cP_s}\right\| < (R_c + R_s)$, and $\overrightarrow{P_cP_s} \cdot \vec{V} \geq \left\|\overrightarrow{P_cP_s}\right\| \cdot \cos\alpha$;

(2) the disk sensing area of s passes across the arc $\overset{\frown}{A_1A_2}$ and the line $\overrightarrow{P_cA_1}$ of the directional sensing area of c at the same time. That is to say the point A_1 is covered by the disk sensing area of s. Thus, we have $\left\|\overrightarrow{P_sA_1}\right\| < R_s$;

(3) the disk sensing area of s passes across the arc $\overset{\frown}{A_1A_2}$ and the line $\overrightarrow{P_cA_2}$ of the directional sensing area of c at the same time. That is to say the point A_2 is covered by the disk sensing area of s. Thus, we have $\left\|\overrightarrow{P_sA_2}\right\| < R_s$;

(4) the disk sensing area of s passes across the line $\overrightarrow{P_cA_1}$ of the directional sensing area of c. That is to say the point B_1 is covered by the sensing area of c. Thus, we have $\left\|\overrightarrow{P_cB_1}\right\| < R_c$, and $\overrightarrow{P_cB_1} \cdot \vec{V} > \left\|\overrightarrow{P_cB_1}\right\| \cdot \cos\alpha$;

(5) the disk sensing area of s passes across the line $\overrightarrow{P_cA_2}$ of the directional sensing area of c. That is to say the point B_2 is covered by the sensing area of c. Thus, we have $\left\|\overrightarrow{P_cB_2}\right\| < R_c$, and $\overrightarrow{P_cB_2} \cdot \vec{V} > \left\|\overrightarrow{P_cB_2}\right\| \cdot \cos\alpha$;

(6) the disk sensing area of s passes across the line $\overrightarrow{P_cA_1}$ and the line $\overrightarrow{P_cA_2}$ of the directional sensing area of c at the same time. That is to say the point P_c is covered by the disk sensing area of s. Thus, we have $\left\|\overrightarrow{P_sP_c}\right\| < R_s$.

3.3 Theorem for Totally Coverage

Theorem 3. In the sensing field which is totally covered by a set of the camera sensor nodes $C = \{c_1, c_2, ..., c_m\}$, if a scalar sensor node s is not the inner-node of any camera sensor node, it must be the fringe-node of some camera sensor nodes.

Proof: For any scalar sensor node s in the sensing field, if it is not the inner-node or the fringe-node of c_1, it must be the outer-node of c_1; Using the same principle, we can get s must be the outer-node of $c_2, c_3, ...c_m$. Thus, the scalar sensor node s in the sensing field is not covered by the set of the camera sensor nodes $C = \{c_1, c_2, ..., c_m\}$ which is contrary to the assumption that the sensing field which is totally covered by a set of the camera sensor nodes $C = \{c_1, c_2, ..., c_m\}$. Therefore, if a scalar sensor node s is not the inner node of any camera sensor node, it must be the fringe-node of some camera sensor nodes.

4 Summary

For energy-constrained wireless multimedia sensor networks (WMSNs), how to minimize the energy consumption while still capturing the events images with low latency becomes a crucial question in the design. A good solution for this question is using two-tier network, in which the camera sensor nodes should only be woken up when an event is detected by the scalar sensor nodes within its sensing field. In this solution, a key step is to know the coverage relationship between the camera sensor nodes and the scalar sensor nodes. Therefore we study the sensing models of these two kinds of sensor nodes and define the coverage relationship between them in this paper. An algorithm to calculate the coverage relationship between these two kinds of sensor nodes is presented. Thus, during the working process of the network, by saving the coverage relationship table in its memory, each camera sensor node can easy to know whether they cove the events or not.

Acknowledgment. This work was supported by the National Science Foundation of PR China under Grant Nos. 60872151, 61003302.

References

1. Akyildiz, I.F., Melodia, T., Chowdhury, K.R.: A Survey on Wireless Multimedia Sensor Networks. Computer Networks 51(4), 921–960 (2007)
2. Newell, A., Akkaya, K.: Distributed Collaborative Camera Actuation for Redundant Data Elimination in Wireless Multimedia Sensor Networks. Ad. Hoc. Networks 9, 514–527 (2011)
3. Chakrabarty, K., Iyengar, S.S., Qi, H., Cho, E.: Grid coverage for surveillance and target location in distributed sensor networks. IEEE Transactions on Computers 51(12), 1448–1453 (2002)
4. Ma, H., Liu, Y.: On coverage problems of directional sensor networks. In: Proceedings of the International Conference on Mobile Ad-Hoc and Sensor Networks, pp. 721–731 (2005)

Cerebellar Model Articulation Controller Applied in Short-Term Wind Power Prediction

Yichuan Shao[1] and Xingjia Yao[2]

[1] School of Information Engineering, Shenyang University, Shenyang 110044, China
[2] College of New Energy Engineering, Shenyang University of Technology, Shenyang China
yichuan.shao@gmail.com, yaoxingjia@sina.cn

Abstract. Wind Power prediction is very important in the wind power grid management. This paper introduces how to use Cerebellar Model Articulation Controller (CMAC) to build a short-term wind power prediction model.CMAC and Back-propagation Artificial Neural Networks(BP) are used respectively to do the short-term prediction with the data from a wind farm in Inner Mongolia. After comparison of the results, CMAC is more stable, accurate and faster than BP neural network with less training data.. CMAC is considered to be more suitable to do the short-term prediction.

Keywords: Wind Power prediction, Cerebellar Model Articulation Controller, Neural network, Back-propagation Artificial Neural Networks.

1 Introduction

As the output power of wind farm is characterized by intermittence and volatility, increasing pressure has been brought to the power dispatching and the management of electricity market in wind-rich areas. Predicting the output power and bringing the wind power into power network dispatching plan is one of the important measures to ensure the stable and economic operation of power network. Power forecasts can be used in economic dispatching, and the output of conventional power unit can be optimized according to the predicted output curve of wind farm, in order to reduce the operating costs.

Continuous prediction method, autoregressive moving average model method, Kalman filter method and intelligent method are distinguished according to different mathematical models that are adopted. Continuous prediction method is the simplest prediction model, this method assumes that the predicted value of wind speed is equal to the sliding average value of several close wind speed values, and it is generally assumed that the wind speed value of the closest point is the predicted value of next point's wind speed. This model has relatively larger prediction errors and unstable prediction results. The ARMA model, vector autoregression model, Kalman filter algorithm or the combination of time series method and Kalman filter algorithm are the improved methods. There are also some intelligent methods, such as artificial neural network methods, among which BP Neural Networks[1] is the most commonly seen method. BP has strong nonlinearity function and mapping capability, and has

D. Jin and S. Jin (Eds.): Advances in FCCS, Vol. 1, AISC 159, pp. 349–354.
springerlink.com

unique advantages in function approximation, pattern recognition, and state forecasting, etc., and at the same time BP is qualified with a certain generalization ability. BP also has a strong non-linear learning ability, and it is a relatively mature method which is widely adopted in the field of international wind power prediction. The training process of neural network is the process that aims at simulating the structure and function of human brain and continuously adjusts the internal network weight and the input-output relationship of the system. After the determination of network structure and algorithm, the performance of network is largely depends on the quality of training samples.

We constructs a prediction model to predict short-term power of different period by adopting CMAC neural network method, and at the same time conducts training and validation by using the actual data of a certain wind farm. Comparing with traditional BP model, CMAC model has faster prediction speed, higher prediction accuracy and a stable output result.

2 Characteristics of CMAC

CMAC is a neural network model which is constructed after the principle of cerebellum controls body movement. When cerebellum commands movement, it makes conditioned reflex without thinking and responds quickly, this conditioned reflex response is a kind of rapid association. As a neural network with association function, it has local promotion ability which is the so called generalization ability. Therefore, similar input will produce similar output, and different inputs will generate separate output. Only a few corresponding weights to neurons have an influence on each output of network, and the neuron which would have an influence on the output is determined by the input. The input and output of A's each neuron is in linear relationship, but it can be seen as a form system that expresses nonlinear mapping in general. As the learning of CMAC[2] is only in the linear mapping part, simple δ algorithm can be adopted, its convergence speed is much faster than BP algorithm, and there is no local minimum problem. This will be good for the short-term forecasting of wind power.

3 The Model Structure of CMAC

The model structure is shown in Figure 1, X represents the input state space of P dimension, which includes the dimensions like wind speed, wind direction, air temperature, humidity, atmospheric pressure, the state of wind turbine and so on. A is a storage area with n units (also known as connected space or concept memory space).Assume that the input vector of CMAC is represented by point Xi $=(x^i_1,x^i_1,\ldots,x^i_1,x^i_p)^T$in the input state space P of P dimension, the corresponding output vector is represented by Yi $=(x^i_1,x^i_1,\ldots,x^i_1,x^i_p)^T$, point Xi of input space will activate N_L elements in A simultaneously, they will be made as 1 at the same time, but most of the other elements are 0, network output y_i is the accumulative total of the corresponding weights of the N_L activated units in A. N_L is called generalization parameter, it reflects the network generalization ability, it also can be seem as feeling size of signal measurement unit. The working process of CMAC generally includes result output calculation, error generation phase and weight adjustment phase.[3]

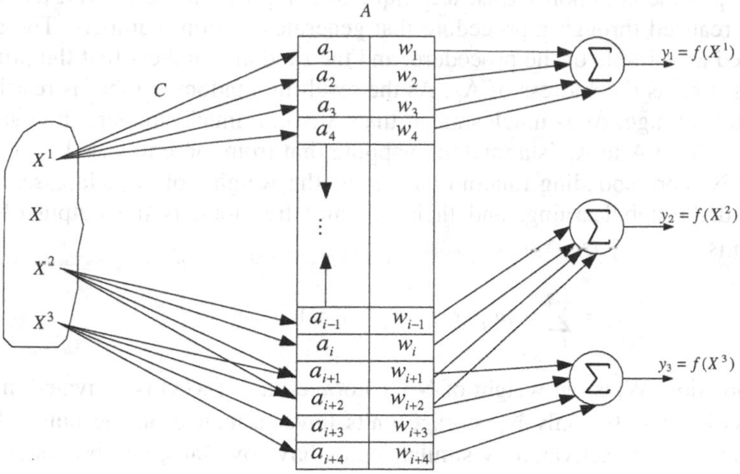

Fig. 1. The Model structure of CMAC

4 Result Output Calculation of CMAC and Error Generation Phase

Assume every component of input vector X can be quantified into q grades, then p components can be combined to q^p possible state X^i , i = 1,2,...,q^p of input state space. Among which, every stateX^i must be mapped to a set A^i in A space storage area, and N_L elements in A^i are all 1.It can be seen from Figure 1 that the mappings of samples X^2 and X^3 which are close in X space are A^2 and A^3 in A,and there is a intersection $A^2 \cap A^3$,which means two of the four corresponding weights are the same. Therefore, the two outputs of weight accumulation and calculation are relatively close, and this feature has generalization effect from the perspective of function mapping.Obviously, the mapping $A^1 \cap A^3$ of samples $X^1$and X^3 that are far from each other in A is an empty set, this kind of generalization has no effect, so it is local generalization. When the input samples is closer in the input space, the elements that are mapped to the corresponding intersection in storage area A are closer to N_L ,and the intersection of corresponding input samples that generated in A has the effect to accumulate similar samples.

In order to make each state of space X to have only one mapping in space A, the number of the units in storage area A should be at least equal to the number of states of spaceX, that is $n \geq q^p$. For the wind power prediction, there are many dimensions of P and q^p is huge, but as most of the learning problems do not contain all the possible input values, q^p storage units do not really need to store the learning weights. A is equivalent to a virtual memory address, each virtual address is corresponding to a sample point of input state space. The address space A with q^P storage units can be mapped into a much smaller physical address connection A_i through Hash coding.

For each input, only N_L units in A are 1, and the rest are 0, so A is a sparse matrix. Hash coding is the commonly used technique to compress sparse matrix, the specific method is realized through a procedure that generates random numbers. The address of A is used as variable of the procedure, and the random numbers that the procedure generate is used as the address of A_i . As the resulting random number is restricted in a small integer range, A_i is much smaller than A ,so far much smaller. Obviously, the compression from A to A_i is a random mapping that from more to less. Every sample in A_i has N_L corresponding random addresses, the weights of N_L addresses storage are attained through learning, and their accumulative total is the output of A. Its expression is :

$$y_i = \sum_{j=1}^{N_L} w_j a_j(x) \qquad i = 1, \cdots, m \tag{1}$$

In the expression, W_j is the weight of No. j storage unit, if $a_j(x)$ is activated, its value is 1, otherwise it is 0, only N_L storage units have influence on the output [3].The storage units that are activated by similar inputs have overlaps, and produces similar outputs. The inputs that are not similar will not produce similar outputs. The corresponding error expression is:

$$\Delta E_i = \overline{y_s} - \sum_{j=1}^{N_L} w_j a_j(x) \qquad i = 1,2,\cdots,m \tag{2}$$

5 Experiments

The data of power and numerical weather prediction of a certain wind farm in a period of time is used as the target of non-linear curve fitting, and the data points will be changed in every 15 minutes, among which 3600 are used as input samples for training, 600 are used to detect the fitting results. Network BP is also used to predict at the same time, in order to compare the forecasting effects. BP adopts a three-tier network (input layer, hidden layer and output layer), and the nodes of each layer are respectively 7, 20 and 1. The node of hidden layer is Tan-Sigmoid function, the node of output layer is linear function [4]; the number of CMAC's input state is 1, and the output number is 1.Four Statistical error indicators are used in the measuring of forecasting effects:

(1) Average absolute error:

$$\delta_{\text{MAE}} = \frac{1}{N} \sum_{k=1}^{N} P_{simu,k} - P_{trag,k} \tag{3}$$

$P_{simu,k}$ and $P_{trag,k}$ represent predicted power and actual power respectively,N is the number of data;

(2) RMS absolute error:

$$\delta_{\mathrm{RMSE}}=\frac{1}{N}\sqrt{\sum_{k=1}^{N}\left(P_{simu,k}-P_{trag,k}\right)^{2}} \tag{4}$$

(3) Mean relative error:

$$\delta_{\mathrm{RAPE}}=\frac{1}{N}\sum_{k=1}^{N}\frac{\left|P_{simu,k}-P_{trag,k}\right|}{P_{trag,k}} \tag{5}$$

(4) RMS relative error :

$$\delta_{\mathrm{RMSE}}=\frac{1}{N}\sqrt{\sum_{k=1}^{N}\left(\frac{\left|P_{simu,k}-P_{trag,k}\right|}{P_{trag,k}}\right)^{2}} \tag{6}$$

Table 1. The Comparison of wind power prediction error between CMAC and BP

Time	δ_{MAE} CMAC	BP	δ_{RMSE} CMAC	BP	$\delta_{RAPE\%}$ CMAC	BP	$\delta_{RMSE\%}$ CMAC	BP
2010.5	1.29	2.81	0.40	0.76	5.61	11.13	1.77	2.91
2010.6	1.41	3.04	0.36	0.91	5.06	10.77	1.30	3.05
2010.7	2.44	2.96	0.67	0.82	7.45	10.04	1.89	2.78

Table 2. The Comparison of computing speed between CMAC and BP

	CMAC	BP
Computing speed/S	1898	9342

Table 1 and Table 2 quantitatively give the wind power prediction error and the computation time of CMAC and BP. It can be seen from table 1 and table 2 that CMAC's forecasting effects are much better than BP's. The prediction error of CMAC and BP in Table 1 also fully illustrates this point. It can be seen from Table 2 that network CMAC has obvious advantages in computing speed. This is very beneficial for short-term wind power prediction. There is only a slight error of forecasting effect in some local areas when the power peak appears. Although BP can achieve better forecasting effects in the power peak sometimes, its predictive value usually has many "glitches", which would affect the overall prediction accuracy greatly.

In addition to the above problems, BP model also has problem of unstable output in the prediction, that is the prediction results may be different or may vary greatly under the same condition.

6 Summary

Through the comparison of short-term wind power prediction results that attained respectively from CMAC and BP models, it can be seen that CMAC is characterized by smaller requirements of training samples, better output stability, fast calculation speed, high prediction accuracy and so on, which is more suitable for the short-term wind power prediction. When BP carries out prediction to those with intense fluctuations like power, the prediction result will be unstable, and it also requires more training samples and has slower calculation speed, therefore, it is not suitable for short-term wind power prediction.

References

1. Sarkav, D.: Methods to Speed up Error BP Learning Algorithm. ACM Computing Survey 27, 519–592 (1995)
2. Albus, J.S.: A New Approach to Manipulator Control. Trans. ASME J. Dyn. Syst. Meas. Contr. 97, 220–227 (1975)
3. Thompson, D.E.: Neighborhood Sequential and Randon Training Techniques for CMAC. IEEE Trans. NN 6, 196–202 (1995)
4. Shuyong, C.: Wind power generation reliability model and its application. Chinese Journal of Electrical Engineering 83-87, 29 (2000)

Analysis of Grain Yield Prediction Model in Liaoning Province

Guang-Hua Yin[1], Jian Gu[1], Zuo-xin Liu[1], Liang Hao[1,2], and Na Tong[1,2]

[1] Institute of Applied Ecology,
Chinese Academy of Sciences,
Shenyang 110016, China
[2] Graduate University,
Chinese Academy of Sciences,
Beijing 100049, China
{ygh006,sygj981,kkuuzhong}@163.com,
liuzuoxin@iae.ac.cn,
haoliang8702@126.com

Abstract. Based on agricultural production data of Liaoning Province from 1980a to 2009a, using stepwise regression method, the gray prediction method, BP neural network, the grain yield prediction model was respectively established in Liaoning Province, China. The grain yield was predicted with these models, and models were compared. The results show that the yield forecasts relative error of the stepwise regression model, gray prediction model, BP neural network model are respectively: 3.41%, 6.59%, and 1.16%. Among the three models, the order of best fit is the BP neural network model, the less is the stepwise regression model, the least is the gray model. It was proved that the BP neural network model is optimum one with high correspondence degreed and high accuracy for food production forecast in Liaoning Province.

Keywords: grain yield, stepwise regression, gray prediction method, BP neural network, forecast.

0 Preface

Agriculture is the foundation of national economy and Guarding food safety is the key role of the development of modern agriculture [1].The primary problem of food security is yield security. How to predict the future changes in grain production is an important work in modern agriculture. Scholars have done a lot of research on food production forecasting. It formed a variety of yield forecasting method models and most predicted methods had regional differences [2]. In this study, we select the agricultural production data affecting the grain yield from 1980a to 2009a in Liaoning Province to apply stepwise regression analysis, gray correlation and prediction model and BP neural network to predict the yield, fit on the forecast and observed data and analysis fitting accuracy, in order to choose a more accurate predictive model of food production.

D. Jin and S. Jin (Eds.): Advances in FCCS, Vol. 1, AISC 159, pp. 355–360.
springerlink.com

1 Study Area and Methods

1.1 Study Area

Liaoning Province (38°43'N~43°26'N and 118°53'E~125°46'E), is located in the south of Northeast China. It has the advantages in the location and agricultural resources, is important grain-producing province in China. The crops are rice, corn, soybeans, sorghum and wheat. The study of agricultural production data are from Statistical Yearbook of Liaoning Province [3].

1.2 Study Methods

1.2.1 Predicted Factors Selection

Selecting eight factors affecting food production analysis: Y is annual grain yield, X_1 is planting area, X_2 is the effective irrigation area, X_3 is the amount of chemical fertilizer, X_4 is total power for agricultural machinery, X_5 is the damage area, X_6 is rural electricity consumption, and X_7 is the grain yield per hectare. In this study, Y is the dependent variables, X_1- X_7 are independent variables.

1.2.2 Model Foundation

Using stepwise regression analysis, gray correlation and prediction, BP neural network to analyze the eight forecast factors, select the appropriate factor to establish the appropriate forecasting model. The grain yield predictive value was calculated based on three kinds of prediction model.

1.2.3 Comparative Analysis

We compared and analyze three kinds of predictive model fitting results and the prediction accuracy. The fitting effect analysis uses correlation analysis method and the evaluation of forecast accuracy uses the relative error calculation method. Through analysis, the optimum yield forecast model of Liaoning Province is selected.

2 Results and Analysis

2.1 The Grain Yield Predicted by Stepwise Regression Analysis

Stepwise regression is a method derived from the basis of multiple linear regressions, which is able to automatically choose the more important variable from a large number of variables in the regression equation [4-5].This study selected grain yield as the dependent variable, seven impact factors as independent variables. Firstly, independent variables are standardized, and secondly, analysis X_5, X_3, X_2 X_7 variable coefficient's significance test t are close to less than 0.01, which indicated that four variables entered into the regression prediction model. The established equation is,

$$Y = 1461.347 - 120.346 X_2 + 173.019 X_3 - 79.894 X_5 + 149.918 X_7$$

The contrast results between grain yields predicted value and observed value are showed in Figure 1. As can be seen from Figure 1, the error between the actual grain yield and predicted value before and after a few years is smaller, in the middle years is greater. The average error is 3.41%. The model has high accuracy.

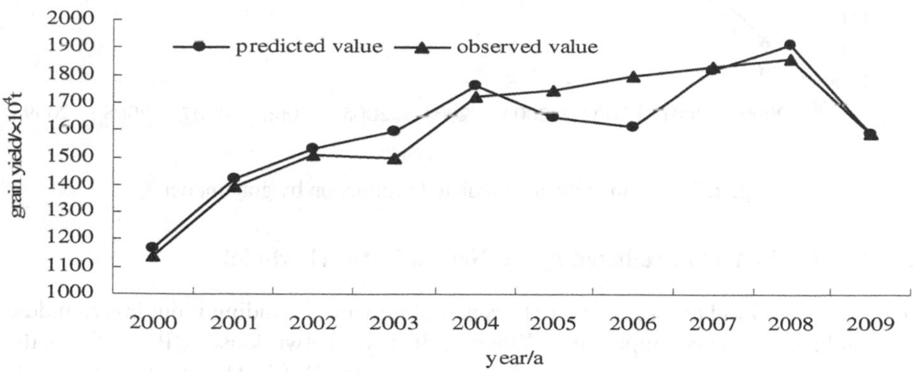

Fig. 1. The grain yield and predicted production by stepwise regression

2.2 The Grain Yield Predicted by Gray Model

Gray system modeling is clarifying the system's key relationships between various factors so as to identify the most influential factor based on the process of the concept through correlation analysis [6]. The commonly used model has GM (1, 1), GM (2, 1), GM (1, N), and so on. In this paper, GM (1, N) model, which corresponds to the first-order differential equations, for a cumulative and variable number is N, is applied [7].

Using data from 1980a to 2009a and the DPS softwarc to establish GM (1, N) model, data analysis results are associated with order: $X_7 > X_3 > X_1 > X_2 > X_5 > X_4 > X_6$, So we selected X_7, X_3, X_1, X_2, X_5 as input factor, Y is the output factors, the establishment of GM (1, N) model is,

$$Y(t+1) = (1221.6 - 0.05945 X_1 + 4.74232 X_2 - 3.60739 X_3 + 0.29562 X_5 - 3.32405 X_7)e^{-1.26532 t} + 0.05945 X_1 - 4.74232 X_2 + 3.60739 X_3 - 0.29562 X_5 + 3.32405 X_7$$

Using the above model to predict grain yield from 2000 to 2009, the results show in Figure 2. As can be seen from Figure 2, the error between the actual yield value and predicted value is relatively larger. The average error is 6.59%. The model has a general accuracy.

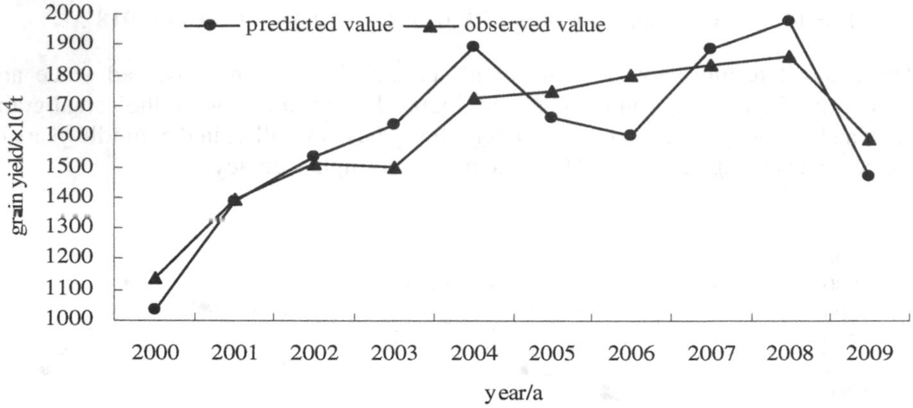

Fig. 2. The grain yield and predicted production by gray model

2.3 The Grain Yield Predicted by BP Neural Network Model

BP network is a kind of three or more tiers neural network including input layer, middle layer (hidden layer) and output layer. When Multi-layer network uses BP algorithm, the forward and reverse transmission includes two stages [8-9]. The prediction selects three-tier structure of the network model. The input layer is the X_1-X_7 variables, the output layer is the Y-grain yield. The network topology is 7:14:1. Training function is the momentum BP algorithm training function, the momentum factor set to 0.9, learning rate set to 0.05, the maximum number of times the training set to 10000, expected error is 0.001.

Using the DPS software and the data from 1980a to 1999a to establish BP neural network software model, after passing the network training, using the model predicted grain yield forecast from 2000a to 2009a, the results show in Figure 3. As can be seen from Figure 3, the error of the actual yield and predicted yield is smaller. The average error of only 1.16%.The model has a very good accuracy.

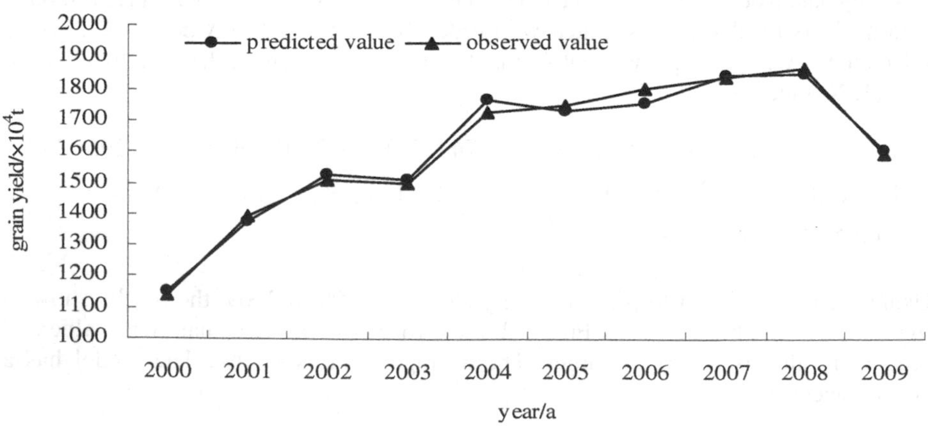

Fig. 3. The grain yield and predicted production by BP neural network model

2.4 Comparison of Three Predictive Models

Forecast and measured values from 2000a to 2009a for trend analysis, combined comparative analysis of the model with the average error of different models and the affecting factors, the results show in Table 1.

Table 1. The comparison of different predicting models

Model name	Trend equation	R^2	Model factors	Average error
Stepwise regression	y=0.8619x+219.5	0.8734	4	3.41%
Gray model	y=1.1003x-163.68	0.8099	5	6.59%
BP neural network	y=0.9678x+51.205	0.9891	7	1.16%

As can be seen from Table 1, the average error of the BP neural network prediction model is lowest and has highest accuracy and consistency of best fit, followed by the stepwise regression model. The fit coefficients of two models are all above 0.85, the average errors are less than 5%, which are higher precision. Prediction model from the point of view to consider the impact factor, stepwise regression model has least impact factor, can be used for long-term forecasts. However, BP neural network model has the more impact factors but the smallest error, which can be used to accurately predict short-term. The error of the Gray model is larger and the trend fitting coefficient is low, only 0.8. Therefore, through comparative analysis, BP neural network prediction model is the optimum method to predict grain yield in Liaoning Province.

3 Conclusions

1) BP neural network prediction model predictions are closer to reality, which is a very good prediction.

2) From the prediction yearly point of view, the stepwise regression model can be considered for long-term forecasts and the BP neural network model can be used short-term forecast.

3) Because of many factors affect food production, the further research that a more comprehensive consideration of factors and the use of more advanced forecasting methods to continuously improve the prediction model.

Acknowledgements. The authors would like to express their sincere thanks to the editor and anonymous reviewer's comments and suggestions for the improvement of this paper.

This research was supported by the program of National Food Production Project (2011BAD16B12), National Science Support Program of China (2012BAD09B02) and Natural Science Foundation of Liaoning Province (20092079).

References

1. Fu, Z.Q., Cai, Y.L., Yang, Y.X., et al.: Research on the relationship of cultivated land change and food security in China. Journal of Natural Resources 4, 313–319 (2001)
2. Mo, X.: The Application of Forecasting Methods in Grain Trade. Jilin University, Changchun (2004)
3. Liaoning Provincial Bureau of Statistics. Liaoning Statistical Yearbook. China Statistics Press, Beijing, China (2010)
4. Liu, Z.Z., Xu, M.: Study on the Forecasting Model of Grain Yield in Sanjiang Plain Region. Transactions of the CSAE 4, 14–18 (1999)
5. Tian, H., Wu, H., Zhao, L.N., et al.: Characteristics and Statistical Model of Road Surface Temperature on Huning Expressway. Journal of Applied Meteorological Science 6, 737–744 (2009)
6. Yang, F., Xiao, L., et al.: Study on cultivated land-grain system balance in Gansu Provence-Based on grey combined mode. Journal of Arid Land Resources and Environment 12, 49–54 (2010)
7. Yao, Z.F., Liu, X.T., Yang, F., et al.: Comparison of several methods in grain production prediction. Agricultural Research in the Arid Areas 4, 264–268 (2010)
8. Zai, S.M., Guo, D.D., Wen, J., et al.: Crop yield forecasting model of BP neural network. Yellow River 9, 71–73 (2010)
9. Zhao, H.Z., Hu, H.Y.: Research on Crop Response to Water Model with BP Neural Network. Journal of North China Institute of Water Conservancy and Hydroelectric Power 2, 17–19 (2006)

An Embedded Image Processing System Based on Qt

Genbao Zhang and Sujuan Pang

School of Electric and Information Engineering,
Shaanxi University of Science and Technology,
Xi'an City, Shaanxi, 710021 China
zhanggb@163.com

Abstract. This paper realized an embedded image processing system based on Qt. On the platform of embedded Linux, the realization method of embedded image processing system is discussed and analyzed. Then, The GUI is designed. After this, the algorithms of image processing are implemented and the code is debugged. The experiments show that using this embedded system has the advantage of high speed, low cost and high expansibility. It also has high practical value for vision detection in some industrial fields.

Keywords: Qt Embedded, image processing, embedded system.

1 Introduction

Embedded system is a computer system that performs some special function. With the development of information technology and the improvement of processing ability of micro-processor, the application areas of embedded system is extended more and more largely [1]. Under the embedded Linux Operating System, some image processing programs can be performed on it now instead of computer. So using embedded system to perform image processing is becoming the hot point at present.

Qt is a development frame for application. It supports several Operating Systems. Its development language is C++. Qt/Embedded is a C++ development collection based on Qt for embedded GUI device and application. It also contains some tools to improve the speed of development using the standard Qt API. So it can be used to develop application programs easily under Windows and Linux. And now, this method is widely used to develop application programs for the embedded devices. The Qt/Embedded uses the mechanism of "signals and slots".

2 The System Overview

2.1 System Hardware

The system's processor is CPU-S3C2440A, produced by SAMSUNG Corporation. Its main working frequency is 400MHz and the highest frequency is 533MHz. So it has strong data processing ability. The hardware board contains eight modules: CMOS

D. Jin and S. Jin (Eds.): Advances in FCCS, Vol. 1, AISC 159, pp. 361–366.
springerlink.com

camera module, power management module, USB interface module, clock module, Ethernet module, serial module, TFT display module and memory module. As shown in fig 1.

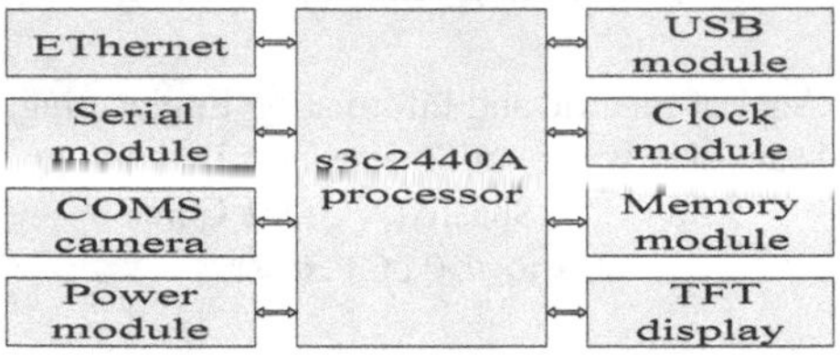

Fig. 1. System hardware block diagram

2.2 System Software

2.2.1 System Software Overview

Generally speaking, the embedded Linux system software structure diagram is shown in fig 2.

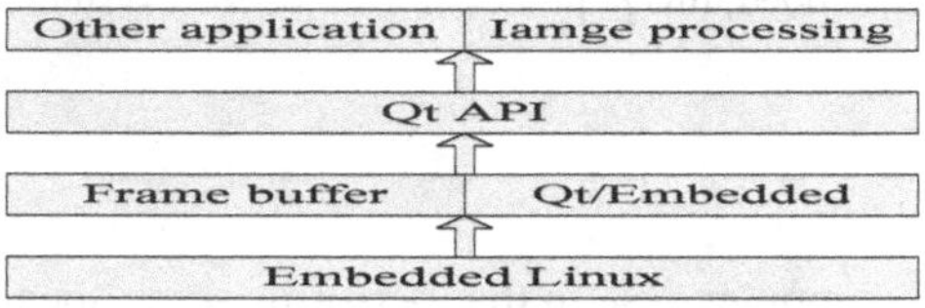

Fig. 2. The overall software structure diagram

Software structure can be divided into four layers: embedded Linux operating system layer, Qt/Embedded and frame buffer layer, Qt API layer and the application layer.

Qt/Embedded and frame buffer layer is used for video and image shows, Qt API layer is used to guarantee the independence of application platform, Application layer mainly includes the users' application program. Operating system, the embedded Linux layer is the embedded Linux 2.6.29 kernel version, the kernel added camera driver, the device driver is the character device drivers. Linux applied in embedded system has another significant advantage which is the program can be simulated using the host environment.

2.2.2 Linux Operation System Transplantation

Transplantation is that a platform of transplanted code running in other platforms. Linux operating system of transplantation [4] includes four parts: BootLoader , Linux kernel and the necessary driver and the file system. Here the vivi which is created by Mizi Company in Korea is used as the target BootLoader; Due to the Linux kernel is very convenient for cutting, it becomes a good choice as embedded operating system, the detail of transplanting the Linux into the ARM platform is in reference [5].

2.2.3 Build QT Cross-Compilation Environment

In order to develop the user Qt program which is running on the target, the first thing need to do is to build the cross-compilation development environment. Cross-compilation environment includes cross-compiling environment based on the host machine and on the development of the target.

Cross-compilation environment based on the host machine is used to cut and compile Linux kernel; that based on the development of the target is used to compile programs running on the target board. The detail of building this environment is in reference [6].

3 System Desgin and Implementation

With a window manager and an application starter (Launcher), allowing developers add their own application program into the Qtopia. The development process of Qtopia is almost the same as Qt[3]. here,the embedded image processing system will be develped on the Qtopia-2.2.0 version platform. the realization process is introduced as follows.

3.1 Graphic Interface Designs

This graphic interface design (GUI) is mainly for image processing operation, including image display, edge detection, gray, binary, filtering algorithm and so on. In order to realize this, the Qt designer is used here. After opening the Qt designer, by placing menu button and other control part and programming it and executes its related operation. Qt Widget class is the base class in all Qt user interface. The QImage class is mainly used for image data processing; so, this QImage class is the base of all images processing class we will develop. QPainter class here is used for image showing.

3.2 Image Processing Algorithm Realization

After the QImage object is initialized, by calling the Bits() function to get the image data pointer and Calling width() and height() function to get the width and height of the image.During the actual operation, we found that image data is stored in the memory of linear, so we can use memcpy() function to copy the image data to a new

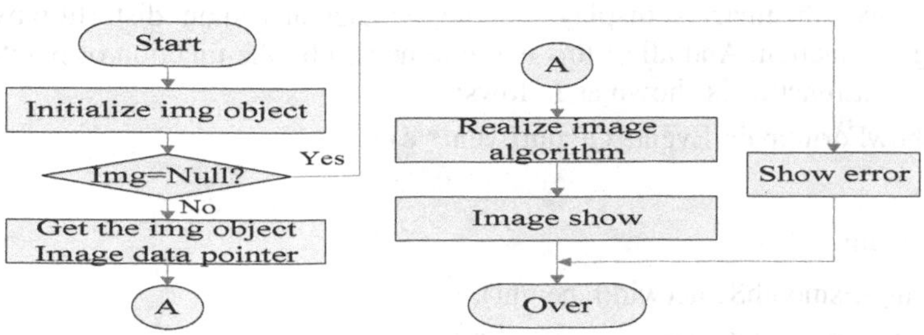

Fig. 3. The main flow of image process

region and write back after processed. Just because the QImage class provides a little functions for user, so it is needed write other image functions by user. The main flow to write new functions in Qtopia is as shown in fig 3.

Due to the image data in memory is stored by bytes, the image data obtained by the CMOS camera is 32 true color images, with 4 bytes representing one pixel and the first three byte represents B, R, G values separately, so we can operate it in our own way. Take the image gray for example, the main code is shown as below, other processing is similar.

```
Void changeImageToGray(QImage & img)
{
   uchar * pointer= img->bits();
numbytes = img->numBytes();
for(i=0;i<numbytes-4;i=i+4)
 {
    B = *(pointer + 0);
    G = *(pointer + 1);
    R = *(pointer + 1);
    Y = B*0.11 +G*0.59 + R*0.30 ;  /*image gray*/
    *(pointer+0) = Y;
    *(pointer+1) = Y;
    *(pointer+2) = Y;
    pointer += 4;
 }
}
```

After realize the image algorithm, it is needed to shown out the processed image, so the image showing is introduced as follows.

3.3 Image Showing

Create a QImage img object, after open an image by QFileDiag. Then create a QPointer object, the image is displayed in the rectangular without distortion by calling drawImage() function. And all of this is implemented by the function of paintEvent(). The painEvent function is shown as follows:

```
void ShowForm::paintEvent( QPaintEvent* e)
{
QImage pim;
pim = img->smoothScale(width,height);
QRect r1( QPoint(1,1), QSize(width,height));
QPainter painter(this);
```

```
painter.setClipRect(e->rect());
setWFlags(getWFlags() | Qt::WRepaintNoErase);
painter.drawImage(QPoint(1,1),pim,r1);
}
```

3.4 Eliminate Flash Processing

In order to show the image in real-time, here, a QTimer is used to get the image data in every 100ms, then calling the paintEvent() function to display images in real-time. But there has a problem that the image has flashing phenomenon. Through analysis, the following two steps are taken to eliminate the flashing:

a) After each frame acquisition, calling the function of video frame redraw update() rather than repaint() in paintEvent(). Because calling the repaint(), Qt immediately erases the background and redrawn the image, due to the two action are time difference. This will cause a severe flashing.While calling update(), the consecutive calling is optimized by Qt, but the repaint() is not allow this.

b) Setting the showing Flags before calling draw function, the method is as follows:

```
setWFlags(getWFlags()|Qt::WRepaintNoErae);
```

By default, the first paintEvent() calls will erase the originally area, and then drawing, in fact drawing twice in every call. So this operation causes serious flashing actually. Setting the mark will draw the image directly without erasing. Experiments results show that: the video display flashing is obvious improvement after the two actions.

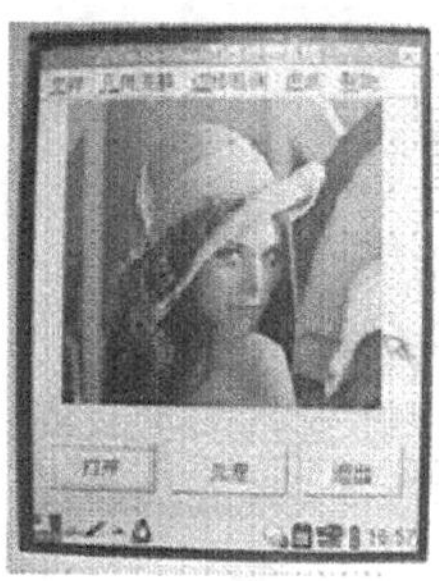

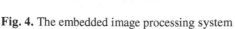

Fig. 4. The embedded image processing system

Fig. 5. The image binary operation

4 Conclusion

This paper realized an embedded image processing system based on Qt/Embedded. On the embedded Linux platform, the realization of image processing algorithms method is discussed. Experiments show that this system can accomplish embedded image processing in 0.5 seconds. Its speed is very high and the cost is low.So it can satisfy the applications of industrial visual inspection requirements.This realization of embedded image processing system has important significance in image processing fields.

Acknowledgments. This paper is supported by projects of the Study on Leather Fast Area Measurement Based on Multi-source Images of Zhejiang province Grant# H20080002 and the System of Fruit Detecting and Grading Based on Computer Vision of Shaanxi province Agricultural Products Processing Research Institute Grant#NYY-090301 and the Graduate Innovation Fund of Shaanxi University of Science and Technology. We would like to thank the support sincerely.

References

1. Ji, J.: Research and Application of Embedded system Based on QT/E. Beijing postal university (2008)
2. Jiang, P.: Research and Realization of video system based on embedded Linux. Tianjin guniversity (2005)
3. Zhang, H.: Automatic Palmprint Identification System Based on Qt/Embedded. Computer Measurement & Control 17(3), 531–533 (2009)
4. Lin, K., Wu, J., Xu, L.: Qt/Embedded Design and Implementation in Wireless Video Surveillance System. Journal of Zhengzhou Institute of Light Industry 23(6), 1–4 (2008)
5. Liu, Z.: The Design and Realization of Image Acquisition System Based on Embedded Linux. Fujian Computer 7(6), 153–154 (2008)

A New Online Soft Measurement Method for Baume Degrees of Black Liquor

Genbao Zhang and Xiuping Li

School of Electric and Information Engineering,
Shaanxi University of Science and Technology
Xi'an City, Shaanxi, 710021 China
zhanggb@163.com

Abstract. The abstract should summarize the contents of the paper and should contain at least 70 and at most 150 words. It should be set in 9-point font size and should be inset 1.0 cm from the right and left margins. There should be two blank (10-point) lines before and after the abstract. This document is in the required format.

Keywords: We would like to encourage you to list your keywords in this section.

1 Introduction

Currently in some control field, some control parameters still cannot be directly measured by sensors, due to the economic and technological conditions. Those parameters must be measured through the soft measurement (indirect measurement). At present there are various soft sensor modeling methods. For example, neural network modeling is a typical method. Although artificial neural network can be arbitrary approximate nonlinear function in theory, as a kind of intelligent technology, but there are still many defects, especially over-fitting, even if the model that training repeatedly only can ensure estimate error of the training sample points minimum, but not able to ensure the best performance of promotion, so it is inevitable that the neural network method makes empirical risk minimization as optimal objective, and limits its application. But in recent years, the support vector machine (SVM) is a new kind of machine learning algorithm based on statistical learning theory. Compared with the traditional neural network, SVM can overcome the local minimum points, and solve the problem which is unavoidable in neural network of the local minimum. The optimal solution of the SVM is based on structural risk minimization, at the same time, according to the limited sample information, it seeks the best compromise between complexities and learning ability, therefore it has stronger generalization ability than other nonlinear function approximation method. Now, with the development of new SVM form, least squares support vector machine (SVM) solves the actual problem rapidly such as a small sample, nonlinear, high dimension, the local minimum points and so on. It is becoming a new hotspot after neural network research, and widely used in pattern recognition, signal processing and so on.

D. Jin and S. Jin (Eds.): Advances in FCCS, Vol. 1, AISC 159, pp. 367–372.
springerlink.com

In paper making industry, the black liquor concentration (Baume degree) in alkali recovery section is a very important parameter, which can not be measured directly by on-line. Due to the obvious correlation between concentration and temperature and pressure of the black liquor, it can build the black liquor concentration (Baume degree) online prediction model using least square support vector machine, by monitoring black liquor temperature and pressure with pressure and temperature sensor. The error of this model is permissible; it can be used for online measurement.

2 Prediction Mode of Baume Degrees

SVM which is put forward by Vapnik based on structural risk minimization principle, makes input map to high dimension space through the nonlinear transform, obtains sparse solution and support vector with global convex quadratic programming optimization problem solving, thus for nonlinear classification and regression. On the premise of SVM characteristic merits, LS-SVM changes inequalities for equality constraints, and simplifies the expression formula, makes global convex quadratic programs for solving linear equation system easily. LS - SVM model principle is in the reference [3,4].

2.1 Prediction Model Structure

Prediction model structure is shown in fig 1. Main parameters including variable x1 that can not be measured, the variable x2 that can be measured, the output variable y which can be measured is the estimated number, the variable Y* is the optimal estimation number and the input variable u, the acquisition data is processed for error processing by using wave-let method. And then the Baume measurement model is built based on LS-SVM. That is $Y^* = f(X, u, y)$.

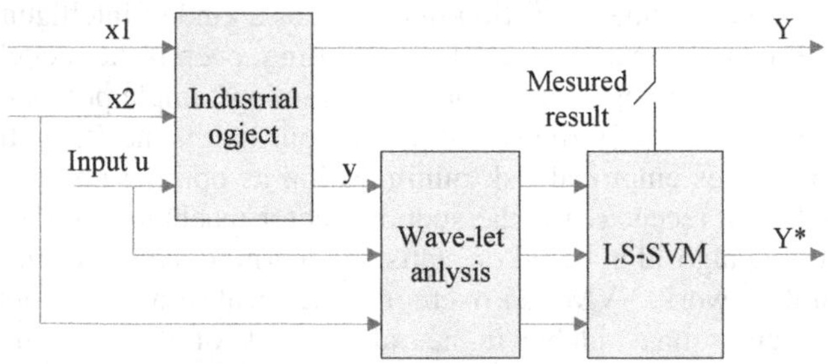

Fig. 1. The structure diagram of prediction model

2.2 The Secondary Variables Selection

In the section of the alkali recovery, the Baume degree of black liquor is a very important optimization of control indicators, and it hard to be measured directly on line. At a certain temperature, it is a linear relationship between Baume degrees of black liquor and its density or the relative density, so using density or relative density usually to represents the Baume. In 1951, Regestad has proposed the practical formula:

$$\rho = 1.007 + 0.006 - 0.000495T_c \tag{1}$$

In the formula, ρ (g/cm3) represents density of black liquor, S(%) represents Solids content, and c represents temperature.

Experimental analysis shows that, for a certain concentration, the relative density d (g/cm3), concentration (S) and temperature (c) has the relation as follows:

$$d = 0.9982 + 0.006s - 0.00054T_c \tag{2}$$

Although type (1) and type (2) are not identical, but it both specified that the density of black liquor or relative density has slight changes with temperature changing. Measuring the density (relative density) directly online is difficult and high cost, so it also must take the indirect measuring method. Using the formula:

$$\Delta P = \rho g \Delta H \tag{3}$$

In the type, ΔP(kPa) represents differential pressure, ρ(g/cm3) represents density, g(9.8 N/kg) represents gravity acceleration, Δ H(m) represents high difference of liquid level. When Δ H is a certain value, it is a linear relationship between ΔP and ρ. So, these shows that the pressure difference ΔP can reflect black liquor density in a certain degree.

2.3 Support Vector Machine Parameter Selection

The modeling uses regression algorithm of LSSVM. The kernel function is selected by debugging the results of the selection index, respectively, the Sigmoid and RBF kernel polynomial function. This paper selects RBF kernel function, as follows (4):

$$K(x, x_i) = \exp\{-|x - x_i|^2 / 2\sigma^2\} \tag{4}$$

Cross validation method is used to select another two parameters (γ, σ^2). When a penalty coefficient γ is 1.854 and σ^2 is 1.108, comprehensive performance(such as running time, the average error of all variances) is the best, when a penalty coefficient γ is 1.854 and σ^2 is 1.108. The following training and validation both use these parameters.

2.4 Wavelet Data Processing

Here, wavelet decomposition is used to eliminate measured bulky error. The bulky error has little rate of appearing. But once exists, it would decrease the quality of data.

And then, all the process of optimization is failure. So, the first thing need to do is to find and eliminate the bulky error.

Wavelet transform has good localization characteristics in the time domain and frequency domain, it makes wavelet transform is more suitable for mutations signal analysis. The measured data which has bulky error can be regarded as a kind of mutations discrete signal. Making wavelet decomposition to the discrete measured data which has bulky error through the analysis of the high frequency signal, the bulky error data points can be determined, and further more, it can be removed by certain threshold [7].

Using wavelet decomposition to eliminate measured bulky error data points has the following steps:

(1) Selecting wavelet function to start the wavelet decomposition. the high frequency signal is as follows.

$$D=(d_1,d_2,d_3,...d_n) \tag{5}$$

(2) Computing threshold: $\delta = t\delta_0$

In this type, $t > 0$ represents adjustment factor;

$$\delta_0 = \sqrt{\frac{d_1^2+d_2^2+...d_n^2}{n}} \tag{6}$$

(3) If $|d_i| > \delta(i=1,...n)$, then the point is recorded as a bulky error location.

(4) Given the integer number N. if the two error data points internal number difference is smaller than N, the data points between two neighbor bulky error data all are bulky error points.

3 The Model Results Analysis

In order to test modeling effect, two performance index is defined as follows [9]:

(1) RMS error: $RMSE = \sqrt{\frac{1}{n}\sum_{i=1}^{n}(y_i - \hat{y}_i)^2}$

(2) Mean absolute error: $MAE = \frac{1}{n}\sum_{i=1}^{n}|y_i - \hat{y}_i|$, in type, y_i and $\hat{y}_i$ are practical value and predicted value.

During this experiment, 55 group data is collected from one paper making factory. 45 group data which has higher accuracy is selected out. Using Pauta criteria, the normalized 30 group data is used to build the LS SVM forecasting model, and another 15 group is used for model generalization inspection. The training and validation line is shown in figure 2 and figure 3. The performance index of eliminating bulky error prediction model is seen in table 1, using Pauta criteria and wavelet decomposition.

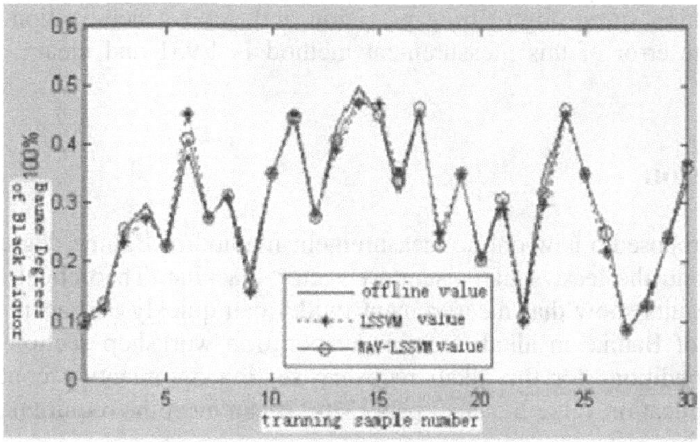

Fig. 2. The training sample result

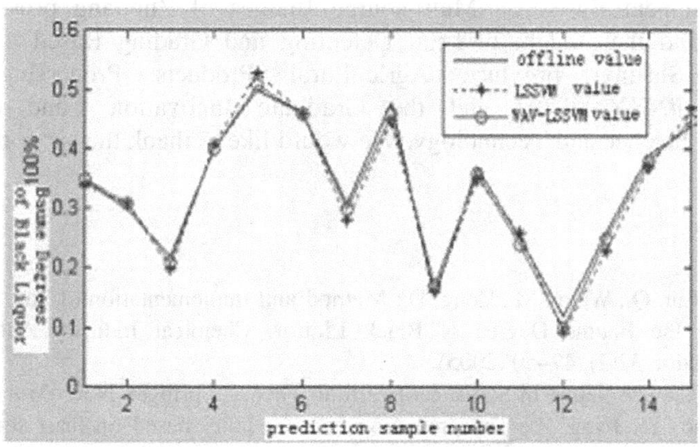

Fig 3. The prediction sample result

Table 1. Different model performance comparing

Model	RMSE	MAE	T(s)
3δ -SVM	2.138	2.907	4.635
3δ -LSSVM	1.982	2.096	0.032
WAV-LSSVM	1.982	1.931	0.015

From the training and curve of forecasting model, it indicates that Baume degrees prediction can follow the change of industrial process of Baume degrees, and it is of high precision and model generalization ability. From table 1, we can see that the deviation between the three kinds of model prediction results and the actual output in production process is allowed (absolute error range no more than 5), but the comprehensive performance is best based on the model prediction LSSVM-WAV, which is of high speed, high fitting precision and good generalization ability. The mean absolute error of this measurement method is 1.931 and mean squared error is 1.218

4 Conclusion

This paper proposed a new online measurement method of Baume degrees based on the wavelet and the least squares support vector machine. Theoretical analysis and simulation results show that measurement model can quickly and accurately predict the changes of Baume in alkali reclaim evaporation workshop section. It provides very good conditions for the alkali recovery section optimization control and has important application value in actual paper site. It can meet the requirements of paper making industry.

Acknowledgments. This paper is supported by projects of the Study on Leather Fast Area Measurement Based on Multi-source Images of Zhejiang province Grant# H20080002 and the System of Fruit Detecting and Grading Based on Computer Vision of Shaanxi province Agricultural Products Processing Research Institute Grant#NYY-090301 and the Graduate Innovation Fund of Shaanxi University of Science and Technology. We would like to thank the support sincerely.

References

1. Tang, W., Liu, Q., Wang, M., Zong, D.: Method and Implementation of Soft Measurement On-line to the Baume Degree of Black Liquor. Chemical Industry Automation and Instrumentation 32(2), 47–50 (2005)
2. Vapnik, V.N.: The Nature of Statistical Learning Theory. Springer, New York (1999)
3. Li, F., Zhao, Y., Jiang, Z.: The prediction of oil quality based on least squares support vector machines and dau2bechies wavelet and mallat algorithm. In: Proceedings of the Sixth International Conference on Intelligent Systems Design and Applications (2006)
4. Zhang, M., Li, Z., Li, W.: Study on least squares support vector machines algorithm and its application. Science press, Beijing (2004)
5. Yan, W., Shao, H.: Application of support vector machines and least squares support vector machines to heart disease diagnoses. Control and Decision 18(3), 358–360 (2003)
6. Vapnik, V.N.: An overview of statistical learning theory. IEEE Trans. Neural Network 10(5), 988–999 (1999)
7. Zhang, L., Zhou, W., Jiao, L.: Wavelet support vector machine. IEEE Transactions on Systems, Man, and Cybernetics, Part B: Cybernetics 34(1), 34–39 (2003)
8. Lin, C.-J.: Formulations of support vector machines:A note from an optimization point of view. Neural Computation 13(2), 307–317 (2001)
9. Varma, A., Jacobson, Q.: Destage Algorithms for Disk Arrays with Nonvolatile Caches. IEEE Trans. on Computers 47(2), 228–235 (1998)

The Anti-electromagnetic Interference Design of the Hot Spot Temperature and Current of Electrical Equipment Online Detection Warning System

Liming Li and Ying Chen

Taiyuan ZhongBo Academy of Information Science and Technology (Company Limited)
Taiyuan, Shanxi, China 030051
{azhaoyang-xf,bhezhijing-xz}@s-hhc.com

Abstract. For running in strong electromagnetic interference environment, we designed the wireless transmission system based Zigbee, used the software of Ansoft Designer to simulate the influence from Electromagnetic to the PCB(Printed Ciruit Board) ,obtained the current plot and near field plot to decide its Electromagnetic Compatibility, then for the area where is high Electromagnetic radiation to redesigned, at last get the optimist anti-electromagnetic design.

Keywords: Zigbee, wireless transmission system, anti-electromagnetic interference, 50Hz.

1 Introduction

With the world's first electromagnetic compatibility specification born in 1944 in Germany, EMC design is becoming increasingly important in modern electronic design [1]. Ordinary 10KV/630KW transformer low-frequency electric field noise radiation at the radiation is generally up to 800V / m, the electromagnetic radiation can up to 30B/μT[2], the wireless transmission module has a very large impact to work in this environment, so it is necessary to anti-electromagnetic interference design for the wireless transmission module.

2 Interference of PCB Design

2.1 The Anti-electromagnetic Interference Design of Hardware

1. Choose the electronic devices of high integration, good anti-interference ability and small power.

2. Good grounding design. For the following work in the 2MHz frequency use the common-ground, that is one point earthing; For more than work in the 10MHz frequency should be used subdivision, that is multi-point grounding. Meanwhile, separate digital ground and analog ground, in the middle with a ferrite bead connect each other. The ground of the senor signal isolate from the earth. All the unused microcontroller IO port, should be grounded and not connected to the supply.

D. Jin and S. Jin (Eds.): Advances in FCCS, Vol. 1, AISC 159, pp. 373–378.
springerlink.com

3. Filtering. The power port of the IC parallel a high-frequency capacitor to reduce the impact of the IC to the power.

4. Design of the PCB: ①Follow the 3W principle, which is adjacent center distance between two lines should be greater than or equal to 3 times the width. With the increase of distance between lines, but also can reduce the crosstalk coupling between lines. ②According to different functions, different modules connected to the corresponding power supply. ③Power and ground as much as width, the direction of current and signal lines keep in the same. ④Crystal oscillator as close as possible to the MCU. ⑤Using the broken line of 45° instead of 90°. ⑥Multilayer design, the power and ground planes in the middle layer, so the benefit of all components of the fast ground, rejecting common mode interference is conducive to anti-electromagnetic interference.

2.2 The Anti-electromagnetic Interference Design of Software

1. Using the watchdog, in which the MSP430 has integrated, then do not go into details.

2. Data redundancy, adding some redundancy to the data bits for parity in the data transmission in noise environment can increase the ability of error detection and correction and have a very significant interference effect.

3 Introduction of the EMI Simulation Software

Electromagnetic interference of numerical simulation process is electromagnetic field numerical calculation, electromagnetic numerical task is based on Maxwell's equations, the establishment of approaching practical problems of continuous-type mathematical model, then using the corresponding numerical method, the discretization, the solution the numerical solution of the model, and then handled the playing field at any point field strength, energy, loss distribution, and other parameters. There are many calculation methods such as time-domain finite element method, frequency domain finite element method, the method of moments, etc., in which the method of moments based on Maxwell's equations in integral equations, high accuracy; frequency domain finite element method based on Maxwell's equations in partial differential equations, have large amount of computation, including the Ansoft Designer is based on frequency-domain finite element method software.

3.1 Principles of Mixed Potential Integral Equation Method

Ansoft Desinger integrated high-frequency, electromagnetic modeling based on physical prototypes, circuit and system simulation and analysis with seamless environment, in order to obtain S-parameters and current density J, with a mixed potential integral equation (MPIE Mixed Potential Integral Equation) method and the method of moments (MoM Method of Moments) were calculated, so that the current density can be obtained through the S-parameters and the radiation field.

In Ansoft, as long as the formation of three-dimensional surface grid, you can use mixed point integral equation (MPIE) analysis:

$$\hat{n} \times (-j\omega A - \nabla\phi) = \hat{n} \times Z_s J$$

Where: $\hat{n}$ is perpendicular to the plane of the grid unit vector; j is imaginary unit; ω is the angular frequency (equal to 2πf); A said magnetic vector; φ a scalar potential; Z_s surface resistance for a single grid; J is the current density on the grid. In the mixed integral, use the method of moments (MoM) to solve the current density on the grid J, to get the grid boundary line to the direction of the current law, the boundary line in the direction of current storage component of the grid, the grid points within each the current boundary line through the grid method to get the difference between current components.

4 PCB EMC Simulation Steps

4.1 PCB Design, Layout and Output

PCB design carried out in Protel 99SE, first drawn in Protel schematics, through error checking, generating the network diagram, and then the wiring. This 4-layer PCB design, signal the bottom and top lines are gone, the middle two layers are ground and power supply, the bottom and top of all ground and power pins are connected to the concrete ground and the shortest to the power supply to ensure the signal integrity and electromagnetic interference. After wiring the circuit board shown in Figure 1:

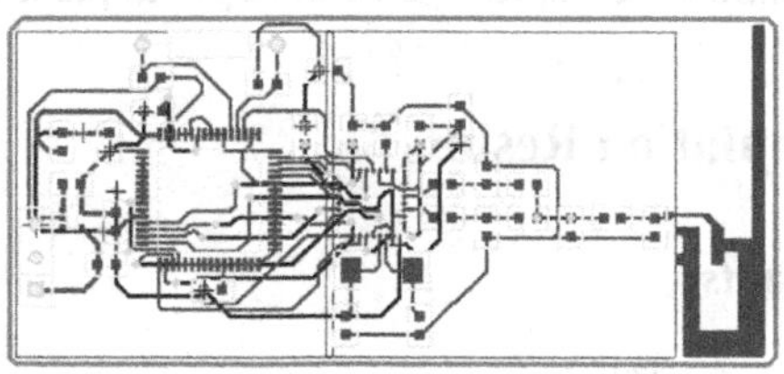

Fig. 1. The placement of PCB

After PCB board design, the output the DXF files.

4.2 Import to the Ansoft Designer and Generate the Model

First create a new Designer Planar EM project, here is a system built using double-sided templates, in order to generate relevant PCB solid model, use the Ansoft Designer's Layout menu, choose Import file listed in items, the Protel exported DXF file into Planar EM project to generate a solid model of the layers of the PCB, in the import process, the units of Protel and Ansoft Designer should be consistent, first, import all the layers, then, modify and adjust according to actual situation of alignment layers on the PCB. PCB dielectric material used is FR-4, to complete the work, the entire PCB solid model simulation can be created in Ansoft Designer. Figure 2:

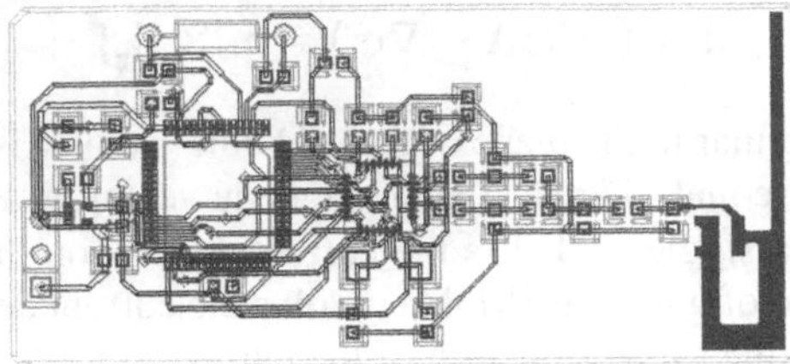

Fig. 2. The model after import in Ansoft

4.3 Set the Resolution Conditions

First, set the excitation. With the selection tool, select the edge you need to add incentives, then right-click to select Add Port. According to the actual situation of the circuit, where the set Port1, Port2, Port3, Port4 a total of four excitation source, current amplitude, respectively 0.1A, 0.2A, 0.1A, 0.3A.

Then set the "Analysis". Right-click the "Adding Solution Setup", the selected initialized grid equal to the "Fixed Mesh" at 60Hz, then in the "Mesh Refinement" labels set the "Refinement Parameters" to default values. Right-click on the increasing "Setup1", selected "Mesh Overlay" and "Dynamic Mesh Update", and then add the "Adding Sweep Frequency", choose the type of "Discrete", re-select the "Generate Surface Current", start frequency of 40Hz, the termination frequency of 60Hz, step 2Hz, a total of 10 frequency points. After setting all the above conditions, also need to set the initial calibration, calibration is complete you can begin to parse the PCB solid model.

5 Analysis of Simulation Results

5.1 The Original Results

Over the Ansoft Designer by 10 of the scanning frequency, we can obtained near-field maps and current maps of the PCB, for example, at 50Hz, analysis the current map, E, H near-field distribution, can be seen in Figure 3,4,5, the circuit appears in four strong field region, distributed in two crystal and RF circuit area, there is a region in which crystal-sensitive components, and just this part of a larger radiation intensity, which the PCB is not appropriate for electromagnetic compatibility.

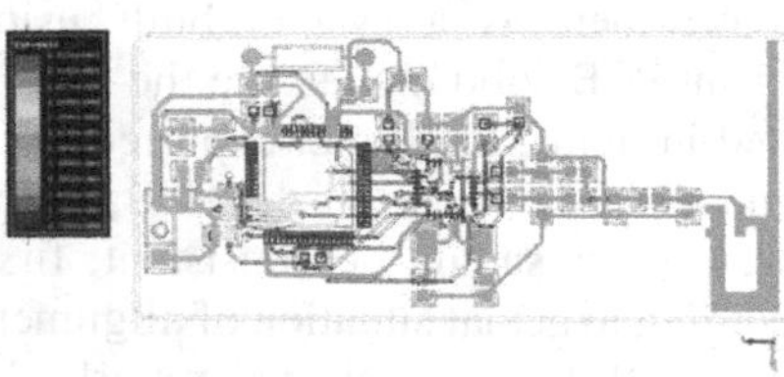

Fig. 3. Current graph at 50Hz

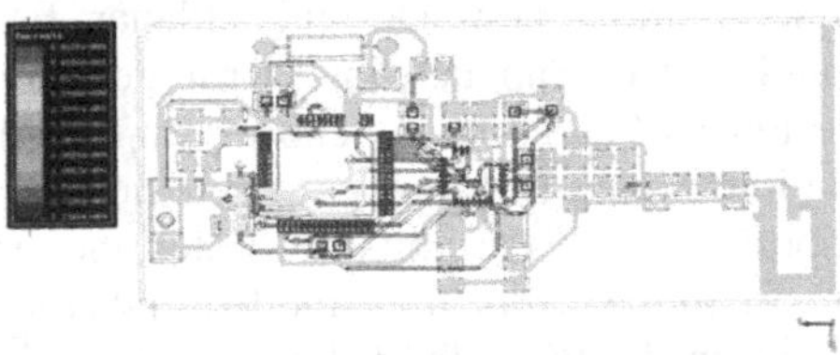

Fig. 6. Current graph at 50Hz after improve

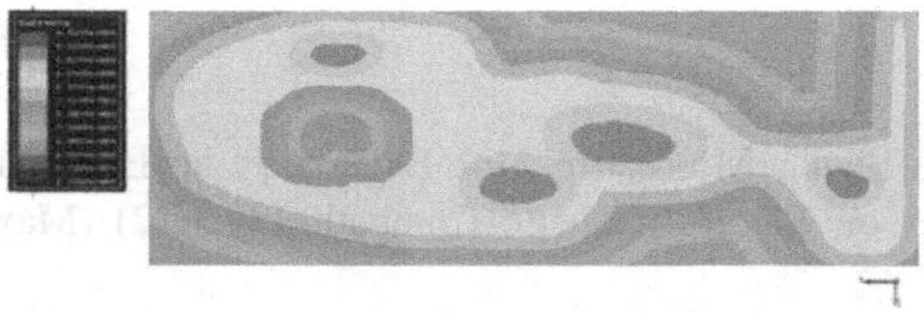

Fig. 4. E near field at 50Hz

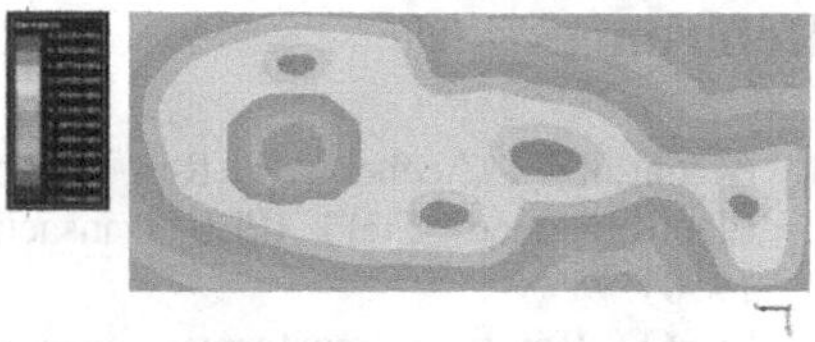

Fig. 7. E near field at 50Hz after improve

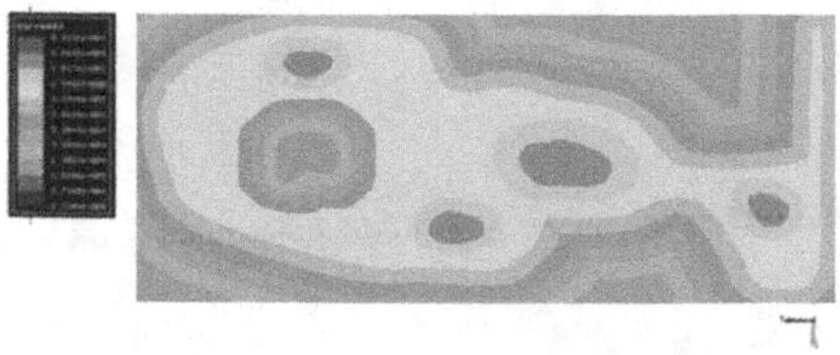

Fig. 5. H near field at 50Hz

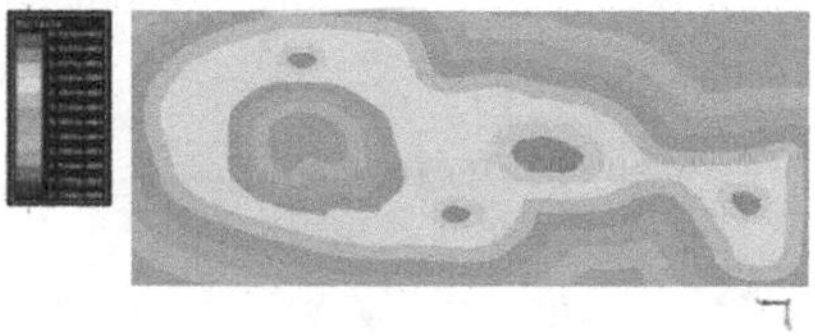

Fig. 8. H near field at 50Hz after improve

In response to these problems, we re-wiring of the PCB, adjust the position above two components of the sensitive area, shorten the distance from the crystal to the microcontroller, the signal into the chip as quickly as possible, smooth corners, as far as possible perpendicular to the inductor and capacitor place, to reduce unnecessary interference.

5.2 The Improved Analysis

We import the improved modal again into ansoft Designer for analysis, get chart 6,7,8, then you can find, previous radiation area smaller or larger deformation zone, a larger crystal area affected intensity decreased, the impact on the surrounding components weakened accordingly, the system of anti-electromagnetic interference capability is improved.

6 Conclusion

In the paper, we analyzing the PCB's electromagnetic compatibility via Ansoft Designer software, according to their current plans and near-field distribution, analyzing electromagnetic compatibility analysis of the PCB, the high area has been redesigned, the index of electromagnetic compatibility, electromagnetic compatibility of the PCB after re-design declined, it is improved by Ansoft Designer verification.

Acknowledgement. This research was financially supported by the International Cooperation and Exchange of Special of China (Grant NO. 2009DFB10570).

References

1. Bernardi, P., Cicchetti, R.: Response of a Planar Microstrip Line Excited by an External Electromagnetic Field. IEEE Transactions on Electromagnetic Compatibility 32(2) (May 1990)
2. Li, Q.: Box-type transformer noise and electromagnetic radiation treatment. Shanghai Railway Technology (1) (2005)

The Quality Inspection of DEM and DOM Based on the ArcGIS Engine

Hong Yu[1,*], HuiMing Li[1], and MingLiang Wang[2]

[1] Nankai University, 94 Weijin Road, Tianjin, 300071, P.R. China
[2] Heilongjiang First Surveying and Mapping Engineering Institutions

Abstract. This paper in order to improve the inspection accuracy and efficiency of digital geographical information production quality, start with check the method of DEM, DOM raster data, aim at the characteristics of large quantity of data, a lot types of data, data format complex of DEM (digital elevation model), DOM (Digital Orthophoto Map), based on component ArcGIS Engine re-development, implement simple, accurate and efficient DEM, DOM quality inspection. Discuss the application and implementation of ArcGIS Engine in DEM, DOM quality inspection.

Keywords: Geographic Information System, ArcGIS Engine, DEM, DOM, quality inspection.

1 Introduction

In recent years with the development of spatial information technologies, the Digital Elevation Model (DEM) and Digital Orthophoto Map (DEM) as geographic information based data is more and more important applied in every field. Meanwhile, related national standards have higher quality requirements to DEM, DOM as achievement of geographical information, so in the process of in surveying and mapping production and inspection required that we can control quality precisely and effectively. To DEM, DOM quality inspection there are lots of different ways, in the current market geographic information system software are not have very complete quality checking, and for quality controllers aspired to use a set of quality inspection software has perfect function and method, to help checking the work, improve the quality and efficiency of inspection. For that reason, the author uses based on component ArcGIS Engine re-development, expand quality inspection function, and implement simple, accurate and efficient DEM, DOM quality inspection.

2 ArcGIS Engine and Digital Geographical Information Production Quality Inspection Analysis

It's well known that ArcGIS is very popular and very mature geographic information software in international, ArcGIS Engine development kit as an extension of the

---

* Corresponding author.

D. Jin and S. Jin (Eds.): Advances in FCCS, Vol. 1, AISC 159, pp. 379–383.
springerlink.com

ArcGIS function is ArcObjects component based on software development products, to construct user-defined GIS and mapping application software. It contains the rich class library which can make users convenient to create a various applications simply and efficiently. In ArcGIS the DEM and DOM have similar data organization format, they are all grid of matrix forms, all can be expressed as Raster data type , so there are some same method to operate. In the past inspection process of DEM and DOM, it's troublesome that the mathematical foundation inspection and the precision inspection of edge match. Especially face to so many kinds of data sources such as *. tif, *. img, *. grd and so on, it's restrictive of the primary software. ArcGIS Engine provides us so many kinds of data sources of support exactly with DataSourcesRaster library, DataSourcesFile library, and SpatialAnalyst, 3 DAnalyst, GeoAnalyst, NetworkAnalysis many kinds of spatial analysis function data library. Through lots of experiments and analysis, we found that realize the DEM and DOM quality check by using ArcGIS Engine re-development, and, method rigorous, operation simply, visual function is strong, can satisfy most of requirements of the data quality inspection.

3 Application Base on the ArcGIS Engine

3.1 Mathematics Foundation Inspection

Mathematics foundation inspection is the most important content to check work, All the check work are begin with mathematics foundation inspection. Check the space reference information of grid data, the geographical location information, X, Y direction pixel size, ranks number, begins and end point coordinates of DEM, DOM and the correctness of the coverage. The above information extraction realized through the IRasterProps (grid attributes) interface, this interface is located under ESRI. ArcGIS. DataSourcesRaster, the interface support such as RasterBand, Raster, MosaicRaster data format, and also can offer spatial reference, data ranks number, pixels size, and the basic information of data scope:

```
IRaster clipRaster = rastDataset.CreateDefaultRaster()
as IRaster;
IRasterProps rasterProps = (IRasterProps)clipRaster;
ISpatialReference spatialReference =
rasterProps.SpatialReference();//spatial reference of
grid data
int dHeight = rasterProps.Height;// row of grid data
int dWidth = rasterProps.Width; // column of grid data
double dX = rasterProps.MeanCellSize().X; //width of
each pixel
double dY = rasterProps.MeanCellSize().Y; //height of
each pixel
IEnvelope extent = rasterProps.Extent; //scope of grid
data
IPoint iPointUupLeft = new PointClass();//top left
corner coordinate
IPoint iPointLowRight = new PointClass();//bottom
right corner coordinate
iPointLowRight = extent.LowerRight;
```

```
iPointUupLeft = extent.UpperLeft ;
rstPixelType pixelType = rasterProps.PixelType;
//current type of grid pixel
```

Obtain the information above, using the convenient method compare grid attribute information with the theoretical value, so can judge which picture the mathematical basis Incorrect, Finally output error information for inspection.

3.2 Data Quality Inspection

Check DEM superposition error by DEM chromatography superimposed contours, is a classical method of evaluating the DEM quality. The rendering interface of IRasterRenderer grid image offer rich rendering method such as the only value, classification and stretching and so on, the classification rendering function realized based on IRasterClassifyColorRampRenderer interface. Set up the rendering grading interval, the output grading chromatography, artificial interactive inspection according to contour interval.

Gray histogram image is function of grayscale, the inspectors often judge some properties of the images through condition of gray histogram. The method of grid DEM classification is also applicable in building DOM gray-level histogram, only change the render way into unique value rendering of a pixel values IRasterUniqueValueRenderer, Then easy to get gray histogram information of each image after numerical statistics, and convenient to evaluate the image quality.

3.3 Precision Testing

Evaluate the precision of the DEM, DOM by mid-error in "The provisions of inspection acceptance and quality evaluation of digital mapping products". Digital elevation model point mid-error calculates by following formula:

$M_h = \pm\sqrt{\dfrac{\sum_{i=1}^{n} \Delta h_i^2}{n-1}}$ in this formula: Δh_i—elevation D-value, n—number of checkpoint.

In this paper the DEM precision statistics by the method of comparing the checkpoint elevation value with DEM inserted elevation value. IFunctionalSurface interface can get elevation value of any coordinate point on the TIN. Get any points of elevation information in DEM by this interface is more concise than that we use various interpolation theory calculate elevation information of any coordinate points, and also can satisfy the requirements of precision statistics. For this, may build applied model by various formats of DEM through GeoProcessing, transfer the data, sum up the accuracy.

```
IFunctionalSurface pFuncSurf= pTinLayer.Dataset
double z = pFuncSurf.Z(x, y) //x,y are testing point
coordinates
```

DOM precision testing method is simple relatively, after import testing point, doing precision statistics by artificial measurement method.

3.4 Checking Seamless Edge-Connecting

Because the maps are separate, complete surface features is partitioned by maps, during process of digital geographical information production, hard to avoid because of various factors make difference of same coordinates pixels of the adjacent map. Although there are no Strict rules of lattice point attribute value must be completely consistent in the same level position in the "The provisions of inspection acceptance and quality evaluation of digital mapping products", but for rigorous of database organization and in-depth research in large area 3D model, the data seamless edge-connecting is particularly important. Because of the uncertainty of the data sources, and the quantity of data of the DEM, DOM are very big, the traditional methods to comparing pixels one by one is unable to complete our needs. Therefore, the use of the IMathOp interface under space analysis is the best choice, the mathematical operation interface is located in ESRI. ArcGIS. SpatialAnalyst. The interface can be quickly realized grid data for addition, subtraction, multiplication, division, absolute value, and logarithmic commonly used mathematical function. As the following:

```
// tempGeodata1, tempGeodata2
IMathOp pMathOP = new RasterMathOpsClass();
IGeoDataset outGetDataset = pMathOP.Minus(tempGeodata1,
tempGeodata2);
```

Then statistical analysis the result of grid operation, judge the error value through the analysis of statistic results. Use IRasterCalcStatsHistogram interface, the grid statistical interface is located in ESRI. ArcGIS. DataSourcesRaster, the interface can realize the maximum, minimum and average values of grid etc.

```
IStatsHistogram pStatsHistogram = new
StatsHistogramClass();
IRasterCalcStatsHistogram pCalStatsHistogram = new
RasterCalcStatsHistogramClass();
pCalStatsHistogram.ComputeFromRaster(outRaster, 0,
pStatsHistogram);
double pMeanVal = pStatsHistogram.Mean;
 if (pMeanVal != 0)
```

Through the two steps simultaneous operation above can easy to realize the comparison of grid image pixel values. Usually there are many DEM data formats, such as *. grd, *. dem, etc. For the data format supported by ArcGIS Engine can be operated directly to load in MapControl, for the data format can't be directly loaded in MapControl, ArcGIS Engine provide plenty of data format transition classes. Such as DEM of ASCII format, using ESRI.ArcGIS.GeoProcessing.Tools toolkit, transfer data by calling the GeoProcessing build application model in the secondary development:

```
Geoprocessor gp = new Geoprocessor();
ESRI.ArcGIS.ConversionTools.ASCIIToRaster
ASCIIToRasterTool = new ASCIIToRaster();
```

4 Conclusion

This paper use Visual Studio 2005 as development platform, with c #.net language based on ArcGIS Engine9.2 re-development. In order to improve the quality inspection accuracy and efficiency of digital geographical information production, from the method of check DEM, DOM grid data, discusses application and implementation of ArcGIS Engine in DEM, DOM quality inspection. Practice has proved that the COM second development technology is simple and efficient, and avoid cockamamie testing process of first development, make inspection program can applied to production quickly, improve the product efficiency, ensure the quality safety. Of course, the above content, cannot solve all the problems of the inspection work, this is the problem we urgently need to continue to think deeply about.

References

1. Qiu, H.: ArcGIS Engine development from approaches to master. Posts and Telecom press, Beijing (2010)
2. Jiang, B.: The design and realization of Plug-in GIS application framework based on c # and ArcGISEngine 9.2. Electronic Industry Press, Beijing (2009)
3. Song, X., Niu, X.: Geographic information system practice tutorial. Science press, Beijing (2007)
4. The fundamental geographic information digital products 1:10 000 1:50 000 production technique rules part 2: digital elevation model (DEM), (CH/T 1015.2-2007)
5. The fundamental geographic information digital products 1:10 000 1:50 000 production technique rules part 3:Digital Orthophoto Map(DOM), (CH/T 1015.2-2007)
6. 1:500 1:1000 1:2000 foundation geographical information digital achievement digital elevation model (DEM), (CH/T 9008.2-2010)
7. 1:500 1:1000 1:2000 foundation geographical information digital achievement Digital Orthophoto Map(DOM), (CH/T 9008.3-2010)
8. The provisions of inspection acceptance and quality evaluation of digital mapping products, (GB/T 18316-2001)

Risk Assessment of Information System Security Based on Fuzzy Multiple Criteria Group AHP

Shengchun Sun[1], Huaiyu Pu[1], and Shuxin Tian[2]

[1] Naval Univ. of Engineering, Wuhan 430033, China
[2] Unit 91829 of the PLA, Dalian 116041, China

Abstract. Based on the fuzzy multiple criteria group AHP, a new method has been proposed for the risk assessment of information system security and then its index system has been established. The least variance priority method of triangular fuzzy number complementary judgment matrix is necessary to obtain the weight of each index. The fuzzy weights of the second-class indexes can be derived from the synthesis of experts' opinions by the fuzzy Delphi method. Combined with the security values of the second-class indexes expressed by the fuzzy language variables, the fuzzy indexes for evaluating information system security are gained by use of the hierarchical integration method. Finally, with the aid of the λ average area measurement method, the security values of the system are determined so that the system can reach its security rating. Furthermore, a practical example given in this paper verifies that the new method is feasible and effective.

Keywords: risk assessment, information system security, fuzzy multiple criteria group AHP, fuzzy weight, fuzzy language variables.

1 Introduction

With information technology advancing by leaps and bounds, national economy and social development rely increasingly on fundamental information networks and essential information systems. Now, information security has become a major problem, for which there is considerable concern. For this reason, the best way to solve this problem is to make an objective and reasonable assessment of information security. The making of a set of scientific analysis and assessment methods or models is theoretically and practically vital for the construction of informationization and the development strategy of information security in our country[1].

Using fuzzy mathematics, grey theory, artificial neural network and so on, scholars at home and abroad have now established risk assessment models and made great achievements[2-4]. Because of complexity in risk assessment of information system security and some ambiguity and uncertainty in human thinking, the adoption of a fuzzy method has gradually called scholars' attention to risk assessment[5-7]. As the information security assessment refers to the fuzzy multiple criteria group AHP, this paper has used triangular fuzzy numbers to conduct a quantitative analysis of fuzzy information.

D. Jin and S. Jin (Eds.): Advances in FCCS, Vol. 1, AISC 159, pp. 385–391.
springerlink.com

2 Risk Assessment Index System of Information System Security

The risk assessment of information system security must be based on a scientific and complete assessment index system, without which the work of assessment cannot be done objectively and smoothly.

According to the characteristics of risk assessment of information system security, the general principles of establishing an index system and the fuzzy Delphi method, the index system for risk assessment of information system security has been set up which is related to physical security, operation security, information security and security management. The hierarchical structure model of the index system is shown in [5], including 4 first-class indexes: physical security (F_1), operation security (F_2), information security (F_3) and security management (F_4), and 23 second-class indexes such as environment safety (C_1), equipment safety (C_2), medium safety (C_3), and backup and recovery (C_4), etc.

3 Fuzzy Multiple Criteria Group AHP

3.1 Preparatory Knowledge

Definition 1: If $a=(a_l,a_m,a_u)$, $0\le a_l\le a_m\le a_u$, $a=(a_l,a_m,a_u)$ is regarded as a triangular fuzzy number.

The rule of operation for triangular fuzzy numbers is as follows:

Suppose $a=(a_l,a_m,a_u)$, $b=(b_l,b_m,b_u)$, $\oplus$ and $\otimes$ represent fuzzy addition and fuzzy multiplication operators respectively, then, $a\oplus b=(a_l+b_l,a_m+b_m,a_u+b_u)$, $a\otimes b=(a_lb_l,a_mb_m,a_ub_u)$.

Definition 2: If the elements in the triangular fuzzy number matrix B satisfy: $b_{ij}=(b_{lij},b_{mij},b_{uij})$, $b_{ji}=(b_{lji},b_{mji},b_{uji})$, and $b_{lij}+b_{uji}=b_{mij}+b_{mji}=b_{uij}+b_{lji}=1$, $b_{nij}\ge b_{mij}\ge b_{lij}\ge 0$, $i,j\in N$, then the matrix is referred to as a triangular fuzzy number complementary judgment matrix[8].

3.2 Quantification of Fuzzy Language Variables

In the fuzzy multiple criteria group AHP model, the security value of each index is assessed by use of the fuzzy language variables which represent the rating set of security assessment adopted by experts as W={very bad, bad, moderate, good, very good}. The quantification of rating set W is expressed by triangular fuzzy numbers. The corresponding relationship between the fuzzy language variables and the triangular fuzzy numbers is shown as Tab. 1:

Table 1. Relationship between fuzzy language variables and triangular fuzzy numbers

number	security assessment rating	triangular fuzzy numbers
1	very bad	(0,0,0.25)
2	bad	(0,0.25,0.5)
3	moderate	(0.25,0.5,0.75)
4	good	(0.5,0.75,1)
5	very good	(0.75,1,1)

3.3 Fuzzy Weights of First-Class Indexes

Assuming that experts $E_1, E_2, \cdots, E_n$ carry out risk assessment of information system security, the degrees of their risk preference are expressed by $\lambda_1, \lambda_2, \cdots, \lambda_n$. Value λ expresses the attitude of each expert towards risk. When $\lambda > 0.5$, the experts are thought to show preference to risk. When $\lambda = 0.5$, it implies that they take a neutral attitude towards risk. When $\lambda < 0.5$, it shows that they take an aversion to risk.

First of all, based on the important assessment values of the first-class indexes given by the experts, it is necessary to establish a fuzzy judgment matrix $R = (\tilde{r}_{ij})_{m\times m}$, among which $r_{ij} = (r_{lij}, r_{mij}, r_{uij})$ is a triangular fuzzy number. r_{lij}, r_{mij} and r_{uij} respectively express the index i given by the experts, which reflects the most pessimistic, the most possible or the most optimistic estimations of the importance of the index j. Assuming that the relative importance of i to j is $r_{ij} = (r_{lij}, r_{mij}, r_{uij})$, that of j to i is $r_{ji} = (1 - r_{uij}, 1 - r_{mij}, 1 - r_{lij})$. From Definition 1, it can be seen that $R = (\tilde{r}_{ij})_{m\times m}$ is a triangular fuzzy number complementary judgment matrix.

Then, according to the experts' risk preference, the least variance priority method for arranging the triangular fuzzy number complementary judgment matrix proposed in [9] is used to arrange the fuzzy judgment matrix $R = (\tilde{r}_{ij})_{m\times m}$ in a sequence so as to obtain the weight vectors of the first-class indexes, namely $w_i = (w_{i1}, w_{i2}, \cdots, w_{im})$. Here, w_{ij} expresses the weight given by expert i to first-class index j.

3.4 Fuzzy Weights of the Second-Class Indexes

It can be learned from the hierarchical structure model for the index system that each individual first-class index is divided into a certain number of second-class indexes. Similarly, a fuzzy judgment matrix needs to be set up first according to every expert' s important assessment of individual second-class indexes corresponding to the first-class indexes. Then, combined with each expert's risk preference, the least variance priority method is adopted to obtain the weight vectors of individual second-class indexes, namely $w_{ip} = (w_{ip1}, w_{ip2}, \cdots, w_{ipr})$ among which w_{ipj} expresses the weight of second-class index j acquired by expert i with regard to first-class index p.

The synthesis of the weights of the first-class and second-class indexes results in the integral weight of the second-class indexes, that's, $W_i = (w_{i1}, w_{i2}, \cdots, w_{it})$.

Here, w_{ij} expresses the integral weight given by expert i to second-class index j. In order to synthesize the opinions of individual experts and reflect the uncertainty in their judgment, this paper has adopted the fuzzy Delphi method, which means that the triangular fuzzy number is applied to integrating the weights obtained by the experts with respect to the second-class indexes. Thus, the triangular fuzzy number weight of each second-class index is determined in such a way as following:

$$W_j = (l_j, m_j, u_j) . \tag{1}$$

In this formula, W_j expresses the fuzzy weight of second-class index j, $l_j = \min_i\{w_{ij}\}$, $m_j = \left|\prod_{i=1}^{n} w_{ij}\right|^{1/n}$, $u_j = \max_i\{w_{ij}\}$, where w_{ij} expresses the integral weight gained by expert i concerning second-class index j.

3.5 Determination of Fuzzy Security Values of Information System

In order to describe the uncertainty of the assessed object and to integrate the experts' opinions conveniently, the security values of the second-class indexes should be determined by the fuzzy language variables shown in Tab.1, marked as $s_{ij} = (\alpha_{ij}, \beta_{ij}, \gamma_{ij})$.

The adoption of average operators, in combination with the experts' opinions, can obtain the average fuzzy security values of the second-class indexes as follows:

$$S_j = \left|\frac{1}{n}\right| \otimes (s_{1j} \oplus s_{2j} \oplus \cdots \oplus s_{nj}) . \tag{2}$$

Where S_j expresses the average fuzzy security values given by the experts to second-class index j, and s_{ij} represents the average fuzzy security values offered by the expert i to second-class index j.

Then, the fuzzy weights and fuzzy security values of the second-class indexes are integrated to obtain the system's fuzzy security assessment index M, which is expressed as following:

$$M = (W_1 \otimes S_1) \oplus (W_2 \otimes S_2) \oplus \cdots \oplus (W_t \otimes S_t) \tag{3}$$

Because both W_j and S_j are triangular fuzzy numbers, M can be expressed as $M = (Y, Q, Z)$, where $Y = \sum_{j=1}^{t} \alpha_j l_j$, $Q = \sum_{j=1}^{t} \beta_j m_j$, $Z = \sum_{j=1}^{t} \gamma_j u_j$, $\alpha_j = \frac{1}{n}\sum_{i=1}^{n} \alpha_{ij}$, $\beta_j = \frac{1}{n}\sum_{i=1}^{n} \beta_{ij}$, $\gamma_j = \frac{1}{n}\sum_{i=1}^{n} \gamma_{ij}$.

The system's fuzzy security assessment value M can be derived from the integration of W_j and S_j.

3.6 Fuzzy Treatment and Security Rating Determination

As the system's fuzzy assessment index $M=(Y,Q,Z)$ is not yet clearly indicative of the state of information system security, the λ average area measurement method[10] is used for the fuzzy treatment of the solution of M.

$$s_{\lambda}(M)=([\lambda Y+Q+(1-\lambda)Z])/2 \tag{4}$$

In this formula, $\lambda\in[0,1]$ is the weight value determined by the experts, which reflects the degree of their aversion to risk, and thus the security value $s_{\lambda}(M)$ is obtained.

As for the value $s_{\lambda}(M)$ obtained, its security rating can be determined according to Tab.2.

Table 2. Subordinate Security Rating of $s_{\lambda}(M)$

$s_{\lambda}(M)$	0-0.2	0.2-0.4	0.4-0.6	0.6-0.8	0.8-1
Rating	low	lower	moderate	high	higher

4 Analysis of an Example

Taking a security communication system for example, the application of the fuzzy multiple criteria group AHP to the risk assessment of the system is as follows:

First, there is a need for establishing an index system for risk assessment of information system security.

Second, It is up to experts to determine the fuzzy weights of the second-class indexes. Three experts, whose preference to risk is respectively expressed by (0.5, 0.4, 0.3), are asked to give the relatively important security value r_{ij} so as to create the fuzzy judgment matrix $R=(\tilde{r}_{ij})_{m\times m}$. From this matrix are derived the weights of the first-class and second-class indexes as well as the integral weights of the second-class indexes on each expert's part, $W_i=(w_{i1},w_{i2},\cdots,w_{it})$, as shown in Tab.3.

Table 3. Integral Weights of the Second-class Indexes Offered by Experts

	C_1	C_2	C_3	C_4	C_5	C_6	C_7	C_8	C_9	C_{10}	C_{11}
E_1	0.029	0.054	0.041	0.029	0.014	0.024	0.033	0.027	0.033	0.020	0.018
E_2	0.019	0.047	0.028	0.042	0.022	0.034	0.034	0.026	0.042	0.031	0.017
E_3	0.032	0.050	0.028	0.033	0.025	0.037	0.041	0.048	0.041	0.022	0.016

C_{12}	C_{13}	C_{14}	C_{15}	C_{16}	C_{17}	C_{18}	C_{19}	C_{20}	C_{21}	C_{22}	C_{23}
0.038	0.036	0.058	0.099	0.036	0.036	0.021	0.053	0.071	0.071	0.063	0.100
0.048	0.041	0.065	0.112	0.041	0.041	0.023	0.049	0.066	0.066	0.043	0.063
0.031	0.039	0.075	0.095	0.037	0.034	0.029	0.034	0.062	0.090	0.042	0.053

In accordance with Formula (1), the synthesis of the experts' opinions leads to the formula $W_j = (l_j, m_j, u_j)$, as shown in Tab.4.

Table 4. Fuzzy Weights of the Second-class Indexes Offered by Experts

	C_1	C_2	C_3	C_4	C_5	C_6	C_7	C_8	C_9	C_{10}	C_{11}
l	0.019	0.047	0.028	0.029	0.014	0.024	0.033	0.026	0.033	0.020	0.016
m	0.027	0.051	0.032	0.035	0.020	0.032	0.036	0.034	0.039	0.024	0.017
u	0.032	0.054	0.041	0.042	0.025	0.037	0.041	0.048	0.042	0.031	0.018
C_{12}	C_{13}	C_{14}	C_{15}	C_{16}	C_{17}	C_{18}	C_{19}	C_{20}	C_{21}	C_{22}	C_{23}
0.031	0.036	0.058	0.095	0.036	0.034	0.021	0.034	0.062	0.066	0.042	0.023
0.039	0.039	0.066	0.102	0.038	0.037	0.024	0.045	0.066	0.076	0.049	0.072
0.048	0.041	0.075	0.112	0.041	0.041	0.029	0.053	0.071	0.090	0.063	0.100

Furthermore, the determination of the fuzzy security indexes of the information system depends on the experts to make a security assessment of the second-class indexes. According to Formula (2), the average fuzzy security values of the second-class indexes offered by the experts are obtained, as shown in Tab.5.

Table 5. Triangular Fuzzy Numbers for Average Security Assessment of the Second-class indexes by Experts

	C_1	C_2	C_3	C_4	C_5	C_6	C_7	C_8	C_9	C_{10}	C_{11}
α	0.50	0.42	0.25	0.50	0.50	0.42	0.50	0.25	0.42	0.50	0.50
β	0.75	0.67	0.50	0.75	0.75	0.67	0.75	0.50	0.67	0.75	0.75
γ	0.92	0.83	0.75	0.92	0.92	0.83	0.92	0.75	0.83	0.92	0.92
C_{12}	C_{13}	C_{14}	C_{15}	C_{16}	C_{17}	C_{18}	C_{19}	C_{20}	C_{21}	C_{22}	C_{23}
0.50	0.42	0.50	0.50	0.50	0.42	0.50	0.50	0.25	0.50	0.42	0.42
0.75	0.67	0.75	0.75	0.75	0.67	0.75	0.75	0.50	0.75	0.67	0.67
0.92	0.83	0.92	0.92	0.92	0.83	0.92	0.92	0.75	0.92	0.83	0.83

The integration of the fuzzy weights and average fuzzy security values of the second-class indexes comes to such a result as M = (0.365, 0.692, 1.020).

Finally, the fuzzy security values of the information system need a fuzzy treatment according to Formula (4). Taking λ=0.4, $s_\lambda(M)$=0.725, and following Tab.2, it is possible to make the information system attain the high security rating. This rating should conform to the result of the multi-class fuzzy comprehensive assessment and consequently the method is proved to be effective and reliable.

5 Conclusions

This paper has proposed a method based on the fuzzy multiple criteria group AHP. The method adopts the triangular fuzzy number complementary judgment matrix to obtain the weight of each hierarchy of the information system, and makes use of the fuzzy Delphi method to integrate the experts' opinions to achieve the fuzzy weights of the second-class indexes. The fuzzy language variables are used to describe the security values of the second-class indexes, which are quantified to obtain the average fuzzy security value of each second-class index. The integration of the fuzzy weights and average fuzzy security values of the second-class indexes results in getting the indexes for assessing the system security. With the aid of the λ average area measurement method for fuzzy treatment of the solutions, the security rating of the system can be determined.

References

1. Li, H., Liu, Y., He, D.: Review on Study of Risk Evaluation for IT System Security. China Safety Science Journal (CSSJ) 16(1), 108–113 (2006)
2. Fu, Y., Wu, X., Ye, Q., et al.: An Approach for Information Systems Security Risk Assessment on Fuzzy Set and Entropy-Weight. Acta Electronica Sinica 38(7), 1489–1494 (2010)
3. Wang, X., Lu, Z., Liu, B.: Algorithm of Information Security Risk Evaluation Based on OCTAVE and Grey System. Journal of Beijing University of Posts and Telecommunications 32(5), 128–131 (2009)
4. Zhao, D., Liu, J., Ma, J.: Risk assessment of information security using fuzzy wavelet neural network. Journal of Huazhong University of Science and Technology (Nature Science Edition) 37(11), 43–45 (2009)
5. Wang, J., Fu, Y., Wu, X.: Research on Security Risk Assessment of Information System Based on Improved Fuzzy AHP. Fire Control & Command Control 36(4), 33–36 (2011)
6. Ma, H., Ma, Q., Fu, G.: Evaluation of supply chain default risk based on fuzzy influence diagram. Journal of Southeast University (English Edition) 23(S1), 111–117 (2007)
7. Liu, X., Jiang, Q., Cao, Y., et al.: Transient security risk assessment of power system based on risk theory and fuzzy reasoning. Electric Power Automation Equipment 29(2), 15–20 (2009)
8. Fu, Y., Wu, X., Song, Y.: Assessment for the performance of logistic department of warship equipment on fuzzy group AHP. Ship Science and Technology 28(5), 99–102 (2006)
9. He, Y., Zhou, D., Wang, Q.: Least variance priority method for triangular fuzzy number complementary judgment matrix. Control and Decision 23(10), 1113–1116 (2008)
10. Li, D.: Fuzzy multiobjective many-person decision makings and games. National Defense Industry Press, Beijing (2003)

A New Mobile Learning Platform Based on Mobile Cloud Computing

Shuqiang Huang[1,*] and Hongkuan Yin[2]

[1] Network and Education Technology Center of Jinan University, Guangzhou 510632, China
hsq2008@vip.sina.com

[2] Shenzhen Ruier information Technology Co., LTD, ShenZhen 518000, China

Abstract. Mobile learning is a new learning mode and can bring learner higher learning efficiency and more flexibility. Aiming the problem of existing development technologies for mobile learning platform, a new mobile learning platform based on mobile cloud computing was proposed. The platform was abstracted several layers such as infrastructure layer, resource layer, service layer, middleware layer and client layer. By middleware layer, the platform can adapt different client access request. The communication between client and sever is finished by instant message server and the task processing is finished by cloud application server. Many key technologies were taken in the cloud platform and the platform will provide a good and practical example for developing mobile learning platform by mobile cloud computing.

Keywords: Mobile learning, Cloud computing, Cloud Application Server, Instant Message Server, Memory database.

1 Introduction

Now a variety of portable wireless mobile terminals and network resources are emerging, more and more network applications based on wireless mobile device have become a tread and played an important role on people's lives. With the development of mobile computing and wireless communications technology, a new mode of learning called mobile-learning has come into being. In the mode, leaner can learn knowledge and course at any time and any place by mobile computing devices. This way can help learner get rid restrictions of time and space, and can bring learner new learning experience and higher learning efficiency and more flexibility. Mobile learning has some features as follow: the basic features of online education, mobility, timeliness, and ubiquitous feature. Because of these features, mobile-learning is an indispensable mode of learning and called as the next generation of learning [1].

Distance education has gone through a long process of development from distance learning to e-learning, from e-learning to mobile learning. Mobile learning is the

* Corresponding author.

D. Jin and S. Jin (Eds.): Advances in FCCS, Vol. 1, AISC 159, pp. 393–398.
springerlink.com

expansion of digital learning, and in the mode the learning environment is mobile, including teaching, student and course. Wireless communication, mobile computing and internet are the main base technologies to realize mobile learning. As a new learning mode, mobile learning has a great potential in future teaching application, but now the theory research and application in practice are still in the stage of early development. There are some problems such as lacking of basic theory and application research, lacking perfect support systems and mobile learning platform, and lacking of mobile learning resources in current mobile learning [2]. How to apply the mobile technology for the teaching has become a hot topic of mobile learning application.

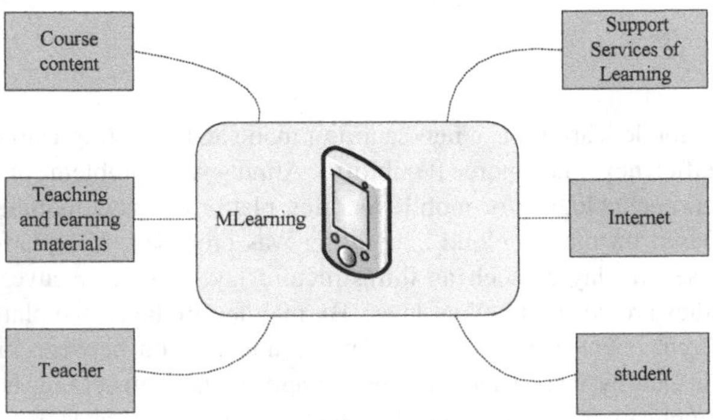

Fig. 1. Mobile learning environment

Short Message Service (SMS), Wireless Application Protocol (WAP) and C/S are the main technologies for developing mobile learning platform [3-4].

SMS is providing text information service of finite length by GSM network, and communication between users, between the user and the server is realized by short message. Short messaging based on phone can send teaching management information, course outline and interactive quiz, but this method is only fit for some simple course and can not provide multimedia information. Compared with traditional WWW communication mode, a gateway is added in WAP and the tasks of protocol conversion, content encoding and decoding can be finished by the gateway. WAP is better then SMS, but WAP can not finish highly interactive task. Developing mobile learning system architecture is feasible by J2EE and J2ME, and the server program can be developed by J2EE architecture and client program can be developed by J2ME architecture. C/S model can finish complex task, but for traditional C/S mode all content must be downloaded to client and processed by client.

2 Combination of Mobile Cloud Computing and Mobile Learning

Cloud computing is a super computing mode based on Internet, in the mode, the shared hardware and software resources and information can be provided to computers and other devices on demand. The core of cloud computing is how to manage and schedule uniformly computing resources connected by network, then constitute a pool of computing resource and provide service for user on demand. The computing resources network is called cloud, and it is a highly scalable computing mode and base technology framework. All the resources are stored in the cloud and user can access the information and services anywhere, anytime from any terminal, for user, the cloud is transparent when using resources.

The application of cloud computing is not limited only to PC, but also to mobile devices and mobile terminals. With the rapid development of mobile Internet, cloud computing services providing for mobile phones and other mobile terminals have emerged. Mobile cloud computing is an application of cloud computing in the mobile internet. In the mode of mobile cloud computing, user can obtain the infrastructure, platform, software and other IT resources service on demand by mobile networks.

By analyzing the features of mobile learning, cloud computing can provide important infrastructure for mobile learning. Because all sources can be stored in cloud and all processing can be finished in cloud, cloud computing can solve the lack of existing platform and technology. Developing mobile learning platform based on cloud computing is of great value and meaning, and all information and learning resources will be moved to cloud. Learner can obtain rich, real-time and accurate learning content and resource by the platform and collaborative learning and situated learning were supported[5]. Platform for mobile learning based on cloud can achieve the following fectures: cloud platform can provide real-time utilization information of resource and allocation of resources on-demand; cloud platform can provide supporting for a variety of applications including computing and storage resources; cloud can provide QoS guarantee for mobile learning system infrastructure; Application and processing are all finished in the cloud. Rich learning resources can be integrated and configure for mobile learning devices was reduced greatly [6].

3 Mobile Learning Platform Based on Cloud Computing

The target of building mobile learning platform is to create a anytime and anywhere learning environment, learner can choose learning content independently and can interact in real-time and non real-time with teacher. When designing mobile learning platform, some factors such as practicality, scalability, adaptability and reliability should be taken into account. The platform should achieve the following objectives: having a rational position, supporting a variety of learning modes, providing real learning environment and realizing unity of online and offline learning.

3.1 System Architecture of Mobile Learning Platform Based on Cloud Computing

The learning platform is made up of infrastructure, resource pool, database and file system, application servers, management servers, middleware and mobile client. The infrastructure includes network、 network storage and security equipment, resource pool is made up of memory, CPU and disk resource. The application servers include CAS (cloud application server) and IMS (Instant Messaging Server), CAS provides mobile cloud computing service and IMS provides instant message service. The management server provides management services as cache management, session Management, log management and message management. The middleware is made up of API, adapter and interceptor. By middleware, authorized mobile client can access cloud application service.

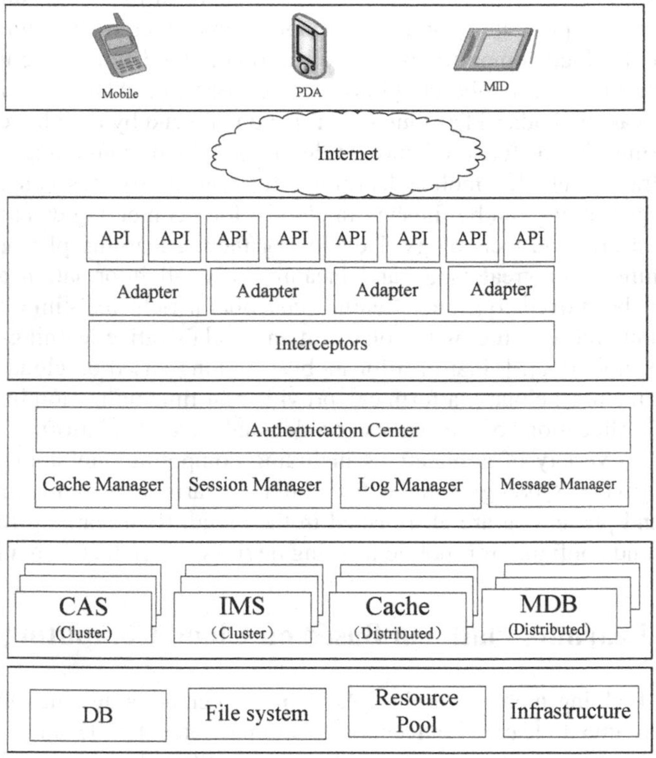

Fig. 2. System Architecture of mobile learning platform

3.2 Architecture of CAS

The CAS is made up of database and proxy, message server, resource accessor, application container, interceptor and adapter. CAS provides API based on http protocol for client, when client sending a request by http to CAS, CAS will process the request by cloud computing programming model and return a response to client. The client only need sent the request and the CAS will finish all the computing and processing, so the client needs only simple processing ability. To cope with a large number of concurrent accesses, the CAS is usually designed by cluster.

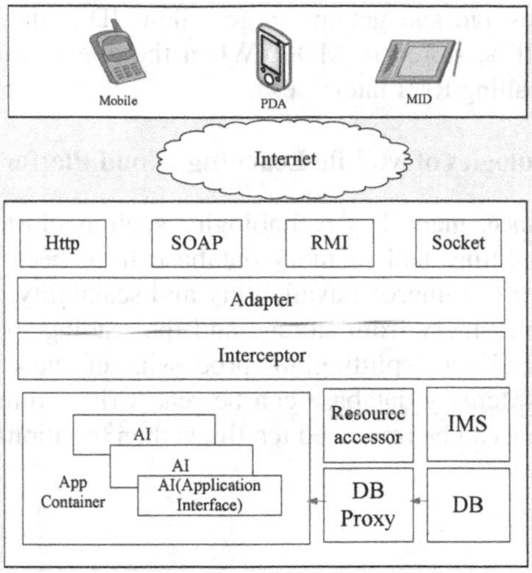

Fig. 3. Architecture of CAS

3.3 Architecture of IMS

The IMS provides communication service between server and client. The session course between client and message server includes server steps as the follow: When the mobile client sending a request to IMS server, the system will return the target server by loading balance algorithm; A long TCP/IP connection between client and IMS is established by reusable channel; Memory database(MDB) is a database stored in memory and the message sent from client to server is stored MDB; IMS addresses

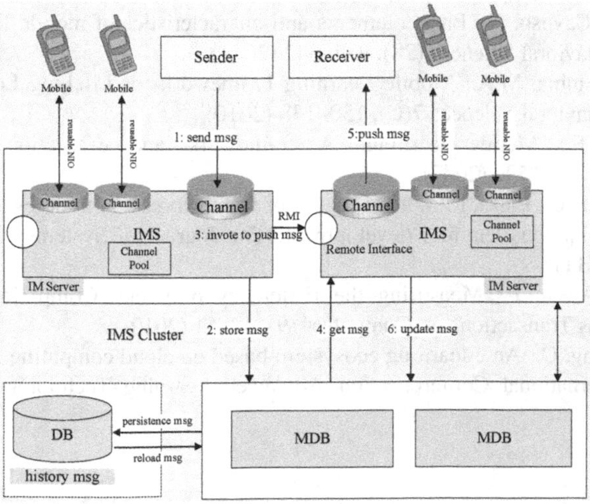

Fig. 4. Architecture of IMS

the client sending message and get the target client ID;If the client is offline, the response message will be stored in MDB. When the client is online, then push the message to client by calling RMI interface.

3.3 The Key Technologies of Mobile Learning Cloud Platform

To improve performance, many key technologies such as cluster, distributed cache, database read/write splitting and memory database have been taken in the mobile cloud platform. Cluster can improve availability and scalability; by distributed cache, the data is called into memory from storage and the reading speed can be improved greatly; by database read/write splitting, the processing efficiency of the database can be improved greatly; Memory database can be read/written directly by user and the processing performance can be improved ten times than traditional database.

Summary

A new mobile learning platform based on mobile cloud computing was proposed and many advanced technologies were taken in the platform, practical learning system based on the platform has been developed and can provide mobile service for mobile phone.

Acknowledgement. The paper was supported by the teaching reform project and education technology project of Jinan University: mobile learning platform based on cloud computing.

References

1. Ozdamlia, F., Cavusb, N.: Basic elements and characteristics of mobile learning. Procedia Social and Behavioral Sciences (28), 937–942 (2011)
2. Nordina, N., Embia, M.A.: Mobile Learning Framework for Lifelong Learning. Procedia Social and Behavioral Sciences 7(C), 130–138 (2010)
3. Motiwalla, L.F.: Mobile learning: A framework and evaluation. Computers & Education (49), 581–596 (2007)
4. Martin, S., Diaz, G., Plaza, I.: State of the art of frameworks and middleware for facilitating mobile and ubiquitous learning development. The Journal of Systems and Software (84), 1883–1891 (2011)
5. Pocatilu, P., Alecu, F.: Measuring the Efficiency of Cloud Computing for E-learning Systems. Wseas Transactions on Computers (9), 42–51 (2010)
6. Dong, B., Zheng, Q.: An e-learning ecosystem based on cloud computing infrastructure. In: 9th IEEE International Conference on Advanced Learning Technologies, pp. 125–127 (2009)

A Survey on MDS-Based Localization for Wireless Sensor Network

Hua Tian, Yingqiang Ding, and Shouyi Yang

School of Information Engineering, Zhengzhou University, Zhengzhou, China
tianhuazzu@163.com, {dyq,iesyyang}@zzu.edu.cn

Abstract. Wireless sensor network (WSN), with broad application prospects, is a novel research area. Due to the data collected by the sensor node, in the monitored area, is useful only when its location information is known, node positioning as a critical technology is the basis of applications of wireless sensor network. Multidimensional scaling has been applied to node localization successfully in recent years. In this paper, several typical MDS-based localization algorithms, such as have been introduced and analyzed.

Keywords: wireless sensor network, node localization, MDS.

1 Introduction

Wireless sensor network (WSN) is composed by hundreds and even thousands of small, battery-powered and wirelessly connected sensor nodes[1]. Wireless sensor networks can be deployed in different scenarios, ranging from military applications to environment monitoring[2]. The relative or actual geographic location of the sensor nodes need to be known for various applications in wireless sensor networks such as event discovery and target tracking. Thus, the node localization is the premise of the applications.

Multidimensional scaling (MDS) as a critical technique is used to extract a set of independent variables from a proximity matrix or matrices[3]. Applications of MDS are found in a wide range of areas, such as data preprocessing, scale development, cybernetics, visualization, localization and so on[4]. MDS is well-suited to node positioning in sensor networks, where the purpose is to determine the coordinates of nodes in a 2-D or 3-D space making use of the distance information between nodes. In recent years, different localization algorithms based on classical MDS have been proposed in the literature. These MDS-based algorithms can achieve a higher accuracy than some other localization algorithms.

2 Localization Problem

In localization, the networks are represented as an undirected graph with vertices V and edges E [5]. The vertices correspond to the nodes, of which there may include $m \geq 0$ special nodes called anchors with known positions. In the connectivity-only case, what nodes are nearby be known only by each node, the edges in the graph

D. Jin and S. Jin (Eds.): Advances in FCCS, Vol. 1, AISC 159, pp. 399–403.
springerlink.com

correspond to the connectivity information. In the case with known distances between neighbors, perhaps with limited accuracy, the edges are associated with values corresponding to the estimated distances. All the nodes be considered in the localization problem are assumed to form a connected graph.

Given a sensor network graph of n nodes and estimated distances P between the nodes, p_{ij} is used to refer to the estimated distance between nodes i and j, the positioning problem is to find the coordinates of the nodes X ($X=X_1, X_2, \ldots, X_n$) make the Euclidean distances between the nodes D , where $d_{ij}=\left\|X_i - X_j\right\|_2$, equal to estimated distances P, i.e., $d_{ij}=p_{ij}$. Usually there is no exact solution to the equations when the estimates p_{ij} are just the connectivity or incorrect distance measurements. For this reason, the localization problem is mostly formulated as an optimization problem that minimizes the sum of squared errors.

when solving the localization problem there are two possible outputs: a relative map or an absolute map. In a relative map, even though accurate absolute coordinates for each node are not necessarily included, correct and useful information can be provided. Some applications such as direction-based routing algorithms only require relative positions of nodes[6]. Sometimes, an absolute map is required to determine the absolute geographic coordinates of all the nodes. It is indispensable in applications such as geographic routing and target discovering and tracking [7,8,9].

3 Classical MDS

Classical MDS, one similarity matrix be used, is the simplest and widely used case of MDS. In the classical MDS, the data is quantitative and the proximities of objects are treated as distances in a Euclidean space [10]. Assume that, n nodes are distributed in a m-D space with coordinate $X_i=(x_{i1}, x_{i2}, \ldots, x_{im})$. The estimated distance p_{ij} is used to denote the similarity between two nodes of i and j. The classical MDS can be divided into three steps. Firstly , compute the square of all elements in the proximity matrix P, i.e., $P^{(2)}=[\,p_{ij}^2\,]\,n\times n$. Secondly, obtain the double centered matrix of $P^{(2)}$, B. Finally, carry out the singular value decomposition on B, i.e., $B = V\,A\,V$, then the solution of classical MDS is $X_0= V\,A^{1/2}$.

4 MDS-Based Localization

Several typical MDS-based localization algorithms proposed in recent years, including centralized pattern and distributed pattern, have been introduced here.

4.1 MDS-MAP(C) Algorithm

MDS-MAP(C) [5] is a centralized algorithm, the earliest usage of MDS in node localization for wireless sensor network, which is proposed by Y. Shang et.. The mean idea of MDA-MAP(C) is to build a global map of considered scenarios using classical MDS. This algorithm can be used for both relative positioning and absolute positioning. There are three steps in MDA-MAP(C) algorithm. At first, compute the shortest paths between all pairs of nodes in the region of consideration, Dijkstra[11] or Floyd[11] algorithm generally be used, to construct the distance matrix for MDS. Then, perform classical MDS on the distance matrix, retaining the first two (or three) largest eigenvalues and eigenvectors to construct a 2-D (or 3-D) relative map of the monitoring region. Finally, transform the relative map to an absolute map based on the absolute positions of anchors.

The result of classical MDS is good in the dense or uniform networks, because the shortest path distance as the estimate of the true Euclidean distance is fine. However, it can not work well in irregular networks. MDS-MAP(C, R)[5] is the method with additional refinement to MDS-MAP(C). The result of MDS-MAP(C, R) is better than that of MDS-MAP(C) in uniform topologies, but in irregular topologies it is much worse than its result on the uniform example.

4.2 MDS-MAP(P) Algorithm

MDS-MAP(P) [12], a distributed pattern, is an improved localization algorithm based on MDS-MAP(C), which is also proposed by Y. Shang et.. In MDS-MAP(P), a local map includes only relatively nearby nodes, such as two-hop, is computed for each node with MDS-MAP(C) algorithm at first. Then two local maps are merged together based on their common nodes to construct a global relative map for the network. Finally, transform the relative map to an absolute map based on the absolute coordinate of anchors. MDS-MAP(P, R)[12], as same, is a method with additional refinement. Both the two distributed algorithm perform better than the methods described above on the accurate of the positioning. However, the consumption of node energy is greatly increased.

4.3 Local MDS Algorithm

Local MDS [13], proposed by X. Ji and H. Zha, is another variant of MDS-MAP(C) with improved performance for irregular topologies. Its main idea is to compute a local map using MDS for each node in that only 1-hop neighboring nodes are consisted, but weights are restricted to be either 0 or 1 in local calculations. Then, patch these local relative maps together to form a global map. Moreover , least square optimization method was introduced in local MDS to refine the local maps.

4.4 dwMDS(G) Algorithm

The dwMDS(G) [14], another distributed improved localization algorithm, in which local communication constraints are incorporated within the sensor network. The mean idea is to comply a weighted (Gauss kernel) cost function and to improve the value of it by a majorization method in which iteration is guaranteed. An adaptive neighbor selection method is introduced to avoid the biasing effects of selecting neighbors based on noisy range measurements. Moreover, arbitrary non-negative weights are allowed which is different from Local MDS and MDS-MAP(P). The local nonlinear least squares problem is solved using quadratic majorizing functions as in SMACOF [14]. The dwMDS(E) algorithm[15], amelioration of dwMDS(G), in which network density, node error and ranging distance are taken into account.

4.5 HMDS Algorithm

HMDS [16] is a localization scheme using clustering algorithm. In HMDS, three shapes are comprised: clustering phase, intra-cluster localization phase, and merge phase. In clustering phase, the whole network is partitioned into multiple clusters by a clustering algorithm. In the following shape, distance measurements from all members of cluster are collected by cluster head and local MDS computation is performed to form a local map. Merge phase is used to calibrate coordinates of members in different clusters and then to compute a global map . HMDS localization algorithm outperforms MDS-MAP algorithm in terms of accuracy. However, there is a drawback as fixing the number of cluster in HMDS. The result of HMDS is inaccurate if holds exist in the sensing field. The CMDS algorithm proposed in reference [17] outperforms HMDS on this problem. The steps of the algorithm proposed in [18] is same as HMDS, while in that the classical Euclidean Algorithm is combined to improve the shortest distance estimation of MDS algorithm.

5 Conclusion

Node localization has been a topic of active research in recent years. The algorithms based on classical Multidimensional Scaling (MDS) have been paid attention extensively due to the higher accuracy and only three or four anchors being required. However, there is no single algorithm can adapt to all scenarios, practical node localization algorithm should be chose according to different applications.

References

1. Akyildiz, I.F., Weilian, S., Sankarasubramaniam, Y., Cayirci, E.: A Survey on Sensor Network. IEEE Communications Magazine 40, 102–114 (2002)
2. Bharathidasan, A., Vijay, A., Sai, P.: Sensor Networks: An Overview. Department of Computer Science. University of California (2002)
3. Shepard, R.N.: The Analysis of Proximities: Multidimensional Scaling with an Unknown Distance Function, I & II. Psychometrika 27, 125–140, 219–146 (1962)
4. Borg, I., Graenen, P.: Modern Mdtidimmional Scaling, Theory and Applications. Spinger, New York (1997)
5. Shang, Y., Ruml, W., Zhang, K., Fromherz, M.: Localization from mere connectivity. In: ACM MobiHoc, Annapolis, MD, pp. 201–212 (2003)
6. Royer, E., Toh, C.: A Review of Current Routing Protocols for Ad Hoc Mobile Wireless Networks. In: IEEE Personal Comm. (1999)
7. Karp, B., Kung, H.T.: GPSR: Greedy Perimeter Stateless Routing for Wireless Networks. In: Proceedings of the Sixth Annual International Conference on Mobile Computing and Networking, pp. 243–254. ACM, New York (2000)
8. Intanagonwiwat, C., Govindan, R., Estrin, D.: Directed Diffusion: A Scalable and Robust Communication Paradigm for Sensor Networks. In: Proceedings of the Sixth Annual International Conference on Mobile Computing and Networking, pp. 56–67. ACM, New York (2000)

9. Chu, M., Haussecker, H., Zhao, F.: Scalable Information-Driven Sensor Querying and Routing for Ad Hoc Heterogeneous Sensor Networks. In: Int'l J. High Performance Computing Applications (2002)
10. Torgerson, W.S.: Multidimensional Scaling of Similarity. Psychometrika 303, 79–393 (1965)
11. Zhan, F.B.: Three Fastest Shortest Path Algorithms on Real Road Networks. Journal of Geographic Information and Decision Analysis 1, 69–82 (1997)
12. Shang, Y., Ruml, W.: Improved MDS-Based Localization. In: Proceedings of the 23rd Conference of the IEEE Communications Society, pp. 2640–2651. IEEE Press, Hong Kong (2004)
13. Ji, X., Zha, H.: Sensor Positioning in Wireless Ad-Hoc Sensor Networks Using Multidimensional Scaling. In: Proceedings of IEEE INFOCOM, pp. 2652–2661. IEEE Press, Hong Kong (2004)
14. Costa, J.A., Patwari, N., Hero III, A.O.: Distributed Multidimensional Scaling With Adaptive Weighting for Node Localization in Sensor Networks. IEEE/ACM Transactions on Sensor Networks 2, 39–64 (2006)
15. Luo, H., Li, J., Zhu, Z., Yuan, W., Zhao, F., Lin, Q.: Distributed Multidimensional Scaling with Relative Error-based Neighborhood Selection for Node Localization in Sensor Networks. In: Proceedings of the 2007 IEEE, International Conference on Integration Technology, pp. 735–739. IEEE Press, Shenzhen (2007)
16. Yu, G., Wang, S.: A Hierarchical MDS-based Localization Algorithm for Wireless Sensor Networks. In: AINA 2008, pp. 748–754. IEEE Press, Okinawa (2008)
17. Shon, M., Choi, W., Choo, H.: A Cluster-based MDS Scheme for Range-free Localization in Wireless Sensor Networks. In: International Conference on Cyber-Enabled Distributed Computing and Knowledge Discovery, pp. 42–47. IEEE Press, Huangshan (2010)
18. Hu, J., Cao, J., Zhao, Y., Mi, X.: A MDS-Based Localization Algorithm for Large-Scale Wireless Sensor Network. In: International Conference on Computer Design and Applications (ICCDA), vol. 2, pp. 566–570. IEEE Press, Qinhuangdao (2010)

The Research of Web Service Composition Reuse Model

GuoFang Kuang and HongSheng Xu

College of Information Technology, Luoyang Normal University, Luoyang, 471022, China
xhs_ls@sina.com

Abstract. This paper discusses the basic concepts of Web services, Web services architecture, Web services protocol stack, analyzes the current status of Web service composition research. In this paper, reusable Web service composition model is put forward. This model can be added to any Web service composition method (either dynamic or static), describes the process of Web service composition reuse. In this paper, the introduction of Web services composition classes and components thought to form the concept of Web services, component class, and the combination of Web services to achieve a certain degree of reuse.

Keywords: web service, reuse model, Web service composition.

1 Introduction

XML allows information transfer from the platform and programming language can be limited, for the various systems on the network provides a common communication standard. SOAP protocol defines a simple messaging service rules, and by the major software providers support. These are to promote Web services applications[1]. Web services have the full interoperability between different platforms, in the ubiquitous network called Web services. The purpose of Web services through the use of Web standards to achieve interoperability between applications.

(1) Web services are self-contained, modular applications that can be in the network (usually the Web) is described, publish, find and call.

(2) Web services are Web-based, distributed modular components, to perform specific tasks, to comply with specific technical specifications that make Web services compatible with other components to interoperate.

(3) Web services is an enterprise release, especially business needs to complete the online application service, or other enterprise application software accessible via the Internet and use the application service.

Various definitions have one thing in common: Web services are packaged into a single entity and published to a collection of functions on the network. Simply put, Web services is a URL resource, the client can get it programmatically request the service, without the need to know how the requested service is implemented. Some people think that Web services and application service provider is somewhat similar, they rent their suppliers on a monthly basis and allows users to run software on the Internet to use it. Although Web services allow some of the new business models, but

D. Jin and S. Jin (Eds.): Advances in FCCS, Vol. 1, AISC 159, pp. 405–410.
springerlink.com

this concern is not the model. Web services is a shared programming method can be used as a Web-COM, but the basic technology is quite different.

A complete Web services in addition to providing some kind of function, it also defines a clear interface. Typically, a Web service interface description and access the content format, the client, a Web service interface description, know that the Web service contains the required function and its call method. A Web service and other Web services can be integrated to form powerful new Web services. Developers can call the remote service, local service or their own hand-writing code to create a new Web service.

Web applications also present defects; it must use more complex methods (such as connection, frame, etc.) and functional integration of different sites. However, different features on the site application are relatively large, as a separate package there is no simple way to separate the user interface and functionality. Existing solutions include: COM / DCOM standards and CORRBA standards, but the two standards are the following problems.

Reuse of Web service composition problem. Currently, you can type, components, Web services reuse, developers are eager to mix in a Web services application of these technical ideas. In this paper, the introduction of Web services composition classes and components thought to form the concept of Web services, component class, a combination of Web services to achieve a certain degree of reuse.

2 Web Services Architecture

Web services architecture shown in Figure 1, the basic architecture consists of three participants and three constitute the basic operation. Three participants include: the service provider (Service Provider), the service requestor (Service Requestor) and the service agent (Service Registry); 3 basic operations: publish (publish), find (find) and binding (bind)[2].

Service providers publish services to a directory service agent, when the service requester need to call the service, it offers the first to use the directory service agent to search for the service, to get information on how to call the service, then that information to call the service providers publish services. When the service requester calls from the service agent to get the information required services, the communication is between service requesters and providers directly, without going through the service agent. Web services standards and protocols using a series of related functions, such as the Web Services Description Language (Web service description language, WSDL) to describe the service, with the Universal Description, Discovery and Integration (universal description, discovery, integration, UDDI) to publish, Find services, and SOAP is used to perform service calls.

In the Web services architecture and the various modules within the module, the message delivered in XML format. The reason is: the message in XML format that is easy to read and understand, because the cross-platform XML documents and loosely coupled with the structural characteristics. XML message format encapsulates the vocabulary can be both internal and external use of the industry; it also has good flexibility and scalability, allowing the use of additional information.

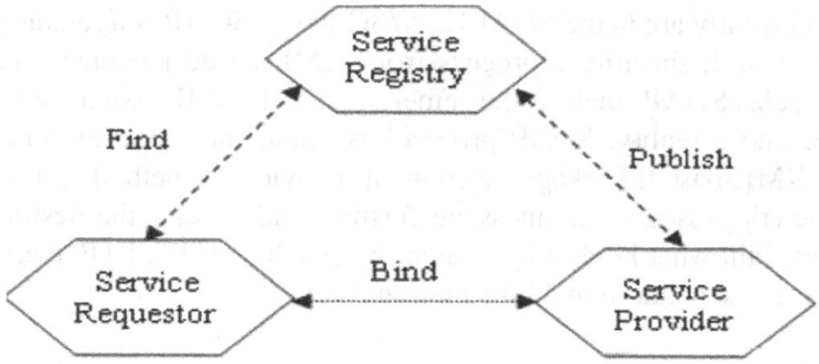

Fig. 1. Web Services Architecture

Web services technology consists of a series of protocols and standard components, including the underlying network transport protocol (such as HTTP, FTP and SMTP, etc.), the message data coding standards (XML, XML Schema), SOAP, WSDL and UDDI[3]. These standards form the basis for Web services, the core standards including SOAP, WSDL and UDDI. Web services technology stack as shown in Figure 2.

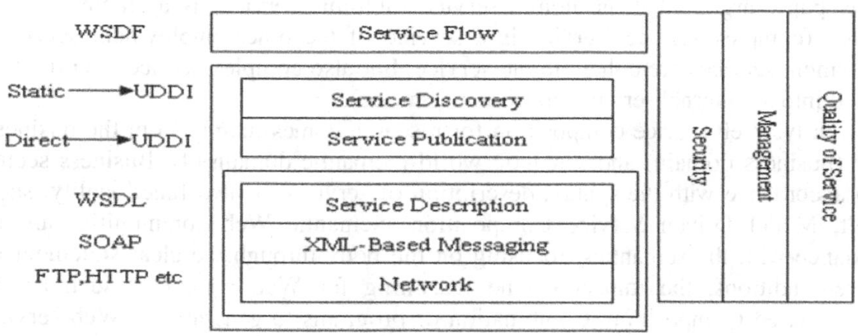

Fig. 2. Web services protocols and standards

XML is a service that all information in the message of unity and the standard way. SOAP provides XML-based information exchange agreements. WSDL is an XML-based language that can be used to describe the service operation. UDDI provides a publish and discover services registry. Currently a variety of WEB application development platform, like Microsoft. NET, IBM Web Sphere Web services standards and so on with varying degrees of support. The formula 1 which calculates the web services nodes of data mining is as follows.

$$\sum_{j=1}^{n+1} f_{i,j} T_{i,j} + \sum_{j=1}^{n} f_{j,i} Rx_i \leq E_i \tag{1}$$

SOAP Simple Object Access Protocol Web services is the core of the agreement, Microsoft, IBM, and other experts in related fields completed in 2000 SOAP1.1

specification, currently are being SOAP1.2 W3C work. SOAP is a remote method call to the XML standard, specifies a process from XML-based message sent to another process approach. SOAP makes the emergence of XML-based remote method invocation become a reality. SOAP protocol is about the message, and specify for sending data (XML-based) package mechanism provides a method and format of the request parameters passed to the message format, and specify the destination of the message and explain what kind of agreement by (such as HTT, FTP, etc.) methods of the SOAP message sent additional information.

3 Web Service Composition

The limited capacity of a single Web service, often can not meet the application requirements. With the rapid development of Web services, Web services, a combination of abstract and more and more important. Web service composition refers to the business process logic to support a set of Web services, which itself can be either the final application, it can be a new Web service. Web service composition is determined by different Web services execution order and the complex interactions between Web services to achieve. The Web service composition problem, there is currently no uniform solution.

This paper argues that an atomic service (automic service) is a single service, complex (complex service) service is a service of the other employment services, employment services are either atomic service, but also complex services, services to be hired into an internal service (constituent service).

Currently, Web service composition for research comes mainly from the business sector (business domain) and semantic world (semantic domain)[4]. Business sector are in accordance with the syntax, description of service-oriented functionality, such as belt, Model Driven Service Composition. Semantic Web communities are in accordance with the semantics, focusing on the body through the clear statement in the pre-conditions, the impact of the reasoning for Web resources, such as Ai planning-based Composition. Combination of programs to generate the Web service method, Web service composition can be divided into two categories: static and dynamic combination of portfolio.

(1) Static portfolio

In a static portfolio, Web service composition is defined in the design phase. Before the implementation of the combined program, you should create a process model.

(2) Dynamic combination

Dynamic combination means that, in the run-time required to select and invoke a Web service to complete a business process. For example, a business travel service system, which through ticket services, hotel services, road services, car rental services, a combination of systems for clients travel arrangements. Such a combination of Web services can not be pre-defined Web services and the required execution order between them, but should rely on the specific application domain, select the appropriate Web services and interact with. The equation 2 which calculates the static portfolio nodes of Web service is as follows.

$$T^0 = \arg\min_{T \subset G(N,A)} \sum_{i=1}^{|N|} (E_{Tx}^i(T) + n_i(T) \times E_{Rx}) \quad (2)$$

Currently, the industry-wide use a single Web service, this does not make full use of the advantages of Web services. In order to build a global Web services market, large-scale reuse in order to strengthen services, more and more people make great efforts to study the combination of Web services.

4 Web Service Composition Reuse Model

For the formation of a new Web service component class is abstract if you can not run through the actual testing of their correctness, accuracy must be checked (see Chapter III), and Web services components can be stored in the corresponding library hierarchical structure. Error for a class of service components into the library the result is serious, such as a Web service component class has a combination of deadlock error, any reuse of such services Web component class has a deadlock error.

This encapsulates the Web service Web service composition component class a total of two forms: class form and SCSL (Service Composition Specification Language) format. Class forms: the Definition, Construction, Prototype, Message Handling, Provider form, this form from the specific combination of Web service composition will control flow, data flow, service providers, the abstract, to form a reuse; with the corresponding form is the SCSL, which is an XML-based language that can be easily carried in the network between the Web service component class transfer. This class is divided into the Web service components: the specific service component class, the abstract service component classes, interfaces, service component class. For the specific service component class, can be directly put into operation, while the other two classes need to implement Web services components can only be put into operation. Fig.3 shows a Web service component class from the interface specification (Interface specification) and structural specification (Construction specification) composed of two parts.

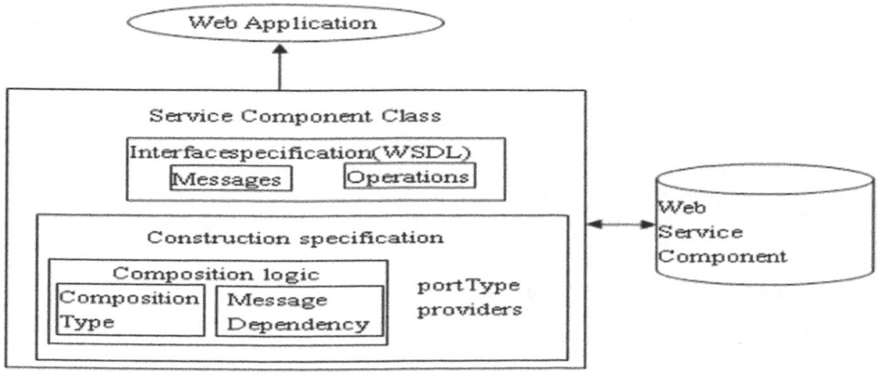

Fig. 3. Elements of a Web service component class diagram

5 Summary

In this paper, the reuse of Web service composition problem. Currently, you can type, components, Web services reuse, developers are eager to mix in a Web services application of these technical ideas. In this paper, the introduction of Web services composition classes and components thought to form the concept of Web services, component class, a combination of Web services to achieve a certain degree of reuse.

References

1. Yang, J., Papazoglou, M.P.: Web Component: A Substrate for Web Service Reuse and Composition. In: Pidduck, A.B., Mylopoulos, J., Woo, C.C., Ozsu, M.T. (eds.) CAiSE 2002. LNCS, vol. 2348, pp. 21–36. Springer, Heidelberg (2002)
2. Yang, J., Papazoglou, M.P.: Service components for managing the life-cycle of service compositions. Elsevier Information Systems 29, 97–125 (2004)
3. Menascé, D.: A QoS Issues in Web Services. IEEE Internet Computing 6(6), 72–75 (2004)
4. Fu, X., Bultan, T., Su, J.: Formal Verification of e-Services and Workflows. In: Bussler, C.J., McIlraith, S.A., Orlowska, M.E., Pernici, B., Yang, J. (eds.) CAiSE 2002 and WES 2002. LNCS, vol. 2512, pp. 188–202. Springer, Heidelberg (2002)

The Research on Fuzzy Adaptive PI Control in Air Conditioning Systems Temperature Control

Li Gaojian

Electronic and Electrical Engineering,
ZIBO VOCATIONAL INSTITUTE, Zibo, 255314

Abstract. The air conditioning system control object multivariable, nonlinear, large time delay, time-varying characteristics of the PI control and fuzzy control combined proposed fuzzy adaptive PI control, and simulation. Simulation results show that this method of control is better than the conventional PI control and fuzzy control to eliminate the shortcomings of steady-state error is greater with short response time, high control precision and good stability.

Keywords: air conditioning systems, temperature, adaptive fuzzy, PI control.

1 Introduction

In recent years, fuzzy control technology in the field of refrigeration and air conditioning has been rapid development, has become a hot research [1]. Fuzzy control does not depend on the mathematical model of the object, through the fuzzy information processing of complex objects can be implemented good control, good robustness and stability, etc. [2, 3]. But do not have the integral part of fuzzy control, run-time steady-state error exists, it is difficult to achieve high control accuracy in the application process still needs further improvement [4, 5]. Currently, PID controller with its simple structure, easy to implement, the steady-state error is small, the control and high accuracy has been widely used in process control, and achieved good control effect, but in the actual control process due to differential effects is not very obvious, and so many practical situations using only PI control [6]. However, air-conditioned room temperature of this multi-variable, nonlinear, large time delay, time-varying systems, the conventional PI control is often difficult to obtain satisfactory results, the need for tuning parameters of PI to get the best control effect. And manual tuning parameters required to run the management staff has extensive experience in project control, so the use of parameter self-tuning PI control method is to ensure that an effective way to effect.

In this paper, the performance requirements of air-conditioning system control, fuzzy control and PI control combined air-conditioning system proposed fuzzy adaptive PI control, the fuzzy rules to achieve adaptive PI parameters online, and through MATLAB for control of air-conditioned room temperature computer simulation.

D. Jin and S. Jin (Eds.): Advances in FCCS, Vol. 1, AISC 159, pp. 411–416.
springerlink.com

2 Mathematical Model of Air-Conditioned Room Temperature

In this paper, a test-bed chamber for the study, test chamber size is 4.8m × 3.8m × 3.5m (length × width × height), the main use of polyurethane foam assembled library board, spray inside and outside surface plate. When the air conditioning system for analysis, and electric heaters, electric humidifier connected to the controller will be treated as air-conditioning system controller. In this paper the mathematical model will only consider the test stand the test chamber temperature control air conditioning system parts, this article refers to the test room air-conditioning systems refer to the test chamber temperature control air conditioning system parts.

Through the use of lumped parameter method established by the mathematical model of air-conditioning system and the load disturbance is a first-order mathematical model of inertia, but because of air-conditioning systems usually have a delay, so the actual mathematical model of the test room air conditioning system load and test chamber interference are first-order mathematical model of the part without delay inertia [7]. Delay part of the first order inertia plus the standard form:

$$G(s) = \frac{K_s \cdot e^{-\tau s}}{T_s s + 1} \tag{1}$$

Which Ks is the amplification factor [°C / kW], Ts is the time constant [min], τ the delay time [min].

In considering the electric heater to control the room temperature test chamber, the only means, the air conditioning system in the mathematical model for the amplification factor, time constant; load disturbance amplification factor of the mathematical model, the time constant. Electric heater which key is the ratio of the output coefficient [kW], the equivalent thermal resistance for the air conditioning system [°C / kW]; C is the equivalent heat capacity of air-conditioning system [kJ / °C]. Interfere with air-conditioning system model and load model pure delay time can be considered consistent, are roughly the delay time can be through the air conditioning system step response experiments. The mathematical model of air-conditioning system and the mathematical model of load disturbance are as follows:

$$G(s) = \frac{Y(s)}{U(s)} = \frac{k_u \cdot e^{-T_d s}}{Cs + 1/R_{eq}} \tag{2}$$

$$G_i(s) = \frac{Y(s)}{Q_i(s)} = \frac{e^{-T_d s}}{Cs + 1/R_{eq}} \tag{3}$$

Figure 1 shows the test room air-conditioning system and the load diagram of the mathematical model of interference, in Figure 1, u (t) is the controller output, the output range of 0% -100%, the controller and electrical heater connected, q (t) is the electric heater output power, it and u (t) proportional relationship exists between, ku electric heater is the scale factor; qi (t) is the amount of load disturbance, y (t) for the test room temperature of the room. Identification algorithm studied in this paper will only be room for testing mathematical models of air-conditioning system, regardless of load disturbances.

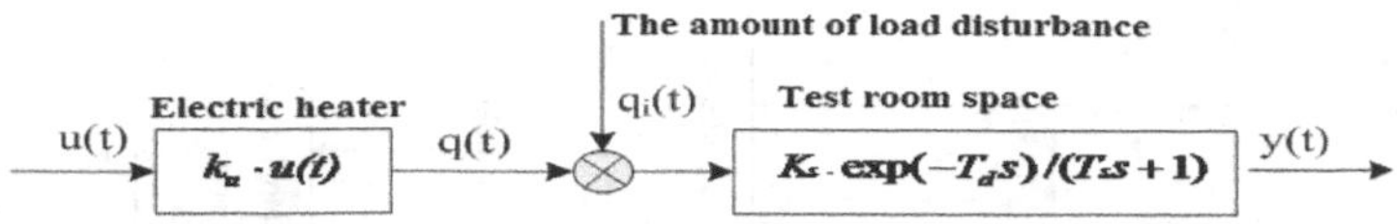

Fig. 1. Mathematical model of air-conditioning system & load disturbance in test room

3 Fuzzy Adaptive PI Controller Design

3.1 PI Controller

Exist in the proportion of PI control, the role of two control points, the conventional control formula is:

$$u = K_p\left(e + \frac{1}{T_i}\int edt\right) \tag{4}$$

Where: respectively, the proportional gain, integration time. To (4) can be written in the form of differential equations:

$$u(n) = K_p e(n) + K_i \sum_{j=0}^{n} e(j) = K_p e(n) + K_i \sum_{j=0}^{n} e(j) \tag{5}$$

Where: The integral coefficient, e (n) for the system error, T is the sampling time.

In PI control process, the first based on the mathematical model to determine the controlled object and then use the error e as input to calculate the amount of control, and drive the appropriate enforcement agency to reduce the error until the charged objects are stable in the permitted range. Figure 2 shows the block diagram of PI control system.

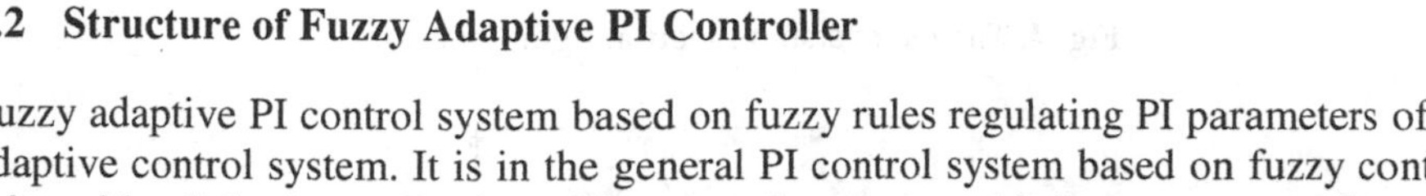

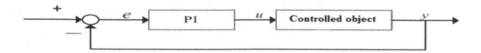

Fig. 2. Diagram of PI control system

3.2 Structure of Fuzzy Adaptive PI Controller

Fuzzy adaptive PI control system based on fuzzy rules regulating PI parameters of an adaptive control system. It is in the general PI control system based on fuzzy control rules with a link, the application of fuzzy set theory to establish parameters and error e and error change rate ec between binary continuous function. Obtained under different conditions by the PI parameters of the reasoning for the results, in order to achieve adaptive PI parameters online to ensure the control effect.Figure 3 shows the fuzzy adaptive PI control system block diagram.

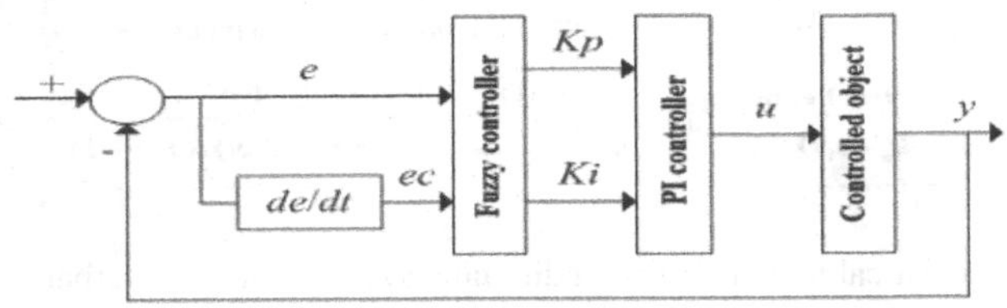

Fig. 3. Diagram of adaptive fuzzy PI control system

According to the actual situation of the control object, where the controller uses two input two output structures. According to air-conditioned room temperature sensor reads the actual sample temperature, set temperature with the calculated value of the deviation e and error change rate ec, e and ec been blurred by the exact amount after the amount of blur into E and EC, After fuzzy reasoning fuzzy decision making, such as: temperature below the set value and the temperature continues to drop, the tone flat is small, then the controller can be adjusted based on the PI parameters of the air-conditioned room temperature to make the necessary adjustments to speed up the response rate and reduce the overshoot.

3.3 Determination of Membership Function

For the fuzzy controller input error e and error change rate ec, output parameters for the PI controller based on the actual situation of the test chamber, the system error e and error change rate ec range of variation is defined as the fuzzy set theory on the domain E , EC = {-3, -2, -1,0,1,2,3}, fuzzy subsets of E, EC = {NB, NM, NS, ZE, PS, PM, PB}, they represent a negative large, negative, negative small, zero, positive small, middle, Chia Tai. Their membership function shown in Figure 4 - Figure 5.

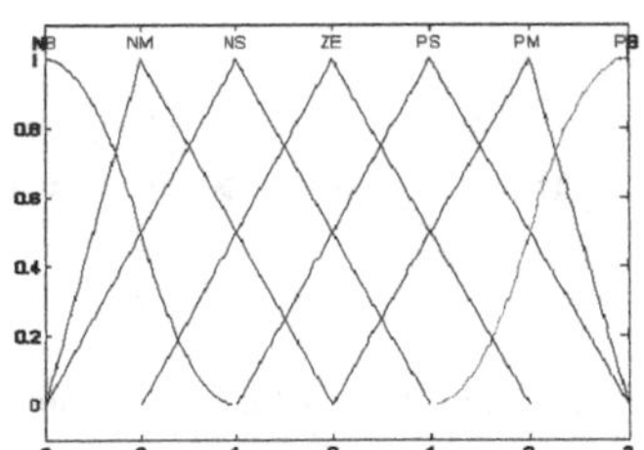

Fig. 4. The membership function figure of error e

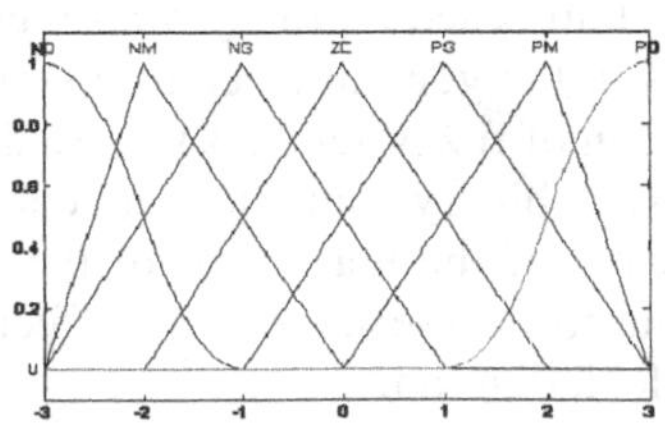

Fig. 5. The membership function figure of error change rate ec

4 Fuzzy Adaptive PID Control System Simulation

4.1 System Simulation Model

U.S. launch of The Mathworks MATLAB software provides a Simulink toolbox and fuzzy logic toolbox. Fuzzy logic toolbox for generating and editing fuzzy inference system (FIS) commonly used utility functions. These functions can be easily applied for membership functions and fuzzy rules for editing. With Simulink toolbox were established adaptive PI control and fuzzy PI control of the simulation model, using the input step response.

Air-conditioned room temperature according to the model set accordingly by step experiments, and according to equation (1), has been delayed time and time constant $\tau = 6\,\mathrm{min}$, amplification factor $T = 60\,\mathrm{min}$, obtained in the simulation program $K = 72^{\circ}C / KW$, the temperature sampling interval $\Delta t = 1\,\mathrm{min}$, the simulation time $200\,\mathrm{min}$.

In order to fully demonstrate the superiority of fuzzy PI control, PI parameters of this initial setting, choose one in the air conditioning system control in the more advanced tuning methods [8], the method for setting the characteristics of air-conditioning system has been taken into account a variety of factors, have a better control effect. After setting the initial value of PI parameters, respectively $K_{p0} = 0.069$, $K_{i0} = 0.00117$ Assume that the initial value at room temperature $0^{\circ}C$, setting $t_{set} = 10^{\circ}C$.

4.2 Analysis of Simulation Results

Model parameters in the system remain unchanged and undisturbed, the conventional PI control algorithm with fuzzy adaptive PI control algorithm of the system response curve shown in Figure 6.

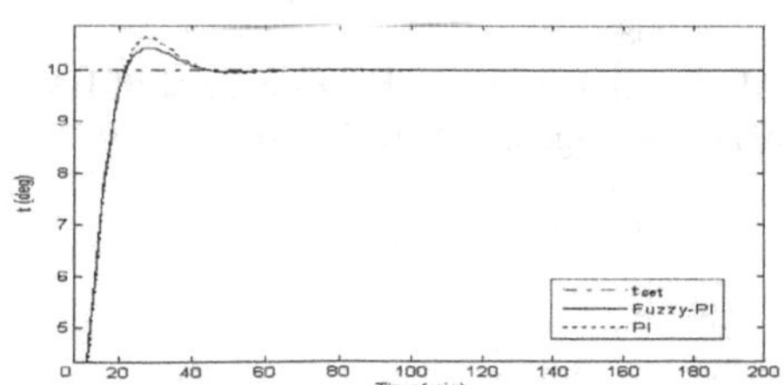

Fig. 6. PI control and fuzzy adaptive PI control system to respond to more conventional

From the simulation results can be seen fuzzy adaptive PI control makes the system dynamic and static performance has been improved. First fuzzy adaptive PI control response speed has increased, the overshoot small compared with conventional PI control about 2%, and fuzzy adaptive PI controller parameter adjustment quickly, can be seen from Figure 6, the fuzzy adaptive PI in $Time = 60\text{min}$ you can reach the steady state, while in the conventional PI control $Time = 100\text{min}$ before reaching the stable fuzzy adaptive PI control slower than 40%.

5 Conclusions

As the air conditioning system control object multivariable, nonlinear, large time delay, time-varying characteristics, makes the conventional PI control is often difficult to obtain satisfactory results. Fuzzy adaptive PI control is simple and easy to use and control by the adaptive process, the controller has better control effect. Simulation results show that the PI control method has the advantages of both fuzzy control and improve the ability to adapt and control system robustness, improve the system dynamic and static quality, more suitable for air-conditioning system control, have better engineering prospects.

References

1. Tanaka, K., Sugeno, M.: Stability Analysis Design of Fuzzy Control systems. Fuzzy Sets and Systems (58), 135–136 (2010)
2. Wu, G., Wang, H., Hu, X., et al.: Of fuzzy control in the field of refrigeration and air conditioning application and development. Refrigeration and Air Conditioning 9(6), 16–18 (2009)
3. Wang, F.H., Mengxiang, Z., Yu, B.: Inverter air conditioner control system simulation. Fluid Mechanics 29(3), 45–47 (2008)
4. Wang, J.W., Wu, Z.: Control Engineering Foundation. Higher Education Press, Beijing (2009)
5. Liu, Y.: Air conditioning and heating automation. Tianjin University Press (1993)
6. Wang, Q.-G., Lee, T.-H., Fung, H.-W., Bi, Q., Zhan, Y.: PID tuning for improved performance. IEEE Transactions on Control Systems Technology 19(4), 456–465 (2011)
7. Yangchun, M., Wang, C.: Room temperature fuzzy control system and simulation. Energy Technology 29(3), 97–99 (2008)
8. Alcala, R., Benitez, J.M., Casillas, J., et al.: Fuzzy control of HVAC systems optimized by genetic algorithms. Applied Intelligence 25(4), 155–177 (2010)

Realization of Oracle Database Expansion and Migration Scheme on AIX Platform

Wenjuan Liu, Dilin Pan, Jinxiang Yang, and Xiaolong Li

Anhui University of Science and Technology Huainan Anhui, China
{liuwj,jxyang}@aust.edu.cn

Abstract. With the development of information technology, the original information system of enterprise gradually can not meet the needs of enterprise development, the enterprise information system expansion is imminent at this time. Taking an enterprise information system in the expansion process, and based on today's common Oracle database, realizing the database expansion and migration as an example, the paper focuses the steps of Oracle database migration and the possible problems and its treatment measures on AIX operating system platform.

Keywords: Oracle, Expansion, Migration, AIX.

1 Introduction

As the rapid development in information technology today, the vast majority of enterprises have implemented information management. And with the rapid development of enterprises, the original information system of some enterprises can not meet the needs of enterprise development, especially large amount of data in the foreground case, the background database is difficult to adapt to the needs of data expansion, at this time the enterprise information system expansion is imminent.

The enterprise information system expansion is focused on improving the database handling ability, enhancing data security, especially for the expansion of data storage capacity[1]. Since the early construction of enterprise information system, generally the amount of data is not large, the data can be stored in the local machine, even one PC can also deal with the enterprise application. But along with the development of enterprise, the increase of data amount, data processing and data security requirement these need to move the database to a larger storage capacity, higher security disk array. If the enterprise originally uses Oracle database, in order to migrate data to the disk array, you need to migrate the data files, control files and the redo log files to the disk array[2],[3].

Taking a enterprise information system in the expansion process , and based on the today's common Oracle database, realizing the database expansion and migration as an example, the paper focuses the steps of Oracle database migration and the possible problems and its treatment measures on AIX operating system platform. If it is HPUX system, the database migration scheme is also applicable, the difference is the VG and file system creation.

D. Jin and S. Jin (Eds.): Advances in FCCS, Vol. 1, AISC 159, pp. 417–421.
springerlink.com

2 Preparing to the Database Expansion and Migration

Before the database expansion and migration, firstly to see the related data of the existing database system, according to the data to analyze the data migration script when needed. When the existing Oracle database in the scheme was installed, the data files and control files used file system, the redo log files used raw device, therefore, after the disk array installation, such as the new file system and raw devices use the same name and store directory, and before new LV and file system creation the data files and control files should be backed up[4],[5].

3 New File System and Raw Device

Firstly to create invg on the disk array. For example, create invg on hiskpower0.

```
#mkvg -y invg hiskpower0
```

3.1 New the LV Raw Device

If LV already exists in the /dev directory, error will be occur when new the same name LV. because the data has already been backed up, so LV can be deleted directly.

```
#rmlv -f lv name
```

Be careful not to delete the raw device for redo log files.

According to the document to create LV that the Oracle database needs in invg. For example, to create lvoracle that is 20G, which ppsize is 256M.

```
#mklv -t jfs2 -y lv01 lvoracle 80
```

Other LV creation method is similar to this.

3.2 New the File System

After LV creation, you can create new file system and mount point. For example, create the jfs2 file system in lvoracle, mount point is /indata/oracle.

```
#crfs -v jfs2 -d lvoracle -m /indata/oracle
```

Other file system creation method is similar to this.

After the disk array creation is finished, all file systems needed to install the database must be mounted on.

```
#mount /indata/oracle
```

Now the data files and control files backed up before can be recovered to the same directory.

4 Recovering the Redo Log Files

Because the raw devices for redo log files are created on the disk array too, so need to rebuild new redo log files[6],[7]. Three log groups were created in Oracle database, each group has one member, the size of each member is 250M. Because two transition log groups are needed at least, you can create any two raw devices for transition log groups. The following is to rebuild the redo log files.

Step 1: Creating the necessary LV needed by the transition log groups

For example, create lv that is 256M, which type is raw, ppsize is 256M.

```
#mklv -t raw -y oraredo04 invg 1
#mklv -t raw -y oraredo05 invg 1
```

Step 2: Opening the database to mount state

```
$sqlplus '/as sysdba'
SQL>startup mount
```

Step 3: Adding new transition log groups

```
SQL>alter database add logfile group4('/dev/ror
aredo04') size 250M;
SQL>alter database add logfile group5('/dev/ror
aredo05') size 250M;
```

Step 4: Switching the current log to new log groups

```
SQL>alter system switch logfile;
SQL>alter system switch logfile;
```

Step 5: Removing the old log groups

```
SQL>alter database drop logfile group1;
SQL>alter database drop logfile group2;
SQL>alter database drop logfile group3;
```

Step 6: Removing the old raw device

After finishing the above operation, the three raw devices created for redo log files on local disk before can be removed

```
#rmlv -f lv oraredo01
#rmlv -f lv oraredo02
#rmlv -f lv oraredo03
```

Step 7: Creating the new LV

After finishing the above delete operation, in order to create new redo log files for preparation, new LV can be created on the disk array. For example, to create lv that is 256M on invg, which type is raw, ppsize is 256M.

```
#mklv -t raw -y oraredo01 invg 1
#mklv -t raw -y oraredo02 invg 1
#mklv -t raw -y oraredo03 invg 1
```

Step 8: Creating new log groups

Firstly you should identify storage directory for new log groups. Because reading and writing the redo log files are connected to the raw device by links, they can be checked in directory /indata/oracle/redolog.

```
redo01.log -> /dev/roraredo01
redo02.log -> /dev/roraredo02
redo03.log -> /dev/roraredo03
```

It can be get that new log groups should be stored in directory /indata/oracle/ redolog. The following is to add three new log groups.

```
    SQL>alter database add logfile group1('/indata/
Oracle/redolog/redo01.log') size 250M;
    SQL>alter database add logfile group2('/indata/
Oracle/redolog/redo02.log') size 250M;
    SQL>alter database add logfile group3('/indata/
Oracle/redolog/redo03.log') size 250M;
```

Step 9: Switching to new log groups

```
SQL>alter system switch logfile;
SQL>alter system switch logfile;
SQL>alter system switch logfile;
```

Step 10: Removing the transition log groups

```
SQL>alter database drop logfile group4;
SQL>alter database drop logfile group5;
```

Step 11: Removing the raw devices for transition log groups

```
#rmlv -f lv oraredo04
#rmlv -f lv oraredo05
```

After finishing the above operation, the database migration will be completed. And then you can execute the following command to open the database, if open successfully without error, it indicates that the database recovery is successful.

```
$sqlplus '/as sysdba'
SQL>alter database open
```

5 Summary

With the above method the database expansion and migration can be realized through migration data files and control files. The scheme can improve the data migration speed and the database expansion efficiency. The cross-platform tablespace transport is provided in Oracle, that makes users to move data more easily[8]. Realization of tablespace migration methods on the same platform in this article adapts to current enterprises oracle database expansion and migration.

References

1. Niemiec, R.J.: Oracle Database 10g Performance Tuning Tips & Techniques. McGraw Hill (2007)
2. Xu, J., Gong, Z., Wan, F., Yang, Z.: Research on Data Recovery of Oracle Database. Computer Engineering 13 (2005)
3. Zhang, Y.-F.: Backup and Recover Strategy of Oracle Database. Computer Engineering 35, 85–87 (2009)
4. Greenwald, R., Stackowiak, R., Stern, J.: Oracle Essentials: Oracle Database 10g, 3rd edn. O'Reilly (2004)
5. Liang, C.-M., Yuan, G.-Z., Qin, Z.-W., Chen, B.: Discussion on how to upgrade or migrate Oracle database. Chinese Medica Equipment Journal 29, 35–37 (2008)
6. He, Y.-R.: Experimental study of repair methods on damaged Oracle s log files. Journal of Computer Applications S2 (2009)
7. Wang, L.: The Analysis of Oracle Database Disaster Recovery Technology. Computer Programming Skills & Maintenance (14), 123–124, 135 (2011)
8. Li, R.: How To Carry Out Tablespace Move in Oracle. Joumal of Wuhan Polytechnic University 27, 57–60 (2008)

[illegible] Oracle, [illegible] more easily[6]. Realization of [illegible]spect [illegible] on [illegible] article adapts to current [illegible] expansion and migration.

References

2. [illegible] Database [illegible]
3. Zhao, Y.: [illegible] Engineering 35 [illegible]
4. [illegible] Database [illegible] 3rd edn. O'Reilly (2010)
5. [illegible] to upgrade oracle [illegible] (2008)
6. [illegible] Oracle [illegible] Journal of [illegible]
7. [illegible]
8. [illegible] Journal of [illegible] 2008

Comparison of Four Fuzzy Control Methods Applied for Electronic Belt Scales

Genbao Zhang and Sujuan Pang

Institute of Electrical and Information Engineering,
Shaanxi University of Technology,
Shaanxi Xian,
China, 710021
zhanggb@163.com

Abstract. One important factor of influencing the electronic belt scale accuracy is that the motor speed can not follow the given speed. To improve the dynamic performance of the motor in electronic belt scale, fuzzy control, fuzzy PI control, self-tuning fuzzy PI control, and fuzzy PID control are introduced into the electronic belt scale control in the paper. It is indicated that fuzzy PID control is the best in those four control methods through simulation and compared analysis. Fuzzy PID method not only control the motor running more precisely and smoothly, but also improve the anti-jamming ability and robustness. The weighing accuracy problem of belt scale can be resolved by applying fuzzy PID control.

Keywords: electronic belt scale, fuzzy control, simulation and comparison, weighing accuracy.

1 Introduction

Electronic belt scale is a measurement equipment of continuous weighing scattered material automatically [1]. It can determine the quantity of the material flow and accumulative quantity of flow in uninterrupted case. It is charactered by continuously and automatically weighing process and independent with human.

With the development of sensor, computer technology and intelligent control theory, weighing material technology progresses rapidly. Weighing apparatus has substantial revolution in digitalization, miniaturization and intelligence. Research and development of weighing apparatus turn into a new stage. Burdening belt scale is a dynamic weighing apparatus, and is extensively applied to the field of metallurgy, coal, chemical engineering and building materials. Adopted new technique, raised the accuracy, and applied intelligence control are the trends of the belt scale currently.

Comparing with traditional control, fuzzy control has many advantages. It does not need accurate mathematics model, and easily understand and learn. So applying fuzzy control in electronic belt scale may obtain better performance.

D. Jin and S. Jin (Eds.): Advances in FCCS, Vol. 1, AISC 159, pp. 423–428.
springerlink.com

2 Paper Preparation

Generally, the model of electronic belt scale is a 3 order system. It transfer function is [2]

$$G(s)=\frac{1}{0.075s(0.017s+0.5)(s+5)} \tag{1}$$

In practice of belt scale, the model is changed because of the motor damaged. So the transfer function is changed to [2]

$$G(s)=\frac{1}{0.075s(0.017s+1)(s+5)} \tag{2}$$

3 Fuzzy Controller Design

Fuzzy control system comprises four component that is fuzzy controller, I/O interface, Object and measurement device in general. Fuzzy controller is the core of fuzzy control system. Two inputs and one output model which are system deviation and deviation rate and the object is adopted in fuzzy controller.

In the control system, variable error e, variable error rate e_c are as the controller inputs. The variable u is as the output. Suppose the domain of linguistic variable e, e_c and u is {-3, -2.5, -2, -1.5, -1, -0.5, 0, 0.5, 1, 1.5, 2, 2.5, 3}. Suppose the linguistic value is {NB,NM,NS,Z,PS,PM,PB}. Suppose the domain of error e is [-3.75, 3.75], the domain of error rate e_c and output u are [-1.5, 1.5]. So the quantization factor $k_e=\frac{3}{3.75}=0.8$, $k_{e_c}=\frac{3}{1.5}=2$, proportion factor $k_u=\frac{3}{1.5}=2$ [3]. The membership function of NM, NS, Z, PS, PM are triangular function. The membership function of NB is zmf curve. The membership function of PB is smf curve.

When building fuzzy control rules table, it should be considered that relations between control variable and error and error rate. When the error is big or slightly big, the chosen control variable should dis-crease the error. When the error is little or slightly little, the chosen control variable should prevent overshoot, and consider the response speed and stability of the system.

3.1 Traditional Fuzzy Control

According the condition forward, the fuzzy controller can be found and simulate by matlab. Fig 1 shows the traditional fuzzy control simulation block.

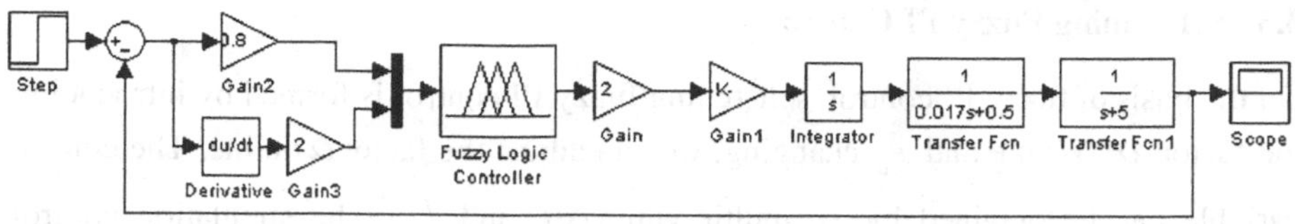

Fig. 1. Traditional fuzzy control block.

Because difference situation needs difference running speed of the belt scale motor, step signal magnitude increasing from 1 to 5 represents the motor speed increase in the paper [4].

With the motor speed increase, the system response is slow down and has stable error. When measuring, the motor of belt scale should run in a constant speed. So it is needed that the system can eliminate stable error and is response rapid. To resolve those problems, fuzzy PI controller is introduced.

3.2 Fuzzy PI Control

The membership function of e, e_c, k_p and k_i are same as traditional fuzzy control. The initial value of k_p and k_i are 1 and 0.05. The proportional factors of k_p and k_i are setted 0.3 and 2. k_p and k_i can tune itself with e and e_c changing. When e is little big, k_p should be little big to increase response. In case of overshoot and integral saturation, the measure of limiting the integral should be taken. So k_i should be little in general. When e and e_c are moderate, to have small overshoot and rapid response, k_p should small and k_i should be proper.. When e is little, to increase the stability for system, k_p and k_i should increase. According those relations, fuzzy PI control rules are found. The simulation block of simulink is shown in fig 2.

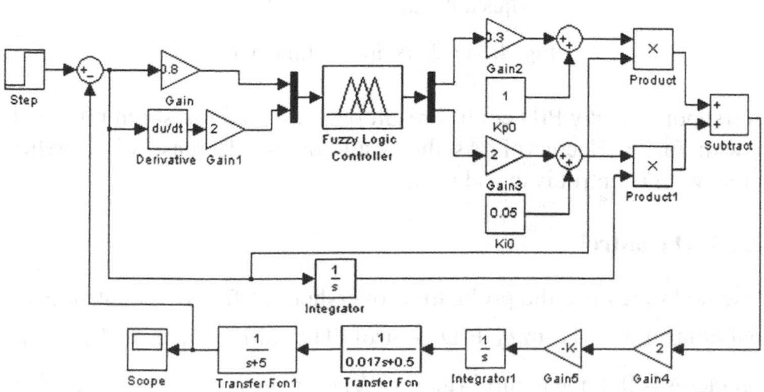

Fig. 2. Fuzzy PI control simulation model

3.3 Self-tuning Fuzzy PI Control

On the basis of fuzzy PI control, self-tuning fuzzy PI control is formed by introducing the factor α. With e and e_c changing, we can adjust the factor α online. The control variable can be obtained by α multiplying Δu and k_u. The simulation control block is shown in fig 3.

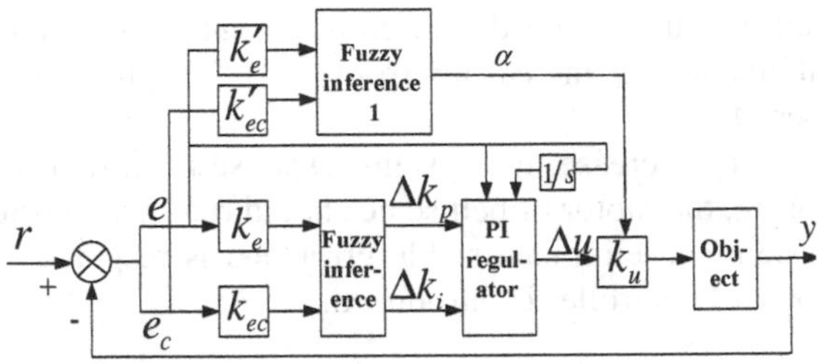

Fig. 3. Self-tuning fuzzy PI control block

The linguistic value of error e and error rate e_c of fuzzy controller 1 set to {N,Z,P}. The linguistic value of α is {S,M,B} and its membership function is shown in fig 4.

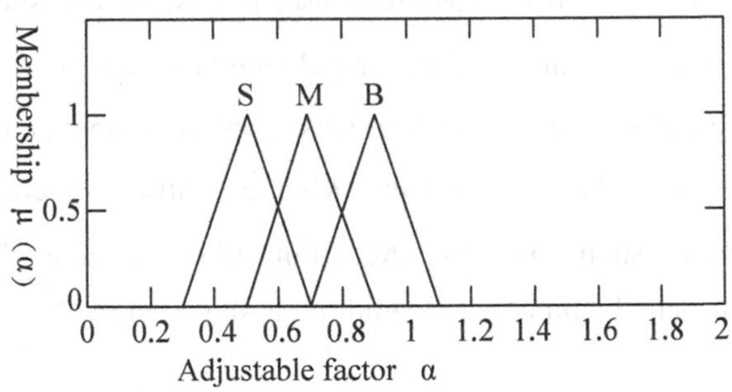

Fig. 4. Membership function of α

The overshoot of fuzzy PID control result decrease and transition process is shorter compare with fuzzy PI control. As the result of the decrease of overshoot is not obvious, fuzzy PID control is introduced.

3.4 Fuzzy PID Control

Another method to resolve the problem of overshoot of fuzzy PI control is introduced differential control, that is fuzzy PID control. The initial value of k_d is 1.2 and the proportion factor is 0.3. the control rules of k_d is same as Δk_p and Δk_i. Fig 5 shows the fuzzy PID control block.

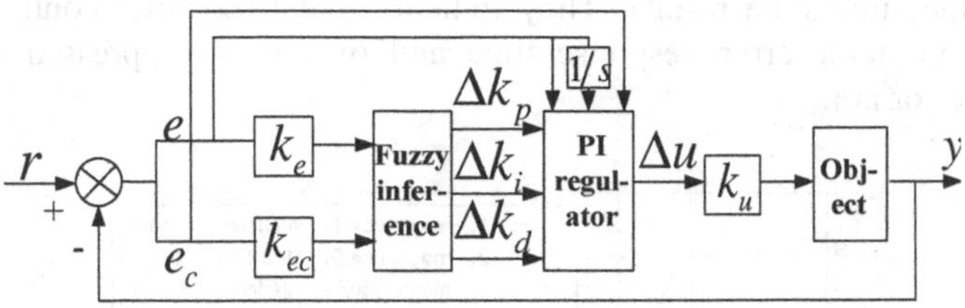

Fig. 5. The fuzzy PID control block

The results of four control methods are shown in fig 6 and fig 7.

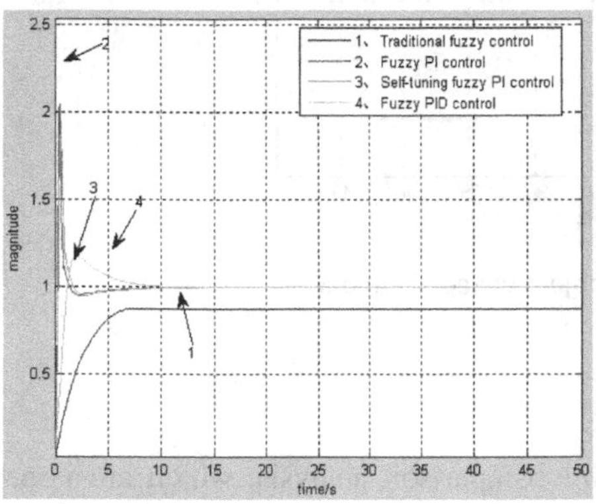

Fig. 6. Step response for magnitude 1

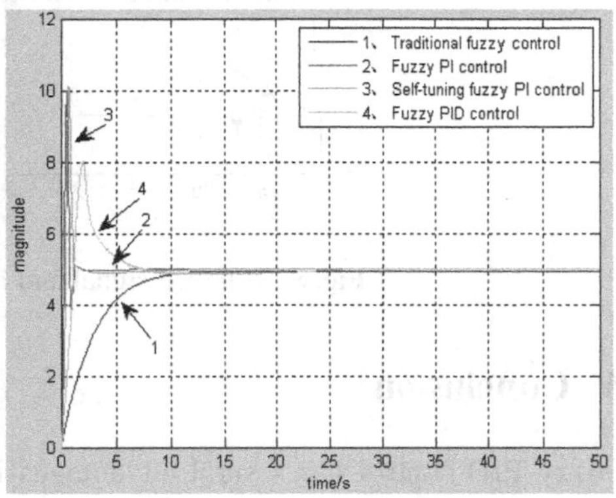

Fig. 7. Step response for magnitude 5

The results of simulation show overshoot of fuzzy PID control decreased obviously comparing with self-tuning fuzzy PI control block. From the comprehensive point, fuzzy PID control is more efficient method than other three fuzzy control methods though the response speed of system is slightly slow.

4 Anti-jamming Ability of Fuzzy PID Control

The simulation results are shown in fig 8, when fuzzy PID control system is disturbed by one signal with magnitude 0.2 and frequency 0.1Hz, and step signal magnitude changed from 1 to 5. the results show that fuzzy PID control has better performance even though there is disturbed signal.

5 Object Changed

In practice, belt scale needs change motor. As the results of motor changed, the model is changed too. Suppose the model changes to expression (2). Line 2 and Line 4 in fig 8 show the simulation results. They indicate that fuzzy PID control has better performance in stable error, response time and overshoot suppressed whether the model changes or not.

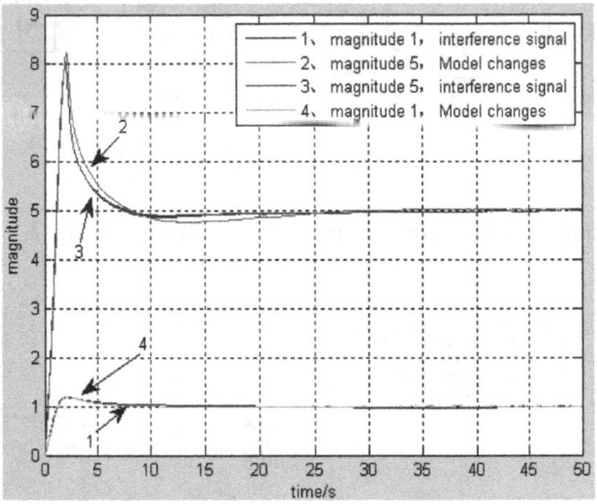

Fig. 8. Disturbed signal and object changed simulation

6 Conclusion

Fuzzy PID control has several advantages by comparison analysis, which are none stable error, high stability, rapid response, small overshoot, strong anti-jamming ability and good robustness. Since those advantages, the fuzzy PID method controls the belt scale motor running in a constant speed more stable and improves the weighing accuracy.

References

1. Li, M.: Intelligent burdening weighing apparatus design and realization based on fuzzy PID control. Taiyuan University of technology, Taiyuan (2008)
2. Liu, M.S., Zhou, Y., Zhang, Z.D., et al.: Research on Application of Fuzzy Control in Electronic Belt Scales. In: IEEE International Conference on Computer Science and Information Technology, pp. 134–138. IEEE Service Center, Beijing (2009)
3. Li, S.: Theory and application of fuzzy control and intelligent control. Press of Haerbin institute of technology, Haerbin (1990)
4. Zhao, Y., Emmanuel, G.C.: Fuzzy PI Control Design for an Industrial Weigh Belt Feeder. IEEE Transactions on Fuzzy Systems 11(3), 311–319 (2003)
5. Tao, Y.: New PID control and its application. Press of mechanical, Beijing (2002)

The Analysis of Wireless Sequence Control Based on Discrete Event System Model

Renshu Wang, Yan Bai, Yuan Ye, Yang Cheng, and Yun Ju

School of Control and Computer Engineering
North China Electric Power University, 102206, Beijing, P.R. China
{wangrenshu_1,baiyan_nce1,yeyuan_123,
chengyang_qw,juyun_asd}@126.com

Abstract. This paper introduces the application of wireless sensor networks (WSNs) in the sequence control of chemical water treatment in power plant. To settle the problem caused by packet loss, the controller and plant are treated as two interacting discrete event systems (DESs) described by automaton. And the wireless channel works as a switch between the DESs, denoting packet loss if the switch is off. Then, the analysis of the state transition is carried out and the solution is proposed with the capability of calculation of wireless nodes utilized. At last, the conclusion is presented.

Keywords: wireless networks, sequence control, automaton, packet loss, discrete event system.

1 Introduction

With the development of wireless technology, it has been a trend that the wireless sensor networks (WSNs) are introduced into the field of industry for the great benefit it will bring [1, 2]. However, there are many problems, such as message delay, packet loss and so on. To a great extent, these factors limit the further development of wireless network in industry field. To solve these problems, some of the researches are carried out and derive remarkable achievement [3]. In these methods, the plant is continuous and described by state equation. However, there are lots of plants which are binary in practical industry field.

As the part of the system of the chemical water treatment in power plant is presented in Fig. 1, the control is carried out by steps according to the requirement of the production process and there are several valves attached with wireless nodes. The controller sends DO to the actuator nodes, attached to the electromagnetic valves and receives DI from sensor nodes, attached to pneumatic valves. In this situation, it is advantageous to model the controller and the plant as two interacting discrete event systems (DESs) respectively in sequence control system [4].

The rest of the paper is organized as follows. In Section 2, the problem caused by packet loss is described in detail while introducing the wireless technology into the sequence control system. Section 3 presents the system model based on DESs. And in Section 4, the state transition is analyzed and the solution is proposed. At last, the conclusion and the further work direction are presented in Section 5.

D. Jin and S. Jin (Eds.): Advances in FCCS, Vol. 1, AISC 159, pp. 429–434.
springerlink.com

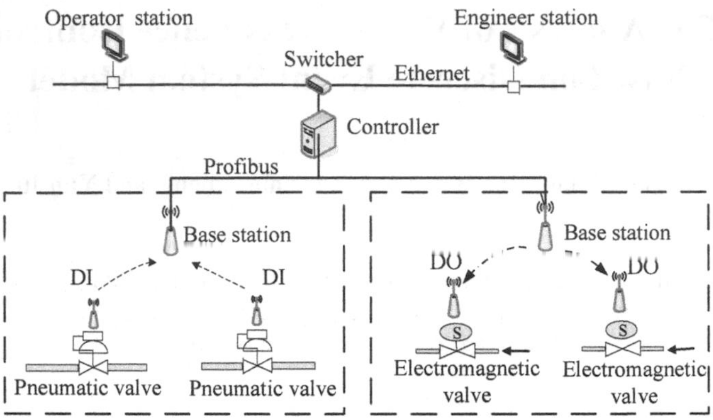

Fig. 1. The architecture of the sequence control system based on WSNs

2 Problem

In this control system, it's necessary that the step of the device in the field coordinates with that estimated by controller. Only that, the controller can generate the right command to ensure the safety of the device and carry out the process of production. The controller estimates the step of device based on the received states of the device through wireless channel. However, for the existing of the wireless networks, there is packet loss which may result that the estimated step in controller disagrees with the practical step of device.

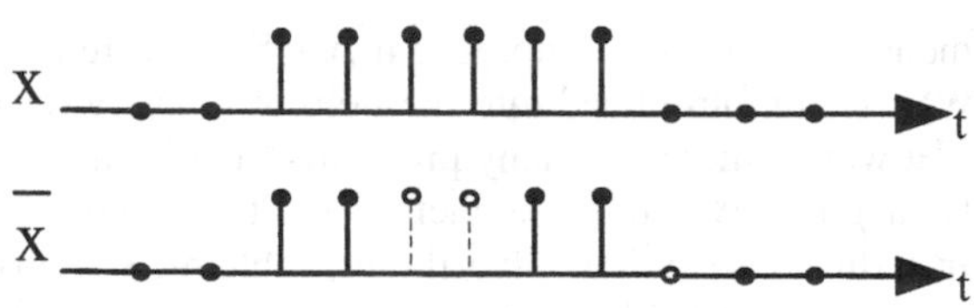

Fig. 2. Signals in device and in controller with packet loss

In Fig. 2, X is the sampled signal in wireless nodes and $\overline{X}$ is the received signal in the controller through wireless channel. The lost signal is indicated with hollow circle. It can be seen that the received signals will be quite different from the sampled signal. Thus, the controller may not know which step that the device stays in and generate wrong command sent to the device.

The situation will be the same when the controller sends the command to the device through the wireless channel. The loss of the packet may let the device entry into the undesired step.

To settle this problem, we model the system as two interacting DESs. And a solution is proposed to keep the estimated step in controller and the practical step of the device staying in consistent.

3 System Model

3.1 Framework of the Model

Although there are pneumatic valves, electromagnetic valves in the workshop of chemical water treatment, they can be taken as an entirety, receiving control command and feeding back the states to the controller. In Fig. 3, it gives the model of the control system. The controller and the plant are treated as two interacting discrete event systems (DESs) described by automaton. Frame A is the model of controller and frame B presents the plant's [5]. For there is wireless path both in the connection between controller and actuators, and in the connection between controller and sensors, the switches between A and B are on behalf of the packet loss through the wireless communication. While the switch is on, it means that the packets are transmitted without loss, else packets lost.

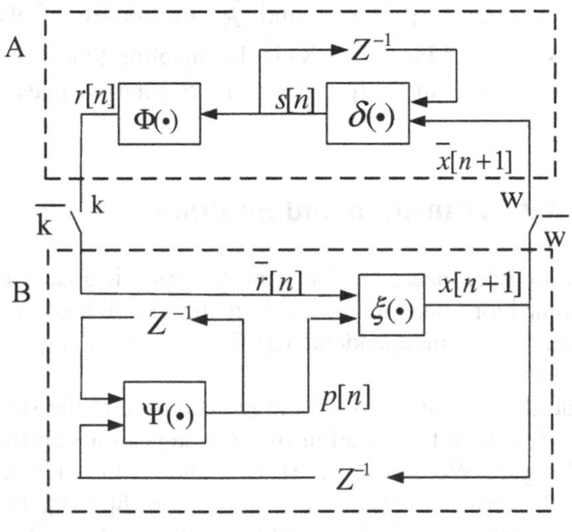

Fig. 3. The model of the control system based on WSNs

3.2 Automaton

Frame A shows the internal structure of the controller. And it can be modeled by quintuple, (S, $\overline{X}$, R, δ, Φ). The behavior of the DES plant is as follows:

$$s[n] = \delta_i(s[n-1], \bar{x}[n]),\ t_i < n < t_{i+1} \tag{1}$$

$$r[n] = \Phi(s[n]) \tag{2}$$

Where s[n] ∈S, the value of s[n] keeps the same in a step but differs in different steps, and can be used to denote the estimated step of the controller based on the collected states from devices and S is the set of steps; $\bar{x}[n] \in \overline{X}$ and $\overline{X}$ is the set of received plant symbols and the number of elements equals to the number of device in field;

$r[n] \in R$ and R is the set of controller symbols. δ: $S \times \overline{X} \rightarrow S$ is the set of state transition function and the function of δ varies with n, denoted as δ_i; Φ: S→R is the output function and $|t_{i+1}-t_i|$ means the duration of a step. The index n is analogous to a time index in that it specifies the order of the symbols in a sequence.

Frame B presents the plant which is also a DES and is similar with the controller. The automaton is specified by a quintuple, (P, X, $\overline{R}$, Ψ, ξ) and the capability of calculation of the wireless nodes can be used for this realization. In this model, we let Ψ=δ. The behavior of the DES plant is as follows:

$$x[n+1] = \xi(p[n], \overline{r}[n]) \tag{3}$$

$$p[n] = \delta_i(p[n-1], x[n]), t_i < n < t_{i+1} \tag{4}$$

Where $p[n] \in P$, the value of p[n] keeps the same in a step but differs in different steps, and can be used to denote the practical step of the devices in the field practically and P is the set of p[n]; $\overline{r}[n] \in \overline{R}$, $x[n] \in X$, X and $\overline{R}$ are the sets of plant symbols and received controller symbols; ξ: $P \times \overline{R} \rightarrow \Omega(X)$ is the enabling function. While receiving command from the controller, the output of the DES plant changes according to the enabling function.

4 Analysis of State Transition and Solution

Some of the assumptions are made: (1) The sample period is greater than the delay of the wireless communication, but much less than the duration of each step; (2) The probability of packet loss is independent; (3) There is no delay in the process of signals in the controller.

In this system, the states of the device change according to the step command. For the sample period is less than the duration of each step, in a step the values of p[n] and s[n] stay unchanged. We introduce H to indicate the step and H has N+1 elements, including N normal steps and one fault step, h[N+1]. For there are two DESs, we use h_a to indicate the step estimated by controller and h_b to indicate the step of the practical devices. The relationships between s[n] and $h_a[k]$, p[n] and $h_b[k]$ are shown respectively where k denotes the step:

$$h_a[k] = \{s[i] \mid s[i] = k, \forall i \in [0, l], l = |S|\} \tag{5}$$

$$h_b[k] = \{p[i] \mid p[i] = k, \forall i \in [0, l], l = |P|\} \tag{6}$$

4.1 State Transition

For the sake of convenience, the subscript of h[k] won't be referred, meaning the practical steps of the devices in the field. In this system, what we focus on is the transition of steps of the devices in the field. Suppose that at the beginning of the control process, the devices in the system are all in the halt and start from Step 1, h[1]. The transition of steps is shown in Fig. 4, with the states of the devices noted aside the transition line. And the state of device is denoted by $x^{a \rightarrow b}$, where a→b means the step transformed from h[a] to h[b]. $\overline{r^q}$ means the command the devices received in Step q. The solid line indicates the right transition according to the requirement of the process. The dashed line means that the step transfers to a fault one with the state marked aside and it should be forbidden.

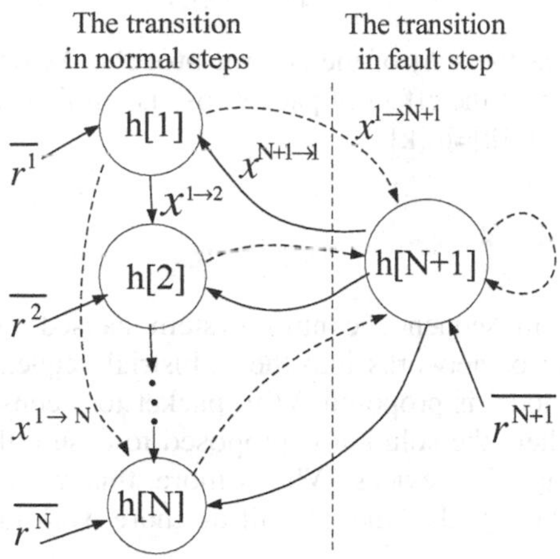

Fig. 4. The transition of steps

4.2 Solution

The communication between controller and plant is interfered by packet loss. The aim is to ensure that $h_a[k]=h_b[k]$. However, with the wireless sensor network introduced, $h_a[k]$ may not equal to $h_b[k]$, for the packet loss. There are two situations that we have to deal with. One is that the devices receive wrong commands. The other is that the controller receives wrong states fed back from devices. Both of these two situations will result in the fault transition, such as the dashed line shown in Fig. 4.

Firstly, we deal with the feedback of the states of the devices. When the packet loss in the feedback, $\overline{x}[n+1]=\mathrm{x}[\mathrm{n}]$. According to Eq. 1 and $t_i<n<t_{i+1}$

$$\mathrm{s}[\mathrm{n}+1]= \delta_{\mathrm{i}}(\mathrm{s}[\mathrm{n}],\overline{x}[n+1])=\delta_{\mathrm{i}}(\mathrm{s}[\mathrm{n}],\mathrm{x}[n]) \tag{7}$$

For the states of the devices are not known by the controller, considering the safety of system, it should not change the output of δ_i. If there are m continual packets lost, we have [6]:

$$\mathrm{s}[\mathrm{n+m}]= \delta_{\mathrm{i}}(\mathrm{s}[\mathrm{n+m-1}],\overline{x}[n+\mathrm{m}]) = \delta_{\mathrm{i}}(\mathrm{s}[\mathrm{n+m-1}],x[n]) = s[n] \tag{8}$$

Then we have

$$\mathrm{r}[n+\mathrm{m}]=\mathrm{r}[\mathrm{n}] \tag{9}$$

It means that when controller receives the states which don't coordinate with the requirement of the process, the controller will send out the same command.

Secondly, when the packet loss happens in the transmission of the command, $\bar{r}[n+m]=r[n]$. And the output of ξ will keep unchanged, so x[n+m]=x[n]. And it can be derived:

$$p[n+m] = \delta_i(p[n+m-1], x[n+m]) = \delta_i(p[n+m-1], x[n]) = p[n] \quad (10)$$

Suppose that p[1]=s[1] at the start of the control process. Associated with Eq. 3 and Eq. 8-10, it can be seen that the effect of packet loss is eliminated. It is easy to derive that p[n]=s[n], then keep $h_a[k]=h_b[k]$.

5 Conclusion

To settle the problem in sequence control system caused by packet loss while introducing wireless sensor networks into the industrial sequence control system, a method of using DES model is proposed. With packet loss considered, the transition of steps is presented. Then, the solution is proposed to ensure the step synchronized between the controller and the devices. What's more, this model can be also used to deal with the delay of the signals. And we will do more work in this direction in our future research.

References

1. Jerry Daniel, J., Panicker, S.T., et al.: Industrial Grade Wireless Base Station for Wireless Sensor Networks. In: 3rd International Conference on Electronics Computer Technology (2011)
2. Min, W., Keecheon, K., et al.: Research and Implementation on the Security cheme of industrial wireless network. In: International Conference on Information Networking, ICOIN (2011)
3. Hespanha, J.P., Naghshtabrizi, P., Xu, Y.: A Survey of Recent Results in Networked Control Systems Proceedings of the IEEE 95(1) (January 2007)
4. Basile, F., Chiacchio, P.: On the Implementation of Supervised Control of Discrete Event System. IEEE Transactions on Control System Technology 15(4), 725–738 (2007)
5. Antsaklis, P.J., Lemmon, M., Stiver, J.A.: Intelligent Control: Theorv and Practice, pp. 28–62. IEEE Press (1995)
6. Fang, X., Wang, J.: Stochastic Observer-based Guaranteed Cost Control for Networked Control Systems with Packet Dropouts. IET Control Theory Application 2(11), 980–989 (2008)

Fuzzy PID Control Strategy Applied in Boiler Combustion System

Yunjing Liu and Deying Gu

Department of Automation Engineering,
Northeastern University at Qinhuangdao, China
yjliu66@163.com

Abstract. With the rapid development of the power industry coal-fired power plants in China have been built more than decades ago. Boiler system is a complex process and has high nonlinearity, large delay, strong coupling and load disturbance. It is crucial parts of most power plants and has been concerned with analyzing power system dynamics in various works. But it is hard to develop practical mathematical model of the control object using the traditional PID control. In order to realize the optimal control of boiler combustion system, this paper presents the fuzzy-PID control Strategy applied in steam temperature control. Simulation results for different cases of boiler combustion system show that the control strategy has better ability.

Keywords: boiler, combustion, fussy PID, MATLAB Simulink.

1 Introduction

With the rapid development of the power industry coal-fired power plants in China have been built more than decades ago. The boiler system is an important part of the generating units. The boiler system in a thermal power plant consists mainly of a steam-water system and combustion system, which produce a high-pressure superheated steam to drive a generator in order to produce power. The superheated steam temperature control and the separator water level control of a steam water system are very important to guarantee operation safety and to improve economic benefits of the power plant. [1, 2] The boiler system performance will directly affect the safe and economic operation of the entire unit. Because boiler system always works under high pressure and temperature its security is very important. With the improvement of the running level of boiler system, large capacity and high parameters units are applied widely. In order to run with safety, stability, economy, high efficiency, and so on, the central task is to adjust the output power to meet system demand while minimizing unwanted pressure and temperature variations. Because boiler combustion system is usually modeled with a multi input multi output (MIMO) nonlinear system it has characteristics with strong coupling, the loads and main steam pressure control loop depend on each other and restrict to each other. The dynamic characteristics of the unit are nonlinear essentially and it has big time-delay of the boiler side. [3,4]The traditional PID control strategy can't acquire satisfied control effects with the changing of the unit loads. In order to improve boiler efficiency a better control strategy need to be studied. [5,6].

D. Jin and S. Jin (Eds.): Advances in FCCS, Vol. 1, AISC 159, pp. 435–439.
springerlink.com

This paper presents the fuzzy-PID control Strategy applied in temperature control. Simulation results for different cases of boiler combustion system show that the control strategy has better ability.

2 Fuzzy PID Control Strategy

Industrial boiler temperature as a control object is a typical multiple-input multiple-output system. It has many features such as high nonlinearity, large delay, strong coupling, load disturbance, inertial lag, time-varying and so on. It is crucial parts of most power plants and has been concerned with analyzing power system dynamics in various works. But it is hard to develop practical mathematical model of the control object using the traditional PID control. Fuzzy PID which combines traditional PID control method with fuzzy theory has a lot of advantages, for example fast response, small overshoot, strong robustness and strong anti-interference ability. Especially it does not require precise mathematical model. [7,8 and 9] Therefore fuzzy PID control is an effective method for non-deterministic system control.

The fuzzy PID controller consists of two parts, one is a conventional PID controller and another is a fuzzy reasoning controller. It is shown in Fig.1 In this controller, the main idea for the PID parameters adjustment is to find the fuzzy relations among three Parameters K_P, K_I, K_D of traditional PID controller, absolute value of an error e

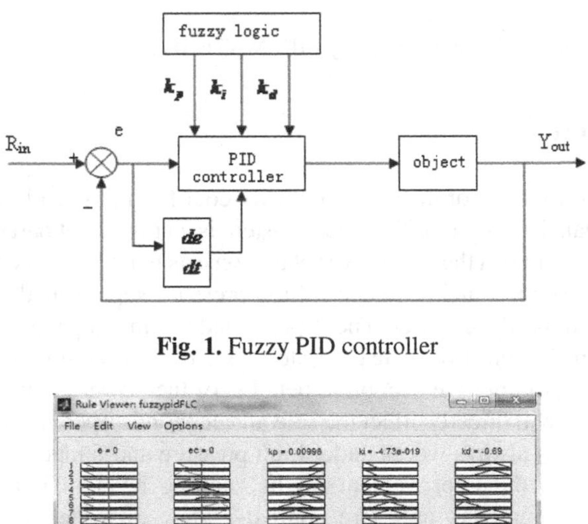

Fig. 1. Fuzzy PID controller

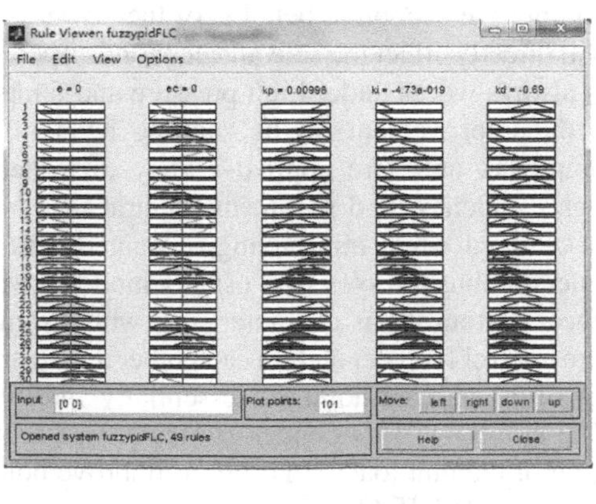

Fig. 2. Fuzzy reasoning rules

and absolute value of error rate. The deviation e and the error rate of the deviation de/dt are used as the input. In practice, with continual testing, three controller parameters are modified on-line to meet the different requirements. The reasoning rules of fuzzy PID are shown as in Fig.2.

A two-input and three-output fuzzy PID controller is designed shown in Fig.3.

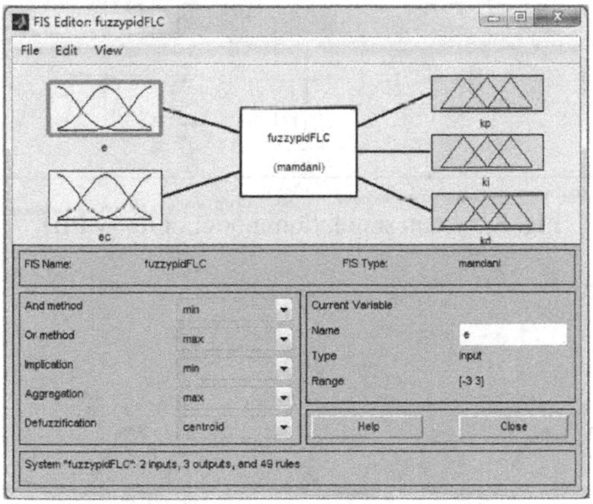

Fig. 3. Fuzzy PID controller

3 Simulation and Results

This controller is simulated in MATLAB Simulink. In order to evaluate and verify the effectiveness of the proposed fuzzy PID controller, and to compare its performance with that of the traditional PID controller, the two model of simulation is given in Fig.4 and Figure.5 based on the control theory of PID , fuzzy-PID and Temperature control system structure described above.

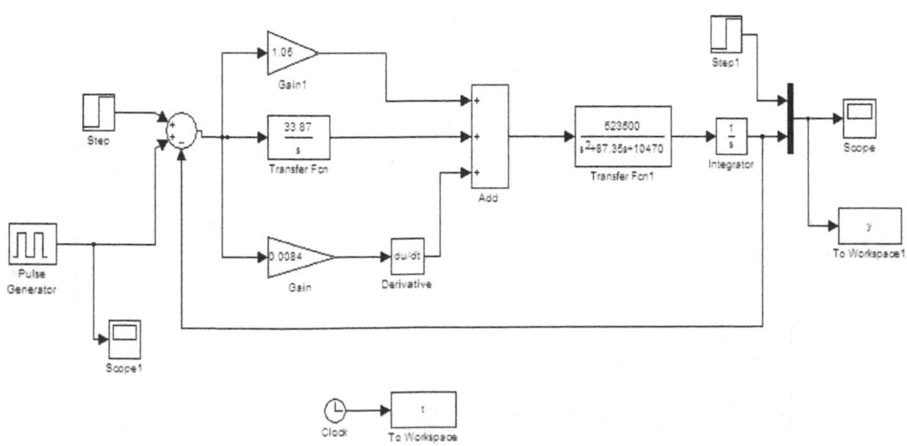

Fig. 4. System simulation model of traditional PID

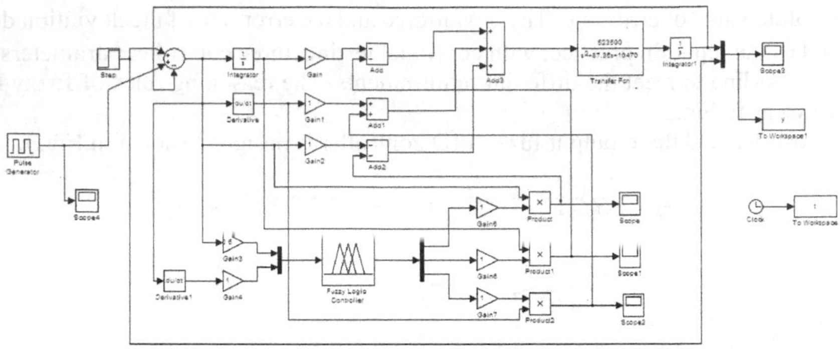

Fig. 5. System simulation model of fuzzy PID

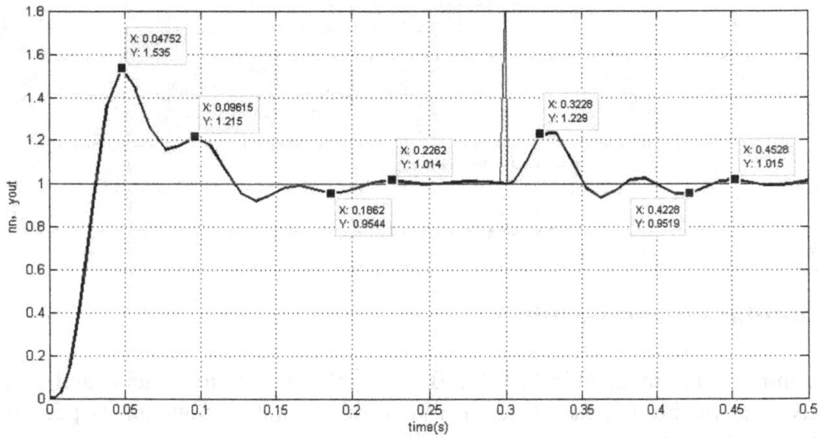

Fig. 6. Step response with disturbance for traditional PID

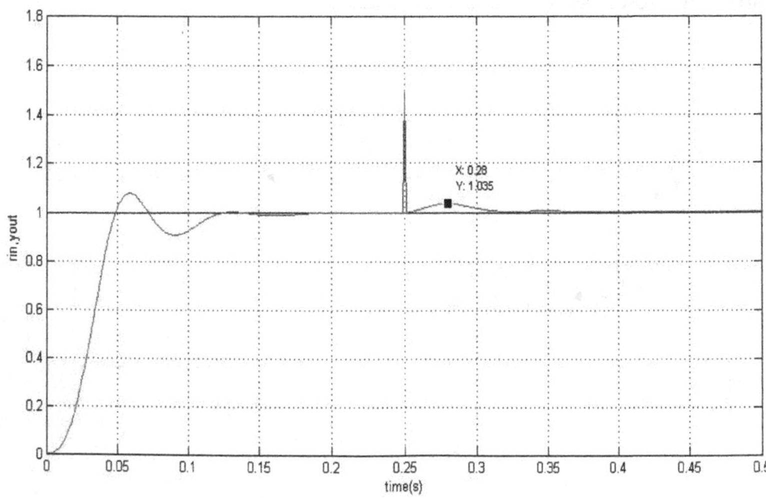

Fig. 7. Step response with disturbance for fuzzy PID

The simulation was run in order to evaluate the behavior of the system when a disturbance occurs. The curves of typical process step response with disturbance are shown in Fig.6 and Fig.7. Simulation results show that the ability of fuzzy PID controller is better than the traditional PID control.

4 Conclusions

This paper analyzes the dynamic characteristics of the boiler combustion system based on the model of traditional PID and fuzzy PID. From the simulations performed, it was found that great improvements can be obtained by using the fuzzy PID controller. The fuzzy PID controller has the quicker and smoother response with a little overshoot in contrast with the traditional PID.

References

1. Li, S., Liu, H., Cai, W.-J., Soh, Y.-C., Xie, L.-H.: A New Coordinated Control Strategy. IEEE Transactions on Control Systems Technology 13(6) (2005)
2. Lam, H.K., Leung, F.H.F.: Stability Analysis of Fuzzy Control Systems Subject to Uncertain Grades of Membership. IEEE Transactions on Control Systems, Man and Cybernetics—Part B 35(6) (2005)
3. Conte, G., Cesaretti, M., Scaradozzi, D.: Combustion control in domestic boilers. Higher Education Press and Springer-Verlag (2009)
4. Seborg, D.E., Edgar, T.F., Mellichamp, D.A.: Process Dynamics and Control. Wiley, New York (1989)
5. Ferrer, J., Rodrigo, M.A., Seco, A., Penya-roja, J.M.: Energy saving in the aeration process by fuzzy logic control. Wat. Sci. Tech. 38(3), 209–217 (1998)
6. Xu, J.-X., Liu, C., Hang, C.C.: Tuning of fuzzy PI controller based on gain/phase margin specifications and ITAE index. ISA Transactions 35, 76–79 (1996)
7. Lin, C.-T., Lee, G.: Neural-network-based fuzzy logic control and decision system. IEEE Transactions on Computers 40(12), 1320–1336 (1991)
8. Lau, C.: Neural networks. Theoretical foundations and analysis. IEEE Press (1992)
9. Lin, Y., Cunningham III, G.A.: A new approach to fuzzy neural system modeling. IEEE Transactions on Fuzzy Systems 3(2), 190–198 (1995)

Adaptive Time-Thresholds for RFID Data Cleaning Algorithm

Jibin Xu[1], Weijie Pan[2,*], Shaobo Li[2], and Jorge Alberto-Reyes-Cruz[1]

[1] School of Computer Science and Information, Guizhou University, 550025, Guiyang, Guizhou, China
[2] Key Laboratory of Advanced Manufacturing Technology, Guizhou University (Ministry of Education), 550003, Guiyang, Guizhou, China
panweijie2009@126.com

Abstract. Traditional RFID data cleaning methods based on event filtering algorithms have their own limitation, which cost too much buffer memory space. In this essay, based on the traditional data cleaning methods, the authors have proposed a new RFID data cleaning mechanism based on event with adaptive Time-thresholds through a deep analyze of the RFID data cleaning methods. According to the current tag information, combining the time-window method with pseudo event filtering method, it adaptively adjust the time-thresholds from the aspect of the tag existing circle. The experimental finding shows that this method has reached theoretical and practical significance to some extent, for it can help reduce the consumption of the cache.

Keywords: RFID, event filtering, sliding window, data cleaning, adaptive time-thresholds.

1 Introduction

In Radio Frequency Identification system, there still three typical undesired scenarios, they are false negative readings, false positive readings and duplicate readings. At the same time, there is a great deal of intermediate data in the RFID system.

At present, the research on the RFID data cleaning technology has made some progress. The literatures [1,2] have addressed issues concerning false negative readings and false positive readings through the way with data cleaning algorithm embedded in RFID readers. But being limited by storage on the RFID readers themselves. The literatures [3,4] use the fixed-length time window cleaning method in the middleware. Though the method is simple, yet it still has its own defect lacking flexibility in data cleaning. The literatures [5] adopt a fixed-length moving time window based on event-driven to process the RFID data. Anyhow, this method still has its lack of flexibility. The literatures [6,7] adopt a new data cleaning algorithm based on pseudo event filtering method which through setting up a fixed-length time-threshold to determine pseudo event so as to throw off the pseudo event data. This

* Corresponding author.

D. Jin and S. Jin (Eds.): Advances in FCCS, Vol. 1, AISC 159, pp. 441–446.
springerlink.com

method solves the duplicate readings problem to a certain extent. The literatures [8] uses the size of the reading area and the velocity of the tags to estimate the length of the time window. However, the size of the reading area is an estimated value, thus there still a major error in the length. The literatures [9] proposes a RFID data cleaning method called SMURF (Statistical smoothing for Unreliable RFID data, SMURF), this literatures makes a statistic and modeling based on the statistical characteristics of RFID data, adjusts the window size in a flexible way.

In order to overcome the shortages of ordinary methods, this paper provides a method called ATDCA (Adaptive Time-thresholds for RFID Data Cleaning Algorithm, ATDCA), combining the time-window method with pseudo event filtering method, to solve the duplicate readings problem and reduce buffer data.

2 Correlative Algorithms

2.1 SMURF Algorithm

IN SMURF Algorithm, when the window size is W , and the epochs number is w_i, the average empirical read rate over these observation epochs is p_i^{avg}.

In the Binomial Sampling Model for Single Tag Readings, The overall probability of reading tag i at least once during W is estimated as:

$$w_i \geq [\frac{\ln(1/\delta)}{p_i^{avg}}] \tag{1}$$

According to the Central limit theorem, SMURF flags a transition for tag i in the current window if the number of observed readings is less than the expected number of readings and the following condition holds:

$$\| S_i | - w_i p_i^{avg} | > 2 \cdot \sqrt{w_i p_i^{avg} (1 - p_i^{avg})} \tag{2}$$

In the adaptive multitag aggregate cleaning, a distribution model called $\pi-estimator$ model is used to solve this problem.

The overall probability of reading tag i at least once during W is estimated as:

$$\pi_i = 1 - (1 - p_i^{avg})^w \tag{3}$$

And the statistically-significant transition in population counting has occurred in the second half of W if the following condition is satisfied:

$$| N_w - N_w' | > 2 \cdot \sqrt{\text{var}(N_w) - \text{var}(N_w')} \tag{4}$$

The following operation is similar to the situation of the Binomial Sampling Model.

2.2 RFID Data Cleaning Method Based on Pseudo Event

In this kind of pseudo event, we defined a threshold as δ_1, when the first time a tag was read, remembered that time as t. If a tag was read more than once between the time [δ_1 , $\delta_1 + t$] , this event would be regarded as a pseudo event, and the tag information would be only output once [10].

3 Adaptive Time-Thresholds for RFID Data Cleaning Algorithm

3.1 Improved Algorithm

Based on pseudo-event filtering algorithm, ATDCA combines sliding-window model and pseudo-event filtering model. This method setting thresholds of event overdue time, in order to filter tag information within the reading range, and it triggers the time-threshold changes its size adaptively based on the number of tag information in the buffer list.

The cleaning filter of the improved method mainly consists of calculator, clock, comparer and buffer list. The processing mechanism is illustrated as in figure 1.

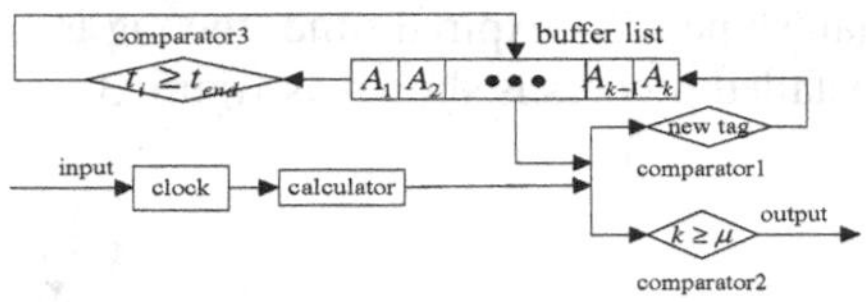

Fig. 1. Structure of the Cleaning Filter

The clock obtains the time t right now. The calculator calculates the size of time window T and gets the expected expired time. the tag Comparer 1 determines if the tag is already in the buffer list, which facilitates its adding to the buffer list. Comparer 2 determines if the tag is a true reading and filters duplicate and false negative readings. Comparer 3 determines if the tag is expired and deletes the redundant information.

The time threshold satisfying certain conditions as:

$$T = [\frac{\ln(1/\delta)}{p_i^{avg}}] \quad (5)$$

δ denote the confidence. The average reading rate of a single reading cycle is satisfied:

$$p_i^{avg} = \sum_{t \in S_i} p_{i,t} / |S_i| \quad (6)$$

$p_{i,t}$ is the number of tags obtained in a single reading cycle. $|S_i|$ refers to the number of tag information in the buffer list. The event of triggering window size changes is defined as:

$$|N' - N| / N \geq \lambda \quad (7)$$

N' is the tag number in buffer list λ is the confidence interval. When the change rate of tag numbers goes beyond the degree of confidence, the time threshold size should be recount. The whole process of adaptive time-threshold adjustment is shown as figure 2.

3.2 The Algorithm Description of ATDCA

Define the unit format of the tag buffer list as: $U^i = \{[UID^i][T_{end}{}^i][N^i]\}$, in which UID is tag ID, T_{end} is expected expired time of tags, N is the times tags are read, and i refers to the serial number of tags.

When the reader read again, the system checks the buffer list and sees if the tag information is in the buffer list. If it is in the queue, then calculate the expected expired time $T_{end} = t + T$, renew related information of the tag and put the tag information into the end of the queue; if not, then directly insert related information into the end of the buffer list and check the queue in order. If the time right now is equal to or even later than some tag's expired time, the tag is considered expired. The tag should be deleted. Detailed process is shown as figure 3.

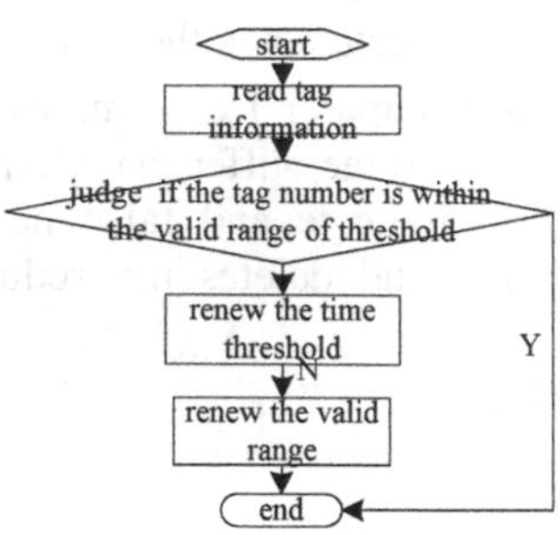

Fig. 2. Single buffer list data filtering

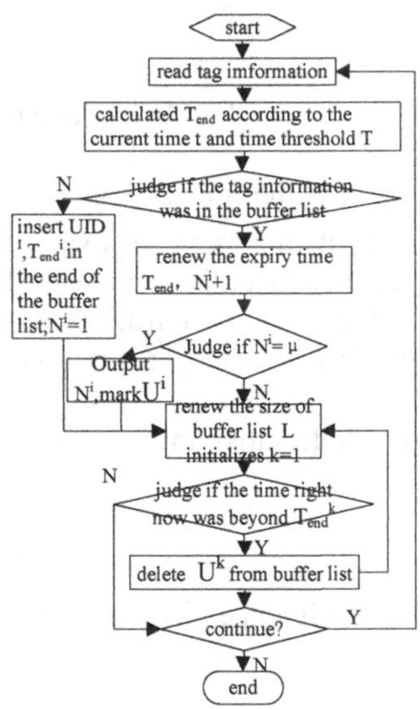

Fig. 3. Process of algorithm

Step 1: The reader runs a new reading and gets tag UID.

Step 2: Determine if the tag numbers in the buffer list are in the valid scope of threshold. If yes, go to step 4; if no, go to step 3.

Step 3: Use $T=[\frac{\ln(1/\delta)}{p_i^{avg}}]$ to calculate new time threshold, and $|N'-N|/N=\lambda$ to renew the scope of threshold.

Step 4: Calculate the expected expired time $T_{end}=t+T$ based on the time when the tag is read t and the time threshold T.

Step 5: Compare the recorded UID with the tag information in the buffer list and determine if the tag is already existed. If yes, go to step 6; if no, go to step 7.

Step 6: The times of the tag's being read N_t plus 1, and renew the expected expired time of the tag and related information. Go to step 8.

Step 7: The times of the tag's being read N_t plus 1, and insert the related information of the tag into the end of the buffer list. Go to step 9.

Step 8: Determine the reading times of the tag's. If they are greater than μ. Output it to the upper layers and mark it. Go to step 9.

Step 9: Renew the buffer list. Count the size of buffer list L and initialize the value $k=1$, which point to the first tag information in the queue. Go to step 10.

Step 10: Compare the current time with the expected expired time of the first tag in the buffer list. If it is equal to or even later than the expected expired time. Delete it from the queue and go to step 9. If not, go to step 11.

Step 11: Determine if the reading process is over. If it is over, end the data cleaning; if no, go to step 1 and continue cleaning.

4 Simulation Result and Analysis

The tag information is generated by Matlab randomly. Take the tag's confidence interval $\delta \le 0.01$ to ensure the probability of the tag's being read, and δ to be fulfilled $1-\delta \ge 0.99$. In each reading cycle, the average reading probability of any tag $p_i^{avg}=0.6$, $\mu=2$. In the test, there are two passing models of the tags, the quick passing model and slow passing model. The difference between the two models is the change rate of the tags within each time window size. In quick passing model, let the change rate $\theta=70\%$. In the slow passing model, let $\theta=10\%$.

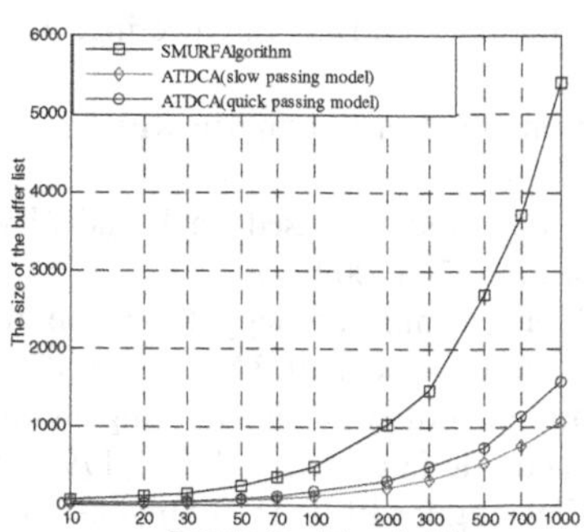

Fig. 4. The comparison between ATDCA and SMURF in buffer list

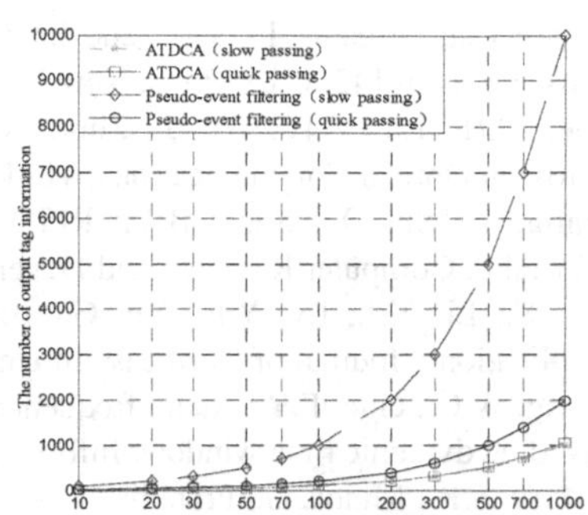

Fig. 5. The comparison between ATDCA and Pseudo-event in output information

As is demonstrated in figure 4, compared with SMURF, the cache space taken up by SMURF increases twice as much as by improved method as the tag number add up. Improved method can thus effectively reduce the cache space cost. As for RFID data cleaning method based on pseudo event, the information load of the tag it generates correlates with the ratio of time-window value T to time threshold T', and the change rate of tag θ .As is shown in figure 5, when time threshold T' is not appropriate, if tag passing slowly, pseudo-event tag cleaning method brings abundant duplicate readings. That severely jeopardizes the performance of the system. When the tags pass quickly, it still generates small amount of duplicate readings. By contrast, the improved methods generally do not produce output of duplicate readings.

5 Conclusion

This essay studies the existing RFID data cleaning technique. Than proposes an improved data cleaning method, which utilizes adaptive time-window size as threshold value to trigger tag output and expiration. Experimental results suggest that the improved algorithm guarantees the accuracy, real-time and simplicity of data. It is more flexible, and solves the problem that inappropriate time threshold generates duplicate readings. In addition, it demands simple hard ware and largely boosts the efficiency of RFID data cleaning.

References

1. Ding, S., Yang, S.L.: Embedded RFID Serviced Component Based on .NET Compact Framework. Computer Engineering 34, 50–52 (2008)
2. Rasmus, J., Karsten, F.N., Petar, P., Torben, L.: Reliable Identification of RFID Tags Using Multiple Independent Reader Sessions. Presented at IEEE RFID 2009 Conference, Orlando, FL, USA, pp. 64–71. IEEE Press, New York (2009)
3. Graham, C., Vladislav, S., Divesh, S., Xu, B.J.: Forward decay: A practical time decay model for streaming systems. In: Proceedings of the 2009 IEEE International Conference on Data Engineering, ICDE 2009, Washington DC, USA, pp. 138–149. IEEE Press, New York (2009)
4. Jeffery, S.R., Gustavo, A., Franklin, M.J.: A Pipelined Framework for Online Cleaning of Sensor Data Streams. In: The 22nd International Conference on Data Engineering, Atlanta, USA, pp. 140–142. IEEE Press, New York (2006)
5. Chen, J.H., Liu, G.H., Wu, J., Zhou, X.: Applications of data filtering in RFID systems. Study on Optical Communications, 41–43 (2009)
6. Wang, Y., Shi, X., Song, B.Y.: RFID Data Cleaning Method Based on Pseudo Event. Journal of Computer Research and Development 46, 270–274 (2009)
7. Gu, Y., Li, X.J., Lv, Y.F., Yu, G.: Integrated Data Cleaning Strategy Based on RFID Applications. Journal of Northeastern University (Natural Science) 30, 34–37 (2009)
8. Wang, W.C., Guo, F.Y.: Radio frequency identification middleware data filtering algorithm based on dynamic time window. Information and Electronic Engineering 7, 177–179 (2009)
9. Jeffery, S.R., Minos, G., Franklin, M.J.: Adaptive Cleaning for RFID Data Streams. In: The First Int'l VLDB Workshop on Clean Databases, Seoul, Korea, pp. 163–174 (2006)

A Design of Hardware-in-the-Loop Simulation System for High Altitude Airship

Zhi Li, Xiao Guo, and Ming Zhu

Bei Hang University, Xue Yuan Road No.37, Hai Dian District, Bei Jing, China

Abstract. This paper introduces a hardware-in-the-loop simulation system for high altitude airship, explains its constitution and how to establish the systems, and also illustrates most involved models and applied software. We have carried out flight simulation experiments in straight line on the system, validated the feasibility and reliability of the hardware-in-the-loop simulation system.

Keywords: high altitude airship, hardware-in-the-loop simulation, simulation model.

1 Introduction

Hardwar-in-the-loop simulation also can be said semi-physical simulation.[1] In the development of a certain high altitude airship, we designed a set of a hardware-in-the-loop simulation system of high altitude airship, which could take simulation experiment of partial system and also the total system.

2 Component of the Simulation System

2.1 Constitution of the System

The hardware of the system includes:4 PC,2 CPCI embedded computers,1 three-axis Rotary tables, 1 Atmospheric data simulation machine,1 network switch,1 optical fiber switch,2 reflect memory cards,1 work Cabinet of Processing System,1 power supply equipment of 28 volts,4 demonstration propellers, 1 semi-physical simulation In interline counter,1 measurement and control station on the ground for simulation.

2.2 System Connection and Subsystem Establishment

The system communicates with all the modules through 3 real-time communication interface in the simulation experiment. The system communicate the physical parts with simulation parts chiefly through Ethernet and reflective memory network assisted by Serial ports. The Ethernet connect the main control table with 2 simulation machines for the download of the program and the communication between the control tables with simulations. Reflective memory network can be communicated fast, operated

D. Jin and S. Jin (Eds.): Advances in FCCS, Vol. 1, AISC 159, pp. 447–451.
springerlink.com

easily, and has high Reliability [2].it execute the data exchange in the system simulation, and connect Processing System, three-axis Rotary table, Atmospheric data simulation machine with the main control table. The Serial Ports connect with station computer, model simulation machine, flight control computer with the main control station, and transmit the information of injected fault and control data.

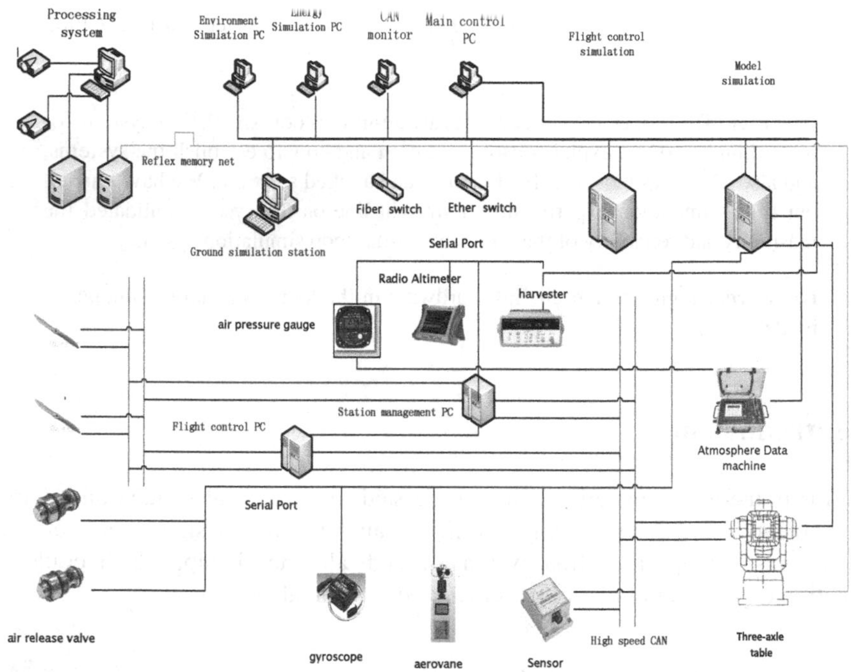

Fig. 1. System Connection

2.3 The Building of the Mathematical Model

The simulation model contains: dynamics model, the model of propeller and electric power, energy system model, environment control system model.

The building of dynamics model bases on certain high altitude airship, and the models set up the motion equation on the body axis system. the velocity on the O_x、O_y、O_z can be described as μ、v、ω, the angular velocity and moment of inertia on the O_x、O_y、O_z can be described as p、q、r、I_x、I_y、I_z. The airship stay in the air based on the buoyancy when it is moving. aerodynamic force includes resultant vector F_a, resultant moment vector M_a. usually F_a can be resolved as: resistance force, side force, lifting power. ρ is air density, v is flight speed, $V^2=\mu^2+v^2+\omega^2$, V is airship volume. [3]

On the stratosphere airship, propeller is pushed by Direct-current Generators which produces thrust T. According to the Dynamics theory [4] and the load-carrying condition, we can get 6 degrees freedom dynamics model of the airship from the 6 degrees freedom balance equation of the airship: [5]

$$m[(\dot{v}-wp+ur)-yG(r^2+p^2)+xG(rp-\dot{p})+xG(qp+\dot{r})]=Y_I+Y_{at}+Y_G+Y_B \tag{1}$$

$$m[(\dot{v}-wp+ur)-yG(r^2+p^2)+xG(rp-\dot{p})+xG(qp+\dot{r})]=Y_I+Y_{at}+Y_G+Y_B \tag{2}$$

$$m[(\dot{\omega}-uq+vp)-zG(p^2+q^2)+xG(rp-\dot{q})+yG(rq+\dot{p})]=Z_I+Z_{at}+Z_G+Z_B \tag{3}$$

$$\begin{aligned}&I_x\dot{p}+(I_x-I_y)qr-I_{xz}(\dot{r}+pq)+I_{xy}(r^2+pq)+I_{xy}(pr-\dot{q})+I_{YZ}(r^2-q^2)\\&+myG(\dot{w}+pv-qu)=L_I+L_a+L_G\end{aligned} \tag{4}$$

$$\begin{aligned}&I_y\dot{q}+(I_x-I_z)rp+I_{xz}(\dot{p}+qr)+I_{yz}(qp-\dot{r})+mzg(\dot{u}+qw-rv)\\&-mxG(\dot{w}+pv-qu)=M_I+M_a+M_G\end{aligned} \tag{5}$$

$$\begin{aligned}&I_x\dot{r}+(I_Y-I_x)pq+I_{xz}(rp-\dot{p})+I_{xy}(q^2-p^2)-I_{yz}(\dot{q}+rp)\\&+mxG(v+r\dot{u}-pw)-myG(\dot{u}+qw-rv)=N_I+N_a+N_G\end{aligned} \tag{6}$$

[6], and then we implement the linearization of the Simulink model, which can be made use of in the simulation experiments, using SISO toolbox [5].

3 The Processing of Hardware-in-the-Loop Simulation

We have taken certain high altitude airship for example, ran the simulation experiment on the platform, simulating the line flight of high altitude airship under the control of flight control law. First, we started the six degree-of-freedom model [6],and made the model working in the condition of floating in the high altitude, then worked the control law of the airship, and made the airship flight straightly.[7]

According to the demand of the system, we set the expecting speed at 6 meters per second on the condition of the injection of wind field at 3 meters per second, and started the straight flight experiment. The pictures below are flight track plots. In the plot, we can know that except a little small disturbance at the beginning time, the airship was always moving straightly, flying toward the directions under the control of the flight control law.

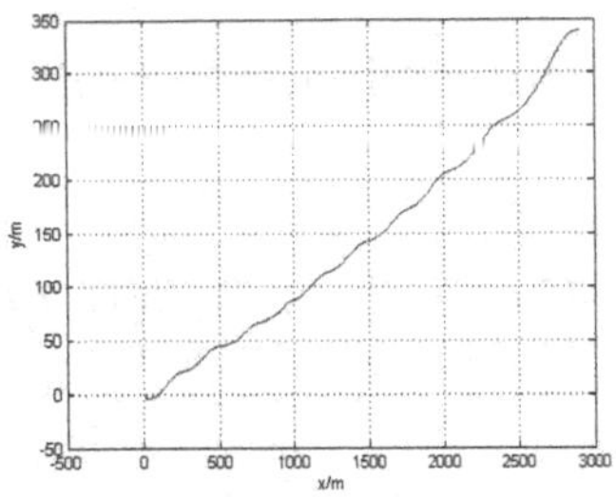

Fig. 2. Fly straight in wind field

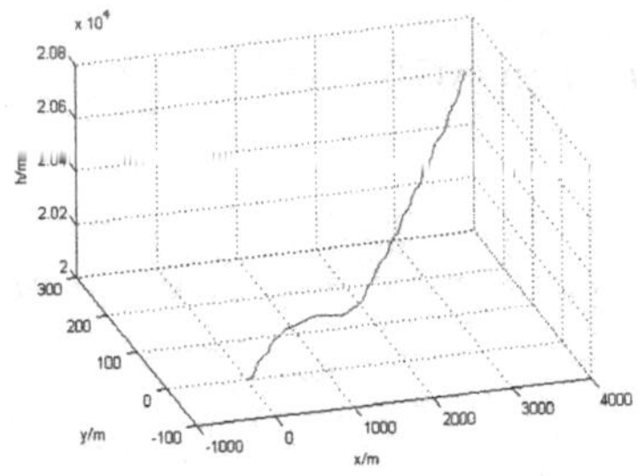

Fig. 3. Fly straight in wind field

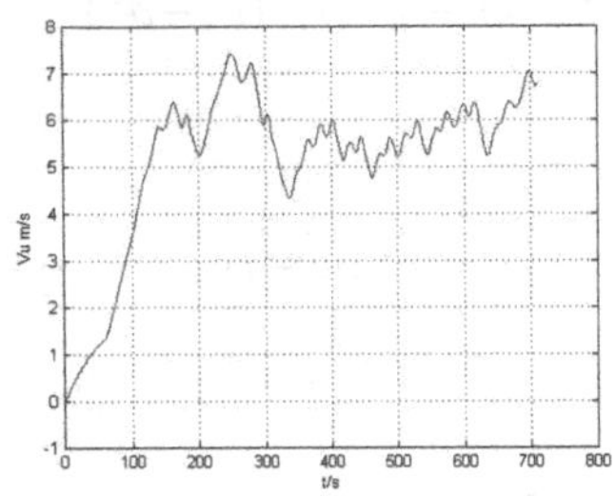

Fig. 4. Speed on X axis direction in wind field

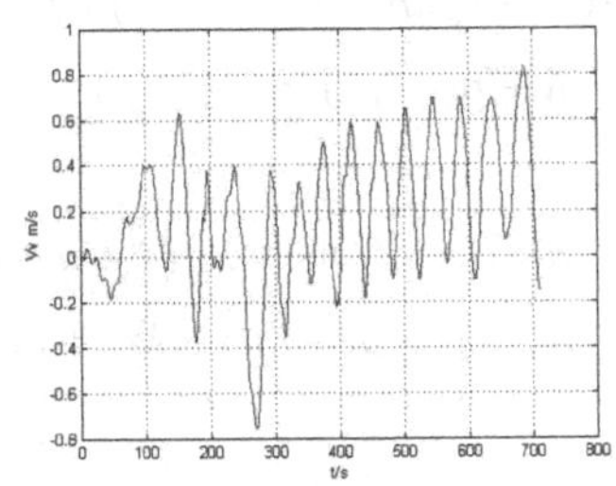

Fig. 5. Speed on Y axis direction in wind field

4 Conclusion

We can know that: under the direction of the flight control law, this simulation system fairly simulated the flight in the high altitude availably, tested the functions of the airship in the condition of cross wind of 3 meters per second on straight line direction. From the simulation results we can see that: under the condition of the cross wind, the body of the high altitude airship can keep dynamic stability on both X and Y axis direction in wind field which is expected and deducted theoretically [8]. we also have taken the test experiments as follow: hovering flight, inertial navigation, wind field Injection, integration test of avionics, the fault of flight control. The simulation system achieved good performance in the test, effectually reflected the truly function of the parts and the software of the airship, which cannot be easily tested and operated in the physical model, helped us a lot in the adjustment.

From the experiments, we can say this simulation system can satisfy the demand of the research of the high altitude airship, and have a certain referential value for the research of other airships.

References

1. He, J.H.: Computer Simulation. Press of University of Science and Technology of China (January 2010) ISBN 978-7-312-02206-7
2. Ryoums: Optimization of data accesses in reflective memory systems. In: IEEE Region 10 Conference, TENCON 2006, pp. 1–4. s,n., Hong Kong (2006)
3. Zhang, M.L.: Flight Control System, pp. 13–50. Press of Aviation Industry, BeiJing (1994)
4. Wei, R.X., Feng, B.Q., Hu, M.L.: Hardware-in-the-Loop Flight Simulate Experimental Platform Designed for UAVs. Flight Dynamics 27(5), 75–78 (2009)
5. Zhao, J.W., Wang, Z.: Theoretical Mechanics. Higher Education Press, BeiJing
6. Jin, O., Qu, W.D., Xi, Y.G.: Stratospheric Verifying Airship Modeling and Analysis. Journal of Shanghai Jiao Tong University 37(6), 956–960, 7788 (2003)
7. Haqp, R., Durrant-Whyte, H.F.: Fuzzy moving sliding mode control with application to robotic manipulators. Automatics 35(4), 607–616 (1999)
8. Jin, O., Dong, Q.W., Geng, X.Y.: Longitudinal Motion Analysis and Simulation for Lighter-Than-Air (LTA) Airship. Journal of Shanghai Jiao Tong University 37(6), 961–963, 56 (2003)

Modeling and Quantifying the Dependability Based on PEPA

Yunchuan Guo[1,3,4], Dongyan Zhang[2], Lihua Yin[1], and Junyan Qian[3]

[1] Institute of Computing Technology, Chinese Academy of Sciences, Beijing, China
[2] Department of Computer Science & Technology, University of Science & Technology Beijing, China
[3] Guangxi Key Laboratory of Trusted Software, Guilin University of Electronic Technology, Guilin, China
[4] Guangxi Key Laboratory of Hybrid Computational and IC Design Analysis, China
{guoyunchuan,qjy2000}@gmail.com

Abstract. Dependability, as a basic problem in computer security is being widely studied, but little work is based on PEPA. In this paper, PEPA are used to model and quantify the dependability of systems. First, backup systems and reparable systems are formally modeled. Then dependability is quantified using PEPA. The result indicates that PEPA can be used to effectively quantify the dependability of systems.

Keywords: Dependability, PEPA, Quantification.

1 Introduction

With the development of information technology, computer network has penetrated to government, military, science and all other fields. Nowadays, people attach more and more importance on the dependability of network and information system. In order to analyze the dependability of systems, one important thing is to formally model it. Generally, methods for analyzing the dependability of systems are divided into two categories[1]: combinatorial methods and state-space methods. Combinatorial methods include Reliability Block Diagrams (RBD), Fault Trees Analysis (FTA), and Fault Mode Effect Analysis (FMEA). State-space methods consist of (1) Markov process, (2) stochastic Petri net, as well as (3) process algebra. In state-space methods, stochastic process algebra has provided a special solution for performance evaluation: operational primitives (called process) are used to describe system components, communication between different processes is adapted to model interactions among processes, and algebra rules are employed to reason process behaviors. Classical stochastic process algebra include: Timed Process and Performance Analysis (TPPA), Performance Evaluation for Process Algebra(PEPA) [2-4], and Extended Markovian Process Algebra (EMPA).

Although dependability is being widely studied, but little work is based on PEPA. In this paper, PEPA are used to model and quantify the dependability of systems. First, backup systems and reparable systems are formally modeled. Then

D. Jin and S. Jin (Eds.): Advances in FCCS, Vol. 1, AISC 159, pp. 453–458.
springerlink.com

dependability is quantified using PEPA. The result indicates that PEPA can be used to effectively quantify the dependability of systems.

2 Service Model for Dependability

Generally, a system is comprised by several sub-systems. We hope that each sub-system may provide requisite functions all the time. However, this is a "saint hope". In fact, sub-system may break down for different reasons (such as an intentional attack). That is, a system at least includes two states: server and failure, and we use $serv$ and $fail$ to denote the two states, respectively. In order to improve the availability of sub-systems, these sub-systems are organized using an appropriate topology [5], for example, series systems, parallel systems, backup systems and reparable systems, etc.. In this paper, only backup systems and reparable systems are models for the limitation of spaces. Next, we model them.

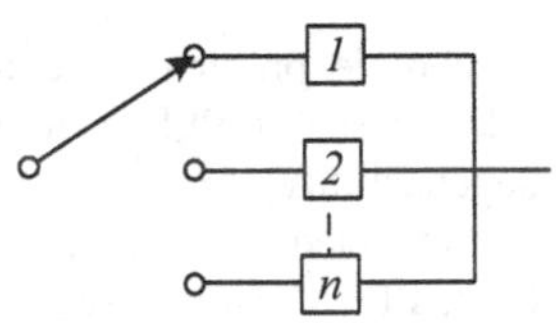

Fig. 1. Backup Systems

2.1 Backup Systems

As shown in Fig. 1, a backup system contains $n(n \geq 2)$ sub-systems. As long as one sub-system provides service, the entire system will work normally. When the system is under running, only one sub-system stays in operation, and the others are sleep. If a sub-system in operation fails, another sub-system will be started immediately. Only all sub-systems break down, will the whole system break down. A backup system can be modeled as follows:

$$Switch = sw_P.sw_V.Switch$$
$$Sub = (SW_P.serv_1.fail_1.SW_V||...||SW_P.serv_n.fail_n.SW_V)$$
$$System = Switch||_{\{SW_P,SW_V\}}Sub$$

Where, $Switch$ contains two temporal events $-sw_P$ and sw_v, with the meaning similar to P event and V event in the two-value mutual exclusion semaphore proposed by E.W. Dijkstra. In Sub, there are n sub-systems, and no synchronization on sw_P and sw_v is required. However, synchronization on sw_P and sw_v is needed in $System$. Here, synchronization ensures that only one sub-system can be switched.

2.2 Reparable Systems

The aforementioned model is not irreparable. In practice, we often have to repair systems from failures. Repairable systems can be modeled as $System = serv.fail.repair.System$

That is, the system will first offer services, and then be repaired if fails. The repaired system is still able to offer services. There are two types of repairable systems: serial repair and parallel repair. Compared with typical serial system, serial repair system is added with the recovery function. Assuming that there are $n(n \geq 1)$ serial components in a serial system, $repair_i$ is used to denote the recovery of the $i^{th}(1 \leq i \leq n)$ component, and $Sub_i = serv_i.fail_i.repair_i.Sub_i$ to denote the ith component, the serial repair system can be modeled as follows:

$$System = f(Sub_1)||_T...||_T f(Sub_n)$$

Where, f is defined as follows:

$$f(x) = \begin{cases} T \ if \ x \in \{serv_1, ..., serv_n\} \\ x \qquad\qquad\qquad otheriese \end{cases}$$

Compared with typical parallel system, parallel repair system is added with the recovery function, and its model is similar.

3 Quantifying the Dependability

Generally, dependability rests on availability, reliability, performability, maintainability and survivability, ect. In this paper, we only take availability as an example to quantify the dependability. Availability refers to the ability of systems' providing services. Formally, availability can be modeled to a five-tuple $(Sys, Res, Env, Event, Serv, Fail)$, where:

(1) Sys represents a system, and is comprised by a groups of sub-systems ;
(2) Res is a set of resources;
(3) Env is a set of outside entities (except for system and resources), such as geographic location, temperature, humidity, voltage, and field intensity;
(4) $Event$ is a set of events related with *Sys*,*Res* and *Env*;
(5) $Serv \subseteq Event$ is a set of services provided by a system or a sub-system;
(6) $Fail \subseteq Event$ denotes system failure.

Generally, events related with availability include: $consumption$ and $update$ of resources, $normal$, $deterioration$ and $restoring$ of environment, $service$, $failure$ and $restoration$ of a system. So, $Event = \{comsume, update, normal, deteriorate, restore, serv, repair, Failure\}$. Assumed that: (1) resources will be updated when after being consumed, and will be consumed after being updated, circulating again and again in this way; (2) the environment is under normal state when the system starts to run, and after a period, the environment deteriorates. Through restoration, the environment returns back to normal; (3) the system is able to offer services if both environment and resources are normal. When system fails, after reparation, it can still offer services normally. So, Sys, Res and Env can be modeled using PEPA as follows:

$res = (comsume, rc).(update, ru).res$

$env = (normal, en).(deteriorate, ed).(restore, er).env$

$Sys = (normal, \infty).(comsume, \infty).(serv, ss).(failure, sf).(repair, sr).Sys$

In order to compute availability, we'll need to model the running of a system. Given Sys, Res and Env, a running of the system is defined to be: $Run = (res||env)||_{\{comsume,normal\}}sys.$

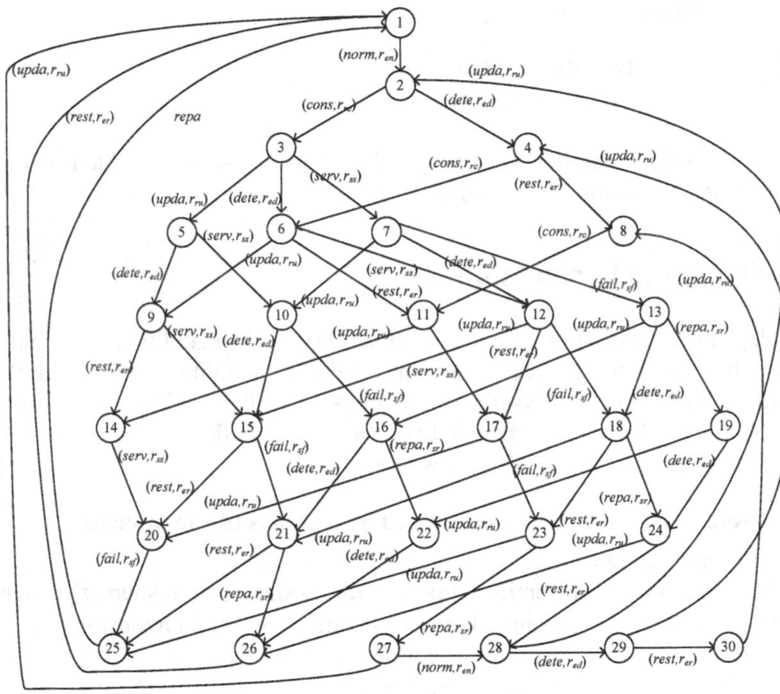

Fig. 2. The derived graph of Run

The above definition requires that Sys and $res||env$ are synchronous on the event $consume$ and $normal$. Its intuitive meaning is that: only when resources are available and the environment is normal, can the system offers services properly. If we assume that the duration time of all events complies with the exponential distribution, then we may make use of PEPA to quantitatively measure availability. Next, some related notions of PEPA are given:

If $P \overset{(ev_1,r_1)}{\longrightarrow} P'$, then, P' is called a one-step derivation of P. If $P \overset{(ev_1,r_1)}{\longrightarrow} \ldots \overset{(ev_n,r_n)}{\longrightarrow} P'$, then, P' is called a derivation of P. Given a component C of $PEPA$, the derivation set $ds(C)$ of C is a minimum set satisfying the following conditions:

(1) If $C \stackrel{def}{=} C_0$, then $C_0 \in ds(C)$;

(2) If $C_i \in ds(C)$ and there exists an event ev such that $C_i \xrightarrow{(ev,r)} C_j$, i.e. $C_j \in ds(C)$.

Given component C of $PEPA$, the derivation graph $D(C)$ of C is a directed graph with label, where, the node set of $D(C)$ is $ds(C)$, the edge set of $\mathcal{D}(C)$ is defined as follows: for any $C_i, C_j \in ds(C)$, $\langle C_i, C_j, (ev, r)\rangle$ is a labeled edge of $\mathcal{D}(C)$, iff $C_i \xrightarrow{(ev,r)} C_j$.

Given component C of $PEPA$, its derivation set $X = ds(C)$ and its event set $Event$, the event source function $evSource : Event \rightarrow 2^X$ is defined as follows: $evSource(ev) = \{x | x \xrightarrow{(ev,r)} y\}$, i.e., $evSource(ev)$ is a set of components, which can be directly derived from event ev. The direct derivation function $evDer : X \rightarrow 2^{Event}$ of a component is defined to be $evDer(x) = \{ev \in Event | x \xrightarrow{(ev,r)} y \wedge y \in X\}$, i.e. $evDer(x)$ is a set of events that the component can directly derived from x via any ev. Now, the availability of a system shall be:

$$Availability = \sum_{C_i \in evSource(serv)} (\Pi(C_i) \times \frac{1/r_{serv}}{\sum_{ev \in evDer(C_i)} 1/r_{ev}})$$

where, $\Pi(C_i)$ is the stable probability of component C_i, and r_x is the speed rate of event x. Next, we discuss how to compute availability.

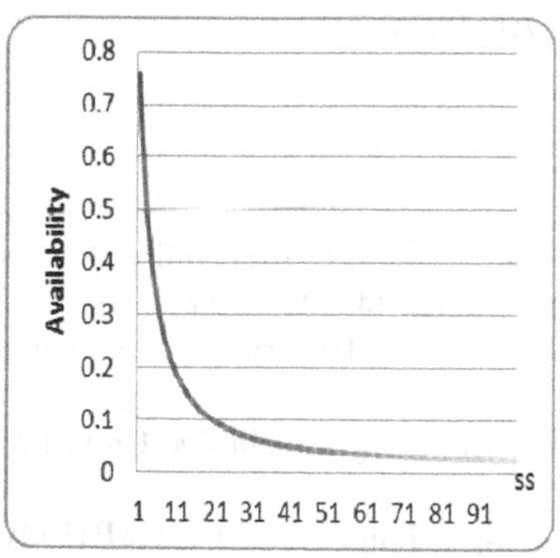

Fig. 3. Availablity changes with ss (whre, rc=10、ru=100、en=10、ed=50、er=30、sf=50 and sr=20)

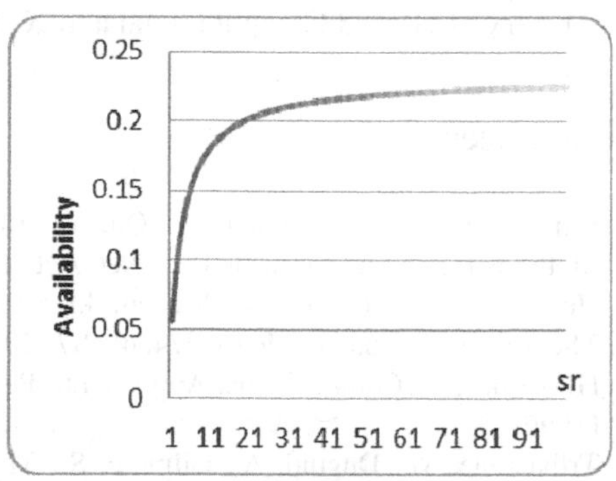

Fig. 4. Availablity changes with sr (whre, rc=10、ru=100、en=10、ed=50、er=30、sf=50 and ss=10)

Fig. 2 shows the derivation graph of Run, where 30 components are derived. For simplification, we only briefly introduce component ① and ②. Component ① and component ② refers to $(res||env)||_{\{comsume,normal\}}sys$ and

$(res||(dete.rest.env))||_{\{comsume,normal\}}(comsume.serv.failure.repair.Sys)$, respectively. The edge between ① and ② means that node ① will be transferred to ② at the rate of r_{en} if the environment is normal. Assume rc=10, ru=100, en=10, ed=50, er=30 and sf=50. Fig. 4 and Figure 5give the PEPA-based result. The result shows that the availability will decrease when ss changes from 1 to 100. The reason is that the growth of ss has reduced the average residence time of service (the average residence time of service state is 1/ss). Similarly, the availability increases with the increase of sr. The reason is that the growth of sr signifies the decrease of average time of reparation. From this it can be seen that, PEPA is adopted to analyze the availability of systems.

5 Conclusion

Computer network has penetrated into almost all fields. Consequently, dependability is becoming increasingly important. It is very necessary to model and quantify dependability. In this paper, PEPA are used to model and quantify the dependability of systems. First, backup systems and reparable systems are formally modeled. Then dependability is quantified using PEPA. The result indicates that PEPA can be used to effectively quantify the dependability of systems. In PEPA, the execution time of events complies with exponential distribution. However, this is not case in some environments. It is worthy to study how to quantify dependability when the execution time of events does not complies with exponential distribution.

Acknowledgements. This work is partially supported by National Natural Science Foundation of China (61070186, 61100186, 61063002, 60903079), Open Foundation of Guangxi Key Lab of Trusted Software, and Open Foundation of Guangxi Key Laboratory of Hybrid Computational and IC Design Analysis.

References

1. Lin, C., Wang, Y.Z., Yang, Y., Qu, Y.: Research on Network Dependability Analysis Methods Based on Stochastic Petri Net. Acta Electronica Sinica 34, 322–332 (2006)
2. Hermanns, H., Herzog, U., Katoen, J.P.: Process Algebra for Performance Evaluation. Theoretical Computer Science 274, 43–87 (2002)
3. Hillston, J.: A Compositional Approach to Performance Modelling. Cambridge Univ. Press (1996)
4. Tribastone, M., Duguid, A., Gilmore, S.: The Pepa Eclipse Plugin. ACM SIGMETRICS Performance Evaluation Review 36(4), 28–33 (2009)
5. Cao, J.H., Cheng, K.: Introduction to Reliability Mathematics. Higher Education Press (2006)

A Improved Industry Data Mining Approach Using Rough-Set Theory and Time Series Analysis

Yang XiaoHua[1] and Lin Na[2]

[1] Shangqiu Medical College
[2] HeNan Industry Commerce Vocational College

Abstract. With rough set and software tools-ROSETTA, the possibility of adopting series timing analysis in the process industry has been able to show in this article. There is some shortcoming for traditional time series analysis adopted in data mining, such as the window method has the limitation on the length of time series which can be analyzed. Because of limitations mentioned above, the direct apply of traditional methods in industrial processes data mining has become unsuitable. The method introduced in this paper adopted a method similar to which operator monitor the actual production process, so which it overcome the limitations of traditional methods.

Keywords: rough set, timing analysis, data mining.

1 Introduction

Rough set theory was first proposed by Zdzislaw Pawlak[1]. The main purpose is to deal with the uncertainties in analysis, calculation and decision-making process. The most significant difference from other theories dealing with the issues of uncertainty such as probability methods, fuzzy set methods and Evidence theory is that it not be required to provide any prior information other than the necessary data collection to address the issue. To solve the problem to a certain acceptable extent, the method that it adopted used the rules the given issue included, even implicit briefly and concisely.

Therefore, in this paper we adopted an improved method for data mining in industry data - based on the time series analysis of the rough set. The main purpose of this article is to discuss possible ways to use the rough set data theory in data mining of industry data.

2 The Rough Set Theory [1]

The basis of Rough set theory is to define the region of a given problem into a number of approximation regions, such as upper approximation and lower approximation or positive region, boundary and negative region.

Definition 1. [Decision system] $S=(U,A,V,f)$ is a knowledge representation system, in which U is a Non-empty finite set of objects, known as the universe; A is a Non-empty finite set of attributes, known as the collection of attributes; V- the range

D. Jin and S. Jin (Eds.): Advances in FCCS, Vol. 1, AISC 159, pp. 459–464.
springerlink.com

of property aϒA, that is, $V=\bigcup_{a\in A}V_a$, V_a is the range of property a: $f:(U,A)\rightarrow V$ is a function of information, given the information value for each attributes of each object, that is, $\forall a\in A, x\in U, f(x,a)\in V_a$ knowledge representation system is also known as information systems.

If A is composed of conditions set C and attributes set D , and CϒD=A, C∩D=φ, S is the decision system.

In a decision system, each element of U is corresponding to a rule, pre-condition is depend on C and its value and post-condition is depend on D and its value. And if V_a and $U\rightarrow V_a$ will not cause confusions, we can also use (U, A) to represent the decision system. For simplicity and facility in analysis , we often use $(U,C\cup\{d\})$ to represent the decision system, that is, the attribute set of the conclusion only contain one element.

Definition 2. [indiscernibility relation] To decision system $S=(U,C\cup\{d\})$, B is a subset of condition attribute set, that is $B\subseteq C$, we said the binary relation $IND(B,\{d\})=\{(x,y)\in U\times U: d(x)=d(y)$ or $\forall a\in B, a(x)=a(y)\}$ is a indiscernibility relation of S, in with x and y is the elements of U.

According to indiscernibility relation $IND(B,\{d\})$, we can divide U into X_1, X_2, …, X_m, X_j(j=1,2,…,m) is indeterministic class.

Definition 3. [upper approximation and lower approximation] Suppose indiscernibility relation R divide U into X_1, X_2, …, X_m, and Y is any subset of U that is Y⊂U , we define the following subset of U the upper approximation and lower approximation of Y under indiscernibility relation R .

$$\underline{R}Y=\{Xj\in U\,|\,Xj\subseteq Y\}\,\overline{R}Y=\{Xj\in U\,|\,Xj\cap Y\neq\varnothing\}$$

Definition 4. [positive region, boundary, negative region]

positive region $POS_R(Y)=\underline{R}Y$;

negative region $NEG_R(Y)=U-\overline{R}Y$

boundary $BN_R(Y)=\overline{R}Y-\underline{R}Y$

The simple explain of these approximate domain is as follow:

$\underline{R}Y$ or POS_R(Y) is a set of elements in universe U which include in Y according to present information (indiscernibility relation R), BN_R (Y) is a set of elements, which may or may not be included in Y according to present information. And $\overline{R}Y$ is a set of elements which may be include in Y (include those elements which include in Y, that is $\underline{R}Y$).

Knowledge reduction is one of the core theories of rough set. Its definition is as follow:

For a given decision system $S=(U,C\cup\{d\})$, the reduction of condition attribute set C is a non-empty subset of C, which we name it C' ,and C' meet the following condition:

$$IND(C',\{d\})=IND(C,\{d\})$$

And not exist $C''\subset C'$ and then $IND(C'',\{d\})=IND(C,\{d\})$

The set of all reduction set of C is RED(C)

Reduction can be interpreted as follow: present the conclusion attribute dependence on and relationship to condition attribute in the simplest way without lose of information. Thus by the calculation of reduction, we can get those simple general rules.

There are a variety of algorithms for knowledge reduction, which we can refer to literature [5] and [6].

3 Problems in Applying Rough Set Theory to Industry Data Mining

Despite the rough set theory suit for dealing with complex systems, and also don't need to address any priori knowledge other than the data collection about the issue which we concern, the direct use of rough set in industry data mining is not appropriate because the main application of rough set theory is the data table in relational database, which is characterized by information have no relevance to time. But for the industry process, time is a very important attribute. The rule obtained using general approach is not much significant for the guide of industry process; we adopt the improved method so that the rough set theory can be qualified to time-series data mining.

Typically, time series can be expressed in two ways:

1) Time series is expressed in accordance with the sequence of events that have occurred, the incident is defined as e = (E, t), of which E is the type of an event, t is that the incident occurred at a time and is a positive integer which can be showed vividly in Fig. 1:

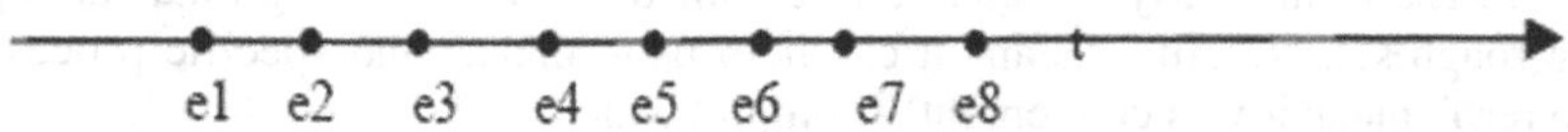

Fig. 1. Time series expression method 1

2) Time series can also be represented as a series of state, and an event can change the status of an object. Fig.2 indicated how event change the status of the object.

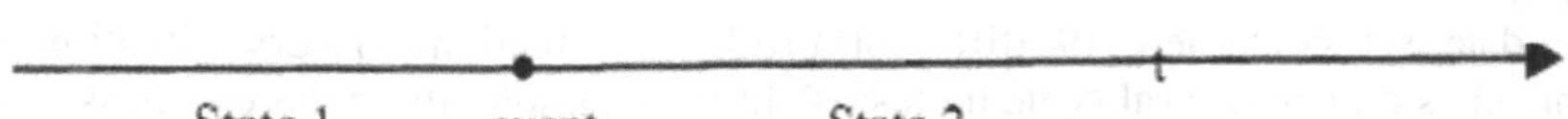

Fig. 2. Time series expression method 2 Event is targeted by the state 1 to the state 2

The following is a simple example of how to describe the time series using the method mentioned above-describe. Envisage describing the atmospheric temperature of Hangzhou in two methods mentioned above; the temperature is measured at the same time every morning. Adopting the first method, the temperature information is representing as a series of temperature (t) which can vividly show in Fig.3:

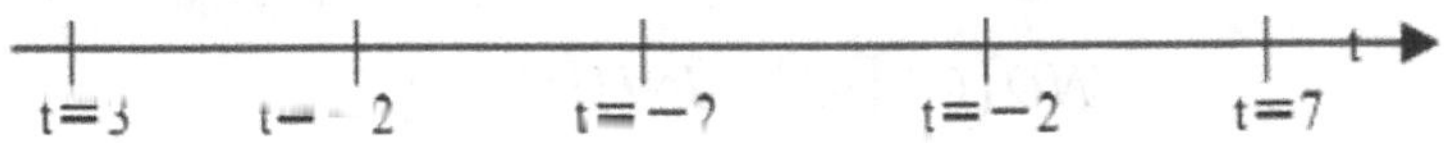

Fig. 3. Describe temperature of Hangzhou in method 1

Adopting the second method, the temperature information is represented as a series of status; Fig. 4 shows how to describe temperature information using two type of status.

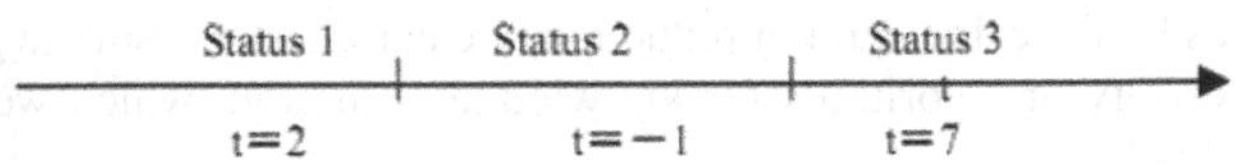

Or

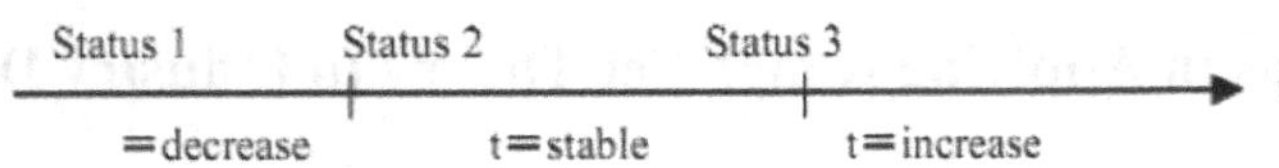

Fig. 4. Describe temperature of Hangzhou in method 2

4 Industry Data Mining Case

The simulation data of Media power station was used in this study of data mining [8]. The process was constituted of water supply system, high-pressure preheater, boiler, turbine steam generator, condenser and Coal-mill. Water which supplied by water supply system and transferred by high pressure preheater converted to high pressure steam in boiler (heated by the coal produced by coal-mill). The high pressure steam produced electrical power with the generator by promoting the turbine. Then the steam was cooled by condenser .thus, one circle was completed.

As mentioned above, there were many methods of time series analysis for the example. But the process industry was more suitably described by state method which indicates the relation of data and time. The purpose of the example was that rulers which were useful and easy to read can be mined from processing data mining by adopting rough set. "Useful" means it can help us to understand specific processing. "Easy to read" mean it was conveniently comprehended.

The original objection of the data set was used to profound a professor system based on case, which helps operators to monitor and operate processing. so the data set contained many possible situations of produce device. And the situations were described by typical variables which were conveniently identified for operators to do appropriate action in Actual use.

The data set contained 49 different produced situations. Every situation was recognized as one or several typical classes. In the example, there were 6 classes such as load variation, the damage valve of preheater, feed pump damage, preheater leak, preheater damage, Coal-mill damage. The table of distribution of 49 situations in 6 classes (as shown table 1):

Table 1. The Classes of Fault Information

Class No.	Class	Number of production situations contained	Same situation in other class
1	load variation	9	6 (class 2)
2	preheater valve damage	7	-
3	Feed pump damage	3	2 (class 1)
4	preheater leak	20	16 (class 2)
5	preheater damage	8	4 (class 2)
6	Coal-mill damage	5	3 (class 1)

Every situation was described by 25 typical process parameters, which select from 157 processing variables, which was similar to actual situation of operators. They not only care change of every process variable, oppositely they care the typical variables which reflect the produced situation. Others would be considered at necessary.

In the example, process parameters are mainly variables tendency. In addition, some state parameters such as valve state are included. Typical variables and classes were shown in table 2.

Table 2. Typical Variable

Parameters class	Typical parameters	Acquisition approach	Variable class
Digital	Electrical power (MW)	Current value	Concrete value
Tendency	Tendency of electrical power	Variable differential	Quickly increase increased invariant decreased quickly decrease
Turbo state	Water supply turbo	State Value	on off
Control valve state	Control valve	State Value	Total on opening above 80% 80% opening below 80% Total off
Safety valve state	Safety valve	State value	Total on opening above 50% opening below 50% Total off

Problems in data set are as follows: for some of the production, one production is attributable to a number of classes at the same time. It is the actual production of the corresponding, because it is common that a number of situations happened at the same time. But if rough data set is used by data mining, every class of production must be unique.

5 Conclusion

It is necessary to mine knowledge from historical data recorded by adopting data mining method for there are a lot of historical data in the chemical production process. Despite experts can derived simple rules and patterns from a large number of data. For more complex models, it would be necessary to adopt data mining technology.

References

1. Zhang, W., Wu, W.: Rough set theory and methods. Science Press (2001 July 2010)
2. Li, Y., Hu, Z.H., Cai, Y.Z., et al.: Feature selection via modified RSBRA for SVM classifiers. In: Proceedings of American Control Conference, vol. 6, pp. 1455–1459. IEEE Press, Portland (2005)
3. Wang, L.W., Zhang, L., Zhan, G.M.: A method of pattern classification based on RS and NCA. In: Proceedings of Machine Learning and Cybernetics, vol. 11, pp. 3090–3094. IEEE Press, Xi'an (2003, 2009)
4. Gregersen, H., Jensen, C.S.: Temporal Entity-Relationship Models–A Survey. IEEE Transactions on Knowledge and Data Engineering 11, 464–497 (1999)
5. Deng, J., Mao, Z., Xu, N.: Attribute classification based on the rough sets theory. Control Theory & Application 25, 591–595 (2008)
6. Deng, J., Mao, Z., Xu, N.: Attribute reduction using attributes partition variation of rough set. Journal of South China University of Technology (Natural Science Edition) 34, 50–54 (2009)
7. Agotnes, T.: Filtering large proportional rule sets while retaining classifier performance. Master's thesis, Norges teknisk-naturvitenskapelige Universitet. (1999)
8. Mannila, H., Toivonen, H.: Discovering generalized episodes using minimal occurrences. In: Second International Conference on Knowledge Discovery and Data Mining, Portland, Oregon (1996)
9. Skourup, C.: Coal-fired power plant: Introduction to the process. Master's thesis, Norwegian University of Science and Technology (1998)

A New Preprocessing Algorithm Used in Color Image Compression

Peng Wu[1], Kai Xie[1,2], Houquan Yu[1,2], Yunping Zheng[3], and Wenmao Yu[1]

[1] School of Electronics and Information, Yangtze University, Jingzhou, China
[2] Key Laboratory of Oil and Gas Resources and Exploration Technology, Yangtze University, Jingzhou, China
[3] School of Computer Science and Engineering, South China University of Technology, Guangzhou, China
peng_wu@189.cn, pami2009@163.com, hq_yu@163.net, zhengyp@scut.edu.cn

Abstract. Still image compression algorithms have been long aiming at gray-scale image compression, in this paper we exploit a new method to improve coding efficiency of color image compression. Firstly, the combination of color space transform and wavelet transform is examined and found out the best one. Then a weighting scheme is proposed to improve objective quality of reconstruction image further. Experimental results show that the proposed scheme not only outperforms JPEG2000 lossy transform scheme and JPEG2000 lossless transform scheme in objective quality by 1–2 db. By conservative estimation, the proposed algorithm outperforms JPEG2000 lossy algorithm from 0.2BPP to 6 BPP and can be used in most practical compression situations.

Keywords: Color Image Compression, Color Space Transform, Wavelet Transform, Color Component Weighting.

1 Introduction

Today, still image compression algorithms have made great progresses and wavelet-based algorithms, such as EZW [1], SPIHT [2], EBCOT [3], have become the mainstream of still image compression algorithms. All these achievements finally result in the breakthrough of JPEG2000 [4] still image compression standard. The color space transform and wavelet transform are crucial parts of JPEG2000, which reduce the correlation between and within components and provide compression potentials.

JPEG2000 Part I includes two kinds of preprocessing algorithms, the lossy and the lossless. The lossy algorithm employs CDF 9/7 wavelet transform and ICT as color space transform, whereas the lossless algorithm uses 5/3 wavelet transform and RCT as color space transform. These two algorithms, as named, are developed in order to make JPEG2000 capable to fulfill the needs of both lossy compression and lossless compression.

In this paper, we have studied these two preprocessing algorithm further and propose a new preprocessing algorithm for lossy compression. The proposed method is based on carefully choosing the combinations of color space transform and wavelet

D. Jin and S. Jin (Eds.): Advances in FCCS, Vol. 1, AISC 159, pp. 465–471.
springerlink.com

transform. The information distribution of color components is exploited to improve objective quality of recovered images. Experiments proved that the proposed algorithm outperforms both the lossy algorithm and the lossless algorithm of JPEG2000 and improves objective quality of reconstruction image. As long as the wavelet coefficients of different components are equally coded, great coding efficiency improvement could be achieved. Since it is a preprocessing algorithm, proposed algorithm can be used in any wavelet based image coding algorithms.

2 The Proposed Algorithm

Firstly, we will give an overview of the proposed algorithm. The part in the dash-box in figure 1 shows the main process of our algorithm.

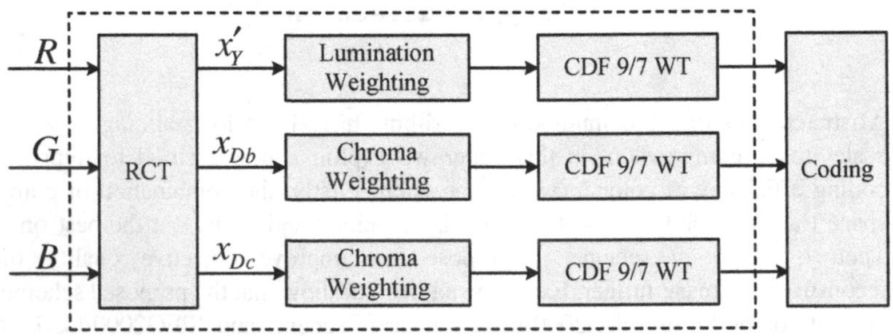

Fig. 1. Diagram of Proposed Algorithm

Our algorithm consists of two parts. The first part is combination of RCT and CDF 9/7 wavelet transform and it outperforms both JPEG2000 lossy combination and JPEG2000 lossless combination in improving objective quality. The second part is weighting the components came from the RCT, which improves objective quality further. In the following part of the paper, we will discuss these parts in detail.

2.1 Selection of Combination of Color Component Transform and Wavelet Transform

The correlation of color components in RGB color space is very large and it is possible to reduce this correlation and improve coding efficiency when transformed to other less correlated color spaces. Researches of years has established ICT(irreversible component transform) and RCT(reversible component transform), defined in JPEG2000 still images compression standard, as the most favorite color space transforms used in compression.

The ICT transforms RGB color space to YC_bC_r color space and as named, it is lossy transform. The ICT has the best compression features among lossy color space transforms. On the other hand, the advantage of RCT is that it precisely reconstructs the original image at the price of a bit lower compressibility features.

The selection of wavelet transforms [5, 6] is also very important. According to JPEG2000, two wavelets transforms, CDF 9/7 wavelet transform and 5/3 wavelet transform, are employed. The first one is better used in lossy compression and the latter one is lossless.

In JPEG2000, lossy transforms are combined together as the lossy preprocessing algorithm and lossless transforms are combined together as lossless algorithm. However, it is very possible that under some range of compression rate combination of lossy transform and lossless transform is able to achieve better coding result. For example, if RCT performs better than ICT when compression rate is above 4 BPP (bit per pixel) and 5/3 wavelet performs better above 2 BPP, then between 2 BPP and 4 BPP combination of ICT and 5/3 wavelet transform could achieve better performance than both lossy scheme and lossless scheme.

In the figure 2, experiment results on Lena and Hell, which are both standard testing images for compression, are showed. Both results show that in a very wide range of compression rate, combination of RCT and CDF 9/7 wavelet transform has the best performances in comparison with other combinations. Experiments on a great number of testing images also come to the same results. The wide range of compression rate where combination of RCT and CDF 9/7 wavelet transform prove that this combination is capable to replace JPEG2000 lossy preprocessing algorithm as usual lossy pre-coding algorithm.

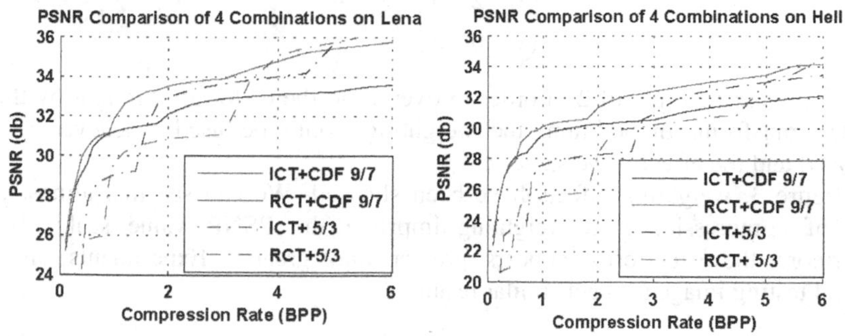

Fig. 2. Coding Efficiency Comparison of Four Different Combinations

2.2 Weighting of Different Component

After color component transform, the standard derivations of each component have been changed which means that information has been partially concentrated to some component. So we should append weights to different color components to distribute bits on different components more reasonably. Since weighting could be considered to be linear transform, we assume the RCT matrix as A and the weighting matrix as:

$$B = \begin{pmatrix} W_{y'} & 0 & 0 \\ 0 & W_{Db} & 0 \\ 0 & 0 & W_{Dr} \end{pmatrix} \tag{1}$$

After RCT, in best situations, the covariance matrix becomes diagonal matrix (because RCT reduce the correlation between components). The aim of weighting is to attribute more bits on the more informative component, which means making none-zero elements of the covariance diagonal matrix become equal. We assume that the covariance matrix Dc1 is:

$$D_{C1} = \begin{pmatrix} d & 0 & 0 \\ 0 & d & 0 \\ 0 & 0 & d \end{pmatrix} \quad (2)$$

Then the covariance matrix of RGB components could be calculated by:

$$D_{RGB} = ((BA)^{-1})^T D_{c1} (BA)^{-1} \quad (3)$$

If we assume the standard derivation of each component of RGB color space to be equal, we can get the following result:

$$W_{Y'} = \sqrt{\frac{48}{11}} W_{Db} = \sqrt{\frac{48}{11}} W_{Dr} \quad (4)$$

Since $\sqrt{48/11} \approx 2$, the actual weighting matrix is established as:

$$\begin{pmatrix} 4 & 0 & 0 \\ 0 & 2 & 0 \\ 0 & 0 & 2 \end{pmatrix} \quad (5)$$

All weights are multiple of 2 in order to overcome the distortion brought by the 5/3 wavelet transform. In addition, the weighting could be used whenever RCT is employed and the effect is the same.

In figure 3, weighting effects have been showed. We can see that almost in all range of compression rate, weighting improve the PSNR value strikingly and furthermore the superiority appears greater and greater. Experiments on other standard testing image also get similar results.

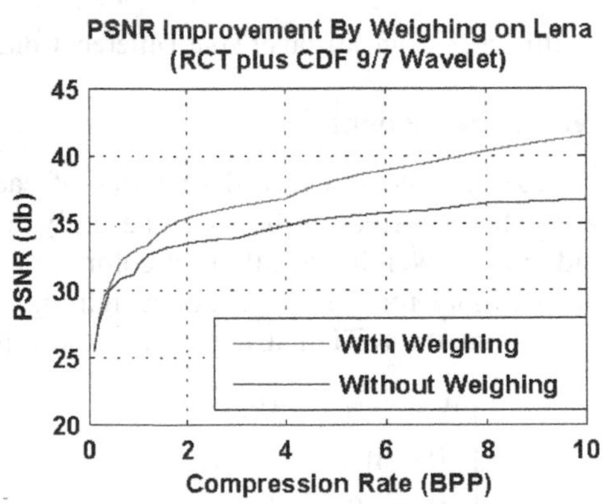

Fig. 3. PSNR Improvement By Weighting

3 Results

In our experiments, SPIHT algorithm is employed as coding algorithm. Since our method is a preprocessing algorithm, any other wavelet-based algorithm will get the similar result.

Objective quality of reconstruction image is measured by PSNR or MSE. The MSE is the mean of the squared differences between the gray-level values of pixels in two pictures or sequences I and I′:

$$MSE = \frac{1}{TXY}\sum_{t}\sum_{x}\sum_{y}\left[I(x,y,t)-I'(x,y,t)\right]^{2} \tag{6}$$

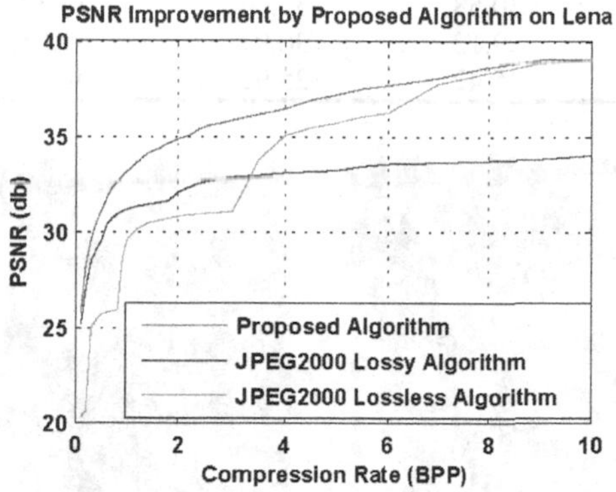

Fig. 4. Objective Quality Evaluation of Different Algorithms

For pictures of size X×Y and T frames in the sequence. The average difference per pixel is thus given by the root mean squared error $RMSE = \sqrt{MSE}$. The PSNR in decibels is defined as:

$$PSNR = 10\log\frac{m^{2}}{MSE} \tag{7}$$

Where m is the maximum value that a pixel can take (e.g. 255 for 8-bit images). Since MSE measures image difference and PSNR measures image fidelity, most traditional compression schemes are aiming at raising PSNR value. The definition of MSE and PSNR could be easily extended to color image compression. In our experiments, PSNR is used as objective quality metric.

We have made experiments on various standard testing images. Firstly, we compute the PSNR value of reconstruction images to evaluate the performance of proposed scheme. In table 1, we could get the conclusion that the proposed algorithm usually outperforms both JPEG2000 lossy scheme and JPEG2000 lossless scheme by 1 to 3 db, which is a great improvement in objective quality of reconstruction images. To give a clearer view about our algorithm, we also compare the PSNR curves (See figure 4) of JPEG2000 lossy algorithm, JPEG2000 lossless algorithm and our algorithm. The superiority of proposed algorithm is obvious.

In addition, the range where proposed algorithm outperforms JPEG2000 lossy algorithm is very wide and even by conservative estimation, it ranges from 0.2 BPP to 6 BPP.

It means that proposed algorithm has better performance than JPEG2000 lossy algorithm in most practical lossy compression situations and proposed algorithm can replace JPEG2000 lossy algorithm as preprocessing algorithm.

In the end of the paper, we will give a comparison of reconstructing image by proposed algorithm and reconstructing image by JPEG2000 lossy algorithm. All of them are compressed in 0.3 BPP.

Table 1. PSNR(db) Improvement by Proposed Algorithm on Lena

BPP	0.4	0.8	1.6	3.0
Proposed Algorithm	30.58	32.64	34.37	35.86
JPEG2000 Lossy	29.12	30.97	31.55	32.89
JPEG2000 Lossless	25.42	25.92	30.62	31.09

Fig. 5. The left image is the original is the reconstruction image by proposed preprocessing algorithm, the right one is the reconstruction image by JPEG2000 lossless preprocessing algorithm.

4 Conclusion

The paper started with a brief introduction of color space transform and wavelet transform, then experiments have been made to select the most outstanding combination of these two transforms. In order to balance the quantity of information in different color component, a weighting matrix is designed which improves objective quality of reconstruction images.

Implemented in the most widely used wavelet coefficients coding algorithm—SPIHT, the experiments have proved that the color image compression algorithm proposed outperform both JPEG2000 lossy algorithm and JPEG2000 lossless algorithm in wide range of compression rate. It is preferable lossy still image compression scheme in most practical situations.

In future, we will work on integrating HVS[7, 8] features, such as masking and so on, into the proposed algorithm and we will also keep on improving objective quality of reconstruction images further.

Acknowledgements. This work has been partially supported by CNPC Innovation Foundation (2010D-5006-0304), Specialized Research Fund for the Doctoral Program of Higher Education of China (20070532077), Natural Science Foundation of Hubei Province of China (2009CDB308), Educational Fund of Hubei Province of China (Q20091211, B20111307).

References

1. Dehkordi, V.R., Daou, H., Labeau, F.: A Channel Differential EZW Coding Scheme for EEG Data Compression. IEEE Transactions on Information Technology in Biomedicine 15, 831–838 (2011)
2. Nikola, S., Sonja, G., Mislav, G.: Modified SPIHT Algorithm for Wavelet Packet Image Coding. Real-Time Imaging 11, 378–388 (2005)
3. Jizheng, X., Zixiang, X., Shipeng, L., Ya-Qin, Z.: Three-Dimensional Embedded Subband Coding with Optimized Truncation (3-D ESCOT). Applied and Computational Harmonic Analysis 10, 290–315 (2001)
4. Skodras, A.N., Christopoulos, C.A., Ebrahimi, T.: JPEG2000: The Upcoming Still Image Compression Standard. Pattern Recognition Letters 22, 1337–1345 (2001)
5. Kaiqi, H., Zhenyang, W., George, S.K.F., Francis, H.Y.C.: Color Image Denoising with Wavelet Thresholding Based on Human Visual System Model 20, 115–127 (2005)
6. Nadenau, M.J., Reichel, J., Kunt, M.: Wavelet-based Color Image Compression: Exploiting the Contrast Sensitivity Function. IEEE Transactions on Image Processing 12, 58–70 (2001)
7. Kai, X., Jie, Y., Yuemin, Z., Xiao-Liang, L.: HVS-based Medical Image Compression. European Journal of Radiology 55, 139–145 (2005)
8. Mullen, K.: The Contrast Sensitivity of Human Color Vision to Red-green and Blue-yellow Chromatic Gratings. J. Physiol. 359, 381–400 (1985)

Performance Simulation of Threaded Type Evaporator

Huifan Zheng and Yaohua Liang

Zhongyuan University of Technology,
450007, Zhengzhou, China
zhenghuifan@163.com

Abstract. A steady distributed parameter method model for predicting the threaded evaporator operation features has been established based on the ejector refrigeration system used R134a as refrigerant. Based on the model to calculating the threaded evaporator best length, the influence of the length on the system capacity and COP are analyzed. The research for the optimization design about heat exchanger provides the theoretical support.

Keywords: Threaded evaporator, HFC134a, length, computer simulation.

1 Introduction

With the growing energy crisis, solar ejector refrigerant system cause the attention of people increasingly, in recent years, many scholars domestic and abroad committed to its research[1~3]. The evaporator is one of the core components, further details about its performance on the basis for structure optimization, is the key measure for improving the solar ejector refrigerant system.

2 Evaporator Model

In this study, evaporator model adopts the steady distributed parameter method, and a computer code is made to calculate the performance of evaporator[3-5]. The following assumption is made:

(1) The flow including refrigerant and water inside the evaporator are steady and one dimension.
(2) Refrigerants and water are all countercurrent.
(3) The radial temperature of wall is same, and the physical property of refrigerant, water, wall is constant.
(4) Two phase area uses homogeneous model.
(5) The influence of gravity can be ignored.
(6) The refrigerant flow of evaporator is even.
(7) The pressure drop of overheated zone is neglected.

Based on the above assumptions, the heat equation of water side can be described as follow literature 7:

$$Q_{e,w} = m_{e,w} c_p (T_{e,win} - T_{e,wout}) = \alpha_{e,w} A_o (T_{e,wa} - T_{wall}) = m_{e,w} (h_{e,win} - h_{e,wout}) \qquad (1)$$

D. Jin and S. Jin (Eds.): Advances in FCCS, Vol. 1, AISC 159, pp. 473–476.
springerlink.com

The heat transfer equation about refrigerant:

$$Q_e = \alpha_e A_i (T_{wall} - T_{e,a}) = m_e (h_1 - h_4) \quad (2)$$

The equation about the heat capacity:

$$Q_{e,w} = \gamma Q_e \quad (3)$$

Where, $Q_{e,w}$, Q_e represent the heat transfer of water side and cooling capacity), kW; $T_{e,win}$, $T_{e,wout}$, $T_{e,wa}$, T_{wall}, $T_{e,a}$ represent import water temperature, the outlet water temperature, the average water temperature, the tube wall temperature and refrigerant average temperature respectively[7], K;

$\alpha_{e,w}$, α_e is the heat transfer coefficient of water side and the refrigerant side, $W/m^2 \bullet k$;

$m_{e,w}$, m_e is the quality of water side and the refrigerant side, kg/s;

c_p express specific heat capacity of water, J/(kg•K) ; A represents the heat transfer area, m^2, and the subscript o represents water side and i represents refrigerants side, m^2;

h_1, h_4, $h_{e,win}$, $h_{e,wout}$ represent the import enthalpy of refrigerant, and the export enthalpy of refrigerant, the import enthalpy of water, and the export enthalpy of water, kJ/kg;

Leakage heat coefficient γ according to the determination of the thermal equilibrium, the value is between 0.92 and 1.0;

The formula of total heat transfer coefficient(Based on heat transfer area of refrigerant):

$$\frac{1}{K} = \frac{1}{\alpha_e} + \frac{d_i}{2\lambda_p} \ln \frac{d_o}{d_i} + \left(\frac{1}{\alpha_{e,w}} + R\right)\frac{d_i}{d_o} \quad (4)$$

where, d_o, d_i represents the outside diameter and inner diameter of spread copper tube, mm;

R represents the additional heat resistance, and its value equals 0.000086, $m^2 \bullet K/W$;

λ_p is the thermal conductivity of copper tube[7], W/m•K.

The calculation formula of length L is:

$$L = \frac{Q_e}{\pi d_i \alpha_e (T_{wall} - T_{e,a})} = \frac{\gamma Q_e}{\pi d_o \alpha_{e,w} (T_{e,wa} - T_{wall})} \quad (5)$$

Choose the Gungor-Winterton as the correlation of boiling heat transfer coefficient calculation in the evaporator. This study only considers two phase area pressure drop, and some related calculation formula are followed[2-6]:

$$\alpha_e = E\alpha_{cv} + S\alpha_{nb} \quad (6)$$

$$\alpha_{cv} = E_{RB}\alpha_{DB} = \alpha_{DB}\left\{1 + \left[2.64\,\mathrm{Re}_l^{0.036}\,\mathrm{Pr}_l^{-0.024}\left(\frac{h}{d_i}\right)^{0.212}\left(\frac{p}{d_i}\right)^{-0.21}\left(\frac{\gamma}{90}\right)^{-0.21}\right]^7\right\}^{1/7} \quad (7)$$

$$E = 1 + 24000Bo^{1.16} + 1.37X_{tt}^{-0.86} \tag{8}$$

$$S = (1 + 0.00000115E^2 \mathrm{Re}_l^{1.17})^{-1} \tag{9}$$

Where, the meaning of some symbol can reference literature 9. The heat transfer coefficient calculation of the water side use Dittus-Boeler correlation:

$$Nu = 0.023\mathrm{Re}^{0.8}\,\mathrm{Pr}^{0.4} \tag{10}$$

On the basis of the above analysis, the evaporator simulation program is designed, its chart flow can reference literature 7 and the input parameters including evaporator program structure parameters, refrigerant inlet temperature and cnthalpy, refrigerant flow, frozen water inlet temperature and chilled water inlet pressure and flow, and the output parameters including the export of refrigerant pressure, temperature, export refrigerant export enthalpy, dry degree, overheat, frozen water export temperature, evaporator heat transfer and refrigerant pressure drop, etc.

3 Result and Discussion

The characteristic of capacity and EER has been researched when the evaporator length is changed. The design cooling capacity is 3 kW, the simulation conditions is followed, the chilled water inlet temperature is 288 K, the mass flow rates of chilled water is 0.27kg/s, the cooling water import temperature is 294K, the mass flow rates of cooling water is 0.69 kg/s, the hot water inlet temperature is 368 K, the mass flow rates of hot water is 0.27kg/s. Using the above calculation program, calculated the performance of ejector refrigerant system when the evaporator length changes and other parameters is constant.

The variation of cooling capacity and COP with evaporator length is shown in figure 2 and figure 3. As can be seen in Fig.2, the cooling capacity of the system increases with the increasing of the evaporator length, and the early show growth faster, when the length of the heat exchanger reach a certain value, the speed of the cooling capacity of the system along with the length of the heat exchanger increases with the increase will slow. The tread of cooling capacity with the change of the evaporator length is almost similar under the condition of different generator and condenser structure size. When evaporator length is between 4.8 and 7.2 m, the system capacity is between 4.35 and 4.65 kW over the range of research.

The COP of the system increases with the increasing of the evaporator length from figure 2, and the early show growth faster, when the length of the heat exchanger reach a certain value, the speed of the COP of the system along with the length of the heat exchanger increases with the increase will slow. The tread of COP with the change of the evaporator length is almost similar under the condition of different generator and condenser structure size. When evaporator length is between 4.8 and 7.2 m, the system COP is between 0.24 and 0.27, and when the length arrive a certain value, even if improving the length of evaporator, the COP of refrigerant system will not increase, that is to say, the length of evaporator has the best value under the condition of design.

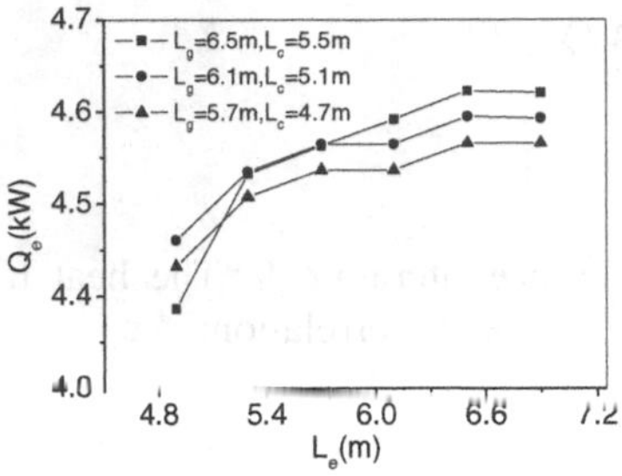

Fig. 2. The effect of length of evaporator on cooling capacity

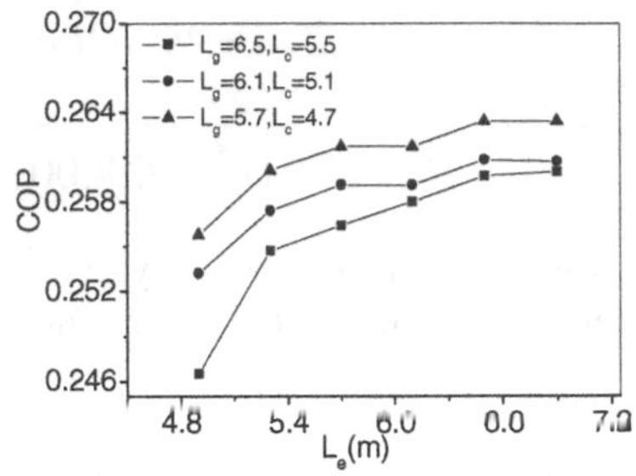

Fig. 3. The effect of length of evaporator on COP

4 Summary

Based on EES software, established the threaded evaporator performance calculation program, the performance of cooling capacity and COP about solar ejector refrigerant system are studied. The results can be summarized as follows:

1. The cooling capacity and COP of the system increases with the increasing of the evaporator length, and the early show growth faster, when the length of the heat exchanger reach a certain value, the increasing speed of the cooling capacity and COP along with the increasing of the heat exchanger length will slow.
2. The length of evaporator heat exchange tube has best value, when it arrives to the best value, even if the heat exchange tube length has been improved, the increasing of cooling capacity and COP will be limited.

References

1. Selvaraju, A., Mani, A.: Analysis of an ejector with environment friendly refrigerants. Applied Thermal Engineering 24(2), 827–838 (2004)
2. Gungor, K.E., Winterton, R.H.S.: A general correlation for flow boiling in tubes and annuli. International Journal of Heat and Mass Transfer 29(3), 351–358 (1986)
3. Kandlikar, S.G.: A general correlation for saturated two-phase flow boiling heat transfer inside horizontal and vertical tubes. ASME Joumal of Heat Transfer 1(112), 219–228 (1990)
4. Yun, R., Kim, Y.: A generalized correlation for evaporation heat transfer of refrigerants in micro-fin tubes. International Journal of Heat and Mass Transfer 45, 2003–2010 (2002)
5. Bandarra Filho, E.P., Saiz Jabardo, J.M.: Convective boiling pressure drop of refrigerant R-134a in horizontal smooth and microfin tubes. International Journal of Refrigeration 27, 895–903 (2004)
6. Zheng, H., Fan, X., Li, A.: Experimental research about solar ejector refrigerant based on the climate condition in the zhengzhou region. Acta Energiae Solaris Sinica 32(8), 1169–1173 (2011)
7. Zheng, H.: The Performance Research of the Solar Ejector/Electric Compression Combined Refrigeration System. Xi'an University of Architecture and Technology (2009)

Iterative Optimization Based Nash Equilibrium for LTE-Femtocell Networks

Hao Chen[1,2], Tong Yang[3], Jianfu Teng[1], and Hong He[4]

[1] School of Electronic and Information Engineering, Tianjin University, Tianjin, 300072, China
[2] Computer Science and Information Engineering College, Tianjin University of Science & Technology, Tianjin, 300222
[3] Tianjin Mobile Communications Co., Ltd., Tianjin 300052, China
[4] Tianjin Key Laboratory for Control Theory and Application in Complicated Systems, Tianjin University of Technology, Tianjin 300384, China

Abstract. To increase LTE (long time evolution) networks spectrum utilization and interference mitigation, a LTE system overlaid with femtocells is studied. The two systems disturb one another with interference. The strategies of the systems are their choices of power allocations in their two bands with regard to individual sum power constraints. Based our prior work, we have derived the Bays-Nash equilibrium a optimal SINR (signal-interference-noise-ratio) threshold. In the paper, with the static non-cooperative power control game, we present a novel distributed power control strategy. That method accelerates the convergence of the Nash game algorithm. The power control scheme can be applied to realistic LTE-femtocell networks to enable robust communication against cross-tier interference thereby obtaining a substantial link quality.

Keywords: LTE, femtocell, Nash equilibrium.

1 Introduction

LTE is being standardized by 3GPP to provide multi-megabit bandwidth, more efficient use of the radio network, latency reduction, improved mobility, and potentially lower cost per bit[1]. Wireless operators are in the process of augmenting the macrocell network with supplemental infrastructure such as microcells, distributed antennas and relays. An alternative with lower upfront costs is to improve indoor coverage and capacity using the concept of end-consumer installed femtocells or home base stations [2].Conventional power control work ties in cellular networks and prior work on utility optimization based on game theory. Results in Foschini et al.[3], Zander [4], Grandhi et al.[5] and Bambos et al [6]. provide conditions for SINR feasibility and/or SIR balancing in cellular systems. Associated results on centralized power control are presented [7]. Our prior research has found optimal SINR threshold with incomplete information. In the paper, we study iterative optimization based the static non-cooperative power control game. The aim is to find suitable functions to mitigate interference and maximum the two systems data throughout. In order to arrive at the aim, we apply results based our prior work to propose a expected utility function at the mobile stations.

D. Jin and S. Jin (Eds.): Advances in FCCS, Vol. 1, AISC 159, pp. 477–482.
springerlink.com

2 System Model

The system consists of a single central macrocell B_0 serving a region C, providing a cellular coverage radius R_c . B_i , $1 \le i \le N$ .The LTE macrocell is underlaid with N co-channel femtocells APs. Femtocell users are located on the circumference of a disc of radius R_f centered at their femtocell AP. Orthogonal uplink signaling is assumed in each slot (1 scheduled active user per cell during each signaling slot), where a slot may refer to a time or frequency resource.During a given slot, let $i \in \{1,2.....N\}$ denote the scheduled user connected to its BS B_i . Designate user $i's$ transmitting power to be p_i Watts. Let σ^2 be the variance of AWGN (Additive White Gaussian Noise) at B_i . Definition 1 The received SINR γ_i of user i at B_i is given as

$$\gamma_i = \frac{p_i g_{i,i}}{\sum_{j \ne i} p_j g_{i,j} + \sigma^2} \ge \Gamma_i \tag{1}$$

Where Γ_i represents the SINR threshold for user i at B_i .The term $g_{i,j}$ denotes the channel gain between user j and BS B_i , but it really is interference term for user i at B_i .The term $g_{i,i}$ can also account for post-processing SINR gains .Definition 2 The term $I_i(p_{i-})$ represents the interference value of user j ($j \ne i$) at B_i .In order to accord with the terms of game theory, $i-$ denotes element sets other than i .

$$I_i(p_{i-}) \triangleq \sum_{j \ne i} p_j g_{i,j} + \sigma^2 \quad , \quad \text{and } \gamma_i = \frac{p_i g_{i,i}}{I_i(p_{i-})} \tag{2}$$

We construct utility function $U_i(p_i, \gamma_i)$ as follows:

$$U_i(p_i, \gamma_i) = R(\gamma_i, \Gamma_i^*) + Q(p_i) = R(\frac{g_{i,i} p_i}{I_i(p_{i-})}, \Gamma_i^*) + Q(p_i) \tag{3}$$

The reward function $R(\gamma_i, \Gamma_i^*)$ denotes the payoff to user i as a function of its individual SINR γ_i .When Γ_i^* has been finalized, $R(\gamma_i, \Gamma_i^*)$ is seemed as increase of p_i brings growth of SINR and improvement of communication quality. The penalty function $Q(p_i)$ is related to the interference that femtocell i give to other femtocell BSs. Based on the physical significance, we give two symbolic functions, without set a specifically analytic function. Because the symbolic function is more general in actual situation.

Assumption 1

$$\frac{\partial R}{\partial \gamma_i} > 0 \quad (\text{if } p_i \in [0, p_{Max}]) \;,\quad \frac{\partial^2 R}{\partial \gamma_i^2} < 0 \tag{4}$$

Eq.(4) ensures that there exist a maximum real value in R function, and in the domain of the definitions, R function is a monotonic increasing function.

Assumption 2

$$\frac{\partial Q}{\partial p_i} < 0 \quad (\text{if } p_i \in [0, p_{Max}]) \;,\quad \frac{\partial^2 Q}{\partial p_i^2} < 0 \tag{5}$$

Eq.(5) ensures that the Q function exists a maximal value in real, and in the domain of definitions Q function is monotonic decreasing function.

Using Eq.(3), taking the second-order derivatives of U_i w.r.t p_i, Eq.(6) is derived as:

$$\frac{\partial^2 U_i}{\partial p_i^{\,2}} = \frac{\partial^2 R}{\partial \gamma^2}\left(\frac{g_{i,i}}{I_i(p_{i-})}\right)^2 + \frac{\partial^2 Q}{\partial p_i^2} < 0 \tag{6}$$

So the utility function exists the maximal value in real.

Lemma 1 [8],[9]and [10]. A Nash equilibrium exists in game $G = \{N, p_i, U_i\}$, if for all $i = 1,2,\ldots\ldots N$.

1) p_i is a nonempty, convex and compact subset of some Euclidean space R^N .
2) $U_i(p_i)$ is continuous in p_i and quasi-concave in p_i .

Following Lemma 1, the optimization problems in Eq.(1) have a Nash Equilibrium. The following theorem derives the SINR equilibrium at each femtocell. As is known from assumption 1 an 2, $\frac{\partial R}{\partial \gamma_i} > 0$, $\frac{\partial Q}{\partial p_i} < 0$. R function is a monotonic increasing function and Q function is monotone decreasing function. So the derivative of U_i surely exists a zero value, that is $\frac{\partial U_i}{\partial p_i} = 0$ in the real number field.

$$p^* = \arg MaxU_i = \arg Max\{R(\frac{g_{i,i} p_i}{I_i(p_{i-})}, \Gamma_i^*) + Q(p_i)\} \tag{7}$$

Note that p^* is not surely in the definition field, that is, $p^* \notin [0, P_{Max}]$

$$\frac{\partial U}{\partial p_i} = 0 \Rightarrow \frac{\partial R}{\partial \gamma} \cdot \frac{g_{i,i}}{I_i(p_{i-})} + \frac{\partial Q}{\partial p_i} = 0 \Rightarrow \frac{\partial R}{\partial \gamma} = -\frac{\partial Q}{\partial p_i} \cdot \frac{I_i(p_{i-})}{g_{i,i}} \tag{8}$$

$$\gamma^* = R^{-1}(-\frac{\partial Q}{\partial p_i} \cdot \frac{I_i(p_{i-})}{g_{i,i}}) \tag{9}$$

Using Eq. (9) and Eq. (3), Eq.(10) can be derived as:

$$\gamma_i^* = \frac{p_i^* g_{i,i}}{I_i(p_{i-})} \Rightarrow p_i^* = \frac{I_i(p_{i-})}{g_{i,i}} \cdot \gamma_i^* = \frac{I_i(p_{i-})}{g_{i,i}} \cdot R^{-1}(-\frac{\partial Q}{\partial p_i} \cdot \frac{I_i(p_{i-})}{g_{i,i}}) \tag{10}$$

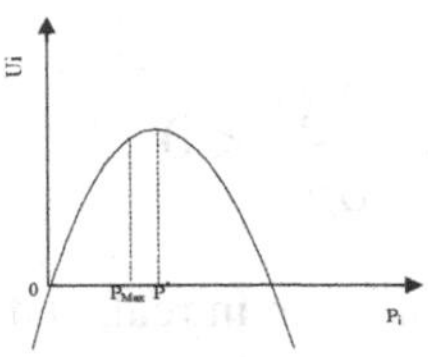

Fig. 1. $p_i^* \notin [0, p_{Max}]$

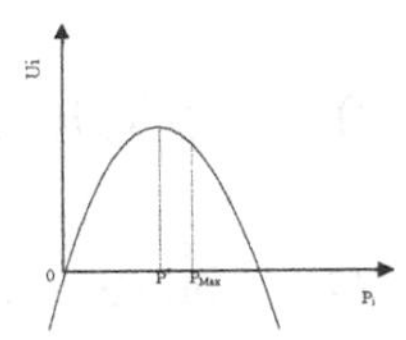

Fig. 2. $p_i^* \in [0, p_{Max}]$

From the Fig.(1)and Fig.(2) , it can be directly seen that if $p_i^* \in [0, p_{Max}]$, the maximum value of utility function exists in the definition field, otherwise has been obtained in p_{Max}. Integrating two kinds of circumstances, Nash equilibrium solution is defined as follows:

$$p_i^{**} = \min(p_i^*, p_{Max}) \Rightarrow p_i^{**} = \min\{\frac{I_i(p_{i-})}{g_{i,i}} \cdot R^{-1}(-\frac{\partial Q}{\partial p_i} \cdot \frac{I_i(p_{i-})}{g_{i,i}}), p_{Max}\} \tag{11}$$

The term p_i^{**} denotes Nash equilibrium solution.

Lemma 2. Yates[11] has proved : If the following three conditions are satisfied , power control can be solved by iterative method .

(1) $T(x) > 0$ (2) if $x_1 > x_2$, then $T(x_1) > T(x_2)$

(3) $\forall a > 1$, $aT(x_1) > T(ax_1)$

then the power control iteration converges to the fixed point(Nash equilibrium point), which is unique. In such a case, $T(x)$ is called a standard interference function. The normal iteration form of power control is given as follows:

$$P^{(k+1)} = T(P^{(k)}) \tag{12}$$

k is the number of iterations.Eq.(11) can be rewritten in Eq.(13) as follows:

$$p_i^{(k+1)} = \min\{\frac{I_i(p_{i-}^{(k)})}{g_{i,i}} \cdot R^{-1}(-\frac{\partial Q}{\partial p_i} \cdot \frac{I_i(p_{i-}^{(k)})}{g_{i,i}}), p_{Max}\} \tag{13}$$

Therefore, the choice of the parameters can entail careful consideration of the trade-offs between the femtocell users' desire to maximize their own data rates and the relative importance of satisfying the LTE cellular users' QoS requirement.

3 Simulation Result and Analysis

In this section, we will simulate in matlab 7.0 environment. We present parameters table, a explicit function and figure results based on two experiments. Simulation result is represented.

Table 1. System Simulation parameters

Variable	Parameter	Value
R	Femtocell Radius	30m
f	Carrier Frequency	2GHz
P_{Max}	Max.Transmission Power femtocell BS	10 W
N	pseudo random cycle number by Monte Carlo method	100
K	Power iterative number	100

Power iterative algorithm is based on some reasonable assumptions and the general formula. In simulation, we will construct an explicit function to adapt the assumptions. The explicit utility function is designed according to Eq.(3) as follows,

$$U_i(p_i)=\log_2(1+\gamma_i)-\beta_i p_i \tag{14}$$

and $\gamma_i = \dfrac{p_i g_{i,i}}{I_i(p_{i-})}$,where β_i is cost factor.

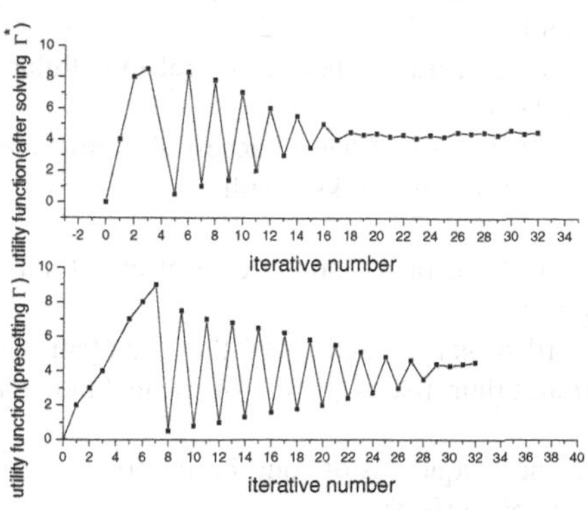

Fig. 3. and **Fig. 4.** Power control by utility function

From experiment results, after obtaining an optimal channel-dependant SINR threshold at each femtocell, power control management will optimize the spectrum utilization and mitigate interference. The power control scheme can be applied to realistic LTE-femtocell networks to enable robust communication against cross-tier interference thereby obtaining a substantial link quality.

4 Conclusion

In this paper, the power control and interference mitigation issues of femtocells in two-tier LTE macro-femto networks is discussed. A novel power control scheme is proposed based on Nash equilibrium to be used into iterative algorithm. The optimal SINR target and optimal transmit power can be derived, which is critical to mitigate interference between neighboring femtocells and improve iterative algorithm efficiency. The simulation results show that, by using the presented scheme, optimal SINR target and optimal transmit power of femtocells can be obtained and sufficient SINR to mitigate interference can be provided. That method can accelerate the convergence of the algorithm.

Acknowledgment. The title selection is mainly originated from Tianjin science and technology innovation special funds project(10FDZDGX00400) and Tianjin Key Laboratory for Control Theory and Application in Complicated Systems, Tianjin University of Technology, Tianjin 300384, China. The name of the project is "the research and development, demonstration and application of new generation mobile communication network coverage key technology".

References

1. Interference Management in UMTS Femtocells, http://www.femtoforum.org
2. Chandrasekhar, V., Andrews, J.G., Gatherer, A.: Femtocell networks: a survey. IEEE Commun. Mag. 46(9), 59–67 (2008)
3. Foschini, G.J., Miljanic, Z.: A simple distributed autonomous power control algorithm and its convergence. IEEE Trans. Veh. Technol. 42(4), 641–646 (1993)
4. Zander, J.: Performance of optimum transmitter power control in cellular radio systems. IEEE Trans. Veh. Technol. 41(1), 57–62 (1992)
5. Grandhi, S.A., Zander, J.: Constrained power control in cellular radio systems. In: Proc. IEEE Veh. Tech. Conf. (1994)
6. Bambos, N., Chen, S.C., Pottie, G.J.: Channel access algorithms with active link protection for wireless communication networks with power control. IEEE/ACM Trans. Networking 8(5), 583–597 (200)
7. Grandhi, S.A., Zander, J.: Constrained power control in cellular radio systems. In: Proc. IEEE Veh. Tech. Conf. (1994)
8. Glicksberg, I.L.: A further generalization of the Kakutani fixed point theorem with application to Nash equilibrium points. Proc. American Mathematical Society 3(1), 170–174 (1952)
9. Rosen, J.B.: Existence and uniqueness of equilibrium points for concave n-person games. Econometrica 33(3), 520–534 (1965)
10. Debreu, G.: A social equilibrium existence theorem. Proc. National Academy of Sciences 38, 886–893 (1952)
11. Yates, R.D.: A framework for uplink power control in cellular radio systems. IEEE J. Select. Areas Commun. 13(7), 1341–1347 (1995)

Handwritten Digit Recognition Based on LS-SVM

Xiaoming Zhao[1] and Shiqing Zhang[2]

[1] Department of Computer Science, Taizhou University,
318000 Taizhou, China
[2] School of Physics and Electronic Engineering, Taizhou University,
318000 Taizhou, China
{tzxyzxm,tzczsq}@163.com

Abstract. In this paper we present a method of handwritten digit recognition based on least squares support vector machines (LS-SVM). Principal component analysis (PCA) is used to extract 10-dimension features from the original digit images for handwritten digit recognition. The performance of LS-SVM on handwritten digit recognition tasks is compared with three typical classification methods, including linear discriminant classifiers (LDC), the nearest neighbor (NN), and the back-propagation neural network (BPNN). The experimental results on the popular MNIST database indicate that LS-SVM obtains the best accuracy of 87.5% with 10-dimension features, outperforming the other used methods.

Keywords: Handwritten digit recognition, Principal component analysis, Least squares support vector machines.

1 Introduction

Handwritten digit recognition is one of the most important problems in computer vision, pattern recognition, etc. There is a great interest in these areas due to its many potential applications, such as optical character recognition, postal mail sorting, bank check processing, form filling, and so on.

Generally, an automatic handwritten digit recognition system consists of two main parts: feature extraction and classification. Various feature extraction methods, such as stroke direction feature [1], the statistical features and the local structural features [2], and celled projection [3], have been developed. Principal component analysis (PCA) [4] is one of the most well-known statistical feature extraction methods. The task of classification is to partition the feature space into regions corresponding to source classes or assign class confidences to each location in the feature space. So far, statistical learning techniques [5], such as linear discriminant classifiers (LDC) and the nearest neighbor (NN), and neural network [6], have been widely used for handwritten digit recognition. Support vector machines (SVM) [7], based on the statistical learning theory of structural risk management and quadratic programming (QP) optimization, has become a popular classification for handwritten digit recognition. In recent years, least squares support vector machine (LS-SVM) [8], as a variant of standard SVM, has been proposed. Like SVM, LS-SVM is based on the

D. Jin and S. Jin (Eds.): Advances in FCCS, Vol. 1, AISC 159, pp. 483–487.
springerlink.com

margin maximization principle that performs the structural risk and it inherits the SVM generalization capacity. Additionally, the training algorithm of LS-SVM is very simplified since a linear problem is resolved instead of the quadratic programming (QP) problem in the SVM case. Motivated by the very limited studies on LS-SVM for handwritten digit recognition, in this work we aim to investigate the potential of LS-SVM on handwritten digit recognition tasks.

2 LS-SVM

LS-SVM [8], as the least squares version of SVM, is readily found with excellent generalization performance and low computational cost, since a set of linear equations can be solved instead of a quadratic programming (QP) problem.

Given a training data points $\{x_i, y_i\}, i = 1,2,\ldots,l$ with input data x_i and output data y_i, in the feature space LS-SVM takes the form

$$y = \operatorname{sgn}\left(w^T \varphi(x) + b\right) \tag{1}$$

where the nonlinear mapping maps the input data into a high-dimensional feature space whose dimension can be infinite. To obtain a classifier, LS-SVM solves the following optimization problem:

$$\min \left\{ \frac{1}{2} w^T w + \frac{\gamma}{2} \sum_{i=1}^{l} e_i^2 \right\} \tag{2}$$

subject to $y_i = w^T \varphi(x_i) + b + e_i$.

Its Wolfe dual problem is

$$\min \left\{ \frac{1}{2} \sum_{i,j=1}^{l} \alpha_i \alpha_j \varphi(x_i)^T \varphi(x_j) + \sum_{i=1}^{l} \frac{\alpha_i^2}{2\gamma} - \sum_{i=1}^{l} \alpha_i y_i \right\} \tag{3}$$

subject to $\sum_{i=1}^{l} a_i = 0$.

We can delete the equality constraint and obtain

$$\min \left\{ \frac{1}{2} \sum_{i,j=1}^{l} \alpha_i \alpha_j \varphi(x_i)^T \varphi(x_j) - \sum_{i=1}^{l} \alpha_i y_i + \sum_{i=1}^{l} \frac{\alpha_i^2}{2\gamma} + b \sum_{i=1}^{l} \alpha_i \right\} \tag{4}$$

The form in (4) is often replaced with a so-called positive-definite kernel function $k(x_i, x_j) = \varphi(x_i)^T \varphi(x_j)$. There are several kernels that can be used in LS-SVM models. These include linear, polynomial, radial basis function (RBF) and sigmoid function. Then we can get the resulting LS-SVM model for function estimation as follows:

$$f(x) = \operatorname{sgn}\left(w^T \varphi(x) + b\right) = \operatorname{sgn}\left(\sum_{i=1}^{l} a_i k(x, x_i) + b \right) \tag{5}$$

For multi-class problem, to construct a set of M binary classifiers, there are mainly four encoding strategies: one-versus-one, one-versus-all, minimal output coding (MOC) and error correcting output codes (ECOC). For the one-versus-one approach, classification is done by a max-wins voting strategy, in which every classifier assigns the instance to one of the two classes, then the vote for the assigned class is increased by one vote, and finally the class with most votes determines the instance classification. For the one-versus-all method, classification is done by a winner-takes-all strategy, in which the classifier with the highest output function assigns the class. Minimum output coding (MOC) uses M outputs to encode up to 2^M classes and thus have minimal M. Error correcting output codes (ECOC) use more bits than MOC, and allow for one or more misclassifications of the binary classifiers.

3 Experiments

3.1 Experiment Setup

To verify the performance of LS-SVM on handwritten digit recognition task, three typical methods, i.e., linear discriminant classifiers (LDC), the nearest neighbor (NN) and the back-propagation neural network (BPNN) as a representative neural network were used to compare with LS-SVM. For BPNN method, the number of the hidden layer nodes is 30. We implement LS-SVM algorithm with the RBF kernel, kernel parameter optimization, one-versus-one strategy for multi-class classification problem. The RBF kernel is used for its better performance compared with other kernels. The kernel parameter of RBF kernel is optimized by using a grid search in the hyper-parameter space.

Fig. 1. Some samples from the MNIST database

The popular MNIST database [9] of handwritten digits, which has been widely used for evaluation of classification and machine learning algorithms, is used for our experiments. The MNIST database of handwritten digits contains a training set of

60000 examples, and a test set of 10000 examples. Each image has 28×28 pixels. In our experiments, for computation simplicity we randomly selected 3000 training samples and 1000 testing samples for handwritten digit recognition. Some samples from the MNIST database are shown in Fig.1.

PCA is applied to economically represent the input digit images by projecting them onto a low-dimensional space constituted by a small number of basis images. These basis images are derived by finding the most significant eigenvectors of the pixel wise covariance matrix, after mean centering the data for each attribute. For simplicity, the feature dimension of the original grey image features (28×28=784) is reduced to 10 as an illustration of evaluating the performance of LS-SVM.

3.2 Experimental Results and Analysis

Table 1 presents the different recognition results of four classification methods including LDC, NN, BPNN and LS-SVM. From the results in Table 1, we can observe that LS-SVM performs best, and achieves the highest accuracy of 87.5%, followed by BPNN, NN and LDC. This demonstrates that LS-SVM has the best generalization ability among all used four classification methods. In addition, the recognition accuracies for BPNN, NN and LDC, are 84.8%, 82.7% and 77.6%, respectively.

Table 1. Handwritten digit recognition results with different methods

Methods	LDC	NN	BPNN	LS-SVM
Accuracy (%)	77.6	82.7	84.8	87.5

Table 2. Confusion matrix of handwritten digit recognition results with LS-SVM

Digits	0	1	2	3	4	5	6	7	8	9
0	87	0	4	0	0	4	0	0	1	0
1	0	111	2	0	0	0	0	0	1	0
2	0	0	79	2	3	0	3	1	1	0
3	0	0	4	106	0	3	0	1	4	3
4	0	0	1	0	75	0	2	1	1	7
5	5	0	1	6	3	83	0	0	0	0
6	4	0	0	0	1	0	80	0	0	0
7	0	2	0	0	0	0	0	103	1	4
8	2	0	2	3	0	3	1	1	70	4
9	0	2	0	0	10	0	0	7	1	81

To further explore the recognition results of different handwritten digits with LS-SVM, the confusion matrix of recognition results with LS-SVM is presented in Table 2. As shown in Table 2, we can see that three digits, i.e., "1", "3" and "7", could be discriminated well, while other digits could be classified poor.

4 Conclusions

In this paper, we present a method of handwritten digit recognition via LS-SVM. The experimental results on the benchmarking MNIST database indicate that LS-SVM achieves the best performance with an accuracy of 87.5% with 10-dimension features, outperforming the other used classification methods, i.e., LDC, NN, and BPNN. It's worth pointing out that in our work we only investigate the performance of all used classification methods, based on the extracted 10-dimension features produced by PCA. Therefore, in our future work, it's an interesting task to further study the performance of other more advanced dimensionality reduction techniques than PCA on handwritten digit recognition tasks.

Acknowledgments. This work is supported by Zhejiang Provincial Natural Science Foundation of China under Grant No.Z1101048 and Grant No. Y1111058.

References

1. Trier, O.D., Jain, A.K., Taxt, T.: Feature extraction methods for character recognition—a survey. Pattern Recognition 29(4), 64–662 (1996)
2. Lauer, F., Suen, C.Y., Bloch, G.: A trainable feature extractor for handwritten digit recognition. Pattern Recognition 40(6), 1816–1824 (2007)
3. Hossain, M.Z., Amin, M.A., Yan, H.: Rapid feature extraction for Bangla handwritten digit recognition. In: International Conference on Machine Learning and Cybernetics (ICMLC), Guilin, China, pp. 1832–1837 (2011)
4. Turk, M.A., Pentland, A.P.: Face recognition using eigenfaces. In: IEEE Conference on Computer Vision and Pattern Recognition (CVPR), Maui, USA, pp. 586–591 (1991)
5. Jain, A.K., Duin, R.P.W., Mao, J.: Statistical pattern recognition: a review. IEEE Transactions on Pattern Analysis and Machine Intelligence 22(1), 4–37 (2000)
6. Kang, M., Palmer-Brown, D.: A modal learning adaptive function neural network applied to handwritten digit recognition. Information Sciences 178(20), 3802–3812 (2008)
7. Vapnik, V.: The nature of statistical learning theory. Springer, New York (2000)
8. Suykens, J.A.K.: Least squares support vector machines. World Scientific Pub. Co. Inc. (2002)
9. LeCun, Y., Jackel, L., Bottou, L., Brunot, A., Cortes, C., Denker, J., Drucker, H., Guyon, I., Muller, U., Sackinger, E.: Comparison of learning algorithms for handwritten digit recognition. In: International Conference on Artificial Neural Networks, Nanterre, France, pp. 53–60 (1995)

The Control System of Twin-Screw Feed Extruder

Fujuan Wang and Feiyun Zhang

College of Electrical & Information Engineering, XuChang University
XuChang, China

Abstract. This paper introduced application of intelligent control system in twin-screw feed extruder. In the control system, S7-200 PLC was used to accomplish data acquisition, signal transmitting and display a monitor interface of control system by touch screen MT506T. The design of hardware and software of the control system were discussed also. The result shows that the control system finished demands of performance. The twin-screw feed extruder realized auto-control with it.

Keywords: control system, twin-screw feed extruder, touch screen, PLC.

1 Introduction

With the development of scale, intensification, specialization of aquiculture, requirements of aquiculture feed become higher. Twin-screw extruding technology is a important part in aquiculture feed production, Existing twin-screw extruder control system has the shortcomings of Immunity being low, control function simple, control System not mature. In order to improve the quality of extruded feed of twin-screw extruder. New type twin-screw extruder control system was studied. The hardware of the control system adopted S7-200 PLC and its expansion modules as the control core. All its input and output interfaces used optically isolated interface board, which avoided the interference signal interference the control system under the industrial field condition. It could adapt the feed industrial hash conditions. It is real-time, safe and reliable. EVIEW MT4500TE touch screen was used to accomplish monitor interface display, it has perfect formula managing system and alert processing mechanism.

2 System Summarize

2.1 Twin-Screw Feed Extruder Equipment

The type of the twin-screw feed extruder equipment is TSE98 manufactured by Beijing Modern Yanggong Machinery S&T development CO., LTD. Its structural representation is shown as in the figure 1.

D. Jin and S. Jin (Eds.): Advances in FCCS, Vol. 1, AISC 159, pp. 489–494.
springerlink.com

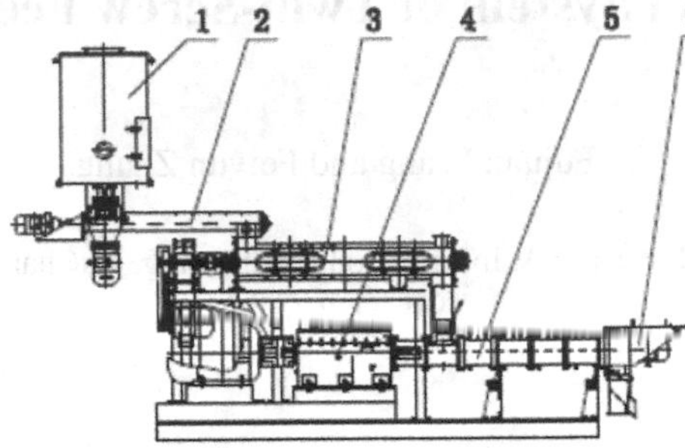

1 flat-bottom bin 2 scroll feeder 3 dual spindle differential speed conditioning tank 4 main generator power transmission system 5 extruder extrusion system 6 cutter

Fig. 1. Twin-screw feed extruder structural representation

2.2 The Basic Requirements and Function of the Control System

The main functions of the control system were to accomplish information acquisition, information procession and sending out control information, so that to accomplish the auto-control of the twin-screw feed extruder. Control process included manual mode and automatic mode. Automatic mode means the parameters, such as motor speed, materiel amount of feed, temperature of the twin-screw extruder and so on, could be auto-controlled according to different feed requirement and control algorithm. Manual control means the extruding parameters could be adjusted on the touch screen artificially to control twin-screw feed extruder and the control system could switch to the manual mode undistributed after the system failure. In addition to basic function, the control system should meet the following demands: good interface to supply simple operation and maintain; strong anti-interference ability, the operation system running reliability to assure machine work normally all the time under the bad conditions of feed production; strong applicability.

3 Hardware Platform of the System

According to the feature and control request of the twin-screw feed extruder, the system hardware is made up of Siemens S7-200 type PLC and EVIEW MT4500TE type touch screen. The hardware diagram of the system is shown as in the figure 2.

The controlled object of the system is the twin-screw extruder, the main measuring devices have temperature sensor, steam flow meter, water flow meter, the actuator of the system are steam valve, inverter, solenoid and electrical machine.

There are amount of control variables in the twin-screw extruder system. Academic study and production experiments showed that the main factors of influencing the feed quality consist of feeding speed of the feeder, adding water and steam speed of the conditioning tank, sleeve temperature, speed of main generator. Then setting up these parameters on the MT4500TE interface, and these variables are monitor by corresponding sensors. For example, the principle of temperature control part is: when the temperature at the outlet of the conditioning tank is higher than the setting up limit value, decrease the opening of the steam valve, that is decrease the amount of adding steam; when the temperature at the outlet of the conditioning tank is lower than the setting lower limit value, increase the opening of the steam valve, that is increase the amount of adding steam. The controller PLC calculates the acquisition data and makes decision in time according to the control algorithm, and forms control command code and output them to control actuator, for accomplishing the auto-control of the twin-screw extruder.

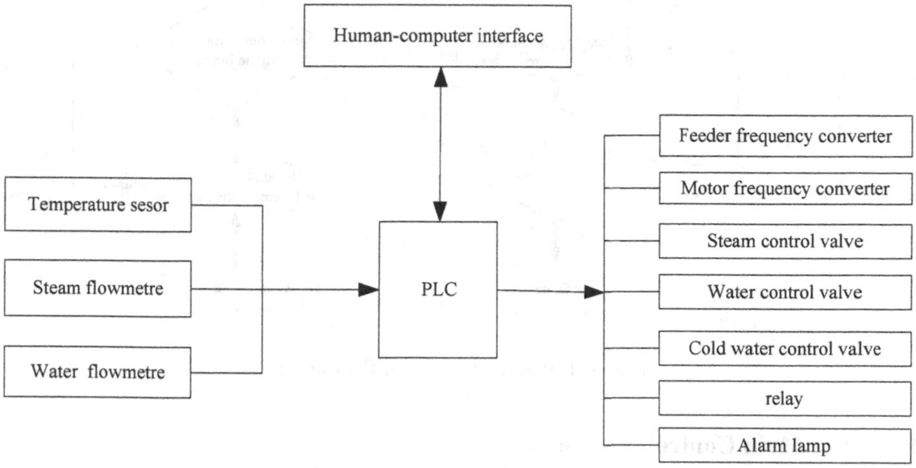

Fig. 2. Hardware diagram of the control system

4 Software Design of the System

The PLC control system hardware platform adopts STEP 7-Micro/WIN as the operating system, and the touch screen type is EVIEW MT4500TE , real time monitoring and test to the working process can be meted through RS-232. The temperature control of the system is showed in figure 3.

4.1 Interface Design

Interface could accomplish human-computer interaction, such as parameters display and parameters input. human-computer interface consist of control interface, real-time curve interface, history curve, alert message interface, data alert interface, parameter input interface, data report interface and help interface.

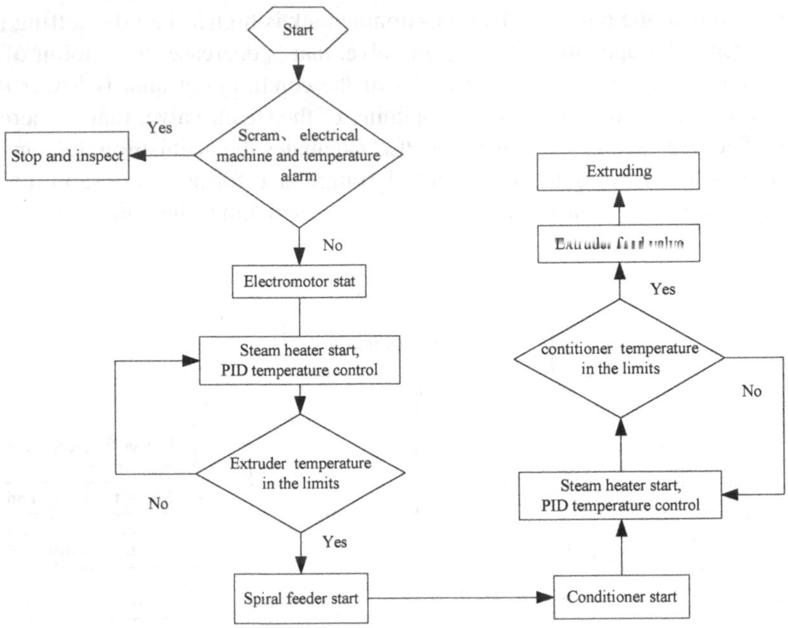

Fig. 3. Temperature control flow chart

4.2 The Main Control Interface

The main control interface is showed as in the figure 4. It displays the whole work frame of the twin-screw extruder system, it dynamic display the running state of multiple parameters of extruder system.

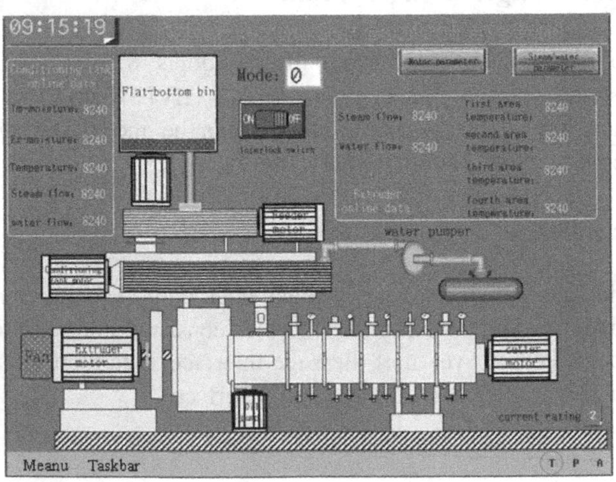

Fig. 4. The main control interface

4.3 Control Menu Interface

The menu of extruding system was shown as in the figure 5, here the extruder parameter configuration could be displayed and the parameters, steam addition, water addition, spindle speed and cutter speed of the conditioning tank could be adjusted according to different formula and different extruding requirement, also the PID parameters and PLC program could be reset. Managing multiple parameters is useful for changing the equipment configuration.

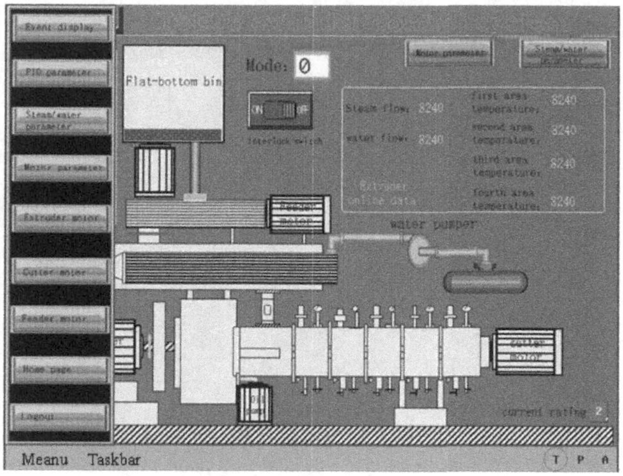

Fig. 5. Control menu interface

4.4 Real-Time Curve and Alert Message Interface

Real-time curve interface displays real-time curve of the temperature, steam addition, water addition, motor speed of the extruding system, through which the dynamic change trend of the parameters could be observed.

Alert function would work when operational error or running fault.

5 Conclusions

Practical application showed that the control system has following feature: the system can run reliably under the bad production conditions, and has high anti-influence; the expanded feed pellet has high dilatation rate, even granule and palatable; during running of the system, the running state of the extruder system could be dynamic analogy displayed, in the same time, the temperature and pressure of the extruder could be displayed. All the curves and data could be printed after stop; The interface is good, it has safety inspect system and alert system. The entry code make non-staff only can look the interface, only the operator could entry into the system and operation is simple.

References

1. Yang, H.: Siemens S7-200PLC application 100 examples. Publishing House of Electronics Industry, Beijing (2009)
2. Ma, C., Lü, J., Wang, H.: Research on Computer Control System of Twin Screw Extrusion Press for Food. Transactions of the CSAE 9(13), 275–278 (1997)
3. Wu, H., Wang, X., Zhang, J.: Study on Speed of Spinning Disc in Model of Spinning Disc Spreader. Journal of Agricultural Mechanization Research 7, 118–119 (2007)
4. Li, H., Hu, G.: Application of PLC and touch screen in the turbo puffing equipment control system. Chinese Agricultural Mechanization 1, 98–101 (2010)

Design and Implementation of Financial Workflow Model Based on the Petri Net

JianBang Chen, Lu Han, DaoYing Xiong, and Jiao Luo

School of Computer and Information Science, Southwest University, Chongqing, China
kfcjb@126.com, hanlu_905@163.com, 419787288@qq.com, 897954496@qq.com

Abstract. In order to ensure the reasonableness, accuracy and reliability of financial workflow nets, this paper analyzes the financial workflow features, and points out the mapping relationship between financial workflow diagram and Petri network. Then, it proposes one modeling method of financial workflow based on the Petri nets. In this paper, the FWF-net model is set up for a financial reimbursement workflow instance and analysis of that shows that the model is correct and reasonable. The model provides a theoretical basis for financial workflow modeling.

Keywords: Petri net, workflow, modeling, financial workflow net.

1 Introduction

Financial workflow is one of the most commonly used workflows in various enterprises and institutions. It is related closely to the economic interests of enterprises and institutions. With the rapid development of enterprises informationization, the diversification of participants and processing mode appears in financial workflow management [1]. The requirements of reliability of financial workflow improve continuously and a modeling method to be represented as a theoretical support is important. Distinguishing strictly the function and execution of activities, the Petri net is the first choice for financial workflow as a modeling tool.

Researchers have made some very meaningful work about financial workflow and Petri nets modeling. Na LiChun and Chen Qingkui [2] came up with financial regulatory data collection model based on concurrent workflow. According to the characteristics of the workflow of medical information system Yan Chungang [3] etc. constructed Petri nets models of medical information system in accord with the standard of IHE and provided theory for medical information system designing and modeling. However, there is still rare research to put financial workflow chart into Petri net model according to the characteristics of financial workflow. After analyzing the characteristics of financial workflow, this paper uses Petri nets modeling method to finance the workflow and comes up with the mapping relationship between the workflow chart and Petri nets. It puts forward the modeling method of the financial workflow net (FWF-net), and constructs the FWF-net model of the financial workflow and analyzes the rationality and validity of the model in the end.

D. Jin and S. Jin (Eds.): Advances in FCCS, Vol. 1, AISC 159, pp. 495–500.
springerlink.com

2 Petri Nets

2.1 Definition of Petri Nets

Definition 1. A Petri net is composed by the four-tuple, namely $PN=(P,T;F,M_0)$, and : (1) Let P be the set of the place node and T be the set of the transition node. (2) Let $P\cap T=\varnothing$, $P\cup T\neq\varnothing$. (3) Let $F\subseteq(P\times T)\cup(T\times P)$ be the set of the directed arcs between the place nodes and transition nodes. (4) Let $M_0:P\rightarrow N$ be the initial marking, where N is the natural number set.

Definition 2. Petri nets are strongly connected graphs if and only if there exits one route from x to y, which are any two nodes in the Petri net.

Definition 3. A Petri net $PN=(P,T,F,M_0)$ modeling is reasonable if and only if [4]: (1) There exits a route from node M, whichever can be reached from the initial node I, to termination node O, namely $\forall_M\left(i\xrightarrow{*}M\right)\Rightarrow\left(M\xrightarrow{*}o\right)$. (2) The termination node O is the only final node that the initial node I reaches, and contains at least one token in the end, namely $\forall_M\left(i\xrightarrow{*}M\wedge M\geq O\right)\Rightarrow(M=O)$. (3) There is no dead transition in PN, namely $\forall_{t\in T}\exists_{M_1,M_2}i\xrightarrow{*}M_1\xrightarrow{t}M_2$.

2.2 Graphic Representation of the Petri Net Basic Structure

Five basic structures in Petri nets are shown in Figure 1: (1) Sequence structure, (2) Parallel split structure, (3) Parallel synchronous structure, (4) Mutex choice structure (5) Simple rendezvous structure.

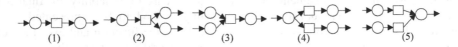

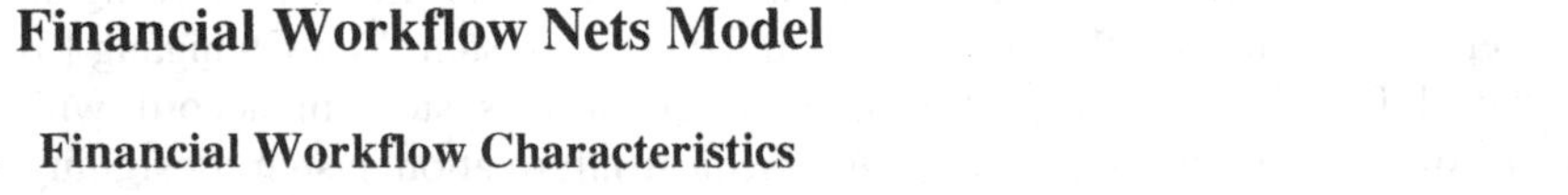

Fig. 1. Five basic structures.

3 Financial Workflow Nets Model

3.1 Financial Workflow Characteristics

As one kind of workflows, financial workflow has its own characteristics: (1) Definiteness. As financial workflow standard is unified, workflow net model have to be unambiguous and definite to ensure that every node is reachable. (2) Flexibleness. Routing in the financial workflow is flexible such as jumping and backing in the financial workflow. For example, there is necessity to back to the workflow node already dealt with or jump some nodes for some reasons. (3) Extensibility. To get more information or ensure the financial work is reliable, the current checker can ask for more related persons in charge of review, namely countersignature. For example, financial executives can ask for one or some persons in charge of countersignature.

3.2 Analysis of Financial Reimbursement Workflow

Here is a financial reimbursement workflow as an example to discuss FWF-net modeling, designing and implementation. In the actual cases, according to the details of the related financial rules, failing to pass audit, requirements of countersigning many times and being rejected and so on would appear. Figure 2 gives workflow chart in the actual case.

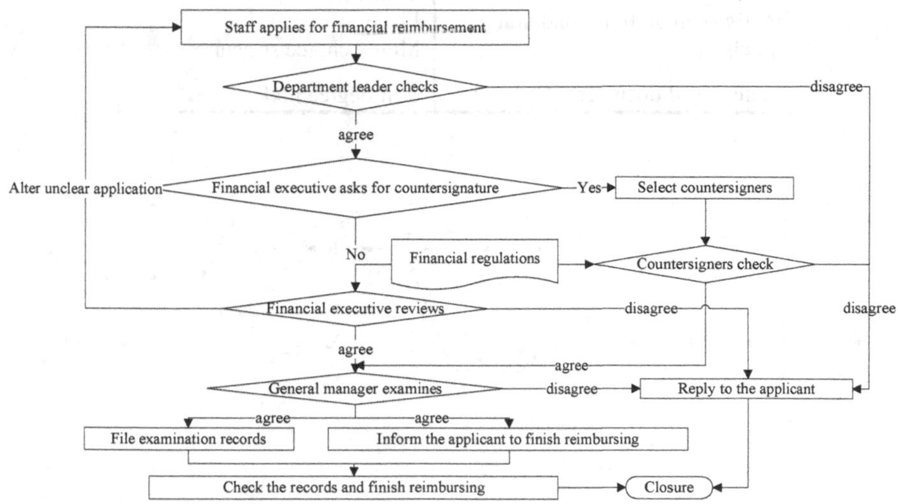

Fig. 2. Detailed financial reimbursement workflow chart.

3.3 Design of FWF-net Model

In the actual financial reimbursement workflow, the application of the staff would be examined by his or her department leader, the financial executive and the general manager in turn. As long as any of them disagrees, the opinions and results will be sent back to the applicant, and the workflow turns to the termination place instead of the following checker, namely the workflow comes to the end. According to the relevant provisions of the financial reimbursement, financial executives can ask for one or more persons in charge of countersigning. If the financial executive chooses somebody to countersign, the workflow turns to the person pointed out by the financial executive. Only when the financial executive and counter-signers both agree, the workflow would turn to the general manager, otherwise the application fails to pass and the result would be sent back to the applicant. If no countersignature, the financial executive checks, and the workflow continues to flow backward: If financial executive agree, it flows to the general manager; otherwise, the result is sent to the applicant. If the financial executive thinks relevant formalities are not complete enough or reasons for application are not clear, it will be returned to the applicant and the workflow will turn back to the initial transition, namely transition T_1 in Figure 3.

According to the mapping relationship between the workflow chart and the Petri net in Table 1, the complex workflow with a variety of routing in Figure 2 will be turned into a correct and reliable FWF-net, as shown in Figure 3 below:

Table 1. Mapping relationship between the workflow chart and the Petri net

Workflow charts	Petri nets
Activity	Transition
Activity migration Condition	Place
Decision node	Mutex choice structure
Referenced document ( )	0-in-degree node ( )

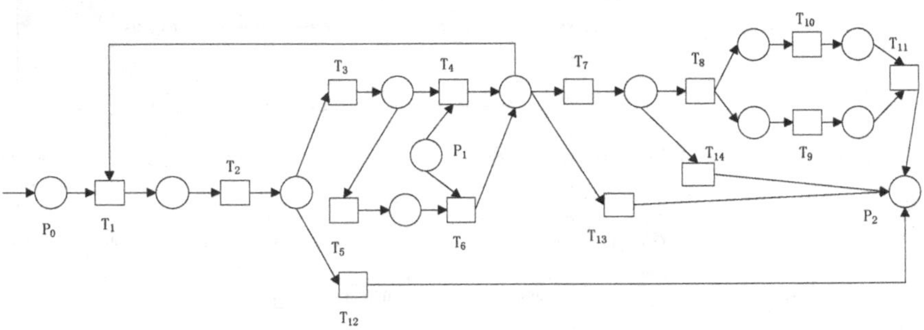

Fig. 3. FWF-net Model Diagram

The meaning of each transition in Figure 3 is shown in Table 2, and the meaning of three key places is shown in Table 3.

Table 2. The meaning of transitions in the FWF-net

Activity unit	Transition	Activity unit	Transition
Reimbursement request	T_1	Agree to reimburse	T_8
Department leader checks	T_2	File examination records	T_9
Financial executive asks for countersignature	T_3	Inform the applicant to finish reimbursing	T_{10}
Financial executive reviews	T_4	Check the records	T_{11}
Financial executive selects countersigners	T_5	Department leader disagrees and replies	T_{12}
Countersigners check	T_6	Financial executive disagrees and replies	T_{13}
General manager examines	T_7	General manager disagrees and replies	T_{14}

Table 3. The meaning of places in FWF-net

Condition	Place
Initial node	P_0
Reimbursement rules node	P_1
Termination node	P_2

4 Verification and Analysis of Rationality of the Model

4.1 Model Simplification

Document [5]~[8] gave some rules of model simplification, especially Document [5] which proved mathematically simplification rules it proposed. According to the simplification rules, a correct and complex Petri net can ultimately be simplified to be a closed-loop net composed by only one place and one transition. Figure 4 gives four basic model simplification rules:

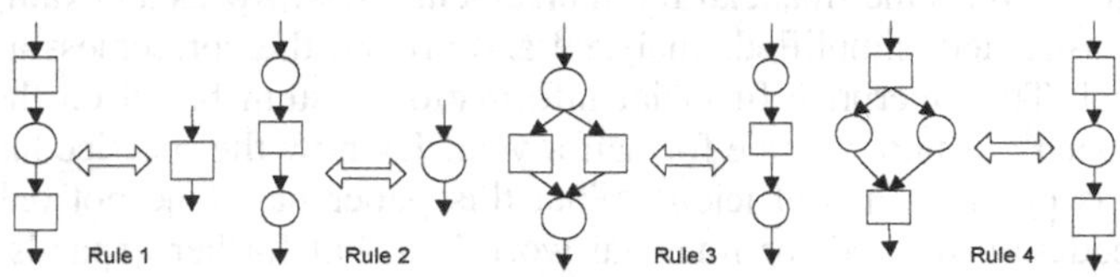

Fig. 4. Simplification Rules

Now simplify the model in Figure 3. The procedure is as follows:(1) Add transition T_0 to form strongly connected graph; (2) Apply Rule 1 to T_1、T_2, and respectively apply Rule 1,2,3,4 to from T_3 to T_6, from T_8 to T_{11}, as shown in Figure 5 (a); (3) Apply Rule 3 and Rule 1 to T_8 and T_{14}, T7 and T_{13} in turn and apply Rule 2 to from P2 to P0, as shown in Figure 5(b); (4) Apply Rule 1, 2, 3 to Figure 5 (b), as shown in Figure 5 (c).

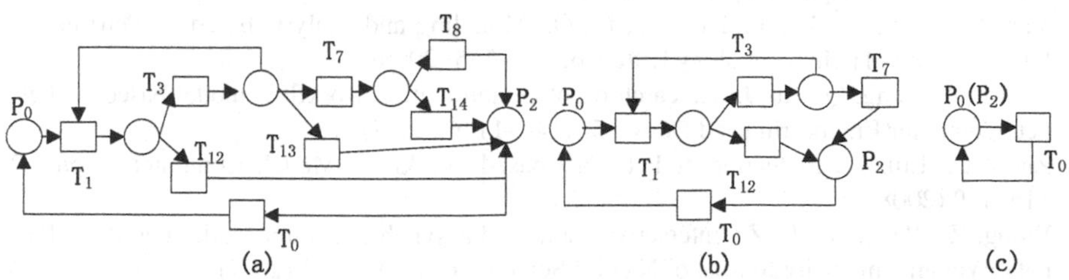

Fig. 5. Simplification procedure

4.2 Analysis of Rationality

To prove the FWF-net model $PN = (P,T,F,M_0)$ in Figure 3 is reasonable, transition t is added between termination place P_2 and initial place P_0, and then the strongly connected graph $PN^{'} = (P^{'}, T^{'}, F^{'}, M_0)$ is constructed. There is one initial state M_0, and $P^{'} = P$, $T^{'} = T \cup \{t\}$, $F^{'} = F \cup \{<P_2, t>, <t, P_0>\}$. After being executed, each transition reaches a new one in $PN^{'}$; and what's more, when transitions are executed, for all of the places in $PN^{'}$, the number of their token is limited. From the two points above, it is clear that workflow net $PN^{'}$ is live and bounded. According to Definition 3 and the theorem in Document [9], one WF net $PN^{'}$ is reasonable if and only if $PN^{'}$ is live and bounded. Based on the above, workflow $PN^{'}$ is reasonable. As a result, the constructed FWF-net model PN is reasonable.

5 Conclusion

Through the analysis of the characteristics of the financial workflow, this paper points out the mapping relationship between the workflow chart and the Petri net and puts forward the modeling method to turn the financial workflow chart into the FWF-net model diagram. With the financial reimbursement workflow as an example, FWF-net model is constructed, simplified, analyzed and proved the correctness and rationality of the model. The enterprise financial information system based on the constructed model has been put into practice for half a year. By now the practice has proved that this model is practical and efficient. What this paper has done not only provides a practical modeling method for financial workflow, but further expands the Petri net modeling application field. The next step of the research is to analyze and research more complex FWF-net model on large scale from multiple perspectives and to provide a method for the performance assessment of this model.

References

1. Wang, K., Zhang, Y., Yang, K., Deng, J., Bai, Z.: Development of workflow engine for OA system. Computer Engineering and Design 19, 4967–4971 (2008)
2. Na, L., Chen, Q.: Gather Model Based on Concurrent Workflow for Finance Monitoring Data. Computer Engineering 10, 62–65 (2009)
3. Yan, C., Jiang, C., Shi, Y., Ding, Z., Li, Q.: Modeling and analysis based on Petri net for IHE workflow. High Technology Letters 6, 551–555 (2006)
4. Wang, G., Yang, Q., Hu, J.: Research on ofice automation workflow model based on Petri net. Computer Engineering and Design 5, 1144–1146 (2007)
5. Zhou, C., Liu, Z.: Reduction of Petri Net-based Workflow Model. Computer Science 2, 115–119 (2008)
6. Wang, Z., Peng, X., Ji, Z.: Interactive multimedia synchronization model based on Petri nets. Wuhan University Journal of Natural Sciences 6, 1019–1023 (2007)
7. Qin, H., Liang, B., Shao, M., Guo, L., Dai, J.: Running management model of corps spatial information system based on Petri nets. Journal of Hubei University (Natural Science Edition) 1, 57–60 (2011)
8. Zhang, L., Yao, S.: Research on Correctness of Workflow Model Based on Petri Nets Reduction Techniques. Computer Engineering 9, 60–61 (2007)
9. Van der Aslst, W., Van Kees, H.C.: Workflow Management: Models, Methods and Systems. Tsinghua University Press, Beijing (2004)

The Merging of Ontology Based on Concept Lattice Merger Technology

HongSheng Xu* and GuoFang Kuang

College of Information Technology, Luoyang Normal University, Luoyang, 471022, China
xhs_ls@sina.com

Abstract. Ontology merging is the process of merging two or more source ontology into target ontology. For the merger of ontology, using traditional editing tools for manually merge time-consuming, strenuous and go wrong easily. These methods are not for the merger to provide a global structure of the ontology to describe the program. The paper offers a methodology for carrying on ontology merging for knowledge sharing and reusing based on concept lattice merger technology.

Keywords: Formal concept analysis, concept lattice, ontology merging.

1 Introduction

At present, there exist defects in the information retrieval and expression of the Internet. For example: the recall ratio, the precision ratio, the retrieval speed and the time of customer response still cannot meet the needs of the users. The main reason is that the purpose of the design is facing users directly reading and processing, but did not provide the semantic information what can be read by the computer, thus limiting the ability of the computer in automatically analysis and process the retrieval information and processing information further intelligent. The ways to solve these problems is the idea of the semantic web, the purpose is to provide the semantic that computer can understand for the information on the Internet. Ontology as the foundation of semantic web is the researched hot spot of the present computer field. The simple introduction of the basic contents described in the ontology and the formal concept analysis.

Ontology merging is the process of merging two or more source ontology into target ontology. For the merger of ontology, using traditional editing tools for manually merge time-consuming, strenuous and go wrong easily. So there are scholars put forward some systems and frameworks to help knowledge engineers in ontology merger, they relied on syntax and semantics matched heuristic method which are taken by ontology engineers when they merge the ontology [1]. For example: (1), Hovy first proposed merger ontology method; (2), Chalupsky put forward OntoMorph system which proposed two kinds of mechanism to transform and merge the ontology; (3), McGuinness put forward Chimaera system; (4), Noy and

* Author Introduce: HongSheng Xu(1979-), Male, lecturer, Master, College of Information Technology, Luoyang Normal University, Research area: concept lattice, ontology.

D. Jin and S. Jin (Eds.): Advances in FCCS, Vol. 1, AISC 159, pp. 501–506.
springerlink.com

Musen put forward the algorithm which merge notology in Protege 2000 system. This algorithm start from distinguishing the matched class name, in this foundation provides repeatedly fold implementation solution for automatically update, find and solve the conflicts when merger.

These tools above, OntoMorph and Chimaera system use a method based on description logic, to describe merger ontology partly. For example: only describe the containment relations between the inspection terms. These methods didn't provide a global structured describing plan for merger ontology.

Ontology merging means that getting several ontology existed in the domain together, eliminating the overlap and not harmonious part. Different ontology merge into a more reasonable concept system and stronger intellectual ability of ontology. The merger of concept lattice is looking for two of the similarities and differences between concept lattices, eliminating repetitive, form one has stronger ability to describe, which has similar functions with ontology merging. In this paper, first, describe and demonstrate dentally which used the FCA technology. On this basis, for the merger of ontology can process based on the principle of concept lattice merging. Here the ontology is merged by concept lattice in accordance with some combining ways of generation, so when the merger is mainly for the merger of concept lattice.

2 The Analysis of Existing Ontology Merging Method

The main purpose of developing body is knowledge sharing and reuse. To achieve knowledge sharing, you need people to use common knowledge representation, such as the common body. But this is clearly unrealistic to domain ontology, for example, even for the same area, different ontology developer may establish different domain ontology, to enable all developers to follow the same methods and rules to build the body are difficult. Therefore, the domain ontology reuse of knowledge among a key issue, then the solution is through the integration and merging from the same field or different fields of knowledge to achieve more than one body, that body merged.

Ontology merging is to merge two or more source ontologies ontology into a target process. For the body of the merger, the uses of traditional hand-editing tools are the merger time-consuming and laborious, and prone to error. Therefore, some scholars have made a number of current systems and framework to help knowledge engineers ontology merging, ontology engineers; they are dependent on the body when taken in the combined syntax and semantics of matching heuristics. The following briefly describes the methods [2].

(1) Hovy the first time the method ontology merging, which describes the heuristic method is different from the body by identifying the appropriate concepts, such as analysis and comparison of the two concepts of names and natural language definitions, and check the concept of hierarchy in the two a concept of similarity relations.

(2) Chalupsky made OntoMorph system two mechanisms proposed to carry out the conversion and merging ontologies. One is: based on syntactic rewriting mechanisms to support knowledge representation language in two different inter-conversions; the other is: Semantic-based rewriting mechanisms for the conversion of inference

provide a solution. The system is explicitly allowed to implement the conversion to a more flexible conversion mechanism; the semantic rules can violate the existing.

(3)McGuinness puts forward the Chimaera system. The system of the body from different sources ontology terms of the merger terms to provide a method. In addition, it can not only help test ontology coverage and accuracy, and can facilitate the maintenance of the body. Although the Chimaera system provides a number of functions, but the current structure on the properties of the underlying ontological assumptions have not made a clear description of.

(4) Noy and Musen was raised conducted in Protégé-2000 ontology merging algorithm. The algorithm is based on identifying the class name that matches the starting point; on this basis to be able to automatically update, find and solve conflicts when combined provide a repeated iteration of the implementation of the program.

These tools, OntoMorph, Chimaera system uses a description logic-based approach, the combined partial ontology to describe, for example: only describes the inclusion relations between the terms inspection. These methods are not for the merger to provide a global structure of the ontology to describe the program.

In this paper, formal concept analysis method has the advantage of ontology merging can reduce the overhead of manual and automated to facilitate the processing, but also for the body of the building, combined to provide a global description of the process.

3 Concept Lattice Operations and Merger

Along with distribution/parallel computer technology matures, many researchers put forward the network technology, especially the rapid development of Internet, the demand of data distributed storage and parallel processing is more and more urgent. The distribution processing thought of concept lattice is: through the split of form context, forming many distributed storage son contexts, and then tectonic corresponding son concept lattice, again by merging the son concept lattices to get required concept lattice. That is, construct required concept lattice by merging several son concept lattices.

Because concept lattice is the form of the relationship between concepts in the formal contexts, it and the corresponding formal context is the one-to-one. Therefore, the distribution processing of concept lattice will inevitably involve to the resolution and merger operations of formal context; it is the premise of concept lattice structure distribution.

Given formal context $K = (G, M, I)$, if formal context $K_1 = (G_1, M_1, I_1)$and $K_2 = (G_2, M_2, I_2)$meet the$G_1 \subseteq G$, $G_2 \subseteq G$, $M_1 \subseteq M$, $M_2 \subseteq M$, then says $K_1$and $K_2$is the same domain formal context, they are all the son formal contexts of K, also says the concept lattice $L(K_1)$ of formal context $K_1$and the concept lattice $L(K_2)$ of formal $K_2$are the same domain concept lattice[3].

Definition 1. For the formal contexts $K_1 = (G, M_1, I_1)$and $K_2 = (G, M_2, I_2)$of the same object domain, if $M_1 \subseteq M$, $M_2 \subseteq M$, $M_1 \cap M_2 = \varnothing$, then says $K_1$and K_2, L (K_1) and L (K_2) were connotation independent; If $M_1 \subseteq M$, $M_2 \subseteq M$, $M_1 \cap M_2 \neq \varnothing$, for any g∈G and arbitrary $m \in M_1 \cap M_2$ meet $gI_1m = gI_2m$, it says K_1 and K_2, L (K_1) and L (K_2) are respectively connotation consistent.

Definition 2. For the formal contexts $K_1 = (G, M_1, I_1)$and $K_2 = (G, M_2, I_2)$of the same attribute domain, if $G_1 \subseteq G$, $G_2 \subseteq G$, $G_1 \cap G_2 = \varnothing$, then says $K_1$and K_2, L (K_1) and L (K_2) were extension independent; If $G_1 \subseteq G$, $G_2 \subseteq G$, $G_1 \cap G_2 \neq \varnothing$, for any $m \in M$ and arbitrary $g \in G_1 \cap G_2$ meet $gI_1m = gI_2m$, it says K_1 and K_2, L (K_1) and L (K_2) are respectively extension consistent.

Below respectively, simply introduce two extension independent domains concept lattice of the vertical combating operation and the two connotation independent domains concept lattice horizontal combating operation. First, the vertical combain operaions of the domain concept lattice have the following definitions.

Definition 3. The two lattice nodes $C_1=(O_1, D_1)$and $C_2=(O_2, D_2)$of concept lattice L, if $D_1=D_2$, says C_1equal to C_2; If $D_1 \subset D_2$, says $C_1$is bigger than C_2, or $C_2$is smaller than C_1.

Definition 4. If L (K_1) and L (K_2) are two extension independent domain concept lattices, define L (K_1) $\cup$L (K_2) is concept lattice L, if L meets:

(1) the maximum lower bound *inf (L (K_1))* of *L (K_1)* and the maximum lower bound *inf (L (K2))* of *L (K2)*, if *inf (L (K_1))* is not equal to, not less than also is no more than the *inf (L (K2))*, then concept $C=(\varnothing, Intent(inf(L(K_1))) \cup Intent(inf(L(K_2)))) \in L$, and if $Extent(inf(L(K_1)))=\Phi$, then $\inf(L(K_1)) \notin L$, if $Extent(inf(L(K_2)))= \Phi$, then $inf(L(K_2)) \notin L$.

(2) one lattice node C1 of L (K_1) and one lattice node C2 of L (K_2), make C3=C1+C2, if any lattice node C'_1 bigger than C_1, has no C'_1 equals to or less than C_3; Likewise, any lattice node C'_2 more than C_2, are not has C'_2 equals to or less than C_3, then $C_3 \in L$.

(3) for any lattice node C_1 of L (K_1), if there is no lattice node equal to or less than C_1 in $L(K_2)$, then $C_1 \in L$;

(4) for any lattice node C_2 of L (K_2), if there is no lattice node equal to or less than C_2 in $L(K_1)$, then $C_2 \in L$;

(5) The lattice node does not belong to L outside above three cases.

Theory 1. If *L* (K_1) and *L* (K_2) are two extension independent domain concept lattice, then has have $L (K_1) \cup L (K_2) = L (K_1+K_2)$.

Definition 5. Lattice nodes $C_1=(O_1, D_1)$ and $C_2=(O_2, D_2)$, if $O_1=O_2$, then says C_1 is equal to C_2; if $O_1 \subset O_2$, says C_1 less than C_2, or C_2 more than C_1

Definition 6. Lattice nodes $C_1= (O_1, D_1)$, $C_2= (O_2, D_2)$ and $C_3=(O_3, D_3)$, if $O_3=O_1 \cap O_2$, $D_3=D_1 \cup D_2$, says C_1+C_2 equals to C_3.

Definition 7. If $L(K_1)$and $L(K_2)$are two connotation independent the same domain concept lattice, define $L(K_1) \cup L(K_2)$ is lattice node L, if L meets:

(1) one lattice node C1 of L (K1) and one lattice node C2 of L (K2), make C3=C1+C2, if any lattice node C'_1 smaller than C_1, has no C'_1 equals to or more than C_3; Likewise, any lattice node C'_2 less than C_2, are not has C'_2 equals to or more than C_3, then $C_3 \in L$;

(2) for any lattice node C_1 of L (K_1), if there is no lattice node equal to or more than C_1 in $L(K_2)$, then $C_1 \in L$;

(3) for any lattice node C_2 of L (K_2), if there is no lattice node equal to or more than C_2 in $L(K_1)$, then $C_2 \in L$;

(4) The lattice node does not belong to L outside above three cases.

Theory 2. If $L(K_1)$and $L(K_2)$are two connotation independent the same domain concept lattice, then has $L(K_1) \cup L(K_2) = L(K_1+K_2)$.

4 Ontology Merging Based on FCA

Ontology merging means that a (within the territory of the existing several Ontology and together, eliminate the overlap and not harmonious part. Different body is for a merger with more reasonable concept system and stronger intellectual ability of ontology. The merger of concept lattice is looking for two of the similarities and differences between concept lattice, eliminating repetitive, form has stronger ability to describe the concept lattice, this and ontology with similar functions merge. This paper has the technology to build to FCA body process demonstration and describe the detailed. On this basis, for the merger of ontology can based on the principle of concept lattice merger for processing.

The body here is to merge the concept lattice generated by a combination of methods, so the merger, the main concept lattice for the merger. In the combined body, the first need to look at different areas of cross-cutting, whether there is some concepts of the semantic overlap, therefore, according to the different areas of the body build has the following two situations: the same domain ontology merging, the combined cross-domain ontology.

According to the merger concept lattice theory, this requires two forms of background and setting, but there will be three special circumstances.

(1), In the context of two forms is-a relationship may exist in two ontology concepts, such as the two concepts m, n, there is m is-a n, and set in the background after only appear in the form of n. That is, if (g; n) ∈I and m is-a n means: (g; n) ∈I.

(2), if the two forms of the background there are two different names but the same sense of ontological concepts such as: a, b, which is due to the different experts in the field taking the concept to be adopted due to the different naming. You can set in and after the formal context to " a\b " to represent this concept, meaning that a or b.

(3), if the two forms of the background there are two names, meanings are different, but they are from the same area of the distribution of documents such as: p, q, describes these documents do not provide sufficient description of the ability to distinguish between these two concepts, then after the form and set the background to $p(q)$to represent, which means p and q.

Then for using Godin algorithm, and, in the end, made in accordance with the combination of ontology and concept lattice ways, from concept lattice into ontology, thus forming the merged body as shown in figure 1.

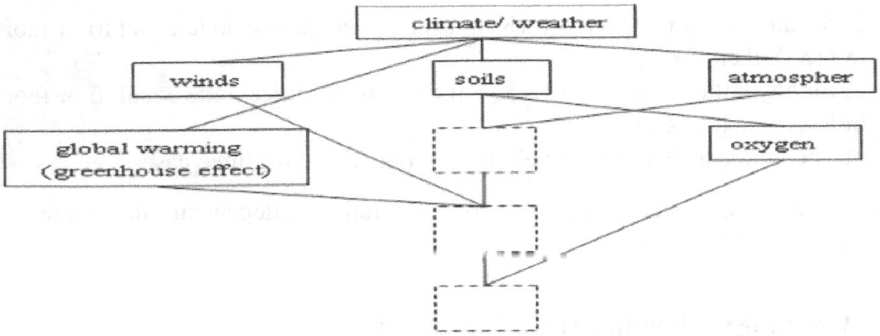

Fig. 1. Ontology merging based on FCA in field combined with the body map

In addition, according to the theorem: the sub-horizontal merger concept lattice and formal context of the background corresponding to the sub-sub-concept lattice is isomorphic to a horizontal merger, the merger of the body then can direct the body corresponding to the two horizontal merger concept lattice. The resulting ontology and the ontology structure obtained is the same, but the drawback of this approach is: do not merge to form the body in the form of background, is not conducive to viewing the document and include the relationship between ontology concepts.

5 Summary

Existing ontology merging method, only the local use of different types of logical method to the whole body can not be combined to provide a unified technology. The article is based on ontology merging FCA technology is the technology used to build the body of the FCA on the basis of cross-cutting areas from the same domain and two aspects, respectively, based on building a body and set the form background and overlay processing and, then after making cell, then the corresponding concept lattice according to some form of switching costs body, the final result of the formation of the combined body.

References

1. Berners-Lee, T., Hendler, J., Lassila, O.: The Semantic Web. Scientific American 5 (2001)
2. Missikoff, M., Navigli, R., Velardi, P.: Integrated Approach for Web Ontology Learning and Engineering. IEEE Computer 35(11), 60–63 (2002)
3. Godin, R., Mili, H., Mineau, G.W., et al.: Design of class hierarchies based on concept (Galois) lattices. Theory and Application of Object Systems 4(2), 117–134 (1998)

Estimation of Fingerprint Orientation Field Based on Improved Structure Tensor

Gong Chen, Mei Xie, and Yu Yu

Image Processing and Information Security Lab, School of Electronic Engineering, UESTC, Chengdu, China

Abstract. The estimation of fingerprint orientation field algorithm is a key step in the whole fingerprint identification algorithm. This paper proposes an accurate and rapid method by using connected component analysis structure tensor to estimate fingerprint orientation field. Compared to the original structure tensor method, this method combines connected component analysis map and Gaussian kernal rather than only use Gaussian kernal in the construction of structure tensor. It can preserve the image local structure better and make the result more reliable. The experiments results show that this method is fast, accurate and can easy be used by subsequent algorithm.

Keywords: fingerprint, fingerprint orientation field, connected component analysis, structure tensor.

1 Introduction

Generally, automatic fingerprint identification system [1] contains many steps like estimation of orientation field, fingerprint image enhance, feature extraction, feature matching,etc. In these steps, the estimation of orientation field is the foundation of the whole algorithm. A fast and accurate method for estimation of orientation field can provide great convenience for subsequent steps.In many algorithms for orientation field estimation [2], Structure Tensor [3] is a simple but effective method to estimate fingerprint orientation field. Just by doing a series of calculation after constructing Structure Tensor, we can get the fingerprint orientation field. Although Structure Tensor is effective, there is still a problem.Generally, Structure Tensor is calculated by component-wise convolving between gradient of fingerprint image and Gaussian kernel, however the Gaussian kernel can not preserve the image structure well.

So in this paper we using a new kernel which combine with Connected Component Analysis [4] map and Gaussian kernel to construct a non-linear Structure Tensor to replace the original linear Structure Tensor [5]. Comparing with original linear Structure Tensor,the Improved Structure Tensor provides more faithful results in the estimation of fingerprint orientation field and only adds little time spent.

D. Jin and S. Jin (Eds.): Advances in FCCS, Vol. 1, AISC 159, pp. 507–512.
springerlink.com

2 The Estimation of Fingerprint Orientation Field and Structure Tensor

2.1 Fingerprint Orientation Field

Fingerprint Orientation Field is the base of fingerprint identification algorithm. It can reflect the overall trend of fingerprint image texture and lay the foundation for the subsequent algorithm.The fingerprint image enhance algorithm need accurate Fingerprint Orientation Field, because the fracture lines need accurate orientation field to connect and the adhesion of lines need accurate orientation field to segment.The binary of fingerprint image also need accurate orientation field.Only we know an accurate estimation of orientation field, we can calculate an accurate mean gray value of lines.It is also necessary for fingerprint feature matching, it provides the direction information of fingerprint features.Of course,fingerprint orientation field also can be use to match as a texture feature.

2.2 The Structure Tensor Algorithm

Structure Tensor can reflect the local structure of fingerprint image well.The natures of Structure Tensor like the Hessian matrix can be used in many place like the analysis of the manifold texture, Angular point and T point detection. Each eigenvalue of this matrix correspond with an eigenvector.This eigenvalue represents the gray-scale contrast in the direction of its corresponding eigenvector.The eigenvector which correspond with minimum eigenvalue represents the minimum wave directation: image texture orientation also known as the Coherence Orientation.

The steps to use Structure Tensor to estimate fingerprint orientation field:consider $u(i, j)$ is a fingerprint image, g_σ and g_ρ are both Gaussian kernel.

$$g_\sigma(i, j) = \frac{1}{2\pi\sigma} exp\left(-\frac{i^2 + j^2}{2\sigma^2} \right) \tag{1}$$

$v(i, j)$ is the enhanced image of $u(i, j)$.

$$v(i, j) = g_\sigma(i, j) * u(i, j) \tag{2}$$

v_x and v_y are the first-order horizontal and longitudinal gradient images of $v(i, j)$ and $S(i, j)$ is the Structure Tensor in (i, j)

$$S(i, j) = \begin{pmatrix} a & b \\ b & c \end{pmatrix} = g_\rho(i, j) * \begin{pmatrix} v_x^2 & v_x v_y \\ v_x v_y & v_y^2 \end{pmatrix} \tag{3}$$

Structure Tensor $S(i, j)$ has two eigenvectors.The one that correspond to the large eigenvalue represent the direction which perpendicular to fingerprint ridge line.The other one represent the direction of fingerprint ridge line.The eigenvalues and eigenvectors of S are calculating below:

$$\lambda_{1,2} = \frac{1}{2}(a + c \pm \sqrt{(a - c)^2 + 4b^2}) \tag{4}$$

$$w_{1,2} = \begin{bmatrix} \dfrac{2b}{\sqrt{(c - a \pm \sqrt{(c - a)^2 + 4b^2})^2 + 4b^2}} \\ \dfrac{c - a \pm \sqrt{(c - a)^2 + 4b^2}}{\sqrt{(c - a \pm \sqrt{(c - a)^2 + 4b^2})^2 + 4b^2}} \end{bmatrix} \tag{5}$$

So the direction of fingerprint ridge line $O(i, j)$ in each pix (i, j) can be calculated by (6):

$$O(i, j) = arctan\left(\frac{w_{21}}{w_{22}}\right) \tag{6}$$

3 The Improved Structure Tensor in Estimating Fingerprint Orientation Field

We know that the Gaussian kernel G, which is used as a weighting kernel of linear structure tensor considers only distance information and ignores local structure information.In some situations,like the fingerprint image is dirty or fuzzy, this linear structure tensor can not perform well.In order to solve this problem,we choose to use Connected Component Analysis Structure Tensor.

At first,Connected Component Analysis Structure Tensor was put forward to cooperate with Coherence Enhancing Diffusion Filtering.By combining with Connected Component Analysis map and Gaussian kernel,we can get a new weighting kernel to calculate structure tensor.This nonlinear structure tensor can consider both local structure information and distance information.We find that this nonlinear structure tensor can also be in the original Structure Tensor algorithm to improve its accuracy.

The steps of Improved Structure Tensor Algorithm is shown below:consider $u(i, j)$ is a fingerprint image,we can divide the image into several blocks size of $W \times W$ .For each block centered on (x_0, y_0),the gradient magnitude of $GM_W(x, y)$ is calculated as follows:

$$GM_W(x_i, y_i) = \sqrt{v_x^2 + v_y^2} \tag{7}$$

Then,binary edge of block can be calculate by (8):

$$BE_W(x_i, y_i) = \begin{cases} 1 & if\left|GM_W(x_i, y_i) - GM_W(x_0, y_0)\right| < th \\ 0 & else \end{cases} \tag{8}$$

the th is threshold for grouping pixels by gradient magnitude of center pixel.Then we scan a block and group pixels into components based on 8-connectivity.Once all the components have been determined,each pixels in the block can be labeled by components.At last,Connected Component Analysis map is constructed as follows:

$$CCA_W(x_i, y_i) = \begin{cases} 1 & label(x_i, y_i) = label(x_0, y_0) \\ 0 & else \end{cases} \tag{9}$$

After constructed the CCA map,we can combine it with Gaussion kernel g_σ to get a new nonlinear weighting kernel $N_\sigma(x_i, y_i)$ as follows:

$$N_\sigma(x_i, y_i) = \frac{1}{C_\sigma} exp\left(-\frac{i^2 + j^2}{2\sigma^2}\right) \times CCA_W(x_i, y_i) \tag{10}$$

The C_σ is normalization operator:

$$C_\sigma = \sum_W exp\left(-\frac{i^2 + j^2}{2\sigma^2}\right) \times CCA_W(x_i, y_i) \tag{11}$$

At last,we get the new nonlinear structure tensor $S_N(i, j)$:

$$S_N(i, j) = \begin{pmatrix} a & b \\ b & c \end{pmatrix} = N_\sigma(x_i, y_i) * \begin{pmatrix} v_x^2 & v_x v_y \\ v_x v_y & v_y^2 \end{pmatrix} \tag{12}$$

After using $S_N(i, j)$ to replace $S(i, j)$ calculating by (3),we use (4) and (5) to calculate the eigenvalues and eigenvectors of $S_N(i, j)$.Finally,we get the direction of fingerprint ridge line $O(i, j)$ by (6).

4 The Experiment Results

In order to inspect the proposed method, we choose various fingerprint images in fvc2004 to test.We experiment the Mark,Structure Tensor and Improved Structure Tensor to estimate fingerprint orientation field respectively.Then we use Gabor filter [6] to enhance these fingerprint images and compare their results.

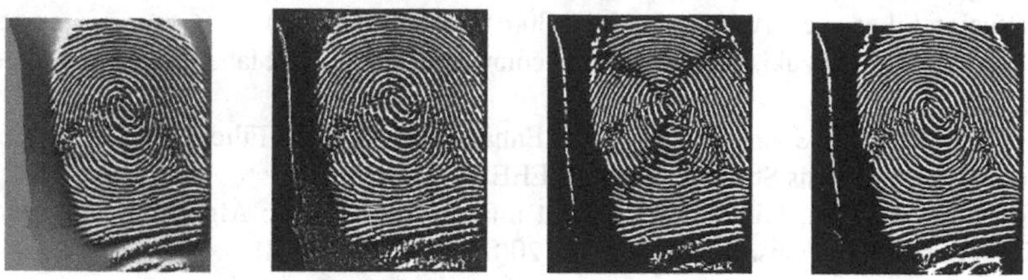

(a) (b) (c) (d)

Fig. 1. The results in enhancing a dirty fingerprint image;(a)Original image.(b)Using Mark to estimate orientation field.(c) Using Structure Tensor to estimate orientation field.(d) Using Improved Structure Tensor to estimate orientation field.

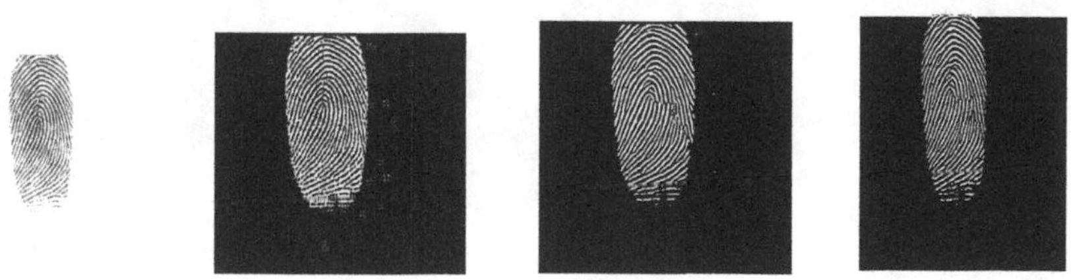

(a) (b) (c) (d)

Fig. 2. The results in enhancing a dry fingerprint image;(a)Original image.(b)Using Mark to estimate orientation field.(c) Using Structure Tensor to estimate orientation field.(d) Using Improved Structure Tensor to estimate orientation field.

5 Conclusions

This paper has shown an Improved Structure Tensor Algorithm for the Estimation of Fingerprint Orientation Field.By using Connected Component Analysis, we construct a nonlinear weighting kernel to calculate Structure Tensor.Comparing to original linear

Structure Tensor,this method perform better in low quality fingerprint images.Though this method has made some progress,its time consumption is big,further work still needs to be done.Our next research is to reduce the time consumption and make this method more robust.

References

1. Tian, J., Yang, X.: Biometric theory and application. Publishing House of Electronics Industry (2005)
2. Chen, D., Ji, X., Fan, F., Zhang, J., Guo, L., Meng, W.: Comparative Analysis of Fingerprint Orientation Field Algorithms. In: IEEE Conferences (2009)
3. Cheng, J., Tian, J., He, Y.: Fingerprint Enhancement Algorithm Based on Nonlinear Diffusion Filter. Acta Automatic Sinica 30(6), 854–862 (2004)
4. He, L., Cao, Y., Suzuki, K., Wu, K.: Fast connected-component labeling. Pattern Recognition (2007)
5. Yoo, H., Kim, B., Sohn, K.: Coherence Enhancing Diffusion Filtering based on Connected Component Analysis Structure Tensor. IEEE (2011)
6. Hong, L., Wan, Y., Jain, A.: Fingerprint image enhancement: Algorithm and performance evaluation. IEEE Transactions on PAMI 20(8), 777–789 (1998)

A New Threshold Algorithm Applied in the Voltage Sag Signal De-noising

Yingjun Sang[1,2], Yuanyuan Fan[1], Qingxia Kong[1], Fei Huang[1], Qi Chen[1], and Bin Liu[1]

[1] Huaiyin Institute of Technology, Meichengstr.1, 223003, Huaian, China
[2] Electrical Engineering School, Southeast University, Sipailoustr.2, 210096, Nanjing, China
sangyingj@163.com

Abstract. In this article, the wavelet analysis method is used to detect and analyze voltage sag signal contaminated by white noise. A new threshold selection method based on the wavelet decomposition layer is proposed, in view of the characteristic of noise coefficients which decrease when the scale increases. This method can obtain the superior wavelet coefficients estimation through adjusting two adjustable parameters. This method has the better ability for reducing noise interference, moreover, better SNR, MSE value are obtained.

Keywords: Sag, wavelet transform, threshold de-noising.

1 Introduction

Electric power quality issues have captured increasing attention in electric engineering in recent years, for its degrading the performance and efficiency of customer loads, especially power electronics loads [1,2]. PQM will be an effective means to detect the power disturbances, such as voltage sags, oscillations, pulses. However, the signal from sensor is often contaminated by noise, so it's necessary to filter the signal with an efficient de-noising method.

Threshold method is common and important for wavelet de-noising, some researches have been carried out. Donoho and Johnstone proposed hard and soft threshold method, which can effectively reduce the noise interference with a certain threshold selection, but each has the drawbacks respectively, so a new threshold selection method is proposed in this article, which is appropriate for the voltage sag and other nonstationary situations of power quality monitoring system. The simulation results proved that the de-noising effect of proposed algorithm is better, compared with the other three algorithms.

2 Discrete Wavelet Transform（DWT）

Discrete wavelet transform was developed by Mallat from fast algorithm based on the conjugate quadratic filters (CQF) [3,4], J-level wavelet decomposition can be computed as follows:

D. Jin and S. Jin (Eds.): Advances in FCCS, Vol. 1, AISC 159, pp. 513–517.
springerlink.com

$$c_{j,k} = \sum_n h_{n-2k} c_{j-1,n} \tag{1}$$

$$d_{j,k} = \sum_n g_{n-2k} c_{j-1,n} \tag{2}$$

Where $c_{j,k}$, $d_{j,k}$ are scale and wavelet coefficients derived from the projection of the signal on the space of scale and wavelet functions respectively. H_n and g_n are the low-pass and high-pass filters respectively, corresponding to the selected wavelet basis, the reconstruction algorithm is shown as following formula:

$$c_{j-1,k} = \sum_n \left[h_{k-2n} c_{j,n} + g_{k-2n} d_{j,n} \right], j = J, J-1, \cdots, 1 \tag{3}$$

3 Applications of Wavelet Analysis in Voltage Sag Signal Processing

Voltage sag is the most common type of voltage disturbances of power distribution system, and is commonly caused by the system short-circuit fault, large motor starting, transformer or capacitor switching, therefore monitoring and detection of voltage sag are necessary for the purpose of improving the power quality of system.

3.1 Threshold-Based Wavelet De-noising

The noise in power quality monitoring system, is usually the Gaussian white noise, and the noisy signal is basically of the following form:

$$y(t) = f(t) + \alpha\varepsilon(t) \tag{4}$$

Where $\varepsilon(t)$ is the standard Gaussian white noise, α is the noise level, $f(t)$ is original pure signal. According to Donoho and Johnstone proposed threshold approach [5], there are two different threshold function, the hard and soft threshold.

The hard threshold function is not a continuous function, and thus the signal reconstruction process will appear oscillations. he soft threshold function is continuous, but when $|w| > T$, which is a constant deviation, and can impact the estimated degree of signal reconstruction, it will result in the loss of useful information, and bring a great distortion to reconstructed signal. Soft and hard threshold function can be expressed in Fig. 1.

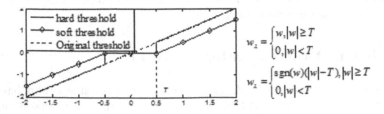

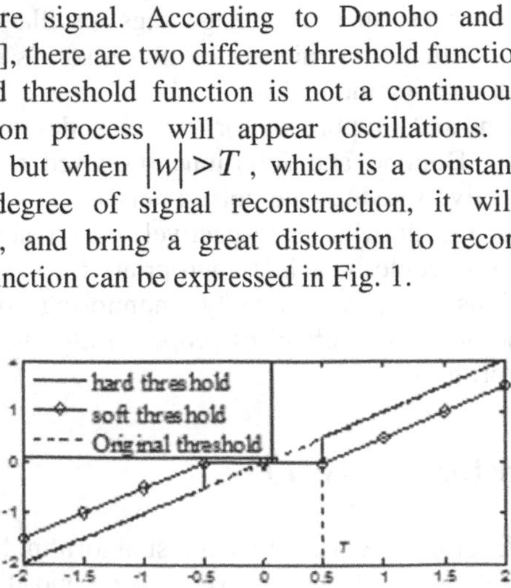

$$w_\lambda = \begin{cases} w, |w| \ge T \\ 0, |w| < T \end{cases}$$

$$w_\lambda = \begin{cases} \operatorname{sgn}(w)(|w| - T), |w| \ge T \\ 0, |w| < T \end{cases}$$

Fig.1. Soft and hard threshold function

Based on the wavelet transform Lyapunov exponents [6], we know that the modulus values corresponding to the signal and noise wavelet coefficients have different characteristics, the magnitude value of the noise decreases gradually with increasing scale, while the magnitude value of the signal is gradually increases or remains unchanged with increasing scale. The scale-based algorithm is proposed for voltage sag signal de-noising, which is described as follows:

$$T = \frac{\sigma\sqrt{2\log(N)}}{S_{L,K}+b}, \quad S_{L,K} = 2^{2L/(2L-K)} \tag{5}$$

Where L indicates decomposition level, K indicates the number of layers of the current threshold, b is an adjustable parameter. Using median estimator to estimate the noise standard deviation, $\hat{\sigma} = \frac{median|x|}{0.6745}$, threshold values is shown in Table 1.

Table 1. The threshold table of four algorithms under noise level 0.125

α	Thresholding Algorithm	Decomposition Level				
		1	2	3	4	5
0.125	Heuristics Sure	0.4382	0.4133	0.3862	0.0633	0.3270
	S-median	0.2388	0.2766	0.3004	0.3139	0.3211
	Penalize Medium	0.3602	0.3602	0.3602	0.3602	0.3602
	proposed	0.4460	0.4307	0.4105	0.3829	0.3434

3.2 Signal De-noising Results and Discussion

The voltage sag fault signal is generated using simulation model, which is performed 5-layer wavelet decomposition in the case of white noise factor of 0.125, choosing daub4 as wavelet function [7], after the decomposition, the low frequency coefficients and high frequency coefficients are separated, and the voltage sag signal is reconstruct to get the voltage sag estimated signal by threshold processing.

As shown in Fig.2, the original voltage sag signal is processed with the proposed threshold de-noising algorithm. The voltage sag signal has an original amplitude of 1, and its amplitude decrease 40% with a continuous interval from 0.4s to 0.9s. The system sampling points is fixed as 1024.

From the noise reduction effect we can see, the noise cancellation signal is very close to the fault signal, the useful low frequency information is remained effectively.

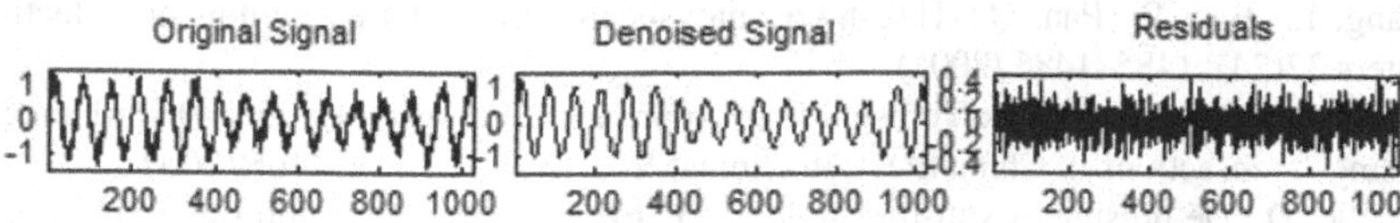

Fig. 2. De-noised result of the voltage sag signal

The purpose of signal de-noising is to minimize the difference between de-noised signal and the ideal signal, which is usually measured by the signal noise ratio (SNR) and the mean square error (MSE), the function is shown as Fig.3.a:

Where $\hat{f}(i)$ is the reconstruction signal amplitude at the point i, and $f(i)$ is the original pure signal amplitude at the point i, N is the length of sample data. The SNR and MSE values after de-noised by the four threshold algorithms under the noise level 0.125 is showed in Fig.3.b. The four algorithms is Heuristics Sure, S-median, Penalize Medium and proposed algorithm. We can see that the larger SNR and smaller MSE are accomplished by the proposed algorithm, which means the de-nosing effect is better than other three algorithms.

$$SNR = 10\lg\frac{\sum_{i=0}^{N}\left|f(i)\right|^2}{\sum_{i=0}^{N}\left|\hat{f}(i)-f(i)\right|^2}$$

$$MSE = \frac{\sum_{i=0}^{N}\left|\hat{f}(i)-f(i)\right|^2}{N}$$

a.

b. 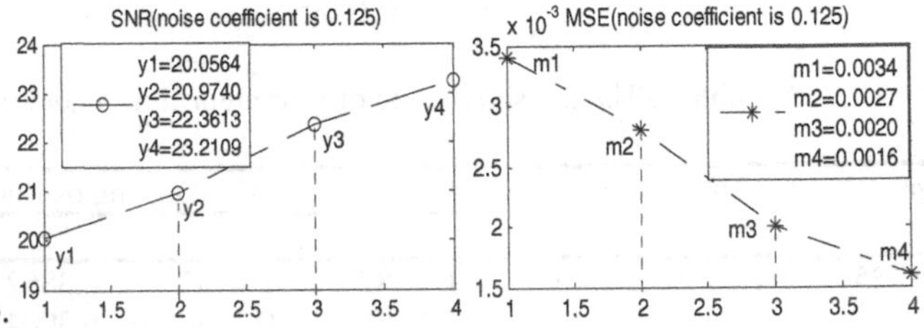

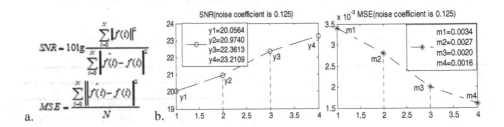

Fig. 3. a. SNR and MSE function **b.** SNR and MSE value

4 Conclusions

In this paper, a new level-dependent proposed noise reduction algorithm based on wavelet transform is proposed. It has improved the effect of power quality signal de-noising for PQM system. A voltage sag signal generated by simulation model is applied for simulation analysis, the simulation results show that the algorithm further improves the SNR, reduces the MSE, and is more effective in de-noising voltage sag signal than other three de-noising algorithms.

Acknowledgements. The authors wish to acknowledge the support of Science Research Foundation (HGC1103) of Huaiyin Institute of Technology.

References

1. Zhang, L., Bao, P., Pan, Q.: Threshold analysis in wavelet-based de-noising. Electronics Letters 37(24), 1485–1486 (2001)
2. Huang, W., Dai, Y.: Block-Thresholding Approach for Power Quality Disturbance De-noising. Transactions of China Electrotechnical Society 22(10), 161–165 (2007)
3. Donoho, D.: De-noising by soft-thresholding. IEEE Transactions on Information Theory 41, 613–627 (1995)
4. Mallat, S., Wen, L.H.: Singularity detection and processing with wavelets. IEEE Transactions on Information theory 38(2), 617–643 (1992)

5. Ji, T.Y., Lu, Z., Wu, Q.H.: Detection of power disturbances using morphological gradient wavelet. Signal Processing 88, 255–267 (2008)
6. Dwivedi, U.D., Singh, S.N.: Denoising techniques with change-point approach for wavelet-based power-quality monitoring. IEEE Trans. Power Del. 24(3), 1719–1727 (2009)
7. Wang, Z., Jiang, H.: Robust incipient fault identification of aircraft engine rotor based on wavelet and fraction. Aerospace Science and Technology 14, 221–224 (2010)

Parallel Clustering of Videos to Provide Real Time Location Intelligence Services to Mobile Users

Alfio Costanzo, Alberto Faro, and Simone Palazzo

Department of Electrical, Electronics and Computer Engineering,
University of Catania, viale A.Doria 6, Catania, Italy
alfioc87@hotmail.it, {afaro,simone.palazzo}@dieei.unict.it

Abstract. Clustering of large datasets has been widely used to retrieve relevant data or to discover novel data associations. Topology preservation of the original dataset is not always mandatory, unless the dataset has a temporal-spatial nature since, in this case, the data in the same cluster as well as the ones of the closest clusters should be related as in the real system. However, only few existing parallel methods preserve the original dataset topology. Also, few times such methods have been used to help the management of real time systems due to their limited parallelism degree. Aim of the paper is to illustrate how such requisites may be fulfilled by using a SOM based parallel clustering method developed recently by the authors. In particular, the clustering of relevant video sequences that helps the management of real time location systems are discussed to point out how this allows us to provide novel location intelligence services to the mobile users.

Keywords: Parallel clustering, neural networks, location intelligence, mobile computing.

1 Introduction

Clustering is a technique widely used not only to study stationary or quasi-stationary data systems such as information retrieval from massive datasets[1] and knowledge discovery from text [2], but also to manage dynamic systems such as decision support systems based on time-evolving business data or the control of time-varying physical systems [3]. Usually, the clustering applications are not in real time unless the applications evolve slowly, because the clustering is a time-consuming computation process especially if it should preserve the topology of the original dataset.

Fortunately, the availability of relatively low cost parallel computation infrastructures such as Grid, Cloud and Cuda, makes possible the parallelization of the clustering techniques to execute topology preserving clustering of time varying datasets in real time. This opens the possibility of processing video sequences of many complex systems with affordable solutions.

Due to the increasing demand of location based services (LBSs) to be provided to mobile users in real time, the paper aims at discussing relevant applications of a fast clustering algorithm previously proposed by the authors that may help the management of real time location based systems such security, logistics and mobility.

D. Jin and S. Jin (Eds.): Advances in FCCS, Vol. 1, AISC 159, pp. 519–527.
springerlink.com

Some corrections of the clustering method is also proposed to increase the accuracy of the clustering without sacrificing the time performance.

Sect.2 briefly recalls the adopted parallel clustering technique. Sect.3 illustrates, by realistic examples, to what extent such techniques may be used effectively to improve the existing LBSs. The concluding remarks briefly illustrate how the fast clustering would allow us to provide the mobile user with Google maps interface.

2 Parallel Neural Clustering of Dynamic Systems

As known, many clustering methods are available in the literature, but if the topology preservation of the original dataset is mandatory, clustering methods based on the SOM [4] are still the best ones [5]. For this reason, the paper adopts the SOM-like based clustering algorithm proposed in [6] where each item of a dataset is represented by a similarity vector $\mathbf{s}_i$ whose general field s_{ij} measures the similarity between the item i and the other items j of the dataset. Such vectors are collected in a single square matrix **S** called similarity matrix. To compute the similarity matrix we have to know the so called feature matrix **F** where each item is represented by a vector $\mathbf{v}_i$ whose general field v_{ik} represents how much a certain attribute (also called feature) k is owned by the item i, e.g., in the image processing domain, v_{ik} could be a measure of the red, green and blue level featuring the pixel i. Matrix **S** may be obtained easily from matrix **F** as follows:

$$s_{ij} = \text{distance}\,(\mathbf{v}_i, \mathbf{v}_j) = [\Sigma_k\,(\mathbf{v}_{ik} - \mathbf{v}_{ik})^2\,]^{0.5} \tag{1}$$

In the adopted method, the cluster of a dataset is obtained by presenting, each item $\mathbf{s}_i$ to a mono dimensional SOM neural network characterized by a number of input neurons equal to the number of data items, and by a number of output neurons corresponding to the number of clusters. At each presentation of an input item, the synaptic weights w_{ir} connecting the input and output neurons are updated following the following formula:

$$w_{jk}\,(t+1) = w_{jk}\,(t) + \eta\,h(k)\,[v_{ij}\,(t) - w_{jk}\,(t)] \tag{2}$$

where η, called learning rate, is used to accelerate or decelerate the convergence of the weight matrix **W** towards the final value, and function h(k) typically ranges between 0 and 1. Here, it is assumed as a function of the distance of the neuron k from the winning neuron according to the following form:

$$h(k) = 1/\text{ord}\,(k) \tag{3}$$

where ord(k) is the position in the distance scale of the neuron k with respect to the winning neuron. The updating of the matrix **W** due to the presentation of all the items is called *learning epoch*. The algorithm terminates after a prefixed number of learning epochs or when the distance between the matrix **W** at epochs n and n+1 is under a prefixed threshold.

To improve the convergence towards a solution that better preserves the topology of the original data set, it is here proposed to apply the updating formula proposed in [5] with a periodic correction that further addresses the weights towards the best solution. Such correction consists in executing after some epochs a learning epoch limited to the items whose vector $\mathbf{s}_i$ activate the output neuron j although the distance between $\mathbf{s}_i$ and the center of the cluster j, also called codebook, is not the minimal one. Following this heuristics, the above formula is applied only to the weights connecting the input neurons to the output neuron related to the cluster whose codebook is at the minimal distance from $\mathbf{s}_i$.

Of course, a suitable high parallelization is requested to apply the above algorithm to cluster massive datasets in a time compatible with the times needed by interactive applications (i.e., in few minutes) or to cluster limited datasets in real time (i.e., in few seconds). Let us recall that two main types of SOM parallel clustering methods are available in the literature: network partition and data partition methods [1]. In the former method the entire dataset is processed by each computation node which is provided with a part of the neural network, whereas in the latter one the data are subdivided in slices that are passed to the computation nodes provided with the same neural network. In [1] it is recalled that the network partition methods are more suitable for highly coupled parallel computation systems (e.g., supercomputer or massive parallel computers), whereas data partition methods are suitable for the low-cost slightly coupled systems (e.g., Grid and Cloud of PCs).

Thanks to the increasing processing power of the PCs and to the diffusion of CUDA PCs provided with many cooperating processing units, the time performance of the latter methods is approaching to the ones of the former methods without sacrificing the accuracy, especially if one applies the mentioned simple correction of the parallel clustering. Also, the data size to be processed by each slave in the data partition method is much less than the one to be processed by the computation nodes of the network partition method. This opens the possibility of clustering the information that supports real time control algorithms by using a cloud of many small systems following our parallel method instead of a central system executing the original sequential SOM.

To fully appreciate how the parallel SOM may be used effectively to cluster datasets for real time applications, let us recall the master-slave parallel algorithm proposed in [1] where the dataset is first subdivided by a master into data slices, e.g., the dataset referring to the pixels of an image is subdivided in slices consisting of sets of adjacent pixels of the image. Then each slave will execute one learning epoch related to its data slice according to the above mentioned sequential SOM. At the end of the learning epoch each slave will send the weights of the proper neural network to the master where the weights are averaged and resent to the slaves for a new learning epoch starting from the same averaged weights. In [1] we have demonstrated that this algorithm converges towards a solution, not necessarily the optimal one. Thus, we propose in this paper that the method will be provided with a measure of topographic error that may cause the execution of the mentioned correction epoch when it is outside a certain threshold, e.g., 10%.

3 Clustering Videos Coming from the Field to Improve LBSs

A relevant field of application of the mentioned clustering method is the one of studying physical systems evolving in time by clustering the consecutive frames of a video related to the process under study. This implies that the time $\Delta_{f,}$ spent to elaborate one frame should be less than the time constant of the evolving process. Considering that the speed of the main physical processes involved the LBSs is moderate, it is enough that $\Delta_{f,}$ is of one hundred of millisecond to obtain from two consecutive frames all the relevant LBS parameters such as the density of the traffic flow in a road or the diffusion rate of a pollutant in the air. Since in [1] we have shown that this time performance is easily satisfied by the clustering method, in the following we illustrate how the method may be used in practice to obtain real time LBSs based on the processing of two consecutive frames. In all the examples the frames consist of images of 30 x 30 pixels. Each image is clustered by 40 nodes working in parallel in few hundred of milliseconds.

3.1 Emergencies

Typically, security LBSs aim at sending the information useful in case of emergencies such as the one that help people for finding safe zones or the escape routes in a disaster (e.g., fires, earthquake, storms), and for avoiding zones affected by car congestions, or that help the city managers to avoid that the urban pollution increases outside the allowed ranges, or to evacuate orderly way a people crowd. All these cases deal with statistical processes and would require very sophisticated mathematical models to be executed by powerful servers in order to inform timely citizens and managers. Since the server may be out of service during the mentioned crisis, simple alert system may be also useful such as the ones that operate on simple rules based systems that on their turn make use of information coming from the field.

3.1.1 Pollution or Storm Evolution

Fig.1 shows how the information about bad weather events may be identified using the proposed approach starting from the videos taken by the camera of the mobiles. In particular, the algorithm derives quantitative measurements of a tornado (e.g., speed, height, width, direction, etc.) able to alert the people located within a suitable area around the tornado so that they may follow timely the safety instructions.

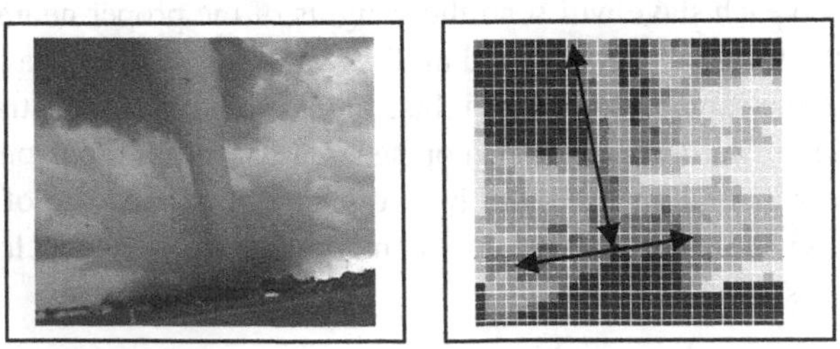

Fig. 1. Sudden formation of a tornado featured by very relevant dimensions.

Fig.2 shows another application of the parallel clustering algorithm to discover from the quantity of the emitted fumes if the emission of an industrial plant is excessive and if the direction of the fumes is not suitable to disperse the powders in a large zone around the industrial site. Also, the grey level of the fumes may be detected to inform the interested people of a potentially dangerous event so that they may follow as soon as possible the relevant safety instructions.

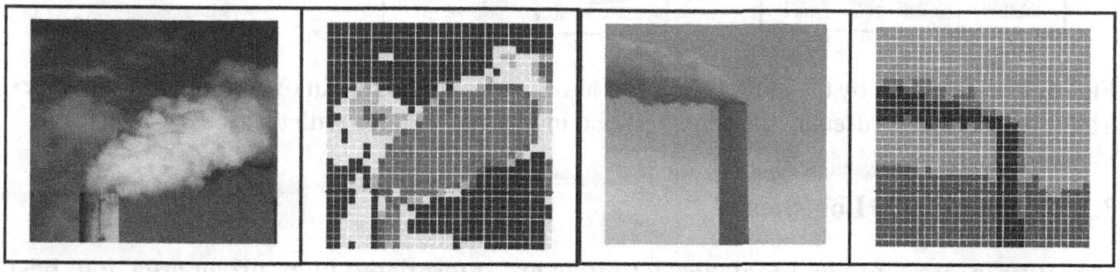

Fig. 2. Air pollution monitoring: a) quasi-vertical emission of medium quantity of powders, b) quasi-horizontal emission of high quantity of powders. The dark zones are colored in blue.

3.1.2 Congested Zones

Traffic congestion due to accidents, sudden increase of car flows, bad weather conditions should be communicated very soon to the drivers so that may change their routes or may decide to postpone the departure. In this case the decision of sending the alarm may be obtained by processing short videos sent from the cameras installed in known critical points of the traffic network. In particular, this information may be obtained by clustering the difference of two frames of the video separated by few seconds. Fig.3 shows that this solution is very effective. Indeed, if the pixels of the difference belong to the same cluster (as in fig.3d), the traffic is blocked and the drivers in the neighborhood should be timely informed.

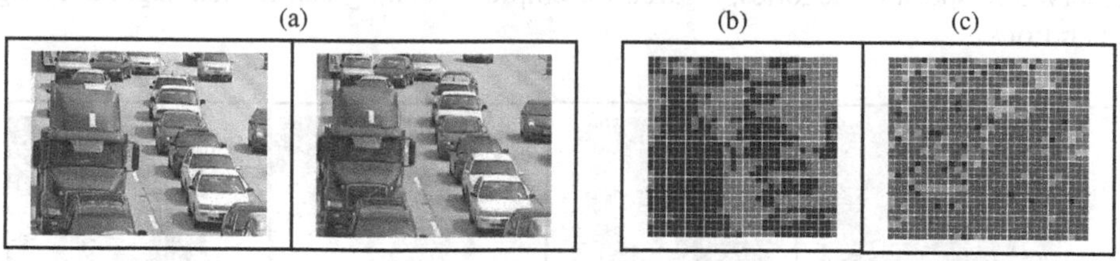

Fig. 3. Identifying traffic congestions: a) Frames separated by 5 seconds. b) clustering of the first frame (cars in blue, asphalt in purple), c) Frame difference (traffic blocked in red, very slow car motion in green).

3.1.3 Escape Routes

Traffic congestions are abnormal situations that may arise in more or less expected points of the traffic network. However, there are other more dangerous events that may appear in any zone of the traffic network suddenly due to less predictable conditions such as the overflowing of a river into a county road. In this case the clustering of the frames of a video received from the camera of a driver carried out by the computing node of the monitoring network closest to the event may identify the relevance of the overflowing and suggest suitable escape routes as shown in fig.4.

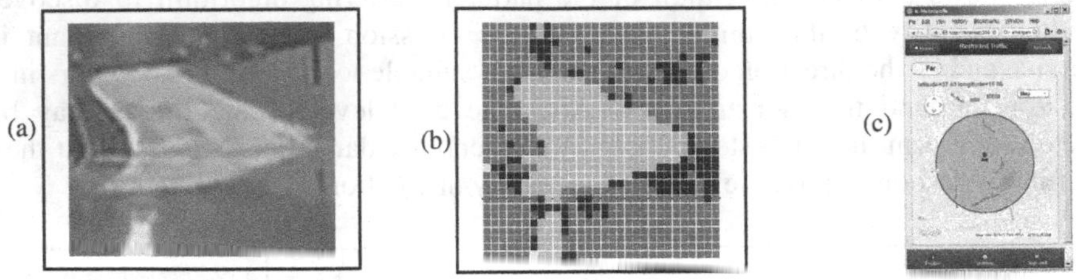

Fig. 4. The detection by the clustering (b) of a road interrupted for an overflowing (a) produces a alert to the mobile users in the neighborhood indicating the area with restricted traffic (c).

3.2 Mobility and Logistics

Mobility information and logistics activities are interrelated in an urban area, e.g. best routes for delivering and collecting goods depend on the traffic conditions. Also, parking info is needed by both citizens for shopping and delivery companies for loading/unloading activities. Both these information systems should be based on that make use of accurate sensing technologies to compute the current traffic parameters or the parking vacancies. Due to the high cost needed to cover the urban area with accurate sensing systems, alternative simple sensing technologies are presented in the following to inform effectively the users with low-cost control systems.

3.2.1 Parking Vacancy

The knowledge of the parking vacancies is very important to avoid useless mobility. Fig.5 shows that also this information may be obtained by clustering a parking area in two main clusters by using black and white cameras installed on the field. The same figure shows that using RGB information we may block the parking in stays reserved to special categories, e.g. red for employers and green for loading/unloading operations.

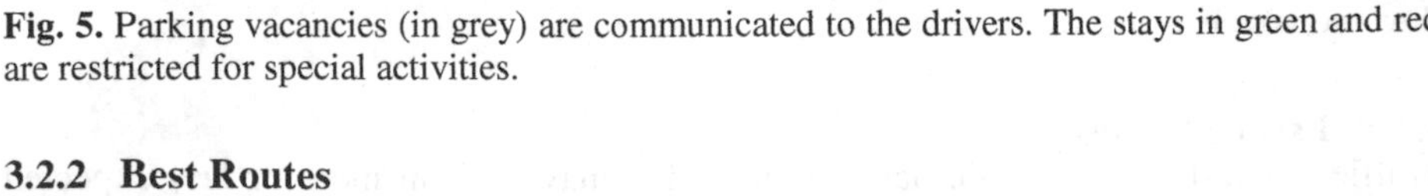

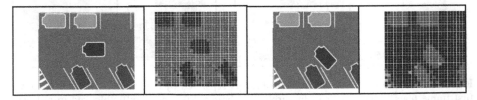

Fig. 5. Parking vacancies (in grey) are communicated to the drivers. The stays in green and red are restricted for special activities.

3.2.2 Best Routes

Better routes to destination should be pointed every time the traffic involving a certain street changes from its normal flowing status. This may arise for two main reasons: a) an increase of the traffic flow beyond a certain threshold, and b) external causes such as heavy rains or disordered parking. Fig.6 shows how the clustering method applied to the difference between the current frame and the background computed in two virtual spires may be used to identify the instant in which a car passes through consecutive spires. For each spire, this instant is given by the moment in which the clustering of the difference between the foreground and background of the frame achieves its maximum value. If the method measures a very low speed, the neighboring drivers should be invited to follow alternative paths to destination. For example in fig.6 the monitoring system measures a speed of 3 km/h (due to the heavy rain) and consequently the neighboring drivers will receive on their mobiles an alert to choose alternative routes to destination according to the approach proposed in [7].

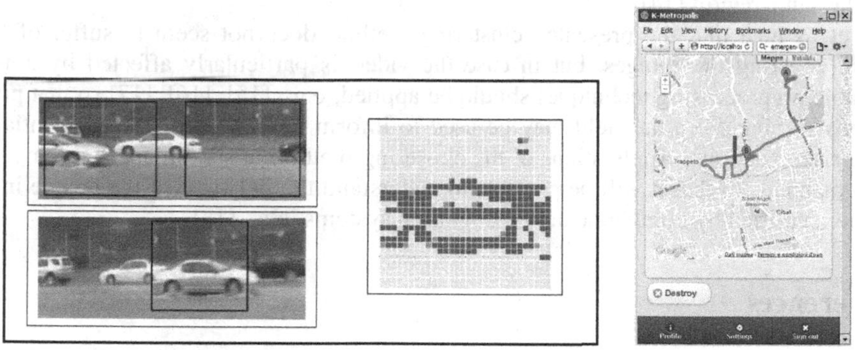

Fig. 6. The passage of a car within the virtual spire is detected when the red pixels reaches the maximum value. The clustering refers to the difference between the foreground and the background. If the traffic is too slow the street is not taken in account to reach the destination and the street affected by the heavy rain is signaled (in red) to the mobile users.

4 Concluding Remarks

The paper has pointed out how the parallel clustering of videos coming from the field may improve the performance of the current LBSs and may open the possibility of novel real time applications that make use of a low cost processing infrastructure and chip videos taken from the cameras of the user mobiles. In particular, a fast parallel clustering that preserves the topology of the original datasets has been illustrated to meet the time performance needed to cluster in real time the frames of physical processes involved in the main LBSs. Also, novel possible applications, especially the ones dealing with emergency management, have been outlined.

However, let us note that all the proposed LBS applications of the clustering algorithm illustrated in the paper are really useful only if the alert information is sent to the user mobiles provided with appropriate interfaces, thus the grphical interface is an integral part of the applications [9]. For this reasons, in the project K-Metropolis led by our University and supported by EC funds we have developed a suitable mobile architecture, named Wi-City, to offer the mentioned SOM based LBSs through a Google Maps interface. In this architecture, the information is sent to the mobile users by a network of computing nodes provided with Ruby on Rails servers [8]. Suitable JQMobile scripts [10] are used to inform effectively the user mobiles about relevant events taking into account their current position as illustrated in fig.4 a dangerous zone around a given critical point and in fig.7 to propose alternativet route to destination depending on the weather conditions.

On the basis of the feasibility study presented in this paper, we plan to start an experimental phase of Wi-City where both time measurements and user feedback will be collected and analyzed to implement a suitable ubiquitous alert system and to activate the mentioned real time mobility-logistics applications for helping the drivers of a metropolitan area. An LBS ontology is also under development inspired by the proposals presented in [11], [12], [13] to make possible the use of the applications in wider areas so that the drivers may be informed in the same way at least in every city of the same region [14].

Let us note that the presented clustering method does not seem to suffer of the noise covering the images, but in case the video is particularly affected by noise, suitable preprocessing techniques should be applied, e.g., [15], [16], [17] so that poor videos coming from the field may be used to inform mobile users about potentially high risk situations. Applications of the clustering method in surveillance systems are also planned so that it will be possible to understand the behavior of the people in a scene identified by complementary recognition systems, e.g., [18].

References

1. Faro, A., Giordano, D., Maiorana, F.: Mining massive datasets by an unsupervised parallel clustering on a GRID: Novel algorithms and case study. Future Generation Computer Systems 27(6), 711–724 (2011)
2. Faro, A., Giordano, D., Maiorana, F., Spampinato, C.: Discovering genes-diseases associations from specialized literature using the grid. IEEE Transactions on Information Technology in Biomedicine 13(4), 554–560 (2009)
3. Denny, G.W., Christen, P.: Redsom: Relative density visualization of temporal changes in cluster structures using self-organizing maps. In: Eighth IEEE Int. Conference on Data Mining, ICDM 2008, pp. 173–182. IEEE (2008)
4. Kohonen, T., Schroeder, M.R., Huang, T.S.: Self-Organizing Maps, 3rd edn. Springer-Verlag New York, Inc., Secaucus (2001)
5. Jin, H., Shum, W.H.K., Leung, K.S., Wong, M.L.: Expanding self-organizing map for data visualization and cluster analysis. Inf. Sci. 163, 157–173 (2004)
6. Faro, A., Giordano, D., Maiorana, F.: Discovering complex regularities by adaptive self organizing classification. World Academy of Science, Engineering and Technology 4, 27–30 (2005)
7. Faro, A., Giordano, D., Spampinato, C.: Evaluation of the traffic parameters in a metropolitan area by fusing visual perceptions and CNN processing of webcam images. IEEE Transactions on Neural Networks 19(6), 1108–1129 (2008)
8. Hartl, M.: Ruby on Rails 3. Addison Wesley (2011)
9. David, M.: Developing Websites with JQueryMobile. Focal Press (2011)
10. Giordano, D.: Evolution of interactive graphical representations into a design language: a distributed cognition account. International Journal of Human-Computer Studies 57(4), 317–345 (2002)
11. Faro, A., Giordano, D., Spampinato, C.: Integrating location tracking, traffic monitoring and semantics in a layered ITS architecture. Intelligent Transport Systems, IET 5(3), 197–206 (2011)

12. Vilches Blázquez, L.M., et al.: Towntology & hydrOntology: Relationship between Urban and Hydrographic Features in the Geographic Information Domain. SCI, vol. 61, pp. 73–84 (2007)
13. Faro, A., Giordano, D., Musarra, A.: Ontology based intelligent mobility systems. In: Proc. IEEE Conference on Systems, Man and Cybernetics, vol. 5, pp. 4288–4293. IEEE (2003)
14. Zhai, J., Jiang, J., Yu, Y., Li, J.: Ontology-based Integrated Information Platform for Digital City. In: IEEE Proc. of Wireless Communications, Networking and Mobile Computing, WiCOM 2008. IEEE (2008)
15. Cannavò, F., Nunnari, G., Giordano, D., Spampinato, C.: Variational method for image denoising by distributed genetic algorithms on grid environment. In: Proceedings of the 15th IEEE International Workshops on Enabling Technologies: Infrastructure for Collaborative Enterprises, WETICE 2006, pp. 227–232. IEEE, Washington, DC (2006)
16. Crisafi, A., Giordano, D., Spampinato, C.: GRIPLAB 1.0: Grid Image Processing Laboratory for Distributed Machine Vision Applications. In: Proc. Int. Workshops on Enabling Technologies: Infrastructure for Collaborative Enterprises, WETICE 2008. IEEE (2008)
17. Spampinato, C., Giordano, D., Di Salvo, R., Chen-Burger, Y.H.J., Fisher, R.B., Nadarajan, G.: Automatic fish classification for underwater species behavior understanding. In: Proceedings of the First ACM International Workshop on Analysis and Retrieval of Tracked Events and Motion in Imagery Streams, ARTEMIS 2010, pp. 45–50. ACM (2010)
18. Faro, A., Giordano, D., Spampinato, C.: An automated tool for face recognition using visual attention and active shape models analysis. In: Proc IEEE Conference in Medicine and Biology Society, EMBS 2006, pp. 4848–4852 (2006)

The Classical Solution for Singular Distributed Parameter Systems

Man Liu[1], XiaoMei Yuan[2], Dan Hang[1], and JinBo Liu[1]

[1] Department of Basic Education, Xuzhou Air Force College, Xuzhou 221000, P.R. China
[2] Department of Aviation Ammunition, Xuzhou Air Force College, Xuzhou 221000, P.R. China
liuman8866@163.com

Abstract. In this paper the classical solution for the singular distributed parameter systems is discussed, we obtain the classical solution of the different system, it is important to the next study for the control of accuracy of the singular distributed parameter systems.

Keywords: Singular distributed parameter systems, Generalized operator semigroup, Classical solution.

1 Introduction

There is an essential distribution between singular and ordinary distributed parameter systems [1-8]. Singular distributed parameter systems are much more often encountered than the distributed parameter systems. They appear in the study of the temperature distribution in a composite heat conductor, voltage distribution in electro-magnetically coupled superconductive circuits, signal propagation in a system of electrical cables, etc. Under disturbance, not only they lose stability, but also great changes take place in their structure, such as leading to impulsive behavior, etc.

One of the most important problems for the study of singular distributed parameter systems is the well-posed. It not only is important to the study of the stability for singular distributed parameter systems, but also is basic to the study of the related optimal control problem.

Let H, H_1 be two separable and infinite dimensional complex Hilbert spaces, $x(t) \in H$, E a bounded linear operator, A a linear operator in Hilbert space H . B a bounded linear operator from H_1 to H ,and $u(t) \in H_1$.

2 Main Definitions and Properties

Definition 2.1. A one-parameter family of bounded linear operators $\{U(t): t \geq 0\}$ on H, E is a bounded linear operator, if $U(t+s) = U(t)EU(s)$, $t \geq 0, s \geq 0$, then $\{U(t): t \geq 0\}$ is called the generalized operator semigroup induced by E , or is called generalized operator semigroup in short.

If the generalized operator semigroup $U(t)$ satisfies $\lim_{t \to 0^+} \|U(t) - U(0)\| = 0$, then $U(t)$ is uniformly continuous.

D. Jin and S. Jin (Eds.): Advances in FCCS, Vol.1, AISC 159, pp. 529–533.
springerlink.com

If the generalized operator semigroup $U(t)$ satisfies $\lim_{t\to 0^+}\|U(t)x-U(0)x\|=0$, $\forall x\in H$, then $U(t)$ is strongly continuous.

Property 2.2. If $U(t)$ is a generalized operator semigroup and strongly continuous on H, then

(i) There exist constants $M>1$, and $\omega\geq 0$, such that $\|U(t)\|\leq Me^{\omega t}$, $t\geq 0$, then $U(t)$ is exponentially bounded.

(ii) $U(t)$ is strongly continuous on H, for $\forall x\in H, t\geq 0$, $\lim_{h\to 0}\|U(t+h)x-U(t)x\|=0$

We assume that the set $\rho_E(A)=\{\lambda\in C: R(\lambda E,A)=(\lambda E-A)^{-1}$ is a bounded operator on H $\}$ is nonempty. which is called the E -resolvent set of operator A ,and the operator $R(\lambda E,A)$ is called the E -resolvent of operator A .

Definition 2.3. Let A be a linear operator and E a bounded linear operator in Hilbert space H . $U(t)$ is a strongly continuous generalized operator semigoup induced by E ,if $(\lambda E-A)^{-1}=\int_0^{+\infty}e^{-\lambda t}U(t)dt$, $\lambda\in C$, $\mathrm{Re}\lambda>\omega$, then A generates the strongly continuous generalized operator semigroup $U(t)$ induced by E .

Property 2.4. Let E be bounded operator, $A: D(A)\to H$ a closed linear operator, when λ, $u\in\rho_E(A)$, we have

(i) $R(\lambda E,A)-R(uE,A)=(u-\lambda)R(\lambda E,A)ER(uE,A)$ (2)

(ii) $\frac{d^n}{d\lambda^n}R(\lambda E,A)=(-1)^n n!R(\lambda E,A)[ER(\lambda E,A)]^n$, $n=1,2,\cdots$

Remark: Identity (2) is called the resolvent identity.

Considering the homogeneous systems of first order singular distributed parameter

systems $\begin{cases} E\frac{dx(t)}{dt}=Ax(t) \\ x(0)=x_0 \end{cases}$ (1)

Definition 2.5. For system (1), $x_0\in H$ .if there exists $x(t)\in H$, $x(t)$ is continuous for $t\geq 0$. $x(t)\in R(\lambda E)H$, $x(t)$ is strongly continuously differentiable ($t>0$), and $x(t)$ satisfies (1), then $x(t)$ is called the classical solution.

Lemma 2.6[7]. For system (1), let E, A be linear operator, E is bounded, A is closed, A is the generator of a strongly continuous generalized operator semigroup $U(t)$ induced by E ,then system (1) is uniformly well-posed on $ER(\lambda)EH$.

Lemma 2.7[7]. For system (1), let A be closed, E be bounded, and the E-resolvent set of operator A non-empty. If A is the generator of a generalized operator semigroup $U(t)$, then

(a) a classical solution $x(t)$ exists for $x_0 \in R(\lambda E)H$ and $x(t) = U(t)x_0$;

(b) the classical solution is unique and uniformly stable for any $T > 0$ and $t \in [0,T]$ with respect to $x_0 \in R(\lambda E)H$.

3 Main Theorems

Considering the following inhomogeneous systems of first order singular distributed parameter systems $\begin{cases} E\dfrac{dx(t)}{dt} = Ax(t) + f(t) \\ x(0) = x_0 \end{cases}$ (3)

Theorem 3.1. Let E be bounded linear operator, A a closed linear operator, and A is the generator of strongly continuous generalized operator semigroup $U(t)$ induced by E, $f(\cdot) \in C^1([0,+\infty),H)$, $f(t)$ and $\dfrac{df(t)}{dt}$ is bounded on $[0,+\infty)$, i.e., there exists $M_1 > 0$, such that $\|f(t)\| \le M_1$, $\left\|\dfrac{df(t)}{dt}\right\| \le M_1$, $t \in [0,+\infty)$, then there exists the unique classical solution of the system (3) for $\forall x_0 \in ER(\lambda)EH$.

Proof. Let $\|U(t)\| \le M e^{\omega t}, \omega > 1$, defining $y(t) = \int_0^t U(t-s)f(s)ds$, then

$$\frac{y(t+h)-y(t)}{h} = \frac{\int_0^{t+h} U(t+h-s)f(s)ds - \int_0^t U(t-s)f(s)ds}{h} \text{ (Letting } h - s = -u \text{)}$$

$$= \frac{\int_{-h}^t U(t-\theta)f(h+\theta)d\theta - \int_0^t U(t-s)f(s)ds}{h} = \frac{\int_{-h}^0 U(t-\theta)f(h+\theta)d\theta}{h}$$

$$+ \frac{\int_0^t U(t-s)[f(h+s)-f(s)]ds}{h}$$

Then $\dfrac{dy(t)}{dt} = \lim\limits_{h\to 0}\dfrac{y(t+h)-y(t)}{h} = U(t)f(0) + \int_0^t U(t-s)\dfrac{df(s)}{ds}ds$

i.e. $\dfrac{dy(t)}{dt} \in C([0,+\infty),H)$

Owing to $\|y(t)\| \le M_1 \int_0 Me^{\omega(t-s)}ds = M_1Me^{\omega t}\left(\frac{1}{\omega}-\frac{e^{-\omega t}}{\omega}\right) = \frac{M_1Me^{\omega t}}{\omega}$

Consequently, for $\operatorname{Re}\lambda > \omega$, we have $\int_0^{+\infty} e^{-\lambda t}y(t)dt = \int_0^{+\infty} e^{-\lambda t}\int_0 U(t-s)f(s)dsdt$

$= \int_0^{+\infty} f(s)\int_s^{+\infty} e^{-\lambda t}U(t-s)dtds$ (Letting $t-s=t'$) $= \int_0^{+\infty} e^{-\lambda t}U(t)dt\int_0^{+\infty} e^{-\lambda s}f(s)ds$

$= (\lambda E - A)^{-1}\int_0^{+\infty} e^{-\lambda t}f(t)dt$

By the unique of Laplace transformation, $y(t) = (\lambda E - A)^{-1}f(t)$,

So $y(t)\in D(A)$,and $\|Ay(t)\| \le \|(\lambda E - A)y(t)\| + \|\lambda Ey(t)\| \le M_1 + \|\lambda E\|\frac{M_1Me^{\omega t}}{\omega}$

Hence,for $\operatorname{Re}\lambda > \omega$, $\lambda\int_0^{+\infty} e^{-\lambda t}Ey(t)dt - \int_0^{+\infty} e^{-\lambda t}Ay(t)dt = \lambda E\int_0^{+\infty} e^{-\lambda t}y(t)dt$

$-A\int_0^{+\infty} e^{-\lambda t}y(t)dt = \int_0^{+\infty} e^{-\lambda t}f(t)dt$ (4)

Be similar to $y(t)$, we can prove for $\operatorname{Re}\lambda > \omega$, $\int_0^{+\infty} e^{-\lambda t}\frac{dy(t)}{dt}dt$ exists.

So $\int_0^{+\infty} \lambda e^{-\lambda t}Ey(t)dt = -e^{-\lambda t}Ey(t)\Big|_0^{+\infty} + \int_0^{+\infty} e^{-\lambda t}E\frac{dy(t)}{dt}dt = \int_0^{+\infty} e^{-\lambda t}E\frac{dy(t)}{dt}dt$ (5)

Combining (5) with (4), and making use of the unique of Laplace transformation, we have $E\frac{dy(t)}{dt} = Ay(t) + f(t)$, $y(0)=0, t\in[0,+\infty)$.

Consequently the solution of system (3) is $x(t) = U(t)x_0 + \int_0 U(t-s)f(s)ds$.

Supposing $x_1(t), x_2(t)$ are the classical solution of system (3), let $y(t) = x_1(t) - x_2(t)$, then $y(t)$ satisfies $E\frac{dy(t)}{dt} = Ay(t)$, $y(0)=0$ (6), By lemma2.6,the system(6) only has the zero solution $y(t)=0$, so $x_1(t) = x_2(t)$.

Considering the following inhomogeneous systems of first order singular distributed parameter systems
$$\begin{cases} E\dfrac{dx(t)}{dt} = Ax(t) + Bu \\ x(0) = Ex_0 \end{cases} \quad (7)$$

Theorem 3.2. Let A be the generator of strongly continuous generalized operator semigroup $U(t)$ induced by E , $u(\cdot)\in C^1([0,+\infty),H)$, $u(t)$ and $u'(t)$ is bounded on $[0,+\infty)$,i.e., there exists $M_1>0$,such that $\|u(t)\|\le M_1$, $\|u'(t)\|\le M_1$, $t\in[0,+\infty)$,then there exists the unique classical solution of the system (3) for $\forall x_0\in(\lambda E-A)^{-1}EH$, and $x(t)=U(t)Ex_0+\int_0^t U(t-s)Bu(s)ds$

Proof. Let $\|U(t)\|\le M\,e^{\omega t},\omega>1$, defining $y(t)=\int_0^t U(t-\tau)Bu(\tau)d\tau$ .Then the following proof is the same as theorem3.1.

References

1. Joder, L., Femandez, M.L.: An implicit difference methods for the numerical solution of coupled system of partial differential equations. Appl. Math. Comput. 46, 127–134 (1991)
2. Lewis, F.L.: A review of 2-D implicit systems. Automatic 28, 245–254 (1992)
3. Hu, Y., Peng, S.G., Li, X.J.: Maximum priciple for optimal control problem of nonlinear generalized systems-infinite dimensional case. Acta. Math. Appl. Sin. 15, 99–104 (1992)
4. Trzaska, Z., Marszalek, W.: Singular distributed parameter systems. IEEE Control Theory Appl. 40, 305–308 (1993)
5. Ahmed, N.U.: Semigroup Theory with Application to System and Control. Longman Scientific and Technical press, New York (1991)
6. Melnikova, I.V., Filinkov, A.I.: Abstract Cauchy problem, pp. 86–89. Chapman & Hall/CRC, London (2001)
7. Ge, Z., Zhu, G., Feng, D.: Generalized operator semigroup and well-posedness of singular distributed parameter systems. Science in China, Ser. A 40(5), 477–495 (2010)
8. Ge, Z.Q., Zhu, G.T., Feng, D.X.: Degenerate semi-group methods for the exponential stability of the first order singular distributed parameter systems. J. Syst. Sci. Complex 21, 260–266 (2008)

Changing the Equipment of Old-Solar Water Heater to Control Water Automatically

XiaoMei Yuan[1], Man Liu[2], Qing Zhang[1], Yan Wei[1], and GuiKao Yang[1]

[1] Department of Aviation Ammunition, Xuzhou Air Force College, Xuzhou, Jiangsu, P.R. China, 221000
[2] Department of Basic Education, Xuzhou Air Force College, Xuzhou, Jiangsu, P.R. China, 221000
liuman8866@163.com

Abstract. Aiming at old-solar water heater has not the function of feeding water automatically, we introduce one kind of simple and convenient equipment can control water automatically. Comparing with the equipment in the market, the equipment we introduce has the characteristic, such as it can be fixed easily, maintain simply and conveniently, has lower breakdown, higher cost performance.

Keywords: water heater of solar energy, tree direct links pipe, electromagnetic valve, automatic water control device.

Introduction

All manuscripts must be in English, also the table and figure texts, otherwise we cannot publish your Supplying heater water has occupied most part of energy consume in the family and business user. In the world most of families consume energy, the requirement of heat water occupied quarter or one third. For example, in Japan consuming heat water had occupied 34 percent in the family in 1999.In recent years, along with the transformation of people's living habits and the improving of living quality, the need of heat water is more and more. At present inland most families consume heat water by means of special water heater in everyday life(such as electronic water heater, gas water heater and so on).Today the condition of energy and environment has become more and more severe, with pollution-free, safety and exhaust never natural energy has many advantage, here solar energy as one kind of representation new energy has been regarded as the hopeful and best of all promising energy, people pay attention to solar energy more and more.

1 The Principle of Automatically Controlling Water Equipment by Solar Energy Water Heater

1.1 The Principle of Solar Energy

The sunshine radiates the glass cover board, then the collected heat board absorbs the sunshine, then the water in the decalescence is heat along the rib slice and pipe wall.

D. Jin and S. Jin (Eds.): Advances in FCCS, Vol. 1, AISC 159, pp. 535–539.
springerlink.com

Then it consist of a system of hot rainbow absorbing, following with the continuously shifting of hot water, and constantly supple the cold water from circulate pipe, in this way it circulates repeatedly, at final the water in the whole box rises to certain temperature.

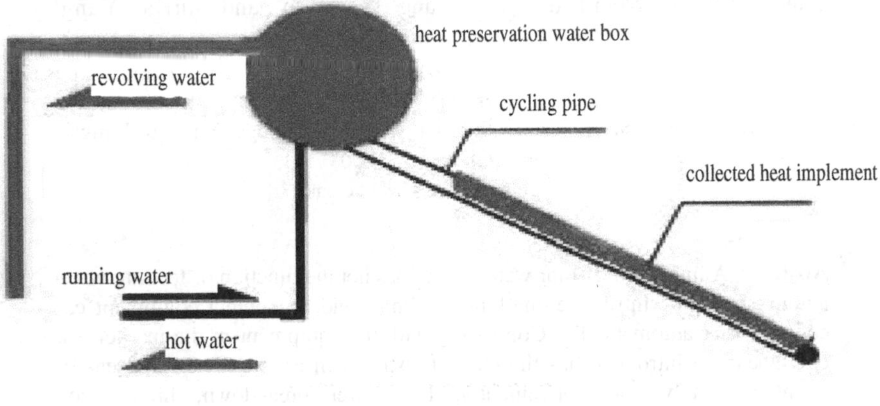

Fig. 1. Schematic diagram of solar water heater structure

1.2 The Details for the Design of the Equipment Automatically Controlling Water

This equipment is that switching on an electromagnetic valve on the original inflowing water with parallel connection, the user push down the button of inflowing water, then the controller launches the electromagnetic valve, when the water is full of the water box, then it will return to the original position, by now the reaction needle will work, this touches off the controller and turns off the electromagnetic valve, and it realizes the function of water controlled automatically.

1.3 The Principle for the Equipment of Water Controlled Automatically

Referring to the Fig. 2, when the water inflows to the water box, we press on the button SQ that sends water, then switch on and put through the controlling electro-circuit. Owing to the break of probe 1,2 the preposition V1 works free, the direct current of +12V goes through R2 and C1, R4 and C2 compose of π-model filter. When the solar energy water box is full of water, the water will overflow from the revolving pipe, two branches probe 1 and 2 break-over when they encounter water. When the preposition pipe V1 works, the compound pipe V2 and V3 stop to work, then the relay switches off, the lamp indicating inflowing water switches off, then the water box stops to inflow water.

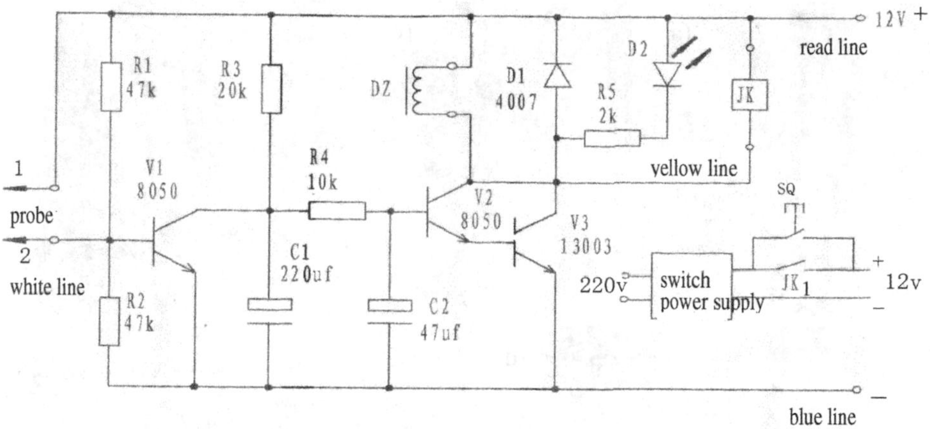

Fig. 2. Circuit diagram of the equipment controlling water automatically

2 Installation and Usefulness for the Equipment of Controlling Water Automatically in the Simple and Convenient Solar Water Heater

2.1 Installation Process of the Equipment of Controlling Water Automatically

At present the material of water pipe generally makes use of aluminum zinc iron pipe, or the PPR pipe (plastic pipe). It is similar to install the two species of materials. Merely the material of water pipe is different. We make use of the different connecting materials. Such as the galvanized zinc iron pipe, we will illustrate the process of installation in detail.

Referring to the Fig.3

(i) At first turn off the total valve of tap-water, at the same time send out the whole surplus water in the solar energy water box.

(ii) Second, take apart (a) and connect (b) with two tree-direct-links pipes at the opening up and down, then install the original water pipe and connect the electromagnetic valve (c) below the tree-direct-links pipes, at last connect (d) with two soft pipes at top and bottom.

(iii) Pay attention to all the connector and make use of raw-material belt to prevent the connector from leaking water.

(iv) Install the response probe (e) at the revolving water pipe and connect the controlling electric circuit box with power supply.

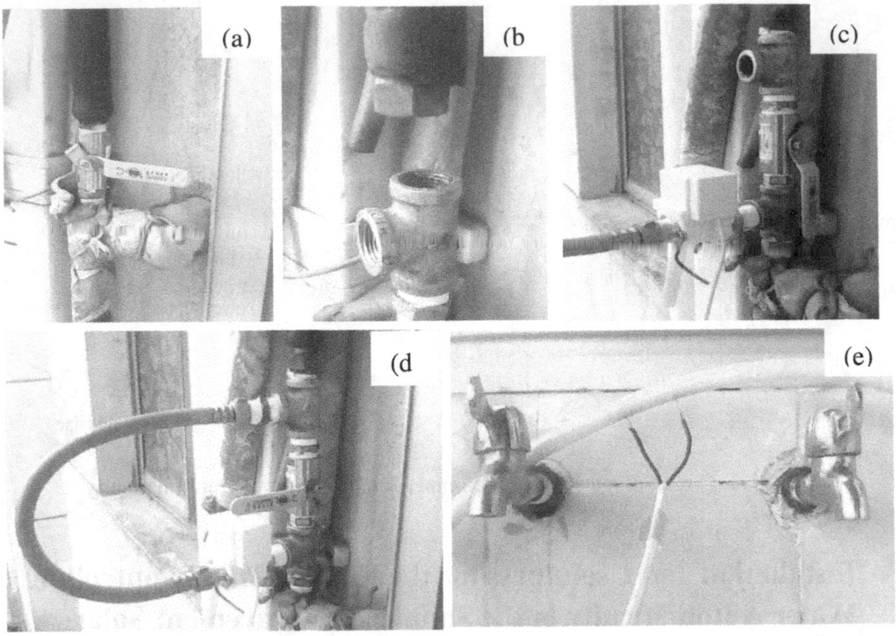

Fig. 3. (a) original inflowing water valve ; (b) take apart (a) and install the tree direct links pipe ;
(c) install the electromagnetic valve ; (d) link the tree direct links pipe up and down with soft pipe ;
(e) install the response probe.

2.2 Process of Using for the Equipment of Controlling Water Automatically

The users need not open the original inflowing water valve, but need only open the button, and cold water can be sent automatically, the operator is so simple and convenient.

Conclusion

In recent year, the solar energy heater water makes use of the sunshine, it has the characteristic such as pollution-free, usefulness-convenient, inputs-low long time, so it fall in love with families. But if the solar energy heater water has not the function of controlling water automatically, it will lead to overflow a lot of water. Consequently it would waste water resource and permeate water at the top of building. The equipment we designed can control water automatically in the solar energy of heater water controller, it is economic and practical, convenient, it is also suitable to popularize.

References

1. Yang, L., Yang, C.J., Wu, S.J., Xie, Y.Q.: Optimal design of an inductively coupled link system for underwater communication. Marine Sciences 35, 9 (2011)
2. Huo, Y.Z., Ma, Y.Q., Yang, Z.: Design of intelligent control system for solar water heater. Coal Technology 30(7), 215–216 (2011)
3. Li, J., Zhang, X.: Introduction to a new kind of solar water heater. Coal Technology 30(6), 212–213 (2011)
4. Zhang, J.J., Yu, M.B., Liu, T.: Construction strategy of using solar water heater in tall apartment building. Renewable Energy Resources 29(1), 90–93 (2011)

Feature Extraction of Iris Based on Texture Analysis

Yufeng He, Zheng Ma, and Yun Zhang

Image Processing and Information Security Lab, Institute of Communication and Information, UESTC, Chengdu, China

Abstract. In general, a typical iris preprocessing system includes image acquisition, quality assessment, normalization and the noise eliminating. This paper focuses on the middle issue and describes a new scheme for iris preprocessing from an image sequence. We must assess the quality of the image sequence and select the clear one from this sequence to the next step. After detecting the pupil coarsely, we get the radius and center coordinate. We can extract local texture features of the iris as our eigenvector, then utilize k-means clustering algorithm to classify the defocused, blurred and occluded image from clear iris image. This method obviously decreases the quality assessment time, especially some people's iris texture are not distinct. Experiments show the proposed method has an encouraging performance.

Keywords: image quality assessment, feature extraction, texture analysis.

1 Introduction

With the development of the current society, personal identification based on biometrics has drawn more and more attention. Iris recognition becomes a hot spot in recent years for its unique advantages.

Iris recognition technology is a new personal identification system based on biometrics since 1990.Nowadays,Iris recognition has its unique advantages of high stability, high reliability and non-intrusive. The people around the world have proposed a lot of different iris recognition algorithms. In this known iris recognition algorithms, John Daugman[1] developed a high grade of mature technology. Of all these biometric features, fingerprint verification is the most mature skill and has been used in law enforcement applications successfully. But when comes to the infectious disease, people don't like to touch the collecting device. Iris recognition is a newly emergent approach to personal identification which has more advantages than disadvantages.

In practical application of iris system, the lens will get series of iris image. It's very important that we adopt a standard to judge whether the images are qualified or not. For example, clarity, blink or shelter from eyelash and eyelid are problems. At present, there are not some particular methods to tackle with this problem. Some common ways such as computing the ratio of high frequency in all the frequency domain of the iris, iris's coefficient of edge kurtosis, iris's gradient energy. Whereas, these methods refered to above all have their defects[7].

D. Jin and S. Jin (Eds.): Advances in FCCS, Vol. 1, AISC 159, pp. 541–546.
springerlink.com

2 Related Work

2.1 Iris Image Quality Assessment

Iris quality assessment is an important content of iris recognition, because the quality of the iris image extremely affects the accuracy of recognition. In today's research, it's still limited in iris quality assessment. Daugman[2] has proposed the high frequency power in the 2-Dimension Fourier spectrum to assess the iris image. However, to do the 2D Fourier transformation is time-consuming. Here, we proposed an effective way to assess the iris image quality by extracting the feature of iris part.

I.local frequency distribution.

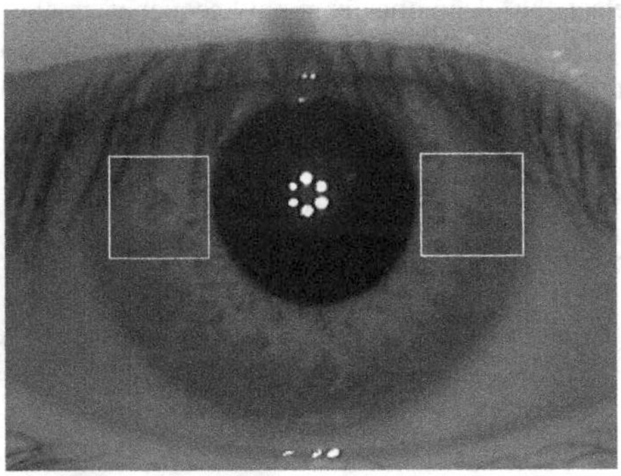

Fig. 1. The rectangle region we choose to compute

As we know, comparing with low quality images, a clear and focused iris image has relatively uniform frequency distribution [5]. In Fig.1.We define the following quality assessment feature:

$$D=[\frac{F_2}{F_1+F_3}]$$
$$F_i=\iint_{\Omega=\left\{(u,v)\middle| f_1^i<\sqrt{u^2+v^2}<=f_2^i\right\}}|F(u,v)|dudv, i=1,2,3. \quad (1)$$

In this formula, $F(u,v)$ is the 2D Fourier spectrum of the rectangle region showed in Fig.1. F_1, F_2 and F_3 are corresponding to the power of low, middle and high frequency. f_1^i, f_2^i are the radial frequency pair and bound the range of the frequency parts respectively. We locate two 64*64 iris regions and compute the descriptor D. In our experiments, we use three frequency pairs of (0,6),(6,22),(22,32).The descriptor D includes two discriminating features. Finally, the mean of the two local descriptors is the ultimate answer that we need. The first one can effectively discriminate clear iris image from occluded one. The second one can effectively discriminate clear iris image from defocused and motion blurred image because clear image have much more middle frequency.

II.Gradient energy.

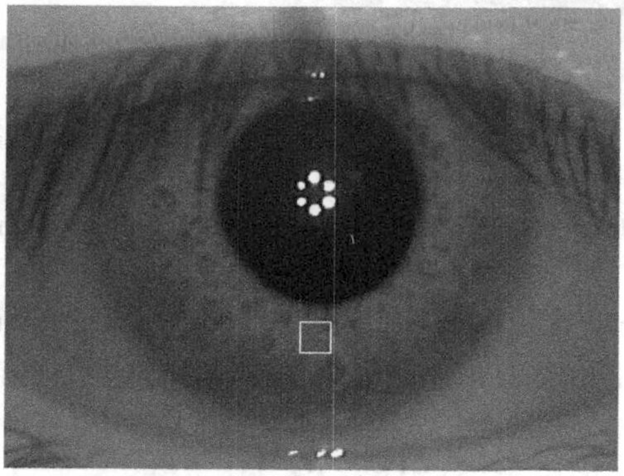

Fig. 2. The rectangle region we choose to compute

Iris's gradient energy describes the iris texture change that we can use it as a feature to express the clarity[6]. In Fig.2.We also take the 20*20 rectangle area as our compute region. We define the following quality assessment feature:

$$E_a = \frac{1}{(M-2)(N-2)}\left[\sum_{i=2}^{M-1}\sum_{j=2}^{N-1}\left(\sum_{k=-1}^{1}(S(i,j)-S(i-1,j+k))^2+\sum_{k=-1}^{1}(S(i,j)-S(i+1,j+k))^2\right)\right] \quad (2)$$

In this formula M=N=20, $S(i, j)$ means the pixel value of coordinate position (i, j). In practice, the quality assessment feature E_a is greater than 100 for clear iris images, while E_a is less than 100 for the defocused, motion blurred and occluded images.

III.Coefficient of edge kurtosis.

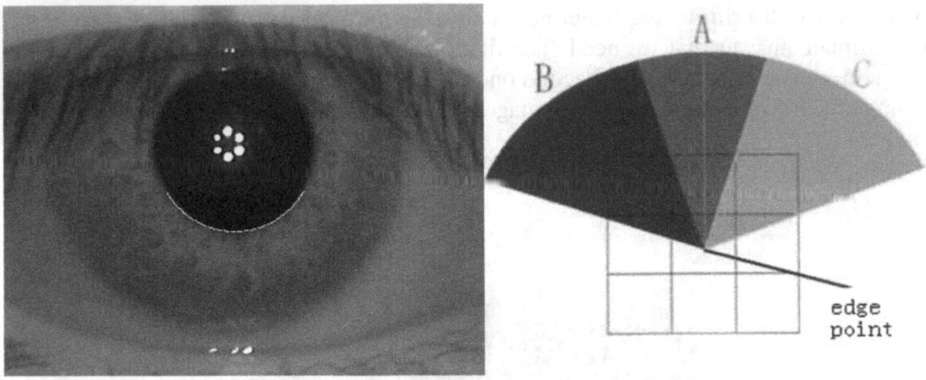

(a) (b)

Fig. 3. (a)The circular boundary we choose to compute. (b)the gradient vector of edge point

Pearson has first proposed this concept in 1905, it is a measurement of information's coefficient of edge kurtosis. It's definition as follows [3]:

$$K = \frac{E[(x-u)^4]}{\sigma^4} \tag{3}$$

In this formula, x means the sample value of random variable, u means the mean value of random variable, σ means the standard deviation of random variable, E means the mathematic expectation of random variable. So if the larger of K, the clearer of the iris image.

In order to reduce the influence of noise of eyelash, we compute the lower edge of the pupil as is showed in Fig.3(a). Digital image is discrete and there are 8 directions for one point. So we should normalize the directions. In Fig.3(b). If the center of pupil is in the A region of edge point, we search for the relevant pixel value in vertical direction to compose a vector. If the center of pupil is in the B region of edge point, we search for the relevant pixel value in top left corner direction to compose a vector. Then the rest can be done in the same manner. We compute the first order gradient value in corresponding direction after done that. Then take the first order gradient into the formula (3), we can get the value of K. In this paper, we choose the points that are under the centre of the pupil left and right 60 degrees as our sample points.

2.2 Feature Classify

For the given quality descriptor D, E_a, K, we take them as 3-dimensions features. We select for these typical features for feature classify so that our system can pass the quality assessment quickly. In our system, we use k-mean clustering algorithm to distinguish whether the iris image is clear. Because defocused, motion blurred and occluded images. Using the algorithm above, one can assess the quality of an iris image precisely and quickly.

3 Experimental Results

In our system, the CASIA Iris Database is adopted [4]. We choose 100 clear images and 90 unclear images (defocused, motion blurred and occluded) to do feature classify. The results are showed in Fig.4.

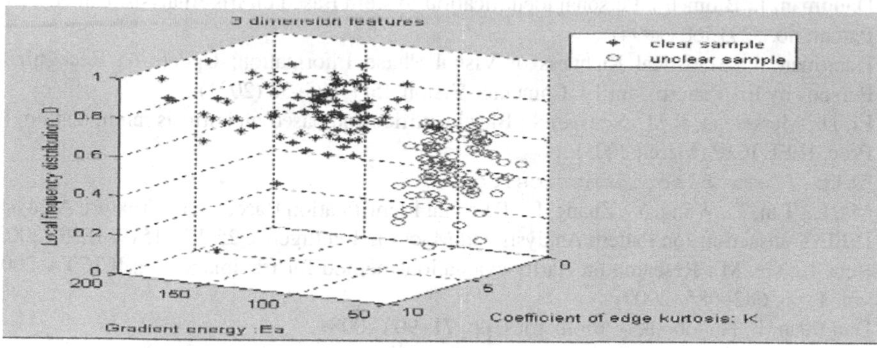

Fig. 4. 3D features of clear sample and unclear sample

From this picture, we can see that there is a clear boundary between clear and unclear sample. So we can classify the clear images from unclear images using k-mean clustering algorithm.

Table 1. The results of image quality assessment

Classification accuracy	Mean time
96.34%	0.06s

We can conclude from Table 1. that our quality assessment algorithm has a high accuracy rate and satisfy the real-time.

4 Conclusion and Future Work

The experiments show that the method has an encouraging performance. Typically iris quality assessment adopts threshold values of the iris image, this is not effective working on people whose iris textures are not distinct. However, our method can overcome the disadvantages above and decrease the quality assessment time, otherwise, this method has a high accuracy rate. Therefore, this method is about to apply in real time iris recognition system. In the near future, we will do experiments to find more iris texture features to design a best classification and improve the accuracy rate.

Acknowledgments. The CASIA Iris Database is available on the web http://www.sinobiometrics.com[4]. Here i would like to pay my thanks to Su Zhang.

References

1. Daugman, J.: Biometric Personal Identification System Based on Iris Analysis. United States Patent, no. 5291560 (1994)
2. Daugman, J.: Statistical Richness of Visual Phase Information: Update on Recognizing Persons by Iris Patterns. Int'l J. Computer Vision 45(1), 25–38 (2001)
3. Li, D., Mersereau, R.M., Simske, S.: Blur identification based on kurtosis minimization. In: Proc. IEEE ICIP (March 2005)
4. http://www.sinobiometrics.com
5. Ma, L., Tan, T., Wang, Y., Zhang, D.: Personal Identification Based on Iris Texture Analysis. IEEE Transactions on Pattern Analysis and Machine Intelligence 25(12), 1519–1533 (2003)
6. Ren, J., Xie, M.: Research on clarity-evaluation-method for iris images. In: ICICTA 2009, vol. 1, pp. 682–685 (2009)
7. Daugman, J.: Handbook of Biometrics, pp. 71–90 (2008)

The Optimization of H.264 Mode Selection Algorithm

Hu Zhang[1] and Lei Meng[2]

[1] Education Science College, Xuchang University 461000, China
[2] College of Computer Science and Technology, Xuchang University 461000, China
{wujix,mengl}@xcu.edu.cn

Abstract. The multi-mode motion estimation algorithm of H.264 standard has the shortcoming of encoding mode complexity and large computation. In order to solve these problems, this paper proposed a fast mode selection algorithm that combined the prediction method with some mode exclusion method. This method can rule out SKIP mode and inter 16 × 16 mode in advance, and divides the rest mode into two parts according to the texture complexity, then rules out the mode contained in each section one by one according to the texture direction, at last, gets the optimal mode. Experimental results show that this method can predict up to 47% of SKIP macroblocks in advance in the case of almost no additional computation, efficiently avoids unnecessary mode selection process, and greatly reduces the encoding complexity and encoding time.

Keywords: H.264, Video encoding, Mode selection, Image difference.

1 Introduction

The latest video encoding standard H.264 [1] adopts the inter prediction mode of high-precision inter prediction mode, its compression efficiency is improved 50% with the comparison of H.263 and MPEG-4, which benefits from the application of multi-mode motion estimation techniques to some extent. The so-called multi-mode motion estimation [2], which divides the 16 × 16 macroblock into 16 × 16,16 × 8,8 × 16,8 × 8,8 × 4,4 × 8,4 × 4, or do the motion compensation with SKIP mode. In the usual video sequence, the adjacent image has the time correlation in general, that, the change of most region of adjacent image is slow. In the region of slow changes, a large proportion of SKIP mode is occupied [3]. Thus, traversing all other modes and identifying the optimal mode before, predicting the SKIP mode in advance can greatly reduce the cost of movement searching for other modes and calculating the rate-distortion.

2 Prediction of SKIP Mode in Advance

As the image sequence has certain continuity and high inter-frame correlation, there are such macroblocks often in the inter mode, whose residual after motion

D. Jin and S. Jin (Eds.): Advances in FCCS, Vol. 1, AISC 159, pp. 547–552.
springerlink.com

compensation is smaller, they become all-zero block or their residuals are very sparse and less than the threshold through DCT and quantization operation. In the H.264 standard, they are called Skipped macroblock [4], the corresponding prediction mode is called SKIP mode [5].

If the current macroblock is a SKIP macroblock, it can be set as all-zero directly without encoding. Therefore, SKIP macroblock mode can reduce the encoding rate significantly in the case of maintaining the image quality [6]. Figure 1 shows the SKIP macroblock decision process of H.264 standard [7].

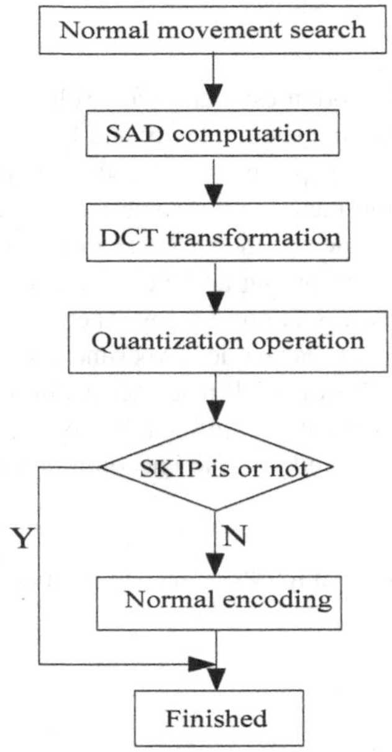

Fig. 1. SKIP decision process

After statistics, there is large proportion of SKIP macroblock in the image sequence [8]. Before the DCT transformation and quantization operation, predicting SKIP macroblock in advance can save the required computation of DCT transformation and quantization operation, therefore can improve the overall encoding rate.

This paper derives the prediction method of SKIP macroblock according to the formula of DCT integer transformation and quantization in H.264. H.264 adopts a similar transformation of 4 × 4 DCT, DCT transformation is $Y=A\times A^T$, the value of A is shown as the formula (1).

$$A=\begin{bmatrix} \frac{1}{2} & \frac{1}{2} & \frac{1}{2} & \frac{1}{2} \\ \sqrt{\frac{1}{2}}cos(\frac{\pi}{8}) & \sqrt{\frac{1}{2}}cos(\frac{3\pi}{8}) & -\sqrt{\frac{1}{2}}cos(\frac{3\pi}{8}) & -\sqrt{\frac{1}{2}}cos(\frac{\pi}{8}) \\ \frac{1}{2} & \frac{1}{2} & \frac{1}{2} & \frac{1}{2} \\ \sqrt{\frac{1}{2}}cos(\frac{3\pi}{8}) & -\sqrt{\frac{1}{2}}cos(\frac{\pi}{8}) & \sqrt{\frac{1}{2}}cos(\frac{\pi}{8}) & -\sqrt{\frac{1}{2}}cos(\frac{3\pi}{8}) \end{bmatrix} \tag{1}$$

The formula of DCT is shown as follows.

$$\begin{gathered} Y(x,y)=C(x)C(y)\sum_{i=0}^{3}\sum_{j=0}^{3}X(i,j)\cos\frac{(2j+1)y\pi}{8}\cos\frac{(2j+1)x\pi}{8} \\ C(x)=\frac{1}{2}(x=0),\ C(x)=\sqrt{\frac{1}{2}}(x>0) \end{gathered} \tag{2}$$

The quantization operation is shown as the formula (3).

$$Z(x,y)=Round\,(\frac{Y(x,y)}{Qstep})=Sign\,(\frac{Y(x,y)}{Qstep})Floor\,(Abs\,(\frac{Y(x,y)}{Qstep})+0.5) \tag{3}$$

In the formula (3), Floor(x) is the largest integer x that less than or equal to x, $Abs(x)=|x|$ and $Sign(x)=1(x>=0), Sign(x)=-1(x<0)$.

Because residuals of SKIP macroblock are zero after DCT and quantization, the quantitative value should satisfy the following formula.

$$|Z(x,y)|=|Floor(Abs(\frac{Y(x,y)}{Qstep})+0.5)|<1 \tag{4}$$

According to the equation (4), the formula (5) can be drawn.

$$|Y(x,y)|<\frac{1}{2}Qstep \tag{5}$$

According to the equation (2), the formula (6) can be drawn.

$$\begin{aligned} |Y(x,y)| &= |C(x)C(y)||\sum_{i=0}^{3}\sum_{j=0}^{3}X(i,j)\cos\frac{(2j+1)y\pi}{8}\cos\frac{(2j+1)x\pi}{8}| \\ &<= |C(x)C(y)||\sum_{i=0}^{3}\sum_{j=0}^{3}X(i,j)| \\ &= |C(x)||C(y)|SAD4 \end{aligned} \tag{6}$$

At different points, C (x) is different. According to the probability distribution, the formula (7) can be drawn.

$$\begin{aligned} |Y(x,y)| &= \frac{1}{4}*\frac{1}{16}*SAD4+\frac{\sqrt{2}}{4}*\frac{6}{16}*SAD4+\frac{1}{2}*\frac{9}{16}*SAD4 \\ &= \frac{25}{64}SAD4 \end{aligned} \tag{7}$$

According to the equation (5) and (7), the formula (8) can be drawn.

$$SAD4 < \frac{32}{25}Qstep \tag{8}$$

Based on the above derivation, if the formula (8) is true, the macroblock can be determined to SKIP macroblock. Thus, each 4 × 4 block requires to judge one time, and a macroblock has 16 sub-blocks of 4 × 4, it would need 16 times conditional judgment. This will generate a lot of jump instructions, which will not be conducive to the implementation of DSP and affect the implementation efficiency of DSP. Therefore, in order to reduce the number of jump instructions, it needs to consider the whole macroblock SAD16.

If the formula (8) is true, the formula (9) is true certainly.

$$SAD16 < \frac{32}{25}*16*Qstep = 20Qstep \tag{9}$$

Although the formula (9) avoids the shortcoming of many jump instructions, it has other weak point. For example, If a macroblock, SAD of some 4 × 4 block is particularly large, and SAD of other 4 × 4 block is very small, the sum of macroblock's SAD also satisfies the formula (9), but it is obviously not the SKIP macroblock. Therefore, it should set a threshold for 4 × 4 block. According to the formula (5) and (6), the equation (10) can be drawn.

$$\begin{aligned} \frac{1}{4}\max(SAD4) &< \frac{1}{2}Qstep \\ \max(SAD4) &< 2Qstep \end{aligned} \tag{10}$$

With the threshold 2Qstep, it can avoid the problem of too large SAD of 4 × 4 block and determine whether a macroblock is SKIP macroblock or not. Therefore, it can avoid DCT, quantization and encoding for SKIP macroblock and effectively improves the overall encoding rate. Figure 2 shows the improved SKIP decision process.

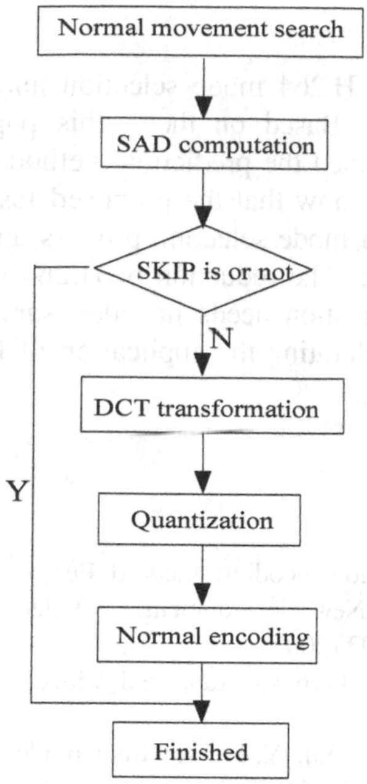

Fig. 2. Improved SKIP decision process

3 Experiment Result and Analysis

Table 1 shows the proportion of SKIP macroblock that predict in advance with the improved decision algorithm. As can be seen from the table, quantization parameter (QP) is greater; the proportion of SKIP macroblock that can predict in advance is higher.

Table 1. The proportion of SKIP macroblock that predict

Sequence	QP=32 SKIP ratio	Detection ratio	QP=36 SKIP ratio	Detection ratio	QP=40 SKIP ratio	Detection ratio
Mobile	89.90%	37.03%	93.58%	41.95%	96.10%	47.55%
Coastguard	79.50%	27.93%	86.88%	30.07%	92.46%	32.76%

In real life, video image sequence has the characteristics of fixed background or slow changes in the background. Therefore, the proportion of SKIP macroblock is higher in unidirectional predicted frames (P frames). The method of improved SKIP decision can predict in advance the highest proportion of up to 47% of SKIP macroblock with virtually no additional computation, greatly reduces the encoding time of H.264.

4 Conclusions

This paper analyzed rules of H.264 mode selection and image texture features of video surveillance sequences. Based on them, this paper proposed a fast mode selection algorithm that combined the prediction method with some mode exclusion method. Experimental results show that the proposed fast mode selection algorithm effectively avoids unnecessary mode selection process, greatly reduces the encoding complexity and encoding time. The reduction of H.264 encoding complexity solves efficiently real-time communication needs in video surveillance and has important practical significance for accelerating the application of H.264 encoding standard in video surveillance system.

References

1. Bi, H.: A new video compression encoding standard. People Post Press, Beijing (2005)
2. Zhou, H.-M., Gong, J.-R.: New development of video compression encoding-H.264. Information Technology, 91–93 (2005)
3. ITU-T Rec.: H.264/ISO/IEC 11496-10, Advanced Video Coding, Final Committee Draft, Document JVTG050 (2003)
4. He, Y.-B., Bi, D.-Y., Ma, S., Xu, Y.-L.: An Inter-mode Decision Algorithm Based on Spatiotemporal Correlation in H. 264. Journal of Image and Graphics, 2456–2462 (2009)
5. Sun, L.-F., Pu, J.-X.: Fast Inter-frame Mode Selection Algorithm Based on H.264 Standard. Computer Engineering, 220–222 (2010)
6. Pei, S.-B., Li, H.-Q., Yu, N.-H.: Research of H.264/AVC intra-prediction mode selection algorithm. Computer Applications, 1808–1810 (2005)
7. Meng, Q., Yao, C., Song, J., Li, W.: Fast selective algorithm of Intra prediction for H.264/AVC. Journal of Beijing University of Aeronautics and Astronautics, 3–5 (2007)
8. Li, H.: The optimization and realization of motion estimation and motion compensation algorithm in H.264 video compression standard. Beijing University of Posts and Telecommunications (2007)

An Investigation of Thermal and Visual Comfort in Classrooms in Ningbo

ZhangKai Li and Jian Yao*

Faculty of Architectural, Civil Engineering and Environment, Ningbo University, China
yaojian@nbu.edu.cn

Abstract. This paper aims to investigate the thermal and visual comfort of classrooms in hot summer and cold winter zone. A survey including objective tests and subjective questionnaires were carried out in a classroom in Ningbo University. Results show that occupant factors have a big influence on indoor comfort and the improvement of the occupant behavior should be enhanced in order to create a comfortable indoor environment for learning.

Keywords: Thermal comfort, visual comfort, building energy.

1 Introduction

Built environment refers to the human-made surroundings that provide the setting for human activity. A good indoor environment is essential for working, learning and living [1-12], especially for students in classrooms. Improvements in school environmental quality can enhance academic performance, as well as teacher and staff productivity and retention. Normally, visual and thermal conditions are two key issues in improving indoor comfort. For visual comfort, it is very important to provide sufficient daylight for classrooms because daylight has been shown to improve study as well as health, awareness and feelings of well-being in classrooms. Thermal comfort is another important factor affecting students' Academic performance. Some students and teachers may experience thermal discomfort in classrooms without mechanical heating, ventilation and air conditioning systems in this hot summer and cold winter zone. Therefore, it is of great importance to investigate the thermal and visual comfortable conditions in classrooms in this region.

2 Methodology

A classroom in Ningbo University (see Fig. 2) in this region was selected to carry out the survey, which includes thermal and visual comfort tests. Indoor and outdoor temperatures and illuminance were recorded on a typical day (December 12, 2011) and then used as objective evaluation indexes, while questionnaires on the thermal and visual comfort (occupants' thermal sensation, which was defined using the ASHRAE Thermal Sensation Scale, a continuous seven-point scale (-3 cold, -2 cool,

* Corresponding author.

D. Jin and S. Jin (Eds.): Advances in FCCS, Vol. 1, AISC 159, pp. 553–558.
springerlink.com

-1 slightly cool, 0 neutral, 1 slightly warm, 2 warm, 3 hot), and according to the seven-point thermal sensation scale, the author also introduced a similar seven-point relative visual sensation scale (-3 too dark, -2 dark, -1 little dark, 0 neutral, 1 little bright, 2 bright, 3 too bright)) were used as subjective evaluation indexes.

Fig. 1. Location of the investigated building

Fig. 2. Location of the investigated classroom

3 Results and Dicussion

Fig. 3 presents the indoor and outdoor temperature fluctuation with time. It can be seen that the indoor temperature has a same trend as outdoor temperature since the windows of the classroom are partly open, and thus the gap between these two temperatures is not big. The indoor temperature is between 10 to 16 °C, which is lower than the comfortable level according to design codes for building energy efficiency.

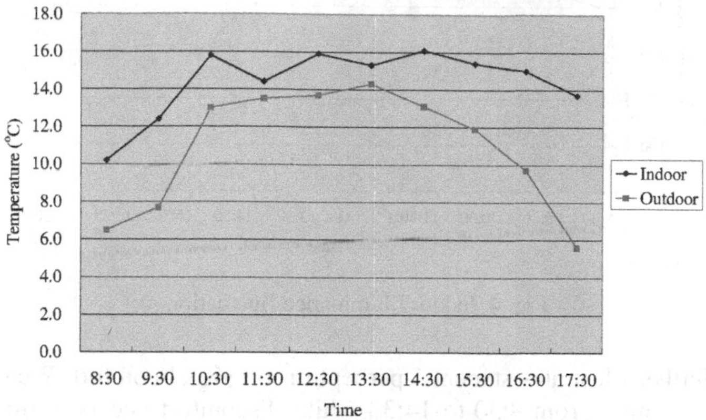

Fig. 3. Indoor and outdoor temperature fluctuation

Table 1 further gives the students' subjective perception on thermal conditions. It can be seen that at 8:30, 9:30 and 17:30 students feel cool because the indoor temperatures at these time points are lower than 14 °C, which is a very low temperature for learning in classrooms. Students' thermal perception changes to slightly cool and neutral when the indoor temperature increases.

Table 1. Students' subjective perception on thermal conditions.

Scale	Thermal comfort									
	8:30	9:30	10:30	11:30	12:30	13:30	14:30	15:30	16:30	17:30
3										
2										
1										
0				√	√	√				
-1			√				√	√	√	
-2	√	√								√
-3										

Fig. 4 gives the indoor illuminance fluctuation with time. In the morning, the illuminance increases from 400 to more than 600 lux, indicating is a suitable visual condition (higher than the normally recognized threshold level of 300 lux). In the afternoon, it drops to lower than 300 lux after 15:00, indicating a discomfortable visual condition for study.

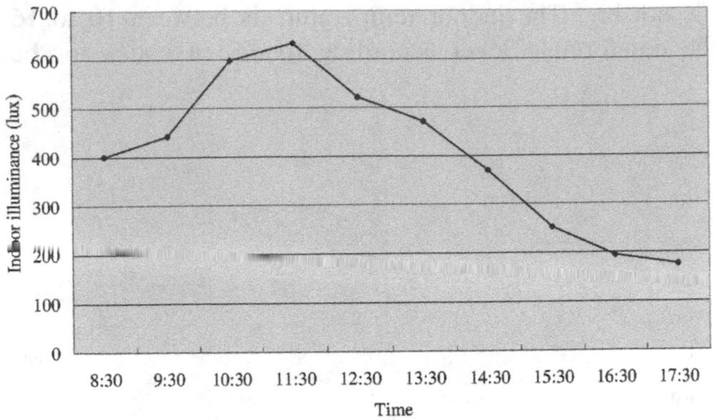

Fig. 4. Indoor illuminance fluctuation

Table 2 further illustrates students' perception on visual comfort. We can see that students feel comfort from 8:30 to 14:30 while discomfort occurs at time points of 15:30, 16:30 and 17:30 because of low illuminance values. To further investigate the cause of visual discomfort, the artificial lighting using was recorded. Fig. 5 listed the percentage of lights being turned on. We can see that 80% of the lights are kept turn on from the beginning of the day to the ending of the day without changing even when the daylight is excessive or insufficient.

Table 2. Students' subjective perception on visual conditions.

Scale	Visual comfort									
	8:30	9:30	10:30	11:30	12:30	13:30	14:30	15:30	16:30	17:30
3										
2										
1										
0	√	√	√	√	√	√	√			
-1								√	√	√
-2										
-3										

If windows are closed when outdoor temperature drops and more lights are turned on, indoor environment conditions will be improved. This means that human behavior is important in maintaining a comfortable indoor thermal and visual condition. And some knowledge should be taught to create suitable indoor environments.

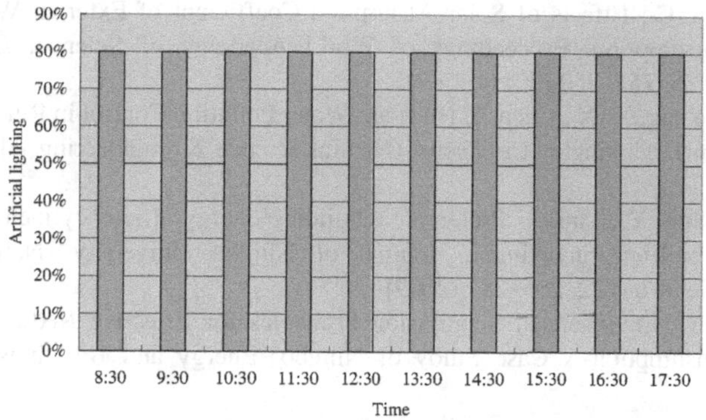

Fig. 5. Application of Artificial lighting

4 Conclusions

This paper investigates the thermal and visual conditions in classrooms in hot summer and cold winter zone. The results show human factors influence these indoor environments and improvement of the occupant behavior should be enhanced.

Acknowledgments. This work was supported by projects in Ningbo University (XYL11004) and (XKL11D2073).

References

1. Koo, S.Y., Yeo, M.S., Kim, K.W.: Automated blind control to maximize the benefits of daylight in buildings. Building and Environment 45, 1508–1520 (2010)
2. Dai, L., Cai, L.: Analysis on the Characteristics of Modern Urban Residence in Ningbo: A Case Study of the Residence at Daici Alley and Deji Alley in Jiangbei District. Journal of Ningbo University (Natural science & engineering edition) 24, 128–132 (2011)
3. Yao, J., Yuan, Z.: Study on Residential Buildings with Energy Saving by 65% in Ningbo. Journal of Ningbo University (Natural science & engineering edition) 23, 84–87 (2010)
4. Yao, J., Zhu, N.: Evaluation of indoor thermal environmental, energy and daylighting performance of thermotropic windows. Building and Environment 49, 283–290 (2012)
5. Wang, W.L.: Master Plan for Small Towns in Perspective of Urban and Rural Planning Act. Journal of Ningbo University (Natural science & engineering edition) 22, 288–292 (2009)
6. Hu, X.-Y., Xu, J., Zhang, J.-Q.: Study on the Modern Rural Residence Planning with Regional Features of Fanshidu Village. Journal of Ningbo University (Natural science & engineering edition) 24, 110–114 (2011)
7. Cai, L., Dai, L.: Analysis on Characteristics of the Residence of Plain Areas in Ningbo: A Case Study on Zoumatang Village. Journal of Ningbo University (Natural science & engineering edition) 22, 430–434 (2009)
8. Yao, J., Yan, C.: Evaluation of The Energy Performance of Shading Devices based on Incremental Costs. Proceedings of World Academy of Science, Engineering and Technology 77, 450–452 (2011)

9. Yao, J., Yan, C.: Effects of Solar Absorption Coefficient of External Wall on Building Energy Consumption. Proceedings of World Academy of Science, Engineering and Technology 76, 758–760 (2011)
10. Gao, H.-S., Yang, X.-X., Chen, Y.H., et al.: Water Pollution Control in Rural Villages: Case Study. Journal of Ningbo University (Natural science & engineering edition) 23, 74–78 (2010)
11. Zhou, Y., Ding, Y., Yao, J.: Preferable rebuilding energy efficiency measures of existing residential building in Ningbo. Journal of Ningbo University (Natural science & engineering edition) 22, 285–287 (2009)
12. Yao, J., Zhu, N.: Enhanced Supervision Strategies for Effective Reduction of Building Energy Consumption-A Case Study of Ningbo. Energy and Buildings 43, 2197–2202 (2011)

Facial Expression Recognition Based on Shearlet Transform

Yan Qu, XiaoMin Mu, Lei Gao, and ZhanWei Liu

School of Information Engineering, Zhengzhou University,
450001 Henan, China
quyanchn@gmail.com, iexmmu@zzu.edu.cn, iegaolei@gmail.com,
iezwliu@zzu.edu.cn

Abstract. In this paper, we explore a facial expression recognition approach based on shearlet transform which is a new image multi-scale time-frequency analysis method. In addition to multi-resolution and time-frequency localization owned by traditional wavelet transform, the shearlet transform also provides directionality and anisotropy. Moreover, the low frequency components in shearlet transform are extracted as features; the SVM (Support Vector Machine) is used for classification. Experimental results show that the proposed approach achieves better recognition rates compared to the traditional approaches on the JAFFE database and Ryerson database.

Keywords: Shearlet transform, expression recognition, support vector machine.

1 Introduction

As the rapid development of the technologies in human-computer interaction (HCI) and affective computing, facial expression recognition has been an actively researched field recently, which plays an essential application on education, entertainment, security and so on [1]. The basic facial expression recognition system generally consists of three stages: face pretreatment, expression feature extraction and expression classification. As a critical step, expression feature extraction has recently received increased attention. Many novel algorithms have been presented to get better performances. Wavelets have been successfully applied to the facial expression recognition due to their time-frequency and multiresolution [2]-[4]. However, it is known that wavelets have the limited ability in expressing directional information [5]. To overcome these limitations, a great number of MGA (Multiscale Geometric Analysis) algorithms, owned good characteristic, such as locality, multiresolution, directionality and anisotropy, are employed for expression feature extraction to offer more discrimination information in the past few years [6]-[8]. Multiscale methods based on shearlets [9] not only have good localization and compactly support in the frequency domain, but also have directionality and anisotropy. With those properties, shearlets can accurately efficiently represent image geometrical information of edges and texture, which are very essential in facial expression recognition. In contrast to other MGA methods such as contourlets [10], ridgelets [11], and curvelets [12], the shearlet framework could provide optimal efficiency and computational efficiency

D. Jin and S. Jin (Eds.): Advances in FCCS, Vol. 1, AISC 159, pp. 559–565.
springerlink.com

when addressing edges [5].The experimental results demonstrate that the proposed approach achieves better performance than the traditional method.

The rest of this paper is organized as follows: Section 2 briefly reviews the theory of Shearlet transform. The proposed facial expression recognition algorithm based on the shearlet transform and its corresponding procedures are explained in Section 3; Section 4 presents the experimental results and performance comparisons. Finally, conclusions are drawn in Section 5.

2 Shearlet Transform

2.1 The Continuous Shearlet

The shearlet transform is a new multiscale geometric analysis tool which has been widely used in image approximation, edge analysis and other fields [5], [13], [14]. The Continuous Shearlet Transform of *f* is defined by

$$SH_f(a,s,t) = \langle f, \psi_{ast} \rangle, \quad a \in R^+, s \in R, t \in R^2 \tag{1}$$

where $\psi_{ast}(x) = a^{-3/4}\psi(M_{as}^{-1}(x-t))$ of three variables, the scale $a \in R^+$, the shear $s \in R$, the transform $t \in R^2$, is called a continuous shearlet system. $M_{as} = (a,s;0,\sqrt{a})$ is the composition of the shear matrices $\mathrm{B} = (1,s;0,1)$ and anisotropic matrices $\mathrm{A} = (a,0;0,a^{1/2})$. For any $\xi = (\xi_1,\xi_2) \in \hat{R}^2, \xi_1 \neq 0$, let

$$\hat{\psi}(\xi) = \hat{\psi}(\xi_1,\xi_2) = \hat{\psi}_1(\xi_1)\hat{\psi}_2(\frac{\xi_2}{\xi_1}) \tag{2}$$

where $\hat{\psi}_1 \in C^\infty(R)$ with $\mathrm{supp}\hat{\psi}_1 \in [-2,-1/2] \cup [1/2,2]$, $\hat{\psi}_2 \in C^\infty(R)$ with $\mathrm{supp}\hat{\psi}_2 \in [-1,1]$ and $\psi_2 > 0$ on $(-1,1)$ .Thus, each function $\hat{\psi}_{ast}$ has frequency support

$$\mathrm{supp}\hat{\psi}_{ast} \subset \left\{ (\xi_1,\xi_2) : \xi_1 \in \left[-\frac{2}{a}, -\frac{1}{2a}\right] \cup \left[\frac{1}{2a}, \frac{2}{a}\right], \left|\frac{\xi_2}{\xi_1} - s\right| \leq \sqrt{a} \right\} \tag{3}$$

As illustrated in Figure.1, each element ψ_{ast} is supported on a pair of trapezoids, oriented along lines of slope *s*. The support becomes increasingly thin as $a \to 0$.That is say the scale of the shearlets controlled by the anisotropic scaling matrices A, while the shear matrices B only control the orientation of the shearlets. Those matrices lead to windows which can be elongated along arbitrary directions and the geometric structures of singularities in images can be efficiently represented and analyzed by using them.

Literature [15] shows that shearlets are localized well and are compactly supported in the frequency domain. Shearlets show highly directional sensitivity and anisotropy. In fact, for two dimension signal, the band-limited shearlets can detect all singular points, and track the direction of singular curve adaptively. Furthermore, along with the parameter changes, shearlets can completely analyze the singular structures of 2-D piecewise smooth functions. Those properties of shearlets are useful especially in image edge and detail information processing.

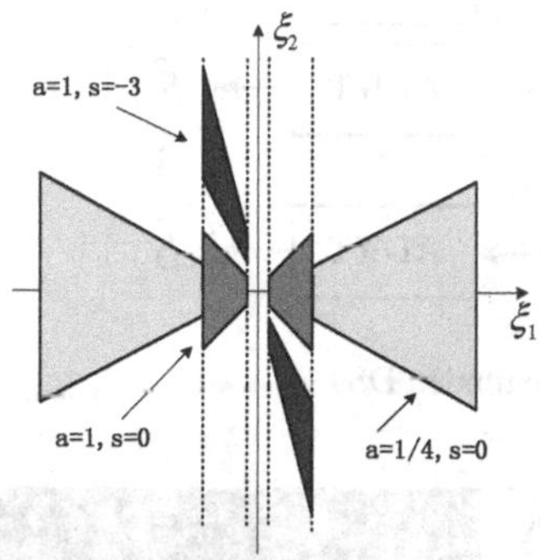

Fig. 1. Support of the shearlets $\hat{\psi}_{ast}$ (in the frequency domain) for different value of a and s

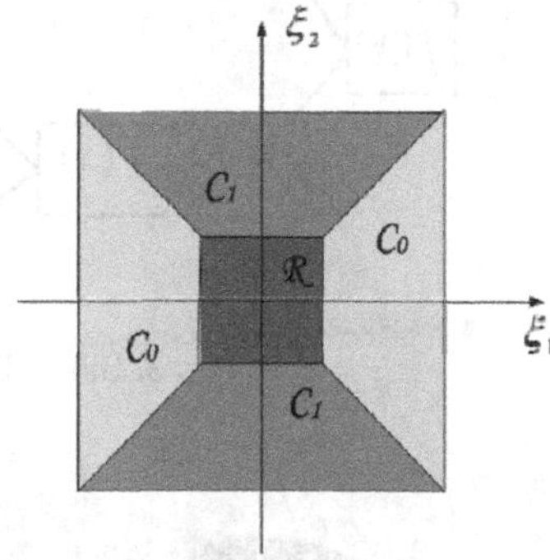

Fig. 2. The horizontal cone C_0 and vertical cone C_1 in the frequency domain

2.2 Discrete Shearlet Transform

The elements of the traditional shearlet can't be separated in the spatial domain, and this property often leads to the difficulty in practically relevant discrete implementation. Based on the above discussion, Wang *et al.* [13] constructed compactly supported shearlets generated by separable functions which are constructed using a MRA (Multiresolution Analysis) ,this lead to a fast DST implementation—the Extended Discrete Shearlet Transform (the extended DST). Figure 3 illustrates the extended DST of the image f. Specifically, the shear matrix B_0^s and B_1^s corresponded to the horizontal cone C_0 and vertical cone C_1 dimensions(see Figure 2), respectively, while the anisotropic scaling matrix A_0 and Λ_1 were offered to construct the anisotropic discrete wavelet basis along the shear direction, and complete the multiscale decomposition. The extended DST is also computationally very efficient, requires $O((2^{M+2}+2)N)$ operations where N is the size of input image and $2^{M+2}+2$ is the number of directions, while the 2D-FRFT costs $O(N(\log N))$, the curvelets transform costs $O(N(\log\sqrt{N})^2)$, the ridgelets transform costs $O(N(\log\sqrt{N}))$ [16].

Using the shearlets, a given image can be analyzed at various resolutions for each direction. The low frequency components—the upper-left corner of the shearlet coefficients matrix—which concentrate most important information and discard the influence of noises and irrelevant parts [2], will be adopted for further analysis in this paper. Thus, the dimensionality of the data is reduced effectively for computation at the next stage. Figure 4 shows the two levels of the extended DST in the horizontal direction for a happy image from the JAFFE database.

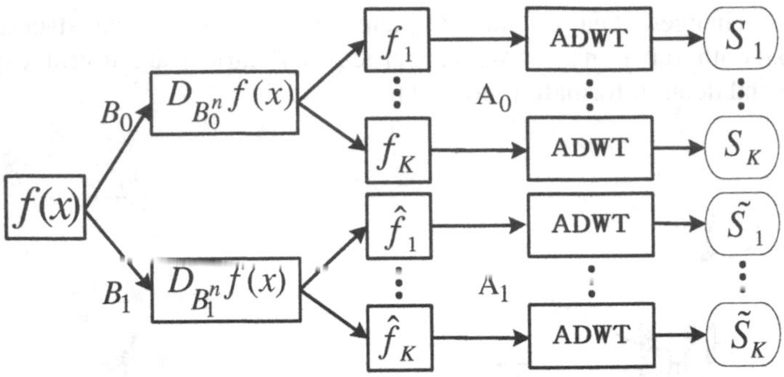

Fig. 3. Block diagram of the extended DST

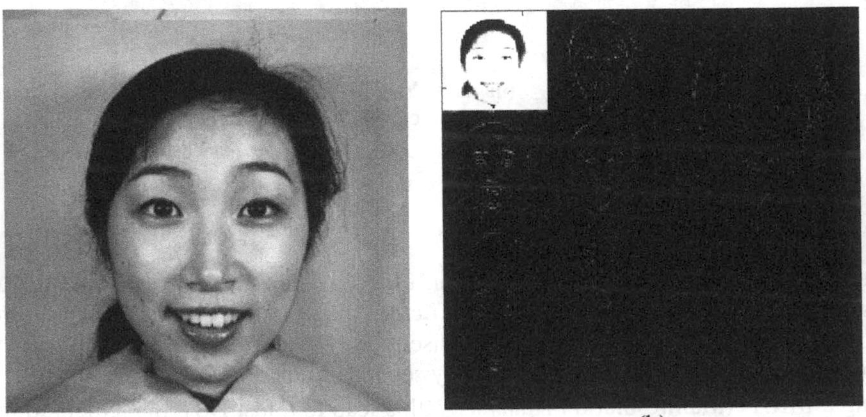

(a) (b)

Fig. 4. Two-level decomposition in the horizontal direction of a happy image (a) the original image. (b) The decomposed subimages.

3 Facial Expression Recognition Based on the Extend DST

In this section, it will implement the new method to facial expression recognition. The algorithm based on the extended DST is implemented with the following steps. Figure 5 illustrate the experimental procedure.

1. Apply histogram equalization to the original image to decrease lighting effects.
2. Detect the central location of eyeballs and mouth of the facial image using the method proposed by Dubuisson *et al.* [17], rotate the face image according to test results, and normalize it to a gray-level image of size 64×64 as the input of the extended DST processing system.
3. Apply the extended DST to the cropped image with 2 level decomposition and 6 directions. Then choose the coefficients of the first 16 columns and the first 16 rows of the shearlet coefficients matrix as the extracted data.
4. At last, use one-against-one RBF SVM for classification.

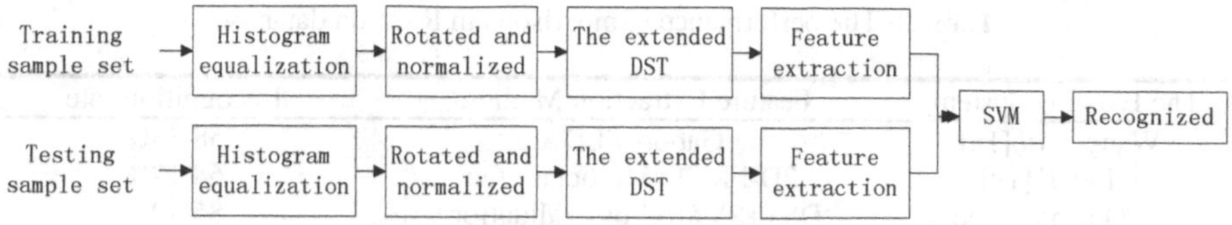

Fig. 5. The experimental procedure based on the extended DST

4 Experimental Results and Discussions

In this section, a series of experiments using two databases separately are carried out to verify the accuracy of the proposed algorithm. The Ryerson database contains 240 images, including 6 basic facial expressions(happy, sad, angry, fear, surprise, disgust) posed by 4 subjects, with 10 images per person in each class. The JAFFE database is a collection of 213 images of 10 Japanese females, including 7 categories expressions (6 basic facial expressions and neutral) and 3～4 images for each expression per person. In the experiments, both leave-one-out approach and cross-validation approach were applied to perform comparisons with other existing algorithms. For leave-one-out approach, each time only one image was tested, while the rest images are used as the training sets, and calculate the final recognition rate by averaging all the recognition rates. For cross-validation approach, the database are randomly divided into ten segments, where one segment is used for test set, and the others are used for training sets, then repeat the same procedure of training and testing for 30 times, while at last average all the 30 recognition rates to obtain the final performance of the proposed system. In the SVM algorithm, set σ^2 to be 1 and C to be 50 for cross-validation approach, σ^2 to be 4 and C to be 11 for leave-one-out approach.

Table 1 compares the performance among the proposed algorithm and the existing algorithms using the same JAFFE database. It investigates that the recognition rate in the proposed system is higher than others using either approach strategy. Furthermore, the performances of the proposed method and classical Gabor wavelet、2D-FRFT approach are compared based on the Ryerson database. The experimental results are shown in Table 2. The results illustrate that the proposed method outperforms Gabor wavelet and 2D-FRFT approach.

Table 1. The performance comparisons in JAFFE database.

The Existing Systems	Feature Extraction Methods	Strategy	Recognition rate
Seyed *et al.* [6]	Contourlet+MRMR+SSIM	Leave-one-out	82.50%
Dubuisson *et al.*[17]	Sorted PCA+LDA		87.60%
Frank *et al.* [2]	Wavelet+2D-LDA+SVM	Cross-validation	95.71%
	Wavelet+2D-LDA+SVM	Leave-one-out	94.13%
Our proposed system	DST+SVM	Cross-validation	96.07%
	DST+SVM	Leave-one-out	97.65%

Table 2. The performance comparisons in Ryerson database

The Existing Systems	Feature Extraction Methods	Recognition rate
Wang *et al.*[18]	Gabor+FLDA	58.33%
QI *et al.*[19]	2D-FRFT+Multiclassifier	54.17%
Our proposed system	DST+SVM(Cross-validation)	85.89%
	DST+SVM (Leave-one-out)	87.92%

The proposed method demonstrates its efficiency through the above mentioned experiments. The experimental results demonstrate that the shearlet transform can capture the expression features of 2-D image more efficiently. Extracting the LF components as features not only improves recognition rate, but also reduces the dimensions of the data and finally reduces the computational complexity. Meanwhile, our proposed system could be an effective approach to practical applications because of its excellent performances.

5 Conclusions

In this paper, a new approach to facial expression recognition based on the extended discrete shearlet transform and SVM is proposed. As the shearlets own properties of multi-resolution, directionality, anisotropy and localization (the each of shearlet elements can be well localized in both the space and the frequency domain), those properties make shearlets more applicable in dealing with facial expression recognition. Furthermore, the extended DST is computationally attractive. The proposed algorithm first applies the extended DST to the image, selects the upper-left matrix of the shearlet coefficients to compress the data effectively, then uses SVM for classification. Compared to traditional methods, the experimental results show outperformance on the accuracy of the proposed system both in JAFFE database and Ryerson database.

References

1. Picard, R.W.: Affective Computing. MIT Press, Cambridge (1997)
2. Frank, et al.: Performance Comparisons of Facial Expression Recognition in JAFFE Database. International Journal of Pattern Recognition and Artificial Intelligence 22(3), 445–459 (2008)
3. Hosseini, I., et al.: Facial Expression Recognition using Wavelet Based Salient Points and Subspace Analysis Methods. In: Canadian Conference on Electrical and Computer Engineering, pp. 1992–1995 (2006)
4. Zhang, W., et al.: Facial Expression Recognition using Kernel Canonical correlation analysis (KCCA). IEEE. Trans. Neural Network 17(1), 233–238 (2006)
5. Yi, S., et al.: A Shearlet Approach to Edge Analysis and Detection. IEEE Trans. Image Processing 18(5), 929–940 (2009)
6. Seyed, et al.: Contourlet Structural Similarity for Facial Expression Recognition. In: The 35th IEEE International Conference on Acoustics, Speech, and Signal Processing, Dallas, Texas, U.S.A (2010)

7. Wu, X., et al.: Curvelet Feature Extraction for Face Recognition and Facial Expression Recognition. In: The Sixth International Conference on Natural Computation, pp. 1212–1216 (2010)
8. Cai, L., et al.: A New Approach of Facial Expression Recogniying Based on contourlet Transform. In: IEEE Conference on Wavelet Annlysis and Pattern Recognition, pp. 275–280 (2009)
9. Guo, K., Labate, D.: Resolution of the wavefront set using continuous shearlets. Transactions of the American Mathe-Matical Society 361(5), 2719–2754 (2009)
10. Do, M.N., et al.: The contourlet transform: An efficient directional multiresolution image representation. IEEE Trans. Image Process. 14(12), 2091–2106 (2005)
11. Candès, E.J., Donoho, D.L.: Ridgelets: A key to higher-dimensional intermittency. Phil. Trans. Roy. Soc. London A 357, 2495–2509 (1999)
12. Candès, E.J., Donoho, D.L.: New tight frames of curvelets and optimal representations of objects with C singularities. Commun. Pure Appl. Math. 56, 219–266 (2004)
13. Lim, W.-Q.: The Discrete Shearlet Transform: A New Directional Transform and Compactly Supported Shearlet Frames. IEEE Trans. Image Process. 19(5), 1166–1180 (2010)
14. Easly, G.R., et al.: Shearlet-Based Total Variation Diffusion for Denoising. IEEE Trans. Image Process. 18(2), 260–268 (2009)
15. Labate, D., et al.: Sparse Directional Image Representations Using the Discrete Shearlet Transform. Applied and Communcational Harmonic Analysis 1(25), 25–64 (2008)
16. Starck, J.-L., et al.: Sparse Image and Signal Processing: Wavelets, Curvelets, Morphological Diversity. Cambridge University Press (2010)
17. Dubuisson, S., Davoine, F., Masson, M.: A Solution for Facial Expression Representation and Recognition. Signal Processing: Image Communication 17, 657–673 (2002)
18. Wang, Y., Guan, L.: Recognizing Human Emotional State from Audiovisual Signals. IEEE Trans. Multimedia 10(5), 936–946 (2008)
19. Qi, L., et al.: Recognizing Human Emotional State Based on the 2D-FrFT and FLDA. In: The 2nd International Congress on Image and Signal Processing, Tianjin, China (2009)

PE Information Technology and College Physical Education

Dong Sheng Zhang[1], He Hong Zhang[2], and Jing Li Yang[1]

[1] Hebei Vocational and Technical College of Building and Materials, China
zdsjcy@163.com, tyxbcg@163.com
[2] Qinhuangdao Technician College, China
tyxxgx@126.com

Abstract. With the continuous development of science and technology, Information technology plays a more and more important role in teaching, which also includes physical education in colleges and universities. This study makes a systematic analysis for physical education in colleges and universities from the perspective of information engineering. At the same time, this study also analyzes the effect of modern information technology on the entire process of physical education and sports teaching in colleges and universities, and the positive impact of modern information technology on physical education teaching innovation.

Keywords: Physical education, information technology, college and university.

1 Introduction

Physical education information formerly known as sports intelligence, it mainly refers to areas such as libraries and information content. With the rapid development of science and technology, the concept of sports intelligence and physical education information is also continuing to expand its areas, sports intelligence is gradually being replaced by Physical education information. The modern concept of PE information refers to the form of text, data, or signals, which to show a variety of interrelated sports phenomena and characteristics collectively through certain delivery and processing.

2 Analysis of Physical Education Teaching System in Universities

Physical education teaching can be consider as sports information systems engineering from the perspective of information technology, which including the subsystem of physical education information acquisition, physical education information processing and arranging subsystem, physical education information output subsystems and physical education feedback subsystem. Information plays a key role for entire physical education teaching process. The modern teaching needs teaching innovation, but innovation should be thought an open process. Teaching innovation is a new review of ago teaching methods, contents and means in the possession of large amounts of

D. Jin and S. Jin (Eds.): Advances in FCCS, Vol. 1, AISC 159, pp. 567–571.
springerlink.com

information, and combining students ' personality traits and definite teaching environment and to seek a breakthrough.

Therefore, the subsystems of information acquisition in teaching innovation of physical education should play a fundamental role in the teaching process. The PE information acquisition mainly refers to teachers of physical education obtained information related to their professional from relevant sources (journals, books, magazines, television, Internet, CD, etc) by some means (bibliography retrieval, computer search, etc). Nowadays, various sports periodicals, magazines, Internet all have a lot of physical education information due to the rapid development of modern communications technology and media technology. At the present time, the key factor affecting public physical teaching innovation is not lack of information related to sport and PE, but lack of a strong consciousness of physical education information acquisition to teachers, and a efficient technology of information acquisition.

3 The Role of Information Technology Subsystem on Physical Education

To have a sense of access to information and certain techniques in order to gets a lot of physical education information is not difficult for Physical education teachers, but all physical education information is not applicable to all teaching and learning environment, teaching innovation should be carry out according to its own existing criteria, including teaching students' characteristics, environment, teaching equipment, customs, and so on. Therefore, on access to information processing and finishing were most important to teaching innovation of course of the public Physical education.

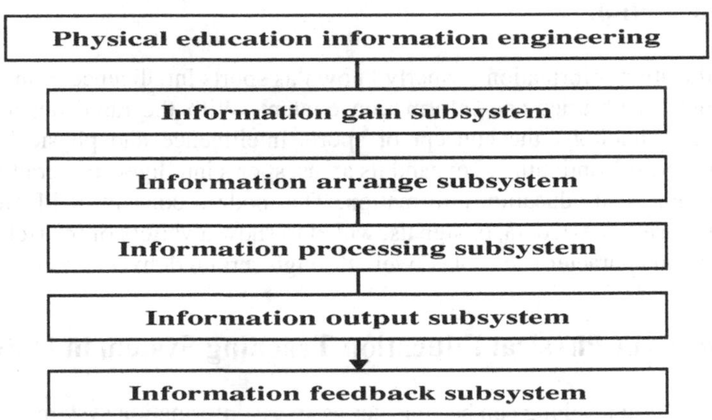

Fig. 1. Physical education information systems

The information feedback subsystem is essential in the whole process of teaching information transmission. The accomplishment of innovative teaching methods and teaching means requires subsystem, which were associated with acceptability of students and innovation of teaching. Teaching innovation subject is teacher; it emphasizes the people-oriented and student-centered, meanwhile, modern PE teaching innovation take physical and mental development of students as center of task. The transfer of the physical education centre is not relaxing its requirements for teachers, but rather put forward higher requirements. The major factor affecting innovation of physical education was the lack of access to information of physical education awareness and ability to information technology, namely the absence of proper physical education information technology. Information technology refers to awareness to importance of information, as well as abilities and overall quality in information activity shown, including information thinking, information ethics, information means.

4 How to Improve Ability Relates to Physical Education Information

From the perspective of physical education, PE information technology can be thought as the ability relates to physical education teachers' awareness of information related to physical education and its characteristics, value and access to application analysis, information, including information consciousness of physical education and physical education teaching information skills, information concept of physical education and sports teaching information ethics and so on. Information consciousness refers to self-consciousness of people's demand for information, which is people's information sensitivity, namely the understanding, experience and evaluation to natural and social phenomena, theoretical perspectives. The information consciousness of physical education including the mental disposition and sensitivity of teacher to get information on physical education. Physical education requires innovation; the innovation premise should hold a wealth of information. Through the analysis of part of the teaching plan for PE teachers and field demonstration, find lesson structure, form, teaching methods, and so more or less, a lesson plans benefit for life, reflects the lack of physical education new, there have been many students "love sports, hates physical education". An innovative physical education teaching point of view is very much, teaching methods, teaching content and teaching methods, teaching methods are dynamic, plasticity is very large, the key is too little attention for the information of physical education teachers, and information sensitivity is too low. The shortages of latest development trend of physical education, and research frontiers, innovation experience not only limit the improvement of its teaching, but also hampered the sustainability of physical education. Information ability refers to teacher's access to information, processing information, the ability to absorb information and innovative information.

To the physical education teaching, mainly gets information about the physical education and physical education teachers for reprocessing, absorption and creating information. The consciousness of physical education information is a prerequisite for teaching innovation, and the ability to physical education information is guaranteed to innovation of the teaching of physical education. It is difficult to physical education gain something if teacher lack of awareness of sensitive information, and the teaching ability is lagging if lack of basic physical information capacity. Nowadays, the pressures of college PE teachers in the teaching, research is very large duo to limited to

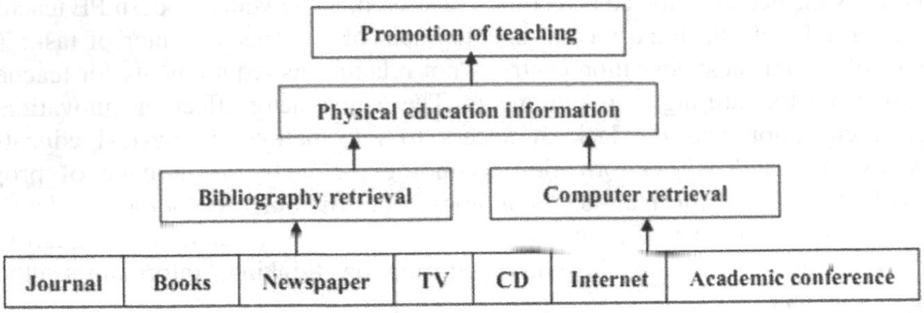

Fig. 2. The access to information for Physical education

the current educational environment, so they is also aware of the importance of information, and want to breakthrough in teaching, bumper win teaching, scientific research, but because the information is less capable, or for physical information could not be found, or face a lot of sports information is at a loss.

5 Physical Education Information Technology and Its Application

The lack of modern information technology application capability is the shackles of innovation in teaching for physical education teachers, especially for the lack of basic knowledge and application of computer ability. According to a survey of some PE teachers in colleges and universities, most teachers get information also remains in the reference room of the "read all over". Teacher gets images of realistic teaching information with the help of information devices such as cameras, scanners. And with the help of computer application software teacher can get a lot of useful information, and easily organize and maintain the physical exchange of information, interpretation, and sports information, and so on. In spite of the importance of perceived characteristics in diffusion research, a literature search indicated that most existing instruments designed to tap these characteristics lacked reliabihty and validity. Furthermore, no comprehensive instrument to measure the variety of perceptions of innovations existed. Such an instrument should be vital to diffusion researchers, and thus the research described in this paper was undertaken to develop one. The reason for focusing on the perceived characteristics of innovations is that the findings of many studies which have examined the primary characteristics of innovations have been inconsistent. As an example of the effect of this division between primary and secondary characteristics, consider the attribute of cost and its influence on buying behavior. Actual cost price is a primary attribute, whereas the perception of cost is a secondary attribute.

6 Conclusion

Physical education teaching in colleges and universities is a complex system of information from the perspective of information engineering, which including 4 subsystems, physical education, sports teaching of information processing, collating information is obtained, the information output of physical education and sports teaching feedback. Physical education information is closely related to quality and teaching innovation of physical education teachers, strengthen information literacy training for physical education teachers of physical education is the inevitable reform of physical education in new century requirements. To obtain physical education information awareness, teachers should use all sources of training physical education and sports for promoting innovation through information. These means including strengthening the information technology training for teachers of physical education, especially computer technology and its application in network capacity training and broaden physical education information access and enhance their physical education information analysis, dissemination, application of skills; and strengthening hardware construction, Department of College physical education teaching and research should be dedicated computer Studios, according to the characteristics of physical education, devoted to the teaching of information acquisition, analysis, and training applications; and hire information professionals to describes the training of application ability of physical education and describes the use of sports applications, production of physical education, access to sports information, analysis and processing of knowledge; and encourage physical education teachers in the management of physical education and sports information technology applications, including multimedia courseware application, use of video recorders, projectors, computer networks, and so on, to change traditional habits of acquisition, analysis, to have a good knowledge of modern information technologies.

References

1. Heim, G.R., Peng, D.X.: The impact of information technology use on plant structure, practices, and performance: An exploratory study. Journal of Operations Management 28(2) (2010)
2. Jonassen, D.H.: Objectivism versus constructivism: Do we need a new philosophical paradigm, vol. (03) (1991)
3. Smith, P.L., Ragan, T.J.: Instructional Design, pp. 59–62 (1992)
4. Reigeluth, C.M.: Instructional Design the ories and Models, pp. 78–82 (1983)
5. Newby, T.J., Stepich, D.A., Lehman, J.D.: Instructional Technology for Teaching and Learning: Designing lnstruction. Integrating Computers and Using Media, 138–141 (2000)
6. Seels, B., Glasgow, Z.: Making instructional Design Decisions, 2nd edn., pp. 121–130 (1998)
7. Rothwell, W.J., Kazanas, H.C.: Mastering the lnstructional Design Process systematic approach, San Francisco, Calif., pp. 125–128 (1998)

The Research of Web Mining Algorithm Based on Variable Precision Rough Set Model

ZhiQiang Zhang* and SuQing Zhang

Henan Occupation Technical College, Zhengzhou, 450046, China
zhangzhiqiang70@sina.com

Abstract. Variable precision rough set model is the extension of rough set model, which introduce theβparameter based on the basic rough set theory, that is, allows a certain degree of misclassification rate existing. This paper presents the web mining algorithm based on variable precision rough set in e-commerce, analyzes the limitation of the reduction methods based on variable precision rough set theory. The experiment shows the CPU Time in the attribute numbers, indicating that VPRS is superior to normal algorithm in web mining in e-commerce. The experiments show the algorithm could flexibly and dynamically web mining and therefore provide a valuable addition to the field.

Keywords: web mining, variable precision rough set model, rough set.

1 Introduction

Rough set (Rough Set, RS) as a new mathematical theory which about dealing with imprecise, uncertain and incomplete data, originally proposed by the Polish mathematician Pawlak.Z in 1982. In 1991, Pawlak published a monograph, "Rough Set - Theoretical Aspects of Reasoning about Data", lays the foundation for the rough set theory[1]. As the most of first study on rough set theory is published in Polish, so it did not cause the several attentions of the international computer science and academic community, the study area has only limited in some countries in Eastern Europe. Until the late 1980s, it eventually caused the attention of scholars from various countries.

Since when rough set theory has many disadvantages when it was used on data mining, main disadvantages include noise data, data lacking and high-capacity data etc, and these disadvantages couldn't be deal well with by classical models, So in 1993, Ziarok published a paper "Variable precision rough set model" in Journal of Computer and System Sciences, which marks the emergence of variable precision model. But the knowledge acquisition of information systems based on variable precision rough set also has no systematic research, such as classification of the variable precision rough set, attribute reduction, decision-making rules and other aspects of theoretical research and application is not mature enough, which need

---

* Author Introduce: ZhiQiang Zhang(1970), male, associate professor, Master, Hehan Occupation Technical College, Research area: web mining, data mining.

D. Jin and S. Jin (Eds.): Advances in FCCS, Vol. 1, AISC 159, pp. 573–578.
springerlink.com

further research. Therefore, the knowledge acquisition of the information systems based on variable precision rough set is a new worthy study both in theory and practice.

Data mining is from the vast amounts of data automatically and efficiently extracts useful knowledge of a new data processing technique. In recent years, the Internet's rapid development and widespread use, making the amount of information on the Web at an alarming rate, current information resources has become an astronomical figure. How people get from vast amounts of data useful for their own data and information, all need to be one kind of resource can be automatically discovered from the Web, access information, and will not get lost in a sea of information technology direction, so Web mining techniques have emerged.

Web data mining originated in data mining, but most of the traditional data mining objects for relational database or data warehouse, data processed with a complete structure. The data has the following characteristics of Web information.

(1) Large-scale mass data. Current information on the Web in order to calculate the magnitude of TB, it is in the rapidly growing and updating.

(2) Information is rich in content. Both related to economic, cultural, educational, news, entertainment, e-commerce and other rich information services, but also implies access the page properties, access path characteristics, characteristics of these potential access time access to information.

Web mining based on variable precision rough set model is proposed based on integrating of variable precision rough set, web mining and data mining, and is used to reduce formal context. Variable precision rough set model is a more effective data mining methods. Rough set method has become an important method of data mining.

2 The Research of Web Data Mining Technology

Web mining is a collection from a large number of Web documents in the discovery of hidden patterns in the process.

Current information on the Web can be divided into three categories: 1) Web page content, including text messages and multimedia messages; 2) Web hyperlinks between pages referenced data; 3) Web server on the user login Web access log data. Thus divided into three categories of Web mining: Web content mining, Web structure mining, and Web access log mining.

Web content mining is a web-based content of Web mining, Web data from a large number of discovered information, the process of extracting useful knowledge. Web content mining is the object of Web documents for information and multimedia information content in terms of its mining, can be divided into text documents on the Web (including Text, HTML and other formats), and multimedia files (including Image, Audio, Video and other media type) of the excavation. Basic Web content mining is a text mining [2]. Web text mining the process shown in Figure 1.

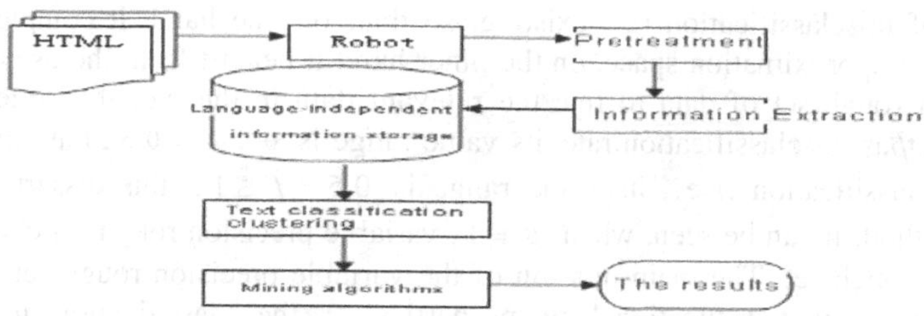

Fig. 1. Web text mining process

Web structure mining is a hyperlink from a web page found in its structure and their mutual relations. By finding a hidden link to a page after the structural model, you can use this model for web page classification, or to establish similarity measure to help users find related topics authority site.

Web usage mining is from user "access signs" to obtain valuable information is the Web log data and related data mining. WWW Each server maintains the access log (Web access log), log on user access and interaction information. Analysis of these data can help understand user behavior, thus improving the structure of the site, or to provide users with personalized service.

The first step: The establishments of the target sample, that is, from the user select the target text, as the characteristics of the user to extract information;

Step two: Extract feature information that, according to word frequency distribution of the target sample, extracted from the statistical dictionary feature vector mining goals and calculate the corresponding weight;

The third step: network access to information, which is to be selected using the search engine site, site collection, recycling collection Robot process static Web page, last accessed the site to obtain the dynamic information network database, generate Web resources index library;

Step four: Information feature matching, that is the source database to extract index information of the feature vector and feature vectors in the target sample to match, will meet the threshold condition information back to the user.

3 Variable Precision Rough Set Model

Rough set theory is built on the basis of the classification system, it states the classification as the equivalence relations on specific space, and equivalence relations constitute the division of space. Rough set theory defines the knowledge as the division of data, each divided collection called (Concept) or category (Category). The main idea of rough set theory is using known database to (approximately) characterize the imprecise or uncertain knowledge [3]. The most notable difference between this theory and other theories dealing with the problem of uncertainty and imprecision is that it does not need to provide any prior information outside of data sets required by the issue.

Variable precision rough set model is the extension of rough set model, which introduce theβ parameter based on the basic rough set theory, that is, allows a certain degree of misclassification rate existing, so that, on one hand, it complements the concept of approximation space, on the other hand it benefit from the use of variable precision rough set of data to fine the relevant data in the thought irrelevant data. Ziarko setβas misclassification rate, its value range is $0 \le \beta < 0.5$, but An[26]setβas correct classification rate, its value range is $0.5 < \beta \le 1$, this dissertation adopt An's method. It can be seen, when $\beta = 1$, variable precision rough set change to the standard rough set. The main mission of the variable precision rough set is to solve the problem of non-functional of properties of the classification in uncertain relationship data.

Set $\{U, R\}$ is the approximation space, U which is the domain, R is the equivalence relationship on U, $X \subseteq U$ then X about R's lower approximation (Lower Approximation) is defined as 1.

$$\underline{R}(X) = \cup\{Y \in U / R : Y \subseteq X\} \tag{1}$$

According to what is known, the collection composed by the object not belongs to collection X is called as the negative domain (Negative Region), written by $NEG_R(X)$, defined as figure 2.

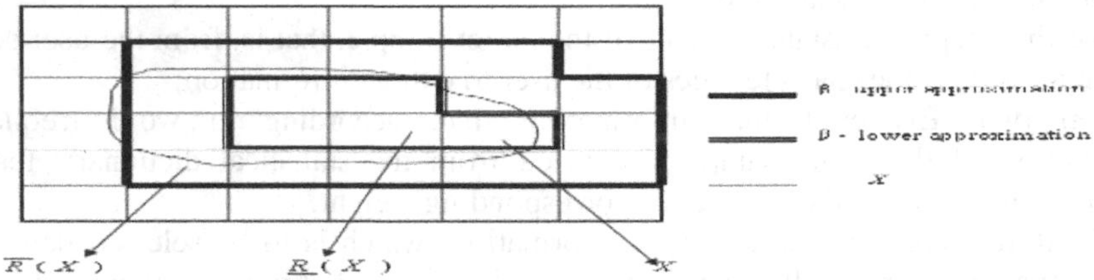

Fig. 2. Sketch map of variable precision rough set.

According to the figure, the classical rough set can not handle the noise date effectively, but the variable precision rough set can handle it well. Such as, X_1 and X_2 are the two equivalence classes under the condition attribute, respectively includes 100 objects. To a decision-making equivalency class Q, X_1 just have one element belongs to Q, but X_2 only own one element not belongs to Q, According to classical rough set, the similar pattern X_1 and X_2 belong to the boundary region of Q, the all can not make certain judgments, but X_2 only own one element not belongs to Q, this maybe leaded by noise.

4 Web Mining Based on Variable Precision Rough Set Model

Web mining techniques are commonly used in Web usage of the specific path analysis techniques commonly used in the field of data mining association rules, sequential patterns, and classification clustering technology.

Web path analysis using data mining techniques, the most commonly used figure. Because the Web can be used to represent a directed graph, is a collection of pages, is a collection of hyperlinks between pages, the page is defined as the vertices in the graph, and hyperlinks between pages is defined as the graph directed edges. Vertices into the edges express reference, a reference to the other side of that page, creating a site structure diagram, from the graph to determine the most frequent access path [4].

Association rule mining technique is mainly used for user access sequences from the sequence database of the sequence database entries to dig out the relevant rules is to dig out the user in a visit (SESSION), from the server to access the page / document the link between these pages may not exist between the direct-reference (RIFERENCE) relationship. The most common is to use APRIOR algorithm, excavated from the largest transactional database of frequently accessed items set, set this item is out of the association rule mining user access patterns.

Set $(U,C\cup D,V,f)$ is the target information system, C is the condition attributes set, $D=\{d\}$ is the decision-making attribute set, set $B\subseteq C$, $U/D=\{D_1,D_2,\cdots,D_r\}$ is as equation 2:

$$\sigma_B^{\beta}=\frac{\Sigma\{|\underline{R_B}^{\beta}(D_j)|: j\leq r\}}{|U|} \tag{2}$$

General approximation algorithm is to gradually examine and delete each condition attribute a_i in decision-making according to the definition of approximation reduction, if the total number of objects with decision-making unhinges after deleting, that is:

$\gamma^{\beta}(C,D)=\gamma^{\beta}(C-\{a_i\},D)$, then it's said that this attribute a_i can be deleted; otherwise it can not be deleted. Therefore, the properties of the matrix condition number of combinations of all the properties of 1 are the nuclear properties. This matrix will identify the corresponding attribute reduction algorithm to improve the basic steps are as follows.

Input: A decision-making table $S=<U,C\cup D,V,f>$ and the classification precision parameter $\beta\in(0.5,1]$, U is the domain, C is the condition attribute set, D and is the decision-making attribute set.

Output: S is all the relative approximation reduction

Steps of algorithm:

Step1 $B=C$;

Step2 $P=B$;

Step3 Implement all condition attributes a_i in the decision-making table:

If $\gamma^{\beta}(C,D)=\gamma^{\beta}(C-\{a_i\},D)$, then $B=B-a_i$;

Step 4 If $P=B$, stop, otherwise, to Step2

Fig.3 shows the detailed comparison results of Variable precision rough set model and normal algorithm.

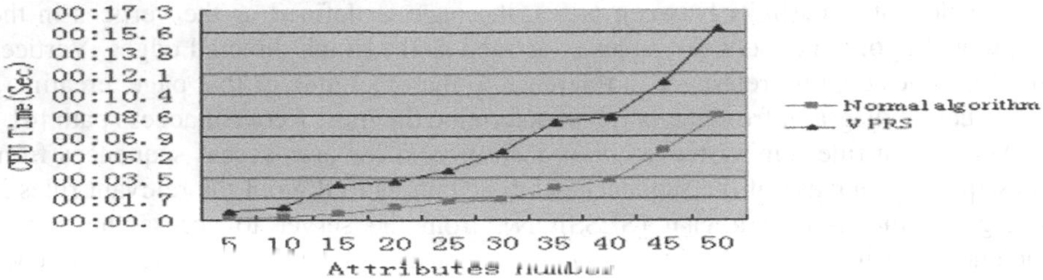

Fig. 3. Web mining method compared result of VPRS and normal algorithm.

5 Summary

This paper presents the web mining methods based on variable precision rough set in e-commerce, analyzes the limitation of the reduction methods based on variable precision rough set theory. Experimental results show that the algorithm ensured the compatibility of the proposition rules, and because of the appropriate given value of β, it significantly increases the credibility of the rule compared with similar algorithms. The experiment shows the CPU Time in the attribute numbers, indicating that VPRS is superior to normal algorithm in web mining in e-commerce.

Acknowledgement. This paper is supported by Education Department of Henan Province, 2011 Natural Science Research Program (2011C520019).

References

1. Mi, J.-S., Wu, W.-Z., Zhang, W.-X.: Approaches to knowledge reduction based on variable precision rough set model. Information Sciences 159, 255–272 (2004)
2. Beynon, M.: Reduces within the variable precision rough sets model: A further investigation. European Journal of Operational Research 134(3), 592–605 (2001)
3. Beynon, M.: An Investigation of -β Reduct Selection within the Variable Precision Rough Sets Model, pp. 114–122. Springer, Berlin (2001)
4. Wu, B., Zhao, L.: Web Mining Model Based on Rough Set Theory. Journal of Southeast University (English Edition) 18(1) (March 2002)

Research on the Development of Vertical Search Engines

XiaoYan Xu and DePing Zhao

Shenyang Jianzhu University, Shenyang City, Liaoning Province,
P.R. China, 110168
qdxxyxlj@126.com

Abstract. With the rapid increase of information on the Internct, vertocal search cngines have become more and more popular. This paper firstly analyzes the development status of vertical search engines, then make comparisons of vertical search engines and horizontal search engines and list the characteristics of vertical search engines. On these basis, it discusses the development necessity of vertical search engines. At last it makes prospect for the development of vertical search engines.

Keywords: Vertical search engine, Horizontal search engine, Internet, Information retrieval.

1 Introduction

Search engines develop from Archie to horizontal search engines represented by Baidu, Google, and they have been greatly improved in the aspects of function and amount of retrieval information, so they are the main channel to get information for users. However, can horizontal search engines really satisfy users' demand for more and more accurate, multiple information? According to the forecast of experts, vertical search engines are the future of search engines. Now many large horizontal search companies begin to develop their own vertical search engine, which indicates that future market of vertical search engines is large and potential. While what is the trend of vertical search engines, and that is what the paper is going to discuss.

2 The Present Situation of Vertical Search Engines

With the development of the Internet, more and more people retrieve useful information by search engines. According to the report published by China Internet Network Information Center (CNNIC) on December 20, 2011 in Beijing which is named the Chinese search engines market research in 2011, as of the third quarter of 2011, the amount of users of search engines scales up to 396 million [1], Chinese search engines market scales to 5.51 billion Yuan, rising 77.8% over the same period, thus it can be seen that the search engines market is rapidly expanding.

At the early development of search engines, the network information resources are less, so it's more easily to make information retrieval, and horizontal search engines can meet the users' demand in their quantity and quality. But along with the network

D. Jin and S. Jin (Eds.): Advances in FCCS, Vol. 1, AISC 159, pp. 579–584.
springerlink.com

information's quantity increasing by geometric series, tens of thousands of results provided by horizontal search engines seem to be good, while it is more and more difficult to meet the requirements of the customers in the accuracy. Therefore the users' concerns shift from how to use search engines to find more information to how to find more accurate information, this is the biggest challenge that horizontal search engines are facing. Vertical search engines produce and develop to solve this problem. Vertical search engines are a class of search engines aimed at a particular area, a specific group or a specific need, providing focused and in-depth information and service; they are the subdivision and extension of horizontal search engines [2], their results have stronger correlation, higher accuracy and can satisfy professional users' requirements in specific areas, thus they get more and more flavors. The report CNNIC published shows that as Chinese Internet users get increasingly mature, they are still using horizontal search engines, while their more subdivided needs begin to be met by vertical search engines, especially in 2011. At present horizontal search engines have a huge customer base, but the searching users of vertical search engines and network application in the station increase quickly [1].

3 The Necessity of the Development of Vertical Search Engines

When anyone enters a medical term in the input box of horizontal search engines, they will return thousands of the same articles no matter who you are. However, most of the search results are not completely related with your retrieval needs or even are quite separate. So it reflects some limitations that the horizontal search engines need to overcome. For example, their retrieval results are poorer in correlation and accuracy; the pattern of retrieval is single. Vertical search engines make up for the lack of horizontal search engines with their "fine, accurate, deep" information service. Their wide applications and developments become a necessary tendency.

3.1 Analysis of the Differences between Vertical Search Engines and Horizontal Search Engines

(1) For horizontal search engines web page is the smallest unit of information search, while vertical search engines firstly extract the web page's unstructured data into structured data, then make searching by the structured data as the minimum unit. This is the biggest difference between them. For example, if you want to search relevant information about MP3, vertical search engines will grab the web pages through web spiders at first, then extract merchandise information, thus the relevant information about MP3 is subdivided into "brand, price, memory, model, and the screen size, ……" The vertical search engines must make deal with the information grabbed before return to the users including information cleaning, reduplicative information removed, analysis, compare, index.

(2) They have the different search fields and service objects. Horizontal search engines provide information retrieval for anyone in any field; while vertical search engines provide retrieval service in a professional field, and service objects are professionals.

(3) Horizontal search engines grab web pages through the fixed links and web spiders take a long time to traverse all the links in the Internet, so it can not update

timely for the current rapidly changing network information; vertical search engine can generate the dynamic data in real time, conveniently for information indexing and retrieval.

3.2 Characteristics of Vertical Search Engines

(1) Vertical search engines can only handle a particular area of data, and satisfy the demand of users in one respect.

Vertical search is a kind of searching behavior targeted to a specific domain, usually associated with industries, at the same time there should be a certain number of information in the industry and the information should be focused. So there are two main data sources for vertical search engines: one is thc industry websites they concerns, the other one is their own platform, for example, GuDu mummy is a vertical search engine for catering information, all of its data comes from the information published by its alliance businesses. The pertinence of their data sources determines that it is impossible to satisfy all demands of users and can only provide the information in the industry they focus. With computers as an example, it can be divided into the purchase of computers, computer repairing, transfer of demand etc. Choosing a certain demand to do deeply and fully meeting the requirements is what vertical search engines should do.

(2) Vertical search engines tend to structured data.

Vertical search engines extract fine, structured data from the unstructured data in the web pages by information extraction technology, and then store them in the database. This can reduce the user's search response time, facilitate data merging and the depth of excavation. For instance, you enter a restaurant name in the input box of a vertical search engine in catering industry, detailed information such as restaurant location, menu, real-time prices and consumer evaluation will be returned.

(3) The data vertical search engines extract is full-scale and in-depth.

Because there are the restrictions of certain areas, vertical search engine can get the collection of related information in the field as possible to comprehensive. Meanwhile, through employing the professionals we can achieve to enrich our knowledge of the industry constantly, mine the contact between information, so as to achieve the effect of "deep" [3].

(4) The quantity of data that vertical search engines need to store is less, so it reduces requirements of the performance of hardware.

4 The Development Trend of Vertical Search Engines

(1) Personalized

The purpose of vertical search engines is to provide satisfactory retrieval service to users, but different users may have different concerns even if they input the same key words. To take a simple example, two users input "vertical search engine" as key words as well, one may hope to get the work principle of vertical search engines; the other one may want to get their development process, so customer satisfaction index will get a great discount. Personalized vertical search engines provide personalized service according to the users' interests, then how to get their interests is the key to vertical search engines. The specific process is that vertical search engines establish the interest models on the basis of key words users input or their browsing history records, of course they will continually adjust to adapt to the changes of users' interests [4]. Personalized vertical search engines take the individuation of users into consideration, so as to provide better service for every customer.

(2) Intelligent

At present, the structured data of vertical search engines is extracted mainly by hand or semi-hand, but this way can not guarantee the information is updated in time. The unstructured information extraction technology has made some progress in intelligent; similarly we can also apply intelligent technology to structured information extraction, which are the intelligence of vertical search engines.

Intelligent vertical search engines are based on natural language processing technology, using Boolean logic operations, these techniques make for better understanding of human language. Users can ask questions by natural language to avoid difficult problems the non professionals may face when they are using vertical search engines. Intelligent vertical search engines also have learning function, so they can make record of all problems users have retrieved and analyze them, so that the same problems will get better answers later on.

(3) Commercial

Compared to horizontal search engines, vertical search engines provide information more in-depth, concentrated and concrete. Now we have entered general electronic business era, the Internet is affecting people's lives in all times and places, and vertical search has attracted many enterprises' attention. Many medium-sized and small enterprises start to use vertical search engines for network marketing, winning huge business opportunities. Users of vertical search engines have clear goals, so the possibility to purchase is relatively large. So it is obviously more accurate and efficient for enterprises to make cooperation in advertising or other aspects aiming at this kind of users [5].

(4) Mobile search

Mobile search is defined to use the search engines to retrieve information through terminal equipments such as mobile phones. With the development of economy, mobile phones have become popular in domestic; China has more mobile phone users than any other country, so mobile search owns a huge market. The applications of vertical search in web have been more and more mature; the horizontal search also turns to vertical search business including Google, and mobile search requires much more vertical search to reduce the amount of search information so as to provide more related information for users [6].

5 Cases of Vertical Search Engines

At present, vertical search engines are in a period of vigorous development at home and abroad, they can be applied to many areas, even all kinds of industries can be subdivided into vertical search engines, such as travel, shopping, video, music and so on. SaDi IT compass launched by SaiDi net is the first Chinese vertical search engine in IT information, also is the first real vertical search engine. It contains most of the network information in the field of IT. There are many other influential vertical search engines, qunar.com, qihu.com, kuxun.cn included.

5.1 Qunar.com

Qunar is founded in February 2005 by FuRui Dai, ChenChao Zhuang and Douglas who form the "ShaWei team". It is currently the largest online Chinese website in the world, and it commits to provide consumers with the depth search of air tickets, hotel, vacation etc. In addition, it provides group purchase of tourism product and other information related to travel to help consumers make better choice. At present, Qunar can search for more than 400 tickets and hotels suppliers' websites, covering more than 30000 hotels and 3000 direct flights and it can offer consumers the price of tourism products and related information as well. Qunar also establishes an objective, intelligent comparable platform to provide the most reasonable suggestions to help consumers make choice. It integrates each distributor's news, so consumers can see related information of the same products in different manufacturers and make a comparison. Qunar establishes various advantages of in-depth service in the aspects of search scope, detailed data, reaction rate, real-time price and so on.

5.2 Qihu.com

Qihu Company issues Qihu.com formally in March, 2008 which aims to provide the service of questions and answers for users. It mainly includes experience search, BBS search, blog search, news search, etc. The community search technology Qihu.com uses is advanced in China, so by virtue of its technological advantage, Qihu makes in-depth analysis of the characteristics of Chinese BBS and formed Qihu BBS known as "the Chinese first portal". Qihu BBS can search for above 95 percent of personalized content like the micro-blog, community websites and others. Qihu BBS has two kinds of functions: one is search, it can retrieve posts in the website; the other one is special topic, that is to say system will generate related topics instantly and automatically according to the specific needs of users and adjust the latest content.

6 Conclusions

As netizen in our country get increasingly mature, their needs for information retrieval are not confined to the entertainment, chat only, search engines are playing more and more important role in people's daily life of food, clothing, shelter and transportation. People used to get information through horizontal search engines like Baidu, Google. However, the nature of horizontal search engines determines that they can't meet requirements of professional people in a specific field for accurate information retrieval. People gradually turn to vertical search engines which are more accurate, professional and personalized. The diverse demand of users determines the search engine market will appear subdivision, which is to say each industry will own vertical search engine in their respective field. It can be said the development of horizontal search engines provides a good market for the appearance of vertical search engines. According to experts' analysis, in the next three years, vertical search engines will occupy a certain share in the Internet market. Search engines will become vertical and personalized; it is the inevitable trend of subdivision in the industry of search engines.

References

1. The Chinese search engines market report in 2011 (2011), http://www.cnnic.cn/dtygg/dtgg/201112/t20111221_23536.html
2. Chen, X.: Analysis of vertical search engine. Modern Information 9, 133–134 (2004)
3. Li, F.: Research and design of vertical search engine. Electronic science and technology University, Chengdu (2009)
4. Zhong, H.: A framework of personalized information service based on vertical search Engine. Journal of Information 1, 118–120 (2008)
5. Ce, L.: Vertical search– a rookie in the field of electronic commerce. Software Guide 6, 57–58 (2006)
6. Zhu, Y.: The current situation of vertical search and its development. Science and Technology Innovation Herald 28, 231 (2011)

A Partial Network Coding Scheme Based on Energy-Balancing in WSN

Guoyin Zhang[1], Lei Wang[1], Chunguang Ma[1,2], and Xiangxiang Li[1]

[1] College of Computer Science and Technology, Harbin Engineering University, 150001, Harbin, China
[2] State Key Laboratory of Networking and Switching Technology, BUPT, 100876, Beijing, China
{zhangguoyin,wanglei24,machunguang,lixiangxiang}@hrbeu.edu.cn

Abstract. This paper provides a new partial network coding scheme suitable for WSN. The sensor nodes of this scheme include four states: Normal, Coding, Suspend and Sleep. Transformation between these states is energy-oriented and relevant scheme are divided into three steps: Initiate coding, Check Period and Balance energy. First, the paper describes the designing process of this scheme, and then describes the coding and decoding process of random linear network coding used in this paper. At last, according to the result of the NS2 virtual experiment, this scheme exceed traditional network coding schemes: increasing the efficiency of network resource usage, decreasing network lag and expanding living period of WSN.

Keywords: partial network coding, WSN, energy-balancing.

1 Introduction

With the process of microelectronic technology, computer technology and wireless communication technology, the low power multi-functional sensor network develop rapidly, so that it can be integrated within a small volume of data acquisition, data processing and wireless communications and so on. Wireless sensor network[1]-[2] (WSN) is a multi-hop self-organizing network system. The aim of WSN is to perceive collaboratively, collect, and process the information of objects perceived in the network coverage area, and send it to the observers.

Although, WSN has its limitations, such as sensor node's energy, communication skills, computing and storage capacity are very limited and so on. The sensors generally work in harsh environments, and can not be supplied when the nodes run out of energy in most cases, which maybe shorten the survival time of the network.

In 2000, Ahlswede[3] etc of the Chinese University of Hong Kong propose the concept of "network coding(NC)", which achieve network multicast capacity limit by allowing the intermediate nodes to encode the information in the input stream before forwarding them. Its core idea is to increase the network throughput, save transmission bandwidth and reduce data transmission delay at the expense of increasing the computational overhead. The emergence of NC provides new ideas to solve some bottlenecks problems for WSN[4]-[5]. In WSN, the nodes consume more

D. Jin and S. Jin (Eds.): Advances in FCCS, Vol. 1, AISC 159, pp. 585–590.
springerlink.com

energy in communication, so reducing the communication time and network delay is the main method to save energy. Therefore, the network coding technology is very suitable for WSN.

2 The Overall System Architecture

2.1 The Process of Node State Conversion

When the network is initialized, all the nodes are in normal state. With the network's operating, some nodes consume more energy because of processing more information, which causes the nodes to change from the ordinary state into the coding state. The nodes in coding state start coding mechanism to encode the information that meet the conditions, start a check mechanism to periodically check coding, and change from the coding state into soft coding state when the coding efficiency is low.

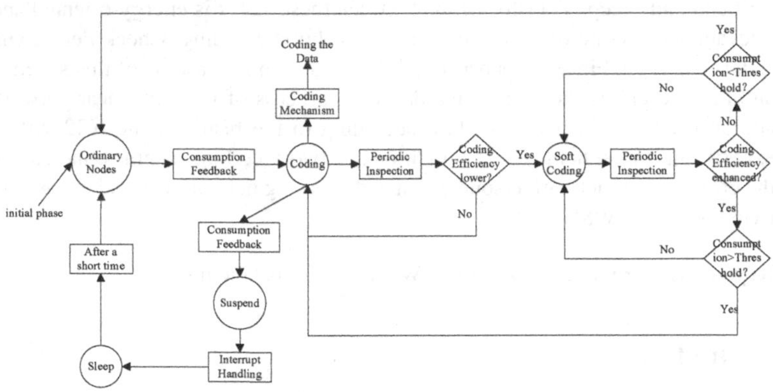

Fig. 1. Transformation process of node states

The soft coding state is a form of coding state, the same with periodic inspection mechanism. When the coding efficiency is increased and energy consumption increase, the nodes change from soft coding state into the coding state, otherwise into the ordinary state. The nodes in coding state may be changed into exceptional states, which is because of excessive consumption. Once extreme reduction of energy is found in a short period, let the nodes find the spare tire, enter the Suspend state, and go into the Sleep state, thus the nodes will avoid the risk of death because of energy depletion. After a short period, the nodes can change from the Sleep state into normal state automatically. Transformation between these states is shown in figure 1.

2.2 Scheme Description

The scheme triggers nodes into coding nodes according to energy consumption. If a node consume great amount of energy in a certain period of time, it means that it processes much information, indicating that it's suitable for high-efficiency coding.

Two threshold values are set for each node: one is the initiation value α and the other is off value β, $\beta<<\alpha$. When the energy consumption in a certain period of time exceeds the initiation value, coding function will be triggered. Then the node starts periodic checking for whether the coding function needs to be shut down. If the power consumption is below the off value, coding function will be turned off. However, if the energy consumption of the node reaches a certain level which is too high, this node will find a replacement for itself in the neighboring nodes to process communication in its spot and temporarily turn into Sleep state. The replacing node should fit the following two conditions: (1) it has more energy than the replaced node; (2) it is able to function as the replaced node. We call the replacement as backup node. The scheme is divided into initiate coding, periodic checking and energy balancing. Not every node has the energy balancing stage but only those in the coding state and has more than 1/10 energy left. When the energy consumption of the node in the time span T_s is the same with the remaining energy, this node turns into Sleep state and in the meanwhile $T_s=T_s/2$. The detailed process is as follows:

Coding initiation stage:

(1)Network initiates in time T.

(2)In T, if a node's energy consumption $\geq\alpha$, it will be triggered into Coding state. Otherwise it stays in Normal state.

(3)Nodes in Coding process will execute coding process on received or stored information as long as they fit the coding requirements.

Periodic checking stage:

(1)Check nodes in the Coding state periodically. The checking period T_p can be adjusted according to the practical conditions of the network.

(2)If the data of a coding node is less than a half of the length of the cache queue for n times, it will get into Soft Coding state, which means doubling the triggering time of the coding state. The value of n can be set according to practical conditions.

(3)When a node in the Soft Coding state detects that the data for coding is less than a half of the cache queue, energy consumption detection will be initiated. If the energy consumption is less than β, coding function of the node will be turned off. Otherwise it stays in Soft Coding state.

(4)Energy consumption detection will be initiated as well when the data for coding is more than a half of the cache queue. If the energy consumption is more than β, the node turns back into Normal state. Otherwise it stays in Soft Coding state.

Energy balancing stage:

(1)For nodes in energy balancing stage, when a node consumes half of its energy in a time $\Delta t \leq T_s$, it seeks for backup node in its neighboring nodes. It will keep searching for backup and when the backup is found, the original node turns into Suspend state, processes interruption for τ and turns into Sleep state for a time of t and will wake up automatically afterwards into Normal state. Suspend state means stop receiving information and process stored information only $\tau << t$.

(2)Backup will communicate in the original node's position. When backup consumes half of its energy in a time $\Delta t \leq T_s$, it searches for backup node in its neighboring nodes. If the search fails, it checks the state of the original node. If the original node is in Sleep state, it will be woke up by force. Then the backup turns into

Suspend state, processes interruption for a time of τ and turns into Sleep state for a time of t. If the original node is in Normal state, the backup will keep searching.

2.3 The Random Linear Network Coding

The Network coding mechanism used in this paper is random linear network coding[6]-[9], which is mapped in a linear form, and select the appropriate weighting factor for each input in a given limited field. In the general case, for a coding node, if the data packet stored in the node is $[X_1, X_2,\ldots,X_n]$, and newly received data packet is $[Y_1, Y_2,\ldots,Y_m]$, the encoded data can be expressed as:

$$Z = \sum_{i=1}^{n} g_i X_i + \sum_{i=1}^{m} f_i Y_i \tag{1}$$

Where g_i and f_i express the corresponding coefficients of coding selected from the finite field F_q randomly, and X_i, Y_i can be a packet of source data or encoded data.

In the network, the data packets transmitted between sensor nodes include two parts: coding vector and information vector, and the encoding vector of source data is corresponded to a unit vector. When the coding node receives one packet, which is stored in the buffer queue temporarily, system will start a timer to set a timeout period. If the packet is encoded, the timer resets. The length of the buffer queue is *q_len*. The system use the following two mechanisms to trigger the coding[10]: (1)any timer of packet in a queue runs out of time; (2) the length of packet in the queue is equal to *q_len*. Then the encoding node broadcasts the encoding packet.

After the cluster head collect a packet, extract vector coding and information vector, and put them into the decoding matrix. Using Gauss elimination method[6] to decode, if the received packet can increase the decoding matrix's rank, update the packet, otherwise neglect it. When the rank of decoding matrix is equal to the number of source packet, decoding is successful. As long as the finite field is large enough, the probability of coding vector's linearly independent will be closed to 1.

3 Virtual and Performance Analysis

Experimented in the NS2 platform, the virtual environment is as follow: 180s of virtual time; sensors using FLOOD protocol with 50 nodes; nodes are deployed randomly in an area of 1200m*1200m; most of the nodes start with 20J of energy while another few with 30J; sensor's sending power is 0.66W and receiving power is 0.395W. Settings: the length of cache queue is 7, overload time is 0.5s, time span T_s=5s, T_p=10s, n=3, the coding array of the random linear network coding in this scheme is based on limited field GF(2^8), α =7×10-2J, β =1.8×10-2J.

We experiment under three different conditions separately: using this scheme; all nodes participate coding and without using NC, as Graph 2, 3 shows.

As graph 2 shows, while using NC, nodes need to collect sufficient data to execute coding process. Compared with not coding, preparation for data consumes more energy. Once the data is ready and gets into coding stage, energy consumption will drop gradually because coding increases transmission efficiency.

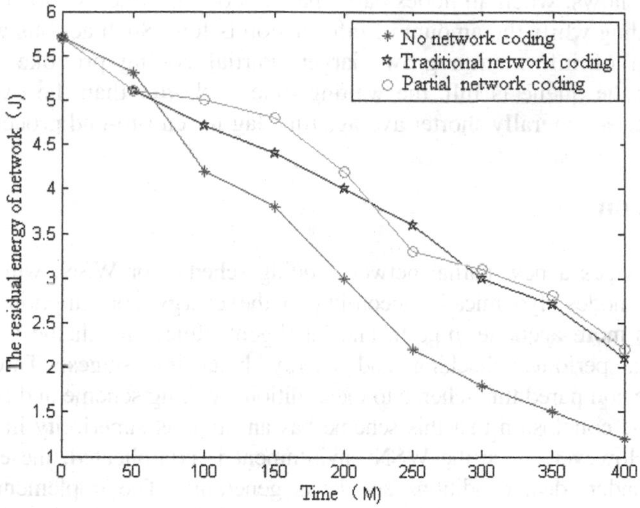

Fig. 2. Comparison of the node residual energy

Partial network coding scheme share almost the same data preparation phase with the full-coding scheme. However, as it proceeds to the coding stage, the extra consumption brought forth by coding makes the partial coding scheme superior than the other one. The tiny fluctuation shown in the graph is due to this reason. As time moves forward and the network stabilize gradually, energy consumption will decrease.

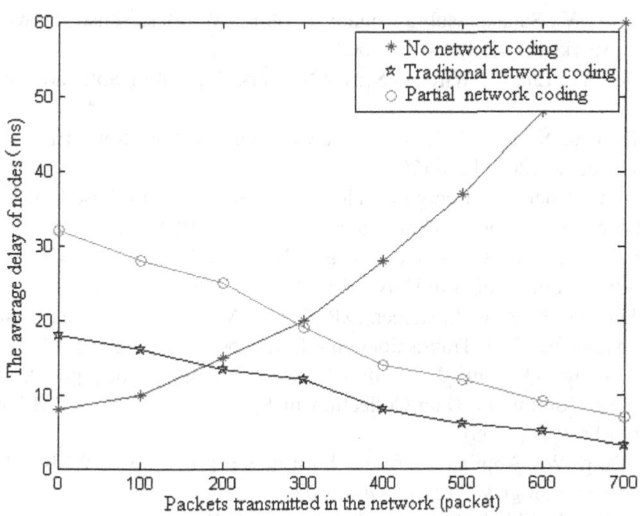

Fig. 3. Comparison of the transmission delay

As graph 3 shows, when all nodes participate in coding, time overload will be used to provoke coding while the amount of information is few. Such actions will decrease as the amount of information grows larger. Partial coding provoke coding only because the cache queue is full, the waiting time is shorter than the overload time, therefore having a generally shorter average time lag for end-to-end process.

4 Conclusion

This paper provides a new partial network coding scheme for WSN, which is able to choose coding nodes dynamically according to the energy consumption, making the coding process more accurate, time-fit and intelligent. Relevant scheme is divided into initiate coding, periodic checking and energy balancing stages. Through virtual experiment, we compared this scheme to the traditional coding scheme and not using NC, and come to the conclusion that this scheme has an obvious superiority in transmission performance. However, since the WSN environment is complicated, the experiment is implemented under ideal conditions and lacks generality. The implementation of the scheme is complex as well, needing support of both hardware and software, which increases the cost of the network. In the next step, we are going to combine technologies from other fields, to complete the design of a simple and cost-efficient NC scheme.

Acknowledgement. This paper is supported by the National Natural Science Foundation of "The research on data distribution mechanism in mobile P2P network"(No. 61073042).

References

1. Akyildiz, I., Su, W., Sanakarasubramaniam, Y., et al.: Wireless sensor networks: A survey. Computer Networks 38(4), 393–422 (2002)
2. Ren, F., Huang, H., Lin, C.: Wireless Sensor Network. Journal of Software 14(7), 1282–1291 (2003)
3. Ahlswede, R., Cai, N., Li, S.Y.R., et al.: Network information flow. IEEE Transactions on Information Theory, 1204–1216 (2000)
4. Lu, W., Zhu, Y., Chen, G.: Energy-efficient Routing Algorithm based on Linear Network Coding in Wireless Network Coding. Electronic Journal 38(10), 2309–2314 (2010)
5. Si, J., Zhuang, B., Cai, A.: Wireless Sensor Network Lifetime Maximization based on Network Coding. Journal of Jilin University 41(3), 822–826 (2011)
6. Ho, T., Medard, M., Koetter, R., Karger, D.R., Effros, M.: A Random Linear Network Coding Approach to Multicast. IEEE Transactions on Information Theory 52(10), 4413–4430 (2008)
7. Wang, D., Zhang, Q., Liu, J.: Partial Network Coding: Concept, Performance, and Application for Continuous Data Collection in Sensor Networks. ACM Transactions on Sensor Networks 4(3) (2008)
8. Wang, W., Tang, W.: Application research on network coding in WSNs. Transducer and Microsystem Technologies 27(2), 68–70 (2008)
9. Ho, T., Medard, M., Shi, J., Efrosand, M., Karger, D.R.: On randomized network coding. In: 41st Annual Allerton Conference on Communication Control and Computing (2003)
10. Qin, J., Yang, B., Li, O.: Cross-layer Research of Random Network Coding Based on WSN. Computer Engineering 36(3), 112–114 (2010)

Information Technology and Its Application in Sports Science

Fei Shen[1], Jun Li[1], and Zhifeng Wang[2]

[1] Hebei Normal University of Science and Technology, 066004 Qinhuangdao, China
[2] Kashi Normal University, 844006 Kashi, China
btterfly@126.com

Abstract. With the economic development and scientific technological progress, information technology raised rapidly, and become the dominant factor of socioeconomic development. Information technology applied in nearly all industries, in the field of sport science, information technology promote the development of sports science, and sports science development cannot be separated from the information technologies; This study analyze the application of information technology in the field of sport science, explain the importance of information technology application to develop sports science, proposed creating a unique means of information technology in sports to make sports science have a better development.

Keywords: Information technology, application, sports science.

1 Introduction

In the emerging information age, sports research and practice development model should meet the times, to make sport toward a "digital sports" direction. Information technology demonstrate a concise form to the regular pattern of complex sports, And can provide strong support to build a digital model of sports. In information technology-driven era, anyone who first mastered the new methods and theory can stand out from the competition. Similarly, in sports, who first mastered the ways to develop sport and cultural resources, the latest scientific training methods and new technology to increase the level of sports can be an invincible position in internal field of sports.

2 Information Technology as a Leader in World Development

2.1 The Rapid Rise of Information Technology

Information technology (computer, technology, sensor, technology, communications technology) were accepted and utilized in society, initially by telephone, radio, etc. With the increasing needs of the community, in 19th century remote sensing, optical fiber communications, computer simulation and other new products appeared. In 20th century, information has been made a huge breakthrough in satellite communications,

D. Jin and S. Jin (Eds.): Advances in FCCS, Vol.1, AISC 159, pp. 591–596.
springerlink.com

software, chip, computer management and support services research. 21st century information technology development have the purpose to the information technology of 20th,to enhance the development of technology, and develop gradually to digital, multi -functional, integrated, intelligent direction. Information technology will be faster, broader and deeper development in the future.

2.2 Information Technology Has Become the Dominant Factor in Social and Economic Development

The development of modern information technology is an important symbol of social information, but also scientific and technological disciplines have highly dynamic and creative process in current world economic, profound changes of social life. For example, information technology's contribution to world economic growth rate reached 14.7% in 1998. Information technology industry will become the pillar industry and the dominant of national economy. Therefore, all the industry regarded information technology as the leader of rapid development. The birth of science and technology will lead to changes in the structure of economic sectors; make the socio-economic development toward the vertical information. As the high innovation, high drive, high multiplier roses of information technology in society enhanced, information technology will be the main driving force for social development.

2.3 Information Technology Has a High Permeability

Now constructions of information technology were speed up as "engine" of social and economic development in all areas. First, the development of information technology has injected new vitality to other industries. For example, the application of information technology in the sports industry, have a sign of importance to raise the level of sports industry, sports and scientific training, to enhance people's physicals and competitive sports. Second, information technology, in combination with other industries constitute make the so-called composite technology, Such as sports information technology, this approach can make the "physical health, mental and physical recreation, exercise the will, strive to" reflect the physical nature of the more obvious features. Finally, information technologies have a significant role of research and development of new technologies, for example, security system in competitive sports; need hardware/software infrastructure solutions, network implementation and some content. As this penetration of information technology applications in sports, sports will develop in the direction toward scientific.

3 Information Technology Applications in the Field of Sports

Information technology has changed the human way of life, and similarly, sports information becomes a necessary requirement to change the sports culture naturally. Information technology use in sport is undergoing tremendous changes from simple application to the depth dependence. Table1 shows that information technology will become a new impetus to the field of sports.

Table 1. List of information technology applied in sports areas

	Sports	use of information technology
Government affair management	Application management software	Management system of equipment acquisition approval
	E-document management	approval for file systems
	Government websites	General Administration of sports sites, sports site
School physical education	Physical education	multimedia, video
	Sports management system	management of physical education curriculum and scheduling
	Physical education in remote	remote physical education curriculum
Competitive sports	Training methods	simulation training technology
	Technology preparation	competitive capacity in data collection, diagnosis
	Information technology	single games and integrated information technology
Mass sports	Virtual information technology	network fitness equipment
	National physical fitness measurement	physical fitness evaluation system, physical fitness testing vehicles
	Management of physical health	fitness guide management system
Sports industry	Management	statistics and analysis of industrial production and marketing
	Equipment	sportswear technology
	Business	many sporting bodies to establish a marketing website
	Media broadcast	television, the Internet sports play
Sports and leisure	Sports Lottery	computer lottery
	Club	full service system technology

3.1 Information Technology Applications in School Sports

Application of information technology, teachers can form various project campaigns, technical difficulty and focused on making courseware for students watching and analysis, through the solution of questions raised by the students, creating exploratory learning environment. First of all, this way so that students from passive to active learning mentality; Secondly can enhance students 'creative thinking, For example, gives a few basketball shooting simulation of lens, so that students' observation, analysis, and then conclude that body posture, carrying arms, shaking when shooting wrist, the amount of technical detail. This combination of information technology teaching mode of sports creates an ideal learning environment and a new way of learning for students. Application of modern information technology will enable school sports culture dissemination and lay the foundations for school sports culture to the level of development.

3.2 Application of Information Technology in Competitive Sports

Look at the development of competitive sports, news on competition venue for timeliness, accuracy of results, standardization of the increasingly high demand, such as, they all depend on the development of information technology. Table 2 shows the application of information technology has become the lifeblood of Olympic economy of supporting elements. For example, in the 26th Olympic Games 100 m final, high-tech computer show champion with Klass de Vries shoulder and torso are a little older than runner-up Otilio leading 0.5cm, semi-cm difference between decide the outcome of this fair judgment, can be achieved only through the high-tech. On information technology training is also of importance, such as the continuous training of athletes with a high speed camera photographed, and then use the computer for analysis and data processing system, or application of simulation technology, establishing commensurate with the players ' identity model, to analysis of the lack of training, and

Table 2. List of information technologies applied to the Olympic Games

Sessions	time	location	application information
The 18th	1964	Tokyo	satellite broadcast Olympic events
The 19th	1968	Mexico	Measuring doping reference information electronics and gender, electronic timing
The 20th	1972	Munich	Electro-optical distance measurement and automatic test technology
The 21st	1976	Montreal	Application of communication technology, the satellite pass the torch
The 22nd	1980	Moscow	Aircraft sprayed with chemicals, creating ideal opening weather
The 23rd	1984	Los Angeles	The computer and infrared optical sensor combinations completed
The 24th	1988	Seoul	Television commentator by monitors to get timing data information
The 25th	1992	Barcelona	Round sports organization and management of the operating system, network system, information system
The 26th	1996	Atlanta	Palace party website, as well as the establishment of the International Olympic Committee Olympic Games website
The 27th	2000	Sydney	Olympic Games, Olympic competition results system information retrieval system and management system.
The 28th	2004	Athens	High-tech security systems-C4I
The 29th	2008	Beijing	IT technology platform supports the Olympic Games

provide scientific training methods. Information technologies can also detect metabolism of athletes for adaptation, for example using combination of human biological rhythms (physical, mental, emotional) software to rationally adjust the training time to maximize the improve players' performance.

3.3 Application of Information Technology in Mass Sports

Since the 21st century, people's living requirements were improved, in order to meet demand for scientific body-building, economic fitness, recreation and fitness, sport must benefit from the development of information technology research and development of new fitness equipment. Combination of information technology, there has been considerable development of mass sports, such as thc application of virtual reality technology in information, to create virtual reality fitness equipment, gives one the feeling is in the scene, For example, networked virtual reality treadmill, digital tennis, digital boxing match, and so on. Through the use of network technology, physical fitness guide Web site first appeared through guidance of fitness experts, Internet users can be summed up in a set of effective fitness on their own ways. Economy of this through information technology, fitness needs of high performance sport culture to intelligent rapid development of physical culture.

4 Summary

Information technology has obtained excellent results in the sports sector, particularly in services, management and other aspects of the application. But for sports technology culture did not conduct a series of carding, systematization and intelligently. Technical culture did not conduct a series of carding, systematization and intelligently. Future developments require the use of information technology, combined with the essence of sports industry and sports, showing brand new sports information technology, and use of information technology to create an ideal learning environment, training and leisure environments. In the process of integration of information technology and physical sciences, from emotional to rational optimization, from irregular to the rule changes from the surface to deep thinking, from structure to function studies. By creating a unique means of information technology and research and development in the sport (that is, application of information technology in physical development), the ultimate sports service, guide the masses to practice sports, school sports, extended era of sports industry for the purpose of intelligent sport.

References

1. Eskofier, B., Oleson, M., DiBenedetto, C., Hornegger, J.: Embedded surface classification in digital sports. Pattern Recognition Letters 30(16), 1448–1456 (2009)
2. Mahmood, M.A., Mann, G.J.: Impact of information technology investment: Anmpirical assessment. Accounting. Management and Information Technologies 3(1), 23–32 (1993)

3. Kaushik, P.D., Singh, N.: Information Technology and Broad-Based Development: Preliminary Lessons from North India. World Development 32(4), 591–607 (2004)
4. Durmuşoğlu, S.S., Barczak, G.: The use of information technology tools in new product development phases: Analysis of effects on new product innovativeness, quality, and market performance. Industrial Marketing Management 40(2), 321–330 (2011)
5. Covell, D., Walker, S., Siciliano, J., Hess, P.W.: Chapter 3-Information technology management and the sports media. Managing Sports Organizations, 2nd edn., pp. 82–115 (2007)
6. Information on, http://www.olympic.org/

A New Fog Removing Method Based on the Degradation Model

Xiao-Guang Li and Wen-Tan Jiao

Department of Electrical Engineering and Automation, Luoyang Institute of Science and Technology, Luoyang, 471023, Peoples R China
lxg@lit.edu.cn, wentanjiao@sina.com

Abstract. Aiming at the degeneration phenomenon of images taken in mist, according to features of the drop quality image, using contrast attenuation of atmosphere scatter model and the mist image has the relation of index with image depth, this paper puts forward a kind of algorithm which estimates scene optical depth from several drop quality images based on degradation model, and then the method is applied to drop quality image fog removing. The experimental results show that the method can effectively improve the fog degeneration phenomenon and improve image clarity, significantly improving drop quality image visual effect.

Keywords: Drop Quality Image, Fog Removing, Degenerate Model, Optical Depth.

1 Introduction

Fog is a common natural phenomenon, and it makes outdoor scene image contrast and color change or degenerate, so in order to guarantee the video monitoring system can work normally, fog removing processing should be done to the video image.

At present, image fog removing methods can be mainly divided into two categories: the non-model algorithm and the algorithm based on the physical model. Typical non-model methods consist of histogram equalization, homomorphic filter method and retinex method, etc; these algorithms can only improve image quality relatively, and cannot achieve the fog removing really. While the algorithm based on the physical model of atmospheric degradation can obtain the relatively ideal fog removing effects [1], but its characteristic is that it is necessary to obtain scenery depth information, and it often needs to resort to radar altimeter to obtain depth information, and then uses the image data and depth information to work out the solution model parameters, which causes difficulties in the actual application.

Recently the fog removing based on a single image has made great progress, and more research is the fog removing algorithm based on the dark primary color priori single image [2].The algorithm is to estimate the fog density directly to achieve high quality fog removing by establishing the priori fog removing model, but this algorithm needs a powerful transcendent or assumption, because dark unbleached prior is a

D. Jin and S. Jin (Eds.): Advances in FCCS, Vol. 1, AISC 159, pp. 597–601.
springerlink.com

statistical rule, for some special images this method might doesn't have good effects, for example, when the scenery is close to the air layer in nature and no shadow covers it, the dark self-colored theory is invalid.

2 The Fog Removing Method Based on Degradation Model

Because the scattering of atmospheric particles makes the mist strong and causes the acquired image to degenerate, the displayed image in the image receiving-end is not the original clear image, therefore, to better display images, degraded image must be processed to restore the realistic primitive clear image.

2.1 Drop Quality Image Degradation Model

Mie scattering theory gives the degradation model of drop quality image in mist [3]. The model describes that the tatal illumination strength of scenery image in mist *E* can be equivalent to the linear superposition of radiation light after atmospheric attenuation E_d and surrounding scenery scattering light E_a, namely

$$E = E_d + E_a = \mathrm{Re}^{-\beta d} + E_\infty(1-e^{-\beta d}) \tag{1}$$

Among them, *R* is the scenery radiation light intensity in fine weather; E_∞ represents the sky radiation light intensity; *d* is called the scenery depth; β is the atmospheric scattering coefficient, and from the angle of atmospheric scattering, Eq. (1) is called atmosphere scatters physical model.

When using monochromatic imaging system on the scene for image acquisition, the degradation model above was simplified as follows form:

$$E = I_\infty \rho e^{-\beta d} + I_\infty(1-e^{-\beta d}) \tag{2}$$

$$E = E_0 e^{-\beta d} + I_\infty(1-e^{-\beta d}) \tag{3}$$

Among them, ρ is the normalization scene radiation degree, and I_d is the sky brightness value, and β and *d* form the optical depth notes that are $Di=\beta_i d$,$E_0=I_\infty \rho$.This model is also known as mono atmosphere scatters physical model.

It can be seen from the model, scene radiation degrees and sky brightness can be obtained by image information, so the problem to achieve normalized radiation degree ρ of each pixel in the image is actually transformed into the computational problem of optical depth βd of each pixel in the image, in order to realize the drop quality image fog removing, we must carry on scene optical depth estimate.

2.2 The Estimate of the Scene Optical Depth Di

According to the threshold value relations between fog visibility and human vision remain given by international illumination committee, Nicolas Hautiere[4] and other people made road surface visibility and scene contrast correspond up to estimate optical depth; Schechner [5] also proposed the method to realize scene optical depth and compute optical depth by using light polarization phenomenon. Based on the influence rule of atmosphere scatter to the image contrast, this paper puts forward a method to estimate scene optical depth from several drop quality images.

Considering getting two drop quality images of one scene in different weathers, and suppose different weather conditions are $(\beta_1, I_{\infty 1})$,$(\beta_2, I_{\infty 2})$,and the observed brightness value of arbitrary point Q_1 in the scene are respectively E_1,E_2, and the scene depth is d. According to the atmosphere scatter model (3), it can be seen observed brightnesses of pixel Q_1 in two images are respectively:

$$E_1 = E_0 e^{-\beta_1 d} + I_{\infty 1}(1 - e^{-\beta_1 d}) \tag{4}$$

$$E_2 = E_0 e^{-\beta_2 d} + I_{\infty 2}(1 - e^{-\beta_2 d}) \tag{5}$$

Then using the atmosphere scatter model and contrast decay rule caused by the atmosphere scattering to achieve the calculation of the optical depth D.

Neatening Eq. (4) and (5), there is:

$$\beta_i d = -\ln(\frac{E_i - I_{\infty i}}{E_0 - I_{\infty i}}) \quad i = 1,2 \tag{6}$$

And therefore, there is:

$$\psi = \frac{D_1}{D_2} = \frac{\beta_1 d}{\beta_2 d} = \ln(\frac{E_1 - I_{\infty 1}}{E_0 - I_{\infty 1}}) \Big/ \ln(\frac{E_2 - I_{\infty 2}}{E_0 - I_{\infty 2}}) \tag{7}$$

In order to find out the optical depth ratio ψ of any point in the two images, we need to select a reference standard point E_0, that is also simply knowing the radiation value E_0 of the fixed reference point in the sunny day, and obtaining the radiation value E_0 of a reference standard point in the scene in clear weather is reasonable and feasible.

In order to further define the optical depth, it needs to use the relation between the image contrast attenuation and scenery depth to get another important relationship equation. According to the contrast attenuation rule proposed by Duntley [6], the definition of each point's contrast in the image is as follows:

$$c = \frac{E_i - \overline{E}_i}{\overline{E}_i} \tag{8}$$

Among the Eq. (8), E_i is the brightness value of image midpoint, and $\overline{E}_i$ is the brightness value average of the background region that midpoint i belongs to. According to the brightness contrast attenuation rule, there are relationships:

$$c_1 = c_0 e^{-\beta_1 d} \tag{9}$$

$$c_2 = c_0 e^{-\beta_2 d} \tag{10}$$

In the Eq. (9) and Eq. (10), c_0 is the contrast value of this point and its region in clear weather. To Eq. (9) Eq. (10), after transformation, it can be obtained:

$$D_1 - D_1 = (\beta_1 - \beta_2)d = -\ln(\frac{c_1}{c_2}) \qquad (11)$$

The combination of the Eq. (9) and Eq. (11) can realize optical depth estimate of arbitrary point in the image.

2.3 The Realization of This Algorithm

Combining the proposed optical depth estimation algorithm, the specific process of using degradation model to realize fog removing is as follows:

Step1: input two distinct obtained drop quality images in different weathers I_1 and I_2, and work out the brightnesses of the sky areas in the two images $I_{\infty 1}$ and $I_{\infty 2}$

Setp2: define the local contrast function, and get the two images corresponding contrast matrixesC_1 and C_2; according to the characteristics of image scene, choose reference standard. Q1, and determine the radiation degree in sunny day E_0;

Step3: use Eq. (7) to obtain each point corresponding ψ, and store them in matrix B; use Eq. (11) to obtain each point correspondingD_1-D_2, and store them in matrix A;

Step4: use matrix A and B to work out each point corresponding optical depth Di=$\beta_i d$, and store them in matrix C, and matrix C is the only depth matrix;

Step5: according to drop quality image degradation model (2), work out each point normalized radiation degree matrix ρ, and stretch to [0,255], then we can get the clear images after fog removing.

3 Experimental Results and Analyses

In order to verify the effectiveness of the proposed algorithm in this chapter, we have done a large number of experiments in MATLAB platform. What needs explanation is that we use simulated images, because the experimental conditions are limited, to get the several drop quality images suitable to the experiment demands, the position of images should be suitable strictly to the requirements, and the filming condition demands are also severe. Another reason of using simulated graph is that experiment can be repeated, makes it possible to test the validity of the algorithm, so we need to make a comparison between the image obtained by the algorithm put forward in this paper and the original image.

(a) Fog added β_1=0.6 (b) Fog added β_1=0.4 (c) New algorithm (d) Original clear image

Fig. 1. Drop quality image recovery results comparison

Fig. 1 (a) and (b) are obtained images under the clear weather, and Fig. 1 (d) is the result of atmosphere scatter model added fog, and f Fig. 1 (a)'corresponding atmospheric scattered coefficient is β_1=0.6. Fig. 1(b)' corresponding atmospheric scattering coefficient is β_1=0.4.Compared with the original clear image in sunny day, the fog added image's overall contrast is bad, and the details are obscure and lost more, Fig. 1 (c) is the fog removing effect of the method put forward in this paper, compared with the two images added fog, the image has very good contrast, and has no obvious difference in details with the original sunny clear image.

4 Conclusions

To the image degeneration phenomenon taken in mist, from features of drop quality image, using atmosphere scatter model and the contrast attenuation have a relation of exponential with scenic spots depth ,study a scenery brightness depth estimate method from several drop quality images, and applied to drop quality image fog removing, the method avoids the high cost of adopting hardware equipment and uses the feature that using an image to estimate depth information is unreliable, can obtain more accurate field attractions optical depth information, and effectively improve the fog degeneration phenomenon and improve image clarity, and significantly improve drop quality image visual effect.

References

1. Narasimhan, S.G., Nayar, S.K.: Contrast reatoration of weather degraded images. IEEE Transactions on Pattem Analysis and Machine Intelligence 25(6), 713–724 (2003)
2. Narasimhan, S.G., Nayar, S.K.: Vision and the atmosphere. International Journal of Computer Vision 48(3), 233–254 (2002)
3. He, K., Sun, J., Tang, X.: Singe Image Haze Removal Using Dark Channel Prior. IEEE Transactions on Pattem Analysis and Machine Intelligence, 1956–1963 (2009)
4. Oakley, J.P.: Improving image quality in poor visibility conditions using a physical model for contrast degradation. IEEE Transactions on Image Processing 7(2), 167–179 (1998)
5. Hautiere, N., Aubert, D.: Contrast Restoration of Foggy Images through Use of an onboard Camera. In: Proceedings of the 8th International IEEE Conference on the Intelligent Transportation Systems, pp. 601–606 (2005)
6. Sehoehner, Y.Y., Narasimhan, S.G., Nayar, S.K.: Instant Dehazing of Images Using Polarization. In: Proeeedings of IEEE Computer Society Conferenee Commuter Vision and Pattem Reeognition, pp. 325–332 (2001)

Dimensions of Harmonious Interaction Design for Mobile Learning

Qinglong Zhan[1] and Xun Deng[2]

[1] School of Information Technology, Tianjin University of Technology and Education, Tianjin, 300222, China,
qlzhan@126.com

[2] Yanzhou Bureau of Education and Sports, Yanzhou, Shandong, 272100, China

Abstract. Mobile learning is learning across a variety of situations or places supported by handheld mobile technologies. To date, few researchers discussed harmonious interaction and its design in m-learning. This paper explains the connotations of harmonious interaction, analyzes dimensions of harmonious interaction design. Harmonious interaction is a whole interaction that learners integrate with mobile learning devices, learning resources and the environment in the learning process according to their needs and goals. Harmonious interaction design is the representation of the effective operation process between the various elements for mobile learning system, and its ultimate goal is to promote learning. Harmonious interaction design in m-learning includes three dimensions: situations and learning resources, device performance, and emotional experience, and reduces learning loads and improves learning efficiency.

Keywords: Mobile learning, harmonious interaction, interaction design.

1 Introduction

Mobile learning (m-learning) is learning across a variety of situations or places supported by handheld mobile technologies. To date, few researchers discussed harmonious interaction and its design in m-learning. This paper explains the meaning of harmonious interaction, and analyzes dimensions of harmonious interaction design.

2 Connotations of Harmonious Interaction

An interaction, grossly speaking, is a transaction between two entities, typically an exchange of information [1]. In m-learning, harmonious interaction is a whole interaction that learners integrate with mobile learning devices, learning resources and the environment in the learning process according to their needs and goals. Immersed in this particular interaction, learners naturally unconsciously ignore all existences of support services, and thus easily achieving to promote learning and acquire high quality experiences. Harmonious interaction mainly includes three levels of meaning.

D. Jin and S. Jin (Eds.): Advances in FCCS, Vol. 1, AISC 159, pp. 603–608.
springerlink.com

2.1 Learner-Centered

Learner centered is the starting point and purpose of harmonious interaction and it is the best interpretation of reflecting the humanistic care and the humanities thinking. Learner-centered requires fully understanding and analyzing the needs of learners at the beginning of design, and focus on learners' feedback in iterative design process, and emphasizes continuous participation for learners to improve the usability of m-learning products.

2.2 Contextualized Intelligence

"Let the computing disappear" is a beautiful picture depicted by intelligent contexts. People are in the transition period currently from computers to no computers. This will undoubtedly depend on the development and promotion of mobile technologies, and harmonious interaction is very important. The concepts of organic unity, user immersion, transparent interface, and unconscious natural interaction advocated by harmonious interaction allow learners to ignore all existing support services unconsciously and focus on the learning process.

2.3 Promoting Learning

Promoting learning can help learners to complete the knowledge acquisition and meaning construction. All tangible and intangible supports are to promote learning and to enhance the learner's abilities. If the learner-centered is the starting point, then the promotion of learning is the end. Therefore harmonious interaction is a procedural definition, and it represents the personal and dynamic behaviors and mental processes.

3 Dimension of Situations and Learning Resources

3.1 Appropriate Learning Resources

Because of physical factors of m-learning devices, not all learning resources are suitable for presentations on them, and appropriate learning resources in harmonious interaction design are mainly reflected in the following aspects.

Learning Resources Are Appropriately Categorized. (1) language learning resources such as English, require smaller capacity of unit resources, and resource modules are relatively independent; (2) learning resources of non-pure language which are basically presented by the texts, such as e-books, electronic dictionaries, short message, social software, courseware and other styles to present micro-learning content; (3) learning resources of images and micro-videos which have lower resolution, size and other parameters.

A Unit of Learning Has Appropriate Capacity. (1) A learning unit contains appropriate learning content; (2) learning time required is proper for completing a full learning process; (3) screen space required is proper for a unit with a full sense.

Learning Units Are Relatively Independent. Because of the mobility, m-learning process may be interrupted and stop at any time, thus resulting in breakpoints of learning process. Therefore, learning resources units are linked each other and relatively independent and learners can restore learning from the "breakpoint" seamlessly and smoothly, and thus maintaining the continuity of the learning process.

3.2 Compatibility of Learning Situations

Learning occurs depending on the situations. Learning environments include learning fields and learning contexts. The field where mobile learning occurs is defined as "big situation", and the learning context is defined as "small situation".

Security. Learning situations cannot be obstacles for conducting the learning activities smoothly. The situation should be able to ensure ongoing learning activities, not because of external interference to excessively transfer the learner's attention, not because of the interrupting learning process impacts on learning efficiency too much.

Compatibility. (1) Compatibility of big situation and small situation. A learning context can be treated as a small situation inside the learning situation, and seamlessly integrated into a big situation while maintaining great harmony. (2) Compatibility of learners, m-learning devices and situations. Building this kind of harmonious interaction environment can better serve the learning process and obtain better learning experience and results.

4 Dimension of Device Performance

4.1 Mainstream Configuration as a Device Standard

Difference is a significant feature of m-learning devices. A design reference in harmonious interaction should select mainstream position of m-learning devices in the screen size, resolution, storage capacity and processing speed. In doing so, the use scope of the design results can be expanded, and can be as a solid foundation for further development. Designers should meet the needs of both mainstream demand and preserve the high-level interfaces to allow for future expansion and upgrades.

4.2 Attention to Constraints of Equipment Space and Capacity

Relatively small screen size of m-learning device is another notable feature, so the design must fully consider the space factor, multi-path focus on research to improve the efficiency of the method of limited space, to meet the shortfall, then advantage. At the same time, computing power of m-learning devices is still subject to certain restrictions, such as handling a small number of tasks, small information capacity, slow feedback and low storage capacity. Therefore, harmonious interaction design can be considered from the following aspects:

Simplicity. (1) Moderate amount of information. The information on the screen should be nothing more, nothing less. Nothing more means that screen space does not accumulate too much information types and information capacity, and means function design cannot be too complex, resulting in the functional load and waste of resources and may be brought to the cognitive load and learners use problems. Nothing less means that the default interface elements and features are self-contained to perform tasks, and means that the graphics, icons, text and other elements are appropriate in size and clarity, not blindly lower standards due to the limited screen.(2)appropriate presentation of information. Human perception of space depends on the psychological experience. Scientific design, reasonable arrangements for the interface elements can improve the small screen, congestion and depression. Specific implementations are: a) a reasonable split-screen, the structure of science; b) classifying information to perform different functions according to similarity and compatibility.

Highlighting Important Information. (1) Information contrast. By contrast of color, size, actual situation, shading, transparency to emphasize key information, weakening the secondary information to give learners more effective stimulation and enhance memory.(2) sorting information. Sort implied 80/20 rule, which is 20% of the functions to meet the 80% of applications. Thus, 20% of the common functions required to be placed prominently sequence menu, the other 80% is not commonly used functions can be included in "more features". (3) Through the progressive disclosure [2] highlighting focused information. Progressive disclosure means that only necessary information and functions are displayed to the learner at a moment. (4) Through constraints focused information. Constraint means to restrict a specific activity. For example, when a function is not available, they will not allow the learner clicks. Constraints include physical constraints and psychological constraints. Psychological constraints also include practices, mapping and symbols to avoid re-learning, re-adaptation process.

Precise Words. Texts are still the main carrier of information. Words should be expressed as precise, easy to understand, simple language, avoiding chaos and confusion for learners.

4.3 Fault Tolerant

Interactive system should be set a certain tolerance limits as indicators of system, the system is always in the learner's control. This can create a comfortable atmosphere to enhance the learner's sense of security and a sense of the subject, so that students in a good attitude and atmosphere to achieve their goals and improve the efficiency of interaction.

4.4 Adaption

Harmonious interaction design can be carried out based on physical characteristics of m-learning devices. Functions are closely attached to the existing m-learning devices will maximize the value-based characteristics. Keypads of m-learning devices using touch technology can act as a touchpad to operate, so the design can be used in gesture interaction methods to take full advantage of this feature. Reasonable set of digital keyboard shortcuts, function keys can reduce the dependence on the learner, to achieve "a key to success", thereby enhancing interaction efficiency.

5 Dimension of Emotional Experience

Based on Maslow's hierarchy of needs [3], combined with the perspective of the learner experience and the availability of the theory, we can construct "harmonious interaction" hierarchy.

5.1 Function

First and foremost task of interactive system is to achieve the basic functions, namely, interactions have to be useful and effective, must support learners to complete the task through various means and methods. Function layer is the starting point and reference points of the entire interactive behavior.

5.2 Safety

Safety belongs to the basic level of interaction, and will bring a sense of security and trust for learners. Safety means to establish mechanisms of helps and fault tolerance to resolve the frustration and support learners to ensure interaction of fluid and reduce learner frustration, access to safe experience. For example, guide the learner to recover from the mistake, the rational use of progressive disclosure, and constraints.

5.3 Efficiency

After usefulness and effectiveness are met in the interactive system, people will have a higher level of physical and psychological demands, that is, how to enhance the interaction efficiency. Improving efficiency is to reduce the loads including cognitive load, memory load, visual load and the physical load [4].

5.4 Pleasure

Pleasure is a human emotion in a positive good and a high level of psychological experience. Good representations of human emotional factors in the design not only help to enhance the value of interactive products, but also create good conditions for learners' high levels of meaning construction. Harmony contrast of Color, size, and brightness, icons with clear meaning, and the notice can bring the feeling of pleasure, the intimacy of the interaction has greatly increased.

5.5 Significance

Significance is the highest level of the system, and it has an extended and deep value. Significance is a representative of high-level cognitive abilities of human. After the process of completing the objectives and tasks, learners not only gained the original meaning of knowledge and gained extended meaning of knowledge.

This process is the construction of meaning. Knowledge of a certain phenomenon, such as the law, we understand the significance of the individual knowledge of the origin, but how to use this rule to analyze the phenomenon of the other groups, to make the abstract, general rules and a common approach is to expand the meaning of knowledge of expression. This process will comprehensively improve learner's cognitive ability, analytical ability, thinking ability of the improvement of the people play a decisive role in sustainable development. Therefore, interaction design should focus on human-computer interaction and beyond human-computer interaction to keeps an eye on improvement of learners' overall strength.

6 Conclusion

Harmonious interaction design in m-learning includes three dimensions: situations and learning resources, device performance, emotional experience. The ultimate goal of harmonious interaction design in m-learning is to promote learning. Harmonious interaction design reduces learning loads and improves learning efficiency.

References

1. Narasimhan, S.G., Nayar, S.K.: Contrast reatoration of weather degraded images. IEEE Transactions on Pattem Analysis and Machine Intelligence 25(6), 713–724 (2003)
2. Narasimhan, S.G., Nayar, S.K.: Vision and the atmosphere. International Journal of Computer Vision 48(3), 233–254 (2002)
3. He, K., Sun, J., Tang, X.: Singe Image Haze Removal Using Dark Channel Prior. IEEE Transactions on Pattem Analysis and Machine Intelligence, 1956–1963 (2009)
4. Oakley, J.P.: Improving image quality in poor visibility conditions using a physical model for contrast degradation. IEEE Transactions on Image Processing 7(2), 167–179 (1998)
5. Hautiere, N., Aubert, D.: Contrast Restoration of Foggy Images through Use of an onboard Camera. In: Proceedings of the 8th International IEEE Conference on the Intelligent Transportation Systems, pp. 601–606 (2005)
6. Sehoehner, Y.Y., Narasimhan, S.G., Nayar, S.K.: Instant Dehazing of Images Using Polarization. In: Proeeedings of IEEE Computer Society Conferenee Commuter Vision and Pattem Reeognition, pp. 325–332 (2001)

An Electronic Commerce Recommendation Approach Based on Time Weight

Huiping Mei

Zhejiang Textile & Fashion College, Ningbo 315211, China
huiping_mei@163.com

Abstract. It has been known that personalized recommendation system is a very important and necessary topic in electronic commerce. Many famous electronic commerce websites employ recommendation systems to convert browsers into buyers. The forms of recommendation include suggesting items to the users, providing personalized service information, summarizing community opinion, and providing society critiques. Collaborative filtering is the most successful technology for building electronic commerce personalized recommendation system and is extensively used in many fields. But traditional collaborative filtering recommendation algorithm does not consider finding the nearest neighbors in different time periods, leading to the neighbors may not be the nearest ones. To solve this issue, an electronic commerce recommendation approach based on time weight is presented. In this method, the time weighted rating is used to search the recommendation items for target users.

Keywords: personalized service, electronic commerce, recommendation approach, collaborative filtering, time weight.

1 Introduction

E-commerce personalized recommendation system is on the Internet to solve the problem of information overload, which is made as an intelligent agent system. It's a lot of information from the Internet and recommending to the user preferences or needs consistent with its interest in the information resources. In personalized recommendation system, one of the most successful methods is the nearest neighbor collaborative filtering recommendation algorithm. Collaborative filtering algorithm has two algorithms. On is user-based collaborative filtering and the other is item-based collaborative filtering. The basic idea is as follows. By calculating the target user and all basic user ratings of the similarity between items, search the collection target user's nearest neighbors, and then score the data from the nearest neighbor to the target user generated recommendations. The target user's score did not score the item through the nearest neighbor to approximate the weighted average rate, resulting in recommendations.

Collaborative filtering recommendation techniques have some problems for waiting to be developed solutions that are more efficient. One of these mainly problems is data sparsity. While the number of items is increase, the ratio of common rated items is decrease so calculating the computations of neighborhood become

D. Jin and S. Jin (Eds.): Advances in FCCS, Vol. 1, AISC 159, pp. 609–614.
springerlink.com

difficult. The other one is scalability which is the performance problem of the existing algorithms on the datasets has large amounts of information. Currently, the traditional disadvantage is the existence of a collaborative filtering algorithm. Looking for the nearest neighbor did not consider the time when the collection of attributes, different users visited the item at different time as equal treatment, which is obviously unreasonable, affecting the accuracy of the algorithm.

Traditional collaborative filtering recommendation algorithm does not consider finding the nearest neighbors in different time periods, leading to the neighbors may not be the nearest ones. To solve this issue, in this paper, an electronic commerce recommendation approach based on time weight is presented. In this method, the time weighted rating is used to search the recommendation items for target users.

2 Collaborative Filtering

Nearest neighbor-based collaborative filtering algorithm is one of the more successful recommendation techniques in recommender systems. The basic idea is similar to the nearest neighbor based on the grading score data to target users have recommended. Collaborative filtering algorithm implementation process can be divided into three steps, namely, the input data, forming neighbors and recommended generation.

Step 1: Enter the data representation

Data indicated that the description of the major finished browsing the data, usually expressed as an m × n user - item rating matrix R, m indicate the number of users, n indicates the number of items, Rij is the first i-j, the user's interest degree.

Step 2: the formation of neighborhood

Neighbors to complete the formation of the main objectives of the user identification of nearest neighbors. Collaborative filtering need to analyze the similarity between users, the formation of the current target user's neighbor set, which recommended the information according to neighbors. Collaborative filtering recommendation system implemented in the core is a need to recommend the service to find the most current target audience similar to the set of nearest neighbors.

Step 3: Generate recommended

The recommended formula of the electronic commerce collaborative filtering in recommendation systems as follows:

$$P_{Target(u,i)} = \overline{R_{Target(u)}} + \frac{\sum_{n \in Neighbor(Target(u))} sim(Target(u,n)) \times (R_{ni} - \overline{R_n})}{\sum_{n \in Neighbor(Target(u))} (|sim(Target(u,n))|)}$$

3 The Electronic Commerce Recommendation Method Based on Time Weight

3.1 The Problem Description

The score of each item the user is not simultaneous, user ratings of the project over time will change. This change reflects the change in user interest. To make the user recommendation system to track the interest changes in time, is generally believed that the relative value of the user's most recent score values than in the past the importance of scoring big. If the user's interest in a different time period equal treatment, the algorithm is difficult to predict the user's current interest. For example, the last time the user plays score higher on humor, but the next time the user interest in the war theater, theater of war also gave a higher score. According to the traditional collaborative filtering algorithm, if the nearest neighbor of the current user is not user rating item scores the same or similar values, the system will likely have humor, drama and theater of war recommended to the user, it is clear that comedy is not the latest interest in the current user. Here is an example of change in user interest, as shown in Table 1.

Table 1. An example of user interest change

item user	item1	item 2	item 3	item 4	item 5	time
user1	1	1	1	5	5	time1
user2	1	1	1	5	1	time2
user3	2	2	2	5	5	time1
user4	2	2	2	5	?	time2

3.2 The Algorithm

By calculating the similarity between users, interest in identifying the current user preferences similar to some of the most similar users. Similar items using these score users to recommend their current user may like the item.

User similarity calculation of two classic functions:
(1) Cosine similarity. Each user is represented in vector form. Cosine similarity of any two vectors can be calculated the angle between the users.

$$sim(x,y) = \frac{\sum_{\{i\in Sx\cap Sy\}} V_{xi} * V_{yi}}{\sqrt{\sum_{\{i\in Sx\cap Sy\}} V_{xi}^2 \sum_{\{i\in Sx\cap Sy\}} V_{yi}^2}}$$

(2) Pearson correlation coefficient. Some users rate the overall high, and some users may be generally low. In order to consider the overall trend rate of each user, use the following Pearson correlation coefficients

$$sim(x,y)=\frac{\sum_{\{i\in Sx\cap Sy\}}(V_{xi}-\overline{V_x})*(V_{yi}-V_y)}{\sqrt{\sum_{\{i\in Sx\cap Sy\}}(V_{xi}-\overline{V_x})^2\sum_{\{i\in Sx\cap Sy\}}(V_{yi}-V_y)^2}}$$

Based on the above similarity measure, the system can draw a prediction derived from the user evaluation of the item. This gives the following definition.

$$P_{ti}=\overline{V_t}+\frac{\sum_{\{u\in U|i\in Su\}}sim(t,u)*(V_{ui}-\overline{V_u})}{\sum_{\{u\in U|i\in Su\}}sim(t,u)}$$

Where: the average of all the ratings a user, defined as

$$\overline{V_t}=\frac{\sum_{\{s\in S_t\}}V_{ts}}{|S_t|}$$

In order to consider the migration of user preferences, the following formula exponential function based on the introduction of a recession function. Exponential function is a widely used class of the decline of function, which can better simulate the impact of historical data over time, weakened the case. Using the exponential function is defined as follows

$$f(h)=e^{-\frac{h}{2}}$$

Can be seen from the above definition, the recession in the range of functions (0,1), and the function value decreases with time variable h, so better to be able to simulate the effect of migration of user preferences. At this point, reached the final migration of the complex combination of user preference function is defined as follows recommended

$$P_{ti}=\frac{\sum_{\{u\in U\}}sim(t,u)*f(h_{ui})*R_{ui}}{\sum_{\{u\in U\}}sim(t,u)*f(h_{ui})}$$

4 Summary

It has been known that personalized recommendation system is a very important and necessary topic in electronic commerce. Many famous electronic commerce websites employ recommendation systems to convert browsers into buyers. The forms of recommendation include suggesting items to the users, providing personalized service information, summarizing community opinion, and providing society critiques. Collaborative filtering is the most successful technology for building electronic commerce personalized recommendation system and is extensively used in many fields. But traditional collaborative filtering recommendation algorithm does not consider finding the nearest neighbors in different time periods, leading to the neighbors may not be the nearest ones. To solve this issue, an electronic commerce recommendation approach based on time weight is presented. In this method, the time weighted rating is used to search the recommendation items for target users.

Acknowledgments. This work was supported by Scientific Research Fund of Zhejiang Provincial Education Department (Grant No. Y201017814) and Ningbo Advanced Textile and Fashion CAD Laboratory (Grant No. 2011ZDSYS-A-004).

References

1. Cho, Y.B., Cho, Y.H., Kim, S.H.: Mining changes in customer buying behavior for collaborative recommendations. Expert Systems with Applications 28, 359–369 (2005)
2. Chee, S.H.S., Han, J., Wang, K.: RecTree: An Efficient Collaborative Filtering Method. In: Kambayashi, Y., Winiwarter, W., Arikawa, M. (eds.) DaWaK 2001. LNCS, vol. 2114, pp. 141–151. Springer, Heidelberg (2001)
3. Min, S.-H., Han, I.: Detection of customer time-variant pattern for improving recommender systems. Expert Systems with Applications 28, 189–199 (2005)
4. Ungar, L.H., Foster, D.P.: A Formal Statistical Approach to Collaborative Filtering. In: Proceedings of Conference on Automated Leading and Discovery, CONALD (1998)
5. Sarwar, B., Karypis, G., Konstan, J., Riedl, J.: Recommender systems for large-scale e-commerce: Scalableneighborhood formation using clustering. In: Proceedings of the Fifth International Conference on Computer and Information Technology (2002)
6. Lee, T.Q., Park, Y., Park, Y.-T.: A time-based approach to effective recommender systems using implicit feedback. Expert Systems with Applications (2007)
7. Gong, S.: A Personalized Recommendation Algorithm on Integration of Item Semantic Similarity and Item Rating Similarity. Journal of Computers 6(5), 1047–1054 (2011)
8. Gong, S.: Privacy-preserving Collaborative Filtering based on Randomized Perturbation Techniques and Secure Multiparty Computation. International Journal of Advancements in Computing Technology 3(4), 89–99 (2011)
9. George, T., Merugu, S.: A scalable collaborative filtering framework based on co-clustering. In: Proceedings of the IEEE ICDM Conference (2005)
10. Conner, M.O., Herlocker, J.: Clustering Items for Collaborative Filtering. In: Proceedings of the ACM SIGIR Workshop on Recommender Systems, Berkeley, CA (August 1999)
11. Kohrs, A., Merialdo, B.: Clustering for Collaborative Filtering Applications. In: Proceedings of CIMCA 1999. IOS Press (1999)

12. Lee, W.S.: Online clustering for collaborative filtering. School of Computing Technical Report TRA8/00 (2000)
13. Resnick, P., Varian, H.R.: Recommender Systems. Special Issue of Communications of the ACM 40(3) (1997)
14. Schafer, J.B., Konstan, J., Riedl, J.: Recommender Systems in E-Commerce. In: Proceedings of ACM E-Commerce 1999 Conference (1999)
15. Sarwar, B.M., Karypis, G., Konstan, J.A., Riedl, J.: Analysis of Recommendation Algorithms for E-Commerce. In: Proceedings of the ACM EC 2000 Conference, Minneapolis, MN, pp. 158–167 (2000)
16. Schafer, J.B., Konstan, J., Riedl, J.: Electronic Commerce Recommender Applications. Journal of Data Mining and Knowledge Discovery 5(1/2), 115–152 (2001)

Research of Collaborative Filtering Recommendation Algorithm in Electronic Commerce

Yibo Huang

Zhejiang Textile & Fashion College, Ningbo 315211, China
yibo_huang1@163.com

Abstract. Recommendation system is becoming an important part of electronic commerce systems. And collaborative filtering is the most hot research topic for building electronic commerce personalized recommendation system and is extensively used in many fields. Collaborative filtering aims at predicting a target user's ratings for new items by integrating other like-minded users' rating information. The user-based approach is a common technique used in collaborative filtering. This method first uses statistical approaches to measure user similarities based on their previous ratings on different items. Users will then be grouped into different neighborhood according to the calculated similarities. Finally, the approach will generate predictions on how a user would rate a specific item by aggregating ratings on the item cast by the identified neighbors of the target user. Collaborative filtering algorithm usually suffers from two fundamental problems: sparsity and scalability. In this paper, the problems of sparsity and scalability are described. And an overview of collaborative filtering recommendation algorithm in electronic commerce is presented.

Keywords: collaborative filtering, electronic commerce, recommendation algorithm, sparsity, scalability.

1 Introduction

With the popularization of Internet, e-commerce has become an important business model and to promote social, economic, and cultural life, an important engine of progress and the tools. The new business environment for enterprises to provide new business opportunities, also posed new challenges enterprises. In the electronic virtual world, how to attract new customers and retain existing customers is to become a major task. On the other hand, the customer facing so many choices from which to pick out the data they really need is very difficult. E-commerce recommendation system is an important way to solve these problems.

E-commerce recommendation system function is to gather information of interest to users, and preferences based on user interest in this initiative to make personalized recommendations for users. This way, when each time a user enters using a user name and password in e-commerce site, according to the user recommendation system it will automatically recommend the level of preference of the user's favorite items. That is, when the system user interest in the item library and information changed, given

D. Jin and S. Jin (Eds.): Advances in FCCS, Vol. 1, AISC 159, pp. 615–620.
springerlink.com

the recommended series will automatically change. The method is convenient for users to browse for goods, but also improve the enterprise level of service.

Collaborative filtering is the most studied, most widely used technique for personalized recommendation. It is based on the user's information neighbor recommended by the target user, a high degree of personalized recommendations. The starting point is the collaborative filtering: users with similar interests may be interested in the same things. So, as long as the maintenance of data on user preferences, it derived from analysis of the user with similar tastes, and then to customer feedback according to their similarity to recommend. Another possible starting point is: the user may have purchased more prefer something with similar products. All things according to the evaluation of the user to determine the degree of similarity between goods, and then recommend those closest to the user interest in commodities. The former thought and the relationship between client and customer as the center, then a train of thought the relationship between the item and the item is the focus. Compared with other methods, collaborative filtering has the following advantages: can cross-type recommendations; do not need domain knowledge, sharing the experience of others; self-adaptability; fully implicit feedback can reduce the amount of user feedback and accelerate individual learning speed.

2 The Problems of Collaborative Filtering

Although collaborative filtering is the most successful recommendation method, but with the growing size of e-commerce system, collaborative filtering recommendation method is also facing many challenges, the most concerns are: data sparsity and scalability.

2.1 Sparsity

Collaborative filtering technology, the user first needs to achieve a matrix of user information, said. Although this is very simple in theory, but in practice many e-commerce recommendation system on a lot of data to be processed, and in general these systems users to buy goods of the total amount of total site about 1% of the amount of goods. Evaluate the resulting matrix is very sparse.

E-commerce sites often have a large number of commodities, and buy or make an assessment of each user only a very small part of it. Collaborative filtering based on the user's system, if two users did not rate the same items, even if the two users have the same interests and hobbies, the system can not come to the similarity between them. Also, because the data is very sparse, in the formation of the target user's nearest neighbor sets of users, they tend to lose information, resulting in reduced effects recommended.

2.2 Scalability

Because collaborative filtering algorithm needs to scan the entire database to calculate the user similarity, so as the user records in the database increases, the computational complexity increases exponentially, resulting in recommendation system performance dramatically. Therefore, the index through the establishment of a number of heuristic pre-filter on the user record will be recommended to minimize the impact of the premise of quality greatly reduce computational complexity. At the same time, global numerical algorithm using the latest information in time for the user a relatively accurate the prediction of user interest or recommended, but the face of the growing number of users, the algorithm scalability problem as a constraint to the implementation of the important factors recommended by the system. Identify the nearest neighbor algorithm for computing the amount of items with the increase of users and greatly increased the number of the millions, the usual method would encounter serious scalability bottlenecks.

Collaborative filtering recommendation system implementation, to obtain nearest neighbor users, must be calculated by a certain similarity between users, and then determine the optimal number of neighbors to form a set of neighbor users. In the process, if all the data sets similar calculation, although straightforward, but computational time and cost are great, can not adapt to the real business system. Therefore, consider using a more effective way to the formation of nearest neighbor users for the application of collaborative filtering is necessary.

3 The Collaborative Filtering Algorithm in E-commerce

With the development of information technology and the expansion of information resources for users to find products of interest to information has become a difficult and expensive to do. Therefore, the e-commerce recommendation system came into being. Personalized e-commerce recommendation system is an important part of the service. Personalized recommendation system includes hot commodity recommendation, New Products, product promotion, and with user groups and recommend the same interest. Recommended system uses data mining technology in e-commerce site to help interested customers to access product information and generate recommendations. These systems expand sales, increase cross sales, improve customer loyalty and other aspects of the larger contribution.

Collaborative filtering technology is the use of the main objectives of the historical evaluation of users of the products to match with other users to predict the users of the evaluation did not evaluate the product, resulting in efficient recommendation. In predicting the target user evaluation of a product before, you first need to score under the user's history and feedback information, select the target user's nearest neighbor set, which are similar to the target users of a collection of user interest. After the selected nearest neighbors, according to a neighbor for some products on the target user's rating score to predict, resulting in recommendations.

3.1 The User-Item Matrix

The use of user behavior and interest in the past with an $m \times n$ matrix R to represent; n is the number of users; m is the number of items in search results, which is shown in table 1.

Table 1. User-item rating matrix

User/Item	I1	I2	…	Ii	…	In
U1	R11	R12	…	R1j	…	R1n
U2	R21	R22	…	R2j	…	R2n
…	…	…	…	…	…	…
Ui	Ri1	Ri2	…	Rij	…	Rm
…	…		…	…	…	…
Um	Rm1	Rm2	…	Rmj	…	Rmn

3.2 Similarity Measurement

There are many similarity measurements between tow users in collaborative filtering methods.

(1) correlation similar: the formula is as following:

$$sim(i,j)=\sum_{z\in I_{ij}}(R_{iz}-\bar{R}_i)(R_{jz}-\bar{R}_j)/\left[\sqrt{\sum_{z\in I_{ij}}(R_{iz}-\bar{R}_i)^2}\times\sqrt{\sum_{z\in I_{ij}}(R_{jz}-\bar{R}_j)^2}\right]$$

(2) cosine similarity: the formula is as following:

$$sim(i,j)=\cos(i,j)=i\times j/(\|i\|\|j\|)$$

(3) the modified cosine similarity: the formula is as following:

$$sim(i,j)=\sum_{z\in I_{ij}}(R_{iz}-\bar{R}_i)(R_{jz}-\bar{R}_j)/\left[\sqrt{\sum_{z\in I_{ij}}(R_{iz}-\bar{R}_i)^2}\times\sqrt{\sum_{p\in I_{ij}}(R_{jz}-\bar{R}_j)^2}\right]$$

3.3 Neighbor Selection

The establishment of similar community collaborative filtering recommendation system is the most important step in order to calculate the degree of similarity between users, as a future recommendation. The idea is shown as figure 1.

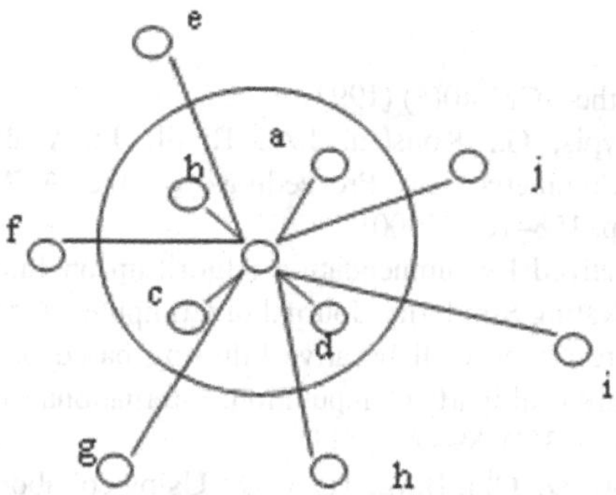

Fig. 1. Neighbor selection

3.4 Producing Recommendation

The producing recommendation formula is shown as following:

$$p_{iz} = \overline{R}_u + \sum_{n\in M_z} sim_{zn} \times R_{i,n} / \sum_{n\in M_z} \left(\left| sim_{zn} \right| \right)$$

4 Summary

Recommendation system is becoming an important part of electronic commerce systems. And collaborative filtering is the most hot research topic for building electronic commerce personalized recommendation system and is extensively used in many fields. Collaborative filtering aims at predicting a target user's ratings for new items by integrating other like-minded users' rating information. The user-based approach is a common technique used in collaborative filtering. This method first uses statistical approaches to measure user similarities based on their previous ratings on different items. Users will then be grouped into different neighborhood according to the calculated similarities. Finally, the approach will generate predictions on how a user would rate a specific item by aggregating ratings on the item cast by the identified neighbors of the target user. Collaborative filtering algorithm usually suffers from two fundamental problems: sparsity and scalability. In this paper, the problems of sparsity and scalability are described. And an overview of collaborative filtering recommendation algorithm in electronic commerce is presented.

Acknowledgments. This work was supported by Scientific Research Fund of Zhejiang Provincial Education Department (Grant No. Y201017814).

References

1. Communications of the ACM 40(3) (1997)
2. Sarwar, B.M., Karypis, G., Konstan, J.A., Riedl, J.: Analysis of Recommendation Algorithms for E-Commerce. In: Proceedings of the ACM EC 2000 Conference, Minneapolis, MN, pp. 158–167 (2000)
3. Gong, S.: A Personalized Recommendation Algorithm on Integration of Item Semantic Similarity and Item Rating Similarity. Journal of Computers 6(5), 1047–1054 (2011)
4. Gong, S.: Privacy-preserving Collaborative Filtering based on Randomized Perturbation Techniques and Secure Multiparty Computation. International Journal of Advancements in Computing Technology 3(4), 89–99 (2011)
5. Goldberg, D., Nichols, D., Oki, B.M., Terry, D.: Using collaborative filtering to weave an information tapestry. Communications of the ACM 35(12), 61–70 (1992)
6. Resnick, P., Iacovou, N., Suchak, M., Bergstrom, P., Riedl, J.: Grouplens: An open architecture for collaborative filtering of netnews. In: Proceedings of the ACM CSCW 1994 Conference on Computer-Supported Cooperative Work, pp. 175–186 (1994)
7. Shardanand, U., Maes, P.: Social information filtering: Algorithms for automating "Word of Mouth". In: Proceedings of the ACM CHI 1995 Conference on Human Factors in Computing Systems, pp. 210–217 (1995)
8. Hill, W., Stead, L., Rosenstein, M., Furnas, G.: Recommending and evaluating choices in a virtual community of use. In: Proceedings of the CHI 1995, pp. 194–201 (1995)
9. Breese, J., Hecherman, D., Kadie, C.: Empirical analysis of predictive algorithms for collaborative filtering. In: Proceedings of the 14th Conference on Uncertainty in Artificial Intelligence (UAI 1998), pp. 43–52 (1998)
10. Balabanovic, M., Shoham, Y.: FAB: Content-based collaborative recommendation. Commun. ACM 40, 3 (1997)
11. Basu, C., Hirsh, H., Cohen, W.: Recommendation as Classification: Using Social and Content-based Information in Recommendation. In: Recommender System Workshop 1998, pp. 11–15 (1998)
12. Billsus, D., Pazzani, M.J.: Learning Collaborative Information Filters. In: Proceedings of ICML 1998, pp. 46–53 (1998)
13. Jin, R., Si, L.: A bayesian approach toward active learning for collaborativefiltering. In: Proceedings of the 20th Conference on Uncertainty in Artificial Intelligence, pp. 278–285. AUAI Press, Banff (2004)
14. Herlocker, J.: Understanding and Improving Automated Collaborative Filtering Systems. Ph.D. Thesis, Computer Science Dept., University of Minnesota (2000)

Research of Intrusion and Monitoring Based on Wireless LAN

ShengBing Che and Shuai Jin

College of Computer & Information Engineering, Central South University of Forestry & Technology, 410004, Changsha, Hunan, China
cheshengbing727@tom.com, jinshuai724@hotmail.com

Abstract. By analyzing the vulnerabilities of IEEE 802.11 protocol, and based on the scheme of capturing, filtering and analyzing frames, the algorithm for cracking WEP protocol key was put forward to intruding the target host in WLAN. Then combined with the key technologies of Trojans such as the covert and anti-killing, a remote monitoring framework based on the mechanism of Trojans was designed. According to practical requirements, some forensics methods were provided in the framework such as screen monitoring and file management. Experiments results show that the forensics system was of high reliability and strong survival ability, and achieved the purpose of real-time, covert and active forensics.

Keywords: WLAN, WEP Protocol, Computer Forensics, Remote Monitoring.

1 Foreword

Traditional computer networks are required to transform from wired/fixed to wireless/mobile ones in order to communicate from anywhere and at anytime. There is no doubt that WLAN will be a new trend of the future development of network technology. The widespread of computer network technology provides a new platform and means for computer criminals. All kinds of new criminal activities that use computer as an object or tool are increasing.

New requirements are presented to case investigation of Public Security Department due to the rampancy of network crimes. How to use WLAN to monitor criminal behaviors in real time and obtain evidences actively has become a new issue to be solved in computer forensics field. In recent years, the security technology called Network Forensics and Analysis is gradually taken seriously[1,2]. However, there is too much uncertainty in these ways of obtaining evidences when analyzing and reverting information content. Moreover the process of analyzing and reverting is too complicated to implement. In addition, it is not difficult to find the following characteristics after comparing all kinds of forensics system at home and abroad[3]. The main ways of forensics are passive rather than active. Beyond this, the main tools of forensics are static rather than dynamic. Further, the main development of forensics system is used to protect the computer instead of monitoring the local computer system.

D. Jin and S. Jin (Eds.): Advances in FCCS, Vol. 1, AISC 159, pp. 621–628.
springerlink.com

Compared with the above wireless network forensics technology and systems, the new forensics idea was put forward. By intruding the specified computer in WLAN, control the external equipments of target host, such as USB camera and hard disk, and then obtain crime evidences actively with Trojan technique. Moreover the target host and its surrounding environment can be monitored.

2 System Design and Implementation

Forensics System Framework. Basic working process of the forensics system is as follows: (1) Set WLAN card to monitoring mode and all types of IEEE 802.11 frame can be captured by the card within its monitoring range. After filtering the captured frames, the frames of specific types can be analyzed. Then obtain key by using the algorithm for cracking WEP protocol key. (2) Connect to WLAN with the correct key. As WLAN may set up SSID broadcast forbidding or MAC address filtering, investigators should provide correct SSID or mask to a valid MAC address. These can be learned by analyzing frame. (3) Implement the process of implanting Trojan through several methods of intrusion. For instance, The Trojans can be implanted by the technology of vulnerability scanning and port intrusion. (4) By analyzing key technologies and basic characteristics of Trojans, the remote monitoring framework based on Trojan mechanism was designed. This framework adopts C/S working model, and is made up of DLL Trojan, Trojan loader and the monitoring program. Trojan loader and DLL Trojan are run on the server side, namely the controlled side. The side is in charge of receiving and dealing with commands of remote monitoring sent by the client, then sends forensics results to the client. The monitoring side implemented according to the programming method of client is mainly responsible for accepting and managing the connections sent by DLL Trojan, sending relevant commands of remote monitoring, then handling and displaying forensics results.

Function Module Partition. Logically speaking, the system can be divided into two parts: WLAN Attack, Real-time and Covert Forensics.

WLAN Attack. It is run on the monitoring side, including four functional modules.

(1) Frame capturing module. WLAN card has a passive working mode, namely listening mode, where the wireless terminal can not access any Basic Service Set (BSS), and the WLAN card can capture all data transmitting in the range of its signal. Set it to monitoring mode and the original MAC frames can be captured from transmission medium. The captured frames are stored in specified memory, waiting to be taken away by the next module.

(2) Frame filtering module. The following frame types should be preserved: Association Request Frames, Probe Response Frames, Beacon Frames, WEP encryption Data Frames and unencrypted Data Frames. Control Frames, other types of Data Frames and Management Frames will be discarded. The WEP encryption Data Frames will be referred in key cracking module.

(3) Frame analysis module. By analyzing the frame header, MAC address of AP and the controlled side can be separated. By analyzing the frame body of Association Request, Probe Response and Beacon Frames respectively, Service Set Identifier (SSID) of the corresponding BSS can be obtained. In addition, IP address can be separated from the unencrypted IP or ARP Data Frames.
(4) Key cracking module. Various attack methods for WEP protocol were designed. Then the key to WEP protocol may be cracked through recursive procedure of cracking. This module will be described in detail in 3rd section.

Real-time and Covert Forensics. After intruding the operating system of target host, DLL Trojan and Trojan loader will be automatically released into specified directory of the server side. The partition of functional module of this part is as follows.

(1) Trojan loading module. It is in charge of the self-start, file recovery and backup, covert running of DLL Trojan. Fig. 1 shows the process of Trojan loading. After running Trojan loader, the existence of self-start record can be checked. If not exist, it means that it is the first time to load DLL Trojan. Otherwise delete the self-start record in order to avoid Trojan trail being found by computer criminals. The self-start mode will be reset when the target host is turned off. Then, detect whether the Trojan and backup files exist, and moreover, both files can be restored to each other when one is deleted. Finally, DLL Trojan will be run covertly by remote thread technology.

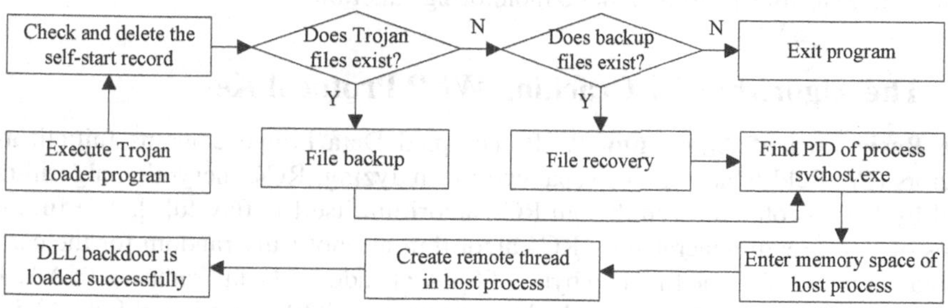

Fig. 1. Process of Trojan loading

(2) Thread guarding module. The structure of thread guarding is as shown in Fig. 2. There are two Trojan threads in system process svchost.exe, and they are called Main Thread and Watch Thread. Watch Thread and Remote Guarding Thread can be created by Main Thread. The operation of shutdown can be watched by hooking the API function ExitWindowEx. As a result, the Trojan will be notified first, and then the default ExitWindowEx can be called after appending the self-start mode. Remote Guarding Thread is run in the process explorer.exe or taskmgr.exe. And the first searched process is selected as the host process. Thus, even if the computer criminals find explorer.exe working abnormally and try to kill it with task manager, a new Remote Guarding Thread will be created again in taskmgr.exe immediately.

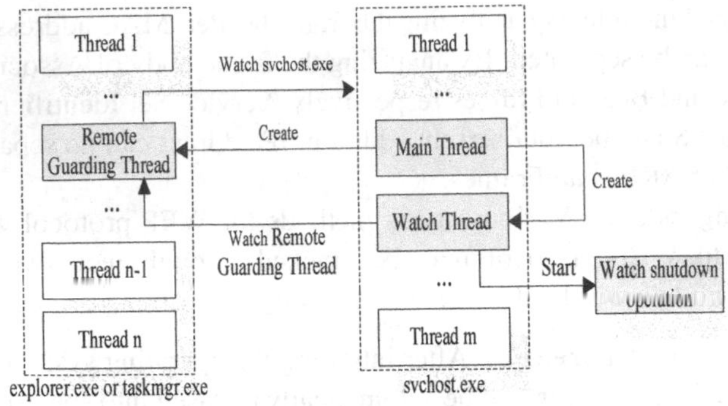

Fig. 2. Structure of thread guarding

(3) Covert communication module. The C/S model based on reverse connection mechanism is adopted in order to traverse the firewall. The connection between them is initiated by the server and the controlled side is implemented through the programming method of client.
(4) Forensics function module. Crime evidences stored in the hard disk of target host can be obtained with the operations of file management function. The remote screen can be monitored and controlled by the mouse and keyboard actively through screen monitoring function. The criminal object and its surrounding environment can be monitored real-timely through video monitoring function.

3 The Algorithm for Cracking WEP Protocol Key

The Basic Idea of Algorithm. WEP encrypted Data Frame contains Initialization Vectors (IV) field which can be separated by analyzing. RC4 encryption algorithm is used by WEP protocol, even though RC4 algorithm itself is flawed[4]. The first few bytes of key stream generated by RC4 algorithm are not truly random for the certain IVs called weak IVs. The first two bytes of SNAP header of Data Frame are 0xAA. The first two outputs of key stream can be known through XOR operation of SNAP header and ciphertext. Since the two outputs are not really randomly generated, the key of WEP protocol can be guessed. Combined with the thought of FMS attack[5], three types of attack method were designed in order to seek more weak IVs.

Assuming that the first $B-1$ bytes of key have been restored, the next step is to crack the byte B of key. The range of possible values of byte B should be $0x00 \sim 0xFF$. According to the conditions of each attack, each WEP encrypted Data Frame can be detected one by one in order to find weak IV towards the certain attack. Assume that the byte affirmed or negated by the certain attack is K_B . The probability of success for this attack can be converted to the corresponding vote value, and then make statistical in K_B . When all WEP encrypted Data Frames are finished the detection process to weak IVs, the possible values of byte B can be sorted by votes in the sequence of descending. The more previous possible value is of greater probability that it is the correct byte of key. After sorting, the possible values can be assumed as an array $\{X_0, X_1, ..., X_{255}\}$. First of all let $K_B = X_0$, then crack the byte $B+1$ of key recursively. After cracking the last byte of key, the correctness of key can be verified. If correct, return success. Otherwise, let $K_B = X_1$, and then continue to implement the above process.

Attack Methods. The known conditions are as follows when detecting weak IVs: (a) Encrypted SNAP header; (b) The first 3 bytes of IV field; (c) The first $B+3$ steps of KSA and the status of S-box after KSA_{B+3}. The first two outputs of key stream are assumed as O_1 and O_2 respectively. Take an example in cracking the byte B of key, and then three types of attack may be described below.

(1) The inverted attack is used to exclude some possible values for byte B of key, as shown in Table 1. The specific analysis of the first inverted attack in Table 1 is as follows.

(a) $KSA_1 \sim KSA_{B+3}$ satisfy the following conditions: $S[1]=0$, $S[0]=1$, $O_1=1$.

(b) Suppose that both $S[0]$ and $S[1]$ are no longer be swapped in $KSA_i (i > B+3)$. If $B=0$, the probability is $P=(1-2/256)^{256-(B+3)}=13.75\%$.

(c) $PRGA_1$: Swap $S[0]$ and $S[1]$, then $S[1]=1$, $O_1 = S[S[1]+S[S[1]]] = S[1] = 1$.

When the above is true, KSA_{B+4} satisfy the following formulas with about 13 percent probability: $j_{B+2}+S[B+3]+K_B \neq 0$, $j_{B+2}+S[B+3]+K_B \neq 1$.

Table 1. Description of each inverted attack

Condition of attack	**Success probability of attack($B=0$)**	**Result of attack (KSA_{B+4})**
$S[1]=0, S[0]=1, O_1=1$	P{Don't swap $S[0]$, $S[1]$ } = 13.75%	$j_{B+2}+S[B+3]+K_B \neq 0, \neq 1$
$S[1]=0$, $S[0]=2$, $O_1=S[2]$	P{Don't swap $S[0]$, $S[1]$, $S[2]$ } = 5.07%	$j_{B+2}+S[B+3]+K_B \neq 0, \neq 1$,
$S[1]=1, O_1=S[2]$	P{Don't swap $S[1]$, $S[2]$ } = 13.75%	$j_{B+2}+S[B+3]+K_B \neq 1, \neq 2$
$S[1]=2, S[2]=0, O_1=2$	P{Don't swap $S[0]$, $S[1]$, $S[2]$ } = 13.75%	$j_{B+2}+S[B+3]+K_B \neq 1, \neq 2$
$S[2]=0, O_2=0$	P{Don't swap $S[2]$ } = 37%	$j_{B+2}+S[B+3]+K_B \neq 2$

(2) If the first $B-1$ bytes of key have been restored, it is such an attack method that K_B can be affirmed only with O_1 and which FMS attack belongs to. This type of attack method is described in Table 2, where $Si[i]$ indicates the position of element i in S-box. The specific analysis of the first attack in Table 2 is as follows.

(a) $KSA_1 \sim KSA_{B+3}$ satisfy the following conditions: $S[1]=B+3$, $O_1=B+3$.

(b) KSA_{B+4}: Swap $S[B+3]$ and $S[Si[0]]$, then $S[B+3]=0$.

(c) Suppose that both $S[1]$ and $S[B+3]$ are no longer be swapped in $KSA_i (i > B+4)$. If $B=0$, the probability is $P=(1-2/256)^{256-(B+4)}=13.86\%$.

(d) $PRGA_1$: Swap $S[1]$ and $S[B+3]$, then $O_1=S[S[1]+S[B+3]]=B+3$.

When the above hypothesis is true, KSA_{B+4} satisfy the following formula with about 14 percent probability: $j_{B+2}+S[B+3]+K_B=Si[0]$.

Table 2. Description of the attack method associated with O_1

Condition of attack	**Success probability of attack(** $B=0$ **)**	**Result of attack (** KSA_{B+4} **)**
$S[1]=B+3, O_1=B+3$	P{Don't swap $S[1]$, $S[B+3]$ } = 13.86%	$j_{B+2}+S[B+3]+K_B=Si$
$Si[O_1]=2$, $S[B+3]=1$	P{Don't swap $S[1]$, $S[2]$ } = 13.86%	$j_{B+2}+S[B+3]+K_B=1$
$S[1]=B+3$, $(1-(B+3)-O_1) \& 0xFF=0$	P{Don't swap $S[0]$, $S[B+3]$ } = 13.86%	$j_{B+2}+S[B+3]+K_B=Si$
$S[B+3]=B+3, S[1]=0$, $O_1=B+3$	P{Don't swap $S[1]$, $S[B+3]$ } = 13.86%	$j_{B+2}+S[B+3]+K_B=1$

(3) If the first $B-1$ bytes of key have been restored, it is such an attack method that K_B can be affirmed with O_1, O_2. This type of attack method is described in Table 3. The specific analysis of the first attack in Table 3 is as follows.

Table 3. Description of the attack method associated with O_1 and O_2

Condition of attack	Success probability of attack	Result of attack (KSA_{B+4})
$S[2]\neq 0$, $S[1]\neq 2$, $O_2=0$, $S[B+3]=0$	P{ $B=0$ && Don't swap $S[1]$, $S[2]$ } = 13.86%	$j_{B+2}+S[B+3]+K_B=2$
$S[1]=2$, $B+3=4$, $O_2=0$	P{ $B=1$ && Don't swap $S[1]$, $S[B+3]$ } = 13.96%	$j_{B+2}+S[B+3]+K_B=S$
$S[1]=2$, $B+3>4$, $Si[O_2]\neq 1$, $Si[O_2]\neq 4$, $S[4]+2=B+3$	P{ $B=2$ && Don't swap $S[1]$, $S[4]$, $S[B+3]$ } = 5.25%	$j_{B+2}+S[B+3]+K_B=S$

(a) $KSA_1 \sim KSA_{B+3}$ satisfy the following conditions: $S[2]\neq 0$, $S[1]\neq 2$, $O_2=0$, $S[B+3]=0$.

(b) KSA_{B+4}: Swap $S[2]$ and $S[B+3]$, then $S[2]=0$.

(c) Suppose that both $S[1]$ and $S[2]$ are no longer be swapped in $KSA_i(i>B+4)$, and let $S[1]=x$. If $B=0$, the probability is $P=(1-2/256)^{256-(B+4)}=13.86\%$.

(d) $PRGA_1$: Swap $S[1]$ and $S[x]$, then $S[x]=x$.

(e) $PRGA_2$: Swap $S[2]$ and $S[x]$, then $S[x]=0$, $S[2]=x$, $O_2=S[S[2]+S[x]]=S[x]=0$.

When the above hypothesis is true, KSA_{B+4} satisfy the following formula with about 14 percent probability: $j_{B+2}+S[B+3]+K_B=2$.

4 Experiment of Key Cracking

Take the cracking process of WEP protocol key with length of 40 bits, for example. Fig. 3 shows the demonstration of key cracking. In the top layer, the crack result is $(71,45)(58,14)(02,13)(21,13)\ldots$, where the unit $(71,45)$ indicates that the possible value $0x71$ has 45 votes. The traversal range can be implied by the number of highlight rectangle units. When reaching the leaf node, the correctness of key will be verified. If the key is error, back to the upper layer and continue recursively. The detailed process of recursive transfer can be revealed by the dashed line with arrow in Fig. 3.

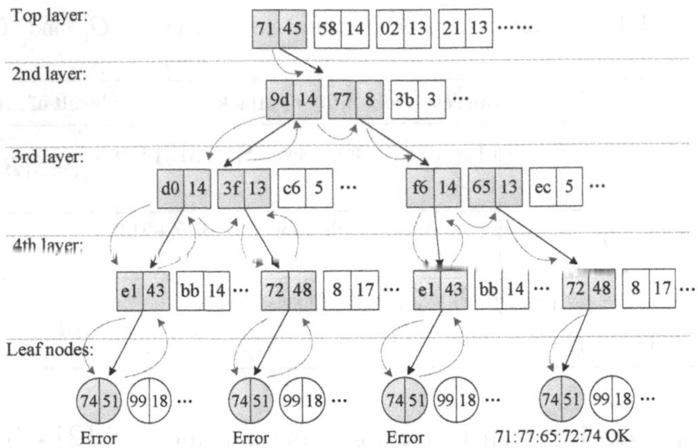

Fig. 3. Demonstration of key cracking

5 Conclusion

According to the vulnerabilities of IEEE 802.11, combining with the basic characteristics and key technologies of current Trojans, it is easy to see that the system framework presented in the paper can be well to implement the real-time, covert and active forensics for crime evidences. According to the results of frame capturing, filtering and analyzing, by designing three types of attack method and the algorithm for cracking WEP protocol key, the key can be cracked successfully byte by byte. And moreover, the Trojan loading and thread guarding module make the forensics system be of strong anti-killed ability, which effectively prevents the computer criminals from terminating or destroying the crime scenes. On the other hand, the usage of DLL and covert Communication technology ensure the concealment performance of this system further.

References

1. Henry, B.W.: Computer forensics. Computers & Security 22(1), 26–28 (2006)
2. Zhang, W.: Research and Implementation of Monitoring System Based WLAN. Northwestern Polytechnical University (2007)
3. Shi, W., Zhang, B., Liu, Y.: Computer monitor and forensics system based on Trojan. Computer Engineering and Design 28(10) (2007)
4. Qing, H.: Analysis on the Security of RC4 Algorithm. Beijing University of Posts and Telecommunications (2009)
5. Stubblefield, A., Ioannidis, J., Rubin, A.: Using the Fluhrer, Mantin, and Shamir Attack to Break WEP. ATT Labs Technical Report-TD4ZCPZZ 45, 1–13 (August 2001)

An IP Design for Keypad Controllers

Liangjun Xiao and Limin Liu

Institute of Embedded Systems IT School, Huzhou University
Huzhou, Zhejiang, 313000, China
sjtzzf@hutc.zj.cn, liulimin@ieee.org

Abstract. As a common device of peripherals, keypads are used popularly. FPGA(Field Programmable Gate Array) devices and VHDL, a hardware discription language, software are common in computer and electronic design. The design is based on FPGA. The simulation and test are taken. With the keypad controller, a microprocessor in SoC will manage keypads more easily and quickly. In this paper, a reconfigurable IP module of keypad controllers is discussed.

Keywords: embedded systems, FPGA, embedded controller, PLD.

1 Introduction

A keypad is an essential part for many applications of embedded systems and is used popularly[1].

FPGA, Field Programmable Gate Arrays, or FPLD, Field Programmable Logic Devices, presents a relatively new development in the field of VLSI, Very Large Scale Integrated Circuit, circuits and is an ideal target technology for VHDL, a hardware discription language, based designs for both prototyping and production volumes. It is different from firmware or micro-controllers that are widely used nowadays. FPGA has more advantages in speed, development time and future modification[2-3]. There are many applications of FPGA in various fields, such as custom reconfigurable processors, telecommunication systems and networks, fault diagnosis, signal processing and so on.

In this paper, FPGA will be employed to design keypad contrl module for a custom MPU, microprocessor unit. The MPU is a reconfigurable processor based on FPGA for a SoC, System on a Chip. The processor has an assembly instruction set, calculation and control unit, address and data bus[4-7].

Actually a reconfigurable kaypad controller is an IP module based on FPGA. On the other hand, the embedded controller is combination of software and hardware in design.

2 Keypad Functional Unit (key_fct)

The development of the key_fct consists of two stages: individual design of key_fct in VHDL and operation of the key_fct together with FLIX, a special MPU.

D. Jin and S. Jin (Eds.): Advances in FCCS, Vol. 1, AISC 159, pp. 629–632.
springerlink.com

2.1 The Outline of key_fct

The chosen keypad is DEVLIN encoded keypad indicated in Appendix 3. It has 16 keys and uses National MM 74C922 encoder. Its input signals concern with *oscillator in* and *clock*. From *oscillator in*, an oscillator signal is supplied for the keypad. The *clock* allows the key code to be gated through to output lines. Normally it can be connected with *data available*. The, *A*, *B*, *C* and *D* are outputs of the keypad. When *data available* is high, keypad gives out key code that represents the key pressed. The value of the key code to the "0" key, for instance, is 1; a hexadecimal number B is for "ENTER" key. The hexadecimal number is composed by *A*, *B*, *C* and *D* 4 bits from bottom to top.

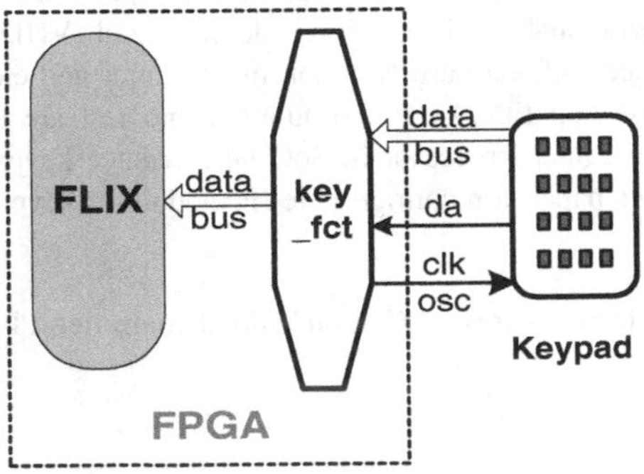

Fig. 1. The Position of key_fct

The key_fct is functional unit designed by VHDL. It is put between FLIX and keypad board and shown in Fig. 1.

The key_fct provides oscillator *osc* and clock *clk* signals to the keypad. It catches data available , *da*, signal from the keypad and has four input lines, *data bus*, to match the *A*, *B*, *C* and *D* signals of the keypad. The key_fct sends key codes with 16 states on 4 lines *data bus* to FLIX.

The connection between keypad and ALTERA UP1 board can be referred to relevant outline. A 14 way cable with a head and a DIP adaptor is used.

2.2 The Structure and Operation of key_fct

As digital circuit design, structure of the circuit is important. The structure can ensure the circuit to work reliably. The reliability is essential for each hardware. The structure of the key_fct is a functional organisation illustrated as Fig.2.

The Reset Circuit is initialised and managed by $\overline{reset}$ and *clk*(system clock). It can produce some initial conditions and constants for the key_fct.

The Clock Regulator may adjust the frequency of system clock and supply a suitable clock, *key_os*, to the output to keypad as oscillator in.

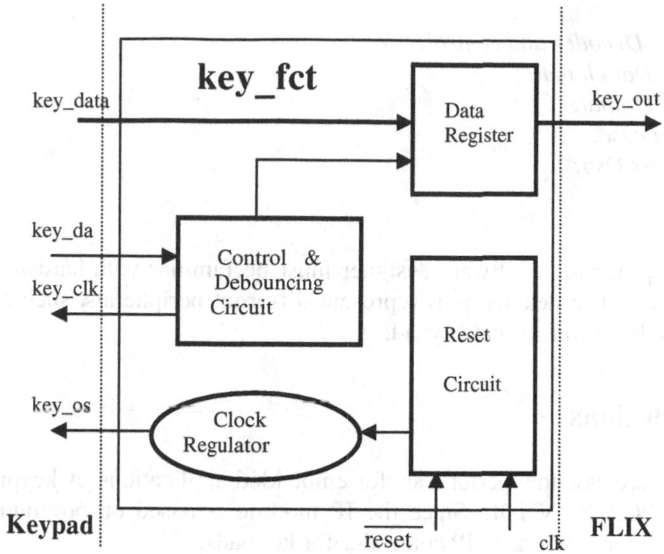

Fig. 2. The Structure of key_fct

Control & Debouncing Circuit monitors the *key_da*, key available signal, from keypad. If the *key_da* is high level, the circuit will control whether Data Register sends input data to output line. So, the key code can be transferred from the key_fct to the keypad. In order to guarantee the reliability of sampling data, a design of a debouncing circuit is required. The operation of the circuit actually is to open Data Register after debouncing the *key_da* signal.

The Data Register receives 4-bit data, *key_data* from keypad. When an output enable signal from Control & Debouncing Circuit is identified, the register sends its data to *key_out*, 4-bit data bus. Then, the data can be read by FLIX through a multiplexer.

3 Design with VHDL

The key_fct is developed by VHDL on FPGA. The VHDL program frame of led_op is as follows.

```
Library
Entity key_fct is
        Port( );
End key_fct;
Architecture struct of key_fct is
 Begin
Dsp: process;
        Case control
```

```
        Decoder and control;
        Data handle;
        Output;
      End case;
  End process Dsp;
End struct;
```

From above program, a software designer must be familiar with hardware for define MPU port pins. The detailed pins represent different peripherals, such as read/write control lines, beep, valves and keypad.

4 Conclusions

The keypads are essential peripherals for embedded applications. A keypad controller is built in FPGA by VHDL. Since the IP module is based on programmable logic devices, it is a reconfigurable IP controller for keypads.

Acknowledgments. This research was supported in part by the National Natural Science Foundation of China under grant 60872057, by Zhejiang Provincial Natural Science Foundation of China under grants R1090244, Y1101237, Y1110944 and Y1100095. We are grateful to NSFC, ZJNSF and Huzhou University.

References

1. Ostua, E., Viejo, J., et al.: Digital Data Processing Peripheral Design for an Embedded Application based on the Microblaze Soft Core. In: Proc. 4th South. Conf. Programmable Logic, San Carlos de Bariloche, Argentina, vol. (3), pp. 197–200 (2008)
2. Kilts, S.: Advanced FPGA Design: Architecture, Implementation, and Optimization. Wiley, New Jersey (2007)
3. Dimond, R.G., Mencer, O., Luk, W.: Combining Instruction Coding and Scheduling to Optimize Energy in System-on-FPGA. In: Proceedings of 14th Annual IEEE Symposium on Field-Programmable Custom Computing Machines, Napa, CA, USA, vol. (4), pp. 175–184 (2006)
4. Liu, L.: A Hardware and Software Cooperative Design of SoC IP. In: Proc. of CCIE 2010, Wuhan, China, vol. (6), pp. 77–80 (2010)
5. Saleh, R., Wilton, S., et al.: System-on-chip: reuse and integration. In: Proceedings of the IEEE, vol. 94(6), pp. 1050–1069 (2006)
6. Sifakis, J.: Embedded systems design - Scientific challenges and work directions. In: Proceedings of DATE 2009, vol. (4), pp. 2–2. IEEE Press, Nice (2009)
7. Liu, L., Luo, X.: The Reconfigurable IP Modules and Design. In: Proc. of EMEIT 2011, Harbin, China, vol. (8), pp. 1324–1327 (2011)

SPN-Based Performance Evaluation for Data and Voice Traffic Hybrid System

Zhiguo Hong[1], Yongbin Wang[1], and Minyong Shi[2]

[1] School of Computer, Communication University of China, Beijing 100024, China
[2] School of Animation and Digital Arts, Communication University of China, Beijing 100024, China
hongzhiguo1977@yahoo.com.cn, ybwang@cuc.edu.cn, myshi@cuc.edu.cn

Abstract. On the basis of analyzing the characteristics of voice traffic, this paper constructed a Stochastic Petri Net (SPN) model for data and ON-OFF voice traffic hybrid system. Then, average time delay of the system was analyzed and model-based simulation was conducted with Stochastic Petri Net Package (SPNP) 6.0. Consequently, variation trends of burst time and idle time on average time delay are derived thereby. The methodology of modeling and simulation in this paper can be further used to analyze the performance of multimedia hybrid traffic system.

Keywords: petri nets, performance evaluation, ON-OFF model, traffic.

1 Introduction

New media usually take media communication mode other than analog television systems, such as digital TV, mobile TV, IPTV, Internet, satellite and etc. With the rapid development of web technologies, recent years have witnessed the convergence of three nets i.e. radio & broadcasting net, Cable TV net and Internet. As a result, integrated services that offer various stream media, voice traffic have been an outstanding trend. Take Internet-based high-definition TV traffic for instance, the bit rate of video stream is considerably high. As a result, to achieve smooth signal network transmission and playback, network bandwidth is also critical for the requirements, usually above 600kbps. In the case of parallel accession by a large number of users, typical broadband network environment is also difficult to satisfy the QoS for all users. Therefore, the construction of a robust, scalable versatile multimedia system should take several factors into account, which are the scale of concurrent users, bottleneck bandwidth of network environment and characteristics of user access behavior etc. Nowadays, methodologies of performance analysis mainly take three ways, which are direct measurement [1], actual experiments and tests for real platform[2][3],mathematic modeling [4], and simulation. However, if performance bottlenecks are detected, adjustment and redeployment should be conducted during running phase of multimedia systems, it would be a waste of manpower and material

D. Jin and S. Jin (Eds.): Advances in FCCS, Vol. 1, AISC 159, pp. 633–636.
springerlink.com

resources. So, how to construct a comparably precise model for multimedia system and evaluate its performance is an urgent and important problem to be solved. In this paper, contraposing to the hybrid system with data and voice traffic, the effect of different parameters' variation on ON-OFF voice traffic system's average time delay is investigated. Firstly, it presented a SPN model for ON-OFF voice traffic system. Then, numerical simulation was carried on to study the average time delay of system with SPNP 6 software [5]. Subsequently, variation trends of burst time and idle time on average time delay are derived thereby.

2 Analysis of ON-OFF Voice Traffic's Characteristics

In order to study the variation of parameters on voice traffic system, we will firstly introduce the classic voice model as follows. A number of voice and data models can be represented as ON-OFF source models. When a source is ON (active), it generates packets with a constant inter arrival time. When the source is OFF (silent), it does not generate any packets. Regarding to voice source modeling, the process of a voice call can be modeled as a two-state Markov chain. The state transition diagram is shown in Fig. 1. If we assume OFF to ON rate to be β1 and ON to OFF rate to be β2, the average lengths of the ON and OFF period are 1/β2 and 1/β1 correspondingly [6].

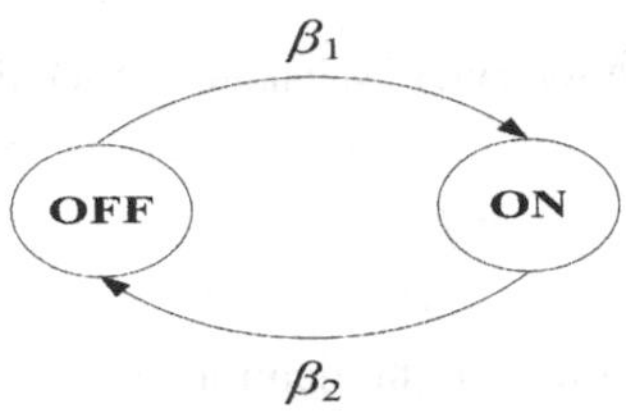

Fig. 1. ON-OFF Model

3 SPN Model and Performance Evaluation for ON-OFF Voice Traffic System

Under network environment, data containing audio, video and text file etc. are transmitted via a shared channel. We concern the whole communication process for voice transmission from sender to receiver via ON-OFF voice traffic system. Further, we make the following assumptions: (1) Voice and data transmitted from sender to receiver follow the Poisson process; (2) Voice and data are transmitted independently in the traffic system. Fig. 2 shows our constructed SPN model for ON-OFF voice traffic system. The objects of the SPN model are listed in Table 1.

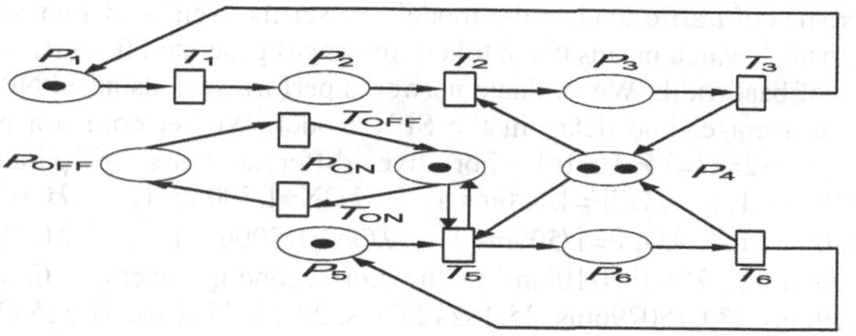

Fig. 2. SPN Model for ON-OFF Voice Traffic System

Table 1. List of Objects of SPN Model for ON-OFF Voice Traffic System

Name	Meaning	Markings	Rate
P1	the number of connected data senders	1	
T1	transmitting data packets		λ1
P2	waiting to access transmission link	0	
T2	delivering data packets to receiver		λ2
P3	receiving voice data	0	
T3	reply answering voice data to sender		λ3
PON	being activity state	1	
TON	altering from activity state to silent one		λON
POFF	being silent state	0	
TOFF	altering from silent state to activity one		λOFF
P4	available bandwidth of shared link	2	
P5	the number of connected voice senders	1	
T5	delivering voice data packets		λ5
P6	receiving voice data	0	
T6	reply answering voice data to sender		λ6

As Fig. 2 shows, PON, TON, POFF and TOFF are constructed to model the behavior of ON-OFF transmission link. That is, activity state and silent state turn to be alternatively by applying TOFF and TON respectively. Two additional arcs between PON and TON are used to control the procedure of voice data's transmission as follows. The firing condition of T2 is that the marking of P2 and that of PON are set to be "1". When ON-OFF voice traffic is in the activity state, it reveals the availability of transmission link so that sender can transmit voice data. For the same reason, once ON-OFF voice traffic is in the silent state, it represents the unavailability of transmission link. Consequently, users couldn't send voice data until ON-OFF voice traffic turn to be active for the next time.

We assume that voice and data packets are allowed to transmit when user obtains enough bandwidth resource. In order to reflect the phenomena of competition in using available bandwidth, the number of tokens in Place P4 is set to be 2, which represents available bandwidth of shared link. Let "#" symbolize the token number of a certain transition. The firing condition of transition T2 is formulated as follows: $(\# P_5 = 1) \wedge (P_{ON} = 1) \wedge (\# P_4 = 2)$. The firing condition of T5 is formulated as follows: $(\# P_2 = 1) \wedge (\# P_4 = 2)$.

Further, we can calculate the average time delay of the constructed SPN model by applying Little's theorem and principle of balance [7]. For presenting the characteristics of traffic load in the model, we set the weight of four arcs connected with P4 to be 2, which means that a token (message) generated from P1 or P5 occupies two units of bandwidth. We evaluate networks performance using SPNP 6.0 software to concern average time delay in the SPN model. We set common parameters as follows: λ1=λ2=λ3=λ5=λ6=0.1. For five different pairs of parameters being λON=1/100ms-1, λOFF=1/900ms-1, λON=1/300ms-1, λOFF=1/700ms-1, λON=1/500ms-1, λOFF=1/500ms-1, λON=1/700ms-1 , λOFF=1/300ms-1, λON=1/900ms-1, λOFF=1/100ms-1, the corresponding average time delays are 20.9731198ms, 23.0750796ms, 25.4824292ms, 28.2674532ms, 31.5254291ms.

4 Summary

In this paper, contraposing to data and voice traffic hybrid system, the performance of system is evaluated by the methodologies of mathematic modeling and simulation validating. Firstly, based on the characteristics of ON-OFF voice traffic, by choosing average time delay as key performance index, SPN-based model for ON-OFF voice traffic system is constructed and performance of the voice traffic system is evaluated by a typical example. On the basis of current work, further investigations on diverse, digital media systems can be carried on, which offers quantification basis and guidance for optimizing the performance of similar hybrid system.

Acknowledgements. The paper is supported by the High Technology Research and Development Program of China (2011AA01A107); Engineering Planning Project for Communication University of China (XNG1126).

References

1. Kambo, N.S., Deniz, D.Z., Iqbal, T.: Measurement-Based MMPP Modeling of Voice Traffic in Computer Networks Using Moments of Packet Interarrival Times. In: Lorenz, P. (ed.) ICN 2001. LNCS, vol. 2094, pp. 570–578. Springer, Heidelberg (2001)
2. Gooding, S.L., Arns, L., Smith, P., et al.: Implementation of a distributed rendering environment for the teraGrid. In: Challenges of Large Applications in Distributed Environments (CLADE), pp. 13–22. IEEE Computer Society, Washington (2006)
3. Zeng, W., Wang, H., Xu, Z.Q., et al.: Raster data cluster cache in network GIS. Journal of Huazhong University of Science and Technology (Natural Science Edition) 37(9), 37–39 (2009) (in chinese)
4. Jin, Y., Zhou, G., Jiang, D.C., et al.: Theoretical mean-variance relationship of IP network traffic based on ON/OFF model. Science in China Series F: Information Sciences 52(4), 645–655 (2009) (in Chinese)
5. Hirel, C., Tuffin, B., Trivedi, K.S.: SPNP: Stochastic Petri Nets. Version 6.0. In: Haverkort, B.R., Bohnenkamp, H.C., Smith, C.U. (eds.) TOOLS 2000. LNCS, vol. 1786, pp. 354–357. Springer, Heidelberg (2000)
6. Chimeh, J.D., Hakkak, M., Alavian, S.A.: Internet Traffic and Capacity Evaluation in UMTS Downlink. In: Proc. of Future Generation Communication and Networking (FGCN 2007), vol. 1, pp. 547–552 (2007)
7. Lin, C.: Stochastic Petri Nets and System Performance Evaluation. Tsinghua University Press, Beijing (2000) (in Chinese)

A Simplified Design of Intelligent Control for MIMO Systems

Liangjun Xiao and Limin Liu

Institute of Embedded Systems
IT School, Huzhou University
Huzhou, Zhejiang, 313000, China
sjtzzf@hutc.zj.cn, liulimin@ieee.org

Abstract. The intelligent control makes automatic regulation of complex processes to be easier. A design of some intelligent control system is not the more complicated than the traditional method. This paper focuses on the simplification of design for automation based on the technology of intelligent fuzzy control. For some MIMO systems, the technology based on artificial intelligent and fuzzy control is more simplified and practical.

Keywords: MIMO systems, intelligent control, fuzzy systems, control simplication.

1 Introduction

A complex industrial system often deals with the problems of multivariable control in which there are more than two controlled variables and acting variables in a system, or to be called as Multiple Input and Multiple Output (MIMO) systems. For some practical MIMO processes, the main task of intelligent control is to manage complicated procedures[1]. The control problems in the complex processes are usually multivariable, closed coupling and ill-structured. In some processes without any amenable mathematical models, purely algorithmic method with mathematics is able to be applied hardly. On the other hand, the design based on the mathematical models is too complicated to be understood by control engineers who are duty to maintain the system in the plants.

Although there are some processes that can not be controlled by regulators, the factories still keep normal operations. Operators, who are with knowledge and experience in system operation, can well manage the production[2-3]. With the development of intelligent control, a controller is entirely able to be instead of a good operator. This is the role of intelligent control in the industrial applications.

Much effort has been put into investigating and implementing intelligent control. Control problems concerned with about ill-structured, uncertain and fuzzy information will be discussed. The intelligent solutions include fuzzy control, neural network and knowledge-based system are three main branches of intelligent control system. The idea of Fuzzy System was first applied to control system in 1974. Since that time, there have been a lot of successful applications based on fuzzy systems [4-6].

D. Jin and S. Jin (Eds.): Advances in FCCS, Vol. 1, AISC 159, pp. 637–640.
springerlink.com

2 Decoupling Control

For MIMO system control, an important problem to be attended to is coupling. The coupling means that changing one acting variable has to concern with several controlled variables. And, the control of closed-loop system may be quite difficult. In order to solve the problem, decoupling algorithm and intelligent control techniques are useful[7].

Here, a typical method to solve multivariable problems is focus of decoupling control. A reason to use the methodology is that decoupling can remove the linkage among related variables and separate a multivariable coupling loop into several single-loop systems. In theory, the way is a path to solve all of multivariable coupling problems through mathematic models. But, in fact, that is impossible. Here is a MIMO system, for instance, with three acting variables and three controlled variables.

The system description in frequency domain is as follows.

$$\begin{bmatrix} Y_1(s) \\ Y_2(s) \\ Y_3(s) \end{bmatrix} = \begin{bmatrix} G_{11}(s) & G_{12}(s) & G_{13}(s) \\ G_{21}(s) & G_{22}(s) & G_{23}(s) \\ G_{31}(s) & G_{32}(s) & G_{33}(s) \end{bmatrix} \bullet \begin{bmatrix} X_1(s) \\ X_2(s) \\ X_3(s) \end{bmatrix} \tag{2.1}$$

where $G_{ij}(s)= Y_i(s)/ X_j(s)=$

$$\prod_{k=1}^{3}(A_k s+C_k)\cdot[\ \prod_{k=1}^{3}(B_k s+D_k)\ ]^{-1} \tag{2.2}$$

In equation (2.2), A_k, B_k, C_k, D_k are coefficients of 3rd order functions in s variable. Decoupling is to find a matrix **D(s)**, and to make **G(s)D(s)** into $\mathbf{G}^*(\mathbf{s})$ that is a diagonal matrix. That is

$$\mathbf{G}^*(\mathbf{s})= \begin{bmatrix} G_{11}(s) & 0 & 0 \\ 0 & G_{22}(s) & 0 \\ 0 & 0 & G_{33}(s) \end{bmatrix}$$

$$= \mathbf{G(s)D(s)} \tag{2.3}$$

And $\quad \mathbf{D(s)}= \mathbf{G}^{-1}\mathbf{(s)}\ \mathbf{G}^*\mathbf{(s)}$ (2.4)

Where $\mathbf{G}^{-1}\mathbf{(s)}= \mathbf{G'(s)}/|\mathbf{G}^*\mathbf{(s)}|$

D(s) is a decoupling matrix of (2.1).

To get the **D(s)** in (2.4), we have to solve a ninth order reverse matrix. For an industrial process, it is too complicated to be utilized. Therefore, the method is not practicable for a real object.

There are several other decoupling methodologies, such as feed-forward and unit-matrix algorithms, based on mathematics, but they are almost not good for complex industrial processes. The complicated structure and special controllers for those algorithms are not easy and convenient in systems testing, trial running, maintaining and improving.

3 Intelligent Fuzzy Control and Simplification

Intelligent fuzzy control actually is a kind of imitation control. In fact, every process, whether simple or complex, in plants can run normally. Otherwise, the plants have to be closed. It is not too much to say that people are universal controllers. Of course, there is various control quality for different operators, some better, others worse. With rapidly growing of computer science and control technology, it is entirely possible that machines are used instead of people to adjust the complex systems. Actually, machines may do better than human sometimes.

Intelligent control solves industrial problems through various paths, like to collect knowledge and experience by expert systems, to express uncertain and imprecise information through fuzzy set, to learn and modify rules of control under neural networks. For some industrial application, the model based on knowledge is relatively simpler than its mathematical model.

The establishment of knowledge model has to depend on practical knowledge and experience of excellent operators. The knowledge model is composed of fuzzy rule sets in fuzzy control system. The intelligent fuzzy control is an extension of fuzzy control.

The difference between them is that the fuzzy control table is replaced by two blocks, analyzing & inferring and evaluating & deciding. Intelligent fuzzy control is different from fuzzy control based on the stationary control table. It applies trace and prevision control of dynamic characteristic to the procedures.

The analyzing in block of analyzing & inferring for intelligent fuzzy control is to analyze operation state of the system, then judge whether its acting variables should be adjusted. The adjusting is made by reasoning which is taken place by an expert system with fuzzy rule sets.

Generally, forward rule based inference (F rule inference) is used because it is the reasoning from facts to results. The inferring conforms to the regularity of industrial control. In practical productions, operators manage the systems always to refer to facts, such as the change of measuring signals, collected from the processes to make decision. The decision in a controller is the task of evaluating & deciding block. The evaluating in block implies to analyze the inferring results and to judge whether they should be passed to the deciding model.

Since industrial processes are conditional and continuous systems, variations of parameters are limited in the special ranges. Therefore, the reasoning in an intelligent fuzzy controller is simpler than a normal expert system. The implementation of the controller is easier. It can be fulfilled by a micro-controller. An intelligent fuzzy controller is built on the multi-level rules and multiple rule sets. The multi-level rules are shown like:

IF <process state>
THEN <intermediate variable 1>

IF < intermediate variable 1>
THEN <intermediate variable 2>
...
IF < intermediate variable n>
THEN <control action>.

The multiple rule sets are as:
Basic Rule Set,
High Level Rule Set,
Tuning Decision Rule Set,
Adapting System Rule Set,
... .

They are designed in different cases.

4 Conclusions

The intelligent method is suitable for the practical applications requested to operate simply, maintain conveniently and modify easily. Intelligent fuzzy control is able to simplify the control design and implementation for some MIMO systems. Actually, the simplification for industrial applications normally means reliable and available.

Acknowledgments. This research was supported in part by the National Natural Science Foundation of China under grant 60872057, by Zhejiang Provincial Natural Science Foundation of China under grants R1090244, Y1101237, Y1110944 and Y1100095. We are grateful to NSFC, ZJNSF and Huzhou University.

References

1. Ling, P., Wang, C., Lee, T.: Time-Optimal Control of T-S Fuzzy Models via Lie Algebra. IEEE Trans. on Fuzzy Systems 17(8), 737–749 (2009)
2. Chang, Y.: Intelligent Robust Tracking Control for a Class of Uncertain Strict-Feedback Nonlinear Systems. IEEE Transactions on Systems, Man, and Cybernetics 39(2), 142–155 (2009)
3. Nazir, M.B., Wang, S.: Optimum robust control of nonlinear hydraulic servo system. In: Proc. of IECON 2008, Orlando, vol. 11, pp. 309–314 (2008)
4. Chu, M., Jia, Q., Sun, H.: Robust Tracking Control of Flexible Joint with Nonlinear Friction and Uncertainties Using Wavelet Neural Networks. In: Proc. of ICICTA 2009, Changsha, China, vol. (10), pp. 878–883 (2009)
5. Szabat, K.: Robust control of electrical drives using adaptive control structures — a comparison. In: Proc. of ICIT 2008, Chendu, China, vol. (4), pp. 1–6 (2008)
6. Hua, C., Wang, Q., Guan, X.: Robust Adaptive Controller Design for Nonlinear Time-Delay Systems via T–S Fuzzy Approach. IEEE Trans. on Fuzzy Systems 17(8), 901–910 (2009)
7. Liu, L.: A Control Based on Rule Updating for Non-Linear Systems. In: Proc. of 2011 CCDC, Mianyang, China, vol. (5), pp. 3094–3097 (2011)

Laplace Inverse Transformation for α -Times Integrated C Semigroups

Man Liu[1], XiaoMei Yuan[2], QingZhi Yu[1,3], and FuHong Wang[1]

[1] Department of Basic Education, Xuzhou Air Force College, Xuzhou 221000, P.R. China
[2] Department of Aviation Ammunition, Xuzhou Air Force College, Xuzhou 221000, P.R. China
[3] Department of Basic Education, University of Electronic Science and Technology of China, Chengdu 610054, P.R. China
liuman8866@163.com

Abstract. By the relationship of α -times integrated semigroups and C semigroups ,the Laplace inverse transformation for α -times integrated C semigroups is obtained,some known results are generalized.

Keywords: α -times integrated C semigroups, Laplace inverse transformation, convolution.

1 Introduction

The theory of operator semigroups has developed during the past decades, many authords[1]-[9] have studied the corresponding conclusions about α -times integrated semigroups and C semigroups, we will discuss the Laplace inverse transformation for α -times integrated C semigroups, and generalize some known results.

Let X be Banach space, $B(X)$ is the space of bounded linear operators from X into X , $D(A)$ denotes the domain of operator A . $R(A)$ denotes the range of operator A , $K(A)$ denotes the ker of operator A , $C\in B(X)$.

For $\alpha\geq 0$, $[\alpha]$, (α) denote the integral part and decimal part of α respectively. $\Gamma(\cdot)$ is Gamma function, and $\Gamma(s)=\int_0^\infty x^{s-1}e^{-x}dx$, $s\Gamma(s)=\Gamma(s+1)$.

For $\beta\geq -1$, the function $j_\beta:(0,\infty)\to R$ is defined as $j_\beta(t)=\dfrac{t^\beta}{\Gamma(\beta+1)}$.

j_{-1} denotes 0 -point Dirac measure δ_0.

For continuous function $f(\cdot)$, $\beta\geq -1$, defined convolution as following

$$(j_\beta * f)(t)=\begin{cases}\int_0^t \dfrac{(t-s)^\beta}{\Gamma(\beta+1)}f(s)ds, & \beta>-1\\ f(t), & \beta=-1\end{cases}$$

D. Jin and S. Jin (Eds.): Advances in FCCS, Vol. 1, AISC 159, pp. 641–645.
springerlink.com

At first we introduce the fractional differential and integral of function.

For arbitrary $\alpha > 0$, α -order differential of function u denotes

$$(D_\alpha u)(t_0) = \omega^{(n-1)}(t_0).$$

For arbitrary $\alpha > 0$, α -times cumulative integral of function u denotes

$$(I_\alpha u) = (j_{\alpha-1} * u)(t).$$

Definition 1.1. Let $\alpha \in R^+$, a strongly continuous family $\{S(t)\}_{t\geq 0} \in B(X)$ is called α -times integrated C semigroups, if

(i) $S(t)C = C\,S(t)$,and $S(0) = 0$;

(ii) $S(t)S(s)x = \dfrac{1}{\Gamma(\alpha)}\left[\int^{s+t} (t+s-r)^{\alpha-1} S(r)Cxdr - \int_0^s (t+s-r)^{\alpha-1} S(r)Cxdr\right], \forall\, t, s \geq 0$

If $\alpha = n \quad (n \in N)$, then $\{S(t)\}_{t\geq 0}$ is called n -times integrated C semigroups.

If $\alpha = n \quad (n \in N)$, and $C = I$,then $\{S(t)\}_{t\geq 0}$ is called n -times integrated semigroups.

If $\alpha > 0$ $S(t)x = 0$ ($t \geq 0$) implies $x = 0$, then α -times integrated C semigroups $\{S(t)\}_{t\geq 0}$ is non-degenerated.

If there exists $M > 0, \omega \in R$, such that $\|S(t)\| \leq M\, e^{\omega t}, t \geq 0$, then $\{S(t)\}_{t\geq 0}$ is called exponentially bounded.

Definition 1.2. Let $\alpha \geq 0$,a strongly continuous family $\{S(t)\}_{t\geq 0} \in B(X)$ is called α -times exponentially bounded integrated C semigroups generated by A ,if $S(0) = 0$,and there exists $M > 0, \omega > 0$,such that $(\omega, \infty) \subset \rho(A)$, $\|S(t)\| \leq M\, e^{\omega t}, t \geq 0$,and for arbitrary $\lambda > \omega$, $x \in X$,we have

$$R_C(\lambda, A)x = (\lambda - A)^{-1}Cx = \lambda^\alpha \int_0^\infty e^{-\lambda t} S(t)xdt$$

Lemma 1.3. [8] Let $\omega \geq 0$, $F(\lambda): (\omega, \infty) \to X$, $F(\lambda)$ is Laplace-type expression: $F(\lambda) = \lambda \int_0^{+\infty} e^{-\lambda t} \alpha(t)dt$, $\alpha(t) = 0$,and $\|\alpha(t+h) - \alpha(t)\| \leq M\, he^{\,\omega(t+h)}$, $t, h \geq 0$,then $\alpha(t) = \dfrac{1}{2\pi i}\displaystyle\int_{\gamma-i\infty}^{\gamma+i\infty} e^{\lambda t} F(\lambda)\frac{d\lambda}{\lambda}, (\gamma > \omega)$

Lemma 1.4. [9] Let $\alpha \geq 0$, then the following conditions are equivalent:

(i) A generates an α -times exponentially bounded integrated semigroups $\{S(t)\}_{t\geq 0}$;

(ii) There exists $\omega > 0$, such that $(\omega, \infty) \subset \rho(A)$,and for all $u > \omega$, A generates an $(u-A)^{-\alpha}$ exponentially bounded semigroups $\{T(t)\}_{t\geq 0}$,and

$$T(t) = (u-A)^{-\alpha}\left(\frac{d}{dt}\right)^{[\alpha]+1}(j_{-(\alpha)} * S)(t).$$

Propersition 1.5. Let A be the generator of an α -times integrated C semigroups $\{S(t)\}_{t\geq 0}$, $\alpha \geq 0$. Then for all $x \in D(A)$, and $t \geq 0$,

(i) $S(t)x \in D(A)$, $AS(t)x = S(t)Ax$, $S(t)x = \dfrac{t^\alpha}{\Gamma(\alpha+1)}Cx + \int_0^t S(s)Axds$

(ii) $\int_0^t S(s)xds \in D(A)$,for all $x \in X$,and $t \geq 0$ and $A\int_0^t S(s)xds = S(t)x - \dfrac{t^\alpha}{\Gamma(\alpha+1)}Cx$

2 Laplace Inverse Transformation

Theorem 2.1. Let A generate an α -times integrated C semigroups $\{S(t)\}_{t\geq 0}$, $\alpha \geq 0$,then

(i) For all $x \in D(A^n), n = 1,2,\cdots,[\alpha], t > 0$,such that $S(t)x \in C^n(R^+, X)$,and

$$S^{(n)}(t)x = S(t)A^n x + \sum_{i=1}^{n} j_{\alpha-i}(t)A^{n-i}Cx$$

(ii)For all $x \in D(A^{[\theta]+1})$, $0 \leq \theta < \alpha$, such that $S(t)x \in C^\theta(R^+, X)$, and

$$(D_\theta S)(t)x = (j_{-(\theta)} * S)(t)A^{[\theta]+1}x + \sum_{j=1}^{[\theta]+1} j_{\alpha-(\theta)+1-j}(t)A^{[\theta]+1-j}Cx$$

(iii) For all $x \in D(A^{[\theta]+1})$, $0 \leq \theta < \alpha$, and $(\theta) = (\alpha)$, $t > 0$,such that $(D_\theta S)(0)x = 0$,and when $x \in D(A^{[\alpha]+1})$, we have $(D_\alpha S)(0)x = Cx$.

Proof. (i) we can differentiate the identity n -times

$$S(t)x = \frac{t^\alpha}{\Gamma(\alpha+1)}Cx + \int_0^t S(s)Axds$$

$$S'(t)x = \frac{\alpha t^{\alpha-1}}{\Gamma(\alpha+1)}Cx + S(t)Ax = j_{\alpha-1}(t)Cx + S(t)Ax$$

$$S''(t)x = j_{\alpha-2}(t)Cx + S'(t)Ax = S(t)A^2x + j_{\alpha-1}(t)ACx + j_{\alpha-2}(t)Cx$$

Therefore $S^{(n)}(t)x = S(t)A^n x + \sum_{i=1}^{n} j_{\alpha-i}(t)A^{n-i}Cx$.

(ii) $\{S(t)\}_{t\geq 0}$ is an α -times integrated C semigroups,
$\lceil\alpha\rceil \leq \alpha < \alpha+1$, and $S(0)=0$

$$(D_\theta S)(t)x = \frac{d^{[\theta]+1}}{dt}(j_{-(\theta)} * [S - S(0)])(t)x = \frac{d^{[\theta]+1}}{dt}(j_{-(\theta)} * S)(t)x$$

So we can obtain $(j_{-(\theta)} * S)(t)$ is an $\alpha-(\theta)+1$ -times integrated C semigroups.

Combining (i) gives (ii).

(iii) Letting $t=0$, the proof is complete.

Next we discuss the Laplace inverse transformation for α -times integrated C semigroups.

Theorem 2.2. Let A be closed linear operator on X , $\rho(A) \neq \Phi$, $\lambda \in \rho(A)$, an α -times exponentially bounded integrated C semigroups $\{S(t)\}_{t\geq 0}$ with infinitesimal generator A ,and $\|S(t)\| \leq M e^{\omega t}$, $\omega \geq 0$, $\gamma > \omega$,then for $\forall x \in D(A)$,

$$\int_0 S(s)ds = \frac{1}{2\pi i}\int_{\gamma-i\infty}^{\gamma+i\infty} e^{\lambda t} \frac{R_C(\lambda, A)x}{\lambda^\alpha}\frac{d\lambda}{\lambda} = \frac{1}{2\pi i}\int_{\gamma-i\infty}^{\gamma+i\infty} (I_\alpha e^{\lambda t} R_C(\lambda, A)x)\frac{d\lambda}{\lambda}$$

Proof. Letting $\alpha(t) = \int_0 S(s)ds$, $F(\lambda) = \frac{(\lambda - A)^{-1}Cx}{\lambda^\alpha}$, $\forall x \in D(A)$

By lemma 1.3 $F(\lambda) = \frac{(\lambda - A)^{-1}Cx}{\lambda^\alpha} = \int_0^{+\infty} e^{-\lambda t} S(t)xdt$

$$= \int_0^{+\infty} e^{-\lambda t} xd \int_0 S(s)ds = \lambda \int_0^{+\infty} e^{-\lambda t}\left(\int_0 S(s)xds\right)dt \quad (\lambda > \omega)$$

So $F(\lambda)$ satisfies lemma 1.3, $\int_0 S(s)xds = \frac{1}{2\pi i}\int_{\gamma-i\infty}^{\gamma+i\infty} e^{\lambda t} \frac{R_C(\lambda, A)x}{\lambda^\alpha}\frac{d\lambda}{\lambda}$, $(\gamma > \omega)$

On the other hand, by lemma 1.4 A generates $(u-A)^{-\alpha}C$ exponentially bounded semigroups $\{T(t)\}_{t\geq 0}$.

So for $\forall x \in D(A)$, we have $\int_0 T(s)xds = \frac{1}{2\pi i}\int_{\gamma-i\infty}^{\gamma+i\infty} e^{\lambda t}(\lambda - A)^{-1}(u-A)^{-\alpha}Cx\frac{d\lambda}{\lambda}$

$$= \frac{1}{2\pi i}\int_{\gamma-i\infty}^{\gamma+i\infty} e^{\lambda t}(\lambda - A)^{-1}R(u, A)^\alpha Cx\frac{d\lambda}{\lambda} = \frac{R(u, A)^\alpha}{2\pi i}\int_{\gamma-i\infty}^{\gamma+i\infty} e^{\lambda t}(\lambda - A)^{-1}Cx\frac{d\lambda}{\lambda}$$

It follows that $S(t)=(u-A)^{\alpha}(I_{\alpha}T)(t)$.

Whence $\int_0^t S(s)xds=(u-A)^{\alpha}\int_0^t (I_{\alpha}T)(s)xds=(u-A)^{\alpha}(I_{\alpha}\int_0^t T(s)xds)$

$$=(u-A)^{\alpha}\frac{R(u,A)^{\alpha}}{2\pi i}\int_{\gamma-i\infty}^{\gamma+i\infty}(I_{\alpha}e^{\lambda t}(\lambda-A)^{-1}Cx)\frac{d\lambda}{\lambda}=$$

$$\frac{1}{2\pi i}\int_{\gamma-i\infty}^{\gamma+i\infty}(I_{\alpha}e^{\lambda t}R_C(\lambda,A)x)\frac{d\lambda}{\lambda}.$$

And the right-hand integral is uniformly convergent on any finite intervals.

3 Conclusion

In this paper we obtain the Laplace inverse transformation for α -times integrated exponentially bounded C semigroups. We use the fomula of convolution and the integration by parts. The discussion of this paper is restricted to the case of X is a Banach space and A is closed linear operator.

Acknowledgment. This work is supported by the Fundamental Research Funds for the Central University (2010LKSXO8).

References

1. Miyadera, I., Tanaka, N.: A remark on exponentially bounded C semigroups. Proc. Japan Acad. Ser. A, 31–34 (1990)
2. Arendt, W.: Vector-balued Laplace transforms and Cauchy problems. Israel J. of Math. 59(3), 327–352 (1987)
3. Arendt, W.: Resolvent positive operators. Proceedings London Mathmatical Society 54(2), 321–349 (1978)
4. Delaubenfels, R.: Existence and uniqueness families for the abstract Cauchy problem. London Math. Soc. 44(2), 310–322 (1991)
5. Wang, S.: Mild integrated C-existence families. Studia Math. 112(3), 251–262 (1995)
6. Gao, M.: Mild integrated C -existence families and abstract Cauchy problem. Northeast Math. J. 14(1), 95–102 (1998)
7. Tanaka, N., Miyadera, I.: Exponentially bounded C semigroups and integrated semigroups. Tokyo J. Math. 12, 99–115 (1989)
8. Mijatovic, M., Pilipovic, S.: α-times integrated semigroup. Journal of Mathematical Analysis and Applications 210, 790–803 (1997)
9. Hieber, M.: Laplace Transforms and α–times Integrated Semigroup. In: Semesterbericht Functional Analysis, Tubingen, Wintersemester, vol. 3, pp. 89–90, 109-126; Forum Math. 3, 595–612 (1991)

Research on the Arbiter and Quantum Memory of Multi-Port

Wangping Xiong[1], Ji-cheng Shu[2], Yao-hui Ye[3,*], and Caiying Peng[1]

[1] School of Computer, Jiang Xi University of Traditional Chinese Medicine, NanChang, JiangXi, China
[2] Key Laboratory of Modern Preparation of Ministry of Education, Jiang Xi University of Traditional Chinese Medicine, NanChang, JiangXi, China
[3] Academic Affairs Office, Jiang Xi University of Traditional Chinese Medicine, NanChang, JiangXi, China
xiaoxiongxwp@126.com

Abstract. Quantum memory offers great potential advantage for solving some problems which are of super-polynomial time complexity by contrasting with classical computer. In order to improve the memory bandwidth for masters accessing external memory in the quantum system, a multi-port memory controller based on the AHB bus is developed. Also, an arbitration strategy for the early arbitration and request waiting priority is proposed. A number of masters are requested to access the external memory through a number of ports. The arbitration selects the highest-priority port in the early arbitration moment and sets the request waiting time for the other ports that are notallowed the access request. The early arbitration moment occurs before the completion for the current read/write operations. When the next early arbitration moment happens, the arbitration arbitrates these timeout portsin preference.

Keywords: Quantum memory, Error-tolerance, Arbiter, Multi-Port Memory.

1 Introduction

Quantum computer becomes a reality, Performance is improving, But quantum computer most only deal with specific problems or algorithms, Computer scientists and physicists are trying to build general-purpose quantum computer, That can not change the case of architecture, types of quantum computation to solve the problem. Can be predicted, If the universal quantum computer was developed, Outstanding issues can be used for a variety of quantum programming language program corresponding quantum, The implementation of the universal quantum computer, So that people could be as easy as operating a classical computer to control the efficient operation of quantum computers. To this end, Need to design corresponding to the universal quantum computer architecture of the universal quantum memory, Need to design corresponding to the universal quantum computer architecture of the universal

* Corresponding author.

D. Jin and S. Jin (Eds.): Advances in FCCS, Vol. 1, AISC 159, pp. 647–652.
springerlink.com

quantum memory, Today the universal quantum computer architecture and memory in the international study of quantum is less.

Memory controller is the core component of the storage system, Its role is to master the right timing to access various types of external memory, Its performance often determines the efficiency of storage systems. However, In the memory controller, Low processing delay and high memory bandwidth are parameters between constraints Which is to support high-bandwidth memory controller needs to improve memory and logic, And the logic is increased processing delay; Conversely, support for low-latency memory controller reduces the logic overhead, But also reduces memory bandwidth support. This problem also occurs in quantum memory.

2 The Architecture of Universal Quantum Computer

Universal quantum computer is a computing device to meet the following conditions:

(1) Accurately simulate any Turing machine;
(2) Can simulate any with arbitrary precision, quantum computer or quantum computation simulation equipment;
(3) Simulate a variety of practical or theoretical physical system, Some of them are classical universal Turing machine T can not be simulated;
(4) Fully simulate random poor physical systems.

The classic Neumann architecture computer classical deterministic Turing machine is based on the calculation model, This is a program stored in the computer architecture as the core. Classical Neumann architecture proposed that the classical deterministic Turing machine is based on the main components. From the following five major parts to consider; Operation, controller, memory, input devices and output devices.Each part correspond to the elements of the classical deterministic Turing machine model,Memory corresponds to the "tape" of Turing machine, Input and output devices corresponds to the "read / write head " of Turing machine. And controller corresponds to the role of state transfer functions, Operation corresponds to pattern changes in uncertainty Turing machines.

Suppose C is the set of complex numbers α: The algorithm of in Polynomial n time, Calculate real and imaginary parts and accurate to within 2^{-n},A universal quantum Turing machine M is defined as a triple (Σ,Q,δ), Them Σ is the finite string of letters that is end by the blank symbol #;Q is the finite identified states that the initial state q0 and the termination of the state $q_f \neq q_0$ logo; δis the quantum state transfer function: δ: $Q\times\Sigma\rightarrow C^{\Sigma\times Q\times\{L,R\}}$. Quantum Turing machine has a direction infinitely long record tape, The subscript on the tape box set Z, A read /write head along the tape to the left (corresponds to the transfer function of L)or right(corresponds to the transfer function of R) move in any. The definition of Initial pattern and termination pattern in Quantum Turing Machine and deterministic Turing machines as defined in identical. Let M is a universal quantum Turing machine, So that the pattern of M S is the finite complex linear combinations of the normalization condition to meet the Euclidean inner product space, Each element of S is a superposition of M. Quantum Turing machine M defines a linear operator UM: S→S, is the time evolution operator M: If M starts with the pattern of C, The current state is p and scans an identifier σ; Next action, Each non-zero α_i corresponds to conversion$\delta(p,\sigma,\tau,q,d)$, c_i is the new pattern that is transformed by c, This operation can be extended to the whole S space from Linear time evolution operator UM.

Reference to the design ideas of classic Neumann computer architecture, We start with the constitute of universal quantum Turing machine, The physical background of Quantum computing is the basis,Make the "parts " that constitute universal quantum computer.

3 Quantum Memory Controller

3.1 Static Quantum Memory Controller

Static quantum memory controller used to generate a variety of functions signals that access memory, These signals not only bring about the basic read / write control of quantum memory and the page mode control,But also achieve the internal address increment control of AHB burst reading. To improve the efficiency of MPMC reading data, When the MPMC Dynamic memory controller is on and not taken the internal bus, Static quantum memory control can activate the read / write access operations, Quantum memory controller static read / write access to the state transition shown in Figure 1.

In the idle state, If the read access request is valid, The first reading of the state (RD) produced memory chip select signal, And based on time so that can wait for a cycle of the quantum memory read enable signal, and then read wait state (RD1) read waiting period for providing read data operation, and in accordance with read access patterns into page mode read status (PG-RD) or continuous reading State (TF-RD). In PG-RD state, except to read the entire outside of the first data page of data, a data read is completed, go to the TF-RD state; in the TF-RD mode, if End of the current AHB burst access is not complete, the resulting address increment enable signal and enters the state to continue to read data RD, burst access is completed, return to the IDLE state. and in IDLE state, if the write access request is valid, First, write state (WR) have the quantum memory chip select signal, and based on that can wait for a write cycle of the quantum memory write enable signal, and then write wait state (WR1) waiting period to achieve the write operation to write data, and finally writing Completion status (WR2) a delayed write cycle to ensure data is correctly written to memory. write access is complete, return to the IDLE state.

3.2 Static Quantum Memory Controller

Dynamic quantum memory controller used to generate various functions of dynamic memory access signals, these signals would not only support adynamic memory controller state machine multi-state machine design, each small state machines only support a dynamic quantum memory Kinds of operations, including dynamic quantum memory controller read / write access to the state transition shown in Figure 2.

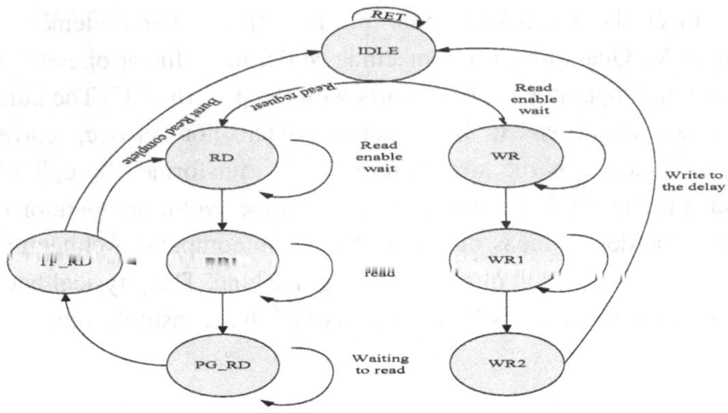

Fig. 1. Static Memory Controller read / write access to the state transition

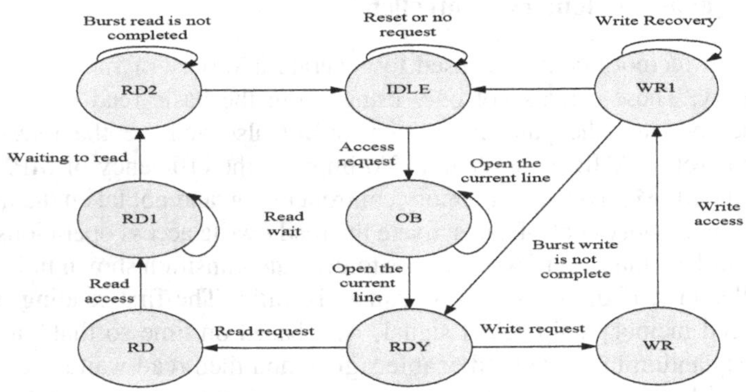

Fig. 2. Dynamic Memory Controller read / write access to the state transition

4 Arbiter

MPMC system contains some port on the main memory and memory device requires large quantities of data between the exchange, that is, the corresponding port should have a higher bus access priority. However, the high priority master port Access to high frequency, low-priority ports easily lead to the master device can not get the bus access time, the arbiter must be reasonable arrangements for the priority of each port and fairness.

Suppose there are m a master device in the ports, their priority from port 0 to port followed by decreasing m-1, the definition of Qf, n the bus for the port is requesting access to the probability of n, $P_{f,n}$ into the request for the port waiting period before the successful The probability of access to the bus, the system reset after the $P_{f,n}$ is:

$$n=0, \quad P_{f,0}=Q_{f,0}; \tag{1}$$

$$n>1, \quad P_{f,n}=Q_{f,n}(1-Q_{f,0})\ldots(1-Q_{f,n-1})=Q_{f,n}\prod_{i-0}^{n-1}(1-Q_{f,i}) \tag{2}$$

By (1) and (2) can be seen in the port waiting period before entering the request, the master device to access the bus through the port the probability of success depends on the level of the port priority, the low priority port can not get the bus resources, namely, the fairness algorithm to take into account enough.

The counter to zero before the port is not responding n the request, then enter the request of the port waiting period, the probability of a successful visit to the difference between . the definition of $Q_{t,n}$ request for the waiting period to enter the port n the probability, $P_{t,n}$ in the request for the port waiting period once the probability of successful access to the bus.

$$n=0, \quad Q_{t,0}=Q_{t,0}-P_{t,0}=0; \tag{3}$$

$$n>1, \quad Q_{t,n}=Q_{t,n}-P_{t,n}=Q_{t,n}[1-\prod_{i-0}^{n-1}(1-Q_{t,i})]; \tag{4}$$

Easy to see: when n is bigger, $Q_{t,n}$ the greater the probability of entering the higher stage of the request to wait. referential (1) and (2), after system reset can be $P_{t,n}$.

$$n=0, \quad P_{t,0}=0; \tag{5}$$

$$n>1, \quad P_{t,n}=Q_{f,n}\prod_{i-0}^{n-1}(1-Q_{f,i})\,Q_{t,n}[1-\prod_{i-0}^{n-1}(1-Q_{t,i})] \tag{6}$$

The priority policies of requesting to wait is based on dynamically adjust the priority level, However,Fixed priority arbitration algorithm to followat all levels remain. Therefore, The request to wait for the arbitration policy priority logic costs less able to achieve higher efficiency of the arbitration, And has a simple algorithm, hardware and simple.

5 Conclusion

Architecture in this study still need to include: quantum memory in quantum error correction system to add it with the existing fault-tolerant quantum wire systems such as quantum error correction with quantum effects to play, and the development of specialized structures for the system High efficiency, low cost of quantum error correcting code.

References

1. Chien, C., Wang, C., Lin, C.: A Low Latency Memory Controller for Video Coding Systems. In: 2007 Proceeding of International Conference on Multimedia and Expo, pp. 1211–1214 (2007)

2. Zhang, H., Zheng, J., Wang, X.: Research on Chinese Proper Nouns Recognition Based on Pattern Matching. Journal of Computational Information Systems 5(6), 1585–1592 (2009)
3. Pu, H.: System-Level Performance Optimization Methodology for SoC Memory Subsystem. Southeast University (2006)
4. Wu, X.: Research on Performance Estimation of SoC On-Chip Bus. Southeast University (2006)
5. Zhang, G., Zhang, S., Li, Y.: FPGA Implementation of a High-throughput Memory-efficient LDPC Decode. Journal of Xidian University 35(3), 427–432 (2008)
6. Wu, Y., Yu, L., Lan, L.: A Coverage driven Constraint Random-based Functional Verification Method of Memory Controller. In: The 19th IEEE/IFIP International Symposium on Rapid System Prototyping, pp. 99–104 (2008)
7. Dai, J., He, Z., Hu, F.: A High Performance Algorithm for Text Feature Automatic Selection Based on Cloud Model. Journal of Computational Information Systems 5(6), 1561–1568 (2009)
8. Lai, Y.P., Chang, C.C.: A Simple Forward Secure Blind Signature Scheme Based on Master Keys and Blind Signatures. In: Proceedings of the 19th International Conference on Advanced Information Networking and Applications, pp. 139–144 (2005)
9. Chow, S.S.M., Hui, L.C.K., Yiu, S.M.: Forward-secure Multisignature and Blind Signature Schemes. Applied Mathematics and Computation 168(2), 895–908 (2005)

Reconfiguration of App Stores for Communications Service Providers

Fei Kang[1], TingJie Lu[1], and ChangChun Lu[2]

[1] E-commerce Research Center, School of Economics and Management, Beijing University of Posts and Telecommunications, Beijing 100876, China
[2] CITC Institute, China Unicom, Beijing 100048, China

Abstract. After the construction wave of app stores which involved almost the whole ICT industry, the communications service providers' (CSPs) app stores have lagged far behind the ones of device manufactures and operating system providers. Based on identifying the platform type and unique network assets of CSPs, a unified strategy framework is proposed for CSPs to choose their appropriate way for operating app stores according to their IT execution and competitive power. Furthermore, for each type of CSPs' app stores, the crucial IT and communications capabilities needed to configure are recommended.

Keywords: app store, platform, charging & billing, APIs, mobile e-commerce.

1 Introduction

Until now, application store (app store) is widely considered as the most successful practice of mobile e-commerce. According to Gartner's forecast, the number of downloads from app stores all around the world will grow by over 13 times to 108 billion per year till 2015 [1]. App store has a simple but efficient B2C business model with low entry barrier and highly simplified business process [2]. App store provider actually operates a service delivery platform, which serves for the two-sided market, but supplies strong technical support to nearly all kinds of developers, while developers and customers can trade on the platform freely based on software development kit (SDK), certification and approval process, charging & billing, and transparent revenue share model, etc.

Combining with smart phones, app stores have been changing consumer behavior on mobile internet and developers' motivation considerably [3]. After the Apple's app store launched in July 2008, a construction wave of app stores had involved almost all of the participants in ICT industry, such as the device manufactures, operating system (OS) provides, chip manufactures, and also communications service providers (CSPs). As for the CSPs, it seems that they have many valuable assets, e.g. network-service application program interface (API), advanced charging & billing mechanisms, etc, which could have been potentially monetized in their own app stores to form differentiated competence. However, after two years of operating, the number of applications of CSPs' app stores had lagged far behind the ones of device manufactures and OS providers. By July 2010, only few mainstream CSPs had more

D. Jin and S. Jin (Eds.): Advances in FCCS, Vol. 1, AISC 159, pp. 653–657.
springerlink.com

than 5,000 applications, while Apple App Store had 100,000 applications and Google Android Market had more than 50,000 applications.

Some strategies for CSPs' app stores have been given out in recent years [2][3][4]. These researches intend to discuss the importance and potential for CSPs' app stores, but the strategies are isolated and scattered. The questions about the crucial assets and capability should have been built for the CSP's app stores are still vague. According to the novel platform theory, a unified strategy framework is proposed for CSPs to choose their appropriate way for operating app stores, based on identifying the type and crucial assets of CSPs' app store-like service delivery platform.

2 Type and Unique Assets of CSPs' Platform

Recently researches on platform theory have started to concern with the different types of platform business models. These researches argue that platforms exhibit strong heterogeneity in terms of the configuration of components. Ballon [5] summarizes them into four main platform components: service brokerage function, creation environment for third-party developers, profile and identity management, and charging and billing modules. Platform provider is just building a business model around these components to lock in customers and further to control over the wider value network. To build these components needs platform to employ most architectural advantages, which in the form of either controlling over the tangible or intangible assets, or controlling over the customers relationship. According to this, Ballon [5][6] proposes a typology of platform models, categorizing by if the assets or customers relationship are controlled over, as shown in Table 1.

Table 1. Typology of platform models [6]

	No control over customers	Control over customers
No control over assets	Neutral Platform Examples: Google search, Paypal	Broker Platform Examples: Facebook, eBay
Control over assets	Enabler Platform Examples: Intel, IMS	Integrator Platform Examples: Apple iPhone, Microsoft OS

This typology of platform hasn't specified the type of CSPs' platform, but it's doubtless that CSPs have control over both of assets and customers relationship, so a CSP's app store-like service delivery platform could be an integrator platform. Same to IT companies, CSPs' platforms largely employ IT-based service delivery platform to build the four platform components, but due to the more hierarchical system than IT companies, their IT capabilities are generally weaker than IT companies, and the less attractiveness of CSPs' app stores would attribute to this problem, as shown in Table 2. One exception here is that the capability of charging & billing of CSPs is strong than IT companies, because CSPs are the only participants who can charge both of network access fees and applications purchase fees in ICT industry.

Table 2. Capability differences of platform between CSP and IT company

Platform components	IT Integrator Platform	CSP's IT capability	CSP's potential communications capability
Service brokerage function	strong	weak	Distribute multi-screen, FMC, enriched applications to customers
Creation environment for 3rd-party developers	strong	weak	Open network APIs to 3rd-party developers to aggregate unique applications
Profile & identity management	medium	medium	Use OSS/BSS data for advanced data mining and customer insight
Charging & billing	weak	strong	Integrate the charging & billing for network access and application payment

Despite of the comparably weak IT capabilities, communications network is the unique assets that CSPs have controlled over. Not only supply network access service to IT companies in ICT industry, but also CSPs will be able to employ their own network assets to supplement their IT capabilities on the four platform components, although they haven't utilized these kinds of assets well until now. More specific information of CSP's potential communications capability has been shown in the last row in Table 2. To summarize, the charging & billing, OSS/BSS data, network APIs, and communications service are the four crucial capabilities that should be relied on by CSPs' platform.

3 Strategy Framework for CSPs' App Store

3.1 Decision Criteria

Generally, CSPs' IT capabilities are disadvantages of CSPs, but it depends on the fact. According to the determinative of the IT capability for platform components, the difference of the level of building and operating of IT capability will result in different strategy and attractiveness of CSPs' app stores. Hence, IT execution will be employed as one of the decision criteria for CSPs' app stores.

The competitive power should be another decision criterion. CSPs' customer market is fragmented comparing to device manufactures and OS providers; moreover, various devices with various OS accessed to CSPs' networks induce incompatibility of applications. These obstacles will decrease the developers' motivation on CSPs' app stores. However, CSPs with scale economics or confronting weak competition from industry chain could overcome these obstacles better than others.

Based on the two decision criteria, IT execution and competitive power, a strategy typology for CSPs' app stores is shown in Table 3.

Table 3. Strategy typology for CSPs' app stores

	Strong competitive power	Weak competitive power
Strong IT execution	Proprietary Store CSP's app store is complete controlled via vertical integration Key capability: CRM, CEM	Aggregation Store CSP's app store is aggregated with other app stores in one place Key capability: search engine, APIs authorization
Weak IT execution	Labeling Store CSP's app store retails applications for other app stores Key capability: charging & billing	Enabler Store CSP's app store wholesales network APIs to other app stores Key capability: APIs portfolio

3.2 Strategy and Crucial Capabilities

The four types of CSPs' app stores can be ranked as proprietary store, labeling store, aggregation store and enabler store according to their most vertically integrated toward the most open to industry chain.

Proprietary store, is a vertical integration strategy that CSPs complete control their own app stores and directly compete with IT companies' app stores. All the components should be well built, but the profile & identity management is the key component to attract customers and developers. Hence, customer relationship management (CRM) system will be crucial to support customer experience management (CEM) and some advanced market functionalities, like customer profiling and in-application advertising.

Labeling store, is a strategy that CSPs will play as retailers or distributors. CSPs may not aggregate application directly, but label on the app stores of solution vendors or third-party internet service providers, or purchase amount of applications from cooperative app stores. With the strong capability of charging and billing, CSPs can retail applications for the app stores providers who don't have effective payment channels. This would be a cost-effective way for some large CSPs to supply applications to customers without developing complete high IT capabilities.

Aggregation store, is a cooperation strategy that CSPs gather other app stores and their own app stores together in one place. By deploying portal linkages, recommending top applications from other app stores, CSPs can aggregate abundant applications for own customers. Vertical search engine will be a crucial capability that enables customers to discover their favorite applications from lots of app stores in one place. Moreover, CSPs still need to maintain the advantages of their own app stores/applications in this broad cooperation platform, so network APIs should not be open to every third-party developers. CSPs should build APIs authorization systems to insure that the APIs can only be opened to partnership or exclusive developers.

Enabler store, is the most open strategy that CSPs will play as a wholesale warehouse, e.g. supplying colocation center service or wholesale distribution service for other app stores, but CSPs may not bother with tangible app stores of themselves, while the most important wholesales will be the network APIs. CSPs ought to open their network APIs to the maximum extent, and build the capability of APIs portfolio management to gain the maximum revenues from openness. For instance, CSPs can charge APIs of some basic network enablers, such as calling, SMS/MMS for low fees, but charge APIs of some advantaged enablers, such as location via base station, network status presence, and especially the OSS/BSS data for high fees.

4 Conclusion

According to novel platform theory, four strategy types of CSPs' app stores are given in this paper. In summary, we tend to recommend the strategy type of aggregation store as the appropriate way for operating app stores, due to the commonly medium IT capability and competitive power of majority CSPs. Despite of this, as for all CSPs' app stores, charging & billing and network APIs openness should always be regarded as the crucial capabilities. As for the future, to utilize CSPs' unique network assets well, CSPs ought to reconfigure app stores again based on IMS. This is because that IMS-based service delivery platform can integrate communications elements well than IT-based service delivery platform [7]. In other words, IMS-based service delivery platform can counterbalance the complexity of building IT capabilities, and can transform some of the IT capabilities to communications capabilities. This will make CSPs cost less to develop IT capabilities. Finally, based on IMS, CSPs can also promote the app stores to all service fields facilely, such as FMC and enterprises.

Acknowledgments. The authors acknowledge the support of the National Natural Science Foundation of China (Research on business model of open service delivery platform. Grant No.: 71172135).

References

1. Baghdassarian, S., Milanesi, C.: Forecast: Mobile Application Stores, Worldwide, 2008-2015. Gartner May report, Gartner (2011)
2. Kimbler, K.: App Store Strategies for Service Providers. In: 14th International Conference on Intelligence in Next Generation Networks, pp. 1–5. IEEE Press, New York (2010)
3. Copeland, R.: Telco App Stores – Friend or Foe? In: 14th International Conference on Intelligence in Next Generation Networks, pp. 1–7. IEEE Press, New York (2010)
4. Goncalves, V., Walravens, N., Ballon, P.: How about an App Store? Enablers and Constraints in Platform Strategies for Mobile Network Operators. In: 9th International Conference on Mobile Business and Global Mobility Roundtable, pp. 66–73. IEEE Press, New York (2010)
5. Ballon, P.: The Platformisation of the European Mobile Industry. Communications & Strategies 75, 15–33 (2009)
6. Ballon, P., Van Heesvelde, E.: ICT Platforms and Regulatory Concerns in Europe. Telecommunications Policy 35, 702–714 (2011)
7. Pavlovski, C.J.: Service Delivery Platforms in Practice: IP Multimedia Systems (IMS) Infrastructure and Services. IEEE Communications Magazine 45(3), 114–121 (2007)

Research on Control System for Single-Crystal Silicon Growth Furnace

Wenpeng Lu

School of Science, Shandong Polytechnic University
Jinan, Shandong, China, 250353
lwp@spu.edu.cn

Abstract. Single-crystal silicon growth furnace is composed with lots of components and peripheral devices, which are difficult to be controlled manually at the same time. In order to satisfy the control requirements of single-crystal silicon growth furnace, the paper designs a control system, which is composed with four modules: control scheme design module, automatic control module, safety monitor module and manual control module. With the control system, operators can control growth furnace more conveniently. The control system can relieve the work of operators greatly.

Keywords: single-crystal silicon, crystal growth, control system, automatic control.

1 Introduction

With the development of electronic technology, because single-crystal silicon is the most important material of integrated circuit (IC), it has been paid more and more attention to[1]. The ability of single-crystal silicon manufacturing is one of the key indicators which represent the science and technology development level of a country[2].

The growth of single-crystal silicon includes a series of complex operations, such as flushing gas, vacuuming gas, heating, shifting and rotating[3, 4]. In order to satisfy the requirements of crystal growth, each stage of growth requires many peripheral devices to work together. If the growth furnace is controlled with manual control form, it is difficult for operators to precisely control all of peripheral devices, which would affect the quality of single-crystal silicon and the safety of operators and growth furnace.

In order to achieve automatic control of single-crystal growth furnace, the paper designs a control system, which is composed with four modules: the module of control scheme design, the module of automatic control, the module of safety monitor and the module of manual control. The module of control scheme design is responsible to help operators design control scheme of single-crystal silicon furnace. Each parameter in each step of control scheme could be set one by one. According to the control scheme of growth furnace, the module of automatic control is responsible to control together multiple devices to satisfy the requirement of each step of growth

D. Jin and S. Jin (Eds.): Advances in FCCS, Vol. 1, AISC 159, pp. 659–664.
springerlink.com

scheme. The module of safety monitor is responsible to monitor the real-time status to assure the safety of growth furnace. When some unexpected accidents happen, safety monitor module would send out alarm or warning messages. The module of manual control allows operators to directly control the devices of growth furnace when automatic control is out of operation or fails to satisfy requirements. The control system can help workers control the operating of single-crystal silicon growth furnace more conveniently, which would greatly reduce their workload.

In the paper, the detailed design of control system of growth furnace is described. The rest of the paper is organized as follow. Section 2 introduces the framework of control system. The detailed design of each child module is described in Section 3. As last, we give the conclusion and future work.

2 Framework of Control System

As is shown in Fig.1, in order to satisfy the requirement of growth furnace, the control system is composed with four modules: the module of control scheme design, the module of automatic control, the module of safety monitor and the module of manual control[5].

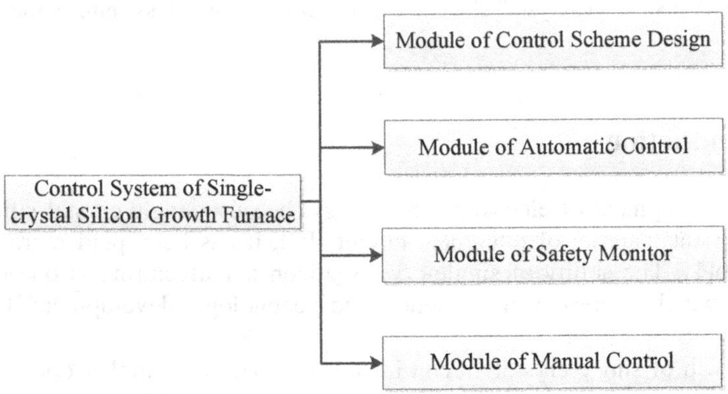

Fig. 1. Framework of Control System

2.1 Module of Control Scheme Design

This module provides an interface for operators to design the control scheme. Operators can preset all kinds of parameters of devices related with growth furnace, which include: time of each stage, status of intake valve and exhaust valve, output power of intermediate frequency supply, work pressure of growth furnace, motion of crucible, et al. The module can perform a preliminary check of control scheme to prevent the mis-operation from causing serious accidents.

2.2 Module of Automatic Control

This module implements the automatic control of single-crystal growth furnace. According to control scheme, the module controls the operation of each components of the furnace to satisfy the requirement of single-crystal silicon growth.

2.3 Module of Safety Monitor

This module is responsible for safety monitor of single-crystal silicon growth furnace. The maximum input power of growth furnace can reach 20KW and internal temperature can reach 2300 degrees centigrade. Once a serious accident happens, its consequence would be disastrous. The module monitors the status of growth furnace. When abnormal phenomena happen, it would inform operators to handle.

2.4 Module of Manual Control

When the furnace needs to be adjusted or tested, operators must control its components manually. The module is responsible to provide an interface for manual control.

3 Design of Each Module of Control System

3.1 Design of Control Scheme Design Module

The growth of single-crystal silicon refers to two types of phrase, referred with *vacuum* and *flush*. For each step of control scheme, there are nineteen parameters which are required to be set, such as *Process, Time, RF, RF (A), Press (mB), Ar, Ar (%), N2, N2 (%), N3, N3 (%), N4, N4 (%), TransDir, TV (mm / h), TS (mm), RotaDir, RV (rpm), Y5*.

Process refers to the type of current growth step, that is, *vacuum* and *flush*.
Time refers to the working time of current growth step.
RF refers to status of intermediate frequency supply, that is, *On* and *Off*.
RF(A) refers to the output current of intermediate frequency supply.
Press(mB) refers to the internal pressure of growth furnace.
Ar refers to the status of intake valve for argon gas, that is, *On* and *Off*.
Ar(%) refers to opening percentage of *Ar* valve.

Except that controlled gases are different, the meanings of *N2, N2(%), N3,N3(%), N4, N4(%)* are same with that of *Ar* and *Ar(%)*.

TransDir refers to the shift direction of the crucible in growth furnace, that is, *Up* and *Down*.

TV(mm/h) refers to the velocity of the crucible.
TS(mm) refers to the shift distance of the crucible.

Rota refers to the rotation direction of the crucible, that is, *clockwise* and *anticlockwise*.

RV(rpm) refers to the rotation velocity of the crucible.
Y5 is the status of total intake valve.

This module provides a friendly interface for operators. When operators need to design a control scheme, for each step of the scheme, the operators could set the nineteen parameters one by one with the interface. Because most parameters of each step are same, in order to conveniently design control scheme, the interface provides a shortcut menu, such as copy current step, paste to current step, delete current step, insert a new step, which would help operators design the scheme more quickly.

This module could check the reasonableness of parameters in each step. When mistakes are found, the module would reject the modification of operators until it was corrected.

3.2 Design of Automatic Control Module

This module is the most important part of control system. According to the requirements of control scheme, the module controls the action of each component and peripheral device of growth furnace, adjusts the status of intake and exhaust valve, adjusts the output power of intermediate frequency supply and the internal pressure of growth furnace, et al.

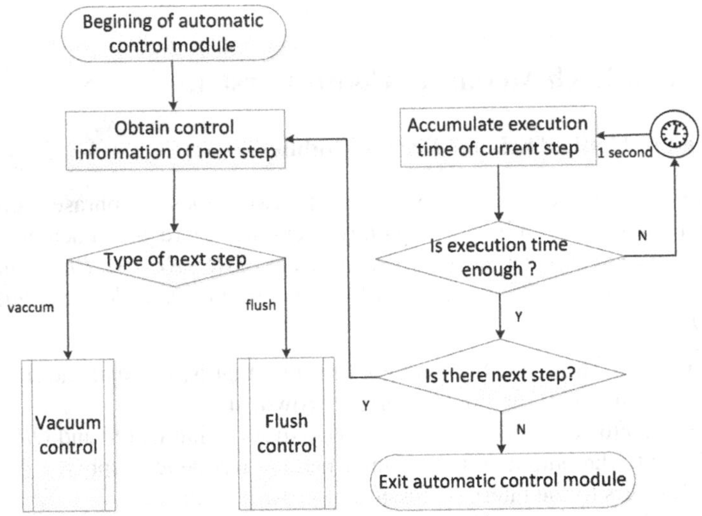

Fig. 2. Program Flow of Automatic Control Module

The program flow of automatic control module is shown in Fig.2. For different type of growth step, the control module would call vacuum control and flush control. There are distinguished differences between vacuum control and flush control, which are introduced in detail as follow.

Vacuum Control

The growth step of vacuum type requires that the furnace should be in vacuum state. The main works of this kind of step include:

Shutdown all of intake valve (*Y1, Y2, Y3, Y4, Y5, Y6, Y7, Y7b*);
Open exhaust valve (*Y10*);
Start mechanical pump;
Start low vacuum pressure gauge (*Pirani*);
Monitor the pressure in growth furnace, when the pressure is lower than 10^{-1} Pa, start molecular pump and high vacuum pressure gauge (*Penning*);
Adjust the output power of intermediate frequency supply.

Flush Control

The growth step of flush type requires that the furnace should work on a certain of air pressure. The main works of this kind of step include:
Stop mechanical pump, molecular pump;
Close *Y7* and *Y7b*;
Shutdown vacuum pressure gauge (*Pirani, Penning*);
Open the intake valve *Y5*;
Adjust the status and opening percentage of four intake valve (*Y1,Y2,Y3,Y4*);
Start pressure gauge (*Barat*) to monitor the real-time pressure in growth furnace;
Control the motion of crucible.

3.3 Design of Safety Monitor Module

When single-crystal furnace is working, its input power can reach 20KW and its internal temperature can reach 2300 degrees. If an unexpected accident happens, it would endanger the safety of growth furnace and operators. The control system should monitor the real-time status of growth furnace to assure that the furnace works in a safe state.

In order to monitor growth furnace, control system considers the following messages:

Water Flow

The backwater flow of flange of growth furnace is monitored. If it exceeds the safety limits, control system would send out alarm message, and shutdown intermediate frequency supply.

Water Temperature

The temperature of cooling water is monitored. If it reaches 45 degrees, control system would send out alarm message, and shutdown intermediate frequency supply.

Furnace Temperature

If the temperature in the growth furnace reaches 2300 degrees, control system would send out alarm message, and shutdown intermediate frequency supply.

Position Limitation

If the crucible moves out of expected range, control system would send out warning message, and stop the motion of crucible.

Other Warnings

When some other accidents happen (such as the failure of opening percentage set of valves, failure of pressure set), control system would send out warning message.

Among the above messages, the first three kinds of messages are alarm message, which mean urgent accidents and should be handled immediately. The last two kinds of messages are warning message, which would be reminded and recorded. Warning message doesn't require to be handled immediately.

3.4 Design of Manual Control Module

This module is responsible to control growth furnace manually. According to requirements, operators can manually control each component and peripheral device. The module is used when operators need to test the growth furnace or particular situations need to be handled.

This module provides a friendly interface in which each component and peripheral device can be controlled separately. Each of them is corresponded with a button or other controls. Operators can manually control each component and device with its button.

4 Conclusions and Future Work

In order to satisfy the control requirement of single-crystal silicon growth furnace, the paper designs a control system. The control system can help operators design control scheme of growth furnace, automatically control the operations of growth furnace, monitor its safety. Besides, it provides the manual control function. This would relieve the work of operators greatly.

The control system can satisfy the basic requirement of growth furnace. In the next work, we would try to improve its interface and try to add a long-range control module.

References

1. Daxi, T.: Research of the Growth and Defects of Czochralski Silicon Crystals. Ph.D Thesis. Zhejiang University, Hangzhou (2010)
2. Jinquan, W.: Tips of the Process Control for Mono-crystal Silicon Foundry. Equipment for Electronic Products Manufacturing, 30–33 (2010)
3. Lixin, L., Ping, L., Chun, L., Hai, L., Xuejian, Z., Ying, Z.: Growth Principle and Technique of Single Crystal Silicon. Journal of Changchun University of Science and Technology 32, 569–573 (2009)
4. Ping, L.: The Principle and Technology of Monocrystall Silicon Growth with Czochralski Methord. Master Thesis, Changchun University of Science and Technology, Changchun (2009)
5. Jianwei, C.: Research on the Key Technology of CZ Silicon Crystal Growth Furnace. Ph.D Thesis, Zhejiang University, Hangzhou (2010)

Unstructured Surface Mesh Generation for Topography Using Interpolating Surface Modeling

Zhiping Deng, Y. Wang, Gang Zhao, and Junfu Zhang*

School of Mechanical Engineering and Automation, Xihua University, Chengdu, China 610039
zhipingdeng@mail.xhu.edu.cn, yu.wang@plymouth.ac.uk,
zhaogang0520@163.com, zhang_junfu@126.com

Abstract. Geological surface mesh generation is a prerequisite for 3D topographical modeling and also is one of the most difficult tasks due to the complication. Interpolating surface modeling is a new paradigm using interpolating grid to model surfaces of any arbitrary topology. In this paper, we combine it with the traditional Delaunay triangulation algorithm to design a new surface mesh generation scheme. Using this scheme can avoid the disadvantage of the traditional analytical model on the discontinuous surface modeling and can establish the underlying geometric model from random scattered measurements, The results of the applications show that it produced a good mesh shape. The ability to deal with the sharp changes also is very good. All of these characters make this scheme have a great potential application in finite element analysis in geological area.

Keywords: Surface mesh, Interpolating subdivision surface, Delaunay triangulation.

1 Introduction

Surface mesh generation is an important prerequisite for shell analysis, volume discretization and 3D free surface flow modelling, but meanwhile it is also among the most difficulties to develop. Particularly, for terrain surface in geological analysis, the complicated and irregular shapes and the multi-composition of the topography make this issue more testing. In the past decades, significant advances have been achieved in the development of surface mesh algorithms by many researchers [1-5]. In general, these algorithms consist of two stages. The first is to represent a surface with an underlying geometric model. Normally analytical models such as B-splines or non-uniformly rational B-spline (NURBS) are utilized for this purpose. In this stage, firstly, a real surface is represented by a collection of surface patches which are usually triangles or quadrilateral patches. Secondly, these patches are represented with the

* Corresponding author. Junfu Zhang, is currently a professor in School of Mechanical Engineering and Automation, Xihua University, China, zhang_junfu@126.com

D. Jin and S. Jin (Eds.): Advances in FCCS, Vol. 1, AISC 159, pp. 665–673.
springerlink.com

analytical spline models. The second stage is the mesh generation itself. Two approaches are employed normally at this stage: (1) using traditional 2D mesh algorithms to generate mesh in a parametric plane, then map the parametric mesh onto surface model [1,2] (Fig. a). (2) using an advancing front method to generate the mesh directly on the surface model [5,6,8].

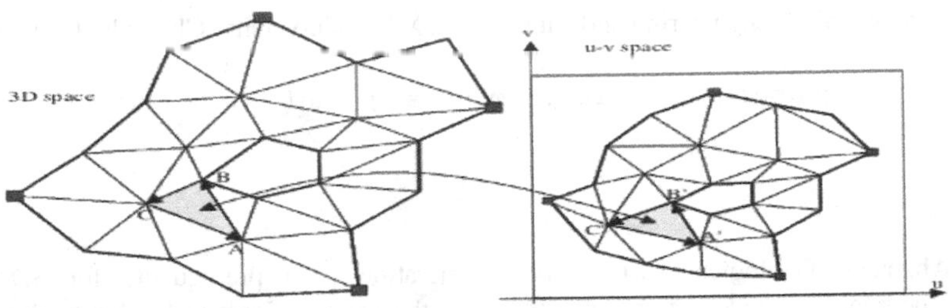

Fig. a. A parametric mapping of a sub-surface [8]

Analytical models are very popular when using CAD packages to define geometric models in engineering design. However, for terrain surfaces, such as a real slope surface (Fig.b), using analytical models is inconvenient because of the complicated and irregular shapes. If we subdivide the slope surface into several patches, to keep the continuity of these patches at their intersection, particularly, is a huge task. In the near decade computer surface modelling using interpolation based on a number of measurement points is a fast developing technique. Many interpolating algorithms have been proposed, such as Inverse Distance Weighted (IDW) and Spline used in GIS and Interpolating Subdivision for arbitrary topology [11-13]. Recently the last one has already been successfully used as an underlying geometric model to help generate surface meshes [7-10]. In these researches, the metric advancing front technique (MAFT) was used to generate triangular mesh directly on a modelling surface constructed using the Interpolating Subdivision. Using a tagging method the proposed scheme also showed a very easy mesh control over the sharp features on surfaces such as cusp nodes and crease lines. MAFT has a good element size and shape control, however the strict mathematical foundation makes it complicated in programming and time consuming in running.

Fig. b. A real slope

In this paper, we take advantage of the traditional 2D Delaunay triangulation algorithm combining it with the interpolating subdivision surface to design a surface mesh generation scheme. Because 2D Delaunay triangulation is a simple, mature and fast technique widely used in finite element analysis, the scheme is quite easy to develop and it has an advantage to cope effectively with the different composition of surfaces, that makes the interface conforming between different components become a easy work without extra concern. In this work, the mesh generation starts from a collection of measurement points. At first these points are connected together to yield an initial mesh which consists of triangle patches. Then for every triangle patch in the initial mesh, we create an associating parametric triangle. After that, a traditional Delaunay triangulation mesh generator is applied on their parametric triangles for all the triangle patches in the initial mesh. Finally the parametric meshes are mapped back on the subdivision surface model respectively to yield the surface mesh. Because Delaunay triangulation method starts from boundary mesh, by adopting a same dividing at common edges of two neighbour patches, this kind of scheme is simple and will keep the conformation of the interface, which is defined by the initial mesh, between different components. However the mesh size control is a critical issue for it. In this paper a non-restrict-mathematical-founded but approximate and simple algorithm is adopted to control the mesh size.

In the following sections, we firstly give a brief review of the interpolating subdivision method that we employ to establish the underlying geometric model (For a detailed review please see references [7,10-13]). A mesh generation algorithm adopted from a 2D Delaunay triangulation method is then discussed in detail and finally we provide some examples to show the performance of the scheme.

2 Modeling Surfaces with Subdivision Methods

Starting from an initial triangular coarse mesh with a sequence of local subdivision, a smooth surface mesh is constructed based on the initial mesh (Fig. 1). A new vertex yielded in subdivision process is defined by the positions of the vertices around it according to a set of rules. The subdivision scheme is fully local, no global equation is needed and the quality of the surface at any point can be improved with local subdivision refinement. There are two subdivision schemes were proposed. They are the Interpolating subdivision scheme (ISS) and Approximating subdivision scheme (ASS) [6,7,8]

(1) ISS is simple to implement, which leads to C1 surfaces. In ISS, the positions of the vertices of the subdivided mesh are fixed, but the position of the new created vertices in the new refined mesh will be computed according to an interpolation rule. Once they are created, these new vertices will be fixed in a further subdivision refinement. Fig. 4(a) shows a one-dimensional example.

(2) In ASS, the position of all the vertices, both the existing ones and the new created ones, will be recalculated (Fig. 4(b)). This scheme will theoretically lead to a C2 continuous surface.

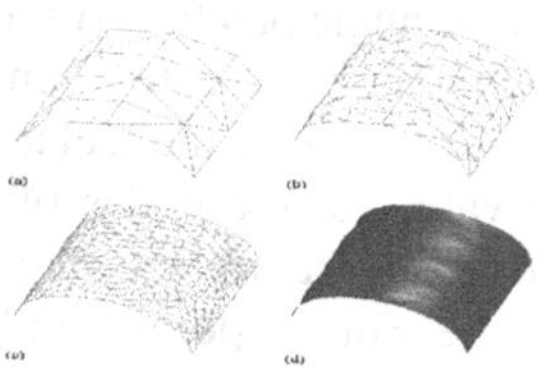

(a)initial mesh; (b) first subdivision:
(c) second subdivision; (d) limiting surface

Fig. 1. Interpolating subdivision surface [7]

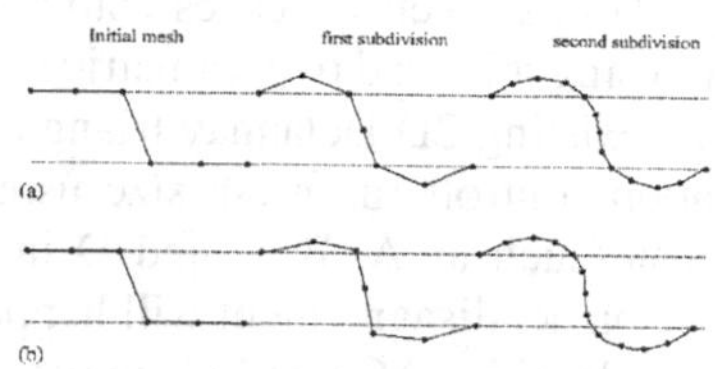

(a) interpolating subdivision;
(b) approximating subdivision.

Fig. 2. Interpolating and approximating subdivision [7]

For our application, because the initial triangle is constructed basing on measurements, ISS is adopted in that it allows us to keep the original measurements unchanged in subdivision process. The interpolation rules of ISS were described in detail in references [6,8].

3 Mesh Generation on Triangular Patches

As mentioned above, an initial triangular mesh is constructed manually at first (Fig. 3). Traditionally, the parametric association can be described as:

$$X = \sum_{i=1}^{3} N_i(\xi,\eta) X_i \qquad (1)$$

Where X represent the physical coordinates of the points in an initial triangular patch. The subscript i indicates three vertices of the patch. ξ and η are the coordinates of the corresponding points in the parametric plane. Ni is the shape function of the three vertices.Eq. (1) describes a one to one correspondence between the points on the triangular patch and that on the parametric plane, only if the area of triangle on parametric plane is not equal to zero. The construction of a parametric domain is not unique and normally a right angle parametric triangle is used as that shown in Fig. 6. For it, the three shape functions are [1]:

$$N_1 = 1 - \xi - \eta, N_2 = \xi, N_3 = \eta \qquad (2)$$

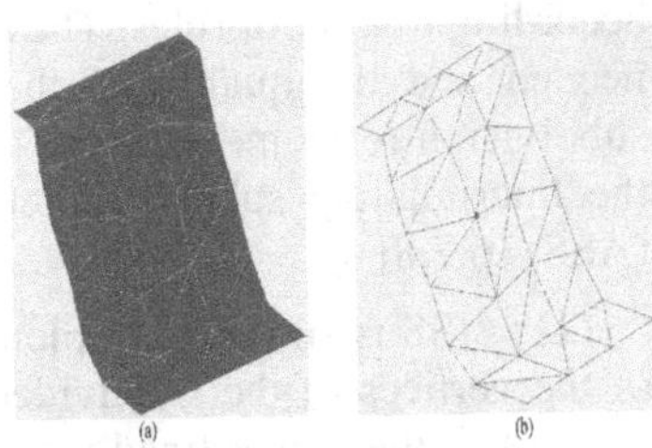

(a) measurement; (b) initial mesh

Fig. 3. A measured slope surface

Different parametric planes, e.g. it may be a vertical side having However this kind of parametric triangle is inconvenient to keep the coherence of mesh nodes along a common edge when the meshes of two neighbored triangles are tiled together if the common edge (e.g. CB in Fig. 4) is at different positions in two unit length in one parametric plane, meanwhile it might be the tilt side whose length is in another parametric plane. Using the same mesh generator, without specific treatment, it is highly along the common edge being completely different. Fig. 4 shows a un-cooperating result after mapping all of parametric meshes onto their respective initial triangle patches. Another problem using this kind of parametric triangle is in controlling the mesh size when we employ an existing 2D Delaunay triangulation mesh generator. Because the Delaunay triangulation controls the mesh size using the predefined values at the nodes of the initial mesh (such as A, B, C and D in Fig. 4), when project these values onto the parametric plane, disagreement will happen at a common node. For the point C in Fig. 5, for example, sides AC and BC are projected onto the unit sides in a same parametric plane. Because the lengths of AC and BC may be different, the projected size value at C will be different when the calculation is based them respectively. To overcome this disadvantage, in this paper we simply use the actual planes which the triangular patches are on as the parametric planes. With this plane the projected parametric triangle has the exactly same shape as its corresponding triangular patch and the disagreed projection of the common edges to different parametric planes is avoided.

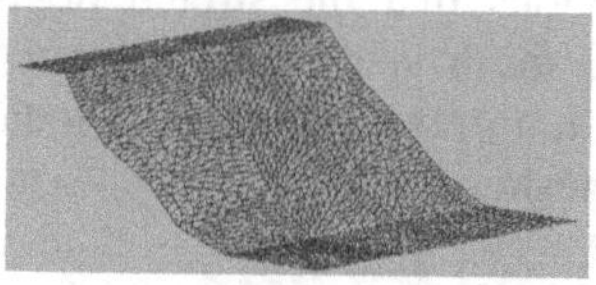

Fig. 4. Un-cooperating patch meshes

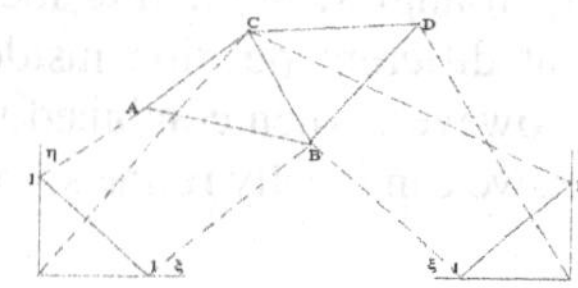

Fig. 5. Parametric triangle and mapping

4 Mesh Size Control

Using Delaunay triangulation to generate meshes on 2D domains, the mesh size distribution is controlled by the predefined values at the domain defining points [8]. For a triangle in a parametric plane, the defining points are its three vertices. Three sides of the triangle are divided into segments at first according to the predefined values, then new nodes are inserted one by one into the triangle and Delaunay algorithm is used to make the connection of all nodes. The new node insertion will not stop until all elements satisfy the size requirement. A local size requirement is calculated by interpolation of the three predefined values. This kind of algorithm is also called boundary controlling. Adopting this algorithm to control the mesh size on a surface, mesh size values are predefined at every measurement point in the initial mesh. These values need to be projected to the parametric plane. However, because the length of the curves on the real surface is not known at first, the first step we needed is to produce the underlying geometric model. Here we use theinterpolating subdivision technique. Based on the interpolated geometric model the length of the surfaces for the example in Figure 3curves corresponding to the sides of the initial patches can be calculated out according to the subdivision mesh. Of course, the result depends on the subdivision level which is determined by the accuracy wanted. Fig. 8 shows the subdivision meshes and their modeling surfaces at different levels.Fig.6. Subdivision meshes and their modeling surfaces for the example in Figure 5Once obtained the length of the curved sides (L1, L2, L3) associated with the sides of the initial triangle patches, we construct the parametric triangle ah of the initial patch by stretch the three curved sides onto a plane as shown in Fig. 7. Because L1, L2

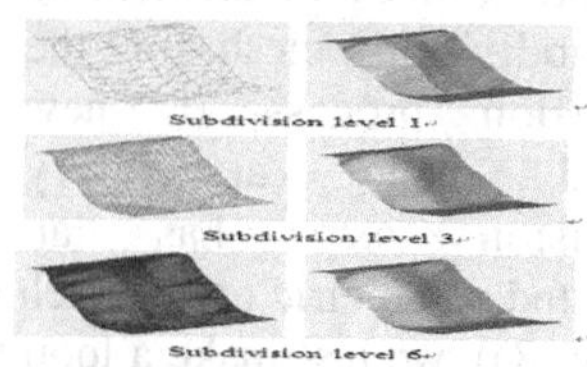

Fig. 6. Subdivision meshes and their modeling

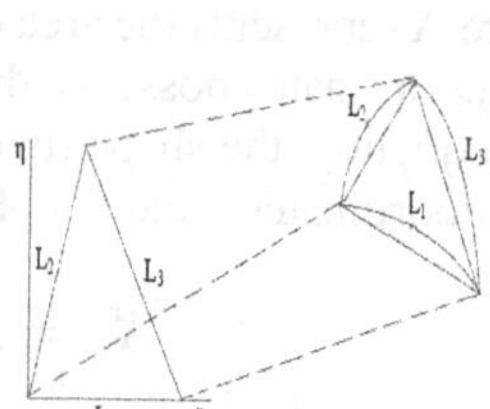

Fig.7. Construction of parametric triangle

and L3 have the full length of their original curves, the predefined mesh sizes can be directly projected onto the corresponding vertices of the parametric triangle without any modification. Then a 2D Delaunay mesh generator is directly applied on the parametric triangles. With this method, it is obvious that the stretch degrees are different at different position inside triangles and mesh size control is not exactly precise. However, when combined this method with a previous proposed local refine technique, we can locally refine some unsatisfied meshes.

5 Mesh Mapping

After the meshing of parametric triangles, the parametric finite element mesh needs to be mapped onto the geometric model. The mapping method is quite simple and straightforward. If we keep the corresponding parametric coordinates of every element of the subdivision mesh in the subdivision process (a new inserted node at every subdivision step has an associated parametric node which is assumed at the midpoint of the side the node lying on (Fig. 8)), we just make a loop to find which subdivision mesh triangle a generated finite element node is in. Once found, the physical coordinates of the finite element node can be interpolated using area coordinates. Assume a node p in a subdivision triangle (Fig. 9), the area coordinates of the point, Li, Lj and Lk, are defined as:

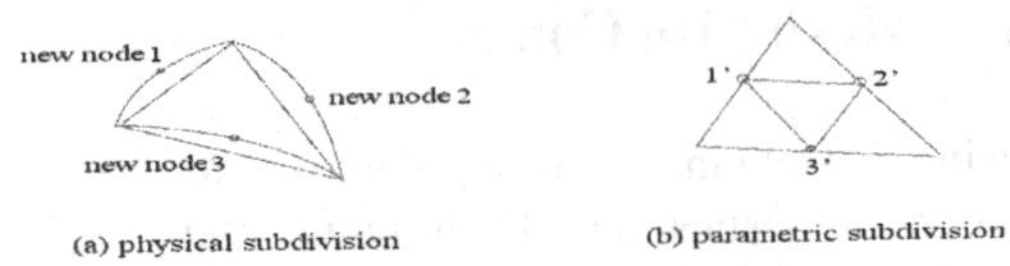

Fig. 8. Physical subdivision and parametric

$$L_i = \frac{A_{pjk}}{A_{ijk}}, L_j = \frac{A_{pki}}{A_{ijk}}, L_k = \frac{A_{pij}}{A_{ijk}} \quad (3)$$

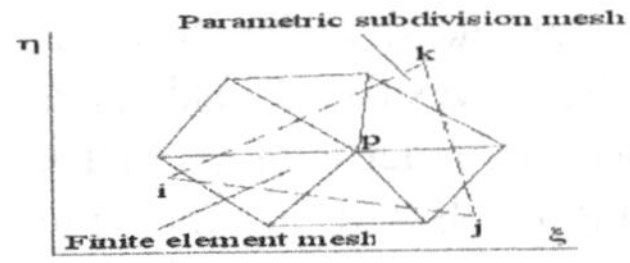

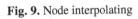

Fig. 9. Node interpolating

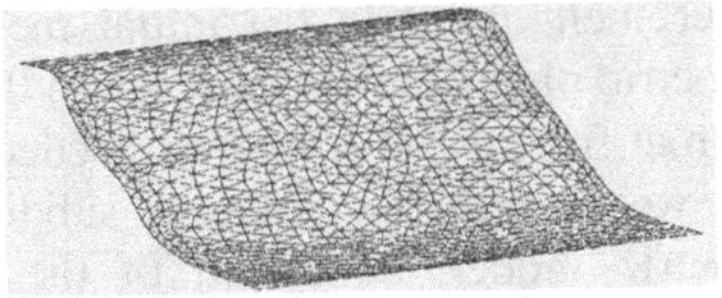

Fig. 10. A final mesh for the example

where A represents the area of the parametric triangle. It is obvious that i+Lj+Lk=1.The area coordinates possess the three properties of C0-continuous shape functions [10], consequently, the shape functions for vertices i, j and k can be given in terms of the area coordinates and the physical coordinates of p is:

$$X_p = L_i X_i + L_j X_j + L_k X_k \quad (4)$$

Finally tile all of the initial patches together, get rid of the over-positioned node and renumber the whole mesh.Fig. 10 shows a final mesh for the example in Fig. 5.

6 Surface Smoothing

Taking advantage of an existing 2D mesh generator, each parametric mesh is subject to additional two treatments:

a) Relaxation by edge swap limits the number of triangles around a point to avoid sharp triangles [9].

b) Laplacian smoothing by moving nodes according to position of the neighbors improves the shape of triangles.

After the parametric mesh has been mapped onto a modelling target surface, an additional global Laplacian smoothing can be applied again for the whole surface. However this has a high possibility to make the material interface moved [1]. So global smoothing is not adopted in this work.

7 Examples

Two examples are designed to demonstrate the performance of the surface mesh scheme proposed. Fig. 11shows a modeling of a complicated slope surface having sharp topographical changes. Fig. 11(a) is the initial mesh based on measurement points. Fig. 11(b) is the subdivision surface constructed at subdivision level 4. Fig. 11(c) is the finite element mesh generated under the condition that all the mesh size values defined at all measurement points are the same. It can be seen that the generated mesh has a reasonable evenly distributed mesh size and a good shape. The calculated ratios of the in-radius to the circum-radius of all the generated triangle elements have a smallest value of 0.218.

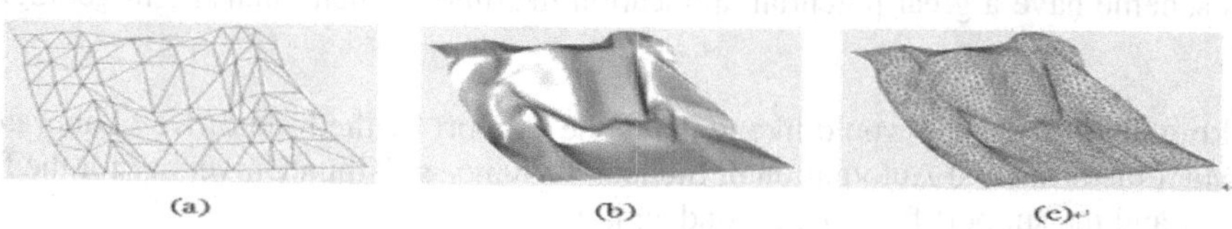

(a) Initial triangle patches; (b) Modelling slope surface; (c) Mesh generated

Fig. 11. Example 1: A complicated slope

Fig. 12 shows an example of a slope surface where there is a drive way built at its middle height. Fig. 12(a) shows the initial mesh constructed. In this example, to model the drive way which causes the break of the continuousness of the surface, the tagging method in references [6] is adopted. As shown in Fig. 12(a) all of

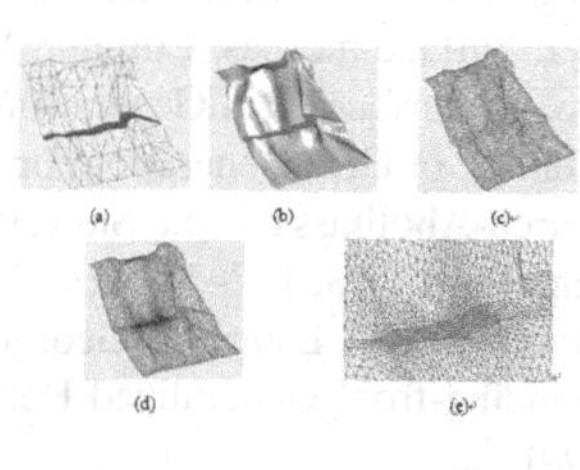

Fig. 12. Example 2: A slope with a drive road

the initial triangular patches which represent the drive way surface are tagged out. In the later subdivision process, all the sides represent the boundary of the drive way are treated as boundary edges and the corresponding interpo- Fig. 14. Example 2: A slope with a drive road lating rules are applied. For the sides which are on the drive way surface, the coordinates of interpolating point generated take the average of the two ends of the drive way, are exactly same as theiroriginal triangle patches .Comparing the mesh on the drive way with that at other places, it is shown that their mesh sizes are quite similar. This proves that the ignorance of the different stretch factors at different position is acceptable and the generated mesh shows a reasonable even distribution of mesh size and a good mesh shape. The result also shows that the mesh representing surface is quite close to the subdivision modelling surface. Fig. 12(d-e) shows the facility of the local refinement of the mesh generator. The refined mesh shows a good controlled mesh distribution and a good mesh shape. The mesh at the sharp places such as the boundaries of the drive way has a good quality as well.

8 Conclusions

In this paper, a simple surface mesh scheme is proposed, which takes advantage of a new development in surface modeling and exploits an existing 2D Delaunay mesh algorithm. Using this scheme can avoid the disadvantage of the traditional analytical model on the discontinuous surface modeling and can establish the underlying geometric model from random scattered measurements, which has a significance in complicated topographical modeling, quite easily. On the other hand, the scheme which employs the widely used 2D Delaunay algorithm has a great flexibility to deal with the multi-component problems. A simple but non-strict mesh size control algorithm was proposed. The results of the applications show that it produced a good mesh shape. The ability to deal with the sharp changes also is very good. All of these characters make this scheme have a great potential application in finite element analysis in geological area.

Acknowledgment. We would like to thank the support by the fund of Key Laboratory of Manufacturing and Automation of Sichuan Province at Xihua University for the first author and the support from the second author.

References

1. Zheng, Y., Lewis, R.W., Gethin, D.T.: Three-dimensional unstructured mesh generation: Part 2. Surface meshes. Comput. Methods Appl. Mech. Eng. 134, 269–284 (1996)
2. Peiro, J.: Surface Grid Generation. In: Thompson, J.F., Soni, B.K., Weatherill, N.P. (eds.) Handbook of Grid Generation, pp. 19.1–19.19. CRC Press, LLC (1999)
3. Samareh-Abolhassani, J., Stewart, J.E.: Surface grid generation in parameter Space. J. Comp. Phys. 113, 112–121 (1994)
4. Borouchaki, H., Laug, P., George, P.L.: Parametric Surface Meshing Using a Combined Advancing-front Generalized Delaunay Approach. Int. J. Numer. Meth. Eng. 49, 233–259 (2000)

5. Ito, Y., Nakahashi, K.: Surface Triangulation for Polygonal Models Based on CAD data. Int. J. Numer. Meth. Fluids 39, 75–96 (2002)
6. Lee, C.K.: Automatic Metric 3D Surface Mesh Generation Using Subdivision Surface Geometrical Model: Part 1: Construction of Underlying geometrical Model. Int. J. Numer. Meth. Eng. 56 (2003)
7. Cirak, F., Ortiz, M., Schroder, P.: Subdivision Surfaces: a New Paradigm for Thin-shell Finite Element Analysis. Int. J. Numer. Meth. Eng. 47, 2039–2072 (2000)
8. Zorin, D., Schroder, P., Sweldens, W.: Interpolating Subdivision for Meshes with Arbitrary Topology. In: Proceedings of Computer Graphic, ACM SIGGRAPH 1996, pp. 189–192 (1996)
9. Rebay, S.: Efficient Unstructured mesh generation by means of Delaunay triangulation and Bowyer-Waston Algorithm. Journal of Computational Physics 106, 125–138 (1993)
10. Stasa, F.L.: Applied Finite Element Analysis for Engineers, CBS International Editions, New York, pp. 249–253, 512-514 (1985)

Effect of Vibration on Imaging Quality of CCD during the CNC Lathe Turning

Gang Zhao, Zhiping Deng*, and Shengjin Chen

School of Mechanical Engineering and Automation, Xihua University,
Chengdu, China 610039
zhaogang0520@163.com, zhiping-deng@163.com, chenniou1314@163.com

Abstract. The CNC lathe will vibrate during turning if it is excited by the off-center force. Vibration is a key factor effecting image quality of CCD in the lathe. One degree of freedom spring-mass-damper mechanical model of the CNC lathe is built. And, the corresponding mathematical model and simulative model are also built. We used Modulation Transfer Function (MTF) to evaluate the influence of vibration on image quality. The effect of linear and angular displacements due to the vibration on the imaging quality is presented. Relationship of image motion with linear and angular displacement is established. It is concluded that the angular displacement is the key factor effecting image quality after having made qualitative analysis on motion image according to practical project.

Keywords: CCD, Modulation Transfer, Function (MTF), vibration, image quality, image motion.

Introduction

With the development of science and technology, there are many changes have taken place in the field of mechanism manufacture technology. The traditional processing equipment has many difficulties in adapting to the market requirements of high-precision high-quality high-efficiency. The future of CNC lathe machining will characterize in high-quality, high precision, high-yield, high-efficiency. And The core of modern manufacturing is numerical control which based on the microelectronics technology; it will combine the traditional mechanical manufacturing technology and modern control technology, sensing measurement technology, information processing technology and network communication technologies in order to form a highly information, flexible, and automated manufacturing system.

Detection technique is one of the main technologies in modern manufacturing industry; it is also the key to guarantee the quality of products. Along with the development of modern manufacturing industry, most of the traditional detection

* Corresponding author. Zhiping Deng, is currently a professor in School of Mechanical Engineering and Automation, Xihua University, China. Her research interests include Mechanical manufacturing and automation, zhiping-deng@163.com

D. Jin and S. Jin (Eds.): Advances in FCCS, Vol. 1, AISC 159, pp. 675–680.
springerlink.com

techniques has cannot satisfy their needs, the modern manufacturing laid great emphasis on real-time, online and non-contact detection. However, advanced detection technology will also has many interference factors to influence the detection results. This essay mainly analyzes the influence of the vibrations caused by the eccentric force which produced in turning of the CNC late on the quality of CCD image and then that provides foundation for effective measures to minimize the potential aberrations.

1 Vibration Source Analysis

No matter what kind of the equipment is or what kind of machine tool it is, if you want to find a wonderful solution to the problem of vibration which produced when the eccentric force is on the correctional level, first and foremost, the work you have to do is to find the causes of vibration.

The vibration source of CNC lathe is the imbalance vibration movement of revolving part, attachment and the free vibration, of which the greatest impact is the imbalance movement of spindle unit and the processed work-piece. Especially, at present time, the direction of the numerical control lathes' development is for high speed, cutting feed parameters, it now s give prominence to the imbalance of spindle unit and the processed work-piece. When the spindle unit and work-piece rotating, due to the decentralization of its center of mass and center of rotating there will produce centrifugal force that around the rotary center, it is the main factor in bring about the vibration and unstable of lathes. The size of the centrifugal force is

$$F_0=(M-m)ew^2 \tag{1}$$

in which

M - *the mass of CNC lathe, kg*

m - *the eccentric mass of parts, kg*

e - *the center of mass for spindle components and parts and the eccentric distance for center of rotating, m*

w - *the angular velocity in parts rotating, rad/s*

F_0 - *eccentric force, N*

From the following formula, we can see that, the magnitude of eccentric force is directly proportional with the mass of rotating parts, offset and the rotating speed. In order to reduce the vibration, when you are processing the great work-piece blank piece which distributed asymmetry and the eccentric work-piece, it will be better in bringing down the rotating speed appropriately, especially in the processing the heavy scale of eccentric piece, the mass balance equipment should be attached to reduce the vibration caused by the decreasing of eccentric force.

2 Mechanics Model

According to the production conditions in reality, we can simplify the system which combined CNC lathe with the foundation to a free spring-mass- damping system, so as to establish a the mechanics model showed in figure 1. The figure 1 shows that, CNC lathe simplified into the rigid body in mass M, the foundation is simplified into K stiffness coefficient of spring and the damper which has a C level in damping coefficient. The eccentric movement of a work-piece can be simplified into a particle with an E degree of throw in offset which will rotate around the rotating center in a speed of W. The eccentric force produced in the swirling process can be break into excitation force of vibration in the vertical point.

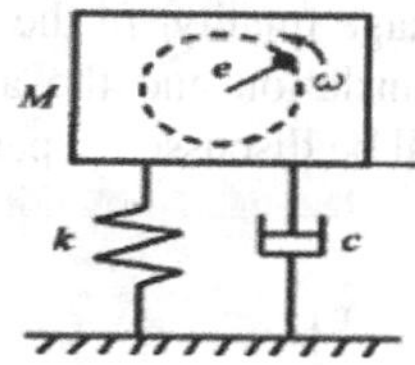

Fig. 1. Mechanical model

Figure 1 is the mechanical model of static equilibrium position for origin, positive direction is upward to establish coordinate system. When the CNC lathe affected by eccentric force, according d'Alembert principle, the system vibration equation is

$$(M\text{-}m)x+m(x+e)+cx^2+kx=0 \quad (2)$$

in which

M - *The mass of CNC lathe, kg;*
m - *The mass eccentricity of parts, kg;*
e - *Offset, m;*
x - *CNC lathe vertical displacement, m*

Based on (1) (2)

$$Mx+rx^2+kx=F_0 \quad (3)$$

3 MTF Analysis

As an evaluation tool in the optical system, test of modulation transfer function(MTF) has been widely used, it can reflect the transitivity relation between the image-forming system and the spatial frequency with different target, target or instrument movement will all have a attenuation effect in the dynamic image of MTF, it's essence is that the response speed to image-forming medium should be limited. In the movement times, the image forward lap, contrast grade is also decreased, and then the MTF decline naturally. Therefore, the use of MTF as an evaluation tool is very effective, simple and intuitive.

3.1 Effect of Linear on Imaging Quality of CCD

In order to analyze the influence of vibration to laser CCD scan images, we can divided the vibration of work-piece into three translation along with axis and three rotation circle around with axis. As shown in figure 2, the vibration state in the coordinate system: laser CCD system and the installation position of work-piece will determines the work-piece in X and Z axis are translation, in Y axis are rotation, what's more, the rotation in X axis will exist no influence on the production of image. On the contrary, the rotation in Z axis and the translation in Y axis all can lead to move and then affect the CCD image quality. In the following part, the influence of linear displacement caused by translation and the angular displacement caused by rotation to the images' forming will be discussed separately:

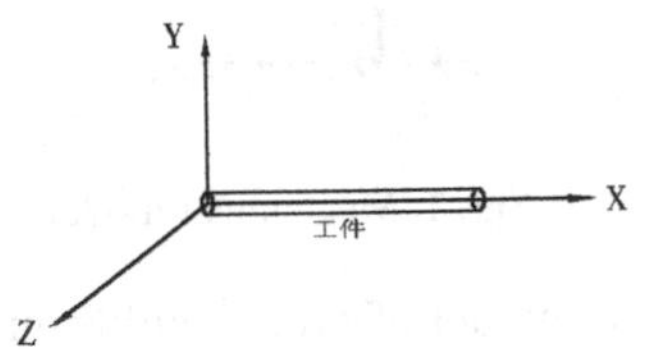

Fig. 2. Workpiece coordinate system used in Vibration Analysis

We assume that the workpiece is placed vertically, the Y-axis direction:

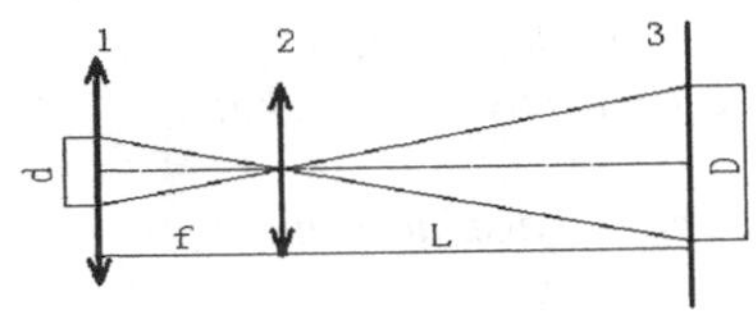

Fig. 3. Imaging situation before Vibration

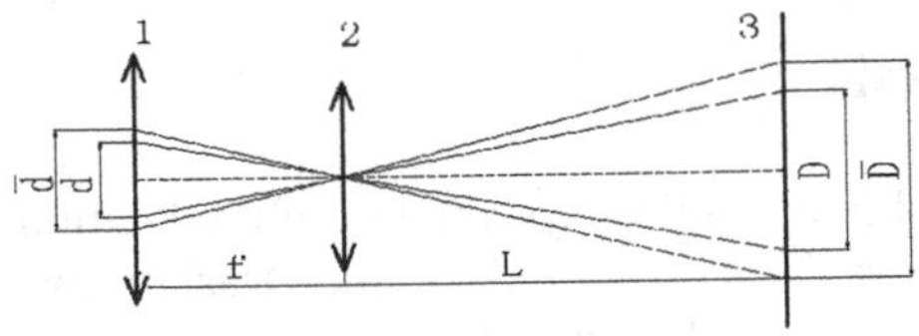

Fig. 4. Image Shift Moving along the Y-axis direction

$$d=D \tag{4}$$

$$= \tag{5}$$

$$= \quad =\Delta D\times \tag{6}$$

in which

D—*The location of the image without vibration;*
—*The location of the image after vibration;*
d—*The location of the image before vibration;*
—*The relative position after vibration;*
ΔD—*Maximum amplitude of CNC lathe;*
f—*focal length;*
*L**The distance between the receiver and focus;*
The image motion for the workpiece moves in the Y-axis

Make the Mechanical models of CNC lathe and mathematical models to establish Simulink simulation model, and ΔD =0.05*mm* f =10 *mm*, L =160*mm*, We will get =3.125*μm*

And MTF= = 1.745302×10^{-2} (7)

in which N=56 *lp*/*mm*

3.2 Effect of Angular Displacements on Imaging Quality of CCD

The effects on image quality cased by angular displacement

= (8)

in which D =30 mm ; a=1'
We will get =4.363 *μm*

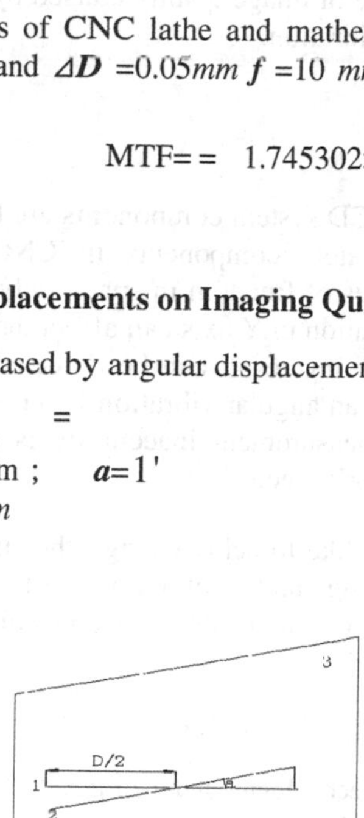

Fig. 5. Image Shift from rotation of Z-axis

1. The location of the image before vibration; 2.The location of the image after vibration; 3.CDD receiving mirror

And MTF= = 1.745277×10^{-2} (9)
in which N=56 *lp*/*mm*

With the method of MTF to analyze the influence created by vibration to the quality of image, it is a simple and direct viewing way which deals with a little bit parameters, in this point of view we can give the quantized value without efforts.

From the values of MTF, it is evident that angular vibration's effect to the quality of image is far more than that caused by linear displacement. The MTF of linear displacement is equal to 1.745302×10^{-2}, is equal to 3.125μm, so the influence of linear displacement is small, in some given conditions it also can be ignored. It is generally acknowledged that if the image drift does not surpass the half of pixel, there will not come out into vague image. But when theis equal to 4.363um, this number is bigger than the half of pixel, as a result, the corresponding measurements must be taken to improve the quality of images. The size of the linear array CCD pixel used in this system is 7um, and also an adoption of optical amplifier, the magnification times is 16, thus the limiting error in this measurement system should be 0.4375 um. To sum up, as for the change of image quality caused by vibration, angular vibrations is more serious than linear vibration.

4 Conclusions

Through the analysis, the CCD system components are being determined to fix on the CNC Lathe. The CCD system components in CNC Lathe make sure that the work-piece in the six degrees of freedom in space only in the rotation vibration in Z axis and the translation vibration in Y axis can affect the image's quality. Through the analysis on MTF, the following result can be detected: as for the change of image quality caused by vibration, an angular vibration is more serious than linear vibration. In the CCD system, the measurement inaccuracy is principally influenced by the image forming quality of work-piece.

Acknowledgment. I would like to acknowledge the support from the university key laboratory of Manufacturing and Automation of Sichuan Province at Xihua University and the support from the Professor Deng Zhiping for the second author.

References

1. Jia, P., Zhang, B.: Critical technologies and their development for airborne opto—electronic reconnaissance platforms. Optics and Precision Engineering 11(1), 82–88 (2003) (in Chinese)
2. Shen, H.H., Liu, J.H., Jia, P.: Overview of image stabilization. Optics and Precision Engineering 9(2), 115–120 (2001) (in Chinese)
3. Sun, H., Zhang, B., Liu, J.H.: Restoration of motion-blurred image based on Wiener filter and its application in aerial imaging system. Optics and Precision Engineering 13(6), 735–740 (2005) (in Chinese)
4. Lim, H., Tan, K.C., Tan, B.T.G.: Edge errors in inverse and Wiener filter restorations of motion—blurred images and their windowing treatment. CVGIP 53, 186–195 (1991) (in Chinese)
5. Li, Z.Y.: Moving image analysis I. National Defence Industry Press, Beijing (1999) (in Chinese)
6. Wang, X.H., Zhao, R.C.: Restoration of arbitrary direction motion—blurred images. Journal of Image and Graphics 5(6), 525–529 (2000) (in Chinese)
7. Xu, P., Huang, C.L.: Effects of satellite vibration on image quality. Journal of Image and Graphics 5(3), 127–130 (2002) (in Chinese)

The Control System of Electric Load Simulator Based on Neural Network

Liu Xia, Yang Ruifeng, and Jia Jianfang

School of Information and Communication Engineering
North University of China
Tai-yuan, Shan-xi, China, 030051

Abstract. The superfluous torque is a key issue on electric load simulator. How to overcome the superfluous force is the critical issue for designing load simulator and improving control system performance. In this paper, the electric load modeling on the basis of the RBF neural network algorithm for system identification of the system model and the PID parameters, real-time updates PID control parameters to ensure that the system dynamic and static characteristics. And through simulation, the controller reduces the excess torque on the system.

Keywords: excess torque, load simulator, RBF neural network, PID.

1 Introduction

Load simulator is a passive torque servo system. In practice the process, due to mechanical reasons connected for load simulator, the aircraft had active exercise its force interference, the output power of the load simulator superposition of a redundant power. As long as the object and the load bearing mechanical connection between the object, then the excess power is inevitable. However, the actual aircraft in flight does not exist in the mechanical connection, it is necessary to eliminate excess force. Excess load simulator performance impact force is a key issue, how to overcome the extra power load simulator is designed to improve the performance of critical system control.

At present, the study for extra torque control strategy from the start, with compensation network, resulting in an additional control loading system to overcome the extra torque. The method is to load the system software from the start, through the inhibition of excess power control methods, including multi-variable decoupling control, auxiliary synchronous compensator control, based on the composite structure invariance principle of control, internal model control based on the compensation method, based on disturbance observer methods of compensation devices [1-4]. Electric load system load, containing a large number of non-linear factors, it makes the control theory based on the traditional controller is very complex and difficult to achieve, so the introduction of neural network control can effectively solve the nonlinear mapping ability and self-learning ability to adapt the problem.

D. Jin and S. Jin (Eds.): Advances in FCCS, Vol. 1, AISC 159, pp. 681–687.
springerlink.com

2 Electric Load System Model

Electric load for the load simulator of the permanent magnet synchronous motor drive components, which uses current feedback PWM inverter drive, permanent magnet synchronous motor from the body, vector control inverter drive and the rotor position detection and processing circuit has three major components. In the magnetic circuit is not saturated, three-phase stator current space generated by the rotor magnetic potential and magnetic flux distribution is sinusoidal shape of the ideal case, without considering the effect of the rotor salient-pole magnetic field, the surface of a rigid body for permanent magnet synchronous motor, usually that does not exist salient effect, $L_d = L_q = L$ 。 Dg ordinary permanent magnet synchronous motor in the coordinate system of differential equations, electrical motor torque equation and the equation of motion were described as (1),(2),(3).

$$\begin{cases} \frac{di_d}{dt} = pi_d = \frac{u_d}{L} - \frac{R_s}{L} i_d - w_e i_q \\ \frac{di_q}{dt} = pi_q = \frac{u_q}{L} - \frac{R_s}{L} i_q - w_e i_d - \frac{w_e \varphi_m}{L} \end{cases} \tag{1}$$

$$T_e = p_n(\varphi_d i_q - \varphi_q i_d) = p_n \varphi_m i_q \tag{2}$$

$$p_n \varphi_m i_q = J\frac{dw_r}{dt} + Dw_r + T_f \tag{3}$$

Here, u_d , u_q , i_d , i_q , φ_d , φ_q is stator voltage, stator current, electron flux d-axis and q-axis , R_s is stator resistance, w_e is the rotor electrical angular velocity($w_e = w_r \cdot p_n$, w_r is rotor speed, p_n is motor pole pairs), p is differential operator. If L_d , L_q is the inductance of d-axis and q-axis, φ_m is the rotor flux that corresponding to the permanent magnet.

To acquire linear equation of state, assume $i_d = 0$, ignoring i_d - u_d loop, analysis i_q - u_q loop, then after the Laplace transform, we get (4).

$$\frac{w_r}{u_q} = \frac{p_n \varphi_m}{JLs^2 + JR_s s + p_n^2 \varphi_m^2} \tag{4}$$

Permanent magnet synchronous motor using current feedback PWM inverter drive, the current regulator can be considered proportional regulator. Set I_m is input current through a proportional controller $K_D(s) = k_p$ then uq, among Kp is the current control gain, Kf current feedback coefficient. A given set of input voltage Vin, Kv from a given input Vin to the ratio between the coefficient of Im, so that $R = R_S + k_f k_p$ Equivalent resistance, which can have permanent magnet synchronous motor current feedback from Vin to wr transfer function (5)

$$G_1(s) = \frac{w_r}{V_{in}} = \frac{k_p k_m k_v}{LJs^2 + RJs + k_e k_m}$$

$K_e = K_m = p_n \varphi_m$, Ke is the potential coefficient, Km is the torque coefficient, then can obtain the transfer function from uq to wr:

3 Torque Sensor Modeling

Torque sensor is connected to the steering gear and load simulator key component of its own inertia and friction are very small, so its allowable range, can be seen as a proportion of links,

$T_f = T_A(\theta_f - \theta)$, among T_A is the torsional rigidity, $\Delta\theta$ is deformation across the sensor for the torque angle difference, T_f is output torque, θ_f is load simulator output angle, θ is steering gear output angle.

From this we can get electric load simulator system block diagram as shown in Figure 1.

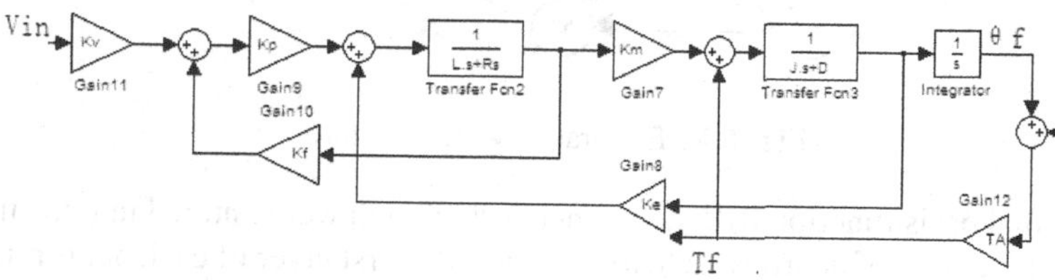

Fig. 1. Block diagram of electric load simulator system

Therefore, the availability of mathematical models for the electric load simulator

$$T_f = P(s)V_{in} - L(s)\theta$$

$$P(s) = \frac{T_A K_p K_m K_v}{s[(R+Ls)(Js+D)+K_e K_m]+T_A(R+Ls)} \tag{5}$$

$$L(s) = \frac{s[(R+Ls)(Js+D)+K_e K_m]T_A}{s[(R+Ls)(Js+D)+K_e K_m]+T_A(R+Ls)}$$

4 RBF Neural Network

RBF neural network is a partial acceptance of the use of the mapping function domain of the artificial neural network, is a hidden layer and output layer consisting of a linear feedforward network structure. The structure shown in Figure 2, the input is the input node of m, n a hidden node, an output node of the three neural networks. The first layer is input layer, the number of nodes from the input signal to determine the dimension of the input layer to hidden layer weights fixed to 1, about the input signal directly to the input of the hidden layer. The second layer is hidden layer nodes as described in the problem needs. From the input space to the hidden layer space transform is nonlinear, activation function of hidden layer unit is a local distribution center of the radial symmetry of the decay of non-negative non-linear function, that is, radial basis function. This hidden layer to the input layer of the incentive will only have a partial response, that is, only when the input into a small space within a given time, the hidden layer neurons will make a meaningful non-zero response. The third layer is output layer, the role of the input patterns respond to the hidden layer to output layer space transform is linear, the weighted coefficient wij, the output layer is a linear combination of the device [3].

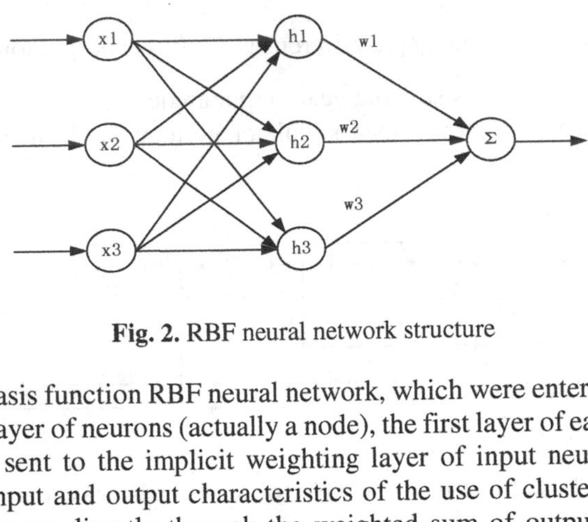

Fig. 2. RBF neural network structure

In the radial basis function RBF neural network, which were entered into the input of each of the first layer of neurons (actually a node), the first layer of each neuron's output without directly sent to the implicit weighting layer of input neurons, hidden layer neurons of the input and output characteristics of the use of clustering, the output of hidden layer neurons directly through the weighted sum of output, the output layer neurons is only a weighted sum, but not non-linearity. Network model can be expressed as:

$Y = W^T H = \sum_{i=1}^{m} w_i h_i$, among : $W = (w_1, w_2, \cdots, w_m)^T$ weight vector for the network ; RBF Network of radial basis vectors $H = [h_1, h_2, \cdots h_m]^T$, among h_j is the Gaussian function $h_j = \exp(-\frac{\|X - C_j\|^2}{2b_j^2}), j = 1,2,\cdots m$, $X = [x_1, x_2, \cdots x_n]^T$ For the network input vector. Network nodes of the j-th center vector $C_j = [c_{j1}, c_{j2}, \cdots c_{ji}, \cdots, c_{jn}]^T, i = 1,2,\cdots,n$; b_i is width parameter of the i th node ; $\|\bullet\|$ is Euler norm; Y is RBF output of the network.

Neural network is used to approximate the loading system transfer function execution unit. The recognition accuracy directly determines the accuracy of electric load control system. After selection of the RBF recognition normalized input vector:

$$X = [x_1, x_2, x_3]^T = [u_k(t-1), T_m(t-1), \dot{\theta}(t-1)]^T$$

Here: Uk is the control signal; Tm torque feedback; $\dot{\theta}$ Steering gear angular velocity.

Identification error is defined as: $e_m(t) = T_e(t) - T_m(t)$

According to gradient descent, node center, width parameter and the output node-based iterative algorithm for the right as follows:

$$\Delta b_i(t) = [T_e(t) - T_m(t)] \cdot w_i(t-1) h_i(t-1) \frac{\|X - C_j\|^2}{b_i^3}$$

$$b_i(t) = b_i(t-1) + \eta \Delta b_i(t) + \alpha(b_i(k-1) - b_i(k-2))$$

$$\Delta C_i(t) = [T_e(t) - T_m(t)] \cdot w_i(t-1) h_i(t-1) \frac{X(t) - C_i(t-1)}{b_i^2(t-1)}$$

$$C_i(t) = C_i(t-1) + \eta \Delta C_i(t) + \alpha(C_i(k-1) - C_i(k-2))$$

$$w_i(t) = w_i(t-1) + \eta[T_e(t) - T_m(t)] h_j + \alpha(w_i(t-1) - w_i(t-2))$$

Among them, α The momentum factor, η is the learning rate.

Set $x_1(t) = u(t)$, When the NNI a good approximation charged objects, can be Tm (t) is approximately Te (t), so you can get the Jacobian matrix algorithm for the NNI:

$$\frac{\partial T_e}{\partial u} \approx \frac{\partial T_m}{\partial u} = \frac{\partial T_m}{\partial x_1} = \sum_{i=1}^{m} w_i h_i \frac{C_{1i} - u}{b_i^2}$$

5 Control Based on RBF Neural Network and PID

RBF neural network combined with conventional PID controller is used as a single-input single-output control system. RBF neural network's primary function is as a reader to accurately track changes in the controlled object, access to online self-tuning PID parameters required for the Jacobian matrix of information (ie, the output of the controlled object input sensitivity control information), in this case, RBF network is generally the single output layer neurons, the network's input layer and hidden layer neurons depends on the complexity of the system.

Radial basis function RBF neural network PID tuning control system structure shown in Figure 3.

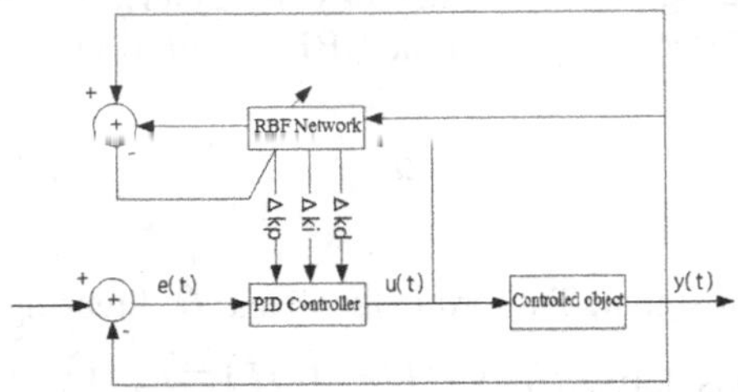

Fig. 3. RBF neural network PID tuning control system structure

Neural network controller using a single neural network structure. Input vector selection:

$$X_C = [x_{C1}, x_{C2}, x_{C3}]^T, W_C = [w_{C1}, w_{C2}, w_{C3}]^T$$

$$x_{c1}(t) = e(t) = T_e(t) - T_m(t)$$

$$x_{c2}(t) = e(t) - e(t-1)$$

$$x_{c3}(t) = e(t) - 2e(t-1) + e(t-2)$$

Control algorithm is:

$$u(t) = u(t-1) + k_i x_{c1}(t) + k_p x_{c2}(t) + k_d x_{c3}(t)$$

Setting targets for the neural network : $E = \frac{1}{2} e(t)^2$

Kp, ki, kd for the PID control parameters, obtained by the gradient descent method to adjust the formula.

6 Conclusions

This chapter will be applied to intelligent control method of controlling electric load simulator, the proposed RBF neural network based tuning of PID control strategy, the use of online real-time RBF neural network PID controller to adjust the three parameters, the object can be any nonlinear followed to achieve good control. Simulation results show that the control method, the system response speed and tracking results were satisfactory, while achieving the PID parameters on-line adjustments.

References

1. Wang, X., Sun, L., Yan, J.: Feedforward composite applications to improve the performance of the experimental study of loading system. Journal of System Simulation 16(7), 1539–1541 (2004)
2. Kawaji, S., Suenaga, Y., Maeda, T., et al.: Control of cutting torque in the drilling process using disturbance observer. In: Proc. of ACC, pp. 723–728 (1995)
3. Profeta, J., Vogt, W., Mickle, M.: Disturbance estimation and compensation in linear systems. IEEE Transaction on Aerospace Electronic Systems 26(2), 225–231 (2002)
4. Fang, Q., Yao, Y., Wang, X.C.: Disturbance observer design for electric aerodynamic load simulator. In: The Fourth International Conference on Machine Learning and Cybernetics, pp. 1316–1321 (2005)
5. Nam, Y.: QFT force loop dsign for the aerodynamic load simulator. IEEE Transactions on Aerospace and Electronic Systems 37(4), 1384–1392 (2001)
6. Shen, D., Hua, Q.: Wang Zhanlin based on neural network for electric load system. Beijing University of Aeronautics and Astronautics 23(6), 525–529 (2002)
7. Jiao, Z.: Huaqing electro-hydraulic load simulator of the RBF neural network control. Mechanical Engineering 39(1), 10–16 (2003)

Realization of Test System for Choice-Question with Authorware

Wenpeng Lu

School of Science, Shandong Polytechnic University
Jinan, Shandong, China, 250353
lwp@spu.edu.cn

Abstract. Based on database technology, a test system of choice-question is realized with Authorware. The test system is composed with three modules: generation module of test, examination module of choice-question and evaluation module of answers. Generation module of test is responsible to randomly select a certain number of questions to form a test paper. Examination module is responsible to display the questions to candidates one by one, and record their answers. Evaluation module is responsible to compare candidates' answers with standard answers, compute their score and generate evaluation information. The test system is developed with database technology. For different tests, the test system only needs to connect corresponding question database without needing to change its source code, which is suitable for large-scale standardized tests.

Keywords: Authorware, Examination System, Database, Standardized Test.

1 Introduction

Authorware is a software developed by Macromedia corporation, which is wildly used to develop multimedia courseware by educators. Authorware provides an icon-based, flow-line method to design a software, and provides many kinds of interactions, which is suitable for develop multimedia courseware and widely applied in education field[1]. With Authorware, courseware can be designed with flowline framework, without requiring any specialized programing knowledge.

For the implementation of test system, if there are only few questions, we can manually design interactions for each question one by one[2, 3]. However, there are a large amount of questions in standardized test system, which needs to randomly generate test paper. This makes that manually design becomes unfeasible[4]. Questions usually exist in a database of questions in standardized test. Authorware needs to access the database to read the selected questions and combine them to generate a test paper.

A method to design a test system for choice-question with Authorware is introduced in the paper. Based on database technology, test system randomly selects a certain number of questions to generate a test paper, display it to candidates and evaluate their answers. The rest of the paper is organized as follow. Section 2

D. Jin and S. Jin (Eds.): Advances in FCCS, Vol. 1, AISC 159, pp. 689–694.
springerlink.com

introduces some basic knowledge about database operation. Detailed implementation of test system is described in Section 3. The summary and future works are mentioned at last.

2 Basic Knowledge of Database Operation

Authorware can access database with open database connectivity (ODBC), manage database with SQL language. The methods to build data source and the functions to operate database are introduced in this section.

2.1 Methods to Build Data Source

In order to access a database with ODBC in Authorware, a data source must be built to connect ODBC with the database. There are three methods to build a data source: to manually create a data source with the manager of "ODBC Data Source", to build a data source with function interface of Authorware, to dynamically link the database. For the convenience of test system, considering its facilitation, the last two methods are commonly adopted.

Authorware can automatically configure its ODBC with tMsDBRegister function within its tMsDSN.u32 file, whose syntax is as follow: Result := tMsDBRegister (dbReqType, dbType, dbList), in which, dbReqType represents the operation type, dbType specifies the driver of ODBC data source, dbList specifies the name, description and file of ODBC data source.

Dynamic Link method is realized with ODBCOpenDynamic function in ODBC.u32 file, which is provided by Authorware 7.0 and higher version. The function can dynamically connect with a database without requiring to create a data source in advance. Its syntax is as follow: ODBCHandle: = ODBCOpenDynamic (WindowHandle, ErrorVar, DBConnString), in which, WindowHandle is window handle provided by Authorware, ErrorVar is error message variable, DBConnString is database connection information.

2.2 Functions to Operate Database

Database operations are implemented with Authorware's functions in its ODBC.u32 file[4]. Three database operation functions are provided in ODBC.u32, which are as follow.

ODBCOpen Function. ODBCOpen() is used to open a database, whose syntax is as follow: ODBCHandle := ODBCOpen (WindowHandle, ErrorVar, Database, User, Password).

ODBCExecute Function. ODBCExecute() is used to execute database operations with SQL, whose syntax is as follow: ODBCdata := ODBCExecute (ODBCHandle, SQLString).

ODBCClose Function. ODBCClose() is used to close the database, whose syntax is as follow: ODBCClose (ODBCHandle).

3 Implementation of Test System

As shown in Fig.1, test system is composed with three modules: generation module of test, examination module of choice-question and evaluation module of answers[4]. Detailed implementation of three modules would be described in this section.

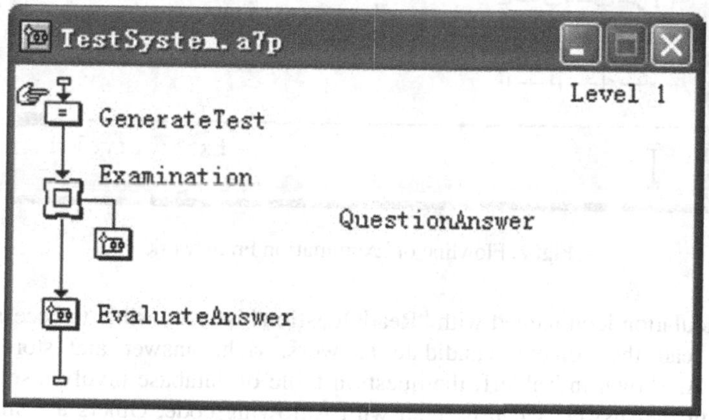

Fig. 1. Main Flowline of Test System.

3.1 Generation Module of Test

A calculation icon named "GeneateTest" is placed on main flowline of test system. The calculation icon is responsible to finish three kinds of works, which are as follow.

1) Initialize global variables of test system, such as database name, table name, array of ID of selected questions, array of right answers, array of candidates' answers, et al.
2) Connect and open question database with the methods mentioned in Section 2.
3) Randomly select the questions that would form test paper. This only needs to generate a random unrepeated array of question IDs. In the paper, the difficulty and knowledge points of questions are not be considered. Once the random array is determined, test paper is generated.

3.2 Examination Module of Choice-Question

As is shown in Fig.1, this module mainly involves two icons: framework of examination and map of question answer.

Framework of Examination

A framework icon named with "Examination" is placed on main flowline, which is responsible to display the question and interact with candidates. As is shown in Fig.2, a calculation icon and an interaction icon are placed on the flowline of "Examination". In the interaction icon, three buttons (last question, next question, hand in) are placed.

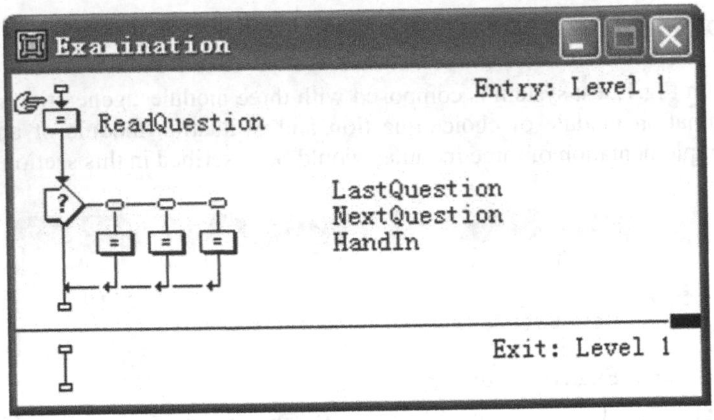

Fig. 2. Flowline of Examination Framework.

The calculation icon named with "ReadQuestion" is responsible to access question database, read the content, candidate answers, right answer and store them in variables. As shown in Table.1, the question table of database involves seven fields. The content of Answer A can be read with following code. Others are similar with Answer A.

SqlStr := "Selection AnswerA from " ^ TableName ^ " Where QuestionID=" ^ No[CurQuesNo]
AnswerA := ODBCExecute(ODBChandle, SqlStr)

Table 1. Table Structure of Choice-Question

Field Name	Data Type	Length/Data Format	Explanation
QuestionID	AutoNumber	Long Integer	ID of the question
QuestionContent	Text	255	Content of the question
AnswerA	Text	255	Content of Answer A
AnswerB	Text	255	Content of Answer B
AnswerC	Text	255	Content of Answer C
AnswerD	Text	255	Content of Answer D
RightAnswer	Numeric	1 byte	ID of right Answer

The calculation icons name with "LastQuestion" and "NextQuestion" is responsible to alter the ID of current question and jump to the calculation icon named with "ReadQuestion". The calculation icon named with "HandIn" is responsible to record the answer of candidate, and jump to the map named with "EvaluateAnswer".

Map of Question Answer

As is shown in Fig.3, a display icon named with "QuestionDisplay" and an interactive icon named with "ChoiceAnswer" are placed on flowline of question answer map. The interaction icon involves four buttons which are corresponding with four candidate answers. The display icon named "QuestionDisplay" is responsible to display the content and candidate answers of questions. The calculation icon of candidate answer is responsible to store the answer of candidate to array variable.

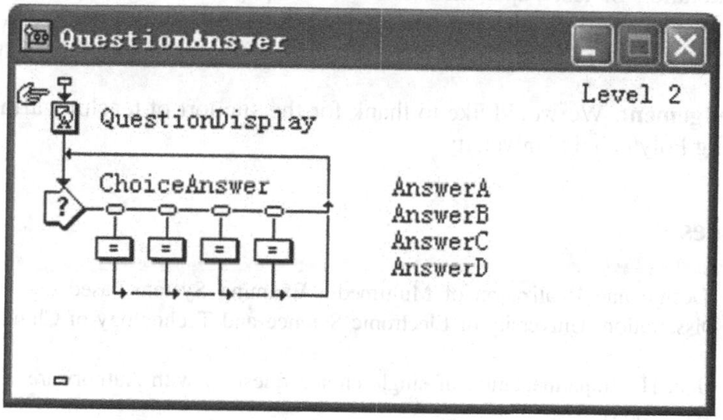

Fig. 3. Flowline of Question Answer Map.

3.3 Evaluation Module of Answers

As is shown in Fig.4, a calculate icon named with "EvaluateAnswer" and a display icon named with "DisplayScoreAndEvaluation" are placed on flowline of "EvaluateAnswer". Comparing the array of right answers and the array of answers of candidate, test system can compute their score and generate feedback information of evaluation.

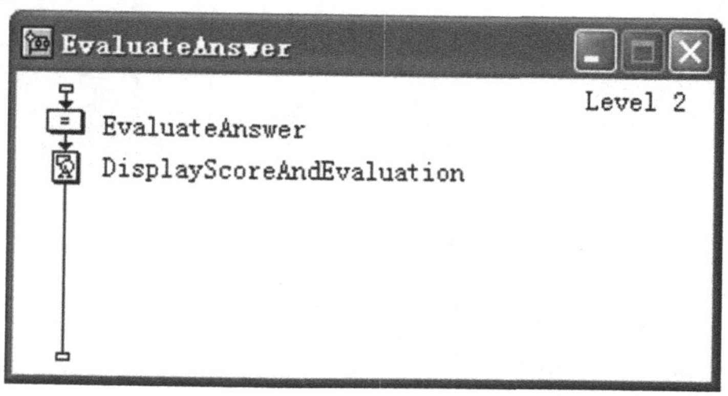

Fig. 4. Flowline of Evaluate Answers Map.

4 Conclusion and Future Work

The paper proposes a method to implement a test system for choice-question with database technology in Authorware. Although only choice-question is implemented in the paper, the method is applicable to other kinds of objective questions. Besides, for different test, the method only needs to change its question database without requiring to modify source code.

The generation of test paper mentioned in the paper is relatively simple, which does not consider the difficulty and knowledge points. In future work, we would like to improve this and try to construct a more practical test system with more functions.

Acknowledgement. We would like to thank for the support of teaching affairs office of Shandong Polytechnic University.

References

1. Lin, J.: Design and Realization of Multimedia Examing System based on Authorware. Master Dissertation, University of Electronic Science and Technology of China, Chengdu (2009)
2. Shi, J., Hua, H.: Implementation of single-choice question with Authorware. China Edu. Info., 51–51 (2003)
3. Zhang, Y.: Design and Implementation of Examination System based on Authorware. China Edu. Info., 37–39 (2005)
4. Lu, W., Jia, Q., Zhao, R.: Realization of random question selection based on ODBC technology in authorware. Journal of Shandong Institute of Light Industry 21, 15–18 (2007)

Author Index